Y0-AAI-925

# LIGHT SCATTERING NEAR PHASE TRANSITIONS

# MODERN PROBLEMS IN CONDENSED MATTER SCIENCES

Volume 5

*Series editors*

V.M. AGRANOVICH

*Moscow, USSR*

A.A. MARADUDIN

*Irvine, California, USA*


*Advisory editorial board*

F. Abelès, Paris, France
N. Bloembergen, Cambridge, MA, USA
E. Burstein, Philadelphia, PA, USA
I.L. Fabelinskii, Moscow, USSR
M.D. Galanin, Moscow, USSR
V.L. Ginzburg, Moscow, USSR
H. Haken, Stuttgart, W. Germany
R.M. Hochstrasser, Philadelphia, PA, USA
I.P. Ipatova, Leningrad, USSR
A.A. Kaplyanskii, Leningrad, USSR
L.V. Keldysh, Moscow, USSR
R. Kubo, Tokyo, Japan
R. Loudon, Colchester, UK
L.P. Pitaevskii, Moscow, USSR
A.M. Prokhorov, Moscow, USSR
K.K. Rebane, Tallinn, USSR

NORTH-HOLLAND PUBLISHING COMPANY
AMSTERDAM · NEW YORK · OXFORD

# LIGHT SCATTERING NEAR PHASE TRANSITIONS

*Volume editors*

H.Z. CUMMINS

*New York, N.Y. 10031, USA*

A.P. LEVANYUK

*Leninskii Prospekt, Moscow B333, USSR*

1983

NORTH-HOLLAND PUBLISHING COMPANY
AMSTERDAM · NEW YORK · OXFORD

UNIVERSITY OF FLORIDA LIBRARIES

© North-Holland Publishing Company, 1983

*All rights reserved. No part of this publication may be reproduced, stored in a retrieval system, or transmitted, in any form or by any means, electronic, mechanical, photocopying, recording or otherwise, without the prior permission of the copyright owner.*

ISBN 0444 86466 0

PUBLISHERS:
NORTH-HOLLAND PUBLISHING COMPANY
AMSTERDAM · NEW YORK · OXFORD

SOLE DISTRIBUTORS FOR THE USA AND CANADA:
ELSEVIER SCIENCE PUBLISHING COMPANY, INC.
52 VANDERBILT AVENUE
NEW YORK, N.Y. 10017

**Library of Congress Cataloging in Publication Data**
Main entry under title:

Light scattering near phase transitions.

(Modern problems in condensed matter sciences ; v. 5)
Bibliography: p.
Includes indexes.
1. Light--Scattering. 2. Phase transformations
(Statistical physics) I. Cummins, Herman Z., 1933-
II. Levanyuk, A. P., 1933- III. Series.
QC427.4.L54 1983 535'.4 83-12103
ISBN 0-444-86466-0 (U.S.)

PRINTED IN THE NETHERLANDS

MODERN PROBLEMS IN CONDENSED MATTER SCIENCES

Vol. 1. SURFACE POLARITONS
V.M. Agranovich and D.L. Mills, *editors*

Vol. 2. EXCITONS
E.I. Rashba and M.D. Sturge, *editors*

Vol. 3. ELECTRONIC EXCITATION ENERGY TRANSFER IN CONDENSED MATTER
V.M. Agranovich and M.D. Galanin

Vol. 4. SPECTROSCOPY AND EXCITATION DYNAMICS OF CONDENSED MOLECULAR SYSTEMS
V.M. Agranovich and R.M. Hochstrasser, *editors*

Vol. 5. LIGHT SCATTERING NEAR PHASE TRANSITIONS
H.Z. Cummins and A.P. Levanyuk, *editors*

Vol. 6. ELECTRON–HOLE DROPLETS IN SEMICONDUCTORS
C.D. Jeffries and L.V. Keldysh, *editors*

*Oh, how many of them there*
*are in the fields!*
*But each flowers in its*
*own way—*
*In this is the highest achievement*
*of a flower!*

*Matsuo Bashó*
*1644–1694*

# PREFACE TO THE SERIES

"Modern Problems in Condensed Matter Sciences" is a series of contributed volumes and monographs on condensed matter science that is published by North-Holland Publishing Company. This vast area of physics is developing rapidly at the present time, and the numerous fundamental results in it define to a significant degree the face of contemporary science. This being so, it is clear that the most important results and directions for future developments can only be covered by an international group of authors working in cooperation.

Both Soviet and Western scholars are taking part in the series, and each contributed volume has, correspondingly, two editors. Furthermore, it is intended that the volumes in the series will be published subsequently in Russian by the publishing house "Nauka".

The idea for the series and for its present structure was born during discussions that took place in the USSR and the USA between the former President of North-Holland Publishing Company, Drs. W.H. Wimmers, and the General Editors.

The establishment of this series of books, which should become a distinguished encyclopedia of condensed matter science, is not the only important outcome of these discussions. A significant development is also the emergence of a rather interesting and fruitful form of collaboration among scholars from different countries. We are deeply convinced that such international collaboration in the spheres of science and art, as well as other socially useful spheres of human activity, will assist in the establishment of a climate of confidence and peace.

The General Editors of the Series,

V.M. Agranovich A.A. Maradudin

# PREFACE

This volume, number 5 in the series Modern Problems in Condensed Matter Sciences, presents an overview of contemporary theoretical and experimental research in this field. In it, several areas of experimental research which were opened up by the advent of lasers twenty years ago, concerning the dynamical properties of crystals, fluids and liquid crystals near phase transitions, are reviewed by authors who have actively contributed to the development of these areas. Similarly, the theoretical concepts which have undergone extensive development during this same period are reviewed by authors who have been responsible for many of the central developments.

The editors wish to express their thanks to all of the authors who have suffered through the complications and long delays inevitably associated with such an international undertaking. We hope that the interested reader will find that by bringing together different scientific and historical perspectives in a single volume, this book will provide a sense of the breadth and interest of this field and of the many remaining questions and areas where additional research is needed.

H.Z. Cummins
New York, USA

A.P. Levanyuk
Moscow, USSR

# CONTENTS

# INTRODUCTION

The field of phase transitions and critical phenomena has experienced a period of extremely rapid growth during the last two decades, both in the development of new theoretical approaches and in the application of new experimental techniques. Since the development of the laser in the early 1960s, light scattering has played an increasingly crucial role in the investigation of many types of phase transitions, and the published work in this field is now widely dispersed in a large number of books and journals. It is our intention in this book to provide a reasonably comprehensive overview of current theoretical and experimental aspects of this field.

Historically, a close association has existed between light scattering and phase transition phenomena since the discovery in 1894 of very intense light scattering (critical opalescence) near the liquid–vapor critical point (cf. Kerker 1969). The early observations of critical opalescence in simple fluids and in fluid mixtures, and subsequent studies of the wavelength and angular dependence of the opalescence, stimulated the development of the theories of Smoluchowski, Einstein, and Ornstein and Zernike which will be discussed below.

The first phase transition in a crystal to be investigated by light scattering was the $\alpha$–$\beta$ transition in quartz at 846 K. Landsberg and Mandelstam (1929) and Raman and Nedungadi (1940) discovered independently that the 207 $cm^{-1}$ Raman line (frequently called a combinational line in the Russian literature) is strongly temperature dependent. As the crystal is heated from room temperature, this line broadens and shifts towards the exciting line and becomes a weak diffuse band as the transition temperature is approached. This was the first case of a soft optic lattice mode associated with a structural phase transition, a phenomenon which has been studied intensively in many crystals since 1960. We also remark that a phenomenon resembling critical opalescence was discovered at the $\alpha$–$\beta$ transition in quartz by Yakovlev et al. (1956a, b).

With the introduction of lasers in the 1960s, light scattering spectroscopy was immediately transformed from a difficult and time consuming experimental *tour de force* to a straightforward experimental technique. Raman scattering studies of the soft modes in $KTaO_3$ and $SrTiO_3$ were first reported by Fleury and Worlock (1967). Subsequently, Kaminow and Damen (1968) observed an overdamped soft mode in the Raman spectrum of $KH_2PO_4$. New Raman scattering studies of soft optic modes soon began to appear regularly. Fleury et al. (1968) first demonstrated the utility of the soft mode concept for cell multiplying phase transitions in their study of the cubic–tetragonal transition in $SrTiO_3$. Similarly, Brillouin scattering (Mandelstam–Brillouin scattering in the Russian literature) was applied to the investigation of acoustic modes near phase transitions, complementing the earlier techniques of acoustic resonance and ultrasonic propagation. A soft acoustic mode in $KH_2PO_4$ was observed by Brillouin scattering by Brody and Cummins in 1968. The very slow fluid density fluctuations responsible for critical opalescence in fluids, although inaccessible with the limited resolution available with Raman or Brillouin scattering spectrometers, also became accessible to experimental investigation with the introduction of intensity fluctuation spectroscopy which allows accurate measurement of extremely narrow spectral lines in the 1 Hz to 1 MHz range. Many investigations of the liquid–vapor critical point, the consulate point of binary fluid mixtures and various liquid crystal transitions have been performed with this technique. For a review of much of the experimental literature on the Rayleigh linewidth in simple fluids and binary mixtures, see Swinney and Henry (1973).

Since 1970, the emphasis in light scattering studies of phase transition has gradually shifted from the simple soft mode analysis to subtler aspects of the spectrum associated with mode coupling, central peaks, modification of classical soft mode behavior due to critical fluctuations, etc. Simultaneously, the range of systems investigated has expanded to include, *inter alia*, cell multiplying transitions, incommensurate transitions, dynamic instabilities and the transition to chaos, etc. In many cases, dramatic and unexpected new phenomena have been discovered and quantitatively explained by light scattering results.

As to the theory of light scattering anomalies near second-order phase transitions, the key idea of the theory goes back to papers by Smoluchowski (1908), Einstein (1910) and Ornstein and Zernike (1918a, b) who dealt with density ($\rho$) fluctuations near the liquid–vapor critical point. Landau (1937) was the first to recognize the enhancement of the order parameter ($\eta$) fluctuations near second-order phase transitions in crystals. The enhancement has the same origin as that of $\rho$ fluctuation near a fluid critical point: loss of the stability of the system. Landau evaluated $\eta$ fluctuations and considered diffuse X-ray scattering near the transition. However, he did not discuss light scattering and not until 1955 was this problem treated by Ginzburg (1955). Ginzburg predicted substantial growth of the scattering intensity for phase transitions close to the point we now call the tricritical point. Unfortunately this prediction proved to be valid for liquids rather than for solids due to the special role of shear strains accompanying $\eta$ fluctuations in solids (Ginzburg and Levanyuk 1974). As to scattering by thermal fluctuations, current theory predicts for most substances a sharp temperature dependence only for relatively weak second-order scattering or for first-order scattering in the critical region which is not usually observed for structural phase transitions. The only exception is for phase transitions now classified as proper ferroelastic (Krivoglaz and Rybak 1957).

Consequently, the strong intensity anomalies observed in crystals by some authors is now attributed to enhancement of light scattering by defects and macroscopic inhomogeneities, or by coexisting domains of two phases within the hysteresis region, a change of view which was first prompted by experimental data for quartz (Shapiro and Cummins 1968). The theory of scattering anomalies due to defects was initiated by Axe and Shirane (1973) who discussed neutron scattering. The same idea was applied to light scattering by Levanyuk et al. (1976).

The main activity in light scattering investigations of phase transitions concerns the spectrum of scattered light. The starting point of the theory of spectral anomalies is also the idea of the loss of stability, i.e., of vanishing of the restoring force for a lattice distortion at the phase transition temperature. This leads to the conclusion that one of the lattice frequencies is expected to become zero at the phase transition, i.e., to the famous "soft mode" concept. This concept was developed primarily by Ginzburg (1949a, b), Cochran (1960, 1961) and Anderson (1960) although several other authors also contributed to its genesis. The effect of a soft mode on the light scattering spectrum was first discussed by Ginzburg and Levanyuk (1960).

The soft mode concept proved to be very useful and a large number of soft modes were studied in various crystals. In many cases, however, a complicated temperature evolution of the spectrum was observed rather than the single temperature dependent feature predicted by the simple soft mode theory. To explain the situation the interaction of the soft mode with other modes must be taken into account. This interaction may be linear or nonlinear and the other modes may be vibrational or relaxational. Linear coupling with vibrational modes was first discussed by Barker and Hopfield (1964). One aspect of nonlinear coupling (interaction with two-phonon excitations) was discussed qualitatively by Scott (1968) and quantitatively by Zawadowski and Ruvalds (1970) and Ruvalds and Zawadowski (1970). Such phenomena as "mode repulsion" and strong temperature dependence of the relative intensity of interacting modes are connected with both linear and nonlinear interactions. The linear interaction of a soft mode with a relaxational mode was considered for the first time by Levanyuk and Sobyanin (1967) and the possibility of a specific temperature evolution of the spectrum including the "three-peak situation" with temperature dependence of the central peak was found. A qualitatively similar picture arises due to nonlinear coupling as shown by Cowley (1969). The experimental observation of the three-peak spectral density distribution prompted a number of theoretical investigations which will be reviewed in the first chapter of this book.

The first theory predicting a three-peak spectrum of scattered light was given by Landau and Placzek (1934) to explain the observation by Gross of a triplet in scattering from fluids where Brillouin and Mandelstam had predicted a doublet. They also noted that it is the intensity of the central component that diverges as the critical point is approached. Coupling of the order parameter to the heat diffusion mode in solids is also one of the mechanisms proposed to explain the phenomenon of central peaks (Pytte and Thomas 1972). Many other mechanisms have also been considered – see chapters 1 and 7.

Of special interest and fundamental importance is the problem of light scattering in the immediate vicinity of the phase transition ($T = T_c$) where the Landau theory often ceases to be valid. The temperature, wavevector and frequency dependence of the correlation function for $\eta$ fluctuations becomes nonanalytic at $T_c$. The development of renormalization group methods for analysis of critical phenomena is one of the most impressive achievements of the modern physics of condensed matter. In this book only limited attention is devoted to this topic. The interested reader can find a number of comprehensive reviews, e.g., in the series of books *Phase Transitions and Critical Phenomena*, edited by C. Domb and M.S. Green (Academic Press, New York).

This book begins with a four part exposition of the theory of light scattering near phase transitions in Part I. The first chapter addresses aspects of light scattering in ideal crystals which are common to a large range of phase transitions. It is written in the context of Landau theory primarily with one dimensional order parameters with only limited emphasis on the breakdown of the theory due to critical fluctuations. Special attention is paid to various mode coupling phenomena including the central peak. The second chapter deals with elastic and inelastic scattering due to defects.

Recently, incommensurate phase transitions have been the subject of many experimental and theoretical investigations. The unusual phonon excitations of incommensurate phases which lack the regular translational symmetry of ideal crystals can be effectively studied by light scattering spectroscopy. The theory of this scattering is the subject of chapter 3.

Chapter 4 is devoted to the theory of light scattering in liquid crystals. Fluctuations in these systems are more complicated than those in simple fluids or crystals owing to the special role of the local orientation of highly anisotropic molecules, the "director". Consequently, the theory of light scattering in liquid crystals is treated separately with a distinct phenomenology.

In Part II, seven areas of experimental investigation are reviewed. Raman scattering spectroscopy, which has played a central role in the investigation of solid state phase transitions is the subject of chapter 5. In view of the large number of transitions already investigated and of other earlier review articles, this chapter is limited to an in depth analysis of a few transitions of current interest. In chapter 6, an exhaustive survey of the less extensive literature of

Brillouin scattering studies of solid state phase transitions is presented, with special emphasis on proper and improper ferroelastic transitions.

The topic of central peaks which has been studied extensively during the last decade and has often been the subject of intense controversy is reviewed in chapter 7. Various theoretical mechanisms which have been proposed to explain central peaks are discussed, and several experimental studies are reviewed which have provided definitive identification of the central peak mechanisms at work in specific cases. It is now clear that no single mechanism can account for all, or even most, of the central peaks associated with structural phase transitions. Rather, a variety of physical phenomena are involved with two or more often contributing at the same transition.

In chapter 8, light scattering investigations of incommensurate phases, both metals and dielectrics are reviewed.

Light scattering from fluids in the critical region is covered in chapter 9. The spectrum of equilibrium systems is described and compared with the predictions of mode–mode coupling theories. A discussion of the theoretical and experimental investigations of the dynamics of phase separation in fluids is also presented which includes a comparison of nucleation and spinodal decomposition.

In chapter 10, the study of liquid crystals by light scattering techniques is reviewed.

As we mentioned earlier, the discovery in 1956 of a scattering anomaly in quartz near the $\alpha$–$\beta$ transition was initially interpreted as critical opalescence, closely related to the same phenomena in fluids. However, this phenomenon has proved to be more complex than initially recognized, and is now understood to be a consequence of optical inhomogeneities of various size scales, although no acceptable theory yet exists. A similar situation exists for ammonium chloride. In the final chapter (11), an historical survey of the experimental work on these two transitions and some recent data are presented with the hope that it will stimulate new theoretical analysis.

Regrettably, a number of important topics which we would have liked to include in the book had to be omitted in order to keep its size from diverging. Among the most interesting experimental areas omitted are superionic conductors, magnetic transitions, hydrodynamic instabilities, melting (in 2 and 3 dimensions) and superfluids. The theoretical topics of dynamical scaling and renormalization group techniques which would also have added significantly to the book have likewise been left out.

We apologize to the reader both for these omissions, and for an inevitable lack of coherence between the chapters. Although we have tried to encourage cross-referencing and continuity, the logistical problems of achieving even a moderate level of success proved to be far more complicated than we had imagined. It is our hope that in spite of the omissions and limited coherence of presentation, the condensation into a single volume of discussions and references covering a broad spectrum of the major areas in the field of light scattering near phase transitions will make the book worthwhile.

A.P. Levanyuk
Moscow, USSR

H.Z. Cummins
Southampton, N.Y., USA

## *References*

Anderson, P.W., 1960, Fizika dielektrikov., ed., G.I. Skanavi. (Akad. Nauk, Moscow) p. 290.
Axe, J.D. and G. Shirane, 1973, Phys. Rev. **B 8**, 1965.
Barker, A.S. and J.J. Hopfield, 1964, Phys. Rev. **A 135**, 1732.
Brody, E.M. and H.Z. Cummins, 1968, Phys. Rev. Lett. **21**, 1263.
Cochran, W., 1960, Adv. Phys. **9**, 387.
Cochran, W., 1961, Adv. Phys. **10**, 401.
Cowley, R.A., 1970, J. Phys. Soc. Jpn. Suppl. **28**, 239.
Einstein, A., 1910, Ann. Phys. **33**, 1275.
Fleury, P.A. and J.M. Worlock, 1967, Phys. Rev. Lett. **18**, 665.
Fleury, P.A., J.F. Scott and J.M. Worlock, 1968, Phys. Rev. Lett. **21**, 16.
Ginzburg, V.L., 1949a, Zh. Eksp. Teor. Fiz. **19**, 36.
Ginzburg, V.L., 1949b, Usp. Fiz. Nauk, **38**, 490.
Ginzburg, V.L., 1955, Dok. Akad. Nauk SSR **105**, 240.
Ginzubrg, V.L. and A.P. Levanyuk, 1960, Zh. Eksp. Teor. Fiz. **39**, 192 (1961, Sov. Phys. JETP, **12**, 138).
Ginzburg, V.L. and A.P. Levanyuk, 1974, Phys. Lett. **47A**, 345.
Kaminow, I.P. and T.C. Damen, 1968, Phys. Lett. **20**, 1105.
Kerker, M., 1969, The Scattering of Light and Other Electromagnetic Radiation (Academic Press, New York).
Krivoglaz, M.A. and S.A. Rybak, 1957, Zh. Eksp. Teor. Fiz. **33**, 139 (1958, Sov. Phys. JETP, **6**, 107.
Landau, L.D., 1937, Zh. Eksp. Teor. Fiz. **7**, 1232 (collected papers (Pergamon Press, Oxford) p. 233).
Landau, L.D. and G. Placzek, 1934, Phys. Z. Sov. **5**, 172.
Landsberg, G. and L. Mandelstam, 1929, Z. Phys. **58**, 250.
Levanyuk, A.P. and A.A. Sobyanin, 1967, Zh. Eksp. Teor. Fiz. **53**, 1024 (1968, Sov. Phys. JETP, **26**, 612).
Levanyuk, A.P., V.V. Osipov and A.A. Sobyanin, 1976, On the Influence of Impurities on Light Scattering at Phase Transitions, in: Theory of Light Scattering in Condensed Matter, eds., B. Bendow, J.L. Birman and V.M. Agranovich (Plenum Press, New York) p. 517.
Ornstein, L. and F. Zernike, 1918a, Phys. Z. **19**, 134.
Ornstein, L. and F. Zernike, 1918b, Phys. Z. **27**, 761.
Pytte, E. and H. Thomas, 1972, Solid State Commun. **11**, 161.
Raman, C.V. and T.M.K. Nedungadi, 1940, Nature, **145**, 147.
Ruvalds, J. and A. Zawadowski, 1970, Phys. Rev. **B 2**, 1172.

Scott, J.F., 1968, Phys. Rev. Lett. **21**, 907.
Shapiro, S.M. and H.Z. Cummins, 1968, Phys. Rev. Lett. **21**, 1578.
Smoluchowski, M., 1908, Ann. Phys. **25**, 205.
Swinney, H.L. and D.L. Henry, 1973, Phys. Rev. **A 8**, 2586.
Yakovlev, I.A., T.S. Velichkina and L.F. Mikheeva, 1956, Dok. Akad. Nauk SSSR, **107**, 675 (Sov. Phys. Dokl. **1**, 215).
Yakovlev, I.A., T.S. Velichkina and L.F. Mikheeva, 1956, Kristallografiya **1**, 123, (Sov. Phys. Crystallogr. **1**, 91).
Zawadowski, A. and J. Ruvalds, 1970, Phys. Rev. Lett. **24**, 1111.

# PART I

# *Theory*

V.A. BELYAKOV
V.L. GINZBURG
V.A. GOLOVKO
E.I. KATS
A.P. LEVANYUK
A.S. SIGOV
A.A. SOBYANIN

CHAPTER 1

# General Theory of Light Scattering Near Phase Transitions in Ideal Crystals

V.L. GINZBURG and A.A. SOBYANIN

*P.N. Lebedev Physical Institute*
*Acad. Sci. USSR, Moscow*
*USSR*

A.P. LEVANYUK

*A.V. Shubnikov Institute of Crystallography*
*Acad. Sci. USSR*
*Moscow*
*USSR*

*Light Scattering near Phase Transitions*
*Edited by*
*H.Z. Cummins and A.P. Levanyuk*

©*North-Holland Publishing Company, 1983*

# Contents

# 1. Introduction

Thermal fluctuations of certain physical quantities and, first of all, those of the order parameter increase near second-order phase transitions. This increase is one of the most prominent features (possibly, even the main feature) of systems which are close to an instability point of their configuration. The second-order phase transition is just an example of such a point. Thermal fluctuations are probed most directly in the measurements of the light, X-ray, and neutron scattering. Hence, investigation of the anomalies in light scattering near phase transitions plays an important role in studying phase transitions as a whole. Light scattering studies provide information on fluctuations manifested not only in scattering but in singularities of diverse thermodynamic functions and of transport (kinetic) coefficients as well.

At the same time fluctuations manifest themselves in light scattering in another manner than, say in neutron scattering. For example, the first-order scattering is known to be mainly due to long-wavelength fluctuations, while the investigation of these fluctuations by means of neutron scattering is a rather difficult problem. Another example is the existence of a symmetry rule forbidding the first-order light scattering by order-parameter fluctuations in a more symmetrical phase for most substances, whereas such a rule is absent in the case of neutron scattering. In other words, light scattering has enough specific features in order that it should be treated separately.

A rather important role in such a treatment is played by the fact that a singular behavior near phase transition points is exhibited only by fluctuations belonging to a restricted range of wavevectors, e.g., only by long-wavelength fluctuations. This fact makes it possible, as far as phase transitions are concerned, to consider a system as a continuous medium whose properties are described by an effective (i.e., dependent on temperature, pressure and so on) Hamiltonian. The latter may be identified, in fact, with the thermodynamic potential used in the well-known Landau phenomenological theory of phase transitions.

What has been said above emphasizes a special role played by a phenomenological (or semiphenomenological) approach in the second-order phase transition theory. This approach proves, as a rule, adequate to the problem, permitting to obtain results by most simple methods and to expose those features of phenomena under consideration which are general for a large class of systems. Nevertheless, it is just the miscroscopic theories that have been used predominantly in the most part of reviews on the light scattering near phase transitions, as well as in original papers. This circumstance has made the clarification of theoretical situation far more complicated.

It seems worth emphasizing in this connection that the first theoretical works on light scattering near phase transitions (Ginzburg 1955, Krivoglaz and Rybak 1957, Ginzburg and Levanyuk 1958, 1960, Levanyuk and Sobyanin 1967) as well as some reviews (Ginzburg 1962, Ginzburg et al. 1980) have been performed in the framework of phenomenological approach. The part of the latter review, which concerns light scattering in ideal crystals, will be widely used in the present chapter. The main attention is payed here to the temperature region where the Landau phase transition theory is applicable (it is often referred to as the classical region) although some specific features of light scattering in critical (scaling) region are discussed too. The emphasis is layed on the classical region because for the structural phase transition in solids, which is the main concern of the present chapter, no critical region has been yet observed with confidence, as we believe. (The situation is quite opposite with superfluid $\lambda$-transition in liquid $^4$He, with critical points in magnetics, and near the "liquid–gas" and consolute critical points in fluids; but such critical points are not considered in detail here; as to the critical points in fluids, see ch. 10). Besides this, the principal features of light scattering in the critical region can be understood by mere generalization of "classical" results, as is frequently done in this chapter.

In sect. 2 some aspects of the phenomenological phase transition theory are presented that will be used further on. Section 3 is devoted to the calculation of total (integrated over frequencies) fluctuations in the order parameter and in other quantities. In sect. 4 elements of the general theory of light scattering are presented. We also discuss there the coupling between fluctuations in order parameter (and other quantities) and those in dielectric permeability tensor which describes optical properties of a crystal. The total (integral) intensity of light scattering is then calculated in sect. 5. Some experimental evidences for an enhancement of light scattering intensity near phase transitions are also treated. In sect. 6 the spectral distribution of scattered light is considered, including such questions as the soft mode conception, the origin of a temperature-dependent central peak appearing in the light scattering spectrum near certain phase transition points, the linear and non-linear mode-coupling effects and so on. The experimental data are mentioned here only as soon as they can serve as an illustration of theoretical results or if they played a stimulating role in the developments of the theory. This is natural, of course, since the second part of this volume is devoted specially to experiment.

It was not our aim to touch upon all theoretical works on light scattering in ideal crystals, although we do cite a rather large number of papers. In any such selection a known subjectivism seems to be inevitable so that we express our apology to all fine researchers whose work is not mentioned.

## 2. *Elements of the phase transition theory*

### 2.1. *Symmetry changes under phase transitions*

A phase transition can be or not be accompanied by a change of symmetry. In the latter case it is referred to as an isomorphic transition. As a well-known example of such transitions we can point to the transformation between gaseous and liquid states. Great attention has been delivered recently to another class of isomorphic transitions, viz, the ones to the state with high ionic conductivity in the so-called superionic conductors.

However, one finds for the most part that symmetry changes under phase transitions in solids. For a large class of systems this change consists in the loss of a part of symmetry elements by a crystal under its transformation from the higher- to the lower-temperature phase. In this case the symmetry group of the new (less symmetrical) phase is a subgroup of the old (more symmetrical) phase. In what follows the more symmetrical phase will be termed for brevity simply the symmetrical one, and the less symmetrical phase will be called non-symmetrical. If, further on, the configuration of atoms (electron density) changes continuously with temperature (second-order transition) or even has a jump at the transition temperature (first-order transition) but a small one, then one can use with efficiency the Landau theory of phase transitions.

### 2.2. *Order parameter*

The introduction of the transition parameter or, as it is more commonly termed, the order parameter is a starting point for the Landau theory (see e.g., Landau and Lifshitz (1976)). It would be the simplest way to define the order parameter as a variable ("inner deformation") that characterizes the magnitude of these atom displacements or the degree of their ordering, which just represent the crystal reconstruction under phase transition. Such a definition is, however, not quite correct, because crystal reconstructions under phase transition may be very diverse, and to describe them, one should consider, in general, not a single but several "inner deformations". Nevertheless, for second-order transitions or for those first-order transitions which are nearly second-order, a dominant role is played by a single inner deformation, namely, by that which is responsible for the loss of stability of the symmetrical phase (at a transition temperature $T_c$ in the case of second-order transitions or at a spinodal temperature $T_{s2}$ in the case of first-order transitions).

The basic formula of the Landau theory is the expression for the thermodynamic potential considered as a function of the order parameter. By minimizing this expression one finds the equilibrium value of the order parameter as well as an increase of the thermodynamic potential over its equilibrium value, the increase determining a probability of corresponding fluctuation, and so on.

Symmetry considerations impose restrictions on the form of the expression for the thermodynamic potential. For the case of one-component order parameters these restrictions are particularly simple and reduce to the requirement that the thermodynamic potential contains only terms with even powers of the order parameter. This is just the case to be first considered.

### 2.3. *Landau theory; transitions with one-component order parameter*

Close to phase transition one can expand the thermodynamic potential density in powers of the order parameter. For systems with one-component order parameter $\eta$ we thus obtain

$$\phi = \phi_0 + \frac{A}{2}\eta^2 + \frac{B}{4}\eta^4 + \frac{C}{6}\eta^6 + \cdots + \frac{D}{2}(\nabla\eta)^2 + \cdots \qquad (1)$$

As to the form of the gradient term, the medium is assumed here to be isotropic or, if we are dealing with a crystal, it has cubic symmetry. The dots in expression (1) denote terms with higher powers of $\eta$ as well as with $\eta$ derivatives of higher degrees and orders as compared to those written out explicitly. If the function $\phi(\eta)$ were an analytic one, then in eq. (1) we could restrict ourselves to terms pointed out, at least, for small enough values of $\eta$. On the other hand, at the second-order phase transition (critical) point (i.e., at $T = T_c$) the function $\phi(\eta)$ is non-analytic, and as $T \to T_c$ more and more terms should be taken into account in eq. (1). In many cases, however, expression (1) with terms written out explicitly can be used even very close to $T_c$, particularly if the temperature dependence of coefficients is not concretized (see, e.g., Luban (1976), Ginzburg and Sobyanin (1976) and below, sect. 2.4). In such cases the coefficients $C$ and $D$ are to be considered positive near $T_c$, as follows from stability conditions, while the coefficient $A$ changes sign at the transition temperature, namely: $A > 0$ for $\tau \equiv (T - T_c)/T_c > 0$, $A < 0$ for $\tau < 0$, and $A = 0$ at $\tau = 0$.

The equilibrium value $\eta_e$ of the order parameter is determined by minimization of the total thermodynamic potential $\tilde{\phi}(\eta) \equiv \int \phi(\eta)\,dV$, i.e., for a spatially homogeneous system it is determined from the equation $\partial\phi(\eta)/\partial\eta = 0$. As a result, we find

$$\eta_e^2 = \frac{1}{2C}\left(-B + \sqrt{B^2 - 4AC}\right) \quad \text{for} \quad \tau < 0\,, \qquad (2)$$

and $\eta_e = 0$ for $\tau > 0$. It is easily seen that (for $\eta = \eta_e$)

$$\left(\frac{\partial^2\phi}{\partial\eta^2}\right)_{\eta=\eta_e} \equiv \phi_{\eta\eta} = 2\eta_e^2(B^2-4AC)^{1/2}, \qquad \tau<0\,, \tag{3}$$

and

$$\phi_{\eta\eta} = A, \qquad \tau>0\,. \tag{4}$$

Neglecting the interaction between (long-wavelength) fluctuations in $\eta$ (which corresponds in particular to the mean or molecular field approximation) one may put (near $T_c$) $A = A_0\tau$, and $B$, $C$, $D$, constant. Such a scheme is usually called the Landau phase transition theory*. In the framework of Landau theory one has $B>0$ for second-order phase transitions, $B<0$ for first-order phase transitions and $B=0$ at the tricritical point ($T=T_{tc}$, $p=p_{tc}$), which separates the second- and the first-order phase transition lines (see fig. 1). Near the

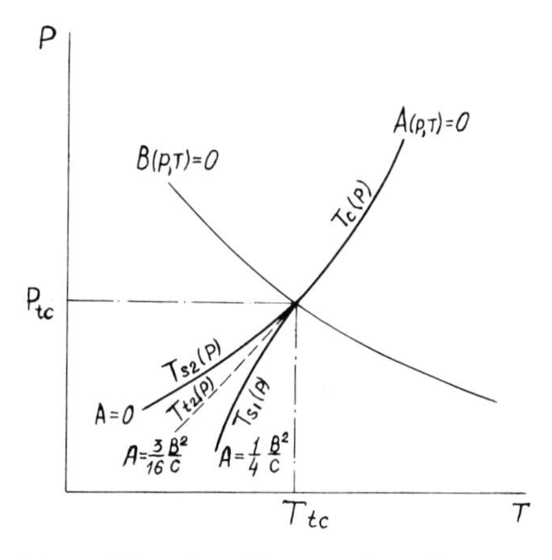


Fig. 1. Schematic picture of the phase diagram of a system near a tricritical point.

tricritical point, $B \propto (T_{tc}-T)$. For $B^2 \gg 4AC$, i.e., for second-order transitions far from the tricritical point,

$$\eta_e^2 \approx -A/B, \quad \tau<0\,, \tag{5}$$

and

$$\phi_{\eta\eta} \approx -2A, \qquad \tau<0\,. \tag{6}$$

For the tricritical point (if approaching it along the direction $B(p,T)=0$),

$$\eta_e^2 = (-A/C)^{1/2}, \qquad \tau<0\,, \tag{7}$$

*We should emphasize that the essence of the Landau theory consists not so much in concretizing the temperature dependence of the coefficients in eq. (1) as in an allowance for symmetry requirements (see, e.g., Landau and Lifshitz (1976)).

and

$$\phi_{\eta\eta} = -4A, \qquad \tau < 0 . \tag{8}$$

In the case of first-order transitions at a "thermodynamic" transition temperature ($T = T_{tr}$), where the thermodynamic potentials of the two phases are equal,

$$A(T_{tr}, p) = \tfrac{3}{16} B^2/C \quad \text{and} \quad \eta_e^2(T_{tr}, p) = -\tfrac{3}{4} B/C . \tag{9}$$

The boundary of stability (the spinodal line $T = T_{s2}(p)$ of the symmetrical phase) at which $\phi_{\eta\eta} = 0$, is determined from the condition (see eq. (4))

$$A(T_{s2}, p) = 0 , \tag{10}$$

and at the spinodal line ($T = T_{s1}(p)$) of the non-symmetrical phase (see eqs. (3) and (2))

$$A(T_{s1}, p) = \tfrac{1}{4} B^2/C, \qquad \eta_e^2(T_{s1}, p) = -B/2C . \tag{11}$$

When approaching the spinodal of the non-symmetrical phase, one has

$$\phi_{\eta\eta} \propto (B^2 - 4AC)^{1/2} \propto (T_{s1} - T)^{1/2} . \tag{12}$$

Note that this power law is different from the linear one, characterizing the $\phi_{\eta\eta}$-temperature dependence near the second-order phase transition or near the spinodal of the symmetrical phase.

### 2.4. *Limits of applicability of the Landau theory. Critical region*

In the immediate vicinity of an instability point, where $\phi_{\eta\eta} \to 0$, the thermal fluctuations in $\eta$ become large, and the Landau theory proves to be not applicable (Levanyuk 1959, Ginzburg 1960). The region, where the Landau theory is not applicable, is termed ordinarily a critical or scaling region. For the temperature width of the critical region from formula (82) one obtains the following estimate:

$$|\tau| \lesssim \left(\frac{3k_B T_c B}{8\pi A_0^2 r_{c0}^3}\right)^2 = \left(\frac{3}{16\pi}\right)^2 \left(\frac{k_B}{\Delta C_p d_0^3}\right)^2 \left(\frac{d_0}{r_{c0}}\right)^6 \equiv \tau_{crit} \tag{13}$$

where $d_0$ is an interparticle (interatomic) distance, $r_{c0} = (D/A_0)^{1/2}$ is the order parameter correlation radius extrapolated to $T = 0$, and the coefficients $A_0 = A/\tau$ and $B$ are those of the Landau theory for a second-order transition far from the tricritical one; $\Delta C_p = A_0^2/2BT_c$ is then the specific heat jump according to this theory (i.e., the difference of specific heat values at the boundaries between the Landau theory and critical region above and below $T_c$). Usually for structural phase transitions $k_B/\Delta C_p d_0^3 \sim 1$, and we see that the width of the critical region depends strongly on $r_{c0}$ (the value of the correlation radius extrapolated to $T = 0$). For example, if $r_{c0} = d_0$, then the width of the critical region $\tau_{crit} \simeq 4 \times 10^{-3}$ and for $r_{c0} = 4d_0$ it is of the order of $\tau_{crit} \sim 10^{-6}$, i.e., it is much smaller already. For different systems the values of $r_{c0}$ may be also quite different. For example, for superconductors $r_{c0} \sim 10^3\, d_0$ and for superfluid helium $r_{c0} \lessapprox d_0$. For magnetic systems and for a 'liquid–gas' critical point $r_{c0}$ is of the order of $d_0$ or even smaller than $d_0$ too. Here in many cases a critical region has undoubtedly been observed (see, e.g., Stanley (1971)). The values of $r_{c0}$ presented in literature for structural phase transitions are usually several times as large as the lattice constant. This fact together with a strong dependence of $\tau_{crit}$ on the ratio $r_{c0}/d_0$ explains possibly why in experiments for structural phase transitions the width of the critical region is probably very small (as far as we know for structural phase transitions the critical region has not yet been reliably observed). Formula (13) does not concern proper ferroelectrics with one axis of spontaneous polarization and proper ferroelastics. In these substances fluctuations of the order parameter are substantially suppressed due to effects of long-range forces. As a result, the Landau theory turns out to be applicable here with a high accuracy up to the phase transition temperature (Levanyuk 1965, Vaks et al. 1966, Levanyuk and Sobyanin 1970, Villain 1970).

In critical region the temperature dependence of the coefficients in formula (1) becomes more complicated. More or less strict conclusions can be drawn here only on the magnitudes of the critical exponents $\alpha$, $\beta$, $\gamma$, . . . (Stanley 1971, Patashinskii and Pokrovskii 1975), which describe the temperature dependence near $T_c$ of different physical quantities, for example,

$$\begin{aligned} &\phi_{\eta\eta}^{-1} \equiv \chi(0) \sim |\tau|^{-\gamma}, \qquad \eta_e(\tau < 0) \propto |\tau|^{\beta}, \\ &C_p = -T(\partial^2\phi_e/\partial T^2)_p \propto |\tau|^{-\alpha}, \qquad r_c = r_{c0}|\tau|^{-\nu}. \end{aligned} \tag{14}$$

According to modern concepts the values of critical exponents (or critical indices) do not depend on details of the microscopic interactions in the system and do depend only on the dimensionality of the system and its symmetry properties, in particular on the number of the order parameter components. Between the values of the critical indices there exist certain relations established by scaling theory (Stanley 1971, Patashinskii and Pokrovskii 1975, Ma 1976), for example,

$$\gamma + 2\beta = 2 - \alpha, \qquad 2 - \alpha = 3\nu\,. \tag{15}$$

The information about the values of critical indices for different systems obtained from experimental data and from numerical calculations for simple model systems is presented, e.g., by Ma (1976) and by Patashinskii and Pokrovskii (1975). Approximate analytical (field theoretical) methods for calculating critical indices have also been recently developed and are exposed in many review papers and books (see, e.g., Wilson and Kogut 1974, Fisher

1974, Ma 1976, Patashinskii and Pokrovskii 1982). Critical indices for quantities in the non-symmetrical phases are sometimes marked by primes (e.g., $\alpha'$, $\gamma'$ etc.). According to the scaling theory for second-order phase transitions the values of primed and non-primed critical indices coincide. It should be noted that in the three-dimensional case one can put approximately $\alpha \simeq 0$ and $\beta \simeq \frac{1}{3}$. From relations (15) we then find $\gamma = \frac{4}{3}$ and $\nu \simeq \frac{2}{3}$.

Although in the critical region one should, strictly speaking, take into account in eq. (1) the whole infinite sequence of terms of the series, it is often practically sufficient to consider only the terms written explicitly, as has been mentioned above. The temperature dependence of the coefficients $\phi_0$, $A$, $B$, $C$ and $D$ can easily be determined by using the above-mentioned definitions of critical indices and also paying attention to the fact that (as follows from the scaling theory) in equilibrium in non-symmetrical phase the terms of expression (1) should all be of the same order in $\tau$. Thus one obtains (for details, see e.g., Luban (1976))

$$A = A_0|\tau|^{\gamma} \operatorname{sgn} \tau, \qquad B = B_0|\tau|^{\gamma - 2\beta}, \qquad C = C_0|\tau|^{\gamma - 4\beta}, \tag{16}$$

where the coefficients $A_0$, $B_0$, $C_0$ can be in general different for $\tau > 0$ and $\tau < 0$.

The temperature dependence of the coefficient $D$ can be found exploiting the relation (see, for example, Landau and Lifshitz 1976, and eq. (31) below)

$$r_c = (D/\phi_{\eta\eta})^{1/2}. \tag{17}$$

From this and eq. (14) one finds $D = D_0|\tau|^{\gamma - 2\nu}$, so that for three-dimensional systems (where $\gamma \approx 2\nu$) the coefficient $D$ remains almost independent of temperature in critical region too. Under such an approach (used, for example, in the phenomenological theory of superfluid helium near the $\lambda$ point; see, Ginzburg and Sobyanin (1976, 1982)) all the expressions used below have formally (before concretizing the temperature dependence of the coefficients) the same form within the Landau theory (or classical region) as well as within the critical region.

As to phase transitions in solids the following important feature has to be borne in mind. In a number of papers (see e.g., Larkin and Pikin 1969, Wilson and Fisher 1972, Wallace 1973, Lyuksyutov and Pokrovskii 1975, Brazovskii and Dzialoshinskii 1975, Natterman and Trimper 1975, Imry 1974) the role of such factors as the shear deformations, crystal anisotropy and the presence of several components of the order parameter was investigated. It was shown that due to these factors phase transitions prove to become, as a rule, first-order transitions, even when these factors are neglected they would be second-order transitions. Therefore, second-order phase transitions in solids could be expected to occur, strictly speaking, only in some exceptional cases (e.g., in proper ferroelastics; Levanyuk and Sobyanin 1970). It should be emphasized, however, that all the above-mentioned factors are essential in the critical region only, which, as has already been said, is often very narrow. So, the corresponding jumps in the thermodynamic quantities, the temperature hysteresis effects, etc., may prove to be very small and difficult to reveal. From this point of view observation of tricritical points for structural phase transitions in some substances, such as $NH_4Cl$ (Garland and Weiner 1971) and $BaTiO_3$ (Clark and Benguigui 1977) is not so surprising. At such conditions the tricritical point may apparently be considered as the point in a phase diagram at which the line of "strong" first-order transitions (described by the Landau theory) passes into the line of "weak" first-order transitions caused by fluctuation effects and treated in the experiment as second-order transitions.

Note finally that the absence of evidence at the present time for the existence of a critical region for structural phase transitions does not exclude, of course, the discovery of such a region in the future with a progress of experimental techniques and improvements of specimen quality. Further on, however, attention will be concentrated mainly on the classical region.

### *2.5. Transitions with multi-component order parameters*

In the case of a multi-component order parameter one cannot write a unified expression of the type (1) for the thermodynamic potential density. The form of the expression depends on the transformational properties of the order parameter under symmetry transformations of the symmetrical phase. These transformational properties determine the form of the invariants composed of the components of the order parameter. These invariants are used in the expansion of the thermodynamic potential, which is also, naturally, an invariant quantity.

In the case of a one-component order parameter, invariants of any order are powers of the quadratic invariant $\eta^2$. For a multi-component order parameter this is not the case, and there exists, generally speaking, a variety of high-order invariants. An insight into the possibilities arising here can be provided by the examples of transitions with a two-component order parmeter.

Let us denote components of the order parameter by $\eta_1$ and $\eta_2$. To find the invariants, we shall first consider the transformational properties of combinations of different powers of the quantities $\eta_1$ and $\eta_2$, which transform by the irreducible representaion of the symmetry group of the symmetrical phase. As is proved in the group theory, using three independent quadratic combinations $\eta_1^2$, $\eta_2^2$ and $\eta_1\eta_2$, one and only one bilinear combination can be composed. With an appropriate choice of $\eta_1$ and $\eta_2$, this invariant combination is $\eta_1^2+\eta_2^2$.

The quantities $\eta_1^2-\eta_2^2$ and $\eta_1\eta_2$ form a two-dimensional representation. For further analysis we shall consider separately the following three possibilities: (i) the above-mentioned representation is reducible and may be divided into two different (i.e., non-equivalent) one-dimensional representations, which corresponds, e.g., to the bases $\eta_1^2-\eta_2^2$ and $\eta_1\eta_2$; (ii) the same as (i) but the two one-dimensional representations are equivalent; (iii) the representation in question is irreducible.

We shall not discuss third-order invariants since according to the Landau theory, which we mainly follow below, if such invariants are admitted by the symmetry, the phase transition will be of first order.

Going over to fourth-order invariants, we note that in the case (i) the quantities $(\eta_1^2 - \eta_2^2)^2$ and $\eta_1^2\eta_2^2$ are invariants (in our example) and, naturally, the quantity $(\eta_2^2 + \eta_2^2)^2$ is also invariant. One of the three above-mentioned invariants can be expressed in the form of a linear combination of the other two, so that there exist, in fact, two independent fourth-order invariants, which can be represented, for example, by $(\eta_1^2 + \eta_2^2)^2$ and $\eta_1^2\eta_2^2$.

In the case (ii) there appears an additional fourth-order invariant: $\eta_1\eta_2(\eta_1^2 - \eta_2^2)$, i.e., there exist three independent fourth-order invariants.

In case (iii), in order to construct a fourth-order invariant, one should take an appropriate quadratic combination of the basis vectors of the indicated two-dimensional representation (in our example $\eta_1\eta_2$ and $\eta_1^2 - \eta_2^2$). Note that if for this quadratic combination we take $(\eta_1^2 - \eta_2^2)^2 + 4\eta_1^2\eta_2^2$ we obtain $(\eta_1^2 + \eta_2^2)^2$; i.e., a definitely invariant quantity, which is the quadratic invariant squared. Since one can obtain only a single invariant quadratic combination on the basis of an irreducible representation, in case (iii) there exists only one fourth-order invariant.

Now let us write the corresponding expressions for the thermodynamic potential. In case (i) we have (disregarding gradient terms)

$$\phi = \phi_0 + \tfrac{1}{2}A(\eta_1^2 + \eta_2^2) + \tfrac{1}{4}B_1(\eta_1^2 + \eta_2^2)^2 + 2B_2\eta_1^2\eta_2^2 . \tag{18}$$

It is convenient to pass over to "polar" coordinates on the plane $(\eta_1, \eta_2)$, i.e., to put

$$\eta_1 = \rho \cos \varphi, \qquad \eta_2 = \rho \sin \varphi . \tag{18a}$$

Then expression (18) takes the form

$$\phi = \phi_0 + \tfrac{1}{2}A\rho^2 + \tfrac{1}{4}(B_1 + B_2)\rho^4 - \tfrac{1}{4}B_2\rho^4 \cos 4\varphi . \tag{19}$$

The equilibrium values of $\rho$ and $\varphi$ are found from the condition of the minimum of $\phi$. Minimizing eq. (19) first over $\varphi$, we obtain

$$\sin 4\varphi_e = 0 . \tag{20}$$

On the plane $(\eta_1\eta_2)$ (see fig. 2(a)) this equation defines 8 directions. Four of these correspond to the minimum value of the thermodynamic potential. Which four directions they are, depends on the sign of the coefficient $B_2$. Minimizing eq. (19) with respect to $\rho$, we find the distance from the reference point to the one corresponding to the equilibrium values of $\eta_1$ and $\eta_2$. Dots stand for the equilibrium values of $\eta_1$ and $\eta_2$ for the case $B_2 > 0$. The other four points (circles) correspond to the non-symmetrical phase if $B_2 < 0$. The set of symmetry elements of this phase is other than that for the case $B_2 > 0$. An arbitrary value $\varphi$ corresponds to the third phase (phase III in fig. 2(b)), whose symmetry is still lower than for the first two (phases I and II). To consider phase

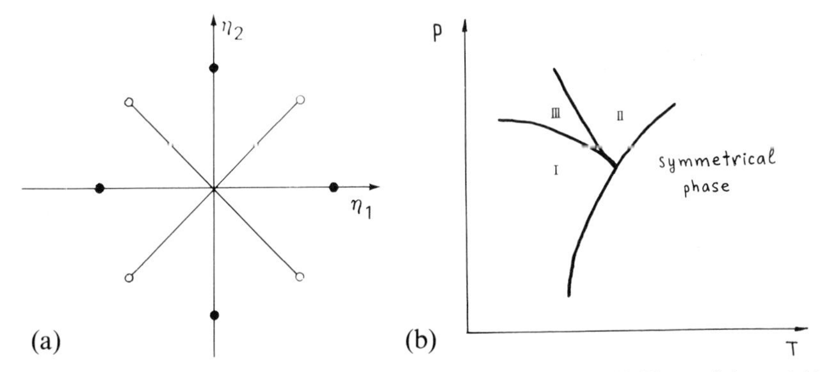


Fig. 2. Phases of a system with a two-component order parameter. (a) Plane of the variables $\eta_1$ and $\eta_2$: dots and circles represent equilibrium values of the order parameter below $T_c$ for the cases $B_2 > 0$ and $B_2 < 0$, respectively. (b) $(pT)$ diagram of the system.

III one should take into account higher-order invariants. We shall not do this here since a second-order transition into this latter phase is possible only at a single point on the $(pT)$ diagram (see Levanyuk and Sannikov 1974).

In case (ii) the thermodynamic potential has the form

$$\phi = \phi_0 + \tfrac{1}{2}A\rho^2 + \tfrac{1}{4}(B_1 + B_2)\rho^4 - \tfrac{1}{4}B_2\rho^4 \cos 4\varphi + \tfrac{1}{4}B_3\rho^4 \sin 4\varphi \,. \tag{21}$$

Minimization of eq. (21) with respect to $\varphi$ now leads to the equation

$$\tan 4\varphi_e = - B_3/B_2 \,. \tag{22}$$

Thus, in this case the polar angle for the point representing an equilibrium state in the $\eta_1\eta_2$-plane is not fixed and can change with temperature since the coefficients $B_3$ and $B_2$ are, in principle, temperature dependent. This means that the phase of one and the same symmetry now corresponds to all the points in the $\eta_1\eta_2$-plane.

In case (iii), the thermodynamic potential does not depend on $\varphi$ at all if only the fourth-order terms are taken into account. Such a dependence does not appear in terms of higher orders either for the so-called degenerate systems. Examples of such systems are superfluid helium, superconductors, and in the case of structural phase transitions – crystals with incommensurate phases, considered in ch. 3.

In the present chapter we deal only with commensurate transitions, for which the dependence of $\phi$ on $\varphi$ certainly appears when higher-order terms are taken into account. Suppose that the dependence of $\phi$ on $\varphi$ appeared in terms of the $n$th order. As above, we may face two possibilities here: there appears only one additional invariant of the form $\rho^n \cos n\varphi$, or two invariants $\rho^n \cos n\varphi$ and $\rho^n \sin n\varphi$. In the former case there exist three non-symmetrical phases, in the latter – only one.

When studying anomalies of thermodynamical characteristics (various generalized moduli) of the systems, one should take into account the dependence of the thermodynamic potential not only on the order parameter, but also on corresponding generalized deformations. Here, along with invariants of the type $(\eta_1^2+\eta_2^2)\xi=\rho^2\xi$, where, like in the case of one-component order parameter, $\xi$ is a totally symmetric generalized deformation, the invariants of the type $\eta_1\eta_2\zeta$ are also possible, where $\zeta$ is a symmetry breaking deformation. Under phase transition both the moduli corresponding to $\xi$ as well as those corresponding to $\zeta$ change in a jump-like way (within the Landau theory).

Now let us touch upon the question of phase stability conditions. For a symmetrical phase it is enough to have only one stability condition, namely the positiveness of the coefficient $A$ in formulae (18), (19) and (21), just like in the case of a one-component order parameter. When investigating a non-symmetrical phase, it is instructive to note that the derivative $\phi_{\rho\varphi}\equiv\partial^2\phi/\partial\rho\partial\varphi=0$ at $\varphi=\varphi_e$, so that the conditions of non-symmetrical phase stability take the form

$$\phi_{\rho\rho}>0 \quad \text{and} \quad \phi_{\varphi\varphi}>0\,. \tag{23}$$

Concrete expressions for $\phi_{\rho\rho}$ and $\phi_{\varphi\varphi}$ are different in different cases, and we shall present them in what follows when necessary.

## 3. Fluctuations near phase transitions

To calculate the light scattering intensity near phase transitions, one should know the mean-square fluctuations of the order parameter as well as of some other quantities, i.e., to have expressions for averages of the type $\langle|\eta(\boldsymbol{q})|^2\rangle$ etc. (see, for example, formula (112)), where $\boldsymbol{q}$ is the transferred wave vector and the brackets $\langle\ldots\rangle$ denote averaging over an appropriate statistical ensemble. On the other hand, when interested in the spectrum of scattered light, one should obtain expressions for averages, like $\langle|\eta(\boldsymbol{q},\Omega)|^2\rangle$, where $\Omega$ is the frequency change due to scattering. This can be done if the state of the scattering medium is known. It will be assumed below that the medium is in equilibrium (or in any case the fluctuations in question are the same as in a thermodynamically equilibrium state including metastable states, such as those of overcooling and overheating). In most cases this assumption holds in experiment. Moreover, fluctuations will be treated as the classical ones, which is admissible under the condition $\hbar\Omega\ll k_BT$ (here $k_B$ is the Boltzmann constant).

In the present section we will be interested in fluctuations (before their spectral decomposition) determining the total (integrated over frequencies) light scattering intensity (see sect. 4). Such fluctuations can be calculated from the expression for an incomplete equilibrium thermodynamic potential $\tilde{\Phi}(\ldots,\eta_\alpha)=\int\phi\,\mathrm{d}V$ (or, specifically, Gibbs free energy), depending on components of the order parameter $\eta_\alpha$ as well as extra variables, such as temperature

$T$, density $\rho$, some inner (optical) lattice deformations, etc., these variables being denoted by dots.

The main role is played below by the mean-square fluctuations $\langle \Delta\xi_i \Delta\xi_j \rangle = \langle \xi_i \xi_j \rangle - \langle \xi_i \rangle \langle \xi_j \rangle$ of a few variables $\xi_1, \xi_2, \ldots$. They are calculated from the standard formula (Leontovich 1944, Landau and Lifshitz 1976)

$$\langle \Delta\xi_i \Delta\xi_j \rangle = k_B T \tilde{\Phi}_{ij}^{-1}, \tag{24}$$

where $\tilde{\Phi}_{ij}^{-1}$ is the tensor inverse to $\tilde{\Phi}_{ij} = \partial^2 \tilde{\Phi} / \partial\xi_i \partial\xi_j$. The quantities $\tilde{\Phi}_{ij}^{-1}$ are referred to as generalized static susceptibilities and are designated by $\chi_{ij}$ or $\chi_{ij}(\boldsymbol{q})$, when one is dealing with Fourier transforms of $\Delta\xi_i$ with different values of the wavevector $\boldsymbol{q}$.

### 3.1. Order parameter fluctuations

Consider first fluctuations in a one-component order parameter $\eta$ in the symmetrical phase. Since in this phase the equilibrium value of $\eta$ is equal to zero and, therefore, the $\eta$ fluctuation, $\Delta\eta = \eta - \eta_e$, coincides with the value of $\eta$ itself, we shall take into account in the expression for the thermodynamic potential density only terms quadratic in $\eta$ (Landau 1937)

$$\Phi = \Phi_0 + \tfrac{1}{2} A \eta^2 + \tfrac{1}{2} D (\nabla \eta)^2 + \ldots . \tag{25}$$

Passing over to Fourier transforms of the function $\Delta\eta(\boldsymbol{r})$

$$\Delta\eta(\boldsymbol{r}) = \sum_{\boldsymbol{k}} \eta(\boldsymbol{k})\, e^{i\boldsymbol{k}\boldsymbol{r}}, \qquad \eta(-\boldsymbol{k}) = \eta^*(\boldsymbol{k}), \qquad \eta(\boldsymbol{k}) = \frac{1}{V} \int \Delta\eta\, e^{-i\boldsymbol{k}\boldsymbol{r}}\, d\boldsymbol{r}, \tag{26}$$

we have for the thermodynamic potential

$$\begin{aligned} \tilde{\Phi} &= \int \Phi\, dV = \tfrac{1}{2} V \sum_{\boldsymbol{k}} A(\boldsymbol{k}) \eta(\boldsymbol{k}) \eta(-\boldsymbol{k}) \\ &= \tfrac{1}{2} V \sum_{\boldsymbol{k}} (A + Dk^2 + \ldots) \eta(\boldsymbol{k}) \eta(-\boldsymbol{k}), \end{aligned} \tag{27}$$

where the dots correspond to terms with higher powers of $k$.

At $A = 0$ (i.e., at the stability loss temperature of the symmetrical phase, including the temperature of a second-order transition) the function $A(\boldsymbol{k})$ is, in general, not analytic and has the form

$$A(k) \sim k^{2-\hat{\eta}}, \tag{28}$$

where $\hat{\eta}$ is one of the critical exponents (this exponent is designated usually by $\eta$, but we take $\eta$ for the order parameter rather than for one of the critical exponents). However, for three-dimensional systems the quantity $\hat{\eta}$ appears to be very small. According to renormalization group calculations, for the Ising model, $\hat{\eta} \simeq 0.04$, while for the Heisenberg model, $\hat{\eta} \simeq 0.03$ (see e.g., Landau and

Lifshitz (1976), Pokrovskii and Patashinskii (1982), Ma (1976)). Neglecting the small distinction of $\hat{\eta}$ from zero, we shall put further on approximately

$$A(k) = A + Dk^2, \tag{29}$$

both in the classical and in the critical regions, but for $k \lesssim r_{c0}^{-1}$.

Now let us consider the $\eta$ fluctuation with a given value of $\boldsymbol{k} = \boldsymbol{q}$. Note that in eq. (27) and in analogous sums the term with $\eta(\boldsymbol{q})\eta(-\boldsymbol{q})$ is met twice since summation is carried out over all $\boldsymbol{k}$'s. With allowance for this fact and using formula (24), we find

$$\langle|\eta(\boldsymbol{q})|^2\rangle \equiv \langle\eta(\boldsymbol{q})\eta(-\boldsymbol{q})\rangle = k_\mathrm{B}T\left[\frac{\partial^2\tilde{\Phi}}{\partial\eta(\boldsymbol{q})\partial\eta(-\boldsymbol{q})}\right]^{-1} = \frac{k_\mathrm{B}T}{V}\frac{1}{A + Dq^2}. \tag{30}$$

As has already been mentioned, for the classical region, $D = \text{const}$. If we assume $\hat{\eta} = 0$, the coefficient $D$ is independent of temperature in the critical region too. So, as follows from eq. (30), the fluctuations with $q \lesssim r_c^{-1}$ have the most pronounced temperature dependence, the latter being determined by the temperature dependence of the coefficient $A \equiv \chi_{\eta\eta}^{-1}(0) \sim |\tau|^{\gamma}$. In the Landau theory, $\gamma = 1$. For the critical region, $\gamma > 1$. According to the results of numerical calculations (see, e.g., Pokrovskii and Patashinskii (1982), Ma (1976)) $\gamma \simeq 1.25$ for a three-dimensional Ising model, $\gamma \simeq 1.43$ for the Heisenberg model, and $\gamma \simeq 1.33$ for the "planar" ("XY") model. On the other hand, for the tricritical transition the temperature dependence of $\chi_{\eta\eta}(0)$ is the same (apart from a logarithmic factor) as in the Landau theory (Bausch 1972, Wegner and Riedel 1973), i.e., for this transition $\gamma = 1$.

What has been said is easily extended to the non-symmetrical phase. For this it is sufficient to replace in eqs. (27)–(30) $A$ by $\phi_{\eta\eta}$ and $A(\boldsymbol{q})$ by $\phi_{\eta\eta}(\boldsymbol{q}) = \phi_{\eta\eta} + Dq^2$. Thus for $\langle|\eta(\boldsymbol{q})|^2\rangle$ we have,

$$\langle|\eta(\boldsymbol{q})|^2\rangle = \frac{k_\mathrm{B}T}{V}\frac{1}{\phi_{\eta\eta} + Dq^2}, \tag{31}$$

where $\phi_{\eta\eta}$ is given by eqs. (6), (8) and (12) for second-order transitions, for the tricritical point, and for the spinodal line $T_{s1}(p)$ in the case of first-order transitions, respectively. Formula (31) is valid for the symmetrical phase as well, in which case $\phi_{\eta\eta} = A$ (see eq. (4)).

### 3.2. *Coupling between order parameter fluctuations and those in other variables; spatially homogeneous fluctuations*

Near the transition point not only fluctuations in the order parmeter increase but also those in other variables as far as these variables are coupled with $\eta$. When calculating the light scattering intensity, these fluctuations should be taken into account too. We shall begin our discussion with the simplest example, considering besides $\eta$ one more fluctuating quantity $\xi$ only, whose origin will not be concretized for the present. We shall abstract ourselves from the $\boldsymbol{q}$ dependence of the corresponding fluctuations on the wavevector $\boldsymbol{q}$ (this dependence, very essential for solids, will be taken into account in the next section 3.3). In other words, it is spatially homogeneous fluctuations with $\boldsymbol{q} = 0$ that will be considered here. At these conditions the change in the density of the thermodynamical potential caused by $\eta$ and $\xi$ fluctuations is determined by the expression (up to the quadratic terms sufficient for our aim)

$$\Delta\phi = \tfrac{1}{2}\phi^{\xi}_{\eta\eta}(\Delta\eta)^2 + \phi_{\eta\xi}\Delta\eta\Delta\xi + \tfrac{1}{2}\phi^{\eta}_{\xi\xi}(\Delta\xi)^2 , \tag{32}$$

where

$$\phi^{\xi}_{\eta\eta} \equiv \left.\frac{\partial^2\phi}{\partial\eta^2}\right|_{\xi,p,T}, \qquad \phi_{\eta\xi} \equiv \left.\frac{\partial^2\phi}{\partial\eta\partial\xi}\right|_{p,T}, \qquad \phi^{\eta}_{\xi\xi} \equiv \left.\frac{\partial^2\phi}{\partial\xi^2}\right|_{\eta,p,T}, \tag{33}$$

and all the derivatives are to be calulated for $\xi = \xi_e$, $\eta = \eta_e$, i.e., in equilibrium.

Proceeding from eq. (32) the mean square fluctuations in $\eta$ and $\xi$ may be calculated by the general formula (24). But for our purposes it is more convenient to take another way (Levanyuk and Sobyanin 1967): namely, represent the quantity $\xi$ as

$$\xi = \xi_0(\eta) + \xi' \tag{34}$$

where $\xi_0(\eta)$ is an "equilibrium" value of $\xi$ corresponding to a given fixed value of $\eta$. The quantity $\xi_0(\eta)$ is found from the equation

$$\left.\frac{\partial\phi}{\partial\xi}\right|_{\xi = \xi_0(\eta)} = 0 .$$

Differentiating both sides of this equality with respect to $\eta$ we find (at $\eta = \eta_e$)

$$\phi_{\eta\xi} + \phi^{\eta}_{\xi\xi}\left.\frac{d\xi_0}{d\eta}\right|_{\eta = \eta_e} = 0 ,$$

whence

$$\left.\frac{d\xi_0}{d\eta}\right|_{\eta = \eta_e} = -\frac{\phi_{\xi\eta}}{\phi^{\eta}_{\xi\xi}} . \tag{35}$$

The $\xi$ fluctuation is expressed through $\eta$ and $\xi'$ fluctuations as follows:

$$\Delta\xi = \left.\frac{d\xi_0}{d\eta}\right|_{\eta = \eta_e}\Delta\eta + \Delta\xi' = -\frac{\phi_{\eta\xi}}{\phi^{\eta}_{\xi\xi}}\Delta\eta + \Delta\xi' . \tag{36}$$

Substituting (36) into (32) one obtains

$$\Delta\phi = \tfrac{1}{2}\phi^{g}_{\eta\eta}(\Delta\eta)^2 + \tfrac{1}{2}(\Delta\xi')^2 , \tag{37}$$

where

$$\phi^{g}_{\eta\eta} = \phi^{\xi}_{\eta\eta} - (\phi_{\eta\xi})^2/\phi^{\eta}_{\xi\xi} . \tag{38}$$

Thus the transformation (34) diagonalizes the quadratic form (32). From eqs. (37) and (24) we obtain immediately

$$\langle(\Delta\eta)^2\rangle = \frac{k_B T}{V}\frac{1}{\phi^g_{\eta\eta}}. \tag{39}$$

It is easily seen that the coefficient $\phi^g_{\eta\eta}$ has the meaning of the derivative $\partial^2\phi/\partial\eta^2$ calculated at a constant (zero) generalized force $g$ conjugate to the variable $\xi$. Indeed, let

$$g \equiv \left.\frac{\partial\phi}{\partial\xi}\right|_{\eta,p,T} = 0\,,$$

then

$$\left.\frac{\partial\phi}{\partial\eta}\right|_{g,p,T} = \left.\frac{\partial\phi}{\partial\eta}\right|_{\xi,p,T} + \left.\frac{\partial\phi}{\partial\xi}\right|_{\eta,p,T}\left.\frac{\partial\xi}{\partial\eta}\right|_{g,p,T} = \left.\frac{\partial\phi}{\partial\eta}\right|_{\xi,p,T}$$

and (in equilibrium)

$$\left.\frac{\partial^2\phi}{\partial\eta^2}\right|_{g,p,T} = \left.\frac{\partial^2\phi}{\partial\eta^2}\right|_{\xi,p,T} + \left.\frac{\partial^2\phi}{\partial\eta\partial\xi}\right|_{p,T}\left.\frac{\partial\xi}{\partial\eta}\right|_{g,p,T} = \phi^{\xi}_{\eta\eta} - (\phi_{\eta\xi})^2/\phi^{\eta}_{\xi\xi} = \phi^g_{\xi\xi}\,.$$

Thus, as to the $\eta$ fluctuations, the account of their coupling with fluctuations of other variables results essentially in imposing an appropriate meaning on the derivative $\phi_{\eta\eta}$ in expression (31). Namely, this derivative should be calculated at constant (zero) generalized forces conjugate to all other variables.

In a similar way, assuming $\eta = \eta_0(\xi) + \eta'$ (cf. eq. (34)), it is easy to find

$$\langle(\Delta\xi)^2\rangle = \frac{k_B T}{V}\frac{1}{\phi^h_{\xi\xi}}, \tag{40}$$

where $\phi^h_{\xi\xi} \equiv \partial^2\phi/\partial\xi^2|_{h,p,T} = \phi^{\eta}_{\xi\xi} - (\phi_{\eta\xi})^2/\phi^{\xi}_{\eta\eta}$, and $h$ is the force conjugate to $\eta$.

Using eq. (36) and taking into account that the variables $\eta$ and $\xi'$ are statistically independent, for $\langle(\Delta\xi)^2\rangle$ one can also obtain the expression

$$\langle(\Delta\xi)^2\rangle = (\mathrm{d}\xi_0/\mathrm{d}\eta)^2_{\eta=\eta_e}\langle(\Delta\eta)^2\rangle + \langle(\Delta\xi')^2\rangle\,. \tag{41}$$

Notice that near the transition point the first term of this formula has the strongest temperature dependence. From eq. (36) for the mixed fluctuation $\langle\Delta\eta\Delta\xi\rangle$ we find

$$\langle\Delta\eta\Delta\xi\rangle = \left(\frac{\mathrm{d}\xi_0}{\mathrm{d}\eta}\right)_{\eta=\eta_e}\langle(\Delta\eta)^2\rangle = -\frac{\phi_{\eta\xi}}{\phi^{\eta}_{\xi\xi}}\langle(\Delta\eta)^2\rangle\,. \tag{42}$$

The variable $\xi$ may have various meanings depending on a concrete situation. For example, in the non-symmetrical phase spatially homogeneous fluctuations in the $\eta$ are always coupled linearly with the fluctuations in the volume deformation $v \equiv u_{ll} = -\Delta\rho/\rho$. This coupling may be taken into account by writing the density of the thermodynamic potential in the form

$$\phi = \phi_0 + \frac{A}{2}\eta^2 + \frac{B_1}{4}\eta^4 + \frac{C}{6}\eta^6 + \frac{r}{2}\eta^2 v + \frac{K}{2}v^2, \tag{43}$$

where $K$ is the modulus of hydrostatic compression (at fixed $\eta$), and we have omitted gradient terms which are insignificant for what follows*. The coupling constant $r$ between $\eta$ and $v$ is equal to

$$r \equiv \left.\frac{\partial A}{\partial v}\right|_{v=0} = \rho \frac{\partial A}{\partial T}\frac{\mathrm{d}T_c}{\mathrm{d}\rho}. \tag{44}$$

Expression (43) comes down to eq. (1) by minimization with respect to $v$. For the equilibrium value of the deformation $v_0(\eta)$ at a given fixed value of $\eta$ one thus obtains

$$v_0(\eta) = -\frac{r}{2K}\eta^2. \tag{45}$$

After substituting this expression in (43) we find that in eq. (1) the coefficient $B$ is

$$B = B_1 - r^2/2K. \tag{46}$$

So, the values of the coefficient $B$ in a "free" ($p = 0$) and "clamped" ($v = 0$) crystal differ by the quantity $r^2/2K$. The expression (31) for fluctuations $\langle(\Delta\eta)^2\rangle$ then still holds (as is seen from eq. (39)) if in eq. (31) one puts $\phi_{\eta\eta} = \phi^p_{\eta\eta}$, i.e., one calculates the derivative $\phi_{\eta\eta}$ not at a constant volume deformation $v$ but at a constant (zero) pressure. To calulate the fluctuations $\langle(\Delta v)^2\rangle$ and $\langle\Delta v\Delta\eta\rangle$ one can use eqs. (33) and (40)–(42) in which one should put now

$$\phi^v_{\eta\eta} = A + 3B_1\eta_e^2 + 5C\eta_e^4 + rv_e = \eta_e^2(2\sqrt{B^2 - 4AC} + r^2/K), \tag{47}$$

$$\phi^\eta_{vv} = K, \qquad \phi_{\eta\xi} = \phi_{\eta v} = r\eta_e. \tag{48}$$

With the aid of these expressions we find

$$\langle(\Delta v)^2\rangle = \frac{k_B T}{V\phi^h_{vv}} = \frac{k_B T}{V}\frac{(B^2 - 4AC)^{1/2} + (r^2/2K)}{K(B^2 - 4AC)^{1/2}}$$

$$= \frac{k_B T}{V}\left(\frac{r^2}{2K^2(B^2 - 4AC)^{1/2}} + \frac{1}{K}\right) = \frac{r^2\eta_e^2}{K^2}\langle(\Delta\eta)^2\rangle + \frac{k_B T}{VK}, \tag{49}$$

$$\langle\Delta v\Delta\eta\rangle = -\frac{r\eta_e}{K}\langle(\Delta\eta)^2\rangle = -\frac{k_B T}{V}\frac{r}{2\eta_e K(B^2 - 4AC)^{1/2}}. \tag{50}$$

In the framework of the Landau theory the coefficients $r$ and $K$ can be taken

*The terms of the form $\frac{1}{4}r_1\eta^4 v$ and $\frac{1}{4}r_2\eta^2 v^2$ are also omitted in eq. (43). Such terms if included would only lead to some renormalization of the coefficient $C$ in the formulae written below.

as temperature independent. It follows then from eq. (50) using eqs. (31) and (2) that $\langle(\Delta v \Delta\eta)\rangle \propto |\tau|^{-1/2}$ for second-order transitions far from the tricritical point, $\langle(\Delta v \Delta\eta\rangle \sim |\tau|^{-3/4}$ for the tricritical point, and $\langle \Delta v \Delta\eta \rangle \sim \langle(\Delta\eta)^2\rangle \sim \langle(\Delta v)^2\rangle \sim (T_{s1} - T)^{-1/2}$ in the region near the non-symmetrical phase spinodal line. At the same time according to eq. (49) the fluctuation $\langle(\Delta v)^2\rangle$ does not increase as $|\tau| \to 0$ for second-order transitions far from the tricritical point, and $\langle(\Delta v)^2\rangle \sim |\tau|^{-1/2}$ near the tricritical point (when approaching it along directions with $B^2 \ll 4AC$).

In the critical region the temperature dependence of the fluctuation $\langle \Delta\eta \Delta v \rangle$ may be found from the following simple coinsiderations. As is clear from eqs (40) and (49),

$$\langle(\Delta v)^2\rangle = \frac{k_B T}{V} \frac{1}{\phi^h_{vv}}, \tag{51}$$

where $(\phi^h_{\eta\eta})^{-1} \equiv -(1/V)(\partial V/\partial p)_T \equiv \beta_T$ is the isothermal compressibility. The critical exponent of this quantity is the same as that of the specific heat (see, e.g., Landau and Lifshitz (1976)), i.e., $\langle(\Delta v)^2\rangle \sim \beta_T \sim |\tau|^{-\alpha}$. On the other hand as follows from eq. (41),

$$\langle(\Delta v)^2\rangle \simeq (\mathrm{d}v_0/\mathrm{d}\eta)^2_{\eta=\eta_e} \langle(\Delta\eta)^2\rangle , \tag{52}$$

and, therefore,

$$(\partial v_0/\partial \eta)^2_{\eta=\eta_e} \simeq \langle(\Delta v)^2\rangle / \langle(\Delta\eta)^2\rangle \sim |\tau|^{\gamma-\alpha} . \tag{53}$$

Thus, in the critical region,

$$\langle \Delta\eta \Delta v \rangle = (\mathrm{d}v_0/\mathrm{d}\eta)_{\eta=\eta_e} \langle(\Delta\eta)^2\rangle \sim |\tau|^{-(\alpha+\gamma)/2} . \tag{54}$$

As is clear from the above example, in the non-symmetrical phase there always exist variables whose fluctuations are linearly coupled with order parameter ones. Besides the density $\rho$ they are represented by lattice deformations corresponding to the normal coordinate of fully symmetric optical phonon modes in the symmetrical phase, by concentrations of mixture components in solid solutions, and so on. In the case of a multi-component order parameter these variables may have zero equilibrium values in the symmetrical phase while their transformation properties can differ from those of the order parameter. They are, for example, polarization and shear deformation in the cases of improper ferroelectrics and improper ferroelastics, respectively.

On the other hand, in the symmetrical phase the situation, when the $\eta$ fluctuations are coupled linearly with those of other variables, is encountered far more rarely. Namely, it is necessary for this that the corresponding variable (or variables) be transformed according to the same irreducible representation of the symmetry group of the symmetrical phase as the parameter $\eta$ itself. Let us consider, for example, two linearly coupled variables $\eta$ and $\xi$. Then the second-order transition temperature $T_c$ (or the spinodal temperature $T_{s1}$ in the case of a first-order transition) is determined from the condition $\phi^g_{\eta\eta} = \phi^\xi_{\eta\eta} - (\phi_{\eta\xi})^2/\phi^\eta_{\xi\xi} = 0$ (see eqs. (38) and (39)). We see that if $\phi_{\eta\xi}$ does not go to zero as $T \to T_c$, none of the coefficients $\phi^\xi_{\eta\eta}$ and $\phi^\eta_{\xi\xi}$ vanishes at the transition point. The bilinear form (32) might be, of course, diagonalized, and the corresponding linear combination of $\eta$ and $\xi$ might be regarded as the order parameter. However, such a diagonalization is not always worthwhile. For example, in the case of $KH_2PO_4$, where one can set $\eta = P_z$ ($z$ component of the polarization vector) and $\xi = u_{xy}$ (component of the strain deformation tensor), it turns out that it is the quantity $\phi^\xi_{\eta\eta} \equiv (\partial^2\phi/\partial P_z^2)_{u_{xy}}$ alone that changes essentially with temperature near the phase transition while the coefficient $\phi^\eta_{\xi\xi} \equiv (\partial^2\phi/\partial u_{xy}^2)_{P_z}$ remains practically constant. For this reason it is just the polarization vector component $P_z$ that is usually taken for the order parameter in $KH_2PO_4$ (Jona and Shirane 1962).

We would like to emphasize finally that the formulae for fluctuations contain not the "bare" coefficients $\phi^\xi_{\eta\eta}$ and $\phi^\eta_{\xi\xi}$ but the "renormalized" quantities $\phi^g_{\eta\eta}$ and $\phi^h_{\xi\xi}$ (see eqs. (38)–(40)) which in the case of $KH_2PO_4$ have the meaning of the inverse dielectric permeability $(\partial^2\phi/\partial P_z^2)_{\sigma_{xy}}$ calculated at constant (zero) stress $\sigma_{xy}$ and the elastic modulus $(\partial^2\phi/\partial u_{xy}^2)_{E_z}$ calculated at constant electric field $E_z$. Since the coefficient $\phi_{\eta\xi}$ does not go to zero as $\tau \to 0$, the temperature dependence of both $P_z$ and $u_{xy}$ fluctuations obey one and the same law below and above $T_c$.

### 3.3. *Spatially inhomogeneous fluctuations; long-range force effects*

Up to now we have considered spatially homogeneous fluctuations. In the absence of long-range forces an extension of the corresponding expressions to the spatially inhomogeneous case is rather evident and consists in adding to the expressions the terms proportional to $q^2$ by analogy with obtaining formula (31). However if fluctuations of the parameter $\eta$ (or variables coupled linearly with it) are accompanied by the appearance of long-range forces fields (electric, magnetic, elastic), passing over to the spatially inhomogeneous case is not trivial. At the same time such fields appear always in solids since in solids (distinct from liquids) any inhomogeneous change of density causes shear deformations which decay slowly with distance.

#### 3.3.1. *Isotropic solid; the role of shear strains*

We discuss the effect of shear strains on the spatially-inhomogeneous $\eta$ and $\rho$ fluctuations by considering the simplest case of an isotropic solid in the presence of a one-component order parameter. The thermodynamic potential density of such a solid is written in the form (Levanyuk 1974, Ginzburg and Levanyuk 1974, 1976)

$$\phi = \phi_0 + \frac{A}{2}\eta^2 + \frac{B_1}{4}\eta^4 + \frac{C}{6}\eta^6 + \frac{r}{2}\eta^2 v + \frac{K}{2}v^2 + \mu(u_{ij} - \tfrac{1}{3}\delta_{ij}v)^2. \qquad (55)$$

Here $u_{ij} = \frac{1}{2}(\partial u_i/\partial x_j + \partial u_j/\partial x_i)$ are components of the strain tensor, $v = \operatorname{div} \boldsymbol{u} \equiv u_{ll}$ is the volume deformation, $\mu$ is the shear modulus, and the sums are taken over the repeated indices. Terms with gradients in $\eta$ and $v$ are omitted in eq. (55). These terms play a negligible role for the long-wavelength fluctuations (with $q \ll r_c^{-1}$), which we are interested in.

Expression (55) differs from that for a spatially homogeneous case (eq. (43)) only by the term $\mu(u_{ij} - \frac{1}{3}\delta_{ij}u_{ll})^2$. Notice now that for a spatially inhomogeneous $v$ fluctuation with the wavevector $\boldsymbol{q}$, it is only a single strain tensor component $u_{zz} = \partial u_z/\partial z \equiv u$ (with the $z$-axis directed along $\boldsymbol{q}$) that is non-zero. With the allowance for the fact, expression (55) passes over to (43) with the replacement in (43) of $v$ by $u$ and of the hydrostatic compression modulus $K$ by

$$\tilde{K} = K + \tfrac{4}{3}\mu\,, \tag{56}$$

where $\tilde{K}$ is the modulus of uniaxial compression without changing the transverse dimensions of the body.

Designating by $\sigma$ the stress tensor component $\sigma_{zz}$ conjugate to $u \equiv u_{zz}$ and calculating the derivatives $\phi^{\sigma}_{\eta\eta}$ and $\phi^{h}_{uu}$, as was done in sect. 3.2 in the calculation of $\phi^{p}_{\eta\eta}$ and $\phi^{h}_{vv}$, but substituting the quantity $\tilde{K}$ for $K$, we find

$$\langle|\eta(\boldsymbol{q})|^2\rangle = \frac{k_B T}{V}\frac{1}{\phi^{\sigma}_{\eta\eta}} = \frac{k_B T}{V}\frac{1}{\phi^{p}_{\eta\eta} + \frac{4}{3}\mu r^2\eta_e^2/K\tilde{K}}\,, \tag{57}$$

$$\langle|u(\boldsymbol{q})|^2\rangle = \frac{k_B T}{V}\frac{1}{\phi^{h}_{uu}} = \frac{k_B T}{V}(\phi^{h}_{vv} + \tfrac{4}{3}\mu)^{-1} = \frac{k_B T}{V}\frac{1}{\tilde{K}} + \frac{r^2\eta_e^2}{\tilde{K}^2}\langle|\eta(\boldsymbol{q})|^2\rangle\,, \tag{58}$$

$$\langle\eta(\boldsymbol{q})u(-\boldsymbol{q}) + \text{c.c.}\rangle = -\frac{2r\eta_e}{\tilde{K}}\langle|\eta(\boldsymbol{q})|^2\rangle\,. \tag{59}$$

We see that expressions (57)–(59) do not coincide with expressions (31), (49) and (50) in the limit $q \to 0$. Thus, spatially inhomogeneous fluctuations of both $\eta$ and $\rho$ differ (are separated by a "gap") from homogeneous fluctuations of these quantities. The origin of this gap for density fluctuations is quite understandable because inhomogeneous $\rho$ fluctuations, as distinct from homogeneous ones, are accompanied by shear deformations, and therefore, as has been shown above, the magnitudes of these fluctuations are determined by different elasticity moduli. (In fact, inhomogeneous density fluctuations in solids can be considered as thermal longitudinal acoustic waves whose velocity square is proportional to the modulus of uniaxial compression $\tilde{K} = K + \frac{4}{3}\mu$.) The appearance of the gap is connected, in other words, with elastic long-range interactions inherent to solids. On the other hand, the presence of a gap (for $\boldsymbol{q} = 0$) in the $\eta$ fluctuations spectrum is due to a linear coupling between $\Delta\eta$ and $\Delta\rho$. From eq. (57) the value of the gap is seen to be proportional to the shear modulus $\mu$ as well as the equilibrium value of the order parameter $\eta_e$.

As to eq. (57) it can be interpreted also in another way. It was shown in the preceding section that an explicit account of the interaction between order parameter and spatially homogeneous density fluctuations leads to the same expression for $\phi_{\eta\eta}$ as without such an account but with a renormalized value of the coefficient $B$ given by formula (46): $B = B_1 - r^2/2K$. To obtain expression (57) one should bear in mind that for the spatially inhomogeneous case renormalization of the coefficient is somewhat different, namely, the modulus $\tilde{K}$ should be substituted for $K$ in eq. (46), for this is just the modulus to describe the response of the system to an inhomogeneous (in the form of a plane wave) change of density.

Now let us discuss the temperature dependence of spatially inhomogeneous fluctuations in the framework of the Landau theory. For a second-order phase transition far from the tricritical point the quantities $\phi^p_{\eta\eta}$ and $\eta_e^2$ depend on temperature in the same way as below $T_c$. Thus, as is seen from eq. (57), the presence of shear deformations does not affect the temperature dependence of $\langle|\eta(\boldsymbol{q})|^2\rangle$ as compared with the spatially homogeneous case but influences only the amplitude of this dependence. Another situation takes place for a tricritical point. Here $\phi^p_{\eta\eta}$ goes to zero (for $|\tau|\to 0$) as $\eta_e^2|\tau|^{1/2}$, i.e., steeper than $\eta_e^2$ (see eqs. (3), (7) and (8)). Therefore near a tricritical point the temperature dependence of $\langle|\eta(\boldsymbol{q})|^2\rangle$ is determined by the temperature dependence of the gap and is the same as that of $\eta_e^2$ (remember that for a tricritical point, $\eta_e^2 \sim |\tau|^{1/2}$; see eq. (7)). In other words, for a tricritical phase transition, long-range elastic forces do affect the temperature dependence of $\eta$ fluctuations below $T_c$ as compared with the spatially homogeneous case.

In the case of first-order phase transitions the quantities $\phi^p_{\eta\eta}$ and $\phi^h_{vv}$ vanish at the spinodal line of the non-symmetrical phase, where $B^2 - 4AC = 0$. Consequently, according to eqs. (31) and (49), the homogeneous fluctuations in $\eta$ and $\rho$ formally diverge here. At the same time spatially inhomogeneous fluctuations of these quantities remain finite at $T = T_{s1}$ due to the gap both for the fluctuations of $\rho$ and $\eta$ (for the latter this takes place because $\eta_e \neq 0$ at the points of the spinodal in question).

In the above considered example long-range forces accompany the $\rho$ and $\eta$ fluctuations in the non-symmetrical phase only. If the order parameter may be identified with polarization, strain and magnetization, i.e., in the cases of proper ferroelectrics, ferroelastics and ferromagnetics, long-range fields accompany $\eta$ fluctuations both below and above the transition temperature (Krivoglaz 1963, Levanyuk 1965, Levanyuk and Sobyanin 1970, Villain 1970, Levanyuk et al. 1968). In certain situations the fields lead to an essential change of the picture of $\eta$ fluctuations. We shall show this at first for proper ferroelectrics.

*3.3.2. Fluctuations near ferroelectric phase transitions; depolarizing field effects*

As is well known, in proper ferroelectrics a spatially inhomogeneous distribution of the polarization $\boldsymbol{P}$ causes a macroscopic electric field $\boldsymbol{E}$ (sometimes referred to as a depolarizing field) which may be found from the equations of electrostatics (retardation being neglected)

$$\operatorname{curl} \boldsymbol{E} = 0\,, \tag{60}$$

$$\operatorname{div}(\boldsymbol{E} + 4\pi \boldsymbol{P}) = 0\,. \tag{61}$$

Free charges are assumed here to be absent. Expressing $\boldsymbol{E}(\boldsymbol{q})$ with the aid of eqs. (60) and (61) in terms of $\boldsymbol{P}(\boldsymbol{q})$ we find

$$\boldsymbol{E}(\boldsymbol{q}) = -4\pi(\boldsymbol{q}\cdot\boldsymbol{P}(\boldsymbol{q}))\boldsymbol{q}/q^2\,. \tag{62}$$

The presence of the field $\boldsymbol{E}$ leads to a change of the free energy in a crystal equal to (Landau and Lifshitz 1957)

$$-\tfrac{1}{2}\int \boldsymbol{E}\boldsymbol{P}\,\mathrm{d}V = 2\pi \sum_{\boldsymbol{q}} \frac{q_i q_j}{q^2} P_i(\boldsymbol{q})P_j(-\boldsymbol{q})\,. \tag{63}$$

The energy should be added to the expression for the thermodynamic potential of the type (1). As formula (62) contains all the components of the polarization vector, all these components are to be taken into account also in other terms of the expression for the thermodynamic potential. As a result we have (for $\tau > 0$)

$$\tilde{\Phi} = \frac{V}{2}\sum_{\boldsymbol{q}}\left(A_{ij} + 4\pi\frac{q_i q_j}{q^2} + D_{ijlm}q_l q_m\right)P_i(\boldsymbol{q})P_j(-\boldsymbol{q})\,. \tag{64}$$

In what follows (as before) we shall neglect for simplicity the anisotropy of $D_{ijlm}$ (which is inessential for us) assuming that $D_{ijlm} = D\delta_{ij}\delta_{lm}$.

If a ferroelectric has three equivalent spontaneous polarization axes, as is the case when the crystal has cubic symmetry in the symmetrical (non-polar, or in other terms, para-electric) phase, then $A_{ij} = A\delta_{ij}$. The bilinear forms in eq. (64) can be diagonalized by introducing appropriate new variables $P_j(\boldsymbol{q})$, the coefficients for the resulting (diagonalized) quadratic combinations have the form

$$\lambda_1(\boldsymbol{q}) = \lambda_2(\boldsymbol{q}) = A + Dq^2, \qquad \lambda_3(\boldsymbol{q}) = A + 4\pi + Dq^2\,, \tag{65}$$

where the 3-axis is directed along $\boldsymbol{q}$.

For mean-square fluctuations one then obtains (Krivoglaz 1963, Levanyuk et al. 1968)

$$\langle|\tilde{P}_{1,2}(\boldsymbol{q})|^2\rangle = \frac{k_\mathrm{B}T}{(A + Dq^2)V}\,,$$

$$\langle|\tilde{P}_3(\boldsymbol{q})|^2\rangle = \frac{k_\mathrm{B}T}{(A + 4\pi + Dq^2)V}\,. \tag{66}$$

Thus, although the order parameter is a three-component parameter $(\{\eta_\alpha\} = \{P_1, P_2, P_3\})$ it is in fact only the fluctuations in the two components $P_1$ and $P_2$, that increase greatly as $A \to 0$ for each direction of the wave-vector $\boldsymbol{q}$. The reason for this is that the inhomogeneous polarization distribution, $\boldsymbol{P}(\boldsymbol{r}) = \boldsymbol{P}_0 \sin \boldsymbol{qr}$, does not cause the appearance of an electric field for $(\boldsymbol{qP}_0) = 0$. Indeed, the bound charge density ($\rho_{\text{Bound}} = -\operatorname{div} \boldsymbol{P}$) is zero in the direction of $\boldsymbol{q}$. The variables $P_i$ are chosen in such a way that for each $\boldsymbol{q}$ two components of the vector $\boldsymbol{P}(\boldsymbol{q})$ were perpendicular to $\boldsymbol{q}$ and the third component was parallel to $\boldsymbol{q}$. Fluctuations of the latter (longitudinal) component are greatly suppressed due to the appearance of an electric field. The difference between magnitudes of longitudinal and transverse polarization fluctuations is, in effect, well known in the theory of lattice dynamics as the splitting between frequencies of longitudinal and transverse infrared active phonons.

Consider now a uniaxial ferroelectric, i.e., one with a single spontaneous polarization axis. We shall restrict ourselves for simplicity to the case of an extremely high anisotropy, when it is sufficient, in spite of the fluctuating electric field, to consider only the component of polarization vector $\boldsymbol{P}$ (say, $P_z$) which is the order parameter. Expression (64) for the thermodynamic potential density then becomes

$$\tilde{\Phi} = \frac{V}{2} \sum_{\boldsymbol{q}} \left\{ A_{zz} + 4\pi \frac{q_z^2}{q^2} + Dq^2 \right\} P_z(\boldsymbol{q}) P_z(-\boldsymbol{q}) \,. \tag{67}$$

From this it follows that

$$\langle |P_z(\boldsymbol{q})|^2 \rangle = \frac{k_{\text{B}} T}{V} \frac{1}{A_{zz} + 4\pi \cos^2\theta + Dq^2}, \qquad \cos\theta = \frac{q_z}{q} \,. \tag{68}$$

Thus we see that in a uniaxial ferroelectric there is a strong suppression of fluctuations: only the $P_z$ fluctuations with wavevectors $\boldsymbol{q}$ perpendicular (or almost perpendicular) to the $z$-axis increase when approaching the phase transition (Levanyuk 1965). This feature (see fig. 3) of uniaxial ferroelectrics (and ferromagnetics) should be displayed, in particular, in a strong angular

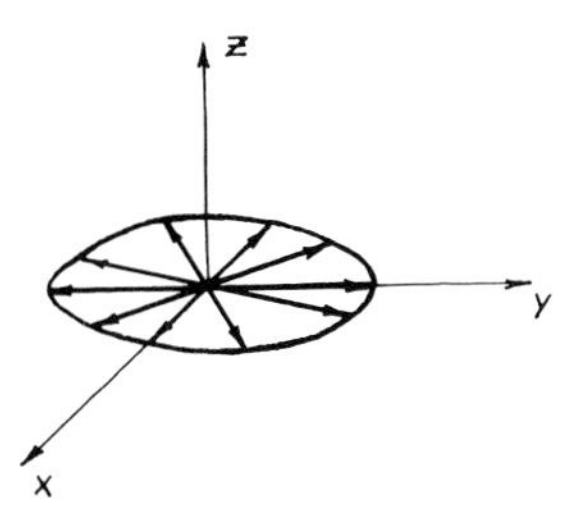


Fig. 3. The set of wavevectors corresponding to the most anomalous (divergent at $T = T_c$) fluctuations in the order parameter in a proper ferroelectric with a single axis of the spontaneous polarization ($z$-axis).

dependence of the anomalous part of the first-order light scattering; besides, critical anomalies of second-order scattering, and of different thermodynamic quantities must be less pronounced here (Levanyuk 1965) and the applicability region of the Landau theory widens greatly (Vaks et al. 1966, Vaks 1973).

It should be stressed that the simple formula (68) holds only for an extremely anisotropic ferroelectric, when the polarizability along $x$ and $y$ directions may be neglected. A more precise consideration includes also an account for fluctuations of $P_x$ and $P_y$ which proves to be essential for $\theta$'s differing markedly from $\pi/2$. The important role of these "non-critical" fluctuations is rather natural as for $\theta \neq \pi/2$ $P_z$ fluctuations are also, strictly speaking, non-critical, i.e., they increase only to a finite amount as $\tau \to 0$ even for $q \simeq 0$.

Note also that in a uniaxial ferroelectric the depolarizing field effects lead to a linear coupling between spatially inhomogeneous fluctuations of variables with different transformation properties ($P_z$, $P_x$ and $P_y$). It is one of the manifestations of the non-analytical behaviour of polarization fluctuations in the vicinity of the point $\boldsymbol{q} = 0$.

In a ferroelectric with two equivalent spontaneous polarization axes an electric field accompanying polarization fluctuations leads, as in the case of a cubic crystal (see eq. (65)), only to a decrease in the number of the order parameter components whose fluctuations increase as $\tau \to 0$. Namely, it is the fluctuations of the $\boldsymbol{P}$-vector component, which is perpendicular to $\boldsymbol{q}$ and lies in the plane of spontaneous polarization vectors, that increase as $\tau \to 0$.

### 3.3.3. *Suppression of critical fluctuations in proper ferroelastics*

To a greater extent than in the above cases fluctuations are suppressed near proper ferroelastic phase transitions when the order parameter(s) may be identified with one (or several) component(s) of the strain tensor $u_{ij}$ (Levanyuk and Sobyanin 1970, Villain 1970). Suppose, for example, that the component $u_{xy}$ corresponds to the order parameter. Then the role of the coefficient $A$ in an expression of the type (64) will be played by the elastic modulus $C_{xyxy}$. It is evident that fluctuations in $\eta$ are in this case thermal (fluctuating) acoustic waves. The mean-square amplitude of the wave is inversely proportional to the square of the wave velocity, i.e., to the corresponding effective elasticity modulus, which is determined by some linear combination of the components of the tensor $C_{ijlm}$, depending on the propagation direction and polarization of the wave. The component $C_{xyxy}$ vanishing as $\tau \to 0$, completely determines the magnitude of the effective modulus only for waves propagating along the $y$-axis and polarized along the $x$-axis, and for waves propagating along the $x$- and polarized along the $y$-axis. For all other waves not only $u_{xy}$ but also other components of the strain tensor are inevitably distinct from zero so that the effective elasticity modulus (and, therefore, the fluctuation wave amplitude) remains finite for $\tau = 0$. Thus, an essential increase of the order parameter fluctuations takes place here only for wavevectors $\boldsymbol{q}$ whose directions are close

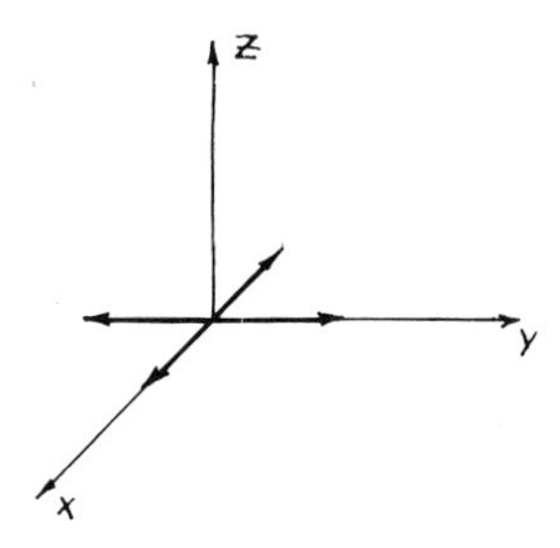


Fig. 4. The set of wavevectors corresponding to the most anomalous (divergent at $T = T_c$) order-parameter fluctuations in a proper ferroelastic of the $KH_2PO_4$ type.

to that of the $x$- and $y$-axes (fig. 4). As the phase volume corresponding to these (large) fluctuations is small, various critical anomalies in the case must be still weaker than in the case of uniaxial ferroelectric, and the applicability region of the Landau theory must extend practically up to the transition temperature (Levanyuk and Sobyanin 1970, Villain 1970, Vaks 1973). For proper ferroelastic transitions with a multi-component order parameter, the situation may occur when the set of the directions of the vectors $\boldsymbol{q}$ corresponding to strongly increasing fluctuations forms a plane, as in a uniaxial ferroelectric (Khmel'nitskii 1974).

For most ferroelastics the change in crystal structure in the phase transition is not reduced to acoustic deformations only but there exists also one or several "optical deformations" arising at the phase transition. Sometimes it is one of these "optical deformations" that is considered as an order parameter, especially if the corresponding "optical" stiffness constant has the strongest temperature dependence. An example is ferroelectrics with piezoelectricity in a non-polar phase ($KH_2PO_4$, Rochelle salt, etc.). Here it is the polarization vector component $P_z$ rather than the strain tensor component $u_{xy}$ (which is equivalent to $P_z$ in its transformation properties) that is taken usually as the order parameter. Another example of this kind is the structural phase transition in $Nb_3Sn$-type superconductors, where at the phase transition along with an "acoustic" deformation displacements appear of certain sublattices ("optical deformations") with the same transformation properties as the acoustic deformation. The linear coupling between these "optical deformations" and elastic strains leads to a non-analytical $\boldsymbol{q}$ dependence also for fluctuations of the "optical deformations". In practice, however, it should be taken into account that the frequency scales of optical and acoustic deformations may be quite different, the latter being usually much slower than the first. In this case the influence of long-range elastic forces will be quite negligible in the range of "optical" frequencies.

### 3.4. Some features of multi-component order parameter fluctuations

Here we restrict ourselves, for simplicity, to the discussion of the spatially homogeneous fluctuations only. We would like to mention, first of all, that above $T_c$ fluctuations in any component of a multi-component order parameter can be treated separately from the fluctuations of its other components. Indeed, the quadratic term in the thermodynamic potential density expansion of the type (1) is written in the form

$$\tfrac{1}{2}A(\eta_1^2+\eta_2^2+\ldots+\eta_n^2), \tag{69}$$

where $n$ in the number of the order parameter components. Therefore, fluctuations in a multi-component order parameter may have some peculiarities as compared with the case of a one-component order parameter in the non-symmetrical phase only.

To illustrate the peculiarities, we consider the case of a two-component order parameter $\eta \equiv \{\eta_1, \eta_2\}$. Like in sect. 2.5, let us introduce the polar coordinates $\rho$ and $\varphi$. This makes the evaluation of fluctuations far more easy, because the derivative $\phi_{\rho\varphi}=0$ in equilibrium, which means that the fluctuations in $\rho$ and $\varphi$ are statistically independent. We shall not discuss the $\rho$ fluctuations, for all, what has been said above of the fluctuations in a one-component order parameter holds for the $\rho$ fluctuations too. So, let us consider the $\varphi$ fluctuations. Since it is the quantity $\rho\Delta\varphi$ that has the dimensionality (and the meaning) of an increment in the order parameter, we shall discuss the temperature dependence of the fluctuation

$$\langle(\rho\Delta\varphi)^2\rangle \simeq \rho_e^2\langle(\Delta\varphi)^2\rangle . \tag{70}$$

We begin with the case when there are two independent invariants of the fourth order formed from the components $\eta_1$ and $\eta_2$. Using eq. (19), one finds the following expression

$$\Delta\phi = 2B_2\rho_e^4(\Delta\varphi)^2 \tag{71}$$

for the change in the thermodynamic potential due to a deviation of $\varphi$ from its equilibrium value; $B_2$ is believed to be positive here. From eq. (71) it follows then that in the classical region

$$\rho_e^2\langle(\Delta\varphi)^2\rangle = \frac{k_B T}{4VB_2\rho_e^2} \propto (T_c - T)^{-1} . \tag{72}$$

Thus, in the case in question the order parameter fluctuations, which are due to changes in the phase $\varphi$, have (in the classical region) the same temperature dependence as the ones due to changes in the amplitude $\rho$. It can be shown easily that the conclusion is valid also for the case when there are three fourth-order invariants.

Now, let us assume that we have only one invariant of the fourth order but that the phase $\varphi$ enters the expression for the thermodynamic potential beginning with the terms of the $n$th order in $\rho$. Equation (71) is replaced then by

$$\Delta\phi \propto \rho_e^n(\Delta\varphi)^2 , \tag{73}$$

and, therefore,

$$\rho^2\langle(\Delta\varphi)^2\rangle \propto \frac{k_B T}{V\rho_e^n} \propto (T_c - T)^{-(n-2)/2} . \tag{74}$$

For example, for $n = 6$

$$\rho^2\langle(\Delta\varphi)^2\rangle \propto (T_c - T)^{-2} . \tag{75}$$

In other words, we see that in the case of a multi-component order parameter there exist situations when fluctuations due to changes in the phase of the order parameter depend much stronger on the distance $(T_c - T)$ than corresponding changes in its amplitude.

### 3.5. Fourth-order fluctuations

When analyzing the light scattering intensity in the symmetrical phase for transitions, which are not proper ferroelastic, one should take into account the order parameter fluctuations of higher orders than second (see sect. 5). Here we discuss the temperature dependence (for $\tau > 0$) of the fourth-order $\eta$ fluctuations,

$$\langle|\eta^2(\boldsymbol{q})|^2\rangle = \sum_{\boldsymbol{k},\boldsymbol{k}'} \langle\eta(\boldsymbol{q}-\boldsymbol{k})\eta(-\boldsymbol{q}-\boldsymbol{k}')\eta(\boldsymbol{k})\eta(\boldsymbol{k}')\rangle . \tag{76}$$

One can obtain exact formulae relating the magnitudes of fluctuations of an arbitrary high order with the thermodynamic potential derivatives with respect to generalized forces conjugate to generalized coordinates whose fluctuations are being considered (see sect. 3.6). However, it is much more convenient to express these fluctuations in terms of derivatives of the thermodynamic potential whose arguments are the generalized coordinates themselves; a potential of just this type enters in the Landau theory. It is essential that the formulae for fluctuations of fourth order include the coefficients at terms no higher than fourth order in the thermodynamic potential expansion. Therefore, to derive the exact expressions for fourth-order fluctuations it is sufficient to write

$$\Phi = \Phi_0 + \tfrac{1}{2}V \sum_{\boldsymbol{k}} A(T;\boldsymbol{k})\eta(\boldsymbol{k})\eta(-\boldsymbol{k})$$
$$+ \tfrac{1}{4}V \sum_{\boldsymbol{k}_1+\boldsymbol{k}_2+\boldsymbol{k}_3+\boldsymbol{k}_4=0} B(T;\boldsymbol{k}_1,\boldsymbol{k}_2,\boldsymbol{k}_3,\boldsymbol{k}_4)\eta(\boldsymbol{k}_1)\eta(\boldsymbol{k}_2)\eta(\boldsymbol{k}_3)\eta(\boldsymbol{k}_4) . \tag{77}$$

Generally speaking, one should have included in this expression also the terms

$\Sigma_{\boldsymbol{k}_1+\boldsymbol{k}_2+\boldsymbol{k}_3=0} B'(T;\boldsymbol{k}_1,\boldsymbol{k}_2,\boldsymbol{k}_3)\eta(\boldsymbol{k}_1)\eta(\boldsymbol{k}_2)\eta(\boldsymbol{k}_3)$, i.e., terms of third order in $\eta(\boldsymbol{k})$. However, for $\tau > 0$, it follows from symmetry requirements that $B'(T;\boldsymbol{k}_1,\boldsymbol{k}_2,\boldsymbol{k}_3) = 0$ for $\boldsymbol{k}_1 = \boldsymbol{k}_2 = \boldsymbol{k}_3 = 0$. Therefore when calculating for $\tau > 0$ the long-wavelength (small $\boldsymbol{q}$) fluctuations (which are the most interesting for light scattering) the third-order terms may be neglected.

We have (Levanyuk 1959, 1976)

$$\begin{aligned}\langle|\eta^2(\boldsymbol{q})|^2\rangle &= 2\left(\frac{k_{\mathrm{B}}T}{V}\right)^2 \sum_{\boldsymbol{k}} A^{-1}(\boldsymbol{k};T)A^{-1}(\boldsymbol{k}-\boldsymbol{q};T)\\ &\quad - 6\left(\frac{k_{\mathrm{B}}T}{V}\right)^3 \sum_{\boldsymbol{k},\boldsymbol{k}'} B(\boldsymbol{k},-\boldsymbol{k}-\boldsymbol{q},-\boldsymbol{k}'+\boldsymbol{q};T)\\ &\quad \times A^{-1}(\boldsymbol{k};T)A^{-1}(\boldsymbol{k}+\boldsymbol{q};T)A^{-1}(\boldsymbol{k}';T)A^{-1}(\boldsymbol{k}'-\boldsymbol{q};T). \qquad (78)\end{aligned}$$

In the approximation corresponding to the Landau theory (and in the absence of long-range forces), $A(\boldsymbol{k};T) = A + Dk^2$, $A = A_0\tau$, $D = \text{const.}$ and $B$ is independent of $\boldsymbol{k}$ and $\tau$. Expression (78) takes then the form

$$\langle|\eta^2(\boldsymbol{q})|^2\rangle = \frac{2(k_{\mathrm{B}}T)^2}{V} S(q)[1 - 3k_{\mathrm{B}}TBS(q)], \qquad (79)$$

where

$$S(q) = \frac{1}{(2\pi)^3}\int_0^\infty \frac{\mathrm{d}\boldsymbol{k}}{(A+Dk^2)[A+D(\boldsymbol{k}-\boldsymbol{q})^2]} = \frac{1}{4\pi D^2 q}\arctan\left(\frac{qr_{\mathrm{c}}}{2}\right), \qquad (80)$$

The upper limit of integration over $\boldsymbol{k}$ has been put here equal to infinity since, for small $\boldsymbol{q}$, the main contribution to the integral is provided by small $\boldsymbol{k}$.

Formula (79) is meaningful, evidently, until its right-hand side is positive, i.e., until

$$3k_{\mathrm{B}}TBS(q) < 1. \qquad (81)$$

This inequality imposes a certain restriction on the thermodynamic potential coefficients. Namely, as the quantity $S(q)$ has a maximum at $q = 0$, one should require that

$$3k_{\mathrm{B}}TBS(0) = \frac{3k_{\mathrm{B}}TB}{8\pi D^{3/2}A^{1/2}} = \frac{3k_{\mathrm{B}}T_{\mathrm{c}}B}{8\pi A_0^2 r_{\mathrm{c}0}^3}\tau^{-1/2} < 1, \qquad (82)$$

where we have used the definition $r_{\mathrm{c}} = r_{\mathrm{c}0}\tau^{-1/2} = (D/A)^{1/2} = (D/A_0\tau)^{1/2}$.

The inequality (82), which is obviously violated close to $T_{\mathrm{c}}$ if the coefficients $A$, $B$, $D$ have the temperature dependence as adopted in the Landau theory, gives the criterion for the applicability of this theory (more precisely, the criterion has, of course, the form (82) with the replacement of $<$ by $\leqslant$; the condition (82) was used in obtaining the estimate (13)). Thus within the Landau theory the second term in brackets in eq. (79) may be omitted.

The inequality (82) (with other temperature dependences for the coefficients) must hold in the critical region too. This means that the quantity $3k_B TBS(0)$ should be in this region independent of $\tau$ or decrease as $\tau \to 0$. One thus arrives at the inequality for critical exponents (Levanyuk 1976)

$$3\nu \geqslant 2 - \alpha\,, \tag{83}$$

which is just the well-known Josephson inequality (see e.g., Stanley 1971).

In the critical region it is only the critical exponent for the quantity $\langle|\eta^2(\boldsymbol{q})|^2\rangle$ that can be found from phenomenological considerations. Let us assume, in accordance with the scaling hypothesis, that

$$A(q;T) = \tau^{\gamma} f(qr_c)\,. \tag{84}$$

Passing over in eq. (78) from summation to integration over $\boldsymbol{k}$ and replacing the integration variable $k$ by $x = kr_c$ we find that the corresponding critical exponent is $3\nu - 2\gamma$. Using then one of the scaling relations, namely, $\gamma = \nu(2 - \hat{\eta})$, we obtain (for $q \to 0$)

$$\langle|\eta^2(\boldsymbol{q})|^2\rangle \sim \tau^{-\nu(1-2\hat{\eta})}\,. \tag{85}$$

Since, as has been mentioned, the index $\hat{\eta}$ is very small for real three-dimensional systems, the temperature dependence of $\langle|\eta^2(0)|^2\rangle$ coincides practically with that of the correlation length $r_c$. As seen from eqs. (79) and (80), these two temperature dependences are the same in the classical region too.

### 3.6. *Some remarks on the calculation of fluctuations and on the role of large fluctuations*

Here we would like to make some general remarks on the methods of the calculation of fluctuations to clarify the meaning of the formulae used in the preceding sections.

Usually (see, e.g., Landau and Lifshitz (1976)) one begins to calculate mean-square fluctuations with the determination of the probability distribution of the fluctuations. Let us be interested in the fluctuations of a generalized coordinate $\xi$. We start from the canonical Gibbs distribution

$$W(\xi, X; p, T) \sim \exp\left\{-\frac{H(\xi, X; p)}{k_B T}\right\}. \tag{86}$$

Here $X$ is the set of all the coordinates of the system but $\xi$. The set $\{X, \xi\}$ characterizes the microscopic state of the system (the volume $V$ is believed to be included in the set so that the pressure $p$ be a parameter of the Hamiltonian). From eq. (86) one finds

$$W_1(\xi) = \int W(\xi, X)\,\mathrm{d}X \sim \exp\left[-\frac{\tilde{\Phi}(\xi; p, T)}{k_B T}\right], \tag{87}$$

where

$$\tilde{\Phi}(\xi; p, T) = -k_B T \ln \int \exp\left[-\frac{H(\xi, X; p)}{k_B T}\right] dX. \tag{88}$$

We call the function $\tilde{\Phi}(\xi; p, T)$ an incomplete thermodynamic potential thus stressing that the integration in eq. (88) is carried out not over all the variables of the system. The minimum of the function $\tilde{\Phi}(\xi)$ corresponds to the most probable value of $\xi$ which can be identified frequently with the observable (equilibrium) value of $\xi$. We will denote this value by $\xi_e$, and assume first that the function $\tilde{\Phi}(\xi)$ has only one minimum.

Passing over to the calculation of the mean-square fluctuation in $\xi$, we expand $\tilde{\Phi}(\xi)$ in powers of $(\xi - \xi_e)$. Using the first non-vanishing term of this expansion only, one obtains

$$\langle(\Delta\xi)^2\rangle = \frac{k_B T}{\tilde{\Phi}_{\xi\xi}} = \frac{k_B T}{V} \frac{1}{\tilde{\phi}_{\xi\xi}}. \tag{89}$$

The corrections to this expression due to an account taken of higher-order terms go to zero as $V \to \infty$.

Recall now that the mean-square fluctuation in $\xi$ may be expressed also through the strict ("complete") thermodynamic potential as used in some (unfortunately, rather rare) text books on statistical thermodynamics (see, e.g., Leontovich (1944), see also Terletskii (1958), Levanyuk (1959)). Let us consider the system in the presence of an external field $g$ conjugate to $\xi$. One has

$$H(\xi, X; g, p) = H(\xi, X; p) - g\xi. \tag{90}$$

Calculate now the thermodynamic potential

$$\Psi(g, p, T) = -k_B T \ln \int \exp\left[-\frac{H(\xi, X; p) - g\xi}{k_B T}\right] d\xi\, dX. \tag{91}$$

As follows from eq. (91), the average value of $\xi$ is

$$\langle\xi\rangle = -\partial\Psi/\partial g. \tag{92}$$

According to the rules of themodynamics, the complete thermodynamic potential depending on $\xi$ is defined as

$$\Phi(\xi; p, T) = \Psi(g, p, T) + g\langle\xi\rangle = \Psi(g, p, T) - g\frac{\partial\Psi}{\partial g}, \tag{93}$$

where $g$ should be expressed through $\langle\xi\rangle = \xi$ with the aid of eq. (92). Using eqs. (91) and (93), one has

$$\langle(\Delta\xi)^2\rangle = \langle\xi^2\rangle - \langle\xi\rangle^2 = k_B T/\Phi_{\xi\xi}. \tag{94}$$

It should be emphasized that formula (94) is a strict one, unlike expression (89), which is valid only if one neglects higher-order derivatives of $\tilde{\Phi}$ with respect to $\xi$. The function $\Phi(\xi)$ differs from $\tilde{\Phi}(\xi)$ in that when obtaining $\Phi(\xi)$ we have performed integration of the partition function over all the variables in contrast to what was done in obtaining $\tilde{\Phi}(\xi)$. That is why we have termed $\Phi(\xi)$ a complete thermodynamic potential.

Comparing eqs. (94) and (89) one sees that both the complete and incomplete potentials give the same results for the fluctuations as $V \to \infty$. In particular, one has in this limit that $\langle \xi \rangle = \xi_e$.

Now we consider the Fourier transform $\eta_0$ of the order parameter $\eta(\boldsymbol{r})$ with $\boldsymbol{q} = 0$ as an example of the variable $\xi$. The condition $\langle \eta_0 \rangle = \eta_{0e}$ evidently holds for $T > T_c$, but the situation is different for $T < T_c$. Indeed, there the function $W_1(\eta_0)$ has two symmetrical maxima (at $\eta_0 = \eta_{0e}$ and $\eta_0 = -\eta_{0e}$), and, therefore, the value of $\eta_0$ averaged over a full statistical ensemble remains equal to zero although $\eta_{0e} \neq 0$. This means, however, that the averaging procedure over a full statistical ensemble is not a good one for our problem. Indeed, in reality one always deals with a single crystal rather than with an ensemble of crystals so that if external conditions are invariable, the uniform value of $\eta$ remains unchanged too. There is, of course, an "academic" probability that $\eta_0$ will change its sign due to a large thermal fluctuation, but such an event is very improbable and one may wait for it much longer than one's lifetime. It is worth mentioning in this connection that statistical mechanics gives values of physical quantities averaged over all the fluctuations, no matter how improbable they are. An attempt to overcome this difficulty for metastable and degenerate systems has been made recently by Ma (1981).

Thus, in the Landau theory (for the spatially homogeneous case) the thermodynamic potential should be considered as incomplete.

One may have an impression that for the calculation of fluctuations near phase transitions it is relevant to use only an incomplete thermodynamic potential. However, the complete thermodynamic potential may be also very useful for the case of spatially-inhomogeneous fluctuations. To show this, we discuss the advantages and shortcomings of the calculation of fluctuations using complete and incomplete thermodynamic potentials. Let us restrict ourselves in the beginning to the case of $T > T_c$. If expression (1) is treated as an incomplete thermodynamic potential, then one finds for the mean square of the Fourier transform of $\eta$ fluctuations with the wavevector $\boldsymbol{q}$ the following expression:

$$\langle |\eta(\boldsymbol{q})|^2 \rangle = \frac{\int \eta(\boldsymbol{q})\eta(-\boldsymbol{q}) \exp[-\tilde{\Phi}/k_B T] \prod_{\boldsymbol{k}} \mathrm{d}\eta(\boldsymbol{k})}{\int \exp[-\tilde{\Phi}/k_B T] \prod_{\boldsymbol{k}} \mathrm{d}\eta(\boldsymbol{k})}, \tag{95}$$

with

$$\tilde{\Phi} = \sum_{\boldsymbol{k}} (A + Dk^2)\eta(\boldsymbol{k})\eta(-\boldsymbol{k}) + \tfrac{1}{2}B \sum_{\boldsymbol{k}_1 + \boldsymbol{k}_2 + \boldsymbol{k}_3 + \boldsymbol{k}_4 = 0,} \eta(\boldsymbol{k}_1)\eta(\boldsymbol{k}_2)\eta(\boldsymbol{k}_3)\eta(\boldsymbol{k}_4)\,.$$

It is essential that the fourth-order term "mixes up" the Fourier transforms of $\eta$ with various $\boldsymbol{k}$ so that the integration in eq. (95) should be carried out over all rather than a single variable $\eta(\boldsymbol{q})$. This cannot be done accurately because the neglect of the fourth-order term leads to an error which does not go to zero as $V \to \infty$, unlike in the case of one variable. This difficulty is crucial in the problem of critical behavior, and in order to overcome it very many efforts have been made for a few past decades.

The difficulty prevents, however, neither the extension to many variables of the method applied above in the case of a single variable, nor the obtaining of exact expressions for the fluctuations in terms of the *complete* thermodynamic potential depending on $\eta(\boldsymbol{q})$. The formulae presented in the preceding sections have just such a sense.

Each method of calculating fluctuations (with the aid of complete and incomplete thermodynamic potentials) has its advantages. Although in the case of incomplete potential no exact expressions for mean fluctuations can be derived, one can have an insight into the temperature dependence of the coefficients in eq. (95). Indeed, when obtaining $\tilde{\phi}(\ldots\eta(\boldsymbol{k})\ldots)$ we have not performed the integration of the partition function over the long-wavelength $\eta$ fluctuations which are responsible for critical anomalies. Therefore the coefficients $A$ and $B$ in eq. (95) may be believed to have an analytic temperature dependence. In particular, they may be expanded in powers of $(T - T_c^*)$, where $T_c^*$ is determined from the equation $A(T) = 0$ and does not coincide generally with the true second-order phase transition temperature. The latter corresponds to the instability point of the system and should be characterized by a divergence of fluctuations, but in this approach they just cannot be calculated precisely!

On the other hand, using complete thermodynamic potential one can obtain precise formulae, such as eqs. (94), (78) and (30), but as for the temperature dependence of the coefficients involved, we can conclude only that $A = 0$ at $T = T_c$ (again from the fact of the loss of stability by the system at the transition temperature). The meaning of formulae (30) and (78) consists, in fact, in expressing the mean-square fluctuations in $\eta$ through the parameters of the equations of state which characterize the response of the system to the fields $h(\boldsymbol{k})$ conjugated to $\eta(\boldsymbol{k})$. The parameters of the equations are still to be determined in this or that way. In the range of applicability of the Landau theory, where critical fluctuations are small enough, the parameters can be expected to have a regular (analytic) temperature dependence (as was assumed repeatedly in preceding sections), but in the critical region this assumption will certainly be invalidated.

Thus formulae (30) and (78) are quite useful only when the equations of state $\eta(\boldsymbol{k}) = f[h(\boldsymbol{k})]$ are known from experiment or if one has some theoretical predictions about them. Therefore, if one develops the theory starting from a given Hamiltonian ("true" or "effective"), then one can first find the equations of state and further on apply formulae (30) and (78), or calculate the mean-square fluctuations immediately.

In the course of direct calculation of fluctuations as well as in the analysis of their dynamics (see sect. 6), the question arises frequently about the role played by large fluctuations. Such fluctuations are neglected usually when average values are considered because an increase in the thermodynamic potential needed for their creation is rather high, and therefore they give a small contribution to the partition function. However, in situations where in the configuration of the system large local changes are possible that are not connected with big changes in the system's energy, large fluctuations should be taken into account along with small ones.

The most evident example of such a situation seems to be provided by a system near an order–disorder phase transition. Let us assume for the sake of illustration that in the disordered phase of such a system some ion has only two possible equally probable positions. At fluctuations of the order parameter redistribution of ions takes place, some ions hop from one position to another, i.e., in a volume of the order of a unit cell the order parameter fluctuations are large. The picture of fluctuations in such a system seems to differ essentially from that of the so-called "phonon-type" fluctuations, which are thought of as weak deviations of ions from an ideal (equilibrium) arrangement. Indeed, in the latter case ion coordinates change continuously, whereas in the case of fluctuations of "order–disorder" type they change in a discrete manner. However, this difference practically disappears when long-wavelength fluctuations are considered, i.e., when we deal with fluctuations in macroscopic volumes but not in a single unit cell. In other words, as far as the phase transition problems are concerned, one considers always values of variables averaged over a great enough volume. This could be the displacement of an ion from a given equilibrium position or the difference in the number of ions located in two possible equilibrium positions. In both cases, if one averages over a great enough volume, the variables change continuously and their fluctuations should be regarded as small and continuous, too. The difference in the order parameter fluctuations for displacive and order–disorder phase transitions manifest themselves only in the kinetics of fluctuations, i.e., in the spectral distribution of scattered light. This question is discussed in sect. 6.

Another example of systems, when large fluctuations seem to be taken into account separately, is the one with a first-order phase transition. Here, along with small fluctuations, the so-called heterophase fluctuations are often considered. Let us explain what this means.

At a temperature somewhat higher than that of the first-order phase transition, the thermodynamic potential as a function of $\eta$ has the form displayed in fig. 5. At the phase transition temperature the curve $\phi(\eta)$ touches the $\eta$-axis at points $\eta = \pm\eta_1$. It seems at first glance that the fluctuations in

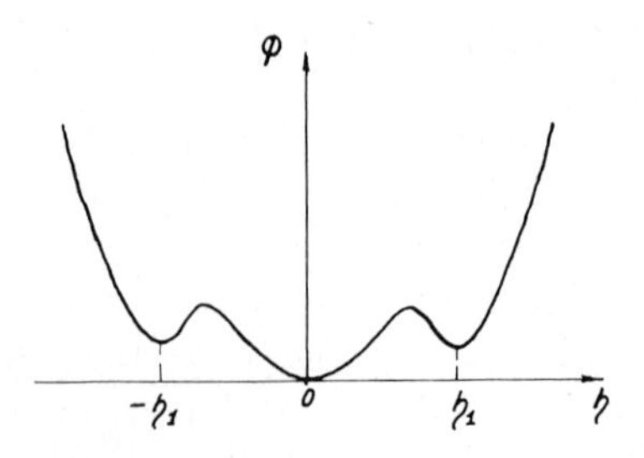


Fig. 5. The thermodynamic potential as a function of the order parameter at a temperature somewhat higher than that of the first-order transition.

which $\eta$ changes from zero to $\pm\eta_1$ are connected with a small change of thermodynamic potential and therefore are quite probable and should be considered on the same footing with small "phonon-like" fluctuations near the point $\eta = 0$.

Note, however, that fig. 5 is for space-homogeneous order parameter variations. At the same time, the appearance of a localized heterophase fluctuation (a change of $\eta$ from zero to a finite value $\eta^*$ in some region, say, inside a sphere of radius $R$) implies the appearance of an extra ("surface") energy. The existence of this energy makes the fluctuation less probable. In particular, at $R \lessapprox r_c$ the contribution from the surface energy will be predominant and will increase monotonically as $\eta^*$ increases. As a result, for each fluctuation the dependence $\tilde{\phi}(\eta^*)$ has no minimum at $\eta^* = \eta_1$ (fig. 6), and the idea of heterophase fluctuations becomes meaningless here. The minima of $\tilde{\phi}(\eta^*)$ at $\eta^* = \pm\eta_1$ can arise for fluctuations in a large enough volume ($R \gg r_c$), but the probability of such fluctuations is very small since they require large energy changes (we mean surface energy); besides, their statistical weight is very small.

The case of solids is worth discussing separately. Here the minima on the curve $\tilde{\phi}(\eta^*)$ are either absent or lie very high even when the surface energy is neglected. The reason is the long-range forces due to the presence of shear modulus in a solid. These forces we discussed in sect. 3.3.1, where it was concluded that they can be effectively taken into account if in the thermodynamic potential $B$ is replaced by $B + r^2\mu/6K\tilde{K}$. The renormalization of the coefficient $B$ due to the presence of the shear modulus is rather substantial for first-order phase transitions close to the tricritical point when $B + r^2\mu/6K\tilde{K} > 0$. Here the function $\tilde{\phi}(\eta^*)$ has no minimum at all, and it is senseless to speak of heterophase fluctuations. A minimum on the curve arises only for phase transitions which are rather far from the tricritical point, but at the phase equilibrium point the distance of this minimum from the $\eta$-axis remains also rather large. Thus, it is quite senseless to speak about heterophase fluctuations in the case of phase transitions in solids.

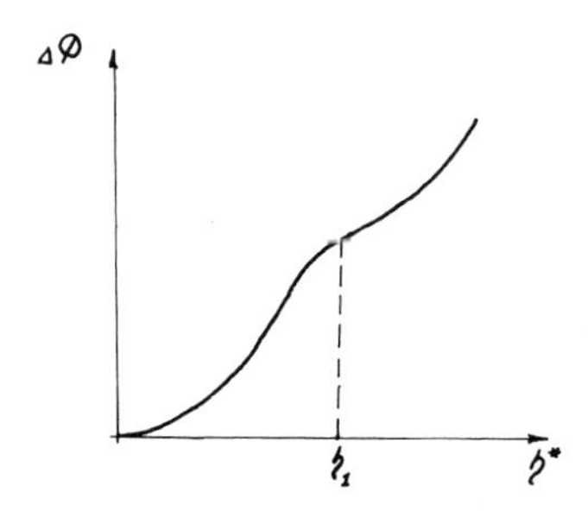

Fig. 6. The thermodynamic potential changes due to the appearance above $T_{tr}$ of a non-zero value ($\eta^*$) of the order parameter within a small volume of a given radius $R \lesssim r_c$.

## 4. *Light scattering by crystals; general relations*

### *4.1. Light scattering intensity*

When formulated in the most general terms the problem of light scattering (to say nothing about that of scattering of electromagnetic waves of all frequencies, as well as of neutrons and other particles) is quite diversified and complicated. The present chapter is dealing only with the scattering of light in ideal systems (as to the light scattering by defects see ch. 2). It is natural then to consider the medium in the absence of (or disregarding) the scattering fluctuations to be optically homogeneous. The second assumption to be made is that of the weakness of the scattering, which allows one to neglect secondary scattering. Even such assumptions are not always justified. So, generally speaking, the medium cannot be considered homogeneous in the presence of domains or interface boundaries which can occur even in perfect crystals, especially near first-order phase transitions (ch. 11). Then, scattering may not be considered weak in the region of critical opalescence (ch. 9). This should be borne in mind, but all the same it is reasonable to begin with the consideration of simpler cases since yearning for maximum generality and accuracy is not always justified and often prevents one from understanding the essence of the matter. Proceeding from such considerations we shall also disregard the light absorption and the spatial dispersion of the dielectric permeability. Finally we shall restrict ourselves (at least in formulae for the scattering intensity) to the case of media whose magnetic permeability (for the range of optical frequencies) may be put equal to unity.

As a result, the medium when light of frequency $\omega$ propagates through it, is characterized by the permeability tensor $\epsilon_{ij}(\omega)$, where $\epsilon_{ij}(\omega)=\epsilon_{ji}(\omega)$, as follows from the Onsager reciprocity principle with a neglect of spatial dispersion and in the absence of an external magnetic field or magnetic structure (see, for example, Agranovich and Ginzburg 1965, 1979). Besides, for a non-absorbing medium $\epsilon_{ij}(\omega) = \epsilon_{ij}^*(\omega)$ and, therefore, under the above-mentioned conditions the tensor $\epsilon_{ij}$ is real. The neglect of spatial dispersion means in particular that the medium is considered to be non-gyrotropic. Extension of all the expressions to gyrotropic and magnetic media is, in principle, no problem but at the present stage we do not represent it for the above-mentioned reason (magnetic media will be, nevertheless, briefly touched upon below).

For a given homogeneous medium the tensor $\epsilon_{ij}$ is spatially independent and will be designated as $\epsilon_{ij}^{(0)}(\omega)$. In such a medium, as is well known from the usual crystal optics, two normal electromagnetic waves* with the refractive indices $n_1(\omega)$ and $n_2(\omega)$ may propagate in each direction. The presence of inhomogeneities (fluctuations or defects) leads to spatial variations of $\epsilon_{ij}$ so that the full tensor is

$$\epsilon_{ij}(\omega, \boldsymbol{r}) = \epsilon_{ij}^{(0)}(\omega) + \Delta\epsilon_{ij}(\omega, \boldsymbol{r}) .$$

It is just the inhomogeneities $\Delta\epsilon_{ij}(\omega, \boldsymbol{r})$ that are responsible for light scattering. We shall be interested here in the situation (fig. 7) when the incident plane wave with a wavevector $\boldsymbol{k}_\mathrm{I}$ propagates in the medium. (The field in the wave is proportional to $\exp(\mathrm{i}\boldsymbol{k}_\mathrm{I}\boldsymbol{r})$.) The scattering is observed in the wave zone where one can single out a plane-scattered wave with the wavevector $\boldsymbol{k}_\mathrm{S}$ (the field is proportional to $\exp(\mathrm{i}\boldsymbol{k}_\mathrm{S}\boldsymbol{r})$).

For normal waves, of course, $k = \omega n_i/c$, where $i = 1, 2$. For inelastic scattering (i.e., always except for the central peak associated with scattering

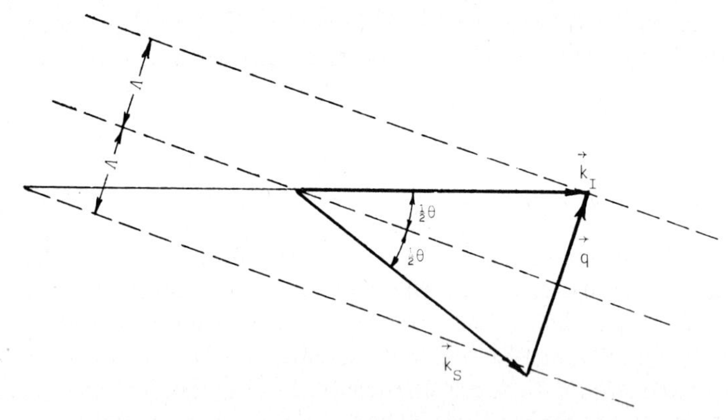


Fig. 7. Light scattering at an angle $\theta$ due to a permeability wave (shown by dotted lines) with the wavelength $\lambda = 2\pi/q$.

*The medium is assumed to be transparent (a non-absorbing medium may be, at the same time, non-transparent as is the case with total internal reflection). Longitudinal waves are not considered (for more details see, e.g., Agranovich and Ginzburg (1965, 1979)).

by static inhomogeneities) the scattered light frequency $\omega_S$ differs from the incident one $\omega_I$. However, if one is interested in scattering with a small change of frequency ($|\Omega| \equiv |\omega_I - \omega_S| \ll \omega_I \approx \omega_S$) only, then electrodynamic calculations are carried out practically in the same way as for scattering by static inhomogeneities (Landau and Lifshitz 1957). The condition $|\Omega| \ll \omega_I$ and, therefore, $\omega_S \approx \omega_I = \omega$ are supposed hereafter to be fulfilled.

In the framework of such a simplified approach the field in the scattered wave is determined by the Fourier transform

$$\Delta\epsilon_{ij}(\boldsymbol{q}) = \frac{1}{V}\int \Delta\epsilon_{ij}(\boldsymbol{r})\, e^{-i\boldsymbol{qr}}\, d\boldsymbol{r}, \qquad \boldsymbol{q} = \boldsymbol{k}_I - \boldsymbol{k}_S\,, \tag{96}$$

where $V$ is the scattering volume. Thus, one can say that scattering in a given direction occurs from a permeability wave with the wavevector $\boldsymbol{q}$. For an isotropic medium $n_1 = n_2 = n$ and, obviously, $k_I \simeq k_S = \omega n(\omega)/c$ and $q = (4\pi n/\lambda_{10})\sin(\theta/2)$, where $\lambda_{10} = 2\pi c/\omega_I$ is the wavelength of the incident light in vacuo and $\theta$ is the angle between $\boldsymbol{k}_I$ and $\boldsymbol{k}_S$, i.e., the scattering angle. In other words, the scattering may be considered as the Bragg reflection from the permeability wave of wavelength

$$\Lambda = \frac{2\pi}{q} = \frac{\lambda_{10}}{2n\sin(\theta/2)} = \frac{\lambda_I}{2\sin(\theta/2)}\,,$$

the wavelength of the incident light in the medium being $\lambda_I = 2\pi c/n\omega_I = \lambda_{10}/n$.

In an optically isotropic medium, $\epsilon_{ij}^{(0)} = \epsilon^{(0)}\delta_{ij}$, but the perturbation $\Delta\epsilon_{ij}$ need not of course be a scalar. Nevertheless in certain cases one can put with a sufficient accuracy $\Delta\epsilon_{ij} = \Delta\epsilon\delta_{ij}$. Then the calculation of the scattered light intensity is especially simple and within the above formulation it goes back to the work by Einstein (1910) and is also exposed, e.g., in the books by Fabelinskii (1965), Landau and Lifshitz (1957), Ginzburg (1975).

The scattered light intensity per unit solid angle and unit intensity of a linearly polarized incident light beam is in this case,

$$\begin{aligned} I(\boldsymbol{q}) &= \left(\frac{V}{4\pi}\right)^2 \left(\frac{2\pi}{\lambda_{10}}\right)^4 \langle|\Delta\epsilon(\boldsymbol{q})|^2\rangle \sin^2\varphi = VQ_{S0}\langle|\Delta\epsilon(\boldsymbol{q})|^2\rangle, \\ Q_{S0} &= \frac{V}{16\pi^2}\left(\frac{\omega_I}{c}\right)^4 \sin^2\varphi\,, \end{aligned} \tag{97}$$

where $\varphi$ is the angle between the incident wave electric vector $\boldsymbol{E}_I$ and the scattered light wavevector $\boldsymbol{k}_S$ (see fig. 8, in which all the vectors are displayed to be lying for simplicity in one plane). Notice that by definition

$$\langle|\Delta\epsilon_{ij}(\boldsymbol{q})|^2\rangle = \frac{1}{V^2}\left\langle \iint \Delta\epsilon_{ij}(\boldsymbol{r}_1)\Delta\epsilon_{ij}(\boldsymbol{r}_2)\exp[i\boldsymbol{q}(\boldsymbol{r}_1 - \boldsymbol{r}_2)]\, d\boldsymbol{r}_1\, d\boldsymbol{r}_2 \right\rangle. \tag{98}$$

In a homogeneous medium the average (98) can depend on the difference

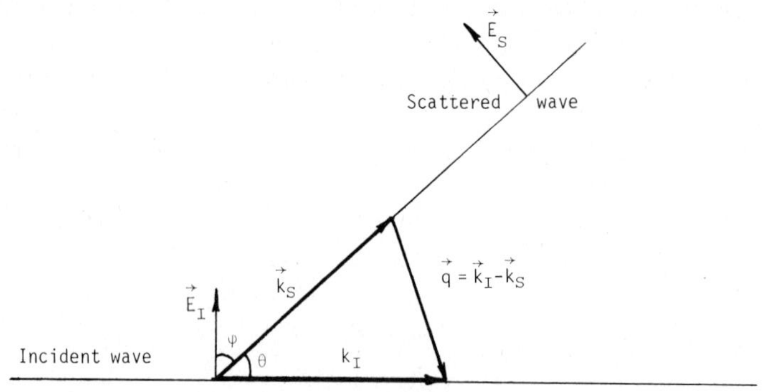


Fig. 8. Schematic of the light scattering in a medium. The field in the scattered wave is linearly polarized, as shown in the figure, only when the fluctuation $\Delta\epsilon_{ij} = \delta\epsilon\delta_{ij}$.

$\boldsymbol{r} = \boldsymbol{r}_1 - \boldsymbol{r}_2$ only. Therefore, going over to the variables $\boldsymbol{r} = \boldsymbol{r}_1 - \boldsymbol{r}_2$ and, say, $\boldsymbol{R} = (\boldsymbol{r}_1 + \boldsymbol{r}_2)/2$, we see that this average is proportional to the inverse scattering volume $V$. As a result in eq. (97) the product $Q_{S0}\langle|\Delta\epsilon(\boldsymbol{q})|^2\rangle$ is independent of $V$ and the intensity $I(\boldsymbol{q})$ is proportional to $V$ as it should be.

Extension of eq. (97) to an anisotropic medium for an incident normal wave with refractive index $n_I$ and unit polarization vector $\boldsymbol{l}_I$ with the formation of a normal scattered wave for which $n = n_S$ and $\boldsymbol{l} = \boldsymbol{l}_S$, respectively, leads to the expression (Motulevich 1950, Fabelinskii 1965)

$$I(\boldsymbol{q}) = \left(\frac{V}{4\pi}\right)^2 \left(\frac{2\pi}{\lambda_{I0}}\right)^4 \frac{n_S\langle|\Delta\epsilon_{ij}(\boldsymbol{q})l_{Ii}l_{Sj}|^2\rangle}{n_I \cos^2\delta_I \cos^2\delta_S} \equiv VQ_S\langle|\Delta\epsilon_{ij}(\boldsymbol{q})l_{Ii}l_{Sj}|^2\rangle\,, \tag{99}$$

where the intensity of an incident beam is considered to be equal to unity and $\delta_I$ and $\delta_S$ are the angles between the electric strength $\boldsymbol{E}$ and electric induction $\boldsymbol{D}$ vectors in the incident and scattered waves, respectively; for a given incident normal wave the total scattered light intensity is equal to the sum of expressions (99) for both normal scattered waves. In a medium that has isotropic optical properties (e.g., in cubic crystals) expression (99) comes down to eq. (97) with the substitution of $\langle|\Delta\epsilon_{ij}(\boldsymbol{q})l_{Ii}l_{Sj}|^2\rangle$ for $\langle|\Delta\epsilon(\boldsymbol{q})|^2\rangle$, the unit polarization vectors $\boldsymbol{l}_I$ and $\boldsymbol{l}_S$ in this case being perpendicular to $\boldsymbol{k}_I$ and $\boldsymbol{k}_S$, respectively. Certainly, the assumption that $\Delta\epsilon_{ij} = \Delta\epsilon\delta_{ij}$ leads already precisely to eq. (97) since the unit vector $\boldsymbol{l}_S$ may be chosen now to lie in the plane $(\boldsymbol{k}_S, \boldsymbol{l}_I)$ and, of course, perpendicular to $\boldsymbol{k}_S$.

Changing the incident light polarization and measuring separately the intensity of both the normal scattered waves (i.e., in essence, determining the polarization of the scattered light), one can obtain information about fluctuations of different components of the tensor $\epsilon_{ij}$. Besides, by varying the scattering angle, we change the vector $\boldsymbol{q} = \boldsymbol{k}_I - \boldsymbol{k}_S$. Different choices of directions and polarizations of waves in scattering experiments are said to correspond to different scattering geometries, and are designated by special symbols. For example, the symbol $x(zz)y$ implies that the incident beam propagates along the crystallographic $x$-axis and is polarized along the $z$-axis, while the scattered light is polarized along the $z$-axis and observed along the $y$-axis. In this case fluctuations in the component $\epsilon_{33} \equiv \epsilon_{zz}$ are studied. In the geometry designated by $x(yz)y$ the fluctuations in the component $\epsilon_{23} \equiv \epsilon_{yz}$ are determined, and so on. Note also that in eq. (99) the factor $n_S/n_I \cos^2\delta_I \cos^2\delta_S$ can be approximated often by unity. For example for quartz this factor differs from unity by 1–2% only (Motulevich 1950, Fabelinskii 1965). Further on we will neglect the distinction of this factor from unity. Under such an assumption eq. (99) is, in fact, equivalent to eq. (97) if $\Delta\epsilon$ in the last equation is supposed to designate the fluctuation of a given component of $\epsilon_{ij}$ or a certain linear combination of these components as specified by the scattering geometry.

The expressions (97) and (99) give the integrated (over frequencies) light scattering intensity. Information on fluctuations of the tensor $\epsilon_{ij}$ obtained from the measurements of the integral intensity provides no insight into their kinetics. To know how the fluctuations change in time and thus to obtain a better insight into their origin, one should carry out spectral measurements. To get expressions for the spectral density of the scattered light, one must expand the fluctuation $\Delta\epsilon_{ij}(\boldsymbol{r}, t)$ in a Fourier series with respect to time, i.e., introduce the quantities

$$\Delta\epsilon_{ij}(\boldsymbol{q}, \Omega) = \frac{1}{2\pi V} \iint \Delta\epsilon_{ij}(\boldsymbol{r}, t) \exp(-\,\mathrm{i}\boldsymbol{q}\boldsymbol{r} + \mathrm{i}\Omega t)\, \mathrm{d}\boldsymbol{r}\, \mathrm{d}t$$
$$= \frac{1}{2\pi} \int_{-\infty}^{+\infty} \Delta\epsilon_{ij}(\boldsymbol{q}, t)\, \mathrm{e}^{\mathrm{i}\Omega t}\, \mathrm{d}t\,. \tag{100}$$

Then the spectral scattered intensity is

$$\mathscr{I}(\boldsymbol{q}, \Omega) = VQ_S\langle|\Delta\epsilon_{ij}(\boldsymbol{q}, \Omega) l_{Ii} l_{Sj}|^2\rangle\,, \tag{101}$$

the integral intensity

$$I(\boldsymbol{q}) = \int_{-\infty}^{+\infty} \mathscr{I}(\boldsymbol{q}, \Omega)\, \mathrm{d}\Omega \tag{102}$$

being determined by eq. (99). In what follows we shall use frequently the simpler expression corresponding to formula (97)

$$\mathscr{I}(\boldsymbol{q}, \Omega) = VQ_{S0}\langle\Delta\epsilon(\boldsymbol{q}, \Omega)|^2\rangle, \qquad I(\boldsymbol{q}) = \int_{-\infty}^{+\infty} \mathscr{I}(\boldsymbol{q}, \Omega)\, \mathrm{d}\Omega\,,$$

$$\Delta\epsilon(\boldsymbol{q}, \Omega) = \frac{1}{2\pi} \int_{-\infty}^{+\infty} \Delta\epsilon(\boldsymbol{q}, t)\, e^{i\Omega t}\, dt\,. \tag{103}$$

A more straightforward meaning of the result obtained, which goes back to Mandelstam and Brillouin (see, in particular, Fabelinskii (1965)), is that the spectral composition of scattered light is determined by changes in time, i.e., by the kinetics of the permeability wave $\Delta\epsilon_{ij}(\boldsymbol{q}, t)$ with the wavevector $\boldsymbol{q} = \boldsymbol{k}_I - \boldsymbol{k}_S$ (see fig. 7). Note that in eq. (96) the fluctuations $\Delta\epsilon_{ij}(\boldsymbol{r})$ also depend on the time $t$, but when we go over to eqs. (97) and (99), the statistical averaging corresponds to time averaging, and therefore the argument $t$ in $\Delta\epsilon_{ij}$ was not written down since it does not enter in the results (the problem is, of course, considered to be homogeneous in time). Equation (101) does not contain the time either, but it is the amplitude squared of the space–time Fourier transform (100) that has been averaged, and thus the spectral density of intensity is obtained.

### *4.2. Coupling of the dielectric permeability with the order parameter*

We have not specified up to now the nature or, if one chooses, the origin of the permeability fluctuations $\Delta\epsilon_{ij}(\boldsymbol{r}, t)$. From a general point of view these fluctuations are caused by fluctuations in different variables describing the state (configuration) of the system. In off-equilibrium the number of such variables is, in general, very large (as distinct from the equilibrium case, when the state of the system is fully determined by fixing the values of a few thermodynamic parameters, e.g., in the case of a one-component fluid by fixing the values of the density $\rho$ and temperature $T$). However, in practice it is enough usually to take into account the dependence of $\epsilon_{ij}$ on a small number of macroscopic variables. For example, when studying light scattering in simple fluids it is enough in the first approximation to take into account the coupling of the $\epsilon_{ij}$ fluctuations with those in the density and entropy. In the next approximation one should also consider the coupling of $\epsilon_{ij}$ with other quantities, e.g., with the so-called anisotropy tensor (Fabelinskii 1965), density gradients (Andreev 1974) etc. In other words, the set of variables whose fluctuations are to be considered is determined by the character of the problem and requirements on the accuracy of the calculations.

In the study of anomalies of light scattering near phase transitions it is natural to take into account first of all the part of the fluctuations in the tensor $\epsilon_{ij}$ caused by fluctuations in the order parameter components $\eta_\alpha$, the latter fluctuations being the first to increase when approaching the transition temperature. Coupling of $\epsilon_{ij}$ with other variables may be also essential but only as far as the fluctuations in these variables are connected with those in $\eta_\alpha$. Fluctuations in the parameters $\eta_\alpha$ manifest themselves in light scattering in different ways depending on the coupling between $\epsilon_{ij}$ and $\eta_\alpha$, which is determined, in turn, by the crystal symmetry and the transformation properties of the parameters $\eta_\alpha$.

Let us discuss first the possibility of a linear coupling between $\epsilon_{ij}$ and $\eta_\alpha$. Such a coupling is allowed evidently only for those components of the tensor $\epsilon_{ij}$ that in the symmetrical phase possess the same transformation properties under symmetry transformations of this phase as the corresponding components $\eta_\alpha$. In particular, these components of $\epsilon_{ij}$ should be equal to zero in equilibrium in the symmetrical phase. Absorption being neglected, the tensor $\epsilon_{ij}$ may be represented in general as the sum of the symmetric tensor $\epsilon_{ij}^{(S)}$, invariant under the time reversal operation $R$, and the antisymmetric tensor $\epsilon_{ij}^{(A)}$ changing its sign under this operation. For a transition accompanied by the appearance of magnetic ordering, we have $R\eta_\alpha = -\eta_\alpha$ and for a non-magnetic structural phase transition $R\eta_\alpha = \eta_\alpha$. Therefore in the case of magnetic transitions it is only the tensor $\epsilon_{ij}^{(A)}$ that may be coupled linearly with $\eta_\alpha$, and in the case of structural transitions it is the tensor $\epsilon_{ij}^{(S)}$.

The tensor $\epsilon_{ij}^{(A)}$ is coupled linearly with $\eta_\alpha$, e.g., in the case of ferromagnetic transitions when the $\eta_\alpha \equiv M_k$ have a sense of certain components of the magnetization vector. Indeed, being an axial vector and an anti-symmetric tensor of rank two, $M$ and $\epsilon_{ij}^{(A)}$ possess the same transformation properties. Thus, in this case we have (see e.g., L'vov (1966) and Moriya (1968))

$$\epsilon_{ij}^{(A)} = \tilde{a}_{ijk} M_k \,. \tag{104}$$

For cubic crystals

$$\tilde{a}_{ijk} = \tilde{a} e_{ijk} \,, \tag{105}$$

where $e_{ijk}$ is the fully antisymmetric unit tensor of rank three. The constant $\tilde{a}$ can be determined from measurements of a magnetic rotation of the polarization plane, i.e., by the Faraday effect (Landau and Lifshitz 1957).

In the case of structural phase transitions the presence of a linear coupling between $\epsilon_{ij}^{(S)}$ and $\eta_\alpha$ means that the order parameter is linearly coupled also with the strain tensor $u_{ij}$ (this tensor, like $\epsilon_{ij}^{(S)}$, is a symmetric tensor of rank two). As the transformation properties of both $\eta_\alpha$ and some components $u_{ij}$ are the same one can choose for the order parameter the corresponding components $u_{ij}$ themselves. The transitions for which the role of $\eta_\alpha$ is played by strains $u_{ij}$ or some other variables whose transformation properties are equivalent to those of $u_{ij}$ (these variables may be, in particular, some components of the polarization vector $\boldsymbol{P}$) are referred to as proper ferroelastic transitions. In the case of ferroelastic transitions fluctuations in $\eta_\alpha$ and in the component (or components) $u_{ij}$ coupled linearly with $\eta_\alpha$ should be treated at the same footing in the light scattering study. For example, for a one-component order parameter coupled linearly with the strain component $u_{xy}$ we have

$$\epsilon_{xy}^{(S)} = \tilde{a}_1 \eta + \tilde{b}_1 u_{xy} \,. \tag{106}$$

For structural phase transitions which are not proper ferroelastic the expansion of $\epsilon_{ij}^{(S)} \equiv \epsilon_{ij}$ in a power series in $\eta_\alpha$ begins with a quadratic term. Besides, since for a magnetic phase transition only bilinear combinations of the components $\eta_\alpha$ may be invariant under time reversal, the dependence of $\epsilon_{ij}^{(S)}$ on $\eta_\alpha$ can also be only quadratic here.

Among bilinear combinations of the components $\eta_\alpha$ there always exists an invariant (scalar) combination $\Sigma_\alpha \eta_\alpha \eta_\alpha \equiv \eta_\alpha \eta_\alpha \equiv \eta^2$. For the components $\epsilon_{lm}$ of the tensor $\epsilon_{ij}$ whose equilibrium values are non-zero in the symmetrical phase this combination is unique so that for the components pointed out, we have

$$\epsilon_{lm} = \epsilon_{lm}^{(0)} + a_{lm}\eta^2 . \tag{107}$$

In the case of a multi-component order parameter one can form also non-invariant bilinear combinations from the components $\eta_\alpha$. Some of them (say $\eta_\alpha \eta_\beta$) may turn out to possess the same transformation properties as a component of the tensor of rank two. Let this component be $\epsilon_{l'm'}$, then one has

$$\epsilon_{l'm'} = a_{l'm'}\eta_\alpha \eta_\beta . \tag{108}$$

Notice that the component of the tensor $\epsilon_{ij}$ is equal to zero in equilibrium in the symmetrical phase. As several non-symmetrical phases may exist in the case of a multi-component order parameter, in a given non-symmetrical phase the equilibrium value of the component $\epsilon_{l'm'}$ may be either zero or non-zero.

Mention also that even for the transition for which some components of $\epsilon_{ij}$ are coupled linearly with $\eta_\alpha$ one can, as follows from eq. (99), by choosing in an appropriate way the scattering geometry, study that part of the scattering which is caused by fluctuations in the components $\epsilon_{ij}$ depending only quadratically on $\eta_\alpha$.

Thus in most cases $\epsilon_{ij}$ is coupled with $\eta_\alpha$ only quadratically. As the coupling is rather weak the contribution of fluctuations in $\eta_\alpha$ to the anomalous part of the light scattering may well turn out to be comparable to that of other variables coupled linearly with $\eta_\alpha$. This may occur even if the fluctuations of these variables increase near the transition much slower than those of $\eta_\alpha$. An example of this kind was first pointed out by Krivoglaz and Rybak (1957). It has been shown by these authors that fluctuations of polarization and concentration are equally important near the tricritical point in a ferroelectric solid solution. A general treatment of the question (Levanyuk and Sobyanin 1967) shows that within the Landau phase transition theory such a situation is, in fact, common to all phase transitions. Besides concentrations the role of the corresponding variables may be played by normal coordinates of fully symmetric lattice vibrations, as well as by the entropy $S$, density $\rho$ etc. Below, for the case of quadratic coupling between $\epsilon_{ij}$ and $\eta_\alpha$, it is fluctuations in the density that are considered in addition to those in $\eta_\alpha$. This is because the permeability $\epsilon_{ij}$ usually depends most strongly on $\rho$ and because of the peculiar long-range character of elastic shear deformations accompanying the $\rho$

fluctuations in solids (see sect. 3.3.1). Instead of $\rho$ fluctuations one can consider also fluctuations in the displacement vector $\boldsymbol{u}(\boldsymbol{r})$ or fluctuations in the volume deformation $v = \operatorname{div} \boldsymbol{u} \equiv u_{ll}$:

$$\Delta v = \Delta u_{ll} = \operatorname{div}[\Delta \boldsymbol{u}(\boldsymbol{r})] = -\Delta\rho/\rho \tag{109}$$

To avoid cumbersome expressions we shall assume as usual that the transition is characterized by a single-component order parameter $\eta$. Then omitting for brevity the tensor indexes and taking into account besides $\eta$ only the quantity $v \equiv u_{ll}$ one can write

$$\Delta\epsilon = a\Delta(\eta^2) + b\Delta v = 2a\eta_e\Delta\eta + a(\Delta\eta)^2 + b\Delta v\,, \tag{110}$$

where $a$ and $b$ are the corresponding coupling constants. We see that in the case under consideration a linear coupling of $\Delta\epsilon$ with $\Delta\eta$ vanishes in the symmetrical phase where $\eta_e = 0$.

We shall be interested in Fourier transforms of the fluctuations

$$\begin{aligned}
&\Delta\eta = \sum_{\boldsymbol{k}} \eta(\boldsymbol{k})\,\mathrm{e}^{\mathrm{i}\boldsymbol{kr}}, \qquad \eta(-\boldsymbol{k}) = \eta^*(\boldsymbol{k}), \qquad \eta(\boldsymbol{k}) = \frac{1}{V}\int \Delta\eta\,\mathrm{e}^{-\mathrm{i}\boldsymbol{kr}}\,\mathrm{d}\boldsymbol{r}\,,\\
&\Delta v = \sum_{\boldsymbol{k}} v(\boldsymbol{k})\,\mathrm{e}^{\mathrm{i}\boldsymbol{kr}}, \qquad v(-\boldsymbol{k}) = v^*(\boldsymbol{k})\,,\\
&\Delta\epsilon = \sum_{\boldsymbol{k}} \Delta\epsilon(\boldsymbol{k})\,\mathrm{e}^{\mathrm{i}\boldsymbol{kr}}, \qquad \Delta\epsilon(-\boldsymbol{k}) = \Delta\epsilon^*(\boldsymbol{k})\,,
\end{aligned} \tag{111}$$

Now we substitute eq. (111) into eq. (110), multiply the expression obtained and its complex-conjugate and carry out a statistical averaging. At the present stage, as before (sect. 4.1), we do not need, however, to consider the state in question to be equilibrium, but only assume that due to averaging all the interference terms vanish, and we have to calculate the quantity $\langle|\Delta\epsilon(\boldsymbol{q})|^2\rangle$. For this we find (c.c. is a complex-conjugate expression)

$$\begin{aligned}
\langle|\Delta\epsilon(\boldsymbol{q})|^2\rangle &= 4a^2\eta_e\langle|\eta(\boldsymbol{q})|^2\rangle + 2ab\eta_e\langle v(\boldsymbol{q})\eta(-\boldsymbol{q}) + \text{c.c.}\rangle\\
&\quad + b^2\langle|v(\boldsymbol{q})|^2\rangle + 2a^2\eta_e\sum_{\boldsymbol{k}}\langle\eta(\boldsymbol{k})\eta(\boldsymbol{q})\eta(-\boldsymbol{q}-\boldsymbol{k}) + \text{c.c.}\rangle\\
&\quad + ab\sum_{\boldsymbol{k}}\langle v(\boldsymbol{q})\eta(\boldsymbol{k})\eta(-\boldsymbol{q}-\boldsymbol{k}) + \text{c.c.}\rangle\\
&\quad + a^2\sum_{\boldsymbol{k},\boldsymbol{k}'}\langle\eta(\boldsymbol{k})\eta(\boldsymbol{k}')\eta(\boldsymbol{q}-\boldsymbol{k})\eta(-\boldsymbol{q}-\boldsymbol{k}')\rangle\,.
\end{aligned} \tag{112}$$

The first three contributions here correspond to the ordinary or the first-order scattering (in the sense the intensity of this scattering $I_1$ is determined by the first-order fluctuations). The last three terms in eq. (112) define, as one says, the second-order scattering. Although its intensity $I_2$ is expressed through the higher-order fluctuations, it may be quite essential especially above the transition point.

As is seen from (110) and (112) the first-order scattering by $\eta$ fluctuations is absent in the symmetrical phase ($\eta_e = 0$) and is always non-zero in the non-symmetrical phase. As to structural phase transitions this conclusion may also be expressed in another language used more frequently in the literature on light scattering. With this aim we recall at first that for a displacive transition the order parameter may be identified with a set of normal coordinates of a soft phonon mode in the symmetrical phase (in the case of a single-component order parameter this mode is non-degenerate). The absence of first-order scattering then means that this mode is not active in the Raman scattering. On the contrary, in the non-symmetrical phase the mode becomes active in the Raman scattering. This is clear both from formula (110) and from the fact that in the non-symmetrical phase the $\eta$ vibrations involve no change of the symmetry of this phase. Hence the $\eta$ vibrations are fully symmetric here. This statement is strictly valid only for a single-component order parameter. For a multi-component order parameter there will be several distinct phonon modes corresponding to $\eta$ vibrations in the non-symmetrical phase (see, for example, Levanyuk and Sannikov 1974). In a given non-symmetrical phase some constraints may exist on the equilibrium values of the components $\eta_\alpha$, e.g., $\eta_{\alpha,e} = \eta_{\beta,e}$ or $\eta_{\alpha,e} \neq 0$, $\eta_{\beta,e} = 0$ etc. Then only those changes in $\eta_\alpha$ which satisfy these constraints will correspond to the fully symmetric vibrations while the activity in the Raman scattering of other modes associated with $\eta$ vibrations should be determined separately in each particular case from relations of the type (108). At the same time, some non-symmetrical phases are possible for which there are no symmetry constraints on $\eta_{\alpha,e}$. In this case all the $\eta$ vibrations will be fully symmetric and active in the Raman scattering. It is also possible that in an expression of the type (108) both components $\eta_\alpha$ and $\eta_\beta$ are zero in equilibrium in a given non-symmetrical phase. Then the fluctuations of the above-mentioned components do not lead to a first-order scattering in the non-symmetrical phase as well, i.e., among the phonon modes corresponding to $\eta$ vibrations there are some which are not active in scattering both above and below $T_c$. From these remarks it is seen already how many possibilities may occur if the order parameter is a multi-component one. We shall not, however, touch upon these possibilities as the most important features of light scattering near phase transitions may be understood by considering the example of a one-component order parameter.

## 5. *Integral intensity of light scattering*

The integral or total (integrated over frequencies) scattered light intensity is determined by permeability fluctuations $\Delta\epsilon_{ij}$ which near phase transitions depend most essentially on fluctuations in the order parameter and other quantities, as far as these quantities are coupled with it (see, for example, eq. (112)). The present section concerns the intensity of light scattering in ideal systems, i.e., the intensity of the so-called molecular scattering (Fabelinskii 1965), scattering by defects being treated in ch. 2.

We begin with discussing first-order scattering and restrict ourselves at first to the temperature region or to the media for which the Landau phase transition theory holds. As mentioned already for all structural phase transitions which are not proper ferroelastic the tensor $\epsilon_{ij}$ is coupled quadratically with the order parameters. Besides, in ferroelastics too a part of the components $\epsilon_{ij}$ depends on $\eta_\alpha$ only quadratically. Therefore first of all we consider this most widespread case. The scattering for a linear dependence of $\epsilon_{ij}$ on $\eta_\alpha$ is discussed in sect. 5.3.2, while the specific features of first-order scattering in critical region will be touched upon in sect. 5.4.

### 5.1. *First-order scattering by order parameter fluctuations (classical region)*

We start our discussion with the consideration of the contribution of order parameter fluctuations to the scattered light intensity, as has been done originally by Ginzburg (1955) and by Ginzburg and Levanyuk (1958). Then proper generalizations will be made connected with the necessity to consider, along with $\eta$ fluctuations, fluctuations in some other variables, long-range forces effects and some extra factors. As a result, the expressions in this and the following subsection are applicable, strictly speaking, merely to fluids.

In the case of a quadratic coupling between $\epsilon_{ij}$ and $\eta$ the first-order scattering takes place for $\tau < 0$ only (see formula (112)). From eqs. (97) and (112) it then follows that

$$I_1(\boldsymbol{q}) = 4Q_{S0}Va^2\eta_e^2\langle|\eta(\boldsymbol{q})|^2\rangle . \tag{113}$$

Using eqs. (31) and (3) we have (Ginzburg and Levanyuk 1958)

$$I_1(\boldsymbol{q}) = \frac{4Q_{S0}a^2\eta_e^2k_BT}{\phi_{\eta\eta} + Dq^2} = Q_{S0}\frac{4a^2\eta_e^2k_BT}{2\eta_e^2(B^2-4AC)^{1/2} + Dq^2} . \tag{114}$$

At the second-order phase transition point $\eta_e = 0$, and the intensity $I_1(\boldsymbol{q})$ vanishes, as expected. Since for light the value of $q \sim 10^5\,\mathrm{cm}^{-1}$ and thus may be considered small, the fall of $I_1(\boldsymbol{q})$ to zero takes place only in a very narrow vicinity of the point $\tau = 0$. Its width $\tau_q$ is given by the condition $Dq^2/\phi_{\eta\eta} = r_c^2q^2 = r_{c0}^2q^2\tau_q^{-1} \sim 1$, whence for structural phase transitions, when $r_{c0} \sim 10^{-7}$–$10^{-8}$ cm, one has $\tau_q \sim 10^{-4}$–$10^{-6}$. For simplicity we shall assume below that $q = 0$ thus disregarding this narrow region.

For second-order phase transitions far from the tricritical point when one may take $C = 0$, the intensity $I_1$, as follows from eq. (114), is practically temperature-independent and is equal to (for $\tau < 0$)

$$I_1(\boldsymbol{q} \simeq 0) = 2Q_{S0}a^2k_BT/B. \tag{115}$$

Thus, the light scattering intensity should undergo a jump at the transition point just equal to expression (115).

On the second-order phase transition line, where $A(T_c) = 0$, formula (115) holds up to the tricritical point, where $B(p, T_c) = 0$ (see fig. 2). Therefore, when approaching the tricritical point along the line of second-order phase transitions (and generally for $B^2 \gg 4AC$) the light scattering intensity increases according to the law

$$I_1(\boldsymbol{q} \simeq 0) \sim (T_{tc} - T)^{-1}, \tag{116}$$

because $B \sim (T_{tc} - T)$ near the tricritical point.

When approaching this point along the directions with $B^2 < 4AC$, the intensity

$$I_1(\boldsymbol{q} \simeq 0) \sim (T_{tc} - T)^{-1/2}, \tag{117}$$

as evident from eq. (114) with $A \sim (T_{tc} - T)$.

On the line of first-order phase transitions, where the relation $3B^2 = 16AC$ holds (see eq. (9)),

$$I_1(\boldsymbol{q} \simeq 0) = 4Q_{S0}a^2k_BT|B|. \tag{118}$$

Hence, when approaching the tricritical point along the line of both first-order and second-order phase transitions, the light scattering intensity increases by one and the same law but with the coefficients differing by a factor two (cf. (115) and (118)).

From eq. (114) it follows that in the case of first-order phase transitions the light scattering intensity increases also when approaching the spinodal of the non-symmetrical phase, for which $B^2 = 4AC$ (see eq. (12) and fig. 2); since here $B^2 - 4AC = \text{const} \times (T_{S1} - T)$, the intensity

$$I_1(\boldsymbol{q} \simeq 0) \sim (T_{S1} - T)^{-1/2}. \tag{119}$$

One can rewrite formula (114) for the light scattering intensity expressing $I_1(\boldsymbol{q} \simeq 0)$ through directly measurable quantities such as the jump in the specific heat $\Delta C_p$ and the temperature derivative of the equilibrium dielectric constant $d\epsilon_e/dT$. One has

$$\frac{d\epsilon_e}{dT} \equiv \left(\frac{\partial \epsilon}{\partial T}\right)_{p,h=0} = \left(\frac{\partial \epsilon}{\partial T}\right)_{p,\eta} + \left(\frac{\partial \epsilon}{\partial \eta}\right)_{p,T} \left(\frac{\partial \eta_e}{\partial T}\right)_{p,h=0}, \tag{120}$$

where $h$ is the generalized force conjugate to the parameter $\eta$. The quantity $(\partial\epsilon/\partial T)_{p,\eta}$ has no singularity at $\tau = 0$ in the framework of the Landau theory and can be found by extrapolating below $T_c$ the temperature dependence of $\epsilon_e$ in the symmetrical phase (away from $T_c$). The derivative $(\partial\eta_e/\partial T)_{p,h=0}$ is found by differentiating the equilibrium equation $(\partial\phi/\partial\eta)_{\eta=\eta_e}=0$ with respect to $T$. We have

$$\left(\frac{\partial\eta_e}{\partial T}\right)_{p,h=0}=-\left(\frac{\partial^2\phi}{\partial\eta\partial T}\right)_{p,h=0}\Bigg/\left(\frac{\partial^2\phi}{\partial\eta^2}\right)_{p,h=0,\,T}\equiv-\phi_{\eta T}/\phi_{\eta\eta}\,. \tag{121}$$

On the other hand, differentiating the equality $(\partial\phi/\partial T)_{p,h=0}=(\partial\phi/\partial T)_{p,\eta}$ valid in equilibrium (when $\partial\phi/\partial\eta=0$), we find

$$\phi^h_{TT}=\phi^\eta_{TT}+\phi_{\eta T}(\partial\eta_e/\partial T)_{p,h=0}=\phi_{TT}-\phi^2_{\eta T}/\phi_{\eta\eta}$$

so that

$$\Delta C_p\equiv-T(\phi^h_{TT}-\phi^\eta_{TT})=T(\phi_{\eta T})^2/\phi_{\eta\eta}=T\left(\frac{\partial\eta_e}{\partial T}\right)^2_{p,h=0}\phi_{\eta\eta}\,, \tag{122}$$

where $\Delta C_p$ is the difference between the specific heats of the phases (or the specific heat "jump") as given by the Landau theory. Therefore, according to eqs. (122) and (31)

$$\left(\frac{\partial\eta_e}{\partial T}\right)^2_{p,h=0}=\frac{\Delta C_p}{T\phi_{\eta\eta}}=\frac{V\Delta C_p}{k_BT^2}\langle(\Delta\eta)^2\rangle,\qquad\langle(\Delta\eta)^2\rangle\equiv\langle|\eta(\boldsymbol{q}=0)|^2\rangle\,. \tag{123}$$

Substituting eq. (123) in eq. (113) with the use of eq. (120) and taking into account that, by definition, $(\partial\epsilon/\partial\eta)_{p,T}=2a\eta_e$ (see, for example, eq. (110)) we find (Levanyuk and Sobyanin 1967)

$$I_1(q\simeq0)=\frac{Q_{S0}k_BT^2}{\Delta C_p}\left[\left(\frac{\partial\epsilon}{\partial T}\right)_{p,h=0}-\left(\frac{\partial\epsilon}{\partial T}\right)_{p,\eta=0}\right]^2. \tag{124}$$

The expression for $I_1(\boldsymbol{q}\simeq0)$ can be written also in another form more convenient when the phase transition is caused by a change in pressure. Proceeding as before we find

$$I_1(\boldsymbol{q}\simeq0)=\frac{Q_{S0}k_BT}{\Delta\beta_T}\left[\left(\frac{\partial\epsilon}{\partial p}\right)_{T,h=0}-\left(\frac{\partial\epsilon}{\partial p}\right)_{T,\eta=0}\right]^2, \tag{125}$$

where $\Delta\beta_T\equiv(\partial^2\phi/\partial p^2)_{T,h=0}-(\partial^2\phi/\partial p^2)_{T,\eta}$ is the difference of the compressibility values below and above $T_c$.

### *5.2. Account of fluctuations in other variables*

Although expressions (124) and (125) were obtained for the case when the light is scattered by fluctuations in one variable only, which is the order parameter itself, they can be extended easily to the case of any number of variables. To this end, one should precise merely the meaning of the derivatives entering formulae (124), (125). Namely, these derivatives should be calculated at fixed (zero) forces conjugate to all generalized coordinates other than the $\eta$.

To show this, consider, for example, beside $\eta$, only one more generalized coordinate $\xi$, i.e., put

$$\Delta\epsilon = \epsilon_\eta \Delta\eta + \epsilon_\xi \Delta\xi \,, \tag{126}$$

where $\epsilon_\eta \equiv (\partial\epsilon/\partial\eta)_{p,T,\xi}$, $\epsilon_\xi \equiv (\partial\epsilon/\partial\xi)_{p,T,\eta}$. From eqs. (97) and (126) it then follows that

$$I_1(\boldsymbol{q} \simeq 0) = Q_{S0} V \{\epsilon_\eta^2 \langle(\Delta\eta)^2\rangle + 2\epsilon_\eta\epsilon_\xi\langle\Delta\eta\Delta\xi\rangle + \epsilon_\xi^2\langle(\Delta\xi)^2\rangle\} \,. \tag{127}$$

Using eqs. (41) and (42) this can be rewritten as

$$I_1(\boldsymbol{q} \simeq 0) = Q_{S0} V \left\{\left[\epsilon_\eta + \epsilon_\xi \left(\frac{\mathrm{d}\xi_0}{\mathrm{d}\eta}\right)_{\eta=\eta_e}\right]^2 \langle(\Delta\eta)^2\rangle + \epsilon_\xi^2 \langle(\Delta\xi')^2\rangle\right\}. \tag{128}$$

It is easily seen that the last term in this expression has a much weaker temperature dependence than the first one. Indeed, let us assume at first that the variable $\xi$ possesses the same transformational properties as $\eta$. Then the derivative $\epsilon_\xi$ is proportional to $\epsilon_\eta$ and both go to zero like $\eta_e$ as $T \to T_c^-$. At the same time the fluctuation $\langle(\Delta\xi')^2\rangle$ is determined by the modulus $\phi_{\xi\xi}^\eta$ (see eq. (37)) which behaves regularly at $T = T_c$. Thus in this case the last term in eq. (128) contributes only to a change of slope of the curve $I_1(T)$ at $T = T_c$, i.e., it has a much weaker anomaly than the first term. On the other hand, if the transformational properties of $\xi$ are different from those of the $\eta$ (as is the case, for example, when $\xi$ is the density $\rho$ or entropy $S$) then both the derivative $\epsilon_\xi$ and the fluctuation $\langle(\Delta\xi')^2\rangle$ depend regularly on $\tau$ and the second term in eq. (128) does not contribute at all to the anomaly of the intensity.

Therefore we neglect the above-mentioned term and take into consideration that the quantity $\epsilon_\eta + \epsilon_\xi(\mathrm{d}\xi_0/\mathrm{d}\eta)_{\eta=\eta_e}$ has the meaning of the derivative of $\epsilon$ with respect to $\eta$ at a fixed generalized force $g$ conjugate to $\xi$. It is seen then that expression (128) coincides with (113) (where, we recall, $2a\eta_e \equiv \epsilon_\eta = (\partial\epsilon/\partial\eta)_{\eta=\eta_e}$) and, with the reasonings analogous to the above (see sects. 5.1 and 3.2), leads to relations (124), (125) where all the derivatives are to be calculated now at $g = \text{const}$. Thus, taking an account of fluctuations of other variables beside $\eta$ is reduced formally to the concretizing in the sense of the derivatives in eqs. (124) and (125). At the same time it should be stressed that the contribution of "the other variables" to the anomalous part of the light scattering intensity may be quite large by itself. This follows already from the fact that eq. (128) still holds even when $\epsilon$ depends, for example, on the density $\rho$ alone, i.e., when $\epsilon_\eta = 0$, $\xi = \rho$. Fluctuations in $\eta$ then manifest themselves in light scattering only due to their coupling with those in $\rho$. In the Landau theory this, however, does not affect the temperature dependence of the intensity $I_1(q \simeq 0)$, since in eq. (128) both the quantities $\epsilon_\eta$ and $\epsilon_\xi(\mathrm{d}\xi_0/\mathrm{d}\eta)_{\eta=\eta_e}$ are proportional to $\eta_e$, and specifically, in the case under discussion (see sect. 3.2)

$$\epsilon_\eta = 2a\eta_e, \qquad \epsilon_\xi\left(\frac{\mathrm{d}\xi_0}{\mathrm{d}\eta}\right)_{\eta=\eta_e} = \left(\frac{\partial\epsilon}{\partial\rho}\right)_{\eta,T}\left(\frac{\mathrm{d}\rho_e}{\mathrm{d}\eta}\right)_{\eta=\eta_e} = -\epsilon_\rho\frac{\phi_{\rho\eta}}{\phi_{\rho\rho}} \propto \eta_e. \tag{129}$$

Notice in conclusion that within the Landau theory the quantities $\Delta C_p$ and $\Delta\beta_T$ either do not depend on $\tau$ (for second-order transitions far from the tricritical point), or increase like $(T_{tc} - T)^{-1/2}$ (for a tricritical transition). Hence, according to eqs. (124) and (125), the anomaly of the light scattering intensity at $T = T_c$ is connected exclusively with the growth of the derivatives $(\partial\epsilon_e/\partial T)_p$ and $(\partial\epsilon_e/\partial p)_T$ as $T \to T_c^-$.

### 5.3. *Long-range field effects*

#### 5.3.1. *Influence of shear stresses*

In solids, distinct from fluids, any spatially inhomogeneous density variation induces long-range shear stresses. As a result, the magnitudes of spatially inhomogeneous fluctuations in $\eta$ and $\rho$ in solids prove to be separated by a "gap" from those of the purely homogeneous (see sect. 3.3.1). This can reduce drastically the scattered light intensity for small but finite values of $\boldsymbol{q}$. To see this, we substitute expression (57)–(59), valid for vanishing but non-zero values of $\boldsymbol{q}$, in the first three terms of formula (112). Instead of eq. (114) (with $\boldsymbol{q} \simeq 0$) one thus obtains

$$I_1(\boldsymbol{q} \simeq 0) = Q_{s0} k_B T \left\{ \frac{[2a - br/\tilde{K}]^2}{2(B^2 - 4AC)^{1/2} + \frac{2}{3}(\mu r^2/K\tilde{K})} + \frac{b^2}{\tilde{K}} \right\}. \tag{130}$$

The right-hand side of expression (130) reaches the maximum for $B^2 - 4AC = 0$, i.e., at the tricritical point and along the spinodal line of the non-symmetrical phase in the case of first-order transitions. However, as distinct from phase transitions in liquids (and, generally speaking, in liquid crystals) the corresponding maximum value of the intensity is finite and does not differ significantly from the intensity of the usual (non-critical) part of scattering away from the transition temperature. Indeed, one can assume usually for estimates that the coefficient $a$ is equal to zero, and the light is scattered by the density fluctuations alone. In this case the ratio of the first to the second term in eq. (130), which is the same in both the phases and corresponds to the "normal" part of scattering, makes up to

$$\frac{3}{2}\frac{K}{\mu}\left[1 + \frac{3K\tilde{K}}{\mu r^2}(B^2 + 4AC)^{1/2}\right]^{-1}. \tag{131}$$

Since $K$ exceeds $\mu$ by no more than a small factor, evidently the light scattering intensity may increase no more than several times as $\tau \to 0$. This circumstance is, of course, very essential for the discussion of experimental data (see sect. 5.7).

Equation (130) can be represented also in a form analogous to eqs. (124) and (125). For this one should take into consideration that, as mentioned in sect. 3.3.1, only the deformation $u_{zz} \equiv u$ is responsible for the density change due to the fluctuation in $\eta$ with the wavevector $\boldsymbol{q}$ (the $z$-axis is directed along $\boldsymbol{q}$).

Therefore all the derivatives in expression (124) should be calculated now for a constant uniaxial stress $\sigma_{zz} \equiv \sigma$ and zero deformations $u_{xx}$ and $u_{yy}$, i.e., for a partially "clamped" crystal that may be deformed only in one direction. Correspondingly, in expression (125) the derivatives with respect to $p$ should be replaced by those with respect to $\sigma$ at fixed $u_{xx}$, $u_{yy}$ and $T$. As a result, instead of eqs. (124) and (125), one obtains

$$I_1(\boldsymbol{q} \simeq 0) = \frac{Q_{S0} k_B T^2}{\Delta C_\sigma} \left[ \left( \frac{\partial \epsilon}{\partial T} \right)_{\sigma, h, u_{xx}, u_{yy}} - \left( \frac{\partial \epsilon}{\partial T} \right)_{\sigma, \eta, u_{xx}, u_{yy}} \right]^2, \tag{132}$$

$$I_1(\boldsymbol{q} \simeq 0) = \frac{Q_{S0} k_B T}{\Delta \tilde{\beta}_T} \left[ \left( \frac{\partial \epsilon}{\partial \sigma} \right)_{h, T, u_{xx}, u_{yy}} - \left( \frac{\partial \epsilon}{\partial \sigma} \right)_{\eta, T, u_{xx}, u_{yy}} \right]^2, \tag{133}$$

where

$$\Delta C_\sigma = C_{h,\sigma,u_{xx},u_{yy}} - C_{\eta,\sigma,u_{xx},u_{yy}} \quad \text{and} \quad \Delta \tilde{\beta}_T = (\partial^2 \phi / \partial \sigma^2)_{h,T,u_{xx},u_{yy}} - (\partial^2 \phi / \partial \sigma^2)_{\eta,T,u_{xx},u_{yy}}.$$

Note that the quantity $(\partial^2 \phi / \partial \sigma^2)_{h,T,u_{xx},u_{yy}} = (K_T + \frac{4}{3}\mu)^{-1}$ can be determined, for example, from measurements of the longitudinal sound velocity while the value of the derivative $(\partial \epsilon / \partial \sigma)_{h,T,u_{xx},u_{yy}}$ can be extracted from studies of the light diffraction by a high-frequency sound wave. For verification of relations of the type (124), (125), (132) and (133) it would be desirable to carry out the whole complex of necessary measurements under identical conditions for one and the same sample. Of course, all the expressions will be more complicated with an account taken of the real anisotropy of elastic properties of the crystal and for an arbitrary direction of the wavevector $\boldsymbol{q}$.

### *5.3.2. Specific features of light scattering in ferroelectrics and ferroelastics*

When considering above shear deformations, we have already taken into account some long-range interactions inherent to a solid. However, long-range effects may manifest themselves still more strikingly when fluctuations in $\eta$ are accompanied by the appearance of macroscopic electric and magnetic fields, or of elastic stresses in some strongly anisotropic bodies (for example, in proper ferroelastics).

It has been shown in sect. 3.3.2 that in proper ferroelectrics with three and two spontaneous polarization axes the presence of long-range forces (macroscopic electric field) leads to an effective decrease of the number of the order parameter components whose fluctuations increase as $\tau \to 0$. The qualitative character of light scattering anomalies remains, however, the same here as in the preceding sections.

In a uniaxial ferroelectric the only fluctuations to increase as $\tau \to 0$ are those corresponding to $q$ perpendicular to the spontaneous polarization axis (see eq. (68)). Therefore the light scattering anomaly in such substances must have a strong angular dependence. Note that in the case of ferroelectrics the quantities $a$ and $\phi_{\eta\eta}$, entering in eq. (114) can easily be measured, because $a$ is there the so-called quadratic electro-optical coefficient and $\phi_{\eta\eta} = 4\pi/(\epsilon_0 - 1)$, where $\epsilon_0$ is the corresponding component of the static dielectric permeability tensor. Taking into account besides $\eta$ the fluctuations in other generalized coordinates comes down as before to a specification of the conditions under which the quantities $a$ and $\epsilon_0$ should be calculated or measured.

In the case of proper ferroelastic transitions, when one can choose for the order parameter a component (or a combination of components) of the strains $u_{ij}$, the light scattering anomaly occurs both in the symmetrical and the non-symmetrical phase (in this case some components of the tensor $\epsilon_{ij}$ are to depend linearly on $\eta$; see sect. 4.2). A strong angular dependence of the light scattering anomaly is to be observed here. In a ferroelastic with a one-component order parameter (say $u_{xy}$) the most pronounced anomaly is expected for wavevectors $\boldsymbol{q}$ along the $x$- or $y$-axis. For such $\boldsymbol{q}$ the intensity of scattering is given by the expression

$$I_1(q_x, q_y) = Q_S\left(\frac{\partial\epsilon}{\partial u_{xy}}\right)^2 \frac{k_B T}{C_{xyxy}}, \tag{134}$$

where $C_{xyxy}$ is the static elastic modulus for the strain component $u_{xy}$, and $\partial\epsilon/\partial u_{xy}$ is the corresponding piezo-optical coefficient. The values of both these quantities are calculated, naturally, at fixed forces conjugate to all the generalized coordinates coupled linearly with $u_{xy}$. For example, in a ferroelectric which is at the same time a ferroelastic (as is the case, in particular, with $KH_2PO_4$ crystals), the quantities $\partial\epsilon/\partial u_{xy}$ and $C_{xyxy}$ should be calculated at a constant electric field $E$. Thus we take implicitly into account in expression (134) the contribution from polarization fluctuations also. As is seen from sect. 6.2.2, formula (134) provides us with the intensity of Mandelstam–Brillouin components if the characteristic frequency of polarization fluctuations is higher than that of strain fluctuations. If not, formula (134) incorporates contributions from both the side bands and the central peak (due to polarization fluctuations).

### 5.4. *First-order scattering in critical region*

The existence of a well pronounced critical region has been established for some second-order magnetic transitions (see, e.g., Stanley (1973)), not to mention the $\lambda$ transition in liquid helium (Ginzburg and Sobyanin 1976, 1982) and critical points in fluids and liquid mixtures (ch. 9). One may expect that the development of experimental techniques and the improvement of specimen quality will permit after all a reliable exposure of this region for structural phase transitions too. A theoretical consideration of light scattering in the critical region is therefore of evident interest.

We shall restrict ourselves, as before, to the case of a one-component order parameter for simplicity. Besides, the dependence of $\epsilon_{ij}$ on $\eta$ will be considered to be quadratic (a linear coupling between $\epsilon_{ij}$ and $\eta$ occurs only for proper ferroelastics, in which case there is no critical region at all; see sect. 3.3.3). Furthermore, consider along with fluctuations in $\eta$ only those in the density $\rho$. The scattering light intensity is given then by eqs. (97) and (112). Designate the contributions to the intensity from the first three terms in eq. (112) by $I_{\eta\eta}$, $I_{\eta\rho}$ and $I_{\rho\rho}$, respectively and analyze the temperature dependence of each of the contributions separately. Using eqs. (112) and (31) and taking as usual first $qr_c \ll 1$ we find (Fleury 1972, Levanyuk 1976)

$$I_{\eta\eta} \propto \eta_e^2 \phi_{\eta\eta}^{-1} \propto |\tau|^{2\beta-\gamma} \propto |\tau|^{2(1-\gamma)-\alpha}, \tag{135}$$

where we used the scaling relation (see, e.g., Landau and Lifshitz (1976))

$$2 - \alpha = \gamma + 2\beta. \tag{136}$$

It is seen from eq. (135) that in the critical region (where $\gamma > 1$ and $\alpha \simeq 0$) the intensity $I_{\eta\eta}$ increases rapidly for $\tau < 0$ when approaching the transition temperature, distinct from the classical region where the intensity is practically independent of $\tau$ (at least for second-order phase transitions far from a tricritical point; see sect. 5.1). With allowance for the possible values of $\gamma$ presented in sect. 3.1 we find

$$I_{\eta\eta} \propto |\tau|^{-\psi}, \qquad \psi = 0.5\text{–}0.8. \tag{137}$$

For $qr_c \gtrsim 1$ one should take into account the term $Dq^2$ in the denominators of eqs. (31) and (114) (more precisely, one should write $Dq^{2-\hat{\eta}}$ for $qr_c \gg 1$; see eq. (28)). It follows then from eq. (114) that the intensity $I_{\eta\eta}$ decreases as $\tau \to 0$ due to the factor $\eta_e^2$. Thus, as $\tau \to 0$, the intensity $I_{\eta\eta}$ for a fixed $\boldsymbol{q}$ increases rather rapidly at first (in contrast to the Landau theory case), has a maximum for $qr_c \sim 1$ and then decreases, vanishing at the phase transition point. We should stress once again that since for light $q \sim 10^5\text{–}10^6\ \mathrm{cm}^{-1}$, the region $qr_c \gtrsim 1$ has not been reached, up till now, for structural phase transitions in solids, although such a region is observable near the critical points in liquids and liquid mixtures (ch. 9) and also near the $\lambda$ transition in helium (Vinen et al. (1974), see also Vinen and Hurd (1978) and references therein).

The mixed fluctuation entering the expression for $I_{\eta\rho}$ has been calculated in sect. 3.2. From eqs. (112) and (54) we find (Levanyuk 1976)

$$I_{\eta\rho} \propto |\tau|^{-\psi_1}, \qquad \psi_1 = 1 - \gamma - \alpha \simeq 0.25\text{–}0.4. \tag{138}$$

The temperature dependence of $I_{\rho\rho}$, as follows from eqs. (112) and (51), is determined by the temperature dependence of the compressibility (which is the same as that of specific heat) i.e.,

$$I_{\rho\rho} \propto |\tau|^{-\alpha}. \tag{139}$$

When the shear modulus is taken into account, i.e., for phase transitions in solids, one should set $\alpha = 0$ in expressions (138) and (139) (see eqs. (58) and (59)).

Notice that although with other temperature dependences for the coefficients $A$, $B$, $C$, $D$ (see sect. 2.2) formulae (114) and (130) as well as relations (124), (125), (132) and (133) hold in the critical region too and can be used for the quantitative interpretation of experimental data.

Thus, in the critical region the order parameter fluctuations may lead to a substantial anomaly of light scattering of the first order even for second-order phase transitions far from the tricritical point. As distinct from the range of applicability of the Landau theory the light scattering due to fluctuations in all other quantities has here a weaker temperature dependence (see eqs. (138) and (139)). Note, however, that the dependence of $\epsilon_{ij}$ on $\eta$ may be, in principle, so weak that it is practically undetectable in experiment. This is the case, in particular, with liquid helium, where near the $\lambda$ transition only a rather slow increase of intensity according to a law which is close to logarithmic, is observed (Vinen and Hurd 1978). The dependence $I_1(\tau)$ for $qr_c \ll 1$ in this case is well described by the Einstein formula (for more details see sect. 5.7):

$$I_1(\tau, \boldsymbol{q} \simeq 0) = Q_{S0} k_B T \rho \left(\frac{\partial \epsilon}{\partial \rho}\right)_T^2 \left(\frac{\partial \rho}{\partial p}\right)_T \propto |\tau|^{-\alpha}. \tag{140}$$

For a tricritical point the exponents $\gamma$ and $\alpha$ coincide with those of the Landau theory, i.e., $\gamma = 1$ and $\alpha = 1/2$. In this case $I_{\eta\eta}$, $I_{\eta\rho}$ and $I_{\rho\rho}$ have the same temperature dependence (see eqs. (135), (138), (139) and sect. 5.1). It should be stressed that such a situation takes place in the non-symmetrical phase only. In the symmetrical phase according to the Landau theory the first-order scattering intensity is practically temperature independent, whereas according to scaling the critical exponent $\alpha$ for the intensity $I_{\rho\rho}$ should have the same values in both phases, i.e., $I_{\rho\rho} \propto |\tau|^{-1/2}$, both for $\tau < 0$ and for $\tau > 0$. For $\tau > 0$ the anomaly of the first-order scattering is entirely due to $\rho$ fluctuations. Increase of the scattering light intensity due to density fluctuations near the tricritical point was observed in mixtures of $^3$He–$^4$He (Watts and Webb 1974). As to solids there are some uncertainties in the question of tricritical points here, as discussed in sect. 2.3, but in any case there cannot be a strong anomaly of $I_1(\tau)$ because of shear deformations (see sect. 5.3.2).

### 5.5. Higher-orders scattering

It has been shown in the preceding sections that, for structural phase transitions which are not proper ferroelastic, the temperature dependence of the first-order scattering intensity in the symmetrical phase is determined by the temperature dependence of the compressibility. The latter is rather weak, at least for second-order transitions far from a tricritical point, and behaves nearly logarithmically even in the critical region. Therefore, the second-order scattering is of particular interest for the symmetrical phase. We calculate the intensity

$I_2$ of this scattering taking into account the fluctuations in $\eta$ only, the contributions of fluctuations in other variables being of less interest here.

According to eq. (112) the temperature depence of the intensity $I_2$ in the symmetrical phase is determined by the quantity $\langle|\eta^2(\boldsymbol{q})|^2\rangle$ calculated in sect. 3.5. Using the circumstance mentioned there that within the Landau theory the second term in formula (79) may be neglected we have from this formula and (112) (for $qr_c \ll 1$):

$$I_2 = 2Q_{S0}a^2(k_B T)^2 S(0) = Q_{S0}\frac{(ak_B T)^2}{4\pi D^2} r_c \equiv I_{20}(r_c/d_0). \tag{141}$$

The factor $I_{20}$ in the order of magnitude is equal to the intensity of the second-order scattering far from the transition point. Thus, the intensity of the second-order scattering increases as $\tau \to 0$ by the same law as $r_c$, i.e., within the classical region as $|\tau|^{-1/2}$.

In the critical region, as discussed in sect. 3.5, both the terms in eq. (79) have the same critical exponents and increase as $\tau \to 0$ like $|\tau|^{-\nu(1-2\hat{\eta})}$. Using the scaling relations $\gamma = \nu(2-\hat{\eta})$ and $3\nu = 2-\alpha$ and comparing formulae (85) and (135) we conclude that the intensities of both first- (for $\tau < 0$) and second-order scattering have the same temperature dependence in the critical region. Furthermore, in this region the intensities $I_2(\tau > 0)$ and $I_1(\tau < 0)$ are of the same order of magnitude. The latter can be seen from a comparison of eqs. (115) and (141) for $3k_B TBS(0) \sim 1$.

For $\tau < 0$ (i.e., in the non-symmetrical phase) the expression for the second-order scattering intensity is rather lengthy (Levanyuk 1976), and will not be presented here. Note only that the above conclusions concerning the coincidence of both the temperature dependence and the orders of magnitude for the intensities of first- and second-order scattering in the critical region remain valid for $\tau < 0$ too. It can be shown that the intensity of scattering of higher orders than the second (considering them in a formula of the type (112) one should take into account terms of higher degrees in $\eta$) does not increase essentially as $\tau \to 0$ (Levanyuk 1976).

Using eqs. (68) and (78) it is easy to show (Levanyuk 1976) that in the case of a uniaxial ferroelectric the intensity of the second-order scattering increases for $\tau \to 0$ as $\ln|\tau|^{-1}$, whereas in the case of a proper ferroelastic $I_2 \propto (c_1 - c_2|\tau|^{-1/2})$.

It is of interest that one can obtain a strict inequality for the total scattering intensity including scattering processes of all orders as well as the coupling of $\eta$ with any number of variables. This inequality, connecting the behaviour of the intensity $I_{tot}(\boldsymbol{q}=0)$ with the intensity of other directly observable quantities, has the form (Sobyanin 1979):

$$I_{tot}(\boldsymbol{q}=0) \geq Q_{S0}\frac{k_B T^2}{C_p - C_{p,0}}\left[\frac{\partial(\epsilon-\epsilon_0)}{\partial T}\right]^2. \tag{142}$$

Here $\epsilon - \epsilon_0$ and $C_p - C_{p,0}$ are singular parts of the specific heat and of the equilibrium value of $\epsilon$, respectively, and the derivative $\partial(\epsilon - \epsilon_0)/\partial T$, as in eq. (124), is taken at fixed $p$ for a phase transition in a liquid; for phase transitions in solids this derivative (as well as the specific heat $C$) should be measured at fixed uniaxial stress (see eq. (132)). If one is interested in pressure rather than in temperature variations, then the right-hand side of eq. (142) should be replaced by that of eq. (125) or eq. (133). Close to $T_c$ the temperature dependence of $I_{tot}$ as given by (142), is connected mainly with that of $\partial\epsilon/\partial T$, so that in rough estimates the quantities $\epsilon_0$ and $C_{p,0}$ may be neglected. Note that within the framework of first-order fluctuation corrections to the Landau theory the inequality (142) becomes an equality if $I_{tot}$ is thought of as an anomalous part of the total scattering intensity. The same seems also to hold in the leading approximation in a critical region. Relation (142) is convenient to use for estimating the anomalous intensity, the temperature dependence of $\epsilon$ being known. For example, on the basis of the experimental data obtained by Smolensky et al. (1977) the scattering light intensity near the antiferromagnetic phase transition point in $KNiF_3$ may be predicted to increase at least one order of magnitude if, of course, the dependence $\epsilon(T)$ reported by these authors is not connected with the presence of defects (see ch. 2).

### *5.6. Specific features of the first-order light scattering in systems with a multi-component order parameter*

In systems with a multi-component order parameter the order parameter fluctuations may have some peculiarities (see sect. 3.4) as compared to the case of a one-component order parameter. These peculiarities can be seen already for a two-component order parameter $\{\eta_1, \eta_2\} = \eta_1 + i\eta_2 = \rho\ \exp(i\varphi)$. This is just the case that has been discussed in sect. 3.4 and will be considered below.

We saw in sect. 3.4 that the mean square of $\varphi$ fluctuations diverges strongly near $T_c$ if the thermodynamic potential density depends on $\varphi$ only through terms of high enough powers in $\eta$. It seems natural then to expect that the contribution of $\varphi$ fluctuations to the light scattering intensity may change much more strongly with temperature when approaching the second-order transition, than the contribution of $\rho$ fluctuations, the latter having the same temperature dependence as in the case of a one-component order parameter. Such a conclusion is, however, wrong, and it will be shown below that the $\varphi$ fluctuations contribute in the intensity no more than the $\rho$ fluctuations, i.e., within the Landau theory (for second-order transitions far from the tricritical point) they contribute no more than to a jump of intensity at $T = T_c$. It is of special interest that the jump of intensity (i.e., a quite marked anomaly) may take place even if the tensor $\epsilon_{ij}$ depends on $\varphi$ through terms of very high powers in $\rho$, that is when the coupling between $\epsilon_{ij}$ and $\varphi$ fluctuations is extremely weak.

Let us suppose at the beginning that there is a bilinear combination of the order parameter components which transforms under symmetry transformations of the symmetrical phase like a component of the $\epsilon_{ij}$ tensor, say, the component $\epsilon_{xy}$. The two possible bilinear combinations of $\eta_1$, $\eta_2$ may be written in the form $\rho^2 \cos 2\varphi$ and $\rho^2 \sin 2\varphi$. Let us assume, for example, that

$$\epsilon_{xy} = b_2 \rho^2 \sin 2\varphi \,. \tag{143}$$

Then, evidently,

$$\langle (\Delta\epsilon_{xy})^2 \rangle = 4 b_2^2 \rho_e^4 \cos^2 2\varphi_e \langle (\Delta\varphi)^2 \rangle \,. \tag{144}$$

Thus, if in a given non-symmetrical phase $\cos 2\varphi_e \neq 0$, the light will be scattered by $\varphi$ fluctuations. As follows from eq. (72),

$$\langle (\Delta\varphi)^2 \rangle \sim \rho_e^{-m} \,, \tag{145}$$

where $m$ is the order of the first $\varphi$-dependent term in the thermodynamic potential expansion. Substitution of expression (145) into (144) shows that an increase of the intensity of light scattered by $\varphi$ fluctuations would take place for $m > 4$. However, in the case under consideration $m$ is equal exactly to four. Indeed, the fourth-order term $\rho^4 \sin^2 4\varphi$ is an invariant in our case and should enter in the thermodynamic potential expansion.

Let us suppose now that the lowest-order combination of the order parameter components which transforms like a component of $\epsilon_{ij}$ tensor is of the $n$th order ($n > 4$), so that one has, for example,

$$\epsilon_{xy} = b_n \rho^n \sin n\varphi \,. \tag{146}$$

There are two possibilities as to the $\varphi$-dependent invariant in the thermodynamic potential expansion. If both the quantities $\rho^n \sin n\varphi$ and $\rho^n \cos n\varphi$ are non-invariant, then $m = 2n$, $\langle (\Delta\varphi)^2 \rangle \sim \rho_e^{-2n}$ and the intensity of the light scattering due to $\epsilon_{xy}$ fluctuations does not depend on $\rho_e$. The result is rather interesting, for it is valid for any $n$. Thus we come to the conclusion that the scattering by $\varphi$ fluctuations may be observed in spite of extremely weak $\varphi$ dependence of $\epsilon_{ij}$ (Golovko and Levanyuk, 1981).

If $\rho^n \cos n\varphi$ or $\rho^n \sin n\varphi$ (or both) are invariant, then $\langle (\Delta\varphi)^2 \rangle \sim \rho_e^{-n}$ and the light scattering intensity is proportional to $\rho_e^n$. This means that there is practically no light scattering due to $\varphi$ fluctuations in this case.

Note that the above considerations are of interest especially for the phase transitions with the formation of long-periodical superstructures since high-order translation-invariant combinations of the order parameter component are only possible in this case. Sequences of such phase transitions are often discussed in connection with the properties of incommensurate phases, where in the case of the so-called devil's staircase there is a sequence of phases with any periods so that any values of $m$ and $n$ may be encountered here. But the above results are valid for any values of $m$ and $n$ only if the scattering at zero angle ($\boldsymbol{q} = 0$) is discussed. If one takes into account that $\boldsymbol{q}$ is never equal to zero, one concludes that only if a unit cell is not greatly increased one may hope to observe the above-discussed jump in the light scattering intensity (Golovko and Levanyuk 1981).

### *5.7. Discussion of some experimental data*

The existing experimental data on the integral intensity of the thermal (molecular) light scattering near phase transition points in solids are neither reliable nor full enough to be compared with the above theoretical predictions. The point is that in direct measurements of the integral intensity it is difficult to distinguish between the scattering by thermal fluctuations and that by static inhomogeneities or defects (the latter scattering may be dominant near phase transitions in real crystals; see below in this section and, especially, ch. 2). On the other hand, in spectral measurements it is not easy to detect properly some frequency regions in the spectrum, which may contribute heavily to the total intensity (in particular, due to a temperature dependent central peak; see sect. 6).

From a few (known to us) direct measurements of the integral intensity of light scattering near phase transitions in solids the first and, possibly, the most striking example is the observation by Yakovlev et al. (1956) of a strong (about $10^4$ times) increase of the scattered light intensity near the point of the structural $\alpha \rightleftarrows \beta$ transition in quartz at a temperature $T_{tr} = 846$ K. However, as will be mentioned in ch. 11, the origin of this anomaly is not yet clear. Furthermore, as shown in recent experiments by Dolino and Bachheimer (1977) there appear to be actually two strong anomalies in quartz at slightly different temperatures: the one for scattering at 90° (it seems to be just the anomaly discovered by Yakovlev et al.) and the other for scattering at small angles. Both these anomalies are mainly of a static nature (Shapiro and Cummins 1968, Shapiro 1969, Dolino and Bachheimer 1977, Shustin et al. 1978) and thus have nothing in common with thermal (molecular) scattering anomalies which are of interest for us here. Unfortunately no attempts have been made up to now to extract the molecular scattering part from the total scattered intensity in quartz. Nevertheless the possible increase of this part of the scattering seems worth estimating.

According to a number of studies (Shapiro and Cummins 1968, Shapiro 1969, Hochli 1970, Axe and Shirane 1970, Bachheimer and Dolino 1975, Dolino and Bachheimer 1977) the $\alpha \rightleftarrows \beta$ transition in quartz is a first-order transition although very close to the tricritical point. As follows from Bachheimer and Dolino (1975), where the results of other experiments are also summarized, the temperature dependence of the equilibrium value of the order parameter in the $\alpha$ phase in a wide range of temperatures is described by a formula of the type (2) with coefficients $B = 166.8$ K, $C = 726$ K and $A = T - T_{s2}$, where $T_{s2} = 837.5$ K*. From eq. (11) it follows then that $T_{s1} = 847$ K, and the maximum possible width of the temperature hysteresis

*The normalization is used here for $\phi$ and $\eta$ in which $dA/dT = A_0 T_{tr} = 1$ and $\eta_e(T = 23°C) = 1$.

region $T_{s1} - T_{s2} = 9.5$ K, which agrees well with the data of neutron experiments in the $\beta$ phase (Axe and Shirane 1970). Note that the reduced width of the hysteresis region is $(T_{s1} - T_{s2})/T_{tr} \simeq 10^{-2}$, i.e., it is very small, and therefore the transition in quartz is indeed very close to the tricritical one (the hysteresis effects are observed in experiments in a still narrower temperature interval of the order of 1.5 K; see ch. 11). It should be mentioned that the possibility also exists (although it seems to us not very probable, particularly in connection with the proportionality between $\eta_e$ and the soft mode frequency observed in the $\alpha$ phase; see sect. 6.1), that the $\alpha \rightleftarrows \beta$ transition in quartz is described in a wide temperature interval not by the Landau but by the scaling theory. In this case a "weak" first-order phase transition could be associated with an influence of shear stresses specific for solids (see Larkin and Pikin (1969) and remarks at the end of sect. 2.4).

It should be emphasized, however, that whichever of these possibilities may correspond to reality, the expected increase of the molecular light scattering intensity in quartz cannot be very large. Indeed, according to eq. (114) this intensity may increase only until the condition $qr_c \lesssim 1$ holds. For the bilinear dependence of $\epsilon_{ij}$ on $\eta$, as in the case of quartz, the intensity $I_1$ (both in the framework of the Landau and scaling theories) increases no faster than $r_c$. Therefore even for phase transitions in liquids at $q \simeq 10^5\,\mathrm{cm}^{-1}$ the intensity $I_1$ may increase near $T_c$ no more than by three orders of magnitude (if we assume that far from $T_c$ the correlation radius $r_c \sim 10^{-8}$ cm). It should be added that for phase transitions in solids the intensities of Raman lines far from $T_c$ make up, as a rule, no more than $10^{-1}$–$10^{-2}$ of the total intensity. Therefore, in the critical region, where the intensity increases mainly due to $\eta$ fluctuations (see sect. 5.5), the maximum expected increase of the intensity near a second-order transition should not actually exceed one or two orders of magnitude. As to the classical behaviour near a tricritical point the increase of intensity is here also limited for solids due to a non-zero value of the shear modulus (sect. 5.3.1), and should not exceed one order of magnitude.

In view of what has been said it is not clear, however, why the integral intensity of light scattering near the transition point in quartz is surprisingly well estimated by formula (132), with the use for the dependence $\epsilon_e(T)$ of the results presented by Baranskii (1953) (see also Yakovlev and Velichkina 1957). At the same time it should be noted that for the anomalous part of the thermal scattering intensity formula (132) includes the derivative $(\partial\epsilon/\partial T)_\sigma$, which (in the case of an isotropic solid) is to be measured not for a "free" crystal but for one clumped from two sides. Therefore, for a verification of eq. (132) the dependence of $\epsilon_e$ on $T$ should be remeasured (or the quantities should be measured, the knowledge of which is necessary for the calculation of $(\partial\epsilon/\partial T)_\sigma$ from the data on $(\partial\epsilon/\partial T)_p$), not to mention the desirability to measure all the quantities that enter eq. (132) on one specimen with the necessary separation in the total intensity of the predominant contribution from the scattering by static inhomogeneities. The same can be said of course about other crystals in which a very strong scattering ("fog") was observed in the transition region (see chs. 2 and 11).

So we see that the light scattering anomaly in the region of the $\alpha \rightleftarrows \beta$ transition in quartz is a very complicated phenomenon. The fact that just this anomaly was the first to be observed gave rise to additional difficulties and misunderstandings in the interpretation of the data on light scattering at phase transitions in solids (in this connection see also Ginzburg and Levanyuk 1974, 1976).

Besides quartz, direct measurements of the integral intensity of light scattering have been carried out also for the well-known structural phase transition point in $NH_4Cl$. This transformation (between cubic phases of different symmetries) occurs at $T_{tr} = 240$ K at atmospheric pressure. An intensity increase of about a factor 8 was observed, when approaching the temperature, by Shustin (1966) as well as by Lasay et al. (1969). Note, however, that the quality of $NH_4Cl$ crystals is much lower than that of quartz crystals, so that the scattering from defects is predominant here even away from $T_{tr}$. Thus if one considered the intensity increase relative to the intensity of the thermal scattering away from $T_{tr}$ (as was done in quartz) then the fractional increase of the scattered light intensity would probably be no less here than in quartz*. It seems quite improbable that such a great anomaly might be connected with scattering from thermal fluctuation for the same reason as in quartz. We conclude therefore that the intensity increase for $NH_4Cl$ is due mainly to scattering by static inhomogeneities too. The evidence for this conclusion is provided also by the presence of a marked "tail" of the intensity anomaly in the symmetrical phase (see fig. 9) and by the discovery (Pique et al. 1977) of light diffraction from some macroscopic structure arising near the phase transition point, this phenomenon being similar here to that in quartz.

At pressures above nearly 1.5 kbar the transition in $NH_4Cl$, which is of first order at low pressures, becomes second-order (Garland and Weiner 1971). When moving along the phase transition line the total intensity of the scattered light has a maximum at the tricritical point (fig. 9). In spite of the fact that the presence of such a maximum is compatible with eq. (130) (see also eqs. (115) and (118) which, however, hold only for fluids), we do not think, in view of the above said, that this maximum is connected with molecular scattering. Thus, extra experiments are needed for extracting the molecular scattering anomaly in $NH_4Cl$ as well.

In a number of works the integral intensity of molecular light scattering was estimated by integrating the light scattering spectral density over frequencies. Although, as mentioned, it is difficult in this way to take into account the light scattering with a small change of frequency, one can obtain in such a way a

*We are indebted to O.A. Shustin who has made this point clear to us.

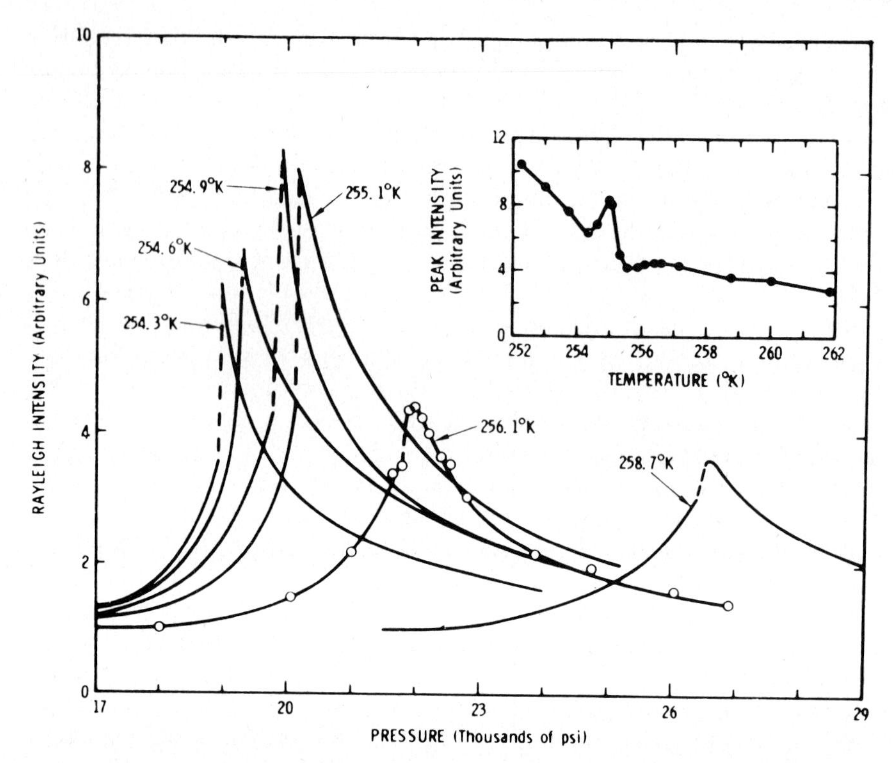


Fig. 9. Pressure dependence of the light scattering integral intensity for $NH_4Cl$ at a number of temperatures. Inset shows the pressure dependence of the intensity at the phase transition line (Fritz and Cummins 1972).

lower bound on the intensity anomaly. In solids which are not proper ferroelastics the most detailed investigations of this kind have been performed on lead germanate ($Pb_5Ge_3O_{11}$) near the second-order ferroelectric transition at $T_c = 451$ K (Hisano and Ryan 1972, Lyons and Fleury 1978). Authors of the works cited observed a comparatively small (about 3 times) increase of the integral intensity in the ferroelectric phase as $|\tau| \to 0$. This increase was proposed by Hisano and Ryan to be due to the closeness of the phase transition to the tricritical point, whereas an attempt was made by Lyons and Fleury to attribute this anomaly to deviations in the corresponding temperature dependences from those predicted by the Landau theory. As lead germanate is a uniaxial ferroelectric for which the range of inapplicability of the Landau theory is expected to be extremely small (sect. 3.3.2), the second proposition seems to us to be far less acceptable.

An increase in the integral intensity due to a broad central maximum was reported by Steigmeier et al. (1971), Penna et al. (1976) and Gorelik et al. (1976)

near ferroelectric phase transitions in SbSI, $LiTaO_3$ and $LiNbO_3$ respectively. According to Hisano and Ryan and to Penna et al., the temperature dependence of the total intensity $I$ is here close to that of the static dielectric permeability $\epsilon_0$ However, such a result is clearly incompatible with what was said in sect. 5.1. Indeed, it follows from eqs. (113) and (114) that $I \simeq I_1(\boldsymbol{q}) \sim P_e^2\epsilon_0$ and therefore if it increases (near a tricritical point) it does this far more slowly than $\epsilon_0$. Thus we do not consider the origin of the light scattering anomaly in these crystals to be well understood.

The temperature dependence of the light scattering intensity reflects directly that of order parameter fluctuations in the case of proper ferroelastic phase transitions only. The most explored example of such transitions is, perhaps, that in $KH_2PO_4$ at $T_c = 122$ K. Measurements of the spectral light scattering intensity near the phase transition have been carried out by many investigators (see, for example, Stekhanov and Popova (1968), Kaminov and Damen (1968), Kaminov (1969), Brody and Cummins (1968, 1969 a, b, 1974), Lagakos and Cummins (1974), Peercy (1973, 1975 a, b), Courtens (1978) and works cited therein). In these works the increase of the scattered intensity was observed both from fluctuations of the polarization vector component $P_z$ and from those of the strain tensor component $u_{xy}$. For pure crystals this increase is well described (Courtens 1978) by expression (134).

The investigations of the temperature dependence of the intensity scattered by a soft acoustic mode aimed at a verification of an expression of the form (134) were performed also by Fleury et al. (1974) (see also Fleury (1975, 1976)) on $PrAlO_3$ near the proper ferroelastic transition at $T_c = 151$ K. In accordance with this expression it was found that the intensity is inversely proportional to the corresponding elastic constant as determined from the soft mode frequency measurements (see sect. 6.1). However, the full complex of measurements including an independent determination of the elastic and piezo-optic constants involved which are necessary for quantitative check of expressions of the form (134) was not carried out in the work cited.

Rather careful and complete investigations of both the integral intensity and the scattered light spectrum have been performed near the $\lambda$ transition in liquid helium (Vinen and Hurd 1978, Greytak 1978). This transition can be regarded as a striking example of a second-order phase transition not described by the Landau theory. The specificity of this transition is that the dielectric permeability may be treated here with great accuracy as being independent of the order parameter $\eta \equiv \Psi$ and coupled with the density $\rho$ only. The integral intensity of the scattered light for $qr_c \ll 1$ is given then by the Einstein formula (Einstein (1910), see also, e.g., Fabelinskii (1965))

$$I_1(\boldsymbol{q} \simeq 0) = Q_{S0}k_BT\rho^2(\partial\epsilon/\partial\rho)_T^2\beta_T, \tag{147}$$

where $\beta_T = \rho^{-1}(\partial\rho/\partial p)_T$ is the isothermal compressibility. This formula has been presented already in sect. 5.4 and follows from eqs. (112) and (51). For

$qr_c \ll 1$ the measured values of the intensity are in good agreement with the expression in question as is clear from fig. 10. Deviations from this expression are observed for temperatures where the order parameter correlation radius $r_c$ becomes comparable with the light wavelength, i.e., for $qr_c \gtrsim 1$. Analysis of the shape of the intensity maximum at $qr_c \gtrsim 1$ is of considerable interest but it has not yet been carried out properly. We would like to mention also the $^3He$–$^4He$ mixtures in which the density and concentration fluctuations increase near the tricritical point (Watts and Webb 1974).

Concluding the discussion of the question of the integral intensity we would like to emphasize once again that measurements of the integral intensity of thermal (molecular) light scattering yield important information on the total magnitudes of fluctuations near the phase transition points. At the same time we do not know any experimental studies of phase transitions in crystals where the whole complex of measurements necessary for a verification of relations of the type (132), (133), (134), (142) has been carried out. Another aspect of the problem is the necessity to separate in the total intensity the contribution of thermal scattering from that of scattering by static inhomogeneities and defects (ch. 2). The latter may appear to dominate in experiments. As to the theory it is important that taking into account variables other than $\eta$ is, as a rule, very essential (particularly if fluctuations of these additional variables are accompanied by the appearance of long-range forces). However, formally (for $qr_c \ll 1$) it is reduced to ascribing the proper meaning to the derivatives in formulae of the type (124), (125), (132), (133). From the physical point of view the matter is that when calculating the anomalous part of the integral scattering intensity

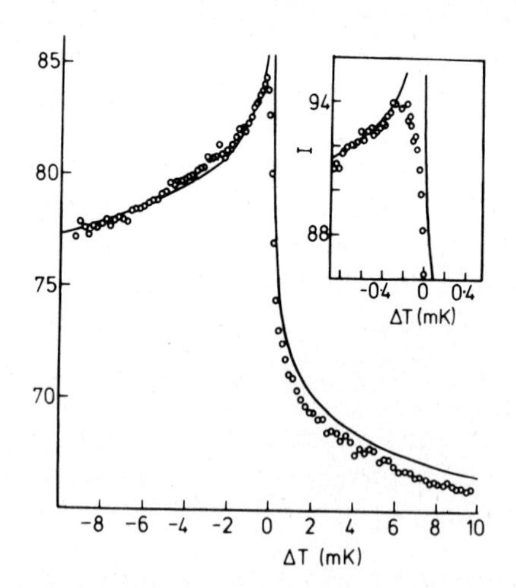


Fig. 10. Integral light scattering intensity $I$ plotted against $\Delta T = T - T_\lambda$ in liquid $^4He$ at a pressure of 18.80 bars (Vinen et al. 1974). Inset: detailed behaviour of $I$ very close to the $\lambda$ line at a pressure of 19.10 bars. The vertical scales are arbitrary and are not the same in the two plots.

one may, in fact, consider that it is the parameter $\eta$ alone that fluctuates, whereas all the other generalized coordinates acquire quasi-equilibrium values corresponding to a given instantaneous value of $\eta$. The rate of reaching these equilibrium values is of no importance for the consideration of the integral scattering intensity.

## 6. *Spectral density of scattered light*

While the measurements of the integral scattered intensity provide information about total (integrated over frequencies) fluctuations of the appropriate variables, measurements of the spectral density yield evidence of fluctuation kinetics. Distinct from the case of the integral intensity, the analysis of the temperature evolution of the light scattering spectral density near phase transitions cannot be reduced, generally speaking, to an effective treatment of the order parameter fluctuations alone. Indeed, the value of the integral intensity is determined by static generalized susceptibilities and therefore, as has already been mentioned, the rates (times) of establishing the equilibrium values of fluctuating quantities are not relevant here. On the contrary, in an evaluation of the spectral density it is just the rates of fluctuations that are of importance. The corresponding characteristic rates may be related to each other differently, and the situation on the whole turns out to be rather complicated. Being not able to characterize it to a full extent, we hope though to help the reader get an idea about the situation, starting from simple cases and proceeding then to more complicated ones.

### 6.1. *First-order scattering; single quasi-harmonic soft mode*

Below, as in the previous sections, we consider mainly a one-component order parameter, and assume $\boldsymbol{q} \simeq 0$. Besides, let us suppose at first that the dielectric permeability of the medium depends on $\eta$ only. Formula (103) for the scattered light spectral density then has the form:

$$\mathscr{I}(0, \Omega) \equiv \mathscr{I}(\Omega) = VQ_{S0}\langle|\Delta\epsilon(0, \Omega)|^2\rangle = VQ_{S0}\left(\frac{\partial \epsilon}{\partial \eta}\right)^2_{\eta = \eta_e} \langle|\eta(0, \Omega)|^2\rangle, \quad (148)$$

where $\langle|\eta(0, \Omega)|^2\rangle \equiv \langle|\eta(\Omega)|^2\rangle$ is the Fourier transform of the two-time $\eta$-correlation function:

$$\langle|\eta(\Omega)|^2\rangle = \iint \langle\Delta\eta(t)\Delta\eta(t')\rangle \exp\{\mathrm{i}\Omega(t' - t)\}\, \mathrm{d}t\, \mathrm{d}t' .$$

To find this Fourier transform (it is also called the spectral density of the $\eta$ fluctuations) one should know the equation of motion for $\eta(\boldsymbol{q} = 0, t) \equiv \eta(t)$. As the simplest version of such an equation (Ginzburg 1949) we take the following (dots denote differentiating with respect to time):

$$m\ddot{\eta} + \gamma\dot{\eta} + \partial\phi/\partial\eta = h(t), \tag{149}$$

where $h(t)$ may refer now either to the external force conjugate to $\eta$ or to a random force (noise source). In the stationary case, eq. (149) reduces, of course, to the usual equilibrium equation $\partial\phi/\partial\eta = h$.

For small deviations of $\eta$ from the equilibrium value $\eta_e$ we put $\eta' = \eta - \eta_e$ and obtain then the equation

$$m\ddot{\eta}' + \gamma\dot{\eta}' + \phi_{\eta\eta}\eta' = h(t). \tag{150}$$

For convenience we shall omit, as a rule, the prime on $\eta'$. Equation (150) was the one used at the first stage of development of the theory under discussion (Ginzburg and Levanyuk 1960).

In spite of their simplicity, eqs. (149) and (150) are quite general in the low-frequency range since their left-hand side may be understood as the first three terms in an expansion of an exact equation of motion for $\eta$ in a time derivatives series. The quantity $\phi_{\eta\eta}$ in eq. (150) is then to be calculated for fixed generalized forces conjugate to all other generalized coordinates and, therefore, has the meaning of an inverse static generalized susceptibility corresponding to the parameter $\eta$. Thus understood, eq. (150), if valid up to optical phonon frequencies, expresses the essence of the so-called soft mode concept (Ginzburg 1949, Cochran 1960, 1961, Anderson 1960).

Indeed, as $\phi_{\eta\eta}(T_c) = 0$ for a second-order phase transition, the frequency $\Omega_0 \equiv (\phi_{\eta\eta}/m)^{1/2}$, which is the soft mode eigenfrequency, goes to zero as $T \to T_c$. Equation (150) includes also two other characteristic frequencies: $\Gamma = \gamma/m$ and $\Omega_R = \phi_{\eta\eta}/\gamma$. The first is a soft mode damping constant, while the second has the meaning of a soft mode relaxation rate when $\Omega_0 \ll \Gamma$ (in this case the $\eta$ fluctuations have a relaxation character, and the soft mode is called overdamped).

Let us stress once more, that the use of eq. (150) means neglecting higher derivatives of $\eta$ with respect to $t$ or, in another language, neglecting a frequency dispersion of the coefficients $m$, $\gamma$ and $\phi_{\eta\eta}$ in this equation. Strictly speaking, a certain frequency dispersion always takes place and may lead to important consequences. Equation (150) is therefore far from being always valid. However, conditions exist when the dispersion of the above-mentioned coefficients is rather small. Besides, the frequency range in question being much higher than the characteristic frequencies of other relevant slow processes, the dispersion can easily be taken into account. To this end, the derivative $\phi_{\eta\eta}$ in eq. (150) must be taken not for fixed generalized forces conjugate to other slowly changing variables but for fixed values of these variables themselves. Below we are discussing this question in more detail, but for the present we accept eq. (150) with $\phi_{\eta\eta}^{-1}$ regarded as the static generalized susceptibility.

To find the quantity $\langle|\eta(\Omega)|^2\rangle$, which is of interest for us (see eq. (148)), we put in eq. (150) $h(t) = h(\Omega)\,\mathrm{e}^{-\mathrm{i}\Omega t}$. Then we obtain

$$\eta(t) = \eta(\Omega)\,\mathrm{e}^{-\mathrm{i}\Omega t} = \frac{h(\Omega)\,\mathrm{e}^{-\mathrm{i}\Omega t}}{m[\Omega_0^2 - \Omega^2 - \mathrm{i}\Gamma\Omega]}, \qquad (151)$$

$$\Omega_0^2 \equiv \phi_{\eta\eta}/m, \qquad \Gamma \equiv \gamma/m,$$

and

$$\langle|\eta(\Omega)|^2\rangle = \frac{\langle|h(\Omega)|^2\rangle}{m^2[(\Omega_0^2 - \Omega^2)^2 + \Gamma^2\Omega^2]}. \qquad (152)$$

For a non-correlated random force $h(t)$ (white noise) the average $\langle|h(\Omega)|^2\rangle$ is frequency-independent and can be found from eq. (148) and the normalization condition (102); thus, as is easily seen,

$$\mathscr{I}_1(\Omega) = I_1 \frac{\Gamma\Omega_0^2}{\pi[(\Omega_0^2 - \Omega^2)^2 + \Gamma^2\Omega^2]}, \qquad I_1 \equiv \int_{-\infty}^{+\infty} \mathscr{I}_1(\Omega)\,\mathrm{d}\Omega. \qquad (153)$$

This result can be derived also from the fluctuation–dissipation theorem (Landau and Lifshitz 1976), according to which (for the case $\hbar\Omega \ll k_\mathrm{B}T$)

$$\langle|\eta(\Omega)|^2\rangle = \frac{k_\mathrm{B}T}{\pi\Omega}\chi''(\Omega), \qquad (154)$$

where $\chi''(\Omega)$ is the imaginary part of the generalized susceptibility corresponding to $\eta$.

Recall that by the definition of the susceptibility $\eta(\Omega) = \chi(\Omega)h(\Omega)$ and therefore it follows from eqs. (151) and (154) that

$$\langle|\eta(\Omega)|^2\rangle = \frac{k_\mathrm{B}T}{\pi m}\frac{\Gamma}{(\Omega_0^2 - \Omega^2)^2 + \Gamma^2\Omega^2}. \qquad (155)$$

To obtain eq. (153) one should evidently substitute eq. (155) into eq. (148) and express, with the aid of eq. (113) or the normalization condition (102), the factor $VQ_{\mathrm{S}0}(\partial\varepsilon/\partial\eta)_{\eta=\eta_\mathrm{e}}$ through $I_1$.

From formula (153) it follows that when $\Omega_0^2 = \phi_{\eta\eta}/m > \Gamma^2/2$, the scattered light spectral density has two maxima at frequencies $\Omega = \pm\Omega_\mathrm{m} = \pm[\Omega_0^2 - (\Gamma^2/2)]^{1/2}$ and a minimum for $\Omega = 0$ (fig. 11). If $\Omega_\mathrm{m} \gg \Gamma$, then near the maxima

$$\mathscr{I}_1(\Omega) \simeq \frac{\Gamma}{4\pi}\frac{I_1}{(\Omega \pm \Omega_0)^2 + (\Gamma^2/4)}. \qquad (156)$$

From this expression it is seen that the full width of the maxima at half height is equal to $\Gamma$.

When approaching the phase transition point the frequency $\Omega_0$ is decreasing, and the peaks go nearer to one another. Within the applicability region of the

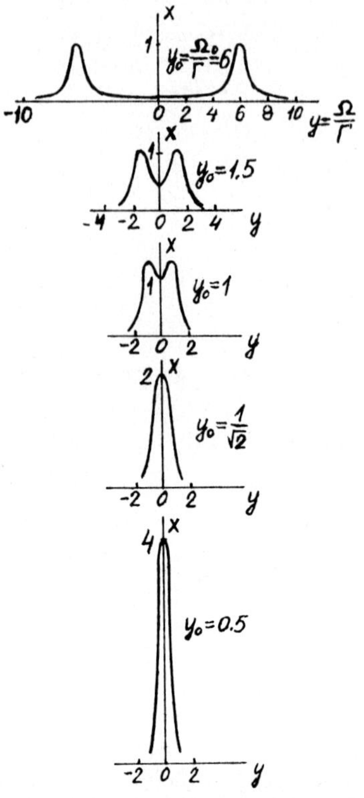


Fig. 11. Spectrum of scattered light $x=\pi\Gamma\mathscr{I}(y)/I$ plotted against $y=\Omega/\Gamma$ for several temperatures (several values of the parameter $y_0=\Omega_0/\Gamma$).

Landau theory the coefficients $\gamma$ and $m$ in eq. (149) may be considered independent of $\tau$, so that here the width of the maxima $\Gamma=\gamma/m$ remains practically constant. The height of the maxima is equal to $I_1/\pi\Gamma$ and for second-order transitions, which are not proper ferroelastic, increases as $\tau\to 0$ only if the transition is close to the tricritical point.

At $\Omega_0^2=\Gamma^2/2$ the side bands merge into a single unshifted (central) line (fig. 11). For $\Omega_0\ll\Gamma$ the intensity distribution in this line is described by the formula

$$\mathscr{I}_1(\Omega)=\frac{\Gamma\Omega_0^2}{\pi}\frac{I_1}{\Omega_0^4+\Gamma^2\Omega^2},\qquad \Omega_0^2=\frac{\phi_{\eta\eta}}{m},\qquad \Gamma=\frac{\gamma}{m}. \tag{157}$$

As follows from this formula, the width of the central line (for $\Omega_0\ll\Gamma$) is $2\Omega_0^2/\Gamma$ and decreases rapidly when approaching the phase transition. The height of the central peak is proportional to $I_1/\Omega_0^2$, and therefore increases dramatically as $|\tau|\to 0$.

In the region $\Omega_0\ll\Gamma$, when $\eta$ fluctuations are of a purely relaxation character, the first term in eq. (150) may be omitted. Going over in eq. (157) to the limit

$m \to 0$, we have

$$\mathscr{I}_1(\Omega) = \frac{\Omega_R}{\pi} \frac{I_1}{\Omega_R^2 + \Omega^2}, \qquad \Omega_R \equiv \frac{\phi_{\eta\eta}}{\gamma}. \tag{158}$$

For order–disorder type transitions, when $m \simeq 0$, formula (158) holds also far from the transition temperature.

In the case of a multi-component order parameter, the picture of the temperature evolution of the spectrum gets more complicated. In particular, below $T_c$ there are now several soft modes (see, e.g., Fleury (1972), Levanyuk and Sannikov (1974)). Some of these modes may be relaxation modes, while the corresponding frequencies and relaxation rates may have different temperature dependences. An example are the $SrTiO_3$-type crystals, where below $T_c$ there exist two soft vibration modes, one of which is two-fold degenerate. At the same time above $T_c$ there is only one triply-degenerate soft mode. Another well-known example is helium II, for which the order parameter $\Psi = \eta \exp(i\varphi)$ is a macroscopic complex wave function with two components (Ginzburg and Sobyanin 1976, 1982). In this case above the $\lambda$ point (i.e., for $T > T_\lambda$) the equation of motion for $\Psi$ is of a relaxation character, whereas for $T < T_\lambda$ the amplitude $\eta \equiv |\Psi|$ is still relaxing but the phase $\varphi$, determining the superfluid velocity $\boldsymbol{v}_S = (\hbar/m)\nabla\varphi$, changes in an oscillating manner (second sound is the soft mode for the phase $\varphi$). Further examples, when there are several optical soft modes in systems with a multi-component order parameter are given, in particular, by Wada et al. (1977) and by Sannikov and Levanyuk (1978).

In spite of the fact that the spectrum picture determined by eq. (153) and shown in fig. 11 is certainly simplified, it has been confirmed up to now in a large number of experiments (see, for example, Scott (1974), Hayes and Loudon (1978), Barta et al. (1976), Makita et al. (1976), as well as conference proceedings edited by Wright (1969), Balkanski (1971), Balkanski et al. (1975), Riste (1974), Bendow et al. (1976), Birman et al. (1979)). The temperature dependence of the soft modes frequencies for some structural phase transitions is presented in figs. 12. Note that even for the displacive-type transitions the soft mode often turns out to be overdamped already far from the transition point (see, e.g., DiDomenico et al. (1968), Fleury and Lazay (1971), Popkov et al. (1971)).

Within the Landau theory the temperature evolution of the scattered light spectral density is determined by the temperature dependence of the quantity $\phi_{\eta\eta}$. If the $\eta$ variations were slowest, the derivative $\phi_{\eta\eta}$ should be calculated at fixed values of forces conjugate to all generalized coordinates other than $\eta$, i.e., for fluids at fixed pressure $p$, whereas for solids it should be calculated at fixed uniaxial stress $\sigma$. But as the soft modes are optical modes (excluding the case of proper ferroelastics) the frequencies of the modes are much higher than the acoustic frequencies for the same wavevectors. So we have to calculate $\phi_{\eta\eta}^u$ at

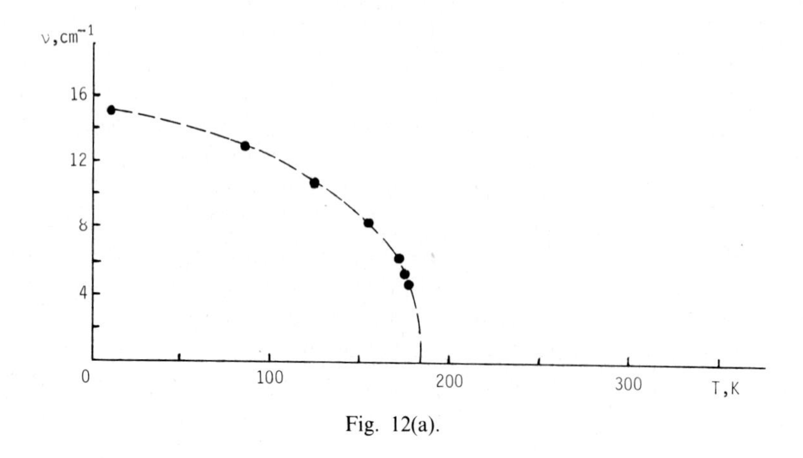


Fig. 12(a).

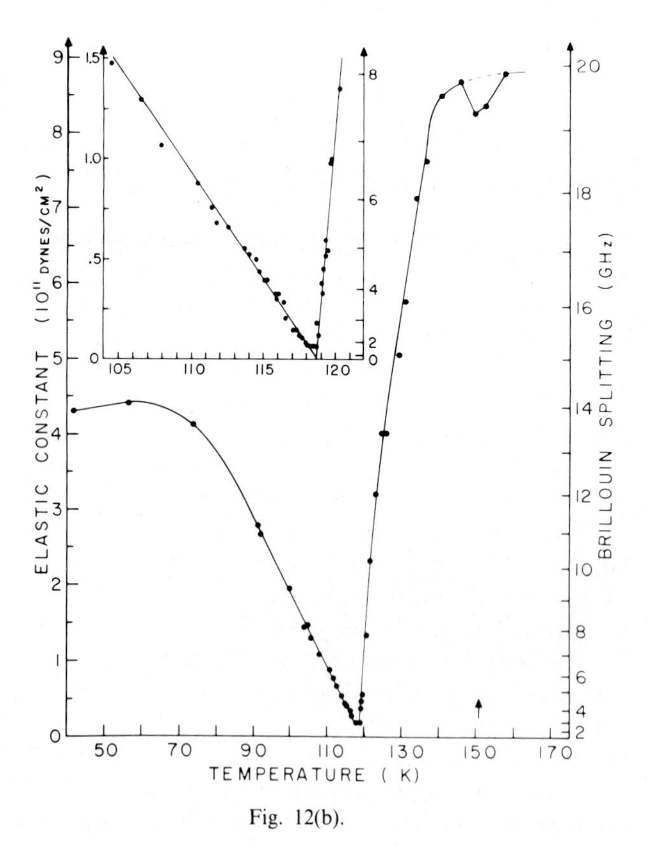


Fig. 12(b).

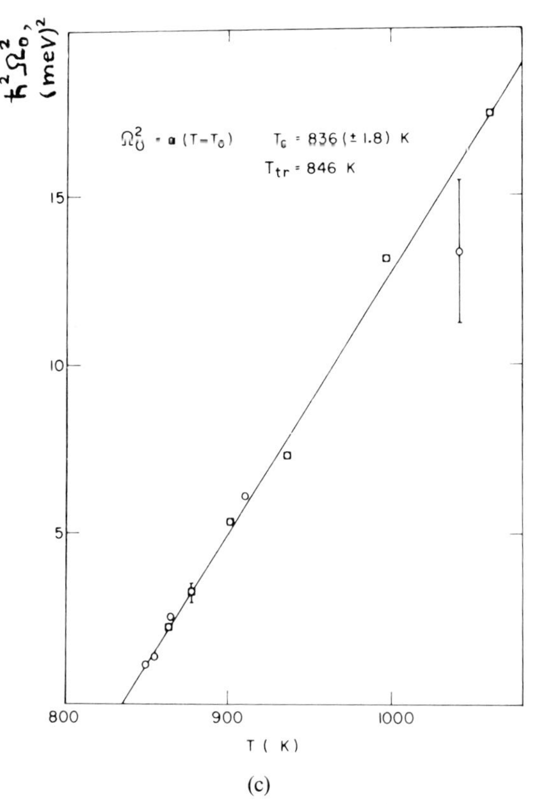


(c)

Fig. 12. Examples of the soft mode temperature dependence: (a) Temperature dependence of the soft mode frequency below the structural phase transition in $Hg_2Cl_2$ (Barta et al. 1976). (b) Temperature dependence of the soft acoustic mode frequency (the right axis) and of the corresponding elastic constant (the left axis) near the proper ferroelastic transition at 118 K in $PrAlO_3$ (Fleury 1975). The inset shows that close to $T_c$ the elastic constant $C_{ij}$ depends linearly on $T - T_c$ in accordance with the simple soft mode concept. (c) Temperature dependence of the square of the soft mode frequency in $\beta$ quartz as determined from neutron measurements (Axe and Shirane 1970).

fixed strain. Using eq. (57) in the case $\mu \to \infty$ which corresponds to the impossibility of any inhomogeneous deformation, one obtains

$$\Omega_0 = (\phi^u_{\eta\eta}/m)^{1/2} = \eta_e[(2\sqrt{B^2 - 4AC} + r^2/K)/m]^{1/2}\,. \tag{159}$$

We see that the soft mode frequency is proportional to $\eta_e$ both for a second-order phase transition far from tricritical point, and for a tricritical transition.

In the critical region alongside the temperature dependence of $\phi_{\eta\eta}$ the temperature dependencies of $\gamma$ and $m$ should, strictly speaking, be taken into account too. As concerns the coefficient $\gamma$, its critical exponent is known to be very small for structural phase transitions (Hohenberg and Halperin 1977), i.e., the temperature dependence of $\gamma$ appears to be very weak. The temperature dependence of the mass $m$ has, as far as we know, not been investigated. Assuming that this dependence is weak too (as seems probable) we see that in the critical region (remember the necessity to distinguish between the coefficient $\gamma$ and the critical exponent $\gamma$!):

$$\Omega_0 = (\phi_{\eta\eta}/m)^{1/2} \sim |\tau|^{\gamma/2} . \tag{160}$$

Since in the critical region the index $\gamma > 1$, then, obviously, the frequency $\Omega_0$ must change more rapidly here than in the classical region, where $\gamma = 2\beta = 1$. Meanwhile, in a number of experiments, deviations have been reported from the linear law for $\Omega_0^2(\tau)$, going to a slower temperature dependence (see Shapiro et al. 1967, Fleury et al. 1968, Steigmeier and Auderset 1973, Barta et al. 1976). The authors of the papers just mentioned emphasize the fact that in a wide temperature range the proportionality $\Omega_0^2(\tau) \sim \eta_e^2(\tau) \sim |\tau|^{2\beta}$ is observed, although taken separately the dependences $\Omega_0^2(\tau)$ and $\eta_e^2(\tau)$ decline from the law $\Omega_0^2 \sim \eta_e^2 \sim |\tau|$, which corresponds to the Landau theory for second-order phase transitions. Hence, the conclusion was drawn that the corresponding experimental data pertain to the critical region. However, such a conclusion contradicts formula (160) with $\gamma > 1$. Thus as the critical region is approached, the proportionality between $\Omega_0(\tau)$ and $\eta_e(\tau)$ must, in fact, be violated (provided proportionality is not re-established due to a strong temperature dependence of the mass $m$, which seems to be quite improbable). The deviations from proportionality of $\Omega_0$ and $\eta_e$ to $|\tau|^{1/2}$ should occur for second-phase transitions close to the tricritical point. In particular, for the tricritical transition itself (when approaching it along the line $B = 0$) $\eta_e \sim |\tau|^{1/4}$ and $\Omega_0 = (\phi_{\eta\eta}^u/m)^{1/2} \sim |\tau|^{1/4}$. Thus, the data of works cited above indicate in our opinion that they refer to the region of applicability of the Landau theory, while the deviations of the dependences $\Omega_0(\tau)$ and $\eta_e(\tau)$ from the law $|\tau|^{1/2}$ are naturally to be ascribed to the proximity of the phase transition to the tricritical point.

Another possibility discussed by Meissner et al. (1981) (Meissner 1981) arises in the case of a multi-component order parameter which, to be precise, corresponds to the case investigated by Barta et al. (1976), Fleury et al. (1968) and Steigmeier and Auderset (1973). There are several modes in non-symmetrical phase now (two for the experimental situation discussed by the authors cited). It proves that if the frequency of one of the modes is higher than that of the other then for the highest mode one has $\Omega_0 \sim \eta_e$ even in critical region. However, both for $Hg_2Cl_2$ (Barta 1976) and for $SrTiO_3$ (Fleury 1968, Steigmeier and Anderset 1973) the difference between frequencies of the modes is rather small compared with the frequencies themselves, so that the theoretical result mentioned seems to be inapplicable to the interpretation of experimental data for these crystals.

In recent years a number of facts, which does not fit into the simple picture described above, have been revealed and have attracted great attention. So, the soft mode side bands not only shift but sometimes increase their width substantially as $\tau \to 0$ (Shapiro 1969, Laulicht et al. 1972, Burns and Scott 1970, Burns 1974, Fleury 1970, Shigenary et al. 1976, Uwe and Sakudo 1976), which should not be the case according to the above simple approach.

Moreover, even for a one-component order parameter not a single but several spectral lines are often observed whose positions and shapes depend essentially on temperature (Scott 1974, Peercy 1975, Slivka et al. 1978).

Particularly great interest was excited by the discovery near some phase transitions of a strongly temperature-dependent central maximum ("central peak", "central line") emerging in addition to the soft mode side bands which may then persist in the spectrum up to $\tau = 0$ (see, e.g., Riste et al. (1971), Riste et al. (1975), Shapiro et al. (1972), Kjems et al. (1973), Axe and Shirane (1974), Scott (1974), Fleury (1975) and ch. 7 in this volume).

There have been many attempts to give a theoretical explanation of these facts. However, the physical meaning of the results obtained is not very transparent, often because of unduly complicated and cumbersome microscopic approaches used. We shall try to discuss the relevant questions in the framework of the simple and general semi-phenomenological theory.

It is worth noticing first of all that eq. (150) is an equation of motion for an oscillator which is placed in a thermal bath formed by all the degrees of freedom of the system except the variable $\eta(\boldsymbol{q} \simeq 0, t)$. The coupling with the bath is reflected approximately in eq. (150) by introducing in it a friction force (the term $\gamma\dot{\eta}$) as well as a random force $h(t)$ and by the temperature dependence of the derivative $\phi_{\eta\eta}$ and, possibly, of the coefficients $\gamma$ and $m$. In the region asymptotically close to the phase transition some probable conclusions regarding the character of a more exact equation of motion for $\eta(\boldsymbol{q}, t)$ or, equivalently, regarding the form of the response function $\chi(\Omega, \boldsymbol{q}, \tau) = \eta(\Omega)/h(\Omega)$ can be made without explicit computations but using the so-called dynamic scaling hypothesis (see, e.g., Stanley (1973), Patashinskii and Pokrovskii (1975, 1982), Hohenberg and Halperin (1977)). This hypothesis is based on the concept of uniqueness of space and time scales for critical $\eta$ fluctuations and permits us to relate, for example, the dependence of the relaxation rate $\Omega_{\mathrm{R}} = \phi_{\eta\eta}/\gamma$ on $\tau$ for $qr_{\mathrm{c}} \ll 1$ (of interest for us) with the dependence of $\Omega_{\mathrm{R}}$ on $\boldsymbol{q}$ for $qr_{\mathrm{c}} \gg 1$ obtained from the neutron scattering data. However, the information which comes from the dynamic scaling is rather limited. On its basis one cannot, for example, describe the merging of the soft mode side bands. In other words, this theory is applicable only to the stage of the evolution of scattered light spectral density, when the side bands have merged already. Thus, to obtain more complete and concrete results concerning the spectral density of $\eta$ fluctuations (not to mention fluctuations in other variables) the coupling between the quantity $\eta(\boldsymbol{q} \simeq 0, t)$ and other degrees of freedom of the system must necessarily be studied in an explicit form. The rest of the section will be devoted, in fact, just to such studies.

### 6.2. *Mode coupling effects: linear coupling*

Let us now consider, along with $\eta$ fluctuations, the fluctuations in some other generalized coordinates. The spectral density of the scattered light is then expressed not only through the correlation function (spectral density) $\langle|\eta(\Omega)|^2\rangle$ but also through correlation functions of other variables (on which the permeability $\epsilon_{ij}$ depends) as well as mixed correlation fuctions. Besides, the coupling between $\eta$ and other variables, affects the very form of the correlation function $\langle|\eta(\Omega)|^2\rangle$. In other terms this means that the interaction between the fluctuations of $\eta$ and those of other variables leads to the appearance of a frequency dispersion of the coefficients in eq. (150).

#### 6.2.1. *Coupling of the soft with a relaxation mode; central peak*

The simplest theory for the frequency dispersion of the coefficients in eq. (150) is the Mandelstam–Leontovich theory (Mandelstam and Leontovich (1936, 1937), see also Landau and Lifshitz (1954)), where it is the linear coupling between $\eta$ and another relaxation variable $\xi$ that leads to the dispersion*. Without concretizing for the present the nature of this extra variable, we shall write the equation of motion for $\eta$ and $\xi$ in the form

$$m\ddot{\eta} + \gamma\dot{\eta} + \phi^{\xi}_{\eta\eta}\eta + \phi_{\eta\xi}\xi = h(t)\,, \tag{161}$$

$$\gamma_{\xi}\dot{\xi} + \phi_{\eta\xi}\eta + \phi^{\eta}_{\xi\xi}\xi = g(t)\,, \tag{162}$$

where the upper index means, as usual, that the derivatives are taken at fixed values of $\eta$ or $\xi$, respectively. To find spectral densities of the fluctuations, one can use, as before, the fluctuation–dissipation theorem, the extension of which to several fluctuating (classical) variables $\xi_i$ has the form (Landau and Lifshitz 1976)

$$\langle\xi_i(\Omega)\xi_j(\Omega)\rangle = \frac{k_B T}{\pi\Omega}\,\mathrm{Im}\,\chi_{ij}(\Omega)\,, \tag{163}$$

where $\chi_{ij}(\Omega)$ are the corresponding generalized susceptibilities defined as coefficients in the relations

$$\xi_i(\Omega) = \chi_{ij}(\Omega)g_j(\Omega) \tag{164}$$

(in eqs. (161) and (162), evidently, $g_1 = h$ and $g_2 = g$). According to this

*The Mandelstam–Leontovich theory was advanced, originally, to describe sound dispersion and attenuation effects irrespective of phase transitions. It was Landau and Khalatnikov (1954) who applied it for an explanation of sound anomalies near the $\lambda$ point in liquid helium. For studies of the light scattering spectrum near phase transition points this theory was used for the first time, as far as we know, by Levanyuk and Sobyanin (1967).

definition, to calculate the susceptibility $\chi_{\eta\eta}(\Omega)$ one should substitute $g(t)=0$, $h(t)=h(\Omega)\,e^{-i\Omega t}$ into the right-hand sides of eqs. (161) and (162) and express $\eta(\Omega)$ in terms of $h(\Omega)$ with the aid of these equations. As a result we come to

$$\chi_{\eta\eta}^{-1}(\Omega) = -m\Omega^2 - i\gamma(\Omega)\Omega + \phi_{\eta\eta}(\Omega), \tag{165}$$

where

$$\begin{aligned} &\gamma(\Omega) = \gamma + m\delta^2\tilde{\Omega}_{R\xi}/(\tilde{\Omega}^2_{R\xi} + \Omega^2), \\ &\phi_{\eta\eta}(\Omega) = \phi^{\xi}_{\eta\eta} - m\delta^2\tilde{\Omega}^2_{R\xi}/(\tilde{\Omega}^2_{R\xi} + \Omega^2) = \phi^{g}_{\eta\eta} + m\delta^2\Omega^2/(\tilde{\Omega}^2_{R\xi} + \Omega^2), \\ &m\delta^2 \equiv \phi^{\xi}_{\eta\eta} - \phi^{g}_{\eta\eta} = \phi^2_{\eta\xi}/\phi^{\eta}_{\xi\xi}, \qquad \tilde{\Omega}_{R\xi} = \phi^{\eta}_{\xi\xi}/\gamma_{\xi}\,. \end{aligned} \tag{166}$$

Here $\phi^{g}_{\eta\eta} \equiv \phi_{\eta\eta}(0)$ has the meaning of the inverse static generalized susceptibility corresponding to $\eta$ or of a generalized elastic modulus for $\eta$ changes at a constant force $g$, $\phi^{\xi}_{\eta\eta} \equiv \phi_{\eta\eta}(\infty)$ is the high-frequency value of this modulus, $m\delta^2$ is the difference of the high-frequency and the static values of the modulus $\phi_{\eta\eta}(\Omega)$, and $\tilde{\Omega}_{R\xi}$ is the relaxation rate (or frequency) of $\xi$ at a fixed value of $\eta$ (here and below non-renormalized values of frequencies, i.e., those without taking into account the coupling between $\eta$ and $\xi$ are marked by the " $\sim$ " sign).

Let us now analyze the picture of the temperature evolution of the spectrum of the scattered light, as given by eqs. (162) and (163). For simplicity we assume that the coefficient $\phi^{\xi}_{\eta\eta}$ alone is temperature-dependent and $\epsilon$ depends, as before, on $\eta$ only. The spectral intensity is then expressed totally through the function $\langle|\eta(\Omega)|^2\rangle$ and can be obtained by using eq. (163) along with expressions (165) and (166) or directly from eq. (153) with the substitution in it of $\Gamma(\Omega) \equiv \gamma(\Omega)/m$ for $\Gamma$ and of $\phi_{\eta\eta}(\Omega)/m$ for $\Omega_0^2$.

Suppose at first that the frequency $\tilde{\Omega}_{R\xi}$ is high as compared to the (non-renormalized) frequency of $\eta$ vibrations: $\tilde{\Omega}_{R\xi} \gg \tilde{\Omega}_{0\eta} \equiv (\phi^{\xi}_{\eta\eta}/m)^{1/2}$; then, as follows from eqs. (166), in the frequency range $\Omega \lesssim \tilde{\Omega}_{0\eta}$ one can put approximately $\Gamma(\Omega) \simeq \Gamma + \delta^2/\tilde{\Omega}_{R\xi} = \text{const}$, $\phi_{\eta\eta}(\Omega) \simeq \phi^{g}_{\eta\eta}$, and we see that the picture of the temperature evolution of the spectrum does not differ significantly from that shown in fig. 11 for the case of a single quasi-harmonic soft mode. Thus the linear coupling of $\eta$ with a quickly relaxing mode leads only to some renormalization of the soft mode damping constant.

The situation is far more interesting in another limiting case when $\tilde{\Omega}_{R\xi} \ll \tilde{\Omega}_{0\eta}$. Here the position and intensity of the side spectral maxima of the soft mode are determined by the high-frequency modulus $\phi^{\xi}_{\eta\eta}$ (which is natural, because $\xi$ does not change essentially for the period of the $\eta$ vibrations), while the total intensity of the first-order light scattering $I_1$ is determined, of course, by the static modulus $\phi^{g}_{\eta\eta}$ (see sects. 5.2 and 5.3). The difference of the intensities, as is easily seen from a simple calculation (Levanyuk and Sobyanin 1967) is contained in a central peak, the fraction of which in the total intensity is given by

$$I_{\text{central}}/I_1 = \delta^2/\tilde{\Omega}^2_{0\eta} = (\phi^{\xi}_{\eta\eta} - \phi^{g}_{\eta\eta})/\phi^{\xi}_{\eta\eta}\,, \tag{167}$$

and goes to unity as $T \to T_c$ if the derivative $\phi^{\xi}_{\eta\eta}$ is finite at $T = T_c$ (see fig. 13).

The presence of the central peak in the $\eta$-fluctuation spectrum reflects evidently the relaxation character of the $\xi$ fluctuations. As far as the $\xi$ fluctuations are slower than those of $\eta$, the changes of $\eta$ follow the $\xi$ ones. Therefore the width of the central peak is determined by the frequency $\Omega_{R\xi} = \phi^{h}_{\xi\xi}/\gamma_{\xi}$ which is proportional to the static generalized modulus

$$\phi^{h}_{\xi\xi} = \phi^{\eta}_{\xi\xi} - (\phi^2_{\eta\xi}/\phi^{\xi}_{\eta\eta})\,, \tag{168}$$

and goes to zero as $\tau \to 0$ if $\phi_{\eta\xi} \neq 0$ at $T = T_c$ (see below). What has been said above is valid when the fluctuations in $\eta$ for fixed $\xi$ is of an oscillatory

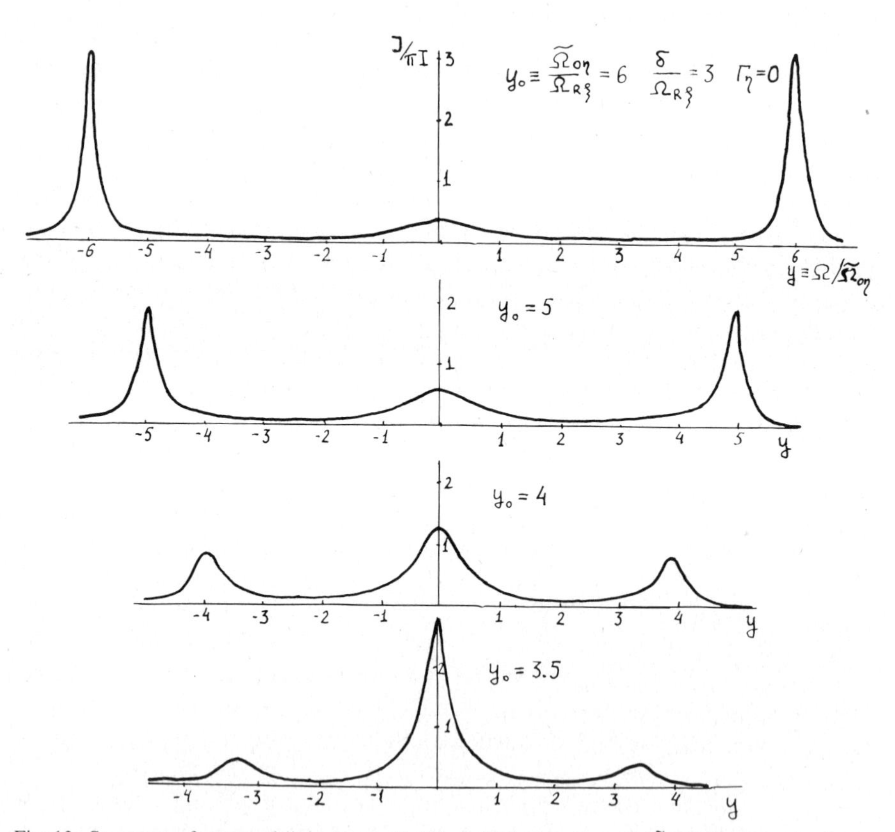


Fig. 13. Spectrum of scattered light $X = \pi\mathscr{I}(y)/I$ plotted against $y = \Omega/\tilde{\Omega}_{R\xi}$ in the case of a linear coupling of the order parameter $\eta$ with a relaxing variable $\xi$ for several temperatures near $T_c$ (several values of the parameter $y_0 = \Omega_0/\Omega_{R\xi}$) and for $\gamma = 0$, $\delta/\tilde{\Omega}_{R\xi} = 3$.

character, i.e., the soft mode is considered to be underdamped. The results presented are, however, easily extended to the case when the soft mode is overdamped, i.e., when $\tilde{\Omega}_{0\eta} < \Gamma_\eta \equiv \gamma/m$. The relaxation of $\xi$ being slower than that of $\eta$, the central peak with the half-width $\Omega_{R\xi}$ will exist against the wide background central maximum with a half-width $\tilde{\Omega}_{R\eta} \equiv \phi^{\xi}_{\eta\eta}/\gamma$ representing the merged components of the soft mode. In what follows we shall for simplicity present the results for the underdamped soft mode only.

The character of the temperature evolution of both side components and the central peak depends essentially on whether or not the coupling coefficient $\phi_{\eta\xi}$ falls to zero at $T \geqslant T_c$.

The coefficient $\phi_{\eta\xi}$ being non-zero at $\tau = 0$ (the case when the transformation properties of $\eta$ and $\xi$ are the same in the symmetrical phase), the modulus $\phi^{\xi}_{\eta\eta}$ determining the frequencies $\tilde{\Omega}_{0\eta}$ and $\tilde{\Omega}_{R\eta}$ is non-zero at the transition point (as distinct from $\phi^{g}_{\eta\eta}$). Thus, these frequencies are non-zero even at $\tau = 0$ (see also sect. 6.2.3). Therefore, if the frequency $\tilde{\Omega}_{0\eta}(\tau = 0) \equiv \delta$ exceeds $\tilde{\Omega}_{R\xi}$ and $\tilde{\Gamma} \equiv \gamma/m$, the side bands are present in the spectrum and do not merge up to the transition point. The width of the side bands in this case is equal to

$$\Gamma \simeq \tilde{\Gamma} + \tilde{\Omega}_{R\xi}\delta^2/\tilde{\Omega}^2_{0\eta}, \tag{169}$$

and increases to some extent as $\tau \to 0$ because $\tilde{\Omega}^2_{0\eta}$ decreases. The fractional integrated central peak intensity, as seen from eq. (167), increases up to unity as $\tau \to 0$. The width of this peak goes to zero as $\tau \to 0$ since, as follows from eqs. (38) and (168), the modulus $\phi^{h}_{\xi\xi} \to 0$ as $\tau \to 0$. The temperature evolution of the spectrum analogous to that described above and presented in fig. 13 has been observed in a whole number of experiments, mainly in neutron but also in light scattering experiments. Much attention is drawn to the fact that the physical origin of the slow relaxation variable remains unclear in many cases. An elucidation of the mechanisms of this slow relaxation, i.e., an establishment of the physical meaning of the variables of the type $\xi$ just makes up the essence of the so-called central peak problem (see sects. 6.2.2, 6.3, 6.4 and ch. 2).

We should emphasize once again that the difference between these results and those obtained on the basis of the simplest soft mode concept presented in sect. 6.1 is due to an account of the frequency dispersion of the generalized modulus $\phi_{\eta\eta}$. Indeed, in sect. 6.1 the high-frequency value of this modulus determining the position of the soft mode sidebands was assumed to coincide with its static value which alone should vanish at the instability point (i.e., at $T = T_c$). In the presence of a linear coupling, once $\xi$ is introduced, the high-frequency and static values of the modulus $\phi_{\eta\eta}(\Omega)$ are different, the high-frequency modulus $\phi^{\xi}_{\eta\eta}$ being non-zero at $T = T_c$ as distinct from the static modulus $\phi^{g}_{\eta\eta}$. That is why the sidebands can be present in the spectrum even at the transition point.

The temperature variation of the intensities of both central peak and the side components depends on the character of the coupling between $\epsilon$ and $\eta$. For a linear coupling, i.e., in the case of proper ferroelastics when $I_1 \propto (\phi^{g}_{\eta\eta})^{-1}$ (see sect. 5.3.2), the intensity of the side components remains finite at the phase transition point, whereas the central component intensity goes to infinite like $|\tau|^{-1}$ as $\tau \to 0$. For a bilinear coupling when $\epsilon = \epsilon_0 + a\eta^2$ and $I_1 \propto \eta_e^2/\phi_{\eta\eta}^g$, (see sect. 5.1) the intensity of the side components is proportional to $\eta_e^2$ and goes to zero as $\tau \to 0$, while the central peak intensity has a finite value at $T = T_c$.

Note that the intensity of some spectral lines, which are not connected with order parameter fluctuations and not active in the Raman scattering in the symmetrical phase, may also change with temperature proportionally to $\eta_e^2$. Indeed, consider the spectral line due to scattering by fluctuations in a variable $\zeta$ such that

$$\epsilon = \epsilon_0 + a_1\eta\zeta , \tag{170}$$

where $a_1$ is a constant. Then the intensity of this line

$$I_\zeta \propto \langle(\Delta\epsilon)^2\rangle \propto \eta_e^2\langle(\Delta\zeta)^2\rangle \propto \eta_e^2 . \tag{171}$$

Let us discuss now the temperature evolution of the spectrum in the case when the coupling coefficient $\phi_{\eta\xi}$ vanishes as $\tau \to 0$. The latter takes place if in the symmetrical phase the variable $\xi$ is bilinearly coupled with $\eta$, whereas below $T_c$ $\xi' \propto \eta'$. Such a coupling occurs when the variable $\xi$ is invariant under operations of the symmetry group of the symmetrical phase. In particular, $\xi$ may be a scalar and have the meaning of the mass density, temperature, component concentration in solid solutions, and so on. In a non-symmetrical phase for second-order phase transitions far from the tricritical point, it follows from eq. (47) that the quantities $\phi_{\eta\eta}^\xi$ and $\phi_{\eta\eta}^g$ have then the same temperature dependence both in the classical and the critical region. (In both cases $\phi_{\eta\eta}^\zeta \sim \phi_{\eta\eta}^g \sim |\tau|^\gamma$, but of course, with different values of $\gamma$.) Therefore, the temperature dependences of the positions and the intensity of the soft mode side bands prove to be the same here, with a precision up to a numerical factor, as if there were no coupling between $\eta$ and $\xi$.

As seen from eq. (167) for a bilinear coupling between $\eta$ and $\xi$ the fractional intensity of the central peak arising in the spectrum when $\tilde{\Omega}_{R\xi} < \tilde{\Omega}_{0\eta}$ is independent of $|\tau|$. Thus for a bilinear coupling, when approaching the phase transition, the intensity of the central component does not increase at the expense of the side ones. The width of the central peak is determined by the relaxation frequency $\Omega_{R\xi} = \phi_{\xi\xi}^h/\gamma_\xi$ which within the Landau theory does not depend on $\tau$, so that the central peak does not get narrower as $T \to T_c$. Since both the relaxation rates $\Omega_{R\xi}$ and $\tilde{\Omega}_{R\xi}$ do not change strongly with temperature while the frequencies $\Omega_{0\eta}$ and $\tilde{\Omega}_{0\eta}$ tend to zero as $|\tau| \to 0$ the condition $\tilde{\Omega}_{R\xi} < \tilde{\Omega}_{0\eta}$ needed for the existence of the central component along with the side ones, will be violated close enough to the transition temperature and the further evolution of the spectrum will be analogous to that described in sect. 6.1 for the case of a single quasi-harmonic soft mode.

Close to the tricritical point (when approached along the paths with

$B^2 < 4AC$, see fig. 1) the temperature dependences of the quantities $\phi^g_{\eta\eta}$ and $\phi^\xi_{\eta\eta}$ do not coincide. Therefore, the intensity of the central component (as compared with that of the side component) increases here as $\tau \to 0$ by the law $I_{\mathrm{central}}/I_{\mathrm{side}} = I_{\mathrm{central}}/(I_1 - I_{\mathrm{central}}) \propto |\tau|^{-1/2}$ which is weaker than for the case of a linear coupling between $\eta$ and $\xi$. For phase transitions in solids, however, (due to shear stresses inherent in solids, see sect. 5.3.1), the ratio $I_{\mathrm{central}}/I_{\mathrm{side}}$ increases near $T_c$, in fact, no more than a few times. Note finally that at spinodal points of the non-symmetrical phase (for first-order transitions) the quantity $\eta_e$, and therefore $\phi_{\eta\xi}$, do not vanish. Hence the character of temperature dependence of the scattered light spectrum must be the same here as in the case of a linear coupling between $\eta$ and $\xi$ (Pytte and Thomas 1972, Levanyuk 1974).

Now let us take into account the dependence of the permeability $\epsilon$ on $\xi$ as well. Neither the form of the spectral density nor the character of its temperature evolution change qualitatively in this case. When the characteristic frequencies of the $\eta$ fluctuations (for fixed $\xi$) and of the $\xi$ fluctuations (for fixed $\eta$) differ considerably, one can write simple formulae for the intensities contained both in the side and in the central components of the spectrum.

If the fluctuations in the parameter $\eta$ are oscillatory and $\tilde{\Omega}_{0\eta} \gg \tilde{\Omega}_{\mathrm{R}\xi}$, it is convenient to divide the $\eta$ fluctuations into "fast" and "slow" parts $\eta = \eta_0(\xi) + \eta'$, as was done for the $\xi$ fluctuations in sect. 3.2. The position and intensity of the side components are determined by the fast fluctuations $\eta'$ which can be considered to proceed at a fixed value of $\xi$. Therefore for the side component intensity we have

$$I_{\mathrm{side}} \simeq Q_{\mathrm{S0}}(\partial\epsilon/\partial\eta)^2_\xi/\phi^\xi_{\eta\eta}\,, \tag{172}$$

i.e., an account of the dependence of $\epsilon$ on $\xi$ proves to be inessential here. The slow fluctuations in $\eta$ follows the $\xi$ fluctuations and contribute, as well as the $\xi$ fluctuations themselves, to the central component, whose intensity is now given by the expression

$$I_{\mathrm{central}} \simeq Q_{\mathrm{S0}}(\partial\epsilon/\partial\xi)^2_h/\phi^h_{\xi\xi}\,, \tag{173}$$

where

$$\left(\frac{\partial\epsilon}{\partial\xi}\right)_h = \left(\frac{\partial\epsilon}{\partial\xi}\right)_\eta + \left(\frac{\partial\epsilon}{\partial\eta}\right)_\xi \frac{\mathrm{d}\eta_0}{\mathrm{d}\xi} = \left(\frac{\partial\epsilon}{\partial\xi}\right)_\eta - \frac{\phi_{\eta\xi}}{\phi^\xi_{\eta\eta}}\left(\frac{\partial\epsilon}{\partial\eta}\right)_\xi. \tag{174}$$

Note that it is the derivative $(\partial\epsilon/\partial\xi)_h$ which takes into account contributions of both the $\xi$ changes and the slow part of changes in $\eta$ which follow the changes in $\xi$.

For $\tilde{\Omega}_{0\eta} \ll \tilde{\Omega}_{\mathrm{R}\xi}$ it is the $\xi$ fluctuations that must be divided into "slow" and "fast" parts. The slow fluctuations in $\xi$ contribute to the side components, the intensity of which is now

$$I_{\text{side}} \simeq Q_{S0}(\partial\epsilon/\partial\eta)_g^2/\phi_{\eta\eta}^g\,, \tag{175}$$

where

$$\left(\frac{\partial\epsilon}{\partial\eta}\right)_g = \left(\frac{\partial\epsilon}{\partial\eta}\right)_\xi - \left(\frac{\partial\epsilon}{\partial\xi}\right)_\eta \frac{\phi_{\eta\xi}}{\phi_{\xi\xi}^\eta}\,. \tag{176}$$

The fast fluctuations in $\xi$ (proceeding at a fixed $\eta$) give a wide background spectral distribution against which the soft mode components are merging. The intensity of this background is contained for the most part in the region $\Omega \gg \tilde{\Omega}_{0\eta}$ and is equal approximately to

$$I_{\text{background}} \simeq Q_{S0}(\partial\epsilon/\partial\xi)_\eta^2/\phi_{\xi\xi}^\eta\,. \tag{177}$$

The results obtained can easily be extended to the case when the fluctuations in $\eta$ have a relaxation character. The intensity $I_{\text{side}}$ in eq. (172) should then be referred to the wide wings of the central peak due to the $\xi$ fluctuations and in eq. (175) to the merged side components associated with the $\eta$ fluctuations.

### *6.2.2. Examples of the central peak situation*

As has already been mentioned, even in the most thoroughly investigated substances it is difficult for the present to unambiguously establish the physical reason for the slow relaxation leading to the appearance in the scattering spectrum of a temperature-dependent central peak. Moreover, in many cases attempts to resolve the width of the central peak have not been successful and therefore it is not clear whether it results from scattering by static inhomogeneities (ch. 2) or from scattering by thermal fluctuations. A few simple examples can be given, however, when the origin of the central peak, i.e., the meaning of the quantities $\eta$ and $\xi$ is clear enough.

Begin, as above, with the case when the variables $\eta$ and $\xi$ are linearly coupled already in the symmetrical phase. This means that the variables $\eta$ and $\xi$ have the same symmetry, so that either of them may be chosen as the order parameter. An example of the kind (mentioned already in sects. 3 and 5) are proper ferroelastics in which the soft mode is one of the zone-center acoustic modes, while the appearance of the spontaneous acoustic deformation below $T_c$ is accompanied by an additional structural reconstruction, e.g., by redistribution of some sort of atoms within the unit cell. The time of this redistribution being large, one may expect that the side components (i.e., in this case the Mandelstam–Brillouin (MB) components) will not merge as $\tau \to 0$, and the intensity increase will proceed at the expense of the central peak growth, as shown in fig. 13.

A well-known example of this type are proper ferroelastics of the $KH_2PO_4$ (KDP) group which are at the same time proper ferroelectrics (Jona and Shirane 1962). The phase transition is usually associated here with ordering in the proton system (Jona and Shirane 1962, Blinc and Žekš 1974, Vaks 1973). However, in KDP crystals the rate of this ordering exceeds considerably the frequency of the MB components at $q \sim 10^5\,\mathrm{cm}^{-1}$ so that the temperature evolution of these components in KDP crystals should be analogous in fact to that described in sect. 6.1. This is just what has been predicted theoretically by Levanyuk and Sobyanin (1967) and verified more recently by Courtens (1978). The latest investigations (Durvasula and Gammon 1977, Courtens 1978) have also shown that a very narrow central peak discovered near the phase transition in KDP by Lagakos and Cummins (1974) is of a purely static origin and is connected in all probability with scattering by defects. A careful investigation of the dynamic central peak in KDP including the observation of its dependence on the direction of scattering wavevector (due to influence of long-range forces) was performed by Sawafuji et al. (1979).

The situation when the proton ordering (polarization) relaxation time exceeds the oscillation period corresponding to MB components has been predicted by Levanyuk and Sobyanin (1967) to occur in Rochelle salt on the ground of the experimental data presented by Yakovlev and Velichkina (1957). This conclusion was confirmed recently in MB-scattering investigations by Sailer and Unruh (1975, 1976) (see also Unruh et al. (1978)) although the spectrum picture of the type shown in fig. 13 was not observed directly in these investigations due to the low scattering by the corresponding transverse acoustic phonon.

Mixed $(\mathrm{KCN})_c(\mathrm{KX})_{1-c}$ crystals, where $\mathrm{X} \equiv \mathrm{Cl}$, Br, represent a very interesting and rather unique example of ferroelastics in which the CN-ordering relaxation frequency can be made either more or less than the MB-components frequency. A KCN crystal in the symmetric phase has a cubic structure similar to KCl with a random orientation of linear molecular CN groups, which substitute for Cl. In the ferroelastic phase these groups are partially ordered, which lowers the crystal symmetry as far as orthorhombic (Bishofberger and Courtens 1975, Courtens 1976a, b). Setting orientational order by rotating CN groups is connected with their hopping over potential barriers produced by the interaction of these groups with nearest neighbours. This leads to a thermo-activation character of the temperature dependence of the coefficient $\gamma_\xi$ in eq. (162). When part of the CN groups is replaced by Cl or Br, the ferroelastic transition temperature lowers, which results in a sharp increase of the CN-ordering relaxation time $\Omega_{\mathrm{R}\xi}^{-1} = \gamma_\xi / \phi^{\eta}_{\xi\xi}$. Thus, by changing the concentration of Cl or Br in mixed crystals of $(\mathrm{KCN})_c\,(\mathrm{KX})_{1-c}$ one can pass over from the spectrum evolution picture with a merging of MB components to the one without their merging but with a strongly temperature dependent central peak appearing in addition to these components (Michel et al. 1978).

In some cases the presence of a slowly relaxing variable $\xi$ connected linearly with $\eta$ (both above and below $T_c$) may be ascribed to the presence in the crystal of defects which may occupy several equivalent positions in the unit cells and hop back and forth to these positions as $\eta$ changes (Halperin and Varma 1976). This mechanism will be discussed in more detail in ch. 2.

It should be noted that in spite of a large number of works concerning the "central peak problem" there are, in fact, very few examples where the dynamic nature of this peak has been established with confidence.

Among the quantities $\xi$, whose fluctuations are linearly coupled with the $\eta$ fluctuations only below $T_c$, the most interesting example is the temperature (Fabelinskii 1965, Balagurov and Vaks 1969, Pytte and Thomas 1972, Wehner and Klein 1972, Levanyuk 1974). In the equations of motion it is however the entropy $S$ rather than the temperature $T$ that is more convenient to use since $S$ has the sense of a generalized coordinate, while $T$ represents the generalized force conjugate to $S$ (sometimes the energy $E$ is also used as a generalized coordinate instead of the entropy $S$, see, e.g., Hohenberg and Halperin (1977)). In eqs. (161), (162) and (166) one should then put

$$\phi^{\xi}_{\eta\eta} \equiv \phi^{S}_{\eta\eta} = \phi^{T}_{\eta\eta}(C_h/C_\eta) = m\tilde{\Omega}^2_{0\eta}, \tag{178}$$

$$\phi_{\eta\xi} = (T/C_\eta)\phi_{\eta T} = TC_\eta \eta_e A'_T, \tag{179}$$

$$\phi^{h}_{\xi\xi} = T/C_\eta, \tag{180}$$

$$\gamma_\xi = T/\kappa q^2, \tag{181}$$

where $A'_T$ is the temperature derivative of the coefficient $A$ in formula (1), $\kappa$ is the heat conductivity and $C_h$ and $C_\eta$ are the specific heats at constant $h$ and $\eta$, respectively. If $\tilde{\Omega}_{RS} \equiv \kappa q^2/C_\eta < \tilde{\Omega}_{0\eta}$ (which holds as a rule for dielectric crystals where $\tilde{\Omega}_{RS} \sim 10^9$–$10^7\,s^{-1}$), the spectrum of scattered light contains, along with the soft mode side bands, also a central maximum whose fractional intensity (see eq. (167)) is given by

$$I_{central}/I_1 = (C_h - C_\eta)/C_h, \tag{182}$$

in accordance with the well-known formula by Landau and Placzek (1934) for fluids in which case $C_h$ and $C_\eta$ are to be replaced by $C_p$ and $C_V$ respectively. Formula (182) is obtained under assumption that $\epsilon$ depends on $\eta$ only. If one takes into account the dependence of $\epsilon$ on other generalized coordinates as well, the expression for the central peak intensity will have the form (see eq. (173))

$$I_{central} = Q_{S0}\left(\frac{\partial\epsilon}{\partial S}\right)^2_h \Big/ \phi^{h}_{SS} = Q_{S0}\left(\frac{\partial\epsilon}{\partial S}\right)^2_h \frac{C_h}{T} = Q_{S0}\left(\frac{\partial\epsilon}{\partial T}\right)^2_h \frac{T}{C_h}. \tag{183}$$

From eqs. (182) and (183) it is clear that for $C_h \gg C_\eta$ (i.e., in the critical or in the classical region when the transition is close to the tricritical point) the main part of the intensity of light scattered by $\eta$ fluctuations is contained in the central peak.

Observations of a dynamical central peak associated probably with entropy (temperature) fluctuations were reported by Schulhof et al. (1970), Birgeneau et al. (1971), Lyons and Fleury (1976, 1978) and Fleury and Lyons (1976) for a number of magnetic and structural phase transitions. Mermelstein and Cummins (1977) observed a rather narrow central peak (with the width of $\sim 50$ MHz) below ferroelectric transition temperature in $KH_2PO_4$. This peak was assumed to originate from entropy fluctuations, the more so as, in contrast with the results of works cited above, the fractional intensity of this peak increased sharply when approaching $T_c$, so that it was observable only within about 0.1 K below $T_c$. Such a strong temperature dependence of the central peak intensity is not surprising since the phase transition in $KH_2PO_4$ is rather close to the tricritical point. The interpretation of this peak as that connected with the entropy fluctuations is, however, queried by Widom et al. (1978).

In solutions the role of a slowly relaxing variable that is linearly coupled with the order parameter for $\tau < 0$ may be played by the concentration of one of the solution's components. This explains in particular the presence of the central peak in the light scattering spectrum in superfluid solutions of $^3$He in $^4$He, as has been predicted theoretically by Gor'kov and Pitaevskii (1957) and observed experimentally by Palin et al. (1971).

Further examples of the central peak situation can be found in sects. 6.3 and 6.4, as well as in ch. 7 of this volume.

*6.2.3. Coupling between the soft and vibration mode*

Before discussing the linear coupling between the soft and an additional vibration mode, we make a remark about some straightforward generalizations of eqs. (161) and (162). In two preceding subsections we used the Mandelstam–Leontovich theory in which the coupling between the two dynamical variables $\eta$ and $\xi$ is assumed to be the same as in the static case, where it is described by some mixed terms in the expression for the thermodynamic potential density. More general equations of motion for $\eta$ and $\xi$ may be obtained by varying the Lagrange function with respect to $\eta$ and $\xi$:

$$
\begin{aligned}
&L = K - U\,,\\
&K = \tfrac{1}{2} m\dot{\eta}^2 + m_{\eta\xi}\dot{\eta}\dot{\xi} + \tfrac{1}{2} m_{\xi}\dot{\xi}^2\,,\\
&U \equiv \Delta\phi = \tfrac{1}{2}\phi^{\xi}_{\eta\eta}\eta'^2 + \phi_{\eta\xi}\eta'\xi' + \tfrac{1}{2}\phi^{\eta}_{\xi\xi}\xi'^2\,,\\
&\eta' = \eta - \eta_e, \qquad \xi' = \xi - \xi_e\,,
\end{aligned}
\tag{184}
$$

together with the dissipation function (see, e.g., Landau and Lifshitz 1976)

$$R = \tfrac{1}{2}\gamma\dot{\eta}^2 + \gamma_{\eta\xi}\dot{\eta}\dot{\xi} + \tfrac{1}{2}\gamma_{\xi}\dot{\xi}^2\,. \tag{185}$$

Two of the three quadratic forms $K$, $U$ and $R$ involved in eqs. (184) and (185) can be simultaneously diagonalized by passing over to new variables in eqs. (184) and (185). In the microscopic theories of lattice dynamics, when $\eta$ and

$\xi$ have the meaning of some atomic displacements (or their linear combinations), it is the Lagrange function (184) that is usually assumed to be diagonalized, which corresponds to introducing normal coordinates. In this case the coupling between normal coordinates is described by mixed terms in the dissipation function. Such a diagonalization is, however, inconvenient for a comparison of the high-frequency data with the results of static and low-frequency experiments. To this end, it is more convenient that the kinetic energy $K$ and the dissipation function $R$ were assumed to be diagonalized. This assumption has been used implicitly above. Of course, it may be unreasonable to diagonalize any of the quadratic forms especially if the variables $\eta$ and $\xi$ in eqs. (184) and (185) have different physical meanings (this is the case, in particular, when $\eta$ is the polarization and $\xi$ is the acoustic deformation; see Sannikov (1962)). All the three above-mentioned approaches (when one diagonalizes $L$, or $K$ and $R$, and when none of the quadratic forms is assumed, in general, to be diagonalized) are used in the literature. The choice of one or another does not, naturally, affect the final results. Remember that within the Landau theory the temperature dependences of all the system's characteristics are determined by the temperature dependence of the coefficients in the expansion of the thermodynamic potential density, while quantities of the type $m$ and $\gamma$ are supposed to be temperature independent. Therefore, below, as above, it will be more convenient for us to use the approach under which the kinetic energy as well as the dissipation function are thought to be represented in a diagonal form, and the coupling between the fluctuations in different variables is taken into account in the thermodynamic potential. For another type of diagonalization the coefficients $m_{ik}$ and $\gamma_{ik}$ may prove to be essentially temperature dependent.

Bearing in mind what has been said above, one can write the equations of motion for two linearly coupled variables $\eta'$ and $\xi'$ (in the case when both the variables are of an oscillatory nature) in the form (used for the first time in the light scattering problems by Barker and Hopfield (1964))

$$m\ddot{\eta}' + \gamma\dot{\eta}' + \phi^{\xi}_{\eta\eta}\eta' + \phi_{\eta\xi}\xi' = h\,, \tag{186}$$

$$m_{\xi}\ddot{\xi}' + \gamma_{\xi}\dot{\xi}' + \phi_{\eta\xi}\eta' + \phi^{\eta}_{\xi\xi}\xi' = g\,. \tag{187}$$

It is obvious that these equations differ from eqs. (161) and (162) by the term $m_{\xi}\ddot{\xi}'$ only.

If the damping of the oscillations of $\eta$ and $\xi$ is small, i.e., if the frequencies $\tilde{\Gamma} \equiv \gamma/m$ and $\tilde{\Gamma}_{\xi} \equiv \gamma_{\xi}/m_{\xi}$ are much lower than all the other characteristic frequencies, then we obtain for the eigenfrequencies of the two normal modes in the system

$$\Omega^2_{\pm} = \tfrac{1}{2}\{\tilde{\Omega}^2_{0\eta} + \tilde{\Omega}^2_{0\xi} \pm \sqrt{(\tilde{\Omega}^2_{0\eta} - \tilde{\Omega}^2_{0\xi})^2 + 4\Delta^4}\}\,, \tag{188}$$

where

$$\tilde{\Omega}^2_{0\eta} = \phi^{\xi}_{\eta\eta}/m_\eta, \qquad \tilde{\Omega}^2_{0\xi} = \phi^{\eta}_{\xi\xi}/m_\xi\,,$$

$$\Delta^4 = \phi^2_{\eta\xi}/m_\eta m_\xi - \delta^2_\eta \delta^2_\xi, \qquad \delta^2_\eta = |\phi_{\eta\xi}|/m_\eta, \qquad \delta^2_\xi = |\phi_{\eta\xi}|/m_\xi\,. \tag{189}$$

The corresponding normal coordinates for these two modes are given by the expression

$$y_\pm = a_\pm \eta + b_\pm \xi\,,$$

where for the upper mode with the frequency $\Omega_+$

$$|a_+/b_+| = \delta^2_\eta/|\Omega^2_+ - \tilde{\Omega}^2_{0\eta}| = |\Omega^2_+ - \tilde{\Omega}^2_{0\xi}|/\delta^2_\xi, \tag{190}$$

and for the lower mode with the frequency $\Omega_-$

$$|a_-/b_-| = \delta^2_\xi/|\Omega^2_- - \tilde{\Omega}^2_{0\xi}| = |\Omega^2_- - \tilde{\Omega}^2_{0\eta}|/\delta^2_\eta\,. \tag{191}$$

Consider first the temperature evolution of the spectrum in the case when for $\tau > 0$ the variables $\eta$ and $\xi$ have the same transformation properties, i.e., when the linear coupling between $\eta$ and $\xi$ takes place in both phases. For simplicity we assume that in eqs. (186) and (187) the coefficient $\phi^{\xi}_{\eta\eta}$ alone is temperature dependent and restrict ourselves to the temperature range above $T_c$, where in the framework of the Landau theory one may put $\phi^{\xi}_{\eta\eta} = A'_T(T - T_{c0})$. It is clear from eqs. (186) and (187) for the static case with $g = 0$ that

$$\chi^{-1}_{\eta\eta}(0) = h/\eta = \phi^{\xi}_{\eta\eta} - (\phi_{\eta\xi})^2/\phi^{\eta}_{\xi\xi} = A'_T(T - T_{c0}) - (\phi_{\eta\xi})^2/\phi^{\eta}_{\xi\xi}\,. \tag{192}$$

We see that the static susceptibility $\chi_{\eta\eta}(0)$ diverges now at a shifted transition temperature $T_c = T_{c0} + (\phi^2_{\eta\xi}/A'_T\phi^{\eta}_{\xi\xi})$ which exceeds the "old" transition temperature $T_{c0}$ (for an uncoupled system) since $A'_T > 0$ and $\phi^{\eta}_{\xi\xi} > 0$.

The temperature dependence of the frequencies $\Omega_+$ and $\Omega_-$, as determined by formulae (188) and (189) is shown under the above assumptions in fig. 14. We see that the interaction between modes leads to a "repulsion" of their frequencies near the "crossing" temperature $T_{cr}$ at which the frequencies of the non-interacting modes coincide. The repulsion of the mode frequencies, responsible for the vanishing of the frequency $\Omega_-$ at a temperature $T_c$ somewhat higher than $T_{c0}$, may be treated as the reason why the transition temperature $T_c$ in a real system is higher than the transition temperature when the interaction between $\eta$ and $\xi$ is neglected.

It follows from eqs. (190) and (191) that the contributions of the $\eta$ and $\xi$ vibrations to the upper and lower modes change with temperature in such a way that above the crossing temperature ($T > T^+_{cr}$) the $\eta$ vibrations contribute mainly to the $\Omega_+$ mode and below this temperature to the $\Omega_-$ mode. For the $\xi$ vibrations the situation is just the opposite. At $T = T_c$ the ratio of the amplitudes of $\eta$ and $\xi$ vibrations in the $\Omega_-$ mode is $|a_-/b_-|_{T=T_c} = \tilde{\Omega}^2_{0\eta}/\delta^2_\eta$ and in the $\Omega_+$ mode this ratio is equal to $|a_+/b_+|_{T=T_c} = \delta^2_\xi/\tilde{\Omega}^2_{0\xi}$. Near the crossing

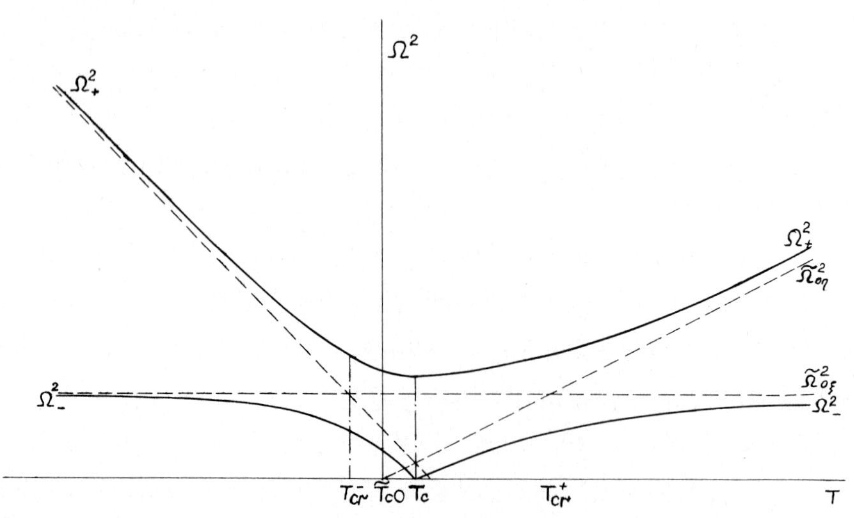


Fig. 14. Temperature dependences of the square of the frequencies of the soft mode and of a vibrating mode linearly coupled with it. The dashed lines are squares of the frequencies of the non-interacting modes.

temperature $T_{\text{cr}}^+$ there occurs a redistribution of the intensity between the $\Omega_-$ and $\Omega_+$ modes. Above $T_{\text{cr}}^+$, when $\tilde{\Omega}_{0\eta} \gg \tilde{\Omega}_{0\xi}$, the ratio of the intensity of the components corresponding to the upper $\Omega_+$ and the lower $\Omega_-$ mode is given by the expression

$$\frac{I_{\Omega_-}}{I_{\Omega_+}} = \left[\frac{(\partial\epsilon/\partial\xi)_\eta}{(\partial\epsilon/\partial\eta)_\xi} - \frac{\delta_\xi^2}{\Omega_{0\xi}^2}\right]^2 \frac{\tilde{\Omega}_{0\eta}^2}{\Omega_-^2}, \tag{193}$$

whereas below $T_{\text{cr}}^+$, when $\tilde{\Omega}_{0\eta} \ll \tilde{\Omega}_{0\xi}$,

$$\frac{I_{\Omega_-}}{I_{\Omega_+}} = \left[\frac{(\partial\epsilon/\partial\eta)_\xi}{(\partial\epsilon/\partial\xi)_\eta} - \frac{\delta_\eta^2}{\Omega_{0\eta}^2}\right]^2 \frac{\tilde{\Omega}_{0\xi}^2}{\Omega_-^2}. \tag{194}$$

These formulae can be obtained in a similar way as expressions (172)–(177) by dividing higher-frequency fluctuations into "slow" and "fast" parts.

For illustration we assume as usual that $\epsilon$ depends on $\eta$ only, i.e. $(\partial\epsilon/\partial\xi)_\eta \simeq 0$. Then it follows from eq. (194) that for $T < T_{\text{cr}}^+$ the ratio $I_{\Omega-}/I_{\Omega+}$ is very large (this means that the mode $\Omega_+$ will not be, in fact, seen in the spectrum), whereas for $T > T_{\text{cr}}^+$ it is equal to $\delta_\xi^4\tilde{\Omega}_{0\eta}^2/\tilde{\Omega}_{0\xi}^4\Omega_-^2$ and is small as far as $\phi_{\eta\xi}^2/\phi_{\eta\eta}^\xi\phi_{\xi\xi}^\eta \ll (\phi_{\xi\xi}^\eta/\phi_{\eta\eta}^\xi)^2$, i.e., in the case of a weak enough coupling. Examples of the temperature dependences of frequencies and intensities of coupled modes as observed experimentally by Peercy (1975c) and by Slivka et al. (1978), are shown in fig. 15 (for more examples see other chapters of this volume). Although it has been assumed above that $\tau > 0$, all the results obtained can be easily extended to the region $\tau < 0$ (see figs. 14 and 15).

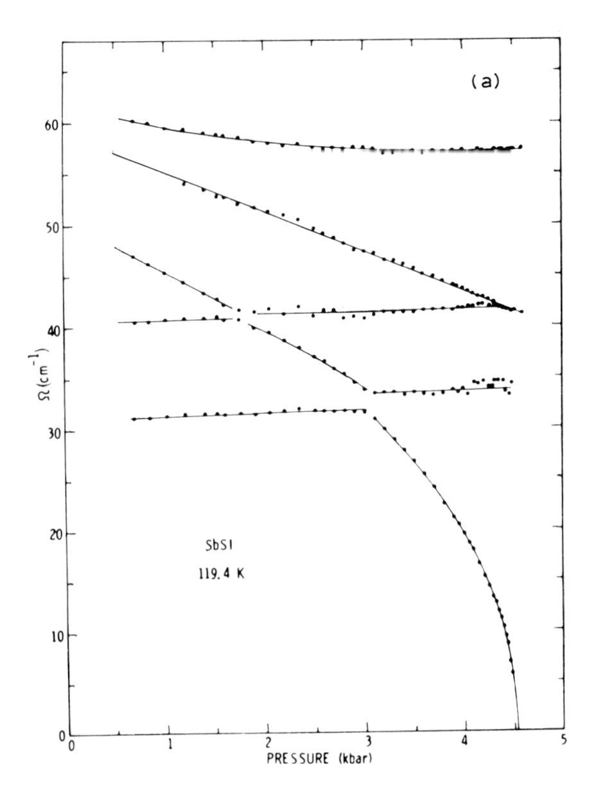


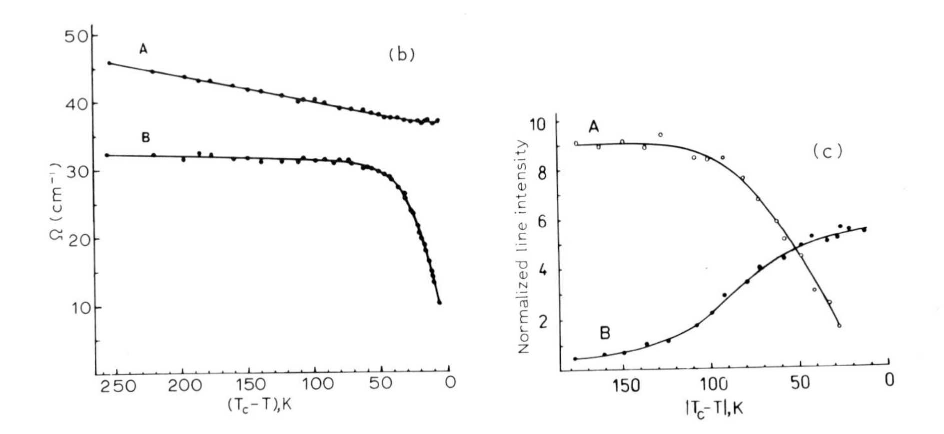


Fig. 15. (a) Pressure dependence of the frequencies of the coupled modes in SbSI (Peercy 1975c); (b), (c) Temperature dependence of the frequencies (b) and the intensities (c) of the two coupled modes in $Sn_2P_2S_6$ (Slivka et al. 1979).

Now consider the situation when the $\eta$ vibrations are coupled linearly with those of $\xi$ only below $T_c$ (in the case of a one-component order parameter this means that for $\tau > 0$ the mode $\xi$ is fully symmetric). The temperature dependences of the frequencies $\Omega_+$ and $\Omega_-$ for this case are shown schematically in fig. 16. Note that since the coupling coefficient $\phi_{\eta\xi}$ vanishes as $|\tau| \to 0$, the interaction of the modes now does not affect the transition temperature. Besides, at $\tau = 0$ the relative contribution of the $\eta$ vibrations to the mode $\Omega_-$ becomes equal to unity, i.e., as $|\tau| \to 0$ the lower mode (which far below the crossing temperature corresponds mainly to the oscillations in $\xi$) at $T = T_c$ corresponds solely to the $\eta$ vibrations. This is just the opposite of the previous case when the $\Omega_-$ mode included both $\eta$ and $\xi$ vibrations even at $T = T_c$. The character of the intensity redistribution between the modes is still described by formulae (193) and (194) and is qualitatively analogous to that considered for the previous case.

In the above analysis the damping of the modes has been assumed to be rather small. Neglect of damping is justified if the frequencies $\tilde{\Gamma}$, $\tilde{\Gamma}_\xi \ll \Delta$, i.e., if the widths of the lines are smaller than the difference of their frequencies at the "crossing" temperature. In the region where this condition is not fulfilled, a more detailed consideration is needed. In particular, the width and the shape of the lines may undergo here considerable changes (Zawadowski and Ruvalds 1970).

### *6.2.4. Further examples and generalizations*

In the preceding section, as well as in sects. 6.2.1 and 6.2.2, we restricted ourselves for the sake of simplicity to the analysis of the linear coupling between the $\eta$ fluctuations and those of only one extra variable $\xi$. In fact, taking into

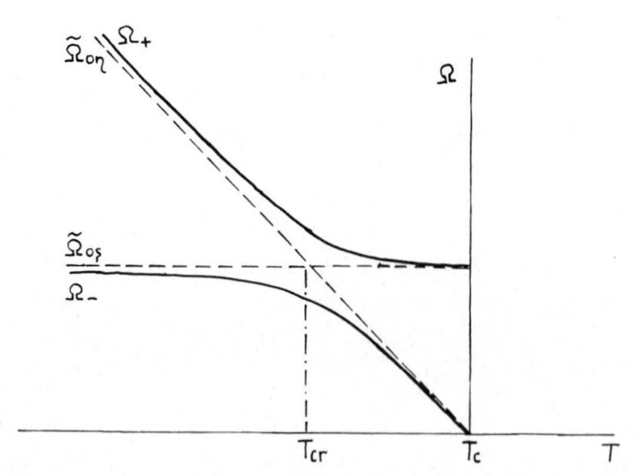


Fig. 16. Temperature dependence of the squares of the frequencies of two "soft" modes interacting only below $T_c$ (when $\phi_{\eta\xi}(T \geqslant T_c) = 0$). The dashed lines are the squares of frequencies of the non-interacting modes.

account a linear coupling between $\eta$ and several oscillation and (or) relaxation variables may, of course, turn out to be necessary. Examples of such an analysis of experimental data may be found, in particular, in Lagakos and Cummins (1974), and Lyons and Fleury (1978).

A further generalization of the theory presented in sect. 6.1 must consist in taking into account the non-linear coupling between fluctuations in the order parameter and fluctuations in other variables. The main characteristic feature of the non-linear case is that the variable $\eta(\boldsymbol{q} \simeq 0)$ is coupled here not with one or several oscillation and relaxation variables, as has taken place in the linear case considered above, but with an infinitely large (as $V \to \infty$, $V$ is the system volume) number of variables. This is the case also for second and higher orders scattering. Moreover, as we shall see in sect. 6.4, these two types of effects are intimately related. Therefore before studying the effects of a non-linear coupling of $\eta$ with other variables, we shall consider the spectral density of second-order scattering by $\eta$ fluctuations.

### 6.3. *Second-order scattering*

To calculate the spectral density of the second-order scattering, we adopt the Gaussian approximation. This implies that the averages of the type $\langle \eta^4 \rangle$ would be replaced by the product of the mean-square fluctuations (with a numerical factor which is equal to the number of ways to make this replacement; in formula (195) this factor is equal to 2) while averages of the type $\langle \eta^3 \rangle$ would be assumed simply to vanish (see also sects. 3.5 and 5.5). The Gaussian approximation is justified within the classical region, when one can restrict oneself to consideration of the first-order fluctuation corrections. With a modification it may be used also for an analysis of the temperature dependence of the spectrum in the critical region (see below). In the above-mentioned approximation the spectral density of the second-order scattering by $\eta$ fluctuations is given by the expression (Levanyuk 1976)

$$\begin{aligned} \mathscr{I}_2(\boldsymbol{q}, \Omega) &= 2Q_{S0}a^2 \int \langle |\eta(\boldsymbol{k}, \omega)|^2 \rangle \langle |\eta(\boldsymbol{q}-\boldsymbol{k}, \Omega-\omega|^2 \rangle \, \mathrm{d}\boldsymbol{k} \, \mathrm{d}\omega \\ &= \frac{8Q_{S0}a^2\gamma^2(k_B T)^2}{(2\pi)^4} \int \chi(\boldsymbol{k}, \omega)\chi(-\boldsymbol{k}, -\omega)\chi(\boldsymbol{q}-\boldsymbol{k}, \Omega-\omega) \\ &\quad \times \chi(\boldsymbol{k}-\boldsymbol{q}, \omega-\Omega) \, \mathrm{d}\boldsymbol{k} \, \mathrm{d}\omega \,, \end{aligned} \tag{195}$$

where the generalized susceptibility

$$\chi(\boldsymbol{k}, \omega) = \frac{1}{\phi_{\eta\eta} + Dk^2 - \mathrm{i}\gamma\omega - m\omega^2} = \frac{1}{m[\Omega_0^2(\boldsymbol{k}) - \mathrm{i}\Gamma\omega - \omega^2]} \,. \tag{196}$$

We will neglect, as before, the finiteness of the light wavevector, i.e., we put $\boldsymbol{q} = 0$ and assume at first that the fluctuations in $\eta$ are of a vibration character, i.e., $\Omega_0(\boldsymbol{k}) > \Gamma/\sqrt{2}$. After integrating in (195) over $\omega$ one finds

$$\mathscr{I}_2(0, \Omega) = \frac{Q_{\mathrm{S0}}}{2\pi^3}\left(\frac{ak_{\mathrm{B}}T}{m}\right)^2 \frac{\Gamma}{\Omega^2+\Gamma^2}\int\left[4+\frac{\Omega^2+4\Gamma^2}{\Omega_0^2(\boldsymbol{k})}\right] \times \frac{\mathrm{d}\boldsymbol{k}}{[4\Omega_0^2(\boldsymbol{k})-\Omega^2]^2+4\Gamma^2\Omega^2}. \tag{197}$$

If we neglect dispersion in the $\eta$-vibration branch, i.e., put $\Omega_0(\boldsymbol{k}) = \text{const.}$, and suppose that $\Omega_0 \gg \Gamma$, then, as is easily seen, the spectral density $\mathscr{I}_2(0, \Omega)$ has three maxima, at frequencies $\Omega = 0$ and $\Omega \simeq \pm 2\Omega_0$. The central maximum is twice as high as the side maxima, the half widths of all the maxima being the same and approximately equal to $\Gamma$.

When expressed in terms of the quantum theory, the side bands in the second-order scattering spectrum correspond to scattering with the emission or absorption of two phonons, while the central peak corresponds to scattering with the emission of one and absorption of another phonon of the same frequency. The widths of the peaks are determined then by the uncertainty in the phonon energy, i.e., by the damping constant $\Gamma$.

Dispersion in the $\eta$-vibration branch will lead to an increase of the width and, in principle, to the disappearance of the side peaks. As to the central peak, it exists always in the spectrum and the dispersion (in this case the dispersion of the coefficient $\Gamma$ is most essential) will only lead to some distortion of its shape. The intensity of this central peak increases as $\tau \to 0$ by the law $I_2 \propto |\tau|^{-1/2}$ and becomes comparable with the intensity of the components in the first-order scattering spectrum when approaching the boundary between the classical and critical regions (see sect. 5.5).

When the $\eta$ fluctuations have a relaxation character ($\Omega_0 \ll \Gamma$), the intensity distribution in the second-order scattering spectrum is given by the expression (Levanyuk 1976)

$$\mathscr{I}_2(0, \Omega) = \frac{Q_{\mathrm{S0}}\gamma(ak_{\mathrm{B}}T)^2}{4\pi^2(\phi_{\eta\eta}r_{\mathrm{c}})^3}\frac{1}{x^2}\left\{\frac{1}{\sqrt{2}}[(1+x^2)^{1/2}+1]^{1/2}-1\right\}, \tag{198}$$

where $x \equiv \Omega^2/\Omega_{\mathrm{R}}^2 = \gamma\Omega^2/\phi_{\eta\eta}$. This expression follows from formula (195) with the substitution $\chi^{-1}(\boldsymbol{k}, \omega) = \phi_{\eta\eta} + Dk^2 - \mathrm{i}\gamma\omega$.

According to eq. (198) the intensity $\mathscr{I}_2(0, \Omega)$ has a single maximum at $\Omega = 0$, the halfwidth of which is about 2.2 $\Omega_{\mathrm{R}}$ and hence changes with temperature according to the same law as the central peak width in the first-order scattering spectrum.

The result of integration over $\boldsymbol{k}$ in eqs. (195) and (197) does not depend on the upper limit, as can be seen, e.g., from eq. (198). Therefore, the corresponding expressions may be used for an analysis of the temperature evolution of the second-order scattering spectrum in the critical region too. As has already been mentioned (see sect. 6.1) in the critical region the coefficient $\Gamma$ may be considered to be still a constant, with the accuracy up to a small critical exponent, while for the quantities $\phi_{\eta\eta}$ and $r_c$ one should use here the modified temperature dependences $\phi_{\eta\eta} \sim |\tau|^{\gamma}$, $r_c \sim |\tau|^{-\nu}$.

For a phase transition in a uniaxial ferroelectric or in a proper ferroelastic the expressions for the second-order light scattering density will be different, since for $\chi(\boldsymbol{k}, \omega)$ in formula (196) it is essential now to take into account the dependence of $\Omega_0^2(\boldsymbol{k})$ on the direction of the wave-vector $\boldsymbol{k}$. It can be shown (Levanyuk 1976) that in a uniaxial ferroelectric the width of the central peak in the second-order scattering spectrum for the case when the order parameter has a relaxation dynamics, decreases like $|\tau|$ for $|\tau| \to 0$, while its height $\mathscr{I}_2(0, 0) \sim |\tau|^{-1} \ln|\tau|^{-1}$. In a proper ferroelastic, $\mathscr{I}_2(0, 0) \sim |\tau|^{-1/2}$, whereas the width of the central peak is proportional to $|\tau|^{1/2}$, so that the total intensity of the second-order scattering turns out to be finite here at $\tau = 0$ (Levanyuk 1976).

A weak Raman scattering line with a temperature-dependent frequency was observed by Barta et al. (1977) and by Benoit et al. (1978) above the point of the structural phase transition (with $T_c = 185$ K) in $Hg_2Cl_2$. Since this transition is not proper ferroelastic, the corresponding soft mode for $\tau > 0$ should not be active in the first-order scattering spectrum, so that the above-mentioned spectral line is naturally ascribed to the second-order scattering by $\eta$ fluctuations as Benoit et al. have proposed based on the observation of the soft optical mode in the neutron scattering spectrum with a frequency about twice as low as that of the line observed in the spectrum of the Raman light scattering. Another possible interpretation of this maximum (as given by Barta et al. and by Ipatova and Klochikhin 1978) is the appearance, for $\tau > 0$, of the soft mode in the spectrum of the first-order Raman scattering due to lattice distortions (see ch. 2). This interpretation seems, however, to contradict the data from neutron experiments. Note that since the intensity of the above-mentioned second-order Raman line is by two orders of magnitude lower than the intensity of the corresponding first-order line (as observed below $T_c$), the width of the temperature region where the Landau theory is not valid is evidently rather small for the phase transition in question. Indeed, as has been just mentioned (see also sect. 5.5), the intensities of first- and second-order scattering should be of the same order on the boundary of applicability of the Landau theory. The fact that the Landau theory is well applicable to the transition in $Hg_2Cl_2$ is confirmed also by the observed proportionality of $\Omega_0(\tau)$ to $\eta_e(\tau)$ below $T_c$ in this crystal (see sect. 6.1).

The works cited (see also Lyons and Fleury (1978), Hosea et al. (1979)) are to our knowledge the first experimental works concerning the spectrum of the second-order scattering by $\eta$ fluctuations. There are indications to the observation of the second-order scattering by $\eta$ fluctuations for a number of other phase transitions (see, for example, Andrews and Harley (1981), Hikita et al. (1981) and papers cited in chs. 7 and 8 in this volume), but the interpretation of these data does not seem to be so definite as in the cases discussed. Further experiments in this direction would be, of course, of great interest.

In the case of a multi-component order parameter, the shape of the second-order scattering spectrum would be generally more complicated and for $T < T_c$ may substantially depend on the geometry of scattering (see, e.g., Courtens (1976)). However we shall not touch upon this question here.

For the discussion of the effects of non-linear interactions of fluctuations the information on the spectra of second-order scattering by entropy and density fluctuations is also needed. This question will be touched upon in sect. 6.4.2. As to higher-order scatterings (higher than the second order), they are of less interest because this part of the scattering is rather weak and does not increase somewhat significantly near $T_c$ (see sect. 5.5). Nevertheless in the following section we shall touch upon one of the higher-order effects.

### 6.4. Non-linear coupling between the soft, and other modes

In the framework of our approach based on the continuum approximation the effects of non-linear interactions between fluctuations in $\eta$ and $\xi$ can be described by taking into account anharmonic terms of the type $\eta^4$, $\eta^2\xi$, $\eta\xi^2$ and $\eta^2\xi^2$, etc., in the expression for the thermodynamic potential density and, therefore, for the quantity $U \equiv \Delta\tilde{\Phi}$ (see eq. (184)). When written in the Fourier representation, the anharmonic terms (in the expression for a change $\Delta\tilde{\Phi}$ of the thermodynamic potential under fluctuations) have the form

$$\begin{aligned} &\sum_{\boldsymbol{k}_1+\boldsymbol{k}_2+\boldsymbol{k}_3+\boldsymbol{k}_4=0} R_{40}\eta(\boldsymbol{k}_1)\eta(\boldsymbol{k}_2)\eta(\boldsymbol{k}_3)\eta(\boldsymbol{k}_4), \\ &\sum_{\boldsymbol{k}_1+\boldsymbol{k}_2+\boldsymbol{k}_3=0} R_{12}\eta(\boldsymbol{k}_1)\xi(\boldsymbol{k}_2)\xi(\boldsymbol{k}_3), \\ &\sum_{\boldsymbol{k}_1+\boldsymbol{k}_2+\boldsymbol{k}_3=0} R_{21}\eta(\boldsymbol{k}_1)\eta(\boldsymbol{k}_2)\xi(\boldsymbol{k}_3), \text{ etc.,} \end{aligned} \tag{199}$$

where $R_{lm}$ are the corresponding coupling constants which may depend, generally, on fluctuation wavevectors so that, for example, $R_{40} = R_{40}(\boldsymbol{k}_1, \boldsymbol{k}_2, \boldsymbol{k}_3, \boldsymbol{k}_4)$, $R_{12} = R_{12}(\boldsymbol{k}_1, \boldsymbol{k}_2, \boldsymbol{k}_3)$, and so on. The equation of motion for the fluctuation $\eta(\boldsymbol{q})$ with a given wavevector $\boldsymbol{q}$ will then contain sums of the type

$$\begin{aligned} &\sum_{\boldsymbol{k}_1+\boldsymbol{k}_2+\boldsymbol{k}_3=\boldsymbol{q}} R_{40}\eta(\boldsymbol{k}_1)\eta(\boldsymbol{k}_2)\eta(\boldsymbol{k}_3), \qquad \sum_{\boldsymbol{k}_1+\boldsymbol{k}_2=\boldsymbol{q}} R_{21}\eta(\boldsymbol{k}_1)\xi(\boldsymbol{k}_2), \\ &\sum_{\boldsymbol{k}_1+\boldsymbol{k}_2=\boldsymbol{q}} R_{12}\xi(\boldsymbol{k}_1)\xi(\boldsymbol{k}_2), \end{aligned} \tag{200}$$

instead of a single term proportional to $\xi(\boldsymbol{q})$ (see eqs. (161) and (186)).

Correspondingly, instead of a set of only two equations for the quantities $\eta(\boldsymbol{q})$ and $\xi(\boldsymbol{q})$, one should now consider an infinite (in the limit $V \rightarrow \infty$) set of coupled equations of motion for the variables $\eta(\boldsymbol{k})$ and $\xi(\boldsymbol{k})$. As a result, instead of one frequency (e.g., $\tilde{\Omega}_{R\xi}(\boldsymbol{q})$) characterizing the frequency dispersion of the coefficients $\phi_{\eta\eta}$ and $\gamma$ there appears a whole set of characteristic frequencies $\tilde{\Omega}_{R\xi}(\boldsymbol{k})$, and the dispersion of these coefficients acquires, generally speaking, a complicated character*.

In the present section when considering the effects of non-linear interactions between fluctuations we shall use perturbation or mode-coupling theory**. Such an approach (proposed originally by Fixman (1962) and Levanyuk (1965b)) is widely used in the study of dynamical critical phenomena (see, e.g.,Kawasaki (1971, 1976), Hohenberg (1971), Hohenberg and Halperin (1977) and the literature cited therein). Although concrete computations even in the framework of perturbation theory often turn out to be rather complicated, the qualitative character of the results obtained under this approach for light scattering problems may be understood from the following simple considerations.

### 6.4.1. *The principle of "interacting spectra"*

We have seen in the preceding sections that a linear coupling between two variables $\eta$ and $\xi$ permit the low-frequency spectral features of fluctuations in one variable to manifest themselves in the power spectrum of fluctuations in another variable. So, the linear coupling between an oscillatory variable $\eta$ and a relaxation variable $\xi$ leads to the appearance of a central peak in the $\eta$-vibration spectrum if the $\xi$-relaxation rate is less than the frequency of $\eta$ vibrations. Analogous conclusions can be drawn when both the $\eta$ and $\xi$ fluctuations are of a vibration character. In this case the spectrum of $\xi$ fluctuations contains two side components which will manifest themselves in the spectrum of $\eta$ fluctuations as soon as the frequency of the $\xi$ vibrations is lower than that of the $\eta$ vibrations. Thus, in the case of a linear coupling between two variables the total picture of the spectrum of a given variable can be understood as a result of the superposition (or hybridization) of spectra for

*In the general case non-linear terms should also be included in the expressions (184) and (185) for the kinetic energy $K$ and the dissipation function $R$. However, for the sake of simplicity this will not be done below.

**This does not mean, however, that we are restricting ourselves to the conventional anharmonic phonon perturbation theory as used in the theories of lattice dynamics (see, e.g., Leibfried and Ludwig 1961, Reissland 1973). In fact, due to temperature dependence of the coefficients and the presence of damping, the macroscopic equations of motion used here can effectively take into account anharmonic phonon effects up to arbitrary high orders. Besides, the entropy or the temperature (i.e., purely thermodynamic quantities) may be considered as a variable $\xi$ interacting with $\eta$.

both uncoupled variables, the necessary condition for this hybridization being the slowness of the dynamics of an extra variable as compared to the dynamics of the variable in question.

This principle of interacting spectra is extended easily to the case of a non-linear coupling (Ginzburg et al. 1980, 1981). Consider, for example, the coupling of the type $\eta\xi^2$ and neglect for simplicity the dependence of the coupling constant $R_{12}$ on wavevectors. The coupling term in the thermodynamic potential (effective Hamiltonian) of the system can then be rewritten in the form

$$\int R_{12}\eta\xi^2 \,\mathrm{d}V = \sum_{q,k} R_{12}\eta(\boldsymbol{q})\xi(\boldsymbol{q}-\boldsymbol{k})\xi(\boldsymbol{k}) = R_{12}\sum_{q}\eta(\boldsymbol{q})\left[\sum_{k}\xi(\boldsymbol{q}-\boldsymbol{k})\xi(\boldsymbol{k})\right]. \tag{201}$$

Designating the second sum in the right-hand side of this expression as

$$\zeta(\boldsymbol{q}) \equiv \sum_{k}\xi(\boldsymbol{q}-\boldsymbol{k})\xi(\boldsymbol{k}), \tag{202}$$

we see that the problem is reduced to the above-considered case of a linear coupling between the variables $\eta(\boldsymbol{q}, t)$ and $\zeta(\boldsymbol{q}, t)$, the role of the second variable being played now by the quantity $\Sigma_k \xi(\boldsymbol{q}-\boldsymbol{k}, t)\xi(\boldsymbol{k}, t)$ which is just the quantity responsible for the second-order scattering by $\xi$ fluctuations. Thus, to reveal the character of the influence of $\eta\xi^2$-type interaction in the $\eta(\boldsymbol{q})$-fluctuations spectrum, one needs to be aware of the spectrum of the second-order scattering by $\xi(\boldsymbol{q})$ fluctuations. The structure of this spectrum can be elucidated using the results obtained above (sect. 6.3) for the second-order scattering by $\eta(\boldsymbol{q})$ fluctuations.

Let us assume, for example, that the $\xi$ fluctuations are of oscillatory nature and the dispersion of the corresponding $\xi(\boldsymbol{k})$-phonon branch is small ($\tilde{\Omega}_{0\xi}(\boldsymbol{k}) \simeq \tilde{\Omega}_{0\xi}(0) = \text{const.}$). Then the spectrum of the second-order scattering by $\xi$ fluctuations will contain two side maxima at frequencies, $\pm 2\tilde{\Omega}_{0\xi}(0)$, corresponding to scattering with the emission or absorption of two phonons, and the central maximum which corresponds to emission of one and absorption of another phonon of the same frequency. Dispersion in the $\xi(\boldsymbol{k})$-oscillation branch will lead to an increase of the width and, in principle, to the disappearance of the side maxima. At the same time, the central component will always exist in the spectrum. The width of this component is determined by the maximum damping constant (inverse lifetime) of the corresponding $\xi(\boldsymbol{k})$ phonons and does not depend on the dispersion of the phonons frequency.

From the above it is clear that if the $\xi(\boldsymbol{k})$-phonon lifetime is greater than the period of $\eta$ vibrations, the central component in the spectrum of the second-order scattering by $\xi$ fluctuations will manifest itself in the spectrum of the first-order scattering by $\eta(\boldsymbol{q}, t)$ fluctuations. Thus, the $\eta\xi^2$-type coupling of $\eta$ vibrations with a weakly damping $\xi(\boldsymbol{k})$ phonon branch leads necessarily to the emergence of a central peak in the power spectrum of $\eta$ fluctuations.

Besides, if $\xi(\boldsymbol{q}, t)$ fluctuations are of low frequency, so that $2\tilde{\Omega}_{\rho\xi} < \tilde{\Omega}_{0\eta}$, then in the first-order spectrum of $\eta$ fluctuations two additional side components will also emerge, which correspond to the components in the spectrum of the second-order scattering by $\xi$ fluctuations. When $\tilde{\Omega}_{0\eta}(\boldsymbol{q})$ becomes close to $2\tilde{\Omega}_{0\xi}(0)$ a 'repulsion' of the frequencies and redistribution of intensity between components in the first- and second-order scattering spectra will take place just as in the case of crossing of two linearly coupled one-phonon bands (sect. 6.2.3).

Based on the principle of interacting spectra, similar qualitative considerations can be presented also for other types of non-linear couplings as will be demonstrated below for a number of concrete cases. Now we consider various non-linear coupling effects in more detail.

*6.4.2. Coupling of the type $\eta\xi^2$ ("Fermi resonance", central peak, etc.)*

Let us discuss at first when a coupling of the type $\eta\xi^2$ is possible*. For one-component parameters $\eta$ and $\xi$ it is allowed by symmetry only below $T_c$. Indeed, the product $\eta\xi^2$ changes the sign under the replacement of $\eta$ by $-\eta$, so that it is not invariant under transformations of the symmetry group of the symmetrical phase and cannot enter in the expression for the thermodynamic potential density. The invariant coupling term of the lowest order in $\eta$ and $\xi$ is in this case $R_{22}\eta^2\xi^2$, which below the phase transition temperature leads to an interaction between fluctuations in $\eta$ and $\xi$ of the form $2R_{22}\eta_e\eta'\xi'^2$ (here $\eta' = \eta - \eta_e$, $\xi' = \xi - \xi_e$; since only fluctuations will be considered, the primes on $\eta'$ and $\xi'$ will be omitted as a rule). Thus, in the case of one-component parameters $\eta$ and $\xi$ the coupling constant $R_{12} = 2R_{22}\eta_e$ vanishes at and above $T_c$. For a two-component quantity $\xi = \{\xi_1, \xi_2\}$ the combination $\eta\xi_1\xi_2$ may be allowed by symmetry and the $\eta\xi_1\xi_2$-type interaction may take place in both phases already. In what follows, interactions of the $\eta\xi_1\xi_2$ and $\eta\xi^2$ type will not be distinguished because at certain conditions (see Note added in proof (1)) the results for both these types of interaction are practically the same.

The equations of motion for the space–time-Fourier transforms of the variables $\eta(\boldsymbol{q} \simeq 0)$ and $\xi(\boldsymbol{k})$ have the form

$$(\chi^{(0)}_{\eta\eta}(\boldsymbol{q}, 0))^{-1}\eta(\boldsymbol{q}, \Omega) + \sum_{\boldsymbol{k},\omega} R\xi(\boldsymbol{k}, \omega)\xi(\boldsymbol{q} - \boldsymbol{k}, \Omega - \omega) = h(\boldsymbol{q}, \Omega)\,, \tag{203}$$

$$(\chi^{(0)}_{\xi\xi}(\boldsymbol{k}, \omega))^{-1}\xi(\boldsymbol{k}, \Omega) + \sum_{\boldsymbol{k}_1,\omega_1} 2R\xi(\boldsymbol{k}_1, \omega_1)\eta(\boldsymbol{k} - \boldsymbol{k}_1, \omega - \omega_1) = g(\boldsymbol{k}, \omega)\,. \tag{204}$$

Here the coupling constant $R \equiv \frac{1}{2}\phi_{\eta\xi\xi} = \frac{1}{2}(\partial^3\phi/\partial\eta\partial\xi^2)_e$ and the generalized susceptibilities $\chi^{(0)}_{\eta\eta}(\boldsymbol{q}, \Omega)$ and $\chi^{(0)}_{\xi\xi}(\boldsymbol{k}, \omega)$ for an uncoupled system will be assumed to be given by usual expressions of the type (196).

*More precisely, we have here in mind the case when the coupling constant $R_{12}$ is distinct from zero at zero values of wavevectors and, henceforth, can be considered to be approximately a constant ($R_{12}(\boldsymbol{k}_1, \boldsymbol{k}_2, \boldsymbol{k}_3) \simeq R_{12}(0, 0, 0)$) in the range of small wavevectors ($R = \text{const}$).

Our aim is to calculate the susceptibility $\chi_{\eta\eta}(\boldsymbol{q}, \Omega)$ for the interacting system. To this end one should express $\xi(\boldsymbol{k}, \omega)$ through $\eta(\boldsymbol{q}, \Omega)$ and $g(\boldsymbol{k}, \omega)$ with the aid of eqs. (204) substitute the expressions obtained into eq. (203) and perform the averaging. As a result one finds

$$\chi_{\eta\eta}^{-1}(\boldsymbol{q}, \Omega) = (\chi_{\eta\eta}^{(0)}(\boldsymbol{q}, \Omega))^{-1} - \Pi(\boldsymbol{q}, \Omega), \tag{205}$$

where the function $\Pi(\boldsymbol{q}, \Omega)$ is called the self-energy. This self-energy function cannot be calculated exactly. In the second order perturbation theory we have

$$\Pi(\boldsymbol{q}, \Omega) = \frac{8k_{\mathrm{B}}T}{(2\pi)^4} \int R^2 \gamma_\xi \chi_{\xi\xi}^{(0)}(\boldsymbol{k}, \omega) \chi_{\xi\xi}^{(0)}(-\boldsymbol{k}, -\omega) \chi_{\xi\xi}^{(0)}(\boldsymbol{q}-\boldsymbol{k}, \Omega - \omega)\, \mathrm{d}\boldsymbol{k}\, \mathrm{d}\omega\,. \tag{206}$$

For classical fluctuations ($h\Omega \ll k_{\mathrm{B}}T$) the expressions (205) and (206) are quite general, especially if one considers the quantities $R$, $\gamma_\xi$ and $\chi_{\xi\xi}^{(0)}$ to be arbitrary functions of the wavevectors. For light scattering problems one can put, with a good accuracy, $\boldsymbol{q} = 0$. Assuming further on that the susceptibility $\chi_{\xi\xi}^{(0)}(\boldsymbol{k}, \omega)$ is given by the expression

$$\chi_{\xi\xi}^{(0)}(\boldsymbol{k}, \omega) = \frac{1}{m_\xi(\boldsymbol{k})[\tilde{\Omega}_{0\xi}^2(\boldsymbol{k}) - \mathrm{i}\tilde{\Gamma}_\xi(\boldsymbol{k})\omega - \omega^2]}, \tag{207}$$

and integrating over $\omega$ in eq. (206), we obtain

$$\Pi(\boldsymbol{q}, \Omega) = \frac{k_{\mathrm{B}}T}{4\pi^3} \int \frac{R^2(\boldsymbol{k})}{m_\xi^2(\boldsymbol{k})\tilde{\Omega}_{0\xi}^2(\boldsymbol{k})} \frac{2\tilde{\Gamma}_\xi(\boldsymbol{k}) - \mathrm{i}\Omega}{\tilde{\Gamma}_\xi(\boldsymbol{k}) - \mathrm{i}\Omega} \frac{\mathrm{d}\boldsymbol{k}}{4\tilde{\Omega}_{0\xi}^2(\boldsymbol{k}) - 2\mathrm{i}\tilde{\Gamma}_\xi(\boldsymbol{k})\Omega - \Omega^2}\,. \tag{208}$$

Suppose in the beginning that $\tilde{\Omega}_{0\xi} \gg \tilde{\Gamma}_\xi$ and neglect the dependence of all the quantities on $\boldsymbol{k}$. Then in the frequency range $\Omega \gg \Gamma_\xi$ we have

$$\Pi(\boldsymbol{q}, \Omega) \simeq \frac{k_{\mathrm{B}}TR^2k_{\max}^3}{3\pi^2 m_\xi^2 \tilde{\Omega}_{0\xi}^2} \frac{1}{4\tilde{\Omega}_{0\xi}^2 - 2\mathrm{i}\Omega\tilde{\Gamma}_\xi - \Omega^2}, \tag{209}$$

where $k_{\max}$ is the upper limit for integration in (208); usually one can assume that $k_{\max} \sim \pi/d_0$, where $d_0$ is the interatomic spacing. It can be shown easily that the self-energy (209) has the same form as in the case of a linear coupling between the soft and another vibration mode (see sect. 6.2.3). The role of that other mode with the frequency $2\tilde{\Omega}_{0\xi}$ and the damping constant $2\tilde{\Gamma}_\xi$ is played now by the two-phonon mode corresponding to the second-order scattering by $\xi$ fluctuations while the linear coupling constant proves to be equal to $(k_{\mathrm{B}}TR^2k_{\max}^3/3\pi^2 m_\xi^2\tilde{\Omega}_{0\xi}^2)^{1/2}$. With the redesignations pointed out all the conclusions concerning the temperature dependence of the $\eta$ fluctuations spectrum presented in sect. 6.2.3 remain valid for the non-linear coupling too. Thus, a non-linear interaction of the type $\eta\xi^2$ leads to a hybridization of the spectrum of the first-order scattering by $\eta$ fluctuations with that of the second-order scattering by $\xi$ fluctuations. Owing to this interaction the spectral features in the second-order scattering by $\xi$ fluctuations manifest themselves in the first-order $\eta$ fluctuation spectrum and may suffer a strong enhancement near $T_{\mathrm{c}}$ as compared to the non-interacting case.

The phenomenon of an anharmonic interaction of two vibrations which is most strongly pronounced when the frequency of one vibration coincides with twice the frequency of the other is referred to as Fermi resonance. In the application to the temperature evolution of the light scattering spectrum near phase transitions in solids a Fermi-resonance phenomenon was first considered by Scott and Porto (1967) (see also Scott 1968 and 1974). A more thorough and accurate analysis of this phenomenon with account taken of the dispersion of the phonon branches was carried out by Ruvalds and Zawadovski (1970) where the quantum case ($\hbar\tilde{\Omega}_{0\xi} \gg k_B T$) was discussed and a Green function formalism was used. Figure 17 shows the temperature dependence of the frequencies of the interacting "one-" and "two-phonon" modes below the point of the structural $\alpha \rightleftarrows \beta$ transition in quartz (Scott and Porto 1967). As is seen from fig. 17 the picture of mode repulsion near the crossing point of their frequencies is quite analogous to the case of the linear interaction of two "one-phonon" modes described in sect. 6.2.3. It should be emphasized that a necessary condition for observing the distinct crossing picture is the smallness of the dispersion in the $\xi$-phonon branch. If this dispersion is pronounced, the two-phonon maxima are widening greatly and may disappear from the spectrum at all.

Let us now analyse the shape of the $\eta$-fluctuations spectrum in a low-frequency range $\Omega \lesssim \tilde{\Gamma}_\xi$. Assume, as above, that the $\xi$ fluctuations represent oscillations with only a small damping, so that $\tilde{\Gamma}_\xi(\boldsymbol{k}) \ll \tilde{\Omega}_{0\xi}(\boldsymbol{k})$. In eq. (208) one can neglect then the terms $2\mathrm{i}\tilde{\Gamma}_\xi(\boldsymbol{k})\Omega$ and $\Omega^2$ as compared to $4\tilde{\Omega}^2_{0\xi}(\boldsymbol{k})$ (recall that we consider the region $\Omega \lesssim \tilde{\Gamma}_\xi(\boldsymbol{k})$). If we neglect also the dependence of $\tilde{\Gamma}_\xi$ on $\boldsymbol{k}$, then the self-energy (208) will have the form

$$\Pi(\boldsymbol{q} \simeq 0, \Omega) = \frac{2\tilde{\Gamma}_\xi - \mathrm{i}\Omega}{\tilde{\Gamma}_\xi - \mathrm{i}\Omega} \frac{k_B T}{8\pi^3} \int \frac{R^2_{12}(\boldsymbol{k})\, \mathrm{d}\boldsymbol{k}}{m^2_\xi(\boldsymbol{k})\, \tilde{\Omega}^4_{0\xi}(\boldsymbol{k})}, \tag{210}$$

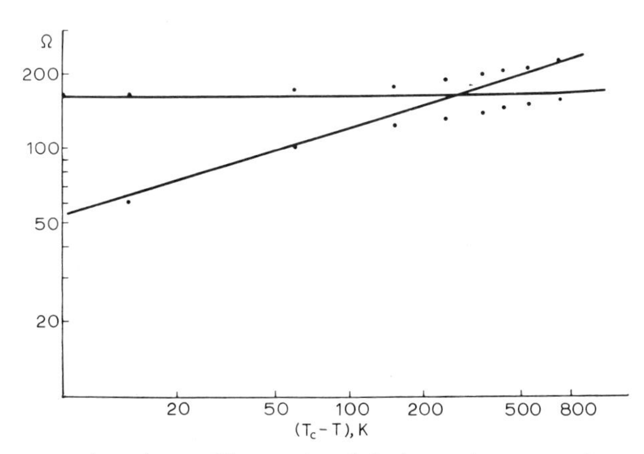


Fig. 17. Temperature dependence of frequencies of the interacting one- and two-phonon modes in $\alpha$ quartz (Scott 1968).

whence for the frequency dispersion of the coefficients $\phi_{\eta\eta}$ and $\gamma$ in a resulting equation of motion for $\eta(\boldsymbol{q} \simeq 0, \Omega)$ we obtain

$$\phi_{\eta\eta}(\Omega) = \phi^{g}_{\eta\eta} + \tfrac{1}{2}m\delta^2\Omega^2/(\tilde{\Gamma}^2_{\xi} + \Omega^2), \tag{211}$$

$$\gamma(\Omega) = \gamma + \tfrac{1}{2}m\delta^2\tilde{\Gamma}_{\xi}/(\tilde{\Gamma}^2_{\xi} + \Omega^2), \tag{212}$$

where

$$\delta^2 = \frac{k_B T}{8\pi^3}\int \frac{R^2_{12}\,\mathrm{d}\boldsymbol{k}}{m m^2_{\xi}(\boldsymbol{k})\tilde{\Omega}^4_{0\xi}(\boldsymbol{k})}, \tag{213}$$

and

$$\phi^{g}_{\eta\eta} = \phi^{\xi}_{\eta\eta} - m\delta^2. \tag{214}$$

Expressions (211), (212) and (214) are to be compared with formulae (166) for the case of a linear coupling between the soft and a relaxation mode. We see that both sets of expressions are almost identical (they differ only by the extra factor $\frac{1}{2}$ in eqs. (211) and (212)). In particular, if the coupling constant $R_{12}$ does not go to zero as $T \to T_c$ (which may be the case for an $\eta\xi_1\xi_2$-type interaction), then eq. (214) determines, as the last one from eqs. (166) does for the case of a linear coupling, a renormalization of the phase transition temperature now due to the coupling between fluctuations in $\eta$ and those in $\xi^2$. Further on, it follows from eqs. (211) and (212) that a central peak will be present in the spectrum of the first-order scattering by $\eta$ fluctuations as soon as the unrenormalized soft mode eigenfrequency $\tilde{\Omega}_{0\eta}$ is less than the sum of the relaxation rates $\tilde{\Gamma} + \tilde{\Gamma}_{\xi}$. The width of this peak far from $T_c$ is determined by the frequency $\tilde{\Gamma}_{\xi}$ and its temperature evolution as $\tau \to 0$ is analogous to that considered in sect. 6.2.1 for the linear coupling case. The appearance of the central peak in the spectrum of the first-order scattering by $\eta$ fluctuations reflects, as has already been mentioned, the presence of a central peak in the spectrum of the second-order scattering by $\xi$ fluctuations. According to formulae (167) and (213) the fractional intensity of this central peak is

$$I_{\text{central}}/I_{\text{side}} = \delta^2/\Omega^2_{0\eta} \sim T_c\Omega^2_a/T_a\Omega^2_{0h} \sim (T_c/T_a)|\tau|^{-1}, \tag{215}$$

where $T_a \sim 10^5$ K is a characteristic ("atomic") temperature, and $\Omega_a \sim 10^{13}\ \mathrm{s}^{-1}$ is a characteristic ("atomic") phonon frequency. The estimate (215) refers to the situations when at $\tau = 0$ the coupling constant $R_{12} \neq 0$ and has a typical "atomic" order of magnitude. It is seen that even in this case the intensity of the central peak becomes comparable with that of the soft mode components for $|\tau| \sim T_c/T_a$ only, i.e., quite near the phase transition point. Note that in the case of a linear coupling between $\eta$ and $\xi$ the small parameter $T_c/T_a$ is absent in the right-hand side of eq. (215).

Fluctuations in the quantity $\xi^2(\boldsymbol{k})$ coupled linearly with $\eta$ may be considered in terms of quantum theory as phonon density fluctuations. Such a terminology is often used in the literature.

Since the acoustic phonons have usually the weakest damping, it is natural in seeking a mechanism for the central peak to consider in the first place the density $\rho$ as a candidate for the variable $\xi$ coupled with the order parameter by means of an iteraction of the type $\eta\xi^2$. Once $\xi$ has the meaning of the density, one should put $m_\xi(\boldsymbol{k}) = \rho k^{-2}$, $\tilde{\Omega}_{0\xi} = c_s k$ and $\tilde{\Gamma}_\xi = \gamma_s k^2$ (where $c_s$ and $\gamma_s$ are the velocity and the damping constant of the longitudinal sound, respectively) in the above formulae beginning with (207). In the approximation under consideration (when the velocity $c_s$ is believed to be independent of $\boldsymbol{k}$) the side components are absent from the spectrum of the second-order scattering by $\rho$ fluctuations due to a strong $\boldsymbol{k}$ dependence of the quantities $\tilde{\Omega}_{0\xi}$ and $\tilde{\Gamma}_\xi$. However, the central peak will be, of course, present in the spectrum. Its width is determined by the maximum relaxation rate $\tilde{\Gamma}_\xi(k_{max})$ of the longitudinal phonons, where $k_{max}$ is of the order of the reciprocal lattice spacing. So, if the soft mode frequency is higher than $\Gamma_\xi(k_{max})$, in the spectrum of light scattered by $\eta$ fluctuations below $T_c$ a central peak should be present, the temperature evolution of which is analogous to that of the thermal diffusion central peak (see sects. 6.2.1 and 6.2.2). The intensity of this peak (far from $T_c$) is, however, low, being $T_c/T_a$ times weaker than the intensity of the soft mode components.

If in a certain region of wavevectors the relaxation rate $\tilde{\Gamma}_\xi$ remains practically constant, in the scattered light spectrum alongside the abovementioned wide central peak a narrower peak may be present with a width determined by the lifetime of the acoustic phonons in this region of wavevectors (Cowley 1970, Cowley and Coombs 1973). This holds in the so-called second-sound regime when $\tau_N$, the characteristic time of N processes (under which the momentum of phonon quasiparticles is conserved) is much less than the characteristic time of U processes (under which the momentum is not conserved). In this case the coefficient $\tilde{\Gamma}_\xi$ turns out to be practically constant for wavevectors $(c_s\tau_U)^{-1} \lesssim k \lesssim (c_s^2\tau_U\tau_N)^{-1/2}$ (Gurevich and Efros 1966, Sham 1967, Enz 1974a). Such a situation may occur in sufficiently pure dielectrics at low temperatures. However, the intensity of the corresponding central peak must be still lower here than in the previous case and proportional to the fraction of the phase volume corresponding to the wavevectors for which $\tilde{\Gamma}_\xi$ is $\boldsymbol{k}$ independent.

Above we neglected, in fact, the dependence of the non-linear coupling coefficient $R_{12}$ on the wavevectors. When such a dependence is taken into account, an interaction of the type

$$\sum_{\boldsymbol{k}_1,\boldsymbol{k}_2} R_{12}(\boldsymbol{k}_1, \boldsymbol{k}_2)\rho(\boldsymbol{k}_1)\rho(\boldsymbol{k}_2)\eta(-\boldsymbol{k}_1 - \boldsymbol{k}_2)\,,$$

where $R_{12}(0, 0) \equiv R_{12} = 2R_{22}\eta_e$, may take place already in both phases.

In the thermodynamic potential density it is the terms with $\eta$- and $\rho$-spatial derivatives which are responsible for the $\boldsymbol{k}$ dependence of the coupling constant. For example, in ferroelectric crystals with a piezoelectricity in the non-polar phase (e.g., in $KH_2PO_4$ crystals) the thermodynamic potential may contain the term (recall that the role of $\eta$ is played here by the component $P_z$ of polarization vector)

$$D_1 \int P_z \frac{\partial \rho}{\partial x} \frac{\partial \rho}{\partial y} \, \mathrm{d}V = \sum_{\boldsymbol{k}_1 \boldsymbol{k}_2} D_1 k_{1x} k_{2y} \rho(\boldsymbol{k}_1) \rho(\boldsymbol{k}_2) P_z(-\boldsymbol{k}_1 - \boldsymbol{k}_2). \tag{216}$$

Thus in this case the full coupling constant for the interaction $\eta\rho^2$ is $R_{12}(\boldsymbol{k}_1, \boldsymbol{k}_2) = 2R_{22}\eta_e + D_1 k_{1x} k_{2y}$. Above the transition point it is only the second term which is distinct from zero and leads to the appearance in the scattered light spectrum of a central peak with a width of the order of $\tilde{\Gamma}_\rho(k_{max})$ as long as the width is less than the distance between the soft mode components. This peak must be rather wide, however, because the relaxation rate $\tilde{\Gamma}_\rho(k_{max})$ is of the order of characteristic (optical) phonon frequencies. An interaction of the type (216) has been considered, in fact, by Cowley (1970) and by Cowley and Coombs (1973), but these authors were concerned mainly with a much more narrow central peak which may appear in the second sound regime when $\tau_U \gg \tau_N$. Note that, as mentioned above, this narrower central peak is contributed only by fluctuations with relatively small $k$'s up to $k \sim (c_s \tau_V \tau_N)^{-1/2}$, and for such values of $k$ the coupling constant $D_1 k_{1x} k_{2y}$ is small. Therefore the peak could have a noticeable intensity only extremely close to the transition temperature. For this reason (to say nothing of the difficulty in meeting the requirement $\tau_U \ll \tau_N$ in real crystals) the narrow central peak is very unlikely to be observed experimentally. As to a wider central peak (with the width of the order of $\gamma_s k^2_{max}$) it could be observed, in principle, even far from $T_c$. An interaction of the type (216) is dominant, probably, also below $T_c$ as compared to the $\boldsymbol{k}$-independent parts of the $\eta\rho^2$ interaction. This is because when calculated at $k_1$, $k_2 \propto k_{max}$, the coupling constant $R_{12}(\boldsymbol{k}_1, \boldsymbol{k}_2)$ does not contain the small parameter $\eta_e/\eta_a \propto (-\tau)^{1/2}$, as distinguished from the case $k_1 = k_2 = 0$.

Note that when in the density of the thermodynamic potential, terms with spatial derivatives are taken into account (or, which is the same, when one includes the dependence of the coupling constant $R_{12}$ on the vectors $\boldsymbol{k}_1$, $\boldsymbol{k}_2$), an interaction of the type $\eta'\rho'^2$ is allowed for a large class of systems. Consider, for example, zone centre structural phase transitions. For such transitions from the components of the vectors $\boldsymbol{k}_1$, $\boldsymbol{k}_2$ one can always compose a combination equivalent in its transformation properties to the order parameter. However, for a real coupling of the $\eta'\rho'^2$ type to be possible, it is necessary that the coupling constant $R_{12}$ be even in $\boldsymbol{k}$'s. Indeed, when this is not so and $R_{12}$ changes sign under the replacement of $\boldsymbol{k}$'s by $-\boldsymbol{k}$'s, the sum $\Sigma_{\boldsymbol{k}} R_{12}(\boldsymbol{k}, \boldsymbol{q} - \boldsymbol{k})\eta(\boldsymbol{q})\xi(\boldsymbol{k}) \times \xi(\boldsymbol{q} - \boldsymbol{k})$ goes to zero as $\boldsymbol{q} \to 0$, because at $\boldsymbol{q} = 0$ terms in the sum with $\boldsymbol{k}$ and $-\boldsymbol{k}$ compensate strictly each other. This important conclusion (which became clear for us in discussions with V.G. Vaks) has been overlooked in the author's preceding publications (Ginzburg et al. 1980, 1981). There the statement was made that "when one includes the dependence of the coupling constant $R_{12}$ in the wavevectors $\boldsymbol{k}_1$, $\boldsymbol{k}_2$, an interaction of the type $\eta'\rho'^2$ is allowed for any transformation properties of the order parameter". Now we see that the statement does not refer, in particular, to proper ferroelectrics with the centre of symmetry in paraelectric phase. By virtue of what has been said, it seems improbable that a central peak with the width of the order of 1 $cm^{-1}$, observed by Fleury and Lyons (1976) and Lyons and Fleury (1978) near the point of ferroelectric transition in $Pb_5Ge_3O_{11}$, is connected with the non-linear mechanism in question (or with an interaction of the $\eta S^2$ type; see below). This is just opposite to an earlier conclusion made by us (Ginzburg et al. 1980, 1981) as well as by the authors of the works cited.

Since energy $U$ or entropy $S$ posseses the same transformation properties as the density $\rho$ (all the quantities are scalars), the coupling of the type $\eta S^2$ or $\eta U^2$ is equally allowed and leads to an interaction of the soft with the entropy (heat diffusional) mode. The appearance of a central peak in the first-order spectrum of $\eta$ fluctuations is then possible once the maximal heat diffusion relaxation rate $\Omega_{RS}(k_{max}) = k^2_{max}/C_\eta$ is less than the eigenfrequency of the soft mode.

Moreover, when one includes the dependence of the coupling constant $R_{12}$ in wavevectors, and the constant is even in $k$'s, an interaction $\eta\xi^2$ becomes possible between the soft and any other phonon modes. If among the modes there are some with a small enough damping (in the whole Brillouin zone up to $k \sim k_{max}$) then a central peak will appear in the spectrum of light scattered by first-order fluctuations.

The first-order spectra of both order parameter and energy fluctuations have been studied numerically by Schneider and Stoll (1973, 1975, 1976, 1978a, b) for a simple model system, in which the soft mode corresponding to the order parameter may interact with itself or with the thermal diffusion mode only. The model and the results of the computer simulations by Shneider and Stoll will be discussed in more detail in sect. 6.5. Here we mention only that in these computer simulations the presence of a central peak has been reported in the $\eta$-fluctuation spectrum for systems of various dimensionality (one, two, three and four). This peak has been observed near the transition temperature even in a purely displacive system. The origin of the peak has been associated by the authors microscopically with the motion of some order parameter domain walls (or clusters). Another interpretation of the peak is that it arises from the non-linear coupling in question (between the soft and thermal diffusion modes) (Ginzburg et al. 1980, 1981). Such an interpretation may be, however, incorrect, since for the model under consideration the order parameter is a vector so that the coupling constant $R_{12}$ should be an odd function of $\boldsymbol{k}$'s (see Note added in proof (2)).

We mention, finally, that the interaction of the type $\eta\xi^2$ leads also to some enhancement of the width of the soft mode components near the phase transition point. This enhancement has the same origin as discussed in sect. 6.2.1 (see eq. (169)) and, in any case, cannot be very large.

*6.4.3. Coupling of the type $\eta^2\xi$*

For a one-component order parameter $\eta$ this type of coupling (with the coupling coefficient $R_{21}$ non-vanishing in the limit of long wavelengths) is allowed for the quantities $\xi$ which transform according to the unit irreducible representation of the symmetry group of a more symmetrical phase. The most well-known examples of such quantities are (again) the density $\rho$, the entropy $S$, coordinates of fully symmetrical phonon modes and so on. To elucidate the effects of this interaction upon the $\eta$-fluctuation spectrum, we shall consider first the spectrum of fluctuations in the quantity $\zeta = \eta\xi \equiv \sum_k \eta(\boldsymbol{k})\xi(\boldsymbol{q} - \boldsymbol{k})$ and apply then the principle of interacting spectra (sect. 6.4.1).

For dispersionless optical phonon branches the $\eta\xi$-fluctuation spectrum contains, generally, two pairs of components at the combinational frequencies $\pm(\tilde{\Omega}_{0\eta} \pm \tilde{\Omega}_{0\xi})$. The inner pair of components merges into a single central line when the distance $|\tilde{\Omega}_{0\eta} - \tilde{\Omega}_{0\xi}|$ is less than the sum of the widths of the corresponding one-phonon lines. Thus, due to the $\eta^2\xi$ coupling a central peak may appear in the $\eta$-fluctuation spectrum only in the case when the difference $|\tilde{\Omega}_{0\eta} - \tilde{\Omega}_{0\xi}|$ is small enough. The latter is possible, evidently, in a limited temperature interval not too close to the transition temperature and is, therefore, of little interest, the more so as in the presence of a dispersion in the $\eta$- and $\xi$-phonon branches the very existence of a central peak in the $\eta\xi$-fluctuation spectrum becomes questionable.

When the $\xi$ fluctuations have a relaxation character, in the spectral density of the $\eta$ fluctuations there is no central peak at all (of course, if the soft mode is underdamped). For this reason the statement (Enz 1974) that the interaction of the type $\eta^2 S$ with the thermal diffusion mode may lead to the appearance of a central peak in the spectrum of $\eta$ fluctuations seems to be incorrect.

A coupling of the $\eta^2\xi$ type, with $\xi$ being an acoustic deformation, has been regarded by Tani (1969) and by Levanyuk and Schedrina (1974) as a possible origin of a significant increase of the soft mode damping constant as $\tau \to 0$. However, it is only a one-dimensional system that has been studied by Tani. At the same time the phase volume corresponding to long-wavelength $\eta$ fluctuations, which contribute mainly to temperature anomalies near phase transitions, is known to depend very essentially on the dimensionality of a system. For this reason the results obtained by Tani for a one-dimensional atomic chain, cannot, in our opinion, be directly compared with experimental data. As to the calculations by Levanyuk and Schedrina, they refer to the three-dimensional case but hold for the frequencies $\Omega$ close to zero. Therefore, a comparison of the results of these calculations with the temperature dependence of the width of the soft mode components is, strictly speaking, questionable too since the value $\gamma(\Omega = \Omega_{0\eta})$ may differ essentially from $\gamma(\Omega = 0)$. Evidences for a strong frequency dispersion of $\gamma$ are given, for example, by the polariton scattering experiments in $BaTiO_3$ and $PbTiO_3$ crystals (Laughman et al. 1972, Heiman and Ushioda 1974, 1978, Heiman et al. 1975).

The most interesting effect of an $\eta^2\xi$-type interaction is the influence of this interaction on the $\xi$-variable dynamics, this influence being the more pronounced, the lower the frequency of the $\xi$ vibrations is. A well studied example of such influence is provided by the so-called critical anomalies in the sound attenuation and its propagation velocity near second-order transitions. These anomalies, considered for the first time by Fixman (1962) and Levanyuk (1965), have been further investigated in a large number of works (for reviews, see, e.g., Garland 1970, Rehwald 1973, Kawasaki 1976). The results of these works are extended easily to a coupling of the $\eta^2\xi$ type with an optical phonon mode. In the latter case, the frequency of $\xi$ vibrations being not very high as compared to the characteristic frequency of $\eta$ changes, the side components due to scattering by $\xi$ fluctuations are widened considerably as $\tau \to 0$ and shift to some extent towards lower frequencies. Widening near $T_c$ of the Raman components, which are not connected with the order parameter, was observed, for example, by Aronov et al. (1977). However, the main contribution to the widening seems to be given by defects (see ch. 2).

*6.4.4. Coupling of the types $\eta^3$ and $\eta^4$*

For second-order transitions a coupling of the $\eta^3$ type (with the coupling constant $R_{30}$, which does not go to zero as $\boldsymbol{k} \to 0$) is allowed in the nonsymmetrical phase merely and is described by the term $\frac{3}{4}B\eta_e(\eta - \eta_e)^3$ in the thermodynamic potential density arising from the term $\frac{1}{4}B\eta^4$ (see eq. (1)). The $\eta^3$-type interaction between $\eta(\boldsymbol{k})$ fluctuations with various wavevectors $\boldsymbol{k}$ can be considered as a particular case of the $\eta\xi^2$-type interaction, the role of the $\xi$ variable being played now by the order parameter fluctuations themselves. Therefore, to consider the effects of this coupling, one may exploit formulae from sect. 6.4.2 setting in them $R_{12} = R_{30} = \frac{3}{4}B\eta_e$ and also $\tilde{\Omega}_{0\eta} = \tilde{\Omega}_{0\xi} = \Omega_0$, $\tilde{\Gamma}_\xi = \Gamma_\eta = \Gamma$ and $m_\xi = m$. It is evident then that the $\eta^3$ interaction leads to the appearance of a central peak in the first-order $\eta$-fluctuation spectrum (when the soft mode is underdamped). This peak reflects the one in the spectrum of the second-order scattering by $\eta$ fluctuations (see sect. 6.3). The temperature dependence of this peak is analogous to that for a linear coupling between the soft and the thermal diffusion mode (sect. 6.2.2).

When one includes the dependence of the coupling constant $R_{30}$ on wavevectors, the $\eta^3$ interaction may occur already in both phases. This is not, however, the case with a one-component order parameter because terms of the type $\int \eta^2 \operatorname{div} \eta \, \mathrm{d}V$ in the thermodynamic potential, which are responsible for the interaction in question, can be expressed then through surface integrals. From this it follows, that the $\eta^3$ interaction does not occur above $T_c$ in the simple model system considered in sect. 6.5.

Another important effect of the $\eta^3$ interaction consists in the fact that it may provide a strongly temperature dependent contribution $\Delta\Gamma$ into the soft mode damping constant. To calculate this contribution, one should substitute

$\tilde{\Omega}^2_{0\xi} = \tilde{\Omega}^2_{0\eta} = (\phi_{\eta\eta} + Dk^2)/m$, $\Gamma_\xi = \Gamma$ in expression (208) and integrate it over $\boldsymbol{k}$. As a result one finds (Levanyuk and Schedrina 1974)

$$\Delta\Gamma = c\Gamma(\tau_{\rm crit}/|\tau|)^{1/2}, \tag{217}$$

where $c \simeq 0.2$ and $\tau_{\rm crit}$ is the temperature width of the critical region, as given by formula (13). It should be emphasized that expression (217) gives a correction to $\Gamma$ at $\Omega = 0$. The width of the soft mode line is characterized, however, by the value of $\Gamma$ at frequencies $\Omega \simeq \Omega_{0\eta}$. The values of $\Delta\Gamma(0)$ and $\Delta\Gamma(\Omega_{0\eta})$ may be, in general, quite different. In the case under consideration, as calculations show (Sobyanin and Schedrina 1979) the ratio $\Delta\Gamma(\Omega_0)/\Delta\Gamma(0)$ does not depend on $\tau$ and is approximately equal to 5.0.

As follows from eq. (217), the correction to $\Gamma$ due to the anharmonic $\eta^3$ interaction is small away from $T_c$ but becomes comparable with $\Gamma$ at the crossover from the classical to the critical region. For $|\tau| < \tau_{\rm crit}$ the result (217), which is, in fact, the first-order correction to the mean field (or Landau) theory, is no longer valid. Nevertheless we can come to some conclusions about the temperature dependence of the soft mode damping constant in the critical region too by means of formula (208) using other (scaling) temperature dependences for the quantities entering this formula, i.e., using expressions (16) and (17) for the coefficients $A$, $B$, $C$, $D$ in eq. (1) for the thermodynamic potential density. In this way one finds easily that in the critical region the contribution $\Delta\Gamma$ is almost temperature independent and its behaviour here is characterized at best by a small critical exponent. This conclusion is supported by more strict theoretical treatments (Hohenberg and Halperin 1977).

A widening of the soft mode components near $T_c$ was observed in a large number of works (see, e.g., Shapiro (1969), Burns and Scott (1970), Burns (1974), Fleury (1970), Laulicht et al. (1972), Shigenari et al. (1976), Uwe and Sakudo (1976)). It is not clear, however, whether this widening is due to the non-linear interaction under consideration or to the presence of defects in crystals (see ch. 2).

Of the higher-order non-linear interactions we shall consider only that of the $\eta^4$ type, since in the symmetrical phase (in the absence of $\eta\xi^2$-type couplings) the $\eta^4$-type interaction is of lowest order leading to considerable temperature anomalies.

It is easily seen that this interaction does not lead to the appearance of a central peak in the spectrum of $\eta$ fluctuations (when these fluctuations are of an oscillatory character). Indeed, as to its influence upon the first-order $\eta$-fluctuation spectrum, this interaction may be considered as linear between the quantities $\eta$ and $\eta^3$. On the other hand the $\eta^3$-fluctuation spectrum may contain peaks near the frequencies $\pm\Omega_0$ and $\pm 3\Omega_0$ only, but not at $\Omega = 0$ (the latter takes place, when $\eta$ fluctuations have a relaxation character).

The most interesting effect of the $\eta^4$ interaction consists, as before, in its contribution to the soft mode damping constant. The corresponding question was investigated by Balagurov et al. (1970) as well as by Levanyuk and Schedrina (1972, 1974) in the framework of the first corrections to the Landau theory. According to these authors the temperature dependence of the coefficient $\Gamma$ at the zero frequency is determined by the expression

$$\Gamma(0) = \Gamma\left[1 + \frac{3}{8}\frac{B\eta_e^2}{\phi_{\eta\eta}}\left|\frac{\tau_{crit}}{\tau}\right|^{1/2} + \frac{1}{6}\left(1 + \frac{2\Omega_0}{\Gamma}\right)\left|\frac{\tau_{crit}}{\tau}\right|\right]. \tag{218}$$

The second summand on this expression is due to the $\eta^3$ interaction. Note that below $T_c$ within the classical region ($|\tau| > \tau_{crit}$) this second summand is the main one (if, of course, the ratio $\Omega_0/\Gamma$ is not too large). On the contrary, above $T_c$ the temperature dependence of $\Gamma$ is determined entirely by the last summand corresponding to the $\eta^4$ interaction. It should be emphasized again that to estimate correctly the width of the soft mode components the damping constant $\Gamma$ should be calculated not for zero frequency but for the frequency of these components. Nevertheless one may hope that this circumstance will not change the results qualitatively.

*6.4.5. Some additional remarks on the non-linear mode-coupling effects*

Above we have restricted ourselves, for simplicity, to the analysis of non-linear interaction effects on the first-order $\eta$-fluctuation spectrum, i.e., on the form of the correlation function $\langle\eta(\Omega)\eta(-\Omega)\rangle$. The scattered light spectrum is determined, however, not only by this correlation function but also by the correlation functions of other variables, which the dielectric tensor $\epsilon_{ij}$ depends on, as well as by mixed correlation functions. Furthermore, even in that hypothetical case, when $\epsilon_{ij}$ depends on $\eta$ only, higher-order scatterings along with the scattering of the first order contribute to the total scattered light intensity. Non-linear interactions couple the scattered light spectra of different orders. Thus, for a complete analysis of effects due to non-linear interactions one needs calculations not only of the correlation function $\langle\eta(\Omega)\eta(-\Omega)\rangle$, but also of the higher-order correlation functions.

For example, in the case of the $\eta^3$ interaction one should know the correlation functions $\langle\eta^2(\Omega)\eta^2(-\Omega)\rangle$ and $\langle\eta(\Omega)\eta^2(-\Omega)\rangle$ in addition to the first-order correlation function $\langle\eta(\Omega)\eta(-\Omega)\rangle$, all these functions enter the expression for the spectral density of the scattered light. Due to the $\eta^3$ interaction the form of the correlation functions $\langle\eta(\Omega)\eta(-\Omega)\rangle$ and $\langle\eta^2(\Omega)\eta^2(-\Omega)\rangle$ may change significantly, whilst the mixed correlation function $\langle\eta(\Omega)\eta^2(\Omega)\rangle$ becomes non-zero. A self-consistent calculation of the $\eta^3$-interaction effects has been carried out by Yacoby et al. (1978) as applied to the light scattering spectrum near the ferroelectric phase transition in lead germanate (Hosea et al. (1979), see also Lyons and Fleury (1978)).

Analogous remarks can be made also as concerns other types of non-linear interactions. So, under the interaction of the $\eta\xi^2$ type a hybridization of the first-order fluctuation spectrum of the order parameter $\eta$ and the second-order one of the variable $\xi$ takes place, as a result of which the correlation functions $\langle\eta(\Omega)\eta(-\Omega)\rangle$ and $\langle\xi^2(\Omega)\xi^2(-\Omega)\rangle$ are changed, and the mixed correlation function $\langle\eta(\Omega)\xi^2(\Omega)\rangle$ becomes non-zero.

Just as the linear mode coupling leads to a redistribution of the intensity among various lines in the first-order scattering spectrum, the non-linear coupling causes intensity redistribution among the lines as well as a change of their shapes in the first- and second-order scattering spectra. Moreover as a result of non-linear coupling the intensity of some lines related genetically to the second-order scattering may become comparable with or even exceed the intensity of first-order lines within the region of applicability of the Landau theory already. In other words, the second-order scattering whose intensity far from the phase transition point is small and is of the order of $T/T_a \sim T/10^5$ K as compared to the first-order scattering intensity increases strongly under certain conditions near $T_c$ due to the non-linear coupling effects. It takes place though in a rather narrow temperature interval $|\tau| = |T - T_c|/T_c \sim T_c/T_a$.

In conclusion we would like to emphasize that although qualitatively the character of the influence of non-linear mode-coupling effects upon the scattered light spectrum is quite comprehensible, a direct comparison of the corresponding theoretical results with experimental data is quite a difficult problem. It is not so much the point of lack of a detailed development of the theory, but the vast information needed for such a comparison. Indeed, once the theoretical formulae include integrals over the $\boldsymbol{k}$ space, it is necessary to know the dispersion laws of the interacting branches of the excitation spectrum, to say nothing of the necessity to know the dependence of the corresponding coupling constants of the wavevectors. One can, however, argue that in this way it is impossible to account for very narrow central peaks (with a width of 4–5 orders less than the characteristic phonon frequencies) observed in some cases. The most possible origin of such peaks is the presence of defects in the crystal (see ch. 2).

### *6.5. On some computer simulations of the order parameter dynamics near structural phase transitions*

The appearance of a central component in addition to the side components in the light scattering spectrum near $T_c$ may be associated, as we have seen already, with a certain type of frequency dispersion of coefficients in the equation of motion for the order parameter. There are, however, many reasons why such a dispersion may occur. Therefore the interpretation of the experimental data concerning the central peak is hampered.

Here we discuss the results of some computer simulations of the order parameter dynamics in a simple model system undergoing a structural phase transition (Onodera 1970, Ishibashi and Takagi 1972, Aubry and Pick 1974,

Schneider and Stoll 1973, 1975, 1976, 1978a, b; Koehler et al. 1975). We will see that even in such a simple system there appears a (wide) central peak in the $\eta$-fluctuation spectrum near $T_c$ but the origin of this peak can be understood only qualitatively.

The system discussed represents a set of particles localized at the sites of a simple cubic ($d$-dimensional) lattice. Each particle is assumed to move in a one-dimensional two-minima anharmonic potential and to interact with neighbouring particles by means of harmonic "springs" (fig. 18). The Hamiltonian of this set of particles has then the form

$$\mathcal{H} = \sum_i \left( -\frac{A}{2} u_i^2 + \frac{B}{4} u_i^4 - h u_i \right) + \frac{C}{2} \sum_{ij}{}' (u_i - u_j)^2 + \sum_i \frac{m_0}{2} \dot{u}_i^2 , \tag{219}$$

where $u_i$ is a displacement of the $i$th particle from the centre of the anharmonic potential, $h$ is the external field conjugate to $u$, the coefficients $A$, $B$, $C$ are taken to be positive, and the prime at the sum over $i, j$ means that the summation is carried out over nearest neighbours only.

Despite its simplicity, the model (219) is capable of describing the structural phase transitions of both the displacive and the order–disorder types. To the former type belong the transitions in systems for which a mean one-particle potential energy $U_1(u)$ (this energy is obtained from (219) by putting there the displacements of all the particles, except a given one, to be equal to zero) has only one minimum at $u = 0$. As is easily seen, this takes place when

$$A < 2QC , \tag{220}$$

where $Q = 2d$ is the number of nearest neighbours. All other systems with

$$A > 2QC , \tag{221}$$

are believed to exhibit the transitions of the order–disorder type. The mean

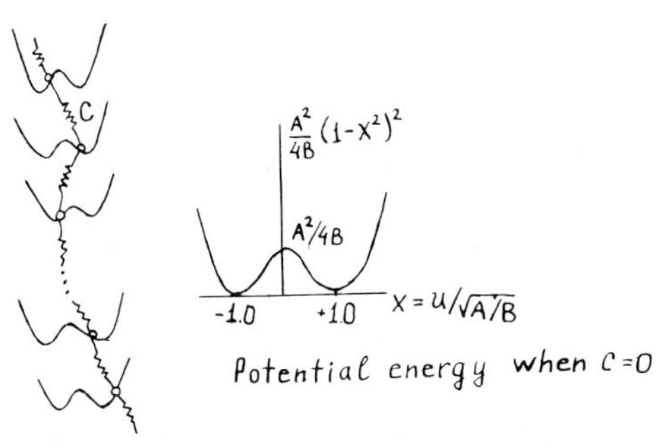


Fig. 18. Schematic of particle interactions for a system described by Hamiltonian (219).

one-particle potential energy $U_1(u)$ has in this case two minima at $u = \pm[(A - 2QC)/B]^{1/2}$.

The difference between the two types of transitions can be demonstrated also in another way. To this end we compare the height of the energy barrier separating the minima in the original anharmonic potential (this height is equal to $A^2/4B$; see fig. 18) with the characteristic particle interaction energy, the latter being equal to $2QCu_0^2$, where $u_0 = (A/B)^{1/2}$ is the distance between the maximum and the minimum in the anharmonic potential. It follows then from inequalities (220) and (221) that

$$A^2/4B < \tfrac{1}{4}(2QCu_0^2) \tag{222}$$

for a displacive system, and

$$A^2/4B > \tfrac{1}{4}(2QCu_0^2) \tag{223}$$

for an order–disorder system.

Since at the phase transition temperature $T_c$ the mean kinetic energy of a particle is of the order of its interaction energy with other particles (i.e., $k_B T_c \sim 2QCu_0^2$), one has

$$A^2/4B \ll k_B T_c \tag{224}$$

for a typical displacive transition, and

$$A^2/4B \gg k_B T_c \tag{225}$$

for a typical order–disorder transition.

Thus, the setting of the zero equilibrium value of the order parameter $\langle \eta \rangle \equiv \langle u \rangle = 0$ at $T \geqslant T_c$ for order–disorder transitions should be connected with thermo-activation jumps of a particle over the barrier, whereas for displacive transitions the presence of such a barrier does not affect, in fact, the particle motion within the anharmonic potential. This is just the reason why the order parameter dynamics for order–disorder transitions is assumed to have mainly a relaxation character and for displacive transitions an oscillatory one.

Computer simulations have shown, however, that even for well pronounced displacive phase transitions, i.e., for $A/C \ll 1$, along with side components in the spectrum of $\eta$ fluctuations there exists a central peak which becomes dominating in the spectrum near $T_c$. Such a picture has been found by Schneider and Stoll in one as well as in two, three and four dimensions. An impression is thus produced that in this model even for well pronounced displacive transitions the dynamics of the $\eta$ fluctuations is not described completely by an oscillatory equation of motion but rather corresponds to the situation described in sect. 6.2.1 when two linearly coupled variables exist possessing the same symmetry as that of the order parameter, the law of motion of one of them being relaxative.

Turning now to the discussion of these results we first of all pay attention to the fact that in the simple model under consideration only one phonon excitation branch exists, namely, the soft optical phonon mode (at $T = 0$ K the frequency of the mode $\Omega_0(0) = (2A/m)^{1/2}$) and one collective excitation branch, which is the thermal diffusion mode. So, if we tried to seek for an explanation of the central peak in the way of a non-linear mode-coupling effect (sect. 6.4), we could expect to find it by taking into account the non-linear coupling of fluctuations in the order parameter $\eta \equiv u$ with fluctuations in the entropy $S$. The corresponding interaction term in the thermodynamic potential density has, for example, the form $S^2 \operatorname{div} \boldsymbol{u}$. That the discussed central peak may actually be attributed to such a non-linear interaction is supported by the fact that the width of this peak not very close to $T_c$ turns out approximately equal to the width of the central line in the energy (or entropy) fluctuation spectrum for wave-vectors of the order of the reciprocal lattice vector. The coupling in question is effective only at finite values of $q$ (see the end of sect. 6.4.2) while for the model under consideration a central peak has been observed also for $q = 0$.

Another qualitative explanation of the appearance of the central peak in the spectrum of $\eta$ fluctuations can be obtained from a direct examination of the model. To this end we recall that for a displacive system the mean one-particle potential energy has only one minimum at $u = 0$. Let us now show that if a joint motion of a group of $N$ particles is considered, the mean potential energy $U_N(u)$ for this motion may have already two minima, like in the case of order–disorder systems. Let us restrict ourselves for the sake of simplicity to the one-dimensional case and assume that two neighbouring particles (e.g., the $i$th and $(i+1)$st particles) move in phase. The mean potential energy $U_2(u)$ corresponding to such a motion is found from expression (219) if one puts in it $u_i = u_{i+1} = u$, the displacements of all the other particles being fixed to zero. As a result we obtain

$$U_2(u) = 2\left(-\frac{A}{2}u^2 + \frac{B}{4}u^4\right) + 2Cu^2 = 2\left(-\frac{A-2C}{2}u^2 + \frac{B}{4}u^4\right). \tag{226}$$

Thus for a displacive-type system with $A/C > 2$ a joint motion of a group of two particles proceeds already in a two-minima potential and, therefore, is expected to have a relaxation character. When considering the joint motion of groups ("clusters") of a larger number of particles the criterion for the presence of two minima in the corresponding mean potential energy $U_N(u)$ will evidently be fulfilled at still lower values of the ratio $A/C$, i.e., for still more displacive systems.

The above considerations allow us to expect that for any displacive-type phase transition, fluctuations in the order parameter will contain, along with oscillations, a certain "relaxation component" connected with joint jumps of particles over a corresponding potential barrier.

Unfortunately to translate these qualitative considerations into a more precise language, i.e., to construct a consistent theory of the order parameter dynamics for displacive-type systems seems to be a very hard problem. Numerous attempts made in this direction (see, e.g., Aubry (1975), Krumhansl and Schrieffer (1975), Varma (1976), Binder et al. (1974), (1975), Binder and Stauffer (1976), Binder (1977)) have not yet provided a full explanation of the situation. Further attempts in this direction are greatly needed.

### 6.6. *On the soft mode concept for the case of a multi-component order parameter*

Here we discuss briefly the peculiarities of the soft mode spectrum temperature evolution which arise in the case of a multi-component order parameter. Consider only the case of a two-component order parameter as we did in sects. 2.5, 3.4 and 5.6. Let us suppose, further on, that the simplest version of the soft mode concept is valid (see sect. 6.1). Then for the frequency of the soft mode corresponding to the $\varphi$ vibrations one has

$$\Omega_\varphi^2 \propto \frac{1}{\rho_e} \frac{\partial^2 \phi}{\partial \varphi^2}. \tag{227}$$

In the case when there are two or three invariants of the fourth order in the Landau thermodynamic potential expansion we find that (see sect. 2.5)

$$\Omega_\varphi^2 \propto \rho_e^2 \propto (T_c - T),$$

i.e., the soft mode frequency corresponding to the $\varphi$ vibrations has the same temperature dependence as that corresponding to the $\rho$ vibrations. Of course, the coefficients at $T_c - T$ are to be different in these two cases so that there is a splitting of the soft mode, which has been double-degenerate in the symmetrical phase, into two non-degenerated soft modes in the non-symmetrical phase. (If the number of order parameter components is greater than 2, in the non-symmetrical phase there may be also degenerate modes.)

It is clearly seen from formula (227) that in the case when a $\varphi$-dependent invariant appears only in a term of $n$th order in the Landau thermodynamic potential one has:

$$\Omega_\varphi^2 \propto \rho_e^{n-2} \propto (T_c - T)^{(n-2)/2}. \tag{228}$$

When $n$ is great enough (it takes place if a long-periodic superstructure arises below the phase transition), the frequency of the $\varphi$ vibrations is very low, not only near but also rather far from the phase transition. At the same time the dissipation function conserves its ordinary form

$$R = \tfrac{1}{2}\gamma(\dot{\eta}_1^2 + \dot{\eta}_2^2) = \tfrac{1}{2}\gamma(\dot{\rho}^2 + \rho^2\dot{\varphi}^2).$$

Remember that the optical phonon damping is usually rather high, the corresponding soft mode damping constant $\gamma$ being only by one or two orders of magnitude less than the characteristic phonon (Debye) frequency. Therefore we expect that for a great value of $n$ the soft "phason" mode is overdamped in a very broad region including the phase transition temperature.

### 6.7. Temperature evolution near $T_c$ of the spectra of rigid modes

It was mentioned already in sect. 6.1 that in the light scattering spectrum below $T_c$ there appear in some cases new lines which are forbidden by symmetry above the transition point and which do not coincide at the same time with the soft mode components. In what follows these new lines will be referred to as those corresponding to "rigid" modes. In the case of a one-component order parameter the intensity of such a rigid mode line is proportional to $\eta_e^2$ (see formula (171)). Here we discuss the temperature evolution of the spectra of these rigid modes in more detail.

The simplest situation occurs for the case of a one-component order parameter. Let us denote, as in sect. 6.2, by $\zeta$ the normal coordinate of a rigid mode. The $\zeta$-dependent part of the thermodynamic potential density may then be written as follows:

$$\phi(\zeta) = a\zeta^2 + b\zeta^2\eta^2 + \dots, \tag{229}$$

where within the Landau theory (which is used below) the coefficients $a$ and $b$ may be thought of as temperature independent. In the symmetrical phase the mode frequency is $\Omega_{0\zeta}^2 = \phi_{\zeta\zeta}/m_\zeta \propto a$, whereas in the non-symmetrical phase,

$$\Omega_{0\zeta}^2 \propto a + b\eta_e^2 . \tag{230}$$

Thus, a jump in the temperature coefficient of the square of the rigid mode frequency is expected at $T = T_c$ (for second-order transitions far from the tricritical point).

In the case of a multi-component order parameter the situation can be more complicated. For example, a $\zeta$-dependent part of the thermodynamic potential density may now have the form

$$\phi(\zeta) = a\zeta^2 + b\zeta\eta_1\eta_2 + \dots, \tag{231}$$

In a non-symmetrical phase (say, in the phase, where $\eta_{1e} \neq 0$ and $\eta_{2e} = 0$) this leads to a linear coupling between fluctuations in $\zeta$ and $\eta_2$, the coupling arises due to the term

$$b\eta_{1e}\Delta\zeta\Delta\eta_2$$

obtained from the second summand in eq. (231). Hence, some linear mode-coupling effects of the type described in sect. 6.2 may occur for the case of a linear coupling between $\eta$ and $\rho$ or $T$ (i.e., between the soft mode and an acoustic or the thermal diffusion mode). However, these mode-coupling effects are expected to be very small usually as $T \to T_c$, due to a great difference between the soft and rigid mode characteristic frequencies near the second-order transition. As a result of this difference the rigid-mode vibrations will take place, in fact, at a fixed value of $\eta$ so that the spectral feature of $\eta$ fluctuations will not manifest themselves in the spectrum of $\zeta$ vibrations, while in the $\eta$-fluctuation spectrum the linear coupling between $\Delta\eta$ and $\Delta\zeta$ will lead only to a renormalization of the parameters which characterize the spectrum. Of course, if the rigid mode in question has an unusually low frequency, then some mode coupling phenomena will be observed not too far from the transition point.

Now we consider the spectral density $\mathscr{I}_{\eta\zeta}(\Omega, \boldsymbol{q}=0)$ of the second-order light scattering with the participation of both the soft and rigid mode. We will be interested in the temperature range above $T_c$ where both the soft and the rigid modes are inactive in the first-order Raman (combinational) scattering. Using eqs. (170) and (103) for the intensity $\mathscr{I}_{\eta\zeta}(\Omega, 0)$ we find,

$$\mathscr{I}_{\eta\zeta}(\boldsymbol{q}=0, \Omega) = VQ_{S0}a_1^2 \sum_{\boldsymbol{k}} \int \langle \eta(\boldsymbol{k}, \omega)\eta(-\boldsymbol{k}, -\omega)\rangle \times \langle \zeta(-\boldsymbol{k}, -\omega+\Omega)\zeta(\boldsymbol{k}, \omega-\Omega)\rangle \, \mathrm{d}\omega . \tag{232}$$

If the $\eta$ fluctuations have very low frequencies and one can neglect the dispersion in the $\zeta$-vibration branch, then the spectral distribution $\mathscr{I}_{\eta\zeta}(0, \Omega)$ proves to be nearly of the same form as the spectrum of the first-order scattering by $\zeta$ fluctuations in the non-symmetrical phase. This is just what one usually bears in mind when speaking about the manifestation of forbidden Raman (combinational) lines in the symmetrical phase due to local violations of symmetry (see, e.g., Whalley and Bertie (1967), Wang and Fleury (1968), Wang (1971), Pick (1972), Sokoloff (1972)). Below we follow Godefroy et al. (1982).

Let us discuss at first the temperature dependence of the integrated intensity of the lines. From the above expression or from eqs. (170) and (97) we have

$$I_{\eta\zeta}(\boldsymbol{q}=0) = VQ_{S0}a_1^2 \sum_{\boldsymbol{k}} \langle \eta(\boldsymbol{k})\eta(-\boldsymbol{k})\rangle\langle \zeta(\boldsymbol{k})\zeta(-\boldsymbol{k})\rangle . \tag{233}$$

Neglecting the dispersion in the "$\zeta$ branch", i.e., putting

$$\langle \zeta(\boldsymbol{k})\zeta(-\boldsymbol{k})\rangle = \langle \zeta^2(0)\rangle = k_B T / V\phi_{\zeta\zeta} ,$$

we find

$$I_{\eta\zeta}(q=0) = \frac{Vk_B T Q_{S0} a_1^2}{\phi_{\zeta\zeta}} \sum_{\boldsymbol{k}} \langle \eta(\boldsymbol{k})\eta(-\boldsymbol{k})\rangle , \tag{234}$$

where the summation over $\boldsymbol{k}$ is carried out up to a cutting wavevector $\boldsymbol{k}_m$, which (we recall) is of the order of the reciprocal lattice spacing. The sum in eq. (234) represents then the average value of $\eta^2$ within a volume with linear dimensions of the order of $k_m^{-1}$, i.e., in fact, within the volume of a single unit cell. Hence, the sum in eq. (234) has a meaning of the "local" average value of $\eta$. This value, which has been calculated many times (see, e.g., Ginzburg (1960)), is known to be finite at the second-order transition temperature, but may have a marked temperature dependence when going away from $T_c$. The latter holds if one uses for the short-wavelength fluctuations in $\eta$ the same expression as for the long-wavelength ones (see eq. (31)).

Such an extrapolation of eq. (31) up to the edge of the Brillouin zone seems reasonable for the displacive-type structural phase transitions but may go wrong for the order–disorder ones. For example, in the "order–disorder limit", i.e., in the Ising-like model when an ordering ion in the symmetrical phase occupies with equal probability only two positions in a unit cell, the distance between them being $2r_0$, the local average value of $\eta^2$ is equal to $r_0^2$ and does not depend on temperature at all. On the contrary, for a pure displacive transition, i.e., in the so-called "displacive limit", one may expect that the lowering of the soft phonon branch by approaching the phase transition proceeds without any deformation of this branch, which means that eq. (31) is valid for any wavevectors. Therefore, in this case we have

$$\begin{aligned}\langle \eta_{\mathrm{loc}}^2 \rangle &\equiv \sum_{k=0}^{k_{\mathrm{m}}} \langle \eta(\boldsymbol{k})\eta(-\boldsymbol{k})\rangle = \frac{k_{\mathrm{B}}T}{2\pi^2}\int_0^{k_{\mathrm{m}}} \frac{k^2\,\mathrm{d}k}{A + Dk^2} \\ &= \frac{k_{\mathrm{B}}T}{2\pi^2 D}\,[k_{\mathrm{m}} - r_{\mathrm{c}}^{-1}\tan^{-1}(k_{\mathrm{m}}r_{\mathrm{c}})] \\ &\simeq \frac{k_{\mathrm{B}}Tk_{\mathrm{m}}}{2\pi^2 D}\left(1 - \frac{\pi}{2k_{\mathrm{m}}r_{\mathrm{c}}}\right),\end{aligned} \tag{235}$$

and

$$I_{\eta\zeta}(\boldsymbol{q}=0) \simeq \frac{VQ_{\mathrm{S0}}(k_{\mathrm{B}}T)^2 a_1^2 k_{\mathrm{m}}}{\phi_{\zeta\zeta}2\pi^2 D}\left(1 - \frac{\pi}{2k_{\mathrm{m}}r_{\mathrm{c}}}\right). \tag{236}$$

Thus, we see that for displacive-type phase transitions the intensity of a "forbidden" line decreases at first when going away from the transition temperature, and is expected to have a weaker temperature dependence for the order–disorder phase transitions. In the literature one can sometimes come across quite an opposite statement. We see no grounds for that though the inverse conclusion obtained above is not reliable due to the utilization of continuum media approximation.

Going over to the discussion of the shape of the $\eta\zeta$ second-order scattering spectrum, we assume, as before, that there is no dispersion in the $\zeta$ branch and moreover that we can neglect damping of the $\zeta$ vibrations. Under such assumptions we have

$$\langle \zeta(\boldsymbol{k},\omega)\zeta(-\boldsymbol{k},-\omega)\rangle = \langle\zeta\rangle^2[\delta(\omega - \Omega_{0\zeta}) + \delta(\omega + \Omega_{0\zeta})]\,. \tag{237}$$

Substituting this into eq. (233), we find (for $\Omega > 0$):

$$\mathscr{I}_{\eta\zeta}(q=0,\Omega>0)=\frac{VQ_{S0}a_1^2k_BT}{2\pi\phi_{\zeta\zeta}}\sum_{k}\langle\eta(\boldsymbol{k},-\Omega_{0\zeta}+\Omega)\eta(-\boldsymbol{k},\Omega_{0\zeta}-\Omega)\rangle, \tag{238}$$

and the same expression, with the replacement of $\Omega-\Omega_{0\zeta}$ by $\Omega+\Omega_{0\zeta}$, for $\Omega<0$.

Now we suppose that the order parameter has a purely relaxation dynamics, this supposition being often valid even for the displacive transitions. The expression (238) may then be rewritten in the form (see, e.g., eq. (155) in the limit $m\to 0$):

$$\mathscr{I}_{\eta\zeta}(\boldsymbol{q}=0,\Omega>0)=\frac{VQ_{S0}(a_1k_BT)^2\gamma}{4\pi\phi_{\zeta\zeta}}\int_0^{k_m}\frac{k^2\,\mathrm{d}k}{A+Dk^2+\gamma^2(\Omega-\Omega_{0\zeta})^2}. \tag{239}$$

Let us discuss first the case when the system is rather far from $T_c$, so that one can neglect the dispersion in the soft mode branch ($A \gg Dk_m^2$). In this case the denominator in eq. (239) does not depend on $\boldsymbol{k}$, and we see that the $\eta\zeta$-vibration spectrum contains two Lorentzian lines with the maxima at $\Omega=\pm\Omega_{0\zeta}$ and with the widths $A/\gamma$. Of course, if we take into account the damping in the $\zeta$ branch, the widths of the forbidden $\zeta$ lines will be equal to about $\Gamma_\zeta+(A/\gamma)$, where $\Gamma_\zeta$ is the $\zeta$-vibrations damping constant (assumed for simplicity to be $\boldsymbol{k}$-independent). Thus, the forbidden $\zeta$ lines become narrower as the phase transition temperature is approached.

Consider now what shape the forbidden $\zeta$ line will have at $T=T_c$. Putting $A=0$ in eq. (239) and performing the integration, we find that the intensity distribution near the centre of the line is given by the expression

$$\mathscr{I}_{\eta\zeta}(q=0,\Omega)\propto\frac{1}{|\Omega-\Omega_{0\zeta}|^{1/2}}. \tag{240}$$

We see that the line under consideration has quite a finite width at the transition point even in the absence of damping in the $\zeta$ branch. With the account taken of the $\zeta$-vibration damping, eq. (240) holds for $|\Omega-\Omega_{0\zeta}|>\Gamma_\zeta$, so that this equation describes the intensity distribution in the wings of the line. Of course, if a dispersion in the $\zeta$ branch were taken into account, the line would be broader.

It should be mentioned that a singular distribution of the intensity of the form (240) is typical of lines in the second-order scattering spectra when the $\boldsymbol{k}$ dependence of frequencies (or damping constants) of the corresponding one-phonon modes is strong enough (see also sects. 6.3 and 6.4).

In the case of a vibration character of the order parameter dynamics the analysis of the $\eta\zeta$-scattering spectrum is more complicated but may be easily carried out by using the general expression (232) and the known results for the $\langle|\eta(\boldsymbol{k},\omega)|^2\rangle$ and $\langle\zeta(\boldsymbol{k},\omega)|^2\rangle$ correlations functions (see sect. 6.1).

## 7. *Conclusion*

For several decades already light scattering has played an outstanding role in the investigation of the structure of molecules, liquids and solids. True, twenty to thirty years ago the possibilities of the method seemed to have been exhausted to some extent, first of all due to the imperfection of the light sources available. The appearance of lasers in the beginning of the sixties changed the situation radically and, particularly, provided the use of much more sensitive apparatus. Thus it became possible to investigate the problems which had been considered insoluble (e.g., the study of the light scattering spectrum in liquid helium or the observation of weak second-order lines in the light scattering spectra in crystals). Simultaneously, though to a considerable extent independently, an increasing attention was paid to certain substances and classes of substances which had been very little or not at all investigated before (some ferroelectrics, ferroelastics, magnetics and many other crystals, polymers, etc.). As a result the study of light scattering in condensed media is now a wide-spread and prosperous field.

On the other hand, in condensed state physics and in solid state physics particularly, one of the central problems is the study of phase transitions and various effects near the points of these transitions. It is quite natural therefore that light scattering near phase transition points is given much attention.

At the same time if we refer to the corresponding literature, it is striking that we come across a predominance of particular results, discussion of individual cases on the basis of model consideration, etc. Meanwhile, it is well known that different phase transitions have much in common, and a general approach to the problem of light scattering near phase transitions would be possible and fruitful. On the whole, such an approach may be called phenomenological (or semi-phenomenological when, for example, the light scattering spectrum is dealt with). There is no need to explain here what we mean because the present paper is entirely based just on the above-mentioned approach. We should emphasize that it is not in the least in contradiction with microtheory and, in fact, facilitates its application when the latter is actually possible. At the present time, of course, the question of the interrelation between phenomenological and microscopic theories is in principle quite clear. In practice, however, we have to come across microscopic derivations of the results which follow much simpler and, above all, in a more general form from the phenomenological (macroscopic) theory. As a typical example we would mention the problem of spatial dispersion in crystal optics (Agranovich and Ginzburg 1965, 1979). The problem of the microtheory consists here in the computation of the permittivity

$\epsilon_{ij}(\omega, \boldsymbol{q})$, whereas the question of light wave propagation in a medium with a given permittivity belongs to the field of macroscopic electrodynamics. Meanwhile, in a large number of papers, from a microtheory (and moreover for particular models), normal waves are directly obtained which can propagate in the medium. This explains, for example, why normal waves in the medium, which were investigated long ago, now in some cases (in a resonance region or in a long wavelength range) are referred to as polaritons. Similarly in the analysis of a whole number of problems concerning light scattering near phase transitions it is reasonable to proceed from the beginning from phenomenological phase transition theory (Landau theory or scaling theory) resorting also to microtheory for heuristic purposes, i.e., to choose more successfully the order parameter, or at the next stage, in the course of a more detailed study of one or another transition, normal oscillation, etc.

What has been said is in general not new, of course, but, as a matter of fact, the situation concerning the theory of light scattering near phase transition points justifies in our opinion the above remarks.

In the present chapter we have tried to elucidate as simply as possible the principal points of the theory but could not, of course, discuss all the questions of interest. We have hardly touched upon the characteristic features of light scattering near magnetic phase transitions and phase transitions in liquids and liquid crystals (see chs. 9 and 4 of this volume). Moreover, even for structural phase transitions in crystals, which have been mainly considered, the presentation is not complete. To be precise, we have almost everywhere restricted ourselves to the simplest case of a one-component order parameter and neglected the optical anisotropy of the medium. Therefore we have not considered anomalies of the polarization properties of scattered light near phase transitions that may also be of interest. Of the number of problems not reflected in this chapter (we always mean a region near a phase transition point) the effects of light scattering from fluctuations of non-linear dielectric permittivities (hyper-Raman scattering; Balagurov and Vaks (1978)), light scattering from and near surfaces (Agranovich 1976, Agranovich and Leskova 1977), scattering in a non-equilibrium medium (in particular, a medium under irradiation, Chanussot et al. (1978)) should be mentioned separately. Light scattering near first-order phase transitions and in the scaling region has not been described in detail (for structural phase transitions, though, this region has not been investigated experimentally). Quantum effects essential in the light scattering at sufficiently low temperatures have not been taken into consideration either, and experimental data have been discussed fragmentarily and only for the purpose of illustration (for details see chs. 5, 6 and 7). Finally, the problem of an inhomogeneous state (structure) strongly scattering the light and appearing under some phase transitions (Shustin 1966, Dolino and Bachheimer 1977, Shustin et al. 1978, Khachaturyan 1974, Roitburd 1974) remains not quite clear and deserves further experimental and theoretical investigations (see ch. 11).

The theory of light scattering near phase transitions has not been sufficiently developed even as regards the questions for the solution of which no essential difficulties seem to arise. At the same time one can point out a number of problems to solve for which new approaches are needed. These problems include the development of a microscopic order parameter fluctuation dynamics not restricted to the framework of perturbation theory. Such a development is necessary for a deeper insight into the nature of the relaxation component in the order parameter spectrum which is present sometimes near the displacive-type phase transitions (see sect. 6.5).

As to experimental problems, we would like to emphasize once again the necessity to perform combined experiments including measurements (on one and the same specimen with a controlled structure, including defects) of both the intensity and polarization of scattered light and different thermodynamic and kinetic parameters of the substance. The investigation of light scattering in a low-frequency range (e.g., by modulation spectroscopy or light scattering by polaritons) deserves special attention in connection with the central peak question discussed above. The fundamental problems include undoubtedly the discovery and investigation (specifically, by the light scattering method) of the scaling region for structural phase transitions.

From the above presentation it follows beyond doubt that the study of light scattering near phase transition points will continue in different directions and yield important results.

## *Notes added in proof*

(1) This is the case, in particular, when the splitting of the double-degenerate mode (with normal coordinates $\xi_1$, $\xi_2$) is small throughout the Brillouin zone and does not exceed the corresponding phonon damping constant. Such a situation, though, is not very typical of real crystals. We are indebted to V.G. Vaks for this comment (see p. 99).

(2) On the other hand, it is supported by the fact that the width of the central peak proves to be close, indeed, to $\Omega_{RS}(k_{max})$ as determined from the computed energy fluctuations spectrum for $k$ of the order of the reciprocal lattice spacing. Further investigations are required before a final conclusion could be made. In particular, the role played by finite size effects seems worthy of more thorough study, because for finite values of $q$ both linear and non-linear interactions between $\eta(\boldsymbol{q})$ and $S$ are allowed by symmetry and may be rather essential. (See p. 105.)

## *References*

Agranovich, V.M., 1976, Pis'ma ZhETF, **24**, 602 (1976, JETP Lett. **24**, 558).

Agranovich, V.M. and V.L. Ginzburg, 1965, Kristallooptica s uchetom prostranstvennoi dispersii i teoriya eksitonov (Nauka, Moscow); Eng. Trans., 1966, Spatial dispersion in crystal optics and theory of exitons (Wiley, London, New York, Sydney); Second ed., 1979 (Nauka, Moscow).

Agranovich, V.M. and T.A. Leskova, 1977, Solid State Commun. **21**, 1065.

Anderson, P.W., 1960, in: Fizika Dielektrikov, ed., G. Skanavi (AN SSSR, Moscow) p. 290.

Andreev, A.F., 1974, Pis'ma ZhETF, **19**, 713 (1974, JETP Lett. **19**, 368).

Andrews, S.R. and R.T. Harley, 1981, J. Phys. **C 14**, L207.

Aronov, A.G., D.N. Mirlin, I.I. Repina and F.F. Chudnovskii, 1977, Fiz. Tverd. Tela, **19**, 193 (1977, Sov. Phys. Solid State II **19**, 110).

Aubry, S., 1975, J. Chem. Phys. **62**, 3217.

Aubry, S. and R. Pick, 1974, Ferroelectrics, **8**, 471.

Axe, J.D. and G. Shirane, 1970, Phys. Rev. **B 1**, 342.

Axe, J.D. and G. Shirane, 1974, Phys. Rev. **B 10**, 1963.

Bacheimer, J.P. and G. Dolino, 1975, Phys. Rev. **B 11**, 3195.

Balagurov, B.Ya. and V.G. Vaks, 1969, ZhETF **57**, 1646 (1970, Sov. Phys. JETP, xxx, 889).

Balagurov, B.Ya. and V.G. Vaks, 1978, Solid State Commun. **25**, 571.

Balagurov, B.Ya., V.G. Vaks and B.I. Shklovskii, 1970, Fiz. Tverd. Tela, **12**, 89 (1970, Sov. Phys. Solid State, **12**, 70).

Balkanski, M., ed., 1971, Light Scattering in Solids (Flammarion Sciences, Paris).

Balkanski, M., R.C.C. Leite and S.P.S. Porto, eds., 1975, Light Scattering in Solids (Flammarion Sciences, Paris).

Baranskii, K.N., 1953, MGU Diploma Work, unpublished.

Barker, A.S. and J.J. Hopfield, 1964, Phys. Rev. **A 135**, 1732.

Barta, Ch., A.A. Kaplianskii, V.V. Kulakov, B.Z. Malkin and Yu.F. Markov, 1976, ZhETF **70**, 1429 (1976, Sov. Phys. JETP, **43**, 744).

Barta, Ch., B.S. Zadokhin, A.A. Kaplianskii and Yu.F. Markov, 1977, Pis'ma ZhETF, **26**, 480 (1977, Sov. Phys. JETP Lett. **26**, 347).

Bausch, R., 1972, Z. Phys. **254**, 81.

Bendow, R., J.L. Birman and V.M. Agranovich, eds., 1976, Theory of Light Scattering in Condensed Matter (Plenum, New York); The Theory of Light Scattering in Solids, Proc. First Sov. Amer. Symp. (Nauka, Moscow, 1976).

Benoit, J.P., Cao Xuan An., Y. Luspin, I.P. Chapelle and J. Lefebre, 1978, J. Phys. **C 11**, L721.

Binder, K., 1977, Solid State Commun. **24**, 401.

Binder, K. and D. Stauffer, 1976, J. Stat. Phys. **15**, 267.

Binder, K., D. Stauffer and H. Muller-Krumbhaar, 1974, Phys. Rev. **B 10**, 3853.

Binder, K., D. Stauffer and H. Muller-Krumbhaar, 1975, Phys. Rev. **B 12**, 5261.

Birgeneau, R.J., J. Skalyo, Jr. and G. Shirane, 1971, Phys. Rev. **B 3**, 1736.

Birman, J.L., H.Z. Cummins and K.K. Rebane, eds., 1979, Light Scattering in Solids (Plenum, New York, London).

Bishofberger, T. and E. Courtens, 1975, Phys. Rev. Lett. **35**, 1451.

Blinc, R. and B. Žekš, 1974, Soft Modes in Ferroelectrics and Antiferroelectrics (North-Holland, Amsterdam); Russian transl. (Mir, Moscow, 1975).

Brazovskii, S.A. and I.E. Dzialoshinskii, 1975, Pis'ma ZhETF, **21**, 360 (1975, JETP Lett. **21**, 164).

Brody, E.M. and H.Z. Cummins, 1968, Phys. Rev. Lett. **21**, 1263.

Brody, E.M. and H.Z. Cummins, 1969a, Phys. Rev. Lett. **23**, 1039.

Brody, E.M. and H.Z. Cummins, 1969b, in: Light Scattering Spectra of Solids, ed. Wright, G.B. (Springer, New York) p. 683.

Brody, E.M. and H.Z. Cummins, 1974, Phys. Rev. **B 9**, 179.
Burns, G., 1974, Phys. Rev. **B 10**, 1951.
Burns, G. and B.A. Scott, 1970, Phys. Rev. Lett. **25**, 167.
Chanussot, G., V.M. Fridkin, G. Godefroy and B. Jannot, 1978, Ferroelectrics, **21**, 615.
Clark, R. and L. Benguigui, 1977, J. Phys. **C 10**, 1963.
Cochran, W., 1960, Adv. Phys. **9**, 387.
Cochran, W., 1961, Adv. Phys. **10**, 401.
Courtens, E.J., 1976a, J. de Phys. Lett. **37**, L21.
Courtens, E.J., 1976b, Phys. Rev. Lett. **37**, 1584.
Courtens, E.J., 1978, Phys. Rev. Lett. **41**, 1171.
Cowley, R.A., 1970, J. Phys. Soc. Jpn. Suppl. **28**, 239.
Cowley, R.A. and G.J. Coombs, 1973, J. Phys. **C 6**, 143.
Di Domenico, M., Jr., S.H. Wemple and S.P.S. Porto, 1968, Phys. Rev. **174**, 522.
Dolino, G. and J.P. Bachheimer, 1977, Phys. Status Solidi (a) **41**, 673.
Durvasula, L.N. and R.W. Gammon, 1977, Phys. Rev. Lett. **38**, 1081.
Einstein, A., 1910, Ann. Phys. **33**, 1275.
Einstein, A., 1966, Collection of Scientific works, Vol. 3 (Nauka, Moscow) p. 216.
Enz, C.P., 1974a, Rev. Mod. Phys. **46**, 705.
Enz, C.P., 1974b, Solid State Commun. **15**, 459.
Fabelinskii, I.L., 1965, Molekularnoe rasseyanie sveta (Nauka, Moskva); Molecular scattering of light (Plenum, New York, 1968).
Fisher, M.E., 1974, Rev. Mod. Phys. **46**, 597.
Fixman, M., 1962, J. Chem. Phys. **36**, 310.
Fleury, P.A., 1970, Solid State Commun. **8**, 611.
Fleury, P.A., 1972, Commun. Solid State Phys. **4**, 149, 167.
Fleury, P.A., 1974, Phys. Rev. Lett. **33**, 492.
Fleury, P.A., 1975, in: Light Scattering in Solids, eds., Balkanski, M., R.C.C. Leihe and S.P.S. Porto (Flammarion Press, Paris) p. 406.
Fleury, P.A., 1976, in: Theory of Light Scattering in Condensed Matter, eds., Bendow, R., J.L. Birman and V.M. Agranovich (Plenum, New York) p. 747.
Fleury, P.A. and P.D. Lazay, 1971, Phys. Rev. Lett. **26**, 1331.
Fleury, P.A. and K.B. Lyons, 1976, Phys. Rev. Lett. **37**, p. 1088.
Fleury, P.A., J.F. Scott and J.M. Worlock, 1968, Phys. Rev. Lett. **21**, 16.
Fleury, P.A., P.D. Lazay and L.G. Van Uitert, 1974, Phys. Rev. **33**, 492.
Fritz, I.J. and H.Z. Cummins, 1972, Phys. Rev. Lett. **28**, 96.
Garland, C.W., 1970, in: Physical Acoustics, ed., Morse, P. (Academic Press, New York) Vol. VII.
Garland, C.W. and B.B. Weiner, 1971, Phys. Rev. **B 3**, 1634.
Ginzburg, V.L., 1949, ZhETF **19**, 36.
Ginzburg, V.L., 1949, Usp. Fiz. Nauk, **38**, 490.
Ginzburg, V.L., 1952, Usp. Fiz. Nauk, **46**, 348.
Ginzburg, V.L., 1955, Doklady Akad. Nauk SSSR, **105**, 240.
Ginzburg, V.L., 1960, Fiz. Tverd. Tela, **2**, 2031 (1961, Sov. Phys. Solid State, **2**, 1824).
Ginzburg, V.L., 1962, Usp. Fiz. Nauk **77**, 621 (1963, Sov. Phys. Uspekhi, **5**, 649).
Ginzburg, V.L., 1975, Theoretical physics and astrophysics (Nauka, Moscow); English transl. (Pergamon Press, Oxford, 1979); Second ed. (Nauka, Moscow, 1981).
Ginzburg, V.L. and A.P. Levanyuk, 1958, J. Phys. Chem. Solids, **6**, 51; for more details see in: Issledovaniya po eksperimental'noy i teoreticheskoy fisike. Sbornik, posvyaschennyy pamyati G.S. Landsberga (Acad. Nauk SSSR, Moscow, 1959) p. 104.
Ginzburg, V.L. and A.P. Levanyuk, 1960, ZhETF, **39**, 102 (1967, Sov. Phys. JETP, **12**, 138).
Ginzburg, V.L. and A.P. Levanyuk, 1974, Phys. Lett. **A47**, 345.
Ginzburg, V.L. and A.P. Levanyuk, 1976, in: Theory of Light Scattering in Condensed Matter, eds., Bendow, R., J.L. Birman and V.M. Agranovich (Plenum, New York) p. 3.

Ginzburg, V.L. and A.A. Sobyanin, 1976, Usp. Fiz. Nauk, **120**, 153, 733 (1976, Sov. Phys. Uspekhi, **19**, 773).
Ginzburg, V.L. and A.A. Sobyanin, 1982, J. Low Temp. Phys. **49**, No. 5/6, p. 507.
Ginzburg, V.L., A.P. Levanyuk and A.A. Sobyanin, 1978, Ferroelectrics, **20**, 97.
Ginzburg, V.L., A.P. Levanyuk, A.A. Sobyanin and A.S. Sigov, 1979, in: Light Scattering in Solids, eds., Birman, J., H.Z. Cummins and K.K. Rebane (Plenum, New York), p. 331.
Ginzburg, V.L., A.P. Levanyuk and A.A. Sobyanin, 1980, Phys. Rep. **57**, 153; for an abbreviated text see also Usp. Fiz. Nauk, 130 (1980) 615.
Ginzburg, V.L., A.A. Sobyanin and A.P. Levanyuk, 1981, J. Raman Spectrosc., **10**, 194.
Godefroy, G., B. Jannot, A.P. Levanyuk and A.S. Sigov, 1983, Izv. AN SSSR, Ser. Fiz. **47**, 688.
Golovko, V.A. and A.P. Levanyuk, 1981, Fiz. Tverd. Tela, **23**, 3179 (1981, Sov. Phys. Solid State, **23**, xxx.
Gorelik, V.S., S.V. Ivanova, I.P. Kucheruk, B.A. Strukov and A.A. Khalezov, 1976, Fiz. Tverd. Tela, **18**, 2297 (1976, Sov. Phys. Solid State, **18**, 1340).
Gor'kov, L.P. and L.P. Pitaievskii, 1957, ZhETF, **33**, 634 (1958, Sov. Phys. JETP, **6**, 486).
Greytak, T.J., 1974, in: Low Temperature Physics – LT13, eds., K.D. Timmerhaus, W.J. O'Sullivan and E.F. Hannuel (Plenum, New York) p. 13.
Greytak, T.J., 1978, in: Quantum Liquids: Lectures in International School on Low Temperature Physics, Erice, 1977 (North-Holland, Amsterdam).
Gurevich, V.L. and A.L. Efros, 1966, ZhETF, **51**, 1693 (1967, Sov. Phys. JETP, **24**, 1146).
Halperin, B.I. and C.M. Varma, 1976, Phys. Rev. **B 14**, 4030.
Hayes, W. and R. Loudon, 1978, Scattering of Light by Crystals (Wiley, New York).
Heiman, D. and S. Ushioda, 1974, Phys. Rev. **B 9**, 2122.
Heiman, D. and S. Ushioda, 1977, Bull. Am. Phys. Soc. **22**, 324.
Heiman, D. and S. Ushioda, 1978, Phys. Rev. **B 17**, 3616.
Heiman, D., S. Ushioda and J.P. Remeika, 1975, Phys. Rev. Lett. **34**, 886.
Hikita, T., K. Suzuki and T. Ikeda, 1981, Ferroelectrics, **39**, 1005.
Hisano, K. and J.P. Ryan, 1972, Solid State Comm. **11**, 1745.
Hochli, U.T., 1970, Solid State Commun. **8**, 1487.
Hohenberg, P.C., 1971, in: Critical Phenomena, Proc. of the Int. Sch. of Phys. Enrico Fermi, Course LI, ed. M.S. Green (Academic Press, New York, London) ch. 5; Russian transl. in: Kvantovaya teoriya polya i fisika fazovykh perekhodov (Mir, Moscow, 1975) ch. 5.
Hohenberg, P.C. and B.I. Halperin, 1977, Rev. Mod. Phys. **49**, 435.
Hosea, T.J., D.J. Lockwood and W. Taylor, 1979, J. Phys. **C12**, 387.
Imry, Y., 1974, Phys. Rev. Lett. **33**, 1304.
Ipatova, I.P. and A.A. Klochikhin, 1978, in: Spectroscopiya kombinatzionnogo rasseyaniya sveta: Materialy II Vsesoyuznoy konferentzii (Moscow) p. 126.
Ishibashi, Y. and Y. Takagi, 1972, J. Phys. Soc. Jpn. **33**, 1.
Jona, F. and G. Shirane, 1962, Ferroelectric Crystals (Pergamon, London); Russian transl. (Mir, Moscow, 1965).
Kaminov, I.P., 1969, in: Light Scattering Spectra of Solids, ed., Wright, G.B. (Springer, New York) p. 675.
Kaminov, I.P. and T.C. Damen, 1968, Phys. Rev. Lett. **10**, 1105.
Kawasaki, K., 1971, in: Critical Phenomena: Proc. of the Int. Sch. of Phys. Enrico Fermi, Course LI, ed., M.S. Green (Academic Press, New York, London) ch. 4; Russian transl. in: Kvantovaya teoriya polya i fizika fazovykh perekhodov (Mir, Moscow, 1975) ch. 4.
Kawasaki, K., 1976, in: Phase Transitions and Critical Phenomena, eds., C. Domb and M.S. Green (Academic Press, New York) Vol. 5a.
Khachaturyan, A.G., 1974, Phase Transition Theory and Solid Solution Structure (Nauka, Moscow).
Khmel'nitzkii, D.E., 1970, ZhETF, **61**, 2110 (1971, Sov. Phys. JETP, **34**, 1125).

Khmel'nitzkii, D.E., 1974, Fiz. Tverd. Tela, **16**, 3188 (1975, Sov. Phys. Solid State, **16**, 2079).
Kjems, J.K., G. Shirane, K.A. Muller and H.J. Scheel, 1973, Phys. Rev. **B 8**, 1119.
Koehler, T.R., A.R. Bishop, J.A. Krumhansl and J.R. Schriffer, 1975, Solid State Commun. **17**, 1515.
Krivoglaz, M.A., 1963, Fiz. Tverd. Tela, **5**, 3437 (1964, Sov. Phys. Solid State, **5**, 2526).
Krivoglaz, M.A. and S.A. Rybak, 1957, ZhETF, **33**, 139 (1958, Sov. Phys. JETP, **6**, 107).
Krumhansl, J.A. and J.R. Schrieffer, 1975, Phys. Rev. **B 11**, 3535.
Lagakos, N. and H.Z. Cummins, 1974, Phys. Rev. **B 10**, 1063.
Landau, L.D., 1937, ZhETF, **7**, 1232 (Phys. Z. Sowjet, **12**, 123); Collected Papers (Pergamon Press, Oxford) p. 233.
Landau, L.D., 1965a, Collected Papers (Pergamon, Oxford) p. 79.
Landau, L.D., 1965b, Collected Papers (Pergamon, Oxford) p. 626.
Landau, L.D., 1969a, Collected Works, Vol. 1 (Nauka, Moscow) p. 4.
Landau, L.D., 1969b, Collected Works, Vol. 2 (Nauka, Moscow) p. 238.
Landau, L.D. and I.M. Khalatnikov, 1954, Dokl. Akad. Nauk SSSR, **96**, 469.
Landau, L.D. and E.M. Lifshitz, 1954, Mechanics of continuous media (Gostekhizdat, Moscow) p. 376; Eng. transl. (Pergamon, Oxford).
Landau, L.D. and E.M. Lifshitz, 1957, Electrodynamics of continuous media (Gostekhizdat, Moskva); Eng. transl. (Pergamon, Oxford).
Landau, L.D. and E.M. Lifshitz, 1976, Statistical Physics (Nauka, Moscow); English transl. (Pergamon, Oxford).
Landau, L.D. and G. Placzek, 1934, Phys. Zs. Sowjet, **5**, 172.
Larkin, A.I. and S.A. Pikin, 1969, ZhETF, **56**, 1664 (1969, Sov. Phys. JETP, **29**, 891).
Laughman, L., L.W. Davis and T. Nakamura, 1972, Phys. Rev. **B 6**, 3322.
Laulicht, I., J. Bagno and G. Schlesinger, 1972, J. Phys. and Chem. Sol. **33**, 319.
Lazay, P.D. and P.A. Fleury, 1971, in: Light Scattering in Solids, ed., M. Balkanski (Flammarion Sciences, Paris) p. 406.
Lazay, P.D., J.H. Lunacek, N.A. Clark and G.B. Benedek, 1969, in: Light Scattering Spectra of Solids, ed., Wright, G.B. (Springer, New York)) p. 593.
Leibfried, G. and W. Ludwig, 1961, Theory of anharmonic effects in crystals, in: Solid State Physics, Vol. 12, eds. F. Seitz, D. Turnbull (Academic Press, New York, London); Russian transl. (I.L., Moscow, 1963).
Leontovich, M.A., 1944, Statistical Physics (Gostekhizdat, Moscow, Leningrad).
Levanyuk, A.P., 1959, ZhETF, **36**, 810 (1959, Sov. Phys. JETP, **9**, 571).
Levanyuk, A.P., 1965a, Izv. AN SSSR, ser, fiz. **28**, 879.
Levanyuk, A.P., 1965b, ZhETF **49**, 1304 (1966, Sov. Phys. JETP, **22**, 901).
Levanyuk, A.P., 1974, ZhETF, **66**, 2256 (1974, Sov. Phys. JETP, **39**, 1111).
Levanyuk, A.P., 1976, ZhETF, **70**, 1253 (1976, Sov. Phys. JETP, **43**, 652).
Levanyuk, A.P. and D.G. Sannikov, 1974, Usp. Fiz. Nauk, **112**, 561 (1974, Sov. Phys. Uspekhi, **17**, 199).
Levanyuk, A.P. and N.V. Schedrina, 1972a, Fiz. Tverd. Tela, **14**, 1204 (1972, Sov. Phys. Solid State, **14**, 1027).
Levanyuk, A.P. and N.V. Schedrina, 1972b, Fiz. Tverd. Tela, **14**, 3012 (1973, Sov. Phys. Solid State, **14**, 2581).
Levanyuk, A.P. and N.V. Schedrina, 1974, Fiz. Tverd. Tela, **16**, 1439 (1974, Sov. Phys. Solid State, **16**, 923).
Levanyuk, A.P. and A.A. Sobyanin, 1967, ZhETF, **53**, 1024 (1968, Sov. Phys. JETP, **26**, 612).
Levanyuk, A.P. and A.A. Sobyanin, 1970, Pis'ma ZhETF, **11**, 540 (1970, JETP Lett. **11**, 371).
Levanyuk, A.P., K.A. Minaeva and B.A. Strukov, 1968, Fiz. Tverd. Tela, **10**, 2443 (1968, Sov. Phys. Solid State, **10**, 1919).
Lockwood, D.J. and B.N. Torrie, 1974, J. Phys. **C 7**, 2729; see also in: Balkanski (1971) p. 147.

Lockwood, D.J., J.W. Arthur, W. Taylor and J.T. Hosea, 1976, Sol. State Commun. **20**, 703.

Luban, M., 1976, in: Phase Transitions and Critical Phenomena, Vol. 5a, eds. C. Domb and M.S. Green (Academic Press, London, New York, San Francisco) p. 35.

L'vov, V.M., 1966, Fiz. Tverd. Tela, **10**, 451 (1966, Sov. Phys. Solid State, **10**, 354).

Lyons, K.B. and P.A. Fleury, 1976, Phys. Rev. Lett. **37**, 161.

Lyons, K.B. and P.A. Fleury, 1978, Phys. Rev. **B 17**, 2403.

Lyuksyutov, I.F. and V.L. Pokrovskii, 1975, Pis'ma ZhETF, **21**, 22 (1975, JETP Lett. **21**, 9).

Ma, S.K., 1976, Modern theory of critical phenomena (Benjamin, Reading, Mass.); Russian transl. (Mir, Moscow, 1981).

Ma, S.K., 1981, J. Statist. Phys. **26**, 221.

Makita, Y., T. Yagi and I. Tatsuzaki, 1976, Phys. Lett. **A 55**, 437.

Mandelstam, L.I. and M.A. Leontovich, 1936, Dokl. Acad. Nauk SSSR, **3**, 111 (1937, ZhETF, **7**, 438).

Meissner, G., 1981a, Ferroelectrics, **37**, 527.

Meissner, G., N. Menyhárd and P. Szépfalusy, 1981b, preprint.

Mermelstein, M.D. and H.Z. Cummins, 1977, Phys. Rev. **B 16**, 2177.

Michel, K.H., J. Naudt and B. de Raedt, 1978, Phys. Rev. **B 18**, 648.

Moriya, T., 1968, J. Appl. Phys. **39**, 1042 (1969, Usp. Fiz. Nauk **98**, 71).

Motulevich, G.P., 1950, Trudy FIAN SSSR, **5**, 11; see also Fabelinskii (1965) ch. 9.

Natterman, T. and S. Trimper, 1975, J. Phys. **A 8**, 11.

Onodera, Y., 1970, Progr. Theor. Phys. **44**, 1477.

Palin, C.J., W.F. Vinen, E.R. Pike and J.M. Vaughan, 1971, J. Phys. **C 4**, L225.

Patashinskii, A.Z. and V.L. Pokrovskii, 1975, Fluctuation theory of phase transitions (Nauka, Moscow); English transl. (Pergamon, Oxford, 1981); 1982, second edition (Nauka, Moscow).

Penna, A.F., A. Chaves and S.P.S. Porto, 1976, Solid State Commun. **19**, 491.

Peercy, P.S., 1973, Phys. Rev. Lett. **31**, 380.

Peercy, P.S., 1975a, Solid State Commun. **16**, 439.

Peercy, P.S., 1975b, Phys. Rev. **B 12**, 2725.

Peercy, P.S., 1975c, in: Light Scattering in Solids, eds., Balkanski, M., R.C.C. Leite and S.P.S. Porto (Flammarion Sciences, Paris) p. 782.

Pique, J.P., G. Dolino and M. Wallade, 1977, J. de Phys. **38**, 1527.

Pick, R.M., 1972, Physics of Impurity Centres in Crystals (Eston. Acad. Sci., Tallin) p. 293.

Popkov, Yu.A., V.V. Eremenko and V.I. Fomin, 1971, Fiz. Tverd. Tela, **13**, 2028 (1971, Sov. Phys. Solid State, **13**, 1701); see also Balkanski (1971) p. 372.

Pytte, E. and H. Thomas, 1972, Solid State Commun. **11**, 161.

Rehwald, W., 1973, Adv. Phys. **22**, 721.

Reissland, J.A., 1973, The Physics of Phonons (Wiley, New York, London); Russian transl. (Mir, Moscow, 1975).

Riste, T., ed., 1974, Anharmonic Lattices, Structural Transitions and Melting (Noordhoff, Leiden).

Riste, T., E.J. Samuelson and K. Otnes, 1971, in: Structural Phase Transitions and Soft Modes, eds., E.J. Samuelson, E. Andersen and T. Feder (Universitetsforlaget, Oslo).

Riste, T., E.J. Samuelson, K. Otnes and J. Feder, 1975, Solid State Commun. **9**, 1455.

Roitburd, A.L., 1974, Usp. Fiz. Nauk, **112**, 69 (1974, Sov. Phys. Uspekhi, **17**, 326).

Ruvalds, J. and Z. Zawadovski, 1970, Phys. Rev. **B 2**, 1172.

Sailer, E. and H.G. Unruh, 1975, Solid State Commun. **16**, 615.

Sailer, E. and H.G. Unruh, 1976, Ferroelectrics, **12**, 285.

Sannikov, D.G., 1962, Fiz. Tverd. Tela, **4**, 1619 (1962, Sov. Phys. Solid State, **4**, 1187).

Sannikov, D.G. and A.P. Levanyuk, 1978, Fiz. Tverd. Tela, **20**, 1005 (1978, Sov. Phys. Solid State, **20**, 580).

Sawafuji, M., M. Tokunaga and I. Tatsuzaki, 1979, J. Phys. Soc. Jpn, **000**, 1860.

Schneider, T. and E. Stoll, 1973, Phys. Rev. Lett. **31**, 1254.
Schneider, T. and E. Stoll, 1975, Phys. Rev. Lett. **35**, 296.
Schneider, T. and E. Stoll, 1976, Phys. Rev. **B 13**, 1216.
Schneider, T. and E. Stoll, 1978a, Phys. Rev. Lett. **41**, 964.
Schneider, T. and E. Stoll, 1978b, Phys. Rev. **B 17**, p. 1302.
Schulhof, M.P., P. Heller, R. Nathans and A. Linz, 1970, Phys. Rev. Lett. **24**, 1184.
Scott, J.F., 1968, Phys. Rev. Lett. **21**, 907.
Scott, J.F., 1974, Rev. Mod. Phys. **46**, 83.
Scott, J.F. and Porto, S.P.S., 1967, Phys. Rev. **161**, 903.
Sham, L.J., 1967, Phys. Rev. **156**, 494.
Shapiro, S.M., 1969, Thesis.
Shapiro, S.M. and H.Z. Cummins, 1968, Phys. Rev. Lett. **21**, 1578; see also in: Wright (1969) p. 705.
Shapiro, S.M., D.C. O'Shea and H.Z. Cummins, 1967, Phys. Rev. Lett. **19**, 361.
Shapiro, S.M., J.D. Axe, G. Shirane and T. Riste, 1972, Phys. Rev. **B6**, 4332; see also in: Riste (1974) p. 23.
Shigenari, T., Y. Takagi and Y. Wakabayashi, 1976, Solid State Commun. **18**, 1271.
Shustin, O.A., 1966, Pis'ma ZhETF, **3**, 491 (1966, JETP Lett. **3**, 320).
Shustin, O.A., T.G. Chernevich, S.A. Ivanov and I.A. Yakovlev, 1978, Pis'ma ZhETF, **27**, 349 (1978, JETP Lett. **27**, 328).
Slivka, V. Yu., Yu.M. Vysochanskii, M.I. Gurzan and D.V. Chepur, 1978, Fiz. Tverd. Tela, **20**, 3530 (1979, Sov. Phys. Solid State **20**, No. 12).
Smolensky, G.A., R.V. Pisarev, P.A. Markovin and B.B. Krichevcov, 1977, Physica, **B86-88**, 1205.
Sobyanin, A.A., 1979, unpublished.
Sobyanin, A.A. and N.V. Schedrina, 1980, unpublished.
Sokoloff, J.B., 1972, Phys. Rev. **B 5**, 4962.
Steigmeier, E.F. and H. Auderset, 1973, Solid State Commun. **12**, 565.
Steigmeier, E.F., G. Harbeke and R.K. Wehner, 1971, in: Light Scattering in Solids, ed. Balkanski, M. (Flammarion Sciences, Paris).
Stenley, H.E., 1971, Introduction to Phase Transitions and Critical Phenomena (Oxford Univ. Press, London); Russian transl. (Mir. Moscow, 1973).
Stekhanov, A.I. and E.A. Popova, 1968, Fiz. Tverd. Tela, **10**, 815 (1968, Sov. Phys. State, **10**, xxx).
Stephen, M.J., 1976, in: Physics of Liquid and Solid Helium, eds., K.H. Benneman and J.B. Ketterson (Wiley, New York) Part 1, ch. 4.
Tani, K., 1969, J. Phys. Soc. Jpn. **26**, 93.
Terletskii, I.P., 1958, Nuovo Cimento, **7**, 3018.
Unruh, H.G., J. Kruger and E. Sailer, 1978, Ferroelectrics, **20**, 3.
Uwe, H. and T. Sakudo, 1976, Phys. Rev. **B 13**, 271.
Vaks, V.G., 1973, Introduction to microscopic theory of ferroelectrics (in Russian) (Nauka, Moscow).
Vaks, V.G., A.I. Larkin and S.A. Pikin, 1966, ZhETF, **51**, 361 (1967, Sov. Phys. JETP, **24**, 240).
Varma, C.M., 1976, Phys. Rev. **B 14**, 244.
Villain, J., 1970, Solid State Commun. **8**, 295.
Vinen, W.F. and D.L. Hurd, 1978, Adv. Phys. **27**, 533.
Vinen, W.F., C.J. Palin and J.M. Vaughan, 1974, in: Low Temperature Physics–LT13, eds., K.D. Timmerhaus, W.J. O'Sullivan and E.F. Hammel (Plenum Press, New York), Vol. 1, p. 524.
Wada, M., H. Uwe, A. Sawada, Y. Ishibashi, Y. Takagi and T. Sakudo, 1977, J. Phys. Soc. Jpn. **43**, 544.
Wallace, D.J., 1973, J. Phys. **C 6**, 1390.
Wang, C.H., 1971, Phys. Rev. Lett. **26**, 1226.

Wang, C.H. and P.A. Fleury, 1968, in: Light Scattering Spectra of Solids, ed. Wright, G.B. (Springer, New York, 1969) p. 631.

Watts, D.R. and W.W. Webb, 1974, in: Low Temperature Physics–LT13, eds., Timmerhaus, K.D., W.J. O'Sullivan and E.P. Hammel (Plenum Press, New York) p. 581.

Wegner, F.J. and R.K. Riedel, 1973, Phys. Rev. **B 7**, 248.

Wehner, R.K. and R. Klein, 1972, Physica, **62**, 161.

Whalley, E. and Berti, J.E., 1967, Jn. Chem. Phys. **46**, 1264.

Widom, A., N.E. Tornberg and R.P. Lowndes, 1978, J. Phys. **C11**, L433.

Wilson, K.G. and M.E. Fisher, 1972, Phys. Rev. Lett. **28**, 240.

Wilson, K.G. and J. Kogut, 1974, Phys. Rep. **12C**, N 2; Russian transl.: K. Wilson and J. Kogut, Renormalizatzionnaya gruppa i $\epsilon$-razlozhenie (Mir, Moscow, 1975).

Wright, G.B., ed., 1969, Light Scattering Spectra of Solids (Springer, New York).

Yacoby, Y., R.A. Cowley, T.J. Hosea, D.J. Lockwood and W. Taylor, 1978, J. Phys. **C 11**, 5065.

Yakovlev, I.A. and T.S. Velichkina, 1957, Usp. Fiz. Nauk, **63**, 411.

Yakovlev, I.A., T.S. Velichkina and L.F. Mikheeva, 1956a, Doklady Acad. Nauk SSSR, **107**, 675 (1956a, Sov. Phys. Doklady, **1**, 215).

Yakovlev, I.A., T.S. Velichkina and L.F. Mikheeva, 1956b, Kristallografiya, **1**, 123 (1956b, Sov. Phys. Crystallogr. **1**, 91).

Zawadowski, A. and J. Ruvalds, 1970, Phys. Rev. Lett. **24**, 1111.

CHAPTER 2

# Light Scattering Anomalies due to Defects

A.P. LEVANYUK

*A.V. Shubnikov Institute of Crystallography*
*Acad. Sci. USSR 117333 Moscow*
*USSR*

A.S. SIGOV

*Institute of Radioengineering, Electronics and Automation*
*117454 Moscow*
*USSR*

A.A. SOBYANIN

*P.N. Lebedev Physical Institute, Acad. Sci. USSR*
*117333 Moscow*
*USSR*

*Light Scattering near Phase Transitions*
*Edited by*
*H.Z. Cummins and A.P. Levanyuk*

© *North-Holland Publishing Company, 1983*

# Contents

## 1. Introduction

At the opening stage of the study of light scattering near phase transitions the temperature anomalies of the scattering had been associated only with the increase of thermal fluctuations of the order parameter when approaching a transition temperature. The scattering from various defects of crystal structure has been retained regardless of the field of action of the investigators and the reason seems to be that the scattering from defects has usually been supposed to be practically temperature-independent (independent of a proximity to a phase transition). The situation has changed, however, after the demonstration by Shapiro and Cummins (1968) that the "critical opalescence" discovered by Yakovlev et al. (1956) near the $\alpha \rightleftarrows \beta$ transition in quartz is mainly no more than elastic scattering, i.e., it is not connected with thermal fluctuations of the order parameter. An increase of intensity of the elastic scattering in the vicinity of structural phase transitions may be caused in principle both by the appearance of macroscopic inhomogeneities, as for instance, inclusions of another phase, and by the increase of contribution to the scattering from lattice defects.

As to the possibility of increase of contribution of microscopic inhomogeneities when approaching a phase transition it has been shown theoretically by Axe and Shirane (1973) with regard to neutron scattering. The analogous mechanism for scattering of light has been considered by Levanyuk et al. (1976). On the other hand, there has been one more reason to concentrate attention on defects: Halperin and Varma (1976) tried to explain anomalies in the low-frequency spectrum of light scattering (the so-called dynamic central peak) by means of an account of the influence of relaxing defects of the type of off-center ions. Thus, there exists a theoretical basis of the possible influence of defects on both dynamic and static central peaks. By now it is known that quite a number of experimental corroborations of a central peak are related with the presence of defects in a crystal (see, for example, Shirane and Axe (1971), Axe et al. (1974), Lagakos and Cummins (1974), Lyons and Fleury (1977), Durvasula and Gammon (1977), Courtens (1977, 1978), Yagi et al. (1977), Mermelstein and Cummins (1977), Prater et al. (1981)). Certain aspects of the problem of central peaks in imperfect crystals have been widely discussed in the review paper by Müller (1979).

It should be mentioned in general, that nowadays the study of light scattering near phase transitions in crystals containing defects is given much attention.

The problems under consideration are, in particular: the appearance of a central peak as well as Raman lines forbidden by the symmetry in the spectrum of thermal light scattering, an increase in light scattering intensity at phase transitions due to defects, widening of different spectral components, etc.

In the present chapter we emphasize the approach of independent (non-interacting) defects being valid if the average distance between defects exceeds the characteristic length of order-parameter distortions near a defect. Just for this case one manages to obtain the results of interest by means of relatively simple theoretical methods. At the same time some results are believed to be valid for systems with large concentrations of defects which are of great interest from the experimental point of view.

In sect. 2 we describe types of the defects under consideration and treat the character of temperature evolution of the defect structure near a phase transition. The main reason for such an evolution is the temperature dependence of the correlation-radius magnitude $r_c$ of the order parameter $\eta$. The correlation radius plays the part of the characteristic dimension of a crystal region distorted due to the influence of a defect. Since the correlation radius in a vicinity of a phase transition substantially exceeds an interatomic distance the crystal and the defect structure is described in the continuous-medium approximation.

Section 3 deals with the elastic light scattering by static (immobile) defects in various crystals and with some experimental results concerning the "static" central peak.

In sect. 4 is analyzed the influence on the light scattering spectrum due to dynamic (hopping) defects which can change their positions in an elementary cell. The presence of such defects may, at certain conditions, give rise to the appearance of the dynamic central peak in the scattered-light spectrum.

In the fifth section we discuss the changes in the light-scattering spectrum caused by the static defects. Such defects are shown to be able to lead to the appearance of new lines, including also the dynamic central peak, to strengthen the attenuation anomalies of acoustic and optic phonon, etc.

Our concluding remarks (sect. 6) are aimed at defining the modern situation in theoretical study of light scattering in crystals with defects near structural phase transitions. The unsolved problems arising from experimental investigations are distinguished.

## 2. *Types of defects and peculiarities of defect structure near phase transitions*

Consider firstly the theoretical classification of the defects which may have strong influence on the physical properties of crystals near a phase-transition temperature $T_c$. In a rather wide vicinity of $T_c$ the correlation radius of the order parameter is much larger than the interatomic distance, therefore the distortions corresponding to $\eta$ may be treated in the continuous-medium approximation. In the simplest case, when the order parameter is a single-component scalar quantity and long-range forces are absent, we describe crystal configuration introducing the Landau expansion for the free energy density (see ch. 1):

$$\Phi[\eta(r)] = \Phi_0 + \tfrac{1}{2}A\eta^2 + \tfrac{1}{4}B\eta^4 + \tfrac{1}{2}D(\nabla\eta)^2 + \ldots - h\eta\,, \tag{1}$$

where $h$ is the generalized field, conjugate to the order parameter $\eta$; the coefficient $A$ depends linearly on temperature, $A = A_0(T - T_c)/T_c \equiv A_0\tau$, and the coefficients $B, D, \ldots$ are independent of temperature in the framework of the Landau theory.

The presence of a defect may lead to local changes in the coefficients $A$, $B$, $D$ of eq. (1) and to the appearance of the local fields conjugate to the order parameter. As a rule, a real defect gives rise to simultaneous changes in a number of the coefficients of eq. (1) but it is convenient in a theoretical treatment to classify defects according to the separate kinds of distortions of a perfect lattice induced by them. A real defect can, in fact, belong to a few types simultaneously. Often considered separately are the defects which give rise to the appearance of local fields $h_d(\boldsymbol{r})$, and the ones which induce changes in the coefficients $A$, $B$, $D$, etc. In the first case one speaks about the defects of "random local field" type and in the second about the defects of "random local transition temperature" type (De Dominicis 1979, Levanyuk et al. 1980, Levanyuk and Sigov 1980).

Study first the defects of "random local field" type. In a microscopic model such a defect can be identified, for example, with the interstitial ion which pushes aside the ferroelectrically active ion and thus induces at the point of its location a nonzero value of the order parameter even in symmetrical (nonpolar) phase. Using the continuous medium approximation and denoting displacement of the ferroelectrically active ion by $\eta$ we describe the crystal configuration by means of the function $\eta(\boldsymbol{r})$ and not by the set of quantities $\eta_i$, where $i$ is the number of a cell. Naturally, the value of $\eta$ differs from zero in all neighbouring cells as well. On account of that, not very close to the interstitial ion the coordinate dependence of $\eta$ is given by the Ornstein–Zernicke function (see, e.g., Levanyuk et al. (1979)):

$$\eta - \eta_e \sim r^{-1}\exp(-r/r_c)\,, \tag{2}$$

where $\eta_e$ is the equilibrium value of the order parameter in a pure crystal and $r_c$ is the correlation radius of the order parameter. The disturbance caused by the interstitial impurity may be considered as being due to the field $h_d(\boldsymbol{r})$ conjugate to the order parameter and acting at the impurity location point. It is convenient to regard fluctuations in the impurity concentration as a random space inhomogeneity of the field $h_d(\boldsymbol{r})$. Naturally, the terms of higher odd powers of $\eta$ as well as the terms with space derivatives of $\eta$ are admissible in the "field-dependent" part of the free energy density due to defects, so that a variety of "random local field" types of defects is possible.

Now consider the defects of the "random local transition temperature" type. The influence of the defects of such a type can be taken into account by ascribing the concentration dependence to the coefficient $A$ in eq. (1):

$$A = A_{\mathrm{id}} + \Delta A_{\mathrm{d}} N(\boldsymbol{r}), \tag{3}$$

where $A_{\mathrm{id}} = A_0(T - T_{\mathrm{c,\,id}})/T_{\mathrm{c,\,id}}$ is the value of the coefficient $A$ in an ideal crystal, $\Delta A_{\mathrm{d}} = A_0(T_{\mathrm{c}} - T_{\mathrm{c,\,id}})/T_{\mathrm{c,\,id}}$ is the change in the coefficient $A_{\mathrm{id}}$ at the point of the defect location, and $N(\boldsymbol{r})$ is a random function having a different form for various systems. Of course, all the rest coefficients $B$, $D$, etc. must also depend on the coordinates but this circumstance does not involve any qualitative changes into results, so the free-energy density of the system may be written as follows:

$$\Phi[\eta(r)] = \tfrac{1}{2}A_{\mathrm{id}}\eta^2 + \tfrac{1}{4}B\eta^4 + \tfrac{1}{2}D(\nabla\eta)^2 + \tfrac{1}{2}\Delta A_{\mathrm{d}} N(\boldsymbol{r})\eta^2 - h\eta\,. \tag{4}$$

An example of the defect of the "random local transition temperature" type is the substitutional impurity ion. Consider the simplest ferroelectric phase transition arising from displacement of one ferroelectrically active ion in each cell of an ideal crystal. Let this ion be substituted by another one of the same valency. If this substitution takes place in the whole number of cells we shall obtain another ideal crystal whose transition temperature will differ from that of the initial crystal. By partial random substituting the crystal will contain regions with various concentration of impurity ions owing to inevitable fluctuations in concentration. The phase transition temperature must be different in these regions. The function $N(\boldsymbol{r})$ (see eqs. (3) and (4)) has the meaning of local concentration just for such a simple example of randomly distributed point defects. Naturally the random dispersion of local temperature of phase transition may also be caused by other reasons. Imagine, e.g., that in a system there are space-inhomogeneous stresses with a characteristic dimension of inhomogeneity considerably exceeding the interatomic distance $a$. The situation is specifically realized in the presence of dislocations. The properties of such systems are characterized by the free energy density function of the form (4) but the quantity $N(\boldsymbol{r})$ must be now understood as being proportional to the elastic stress, i.e., the cause of local dispersion of the phase transition temperature is its dependence on stress.

We mention that there is no impenetrable boundary between the defects of "random local field" and "random local transition temperature" types. Indeed, it seems reasonable to expect that when the increase in the transition temperature in a certain region whose dimension is less than $r_{\mathrm{c}}$ is large enough, one can treat the "random local transition temperature" system as a "random local field" (Schmidt and Schwabl 1977, 1978, Levanyuk et al. 1979, 1981). This is due to the fact that the static local lattice distortions corresponding to the order parameter would now arise near defects far above the phase transition temperature $T_c$. The phenomenon of the appearance of such distortions as a result of changes in temperature or in other external parameters is termed "local phase transition". The "local phase transition" is discussed below. In this chapter the defects are termed as ones of the "random local transition temperature" type if the "local phase transition" does not occur, in fact, up to $T_c$. Thus in symmetrical phase the defects of the "random local transition temperature" type mainly influence the order parameter fluctuations. So the manifestation of the defects of the "random local transition temperature" type is less marked than the manifestation of those of the "random local field" type which induce in a crystal lattice the static distortions corresponding to the order parameter even in symmetrical phase.

In a system with low defect concentration and not too close to the phase-transition temperature $T_c$ one can evaluate the defect contribution to the anomalies of various physical properties (including the light scattering) assuming the defects to be independent (noninteracting). Using the approach of independent defects to find the free energy of imperfect crystals one must calculate the contribution of a single isolated defect to the free energy and multiply the result by the whole number of the defects (or by their concentration $N$ to evaluate the specific free energy). So the independent defect approximation makes it possible to consider defect contributions as additive ones. Such an approach can be expected to fail over that range of temperature and concentration where the mean distance between defects is less than the correlation radius $r_c$, this radius being a scale of the order-parameter space inhomogeneity. In the range where $Nr_c^3 \gg 1$ the interaction between the defects cannot be neglected and one should use, for example, the procedure similar to the mean-field approximation.

In the rest of the section we discuss an isolated defect near a second-order phase transition. The aim of the consideration is to obtain a more precise formula for $\eta(\boldsymbol{r})$ than given by eq. (2), in particular, to investigate the temperature dependence of the coefficient in the Ornstein–Zernicke function, to take into account the influence of the long-range forces which arise due to space inhomogeneity in the order parameter in ferroelectrics and ferroelastics. The results obtained form the basis of the further consideration. It should be stressed that the results may be used for the treatment of both light scattering near $T_c$ and anomalies in thermodynamic quantities and kinetic coefficients (see, e.g., Levanyuk et al. (1979)).

We describe the substance in the continuum approximation, and the presence of defects is taken into account in the boundary conditions to the equations of the continuous medium (Levanyuk et al. 1976, 1979). These conditions are specified at the "core" of the defect, i.e., on the boundary of a spatial region with characteristic dimension $d$ where the continuum approximation is not valid (usually $d \sim r_{c0} \gtrsim a$). We emphasize that in the continuum theory the distortions introduced by the defect are usually taken to be the elastic ("acoustic") deformations only (see, e.g., Eshelby (1956), Kosevich (1972), Leibfried and Breuer (1978)), since the changes of the other internal parameters (such as "optical" deformations of a crystal) fall off as a rule approximately within interatomic distances. However, as already noted, near a phase transition the correlation radius of the order parameter $\eta$ increases substantially, so that these distortions can also be described in the continuous-medium approximation (Levanyuk et al. 1976, 1979). These are precisely the distortions of our interest here.

Consider first the point defect of "random local field" type and assume the boundary of its core to be spherical. The presence of a local field at the point of defect location is equivalent to the specification of the function $\eta(\theta, \varphi; d)$ on a sphere of radius $d$. Expanding the function $\eta(\theta, \varphi; d)$ in spherical functions one can formulate the simplest kinds of boundary conditions at the defect core assuming only one of the terms of this series to differ from zero. Imposing the first term is equivalent to fixing the order-parameter value $\eta = \eta_0$ at $r = d$ and imposing the second term is equivalent to fixing the first space derivative of the order parameter $\partial\eta/\partial z = (\eta_1 \cos\theta)/d$ at $r = d$. Considering these two cases we shall speak of the defect in the states $s$ and $p$ respectively. Of course, the procedure may be prolonged, i.e., the $d$, $f$, etc. states may be introduced in the same way. For our purposes it suffices to consider the first two terms of the expansion. Naturally, the amplitudes $\eta_0$ and $\eta_1$, can differ from zero simultaneously (defect in the state $sp$). The schemes of microscopic models of the defect in the states $s$ and $p$ are illustrated in fig. 1 with the displacive ferroelectric phase transition as an example. Similarly formulated are the boundary conditions for linear (one-dimensional) defects such as dislocations, as well as for planar (two-dimensional) defects such as twin boundaries or crystal grain boundaries and crystal surfaces. In the former case one imposes the function $\eta(\varphi; d)$ on a cylindrical surface of radius $d$ and in the latter the values of $\eta_1$, and $\eta_2$ on the planes bounding the core of the defect. The states $s$ and $p$ of the linear defect correspond, respectively, to the first ($m = 0$) and second ($m = 1$) terms of the expansion of the function $\eta(\varphi; d)$ in $\cos(m\varphi)$. The state $s$ of the planar defect corresponds to the boundary condition $\eta_1 = \eta_2 = \eta_0$ and the state $p$ is defined by the condition $\eta_1 = -\eta_2 = \eta_0$. It is obvious that the planar defect can be only in the states $s$, $p$ or $sp$.

The proposed classification of states of the defects of "random local field" type can easily be extended to the defects of "random local transition temperature" type as it will be shown below. Mention that the classification is simple and useful for theoretical analysis of imperfect crystal problems, but it is not absolutely universal. For example the charged defect in a ferroelectric is difficult to be described within the framework of the given classification. We assume for the beginning that the defect is in the state $s$. Submit the free energy of the substance with such a defect in the form

$$F = \int_{V-V_0} \Phi[\eta(\boldsymbol{r})]\,\mathrm{d}V + [f(\eta_0) - h\eta_0]V_0, \tag{5}$$

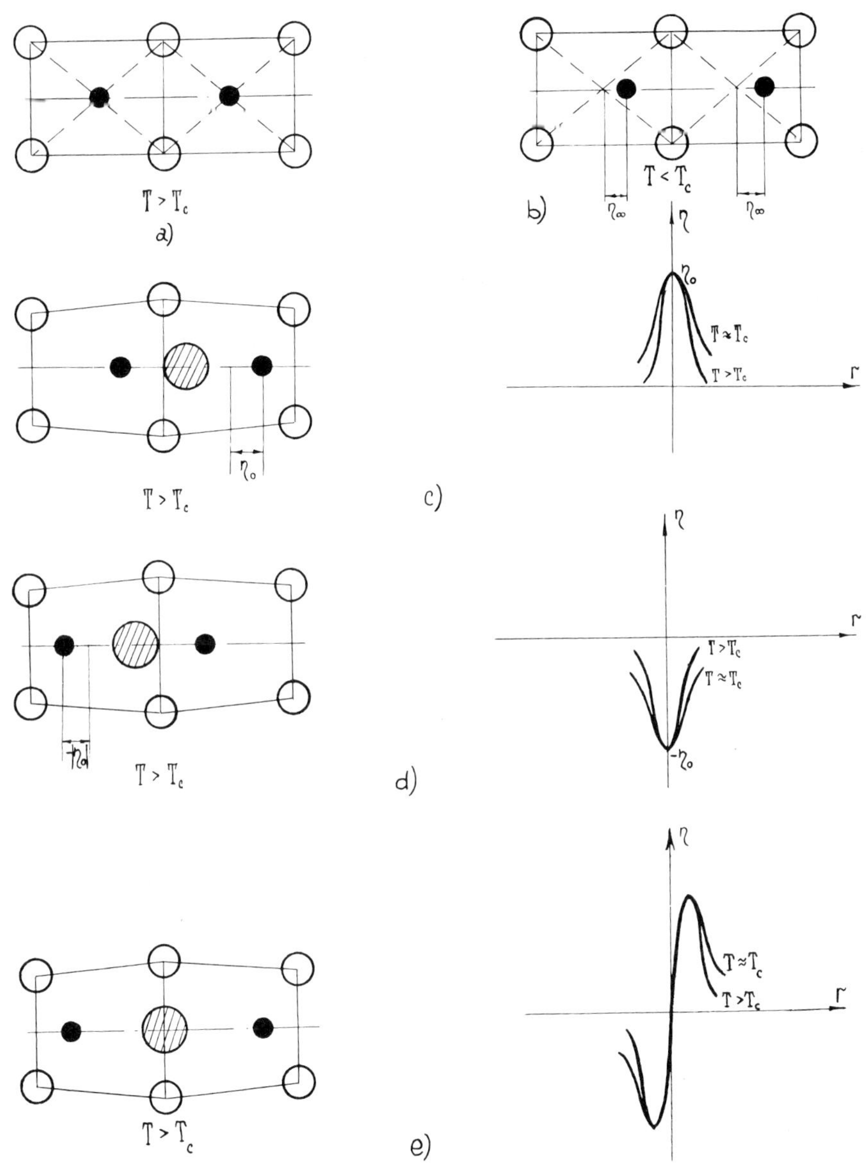


Fig. I. Schematic plane picture of crystal distortions in the vicinity of point defects of "random local field type" (interstitial atoms): (a) an ideal unit cell of symmetrical phase, $T > T_c$; (b) an ideal unit cell of nonsymmetrical phase, $T < T_c$; (c) (on the left) a defect in the state $s$ with $\eta_0 > 0$ and (on the right) the coordinate dependence of the defect-induced displacement of ferroelectrically active ion, obtained in the continuum approximation; (d) (on the left) a defect in the state $s$ with $\eta_0 < 0$ and (on the right) the coordinate dependence of the defect-induced displacement of ferroelectrically active ion, obtained in the continuum approximation; (e) (on the left) a defect in the state $p$ and (on the right) the coordinate dependence of the defect-induced displacements of ferroelectrically active ions, obtained in the continuum approximation. We see that defects of "random local field" type give rise to local appearance of the order parameter both in symmetrical and nonsymmetrical phase.

where the integrand representing the free-energy density of the host matrix is expressed by eq. (1) and the second term is the free energy of the defect core of volume $V_0$.

The form of the function $f(\eta_0)$ is determined from symmetry considerations. We write the function $f(\eta_0)$ in the form of a series in powers of $\eta_0$, i.e., in the form of an expansion about the point $\eta_0 = 0$ which usually corresponds to the most symmetrical state of the defect (an example of such a state is shown in fig. 1(e) (left part)). In the symmetrical phase the function $f(\eta_0)$ must be symmetrical with respect to the transformations belonging to the symmetry group of the crystal with the defect. If this group contains transformations altering the sign of $\eta_0$, then the expansion of $f(\eta_0)$ includes only even powers of $\eta_0$:

$$f(\eta_0) = f_0 + \tfrac{1}{2}\alpha\eta_0^2 + \tfrac{1}{4}\beta\eta_0^4 + \dots. \tag{6}$$

It is just the form which is appropriate also for defects of the "random local transition temperature" type because the coefficient $\alpha$ may be identified with the quantity $\Delta A_d$ in formula (4). For other groups the expansion will also include the summands with odd powers of $\eta_0$ and the sign of the coefficients at these terms can differ for defects of different localization:

$$f(\eta_0) = f_0 \pm |\alpha_1|\eta_0 + \tfrac{1}{2}\alpha_2\eta_0^2 + \dots. \tag{7}$$

Such defects are of course of the "random local field" type because the coefficient $\alpha_1$ may be identified with the extra field $h_d$ induced by the defect at the place of its localization.

We explain the meaning of eqs. (6) and (7) by an example of structural transition in a crystal whose symmetry elements that reverse the sign of $\eta$ are screw axes. Obviously, these symmetry elements cannot be conserved in the presence of any point defect in an arbitrary site of the crystal, i.e., the symmetry of the "crystal + defect" system admits the onset of a nonzero order parameter. In our language this means that $\eta_0 \neq 0$. We now subject the "crystal + defect" system to a screw rotation that belongs to the symmetry group of the ideal crystal. The defect then turns out to be located in another site and the sign of $\eta_0$ (and, consequently, the sign of the coefficients at terms with the odd powers of $\eta_0$) changes. In the case of a multi-component order parameter there exists, surely, the greater variety of types of the function $f(\eta_0)$.

Revert now to eq. (5). Varying the functional $F$ with respect to $\eta(\boldsymbol{r})$ at a given $\eta_0$ we obtain an equation describing the distribution $\eta(\boldsymbol{r})$,

$$D\nabla^2\eta = A\eta + b\eta^3 + \dots - h, \tag{8}$$

with boundary conditions,

$$\eta|_{r=d} = \eta_0, \quad \eta|_{r\to\infty} = \eta_\infty, \tag{9}$$

where the quantity $\eta_\infty$ is determined from the equality,

$$A\eta_\infty + B\eta_\infty^3 + \ldots = h\,. \tag{10}$$

The first of the conditions (9) corresponds to a point defect in the state $s$ Equation (8) is nonlinear and cannot be solved exactly in three- and two-dimensional cases. Linearizing it in the vicinity of $\eta = \eta_\infty$ one sees that for point defects in the state $s$ the solution is written as the well-known Ornstein–Zernike function,

$$\eta(\mathbf{r}) - \eta_\infty = (\eta_0 - \eta_\infty)(d/r)\exp[-(r-d)/r_c], \quad r > d\,, \tag{11}$$

$$r_c = \begin{cases} \sqrt{D/A} & \text{at } T > T_c\,, \\ \sqrt{D/2|A|} & \text{at } T < T_c\,. \end{cases}$$

By iteration method we can calculate the nonlinear corrections to solution (11) and therefore to the free energy $F^{(0)}$ found in the linear approximation as well (Levanyuk et al. 1979):

$$F^{(0)} = 2\pi Dd(1 + d/r_c)(\eta_0 - \eta_\infty)^2\,, \tag{12}$$

$$F^{(1)} = 4\pi(\eta_0 - \eta_\infty)^3 B\eta_\infty d^3 \ln(d/r_c) + \pi(\eta_0 - \eta_\infty)^4 Bd^3\,. \tag{13}$$

Comparison of the values $F^{(0)}$ and $F^{(1)}$ yields a criterion for the applicability of the linear approximation,

$$\left(\frac{Bd^2\eta_0^2}{D}\ln\frac{d}{r_c}\right)^{1/2} \approx \eta_0/\eta_{at} \ll 1\,. \tag{14}$$

It should be mentioned that in the region of applicability of the Landau theory the quantity $\eta_{at} \equiv (D/Bd^2)^{1/2}$ has the meaning of the atomic (maximum possible) value of the order parameter corresponding to total ordering for "order–disorder" phase transitions, to displacement of the sublattices by an atomic distance for displacive transitions, etc. Therefore, the linear approximation (11) appears to be valid even for relatively "strong" defects.

The $\eta(\mathbf{r})$ distribution near a point defect in the state $p$ can be obtained by differentiating eq. (11) with respect to $z$ and satisfying the boundary condition $\eta(\theta, r = d) = \eta_1 \cos\theta$. As a result we have

$$\eta(\mathbf{r}) = \eta_\infty\left[1 - \frac{d}{r}\exp\left(\frac{d-r}{r_c}\right)\right] - \eta_1\frac{d^2\cos\theta}{1 + d/r_c}\frac{\partial}{\partial r} \times \left[\frac{1}{r}\exp\left(\frac{d-r}{r_c}\right)\right]. \tag{15}$$

We see that the distribution $\eta(\mathbf{r})$ for a $p$ defect below $T_c$ has always an admixture of the $s$ state.

For a linear defect (dislocation) in the state $s$ it is also easy to obtain the solution of the linearized equation (8):

$$\eta = \eta_\infty + (\eta_0 - \eta_\infty)\frac{K_0(\rho/r_c)}{K_0(d/r_c)}\,. \tag{16}$$

Here $\rho$ is a distance to the axis of a defect, $K_0(z)$ is a cylindrical function of imaginary argument. We have $K_0(z) \approx (\pi/2z)^{1/2} \exp(-z)$ as $z \to \infty$ and $K_0(z) \approx -\ln(z/2)$ as $z \to 0$.

A criterion for the applicability of the linear approximation has here the form:

$$(Br_c^2 \eta_0 \eta_\infty / D)^{1/2} \approx (\eta_0/\eta_\infty)^{1/2} \ll 1 \,, \tag{17}$$

i.e., the given approximation is valid only for sufficiently "weak" linear defects. The criterion (17) is likewise valid in the case of planar defects (twin boundary or crystal-grain boundary) but now the one-dimensional solution $\eta(x)$ can be calculated exactly. At $d = 0$ we find:

$$\eta = \begin{cases} \sqrt{2A/B}\, \sinh^{-1}\left(\dfrac{x - x_0}{r_c}\right), & T > T_c\,, \\ \sqrt{|A|/B}\, \coth\left(\dfrac{x - x_0}{r_c}\right), & T < T_c\,, \end{cases} \tag{18}$$

where

$$x_0 = \begin{cases} -\dfrac{r_c}{2} \ln\left|\dfrac{\sqrt{1 + (B/2A)\eta_0^2} + 1}{\sqrt{1 + (B/2A)\eta_0^2} - 1}\right|, & \text{at} \quad T > T_c\,, \\ \dfrac{r_c}{2} \ln\left|\dfrac{\eta_0 - \sqrt{|A|/B}}{\eta_0 + \sqrt{|A|/B}}\right|, & \text{at} \quad T < T_c\,. \end{cases}$$

Equations (16) and (18) make it possible to write down the order parameter distribution for linear and planar defects in the state $p$. However, for shortness we do not adduce them here.

In proper ferroelectrics and proper ferroelastics as well as in ferromagnets a change in $\eta(\boldsymbol{r})$ is accompanied by the appearance of long-range forces – electric, elastic or magnetic, respectively. For ferromagnets this circumstance can be disregarded down to very small values of $|\tau|$, but for uniaxial ferroelectrics and ferroelastics it influences the results strongly. In the simplest case of one-component order parameter whose alterations do not induce the long-range forces the space region disturbed by a point defect can be considered to have a spherical form. The long-range forces cause essential changes in a form of the region. For instance, in the case of an extremely anisotropic ferroelectric with one $z$-axis of spontaneous polarization ($\eta \equiv P_z$) the linearized $P_z(\boldsymbol{r})$ distribution for a point defect in the state $s$ has the form of an oblong ellipsoid of revolution with its major axis parallel to spontaneous polarization. The semimajor axis of the ellipsoid equals approximately to $r_c$, the semiminor one is of the same order of magnitude as $d$. As $T \to T_c$ the semimajor axis elongates and the semiminor one does not change its length. Using the Fourier transform one can easily obtain the function $P_z(\boldsymbol{k})$ (Levanyuk et al. 1978):

$$P_z(\boldsymbol{k}) = (P_{z0} - P_{z\infty})(A + Dk^2 + 4\pi k_z^2/k^2)^{-1}\,, \tag{19}$$

but the solution $P(\boldsymbol{r})$ cannot be represented in an explicit form. Of analogous character is the picture of perturbances due to a defect in the state $s$ in ferroelastics with a two-component order parameter.

In a proper ferroelastic with one-component order parameter the point defect induces perturbances of even more complicated form. Indeed, in the continuum theory of anysotropic elastic medium the space distribution of deformations near a defect has been obtained in an explicit form only for a lattice with hexagonal symmetry, but a ferroelastic phase transition in hexagonal crystal is described by the two-component order parameter. Assume $\eta \equiv u_{xy}$ and fix the value of this component of strain tensor on the defect core. Supposing the expression for the energy corresponding to all remaining components of elastic strain tensor to be the same as in the isotropic case we have (Levanyuk and Sigov 1979):

$$\eta(\boldsymbol{k}) = C(\eta_0 - \eta_\infty)\frac{[(n_x^2 - n_y^2)^2 + n_z^2(n_x^2 + n_y^2)](\lambda + 2\mu) + \mu(n_x^2 + n_y^2)}{A + Dk^2 + 4\mu n_z^2 + 4\mu(\lambda + \mu)n_x^2 n_y^2/(\lambda + 2\mu)}, \tag{20}$$

where the constant $C$ is determined from the condition $\eta_0 - \eta_\infty = \int \eta(\boldsymbol{k})\,\mathrm{d}\boldsymbol{k}$, $\boldsymbol{n} = \boldsymbol{k}/k$, $\lambda$ and $\mu$ are the Lame coefficients, and $A$ denotes the elastic modulus corresponding to the component $u_{xy}$.

Let us now take into account the dependence of $\eta_0$ on temperature (on proximity $\tau$ to a phase transition). The temperature dependence of the value of $\eta_0$ can be found from the condition of minimum of the free energy of a substance with the defect as a function of $\eta_0$. Such an approach is well known for plane defects (Sobyanin 1971, Kaganov and Omelyanchuk 1971, Kaganov 1972, Mills 1972, Bray and Moore 1977). Here we consider another simple example of a structural phase transition in the absence of long-range forces in a system with point defect in the state $s$. Take at first the free energy of the defect core to be characterized by the function (7). Minimizing the free energy $F(\eta_0)$ and taking into account eqs. (12) and (7) we get to the equilibrium value $\eta_{0\mathrm{e}}$:

$$\eta_{0\mathrm{e}} = \mp\frac{|\alpha_1|}{\alpha_2}\frac{\xi}{1 + \xi + d/r_\mathrm{c}} + \eta_\infty\frac{1 + d/r_\mathrm{c}}{1 + \xi + d/r_\mathrm{c}}, \tag{21}$$

where $\xi \equiv \alpha_2 d^2/3D$ is a parameter that characterizes the "stiffness" of the defect core. For very "stiff" defects ($\xi \to \infty$) the value of $\eta_{0\mathrm{e}}$ is equal to $\mp|\alpha_1|/\alpha_2$ and is independent of $\tau$. Close to a phase transition ($\tau \to 0$) the expression for $\eta_{0\mathrm{e}}$ can be written in the form

$$\eta_{0\mathrm{e}} = \pm\,\eta_{00}\left(1 - \frac{d/r_\mathrm{c}}{1 + \xi}\right) + \frac{1}{1 + \xi}\eta_\infty, \tag{22}$$

where $\eta_{00} \equiv (|\alpha_1|/\alpha_2)\xi/(1 + \xi)$ is the value of $\eta_{0\mathrm{e}}$ at $T = T_\mathrm{c}$. It follows from eq. (22) that in a symmetrical phase the absolute value of $\eta_{0\mathrm{e}}$ increases as $\tau \to 0$ in proportion to (const $- \tau^{1/2}$). Such a dependence of $\eta_{0\mathrm{e}}$ should, in particular, be observed in magnetic resonance experiments. In a nonsymmetrical phase the temperature dependence of $\eta_{0e}$ is determined mainly by the behaviour of the last term in formula (22) and is different for defects of different polarization.

If we deal with defects to which expression (6) corresponds, the value of $\eta_{0e}$ is determined from the equation:

$$\eta_0[\alpha + \Delta\alpha(1 + d/r_c)] + \beta\eta_0^3 = \Delta\alpha(1 + d/r_c)\eta_\infty \,, \qquad (23)$$

where $\Delta\alpha \equiv 3D/d^2$. At $\alpha + \Delta\alpha > 0$ (we assume the coefficient $\beta$ to be always positive) the equilibrium value of $\eta_0$ is zero in a symmetrical phase, while in nonsymmetrical phase:

$$\eta_{0e} \approx \frac{1 + d/r_c}{1 + \xi + d/r_c}\eta_\infty \,, \qquad (24)$$

where, as before, $\xi \equiv \alpha/\Delta\alpha = \alpha d^2/3D$ is the parameter of the "stiffness" of the defect. We see that defects with $\alpha + \Delta\alpha > 0$ contribute to anomalies of physical properties of the substance near $T_c$ mainly in nonsymmetrical phase. For $\alpha + \Delta\alpha < 0$ the value of $\eta_{0e}$ is finite at $T = T_c$ and is equal to

$$\eta_{0e} \equiv \pm\eta_{00} = \pm\left(-\frac{\alpha + \Delta\alpha}{\beta}\right)^{1/2} . \qquad (25)$$

As before, the temperature dependence of $\eta_{0e}$ near phase transition for such strong defects is given by formula (22) with $\xi = (\alpha + 3\beta\eta_{00}^2)d^2/3D$.

It can be seen from eq. (23) that at conversion of the coefficient at the first power of $\eta_0$ into zero a distinctive "phase transition" in the defect must take place, accompanied by the appearance, around the defect, of distortions corresponding to the order parameter. The temperature $T_{cd}$ of this "local phase transition" is determined from the condition

$$\alpha + \Delta\alpha(1 + d/r_c) = 0 \,. \qquad (26)$$

The phenomenon of such a "local phase transition" has been theoretically considered in a number of papers (Nagaev 1968, Suhl 1975, Lyubov and Solov'ev 1965, Andreev et al. 1977, Höck and Thomas 1977, Schmidt and Schwabl 1977, 1978, Kristoffel 1979a, b, Bulaevskii et al. 1978) referring to concrete examples of various systems, and apparently it has also been observed experimentally (Blinc 1977, Müller et al. 1976). Within the framework of the analysis presented above this "transition" is of second order, since the quantity $\eta_{0e}$ is a continuous function of temperature $\eta_{0e} \sim (T_{cd} - T)^{1/2}$. Of course, a "phase transition" of the first order is also possible in the vicinity of the defect, when the onset of $\eta_{0e}$ takes place jumpwise. The coefficient $\beta$ in expression (6) is then negative and terms of sixth order must be taken into account in the expansion of the function $f(\eta_0)$.

According to condition (26) the magnitude of $T_{cd}$ is influenced by the temperature-dependent value of the correlation radius $r_c$, i.e., by the coefficient

$A$ in expansion (1). Hence, there exists some relation between the "local phase transition" and the phase transition in the matrix (Höck and Thomas 1977, Schmidt and Schwabl 1977, Bulaevskii et al. 1978, Kristoffel 1979a), although the quantities $T_{cd}$ and $T_c$ can differ substantially. Naturally, the "local phase transition" could in principle be connected with the temperature dependence of the coefficients $\alpha$ and $\beta$ irrespective of the temperature dependence of the matrix parameters (Kristoffel (1979b) as well as the other authors cited above).

We emphasize that the term "phase transition" as applied to a restructuring of a defect should be understood only figuratively. Actually, this "transition" affects only a limited number of atoms and the anomalies connected with a phase transition are well known to become smoothed out in a finite system. This smoothing can be taken into account after the analogy of the procedure used, e.g., by Schmidt (1966) and Bray (1974). When calculating the contribution of defects to the free energy of a system we now use the exact formula

$$\Delta\tilde{F}_d = -k_B T \ln\left\{\int \exp[-\Delta F_d(\eta_0)/k_B T]\, d\eta_0\right\}, \tag{27}$$

instead of minimization with respect to $\eta_0$. Then we obtain for the temperature width of the smoothing region,

$$\frac{\Delta T}{T_{cd}} \sim \left(\frac{8}{3\pi}\frac{T_{cd}-T_c}{T_{cd}}\frac{k_B T_c \beta}{D d A_0}\right)^{1/2}. \tag{28}$$

Estimating the value of the ratio $\Delta T/T_{cd}$ one sees that it can be actually less than unity only in the case, when the defect core radius $d$ considerably exceeds an interatomic distance. Thus, it is hardly possible to discern the local phase transition at the defect through the instrumentality of accompanying anomalies of physical quantities of a system. Note, however, that for defects in a uniaxial ferroelectric or in a proper ferroelastic the temperature interval of the smoothing may turn out to be much smaller, since the coefficient of $\eta_0$ in eq. (23) has, far from $T_c$, a stronger temperature dependence of the type $c_1 + c_2\tau$.

In nonsymmetrical phase, as seen from eq. (12), the defect states corresponding to different signs of $\eta_0$ differ in energy. Far enough from a phase transition, at

$$\eta_0 > \frac{2}{3}\frac{|\alpha + \Delta\alpha(1+d/r_c)|^{3/2}}{\Delta\alpha(1+d/r_c)\sqrt{3\beta}}, \tag{29}$$

one of the minima of the function $F(\eta_0)$ vanishes and all the defects acquire the same sign of $\eta_0$. Naturally, the decrease of the number of defects in the metastable state due to thermal and quantum fluctuations takes place before reaching the point of loss of stability of this state as well. Lowering the temperature below the point of stability loss of the metastable state can serve as a method for obtaining a system of fully polarized defects in crystals with structural phase transitions.

The existence of a local order near a defect in symmetrical phase, i.e., the fact that at $\alpha + \Delta\alpha < 0$ the equilibrium value of $\eta_0$ can differ from zero and has a comparatively weak dependence on temperature in a vicinity of phase transition giving rise, when obtaining qualitative results and estimating the anomalies of physical properties near $T_c$, to assume the order-parameter value at the defect core to be fixed and equal to $\eta_0 \equiv \eta_0(T_c)$.

The phenomenological approach to an investigation of a state of immobile defects near structural phase transitions discussed above is, of course, rather simplified, but it has such advantages, as straightforwardness, obviousness and necessity of only a few inherent parameters (the dimension $d$ of the core of the defect, the "strength" $\eta_{00}$ of the defect, the "stiffness" $\xi$ of the defect, as well as the concentration $N$ of defects). These advantages provide us with ample opportunities to apply this approach to study the influence of defects on various anomalies of physical properties of a crystal.

We now turn to the consideration of light-scattering anomalies in imperfect crystals.

## 3. *Elastic scattering from immobile defects*

As a rule an average distance between the defects ($N^{-1/3}$) is small in comparison with a light wavelength. Therefore, when considering light scattering from defects, one must assume the space inhomogeneity of components of the dielectric permeability tensor $\epsilon_{ij}$ to be caused by fluctuations in the defect concentration. For the overwhelming majority of crystals with structural phase transitions there is practically no diffusion of the defects in the temperature interval of interest to us and the fluctuations in $N$ can be regarded as independent of temperature. At the same time, the fluctuations in the components $\epsilon_{ij}$ increase as $\tau \to 0$, owing to the temperature dependence of the contribution made to $\epsilon_{ij}$ by an individual defect (Levanyuk et al. 1976). The increase of space dimensions of the crystal region distorted by the defect leads to an increase of the cross section for scattering of light by a single defect and, consequently, to an increase of the intensity of scattering due to fluctuations in the defect concentration (Levanyuk et al. 1976).

It is well known that the spectral intensity of scattered light $I(\boldsymbol{q}, \Omega)$ is determined by the Fourier transform of the correlation function of space–time variations in $\epsilon_{ij}$ (see ch. 1):

$$I(\boldsymbol{q}, \Omega) = VQ_S\langle \Delta\epsilon_{ij}(\boldsymbol{q}, \Omega)\Delta\epsilon_{ij}(-\boldsymbol{q}, -\Omega)\rangle \,. \tag{30}$$

In formula (30) the brackets $\langle \ldots \rangle$ designate a statistical averaging,

$\boldsymbol{q} = \boldsymbol{k}_{\mathrm{I}} - \boldsymbol{k}_{\mathrm{S}}$, $\Omega = \omega_{\mathrm{I}} - \omega_{\mathrm{S}}$, where $\boldsymbol{k}_{\mathrm{I}}, \omega_{\mathrm{I}}$ and $\boldsymbol{k}_{\mathrm{S}}, \omega_{\mathrm{S}}$ are, respectively, wavevectors and frequencies for incident and scattered light waves within a crystal, $V$ is the scattering volume, $Q_{\mathrm{S}}$ is the geometric factor, written explicitly in ch. 1.

Let us discuss in the beginning the most widespread case of quadratic coupling between an appropriate component of the tensor $\epsilon_{ij}$ and the order parameter $\eta$ (a linear coupling between certain components of $\epsilon_{ij}$ and $\eta$, in the case of structural phase transitions, occurs only in proper ferroelastics, which will be considered separately). Omitting for brevity the tensor indices we have

$$\Delta\epsilon(\boldsymbol{r}, t) \equiv \epsilon(\boldsymbol{r}, t) - \epsilon^0 = a\eta^2(\boldsymbol{r}, t)\,, \tag{31}$$

where $\epsilon^0$ is a spatially independent equilibrium value of $\epsilon$ for a given homogeneous medium. For a crystal with an isolated defect one can write

$$\eta(\boldsymbol{r}, t) = \eta_{\mathrm{e}}(\boldsymbol{r}) + \eta'(\boldsymbol{r}, t)\,. \tag{32}$$

where $\eta_{\mathrm{e}}(\boldsymbol{r})$ is the equilibrium value of the order parameter and $\eta'(\boldsymbol{r}, t)$ is the fluctuating part of $\eta$. Substituting eqs. (31) and (32) into formula (30) we obtain

$$\begin{aligned} I(\boldsymbol{q}, \Omega) = VQ_{\mathrm{S}}a^2 \Big\{ & \sum_{\boldsymbol{k}, \boldsymbol{k}'} \eta_{\mathrm{e}}(\boldsymbol{k})\eta_{\mathrm{e}}(\boldsymbol{q} - \boldsymbol{k})\eta_{\mathrm{e}}(\boldsymbol{k}')\eta_{\mathrm{e}}(-\boldsymbol{q} - \boldsymbol{k}')\delta(\Omega) \\ & + 4\sum_{\boldsymbol{k}} \eta_{\mathrm{e}}(\boldsymbol{q} - \boldsymbol{k})\eta_{\mathrm{e}}(\boldsymbol{k} - \boldsymbol{q})\langle \eta'(\boldsymbol{k}, \Omega)\eta'(-\boldsymbol{k}, -\Omega)\rangle \\ & + 2\sum_{\boldsymbol{k}} \int \mathrm{d}\Omega' \langle \eta'(\boldsymbol{k}, \Omega')\eta'(-\boldsymbol{k}, -\Omega')\rangle \\ & \times \langle \eta'(\boldsymbol{q} - \boldsymbol{k}, \Omega - \Omega')\eta'(\boldsymbol{k} - \boldsymbol{q}, \Omega' - \Omega)\rangle \Big\}\,. \end{aligned} \tag{33}$$

The first sum in this expression corresponds to the elastic scattering of light by static space inhomogeneities of the order parameter, induced by defects. The second term in eq. (33) determines the intensity of the first-order scattering by $\eta$ fluctuations in a crystal containing defects, and the third term describes the second-order scattering by $\eta$ fluctuations. When writing down the last term in eq. (33) we neglect the non-Gaussian character of the fluctuations (for details see ch. 1).

In the present section we discuss only the light scattering by static inhomogeneities of the order parameter. The spectral distribution of the intensity of such a scattering takes the form of a central peak (it becomes evident from eq. (33)) whose width is infinitely small. In order to evaluate the integrated intensity of elastic scattering from defects we find first the contribution of an isolated defect to the dielectric permeability of a crystal.

Taking into account the quadratic coupling between the values $\Delta\epsilon$ and $\eta$,

$$\Delta\epsilon_{\mathrm{d}} = (a/V) \int [\eta^2(\boldsymbol{r}) - \eta_\infty^2]\,\mathrm{d}V\,. \tag{34}$$

The local variation of $\epsilon$ due to defect concentration fluctuations in the volume $V$ equals

$$\Delta\epsilon(\boldsymbol{r}) = V\Delta\epsilon_{\mathrm{d}}\Delta N(\boldsymbol{r})\,, \tag{35}$$

and the integral intensity of elastic light scattering by defects has the form

$$I_{\mathrm{d}}(\boldsymbol{q}) = Q_{\mathrm{S}}V(\Delta\epsilon_{\mathrm{d}}V)^2\langle|\Delta N(\boldsymbol{q})|^2\rangle\,. \tag{36}$$

As well as in the previous section we consider here noninteracting ($Nr_{\mathrm{c}}^3 < 1$) chaotically distributed defects. Hence the correlation between positions of defects is absent, and

$$\langle|\Delta N(\boldsymbol{q})|^2\rangle = \langle N\rangle/V \equiv N/V\,,$$

i.e., the integrated contribution of the system of defects is an additive quantity:

$$I_{\mathrm{d}} = Q_{\mathrm{S}}(\Delta\epsilon_{\mathrm{d}}V)^2N\,. \tag{37}$$

We stress once more that the simple result obtained is only valid within the framework of applicability of the independent-defect approximation: $Nr_{\mathrm{c}}^3 < 1$. It is natural to expect that, starting with $Nr_{\mathrm{c}}^3 \sim 1$, the growth of the intensity will become weaker or will even cease to exist due to overlapping of the crystal regions disturbed by defects. Using eq. (34) and the corresponding formulae for the distribution of the order parameter given in the previous section one can easily get to the values of $\Delta\epsilon_{\mathrm{d}}$ in the case of point defects in the state $s$ (Levanyuk et al. 1979):

(a) in the absence of long-range forces

$$\Delta\epsilon_{\mathrm{d}} \approx (2\pi a d^2/V)(\eta_0 - \eta_\infty)^2 r_{\mathrm{c}}\,, \tag{38}$$

(b) for proper ferroelectrics with three equivalent axes of spontaneous polarization

$$\Delta\epsilon_{\mathrm{d}} \approx (\pi a d^2/V)(P_{z0} - P_{z\infty})^2 r_{\mathrm{c}}\,, \tag{39}$$

(c) for uniaxial proper ferroelectrics

$$\Delta\epsilon_{\mathrm{d}} \approx (ad^2/2V)(\pi D)^{1/2}(P_{z0} - P_{z\infty})^2|\ln(Ad^2/16\pi D)|\,, \tag{40}$$

(d) for proper ferroelastics with the one-component order parameter

$$\Delta\epsilon_{\mathrm{d}} = a(\eta_0 - \eta_\infty)^2(c_1 - c_2 r_{\mathrm{c}}^{-1}), \quad c_1, c_2 = \mathrm{const}\,, \tag{41}$$

(e) for proper ferroelastics with the two-component order parameter

$$\Delta\epsilon_{\mathrm{d}} = a(\eta_{10} - \eta_{1\infty})^2 c\,|\ln(Ad^2/D)|, \quad c = \mathrm{const}\,. \tag{42}$$

In the case of point defects in the state $p$ in systems without long-range forces accompanying the order parameter variations, the contribution $\Delta\epsilon_d$ displays the same temperature behaviour as follows from eq. (41).

Substituting eqs. (38)–(42) into formula (37) one can easily evaluate the temperature dependence of the defect contributions to the integral intensity at $T > T_c$:

$I_d \sim \tau^{-1}$ for the cases (a) and (b);

$I_d \sim \ln^2 \tau$ for the cases (c) and (e);

$I_d \sim (c_1 - c_2\tau^{1/2})^2$ for the case (d).

The last result characterizes also the temperature dependence of the contribution due to point defects in the state $p$ for a system without long-range forces accompanying space variations of $\eta$.

The contributions of linear and plane defects can be treated in an analogous way. It is clear from eqs. (16) and (18) that linear and plane defects of appropriate orientation must in any case give rise to more strong anomalies of the elastic scattering of light than those due to the presence of point defects in the same state. As to the influence of long-range forces caused by space variations of the order parameter, they are obvious to lead to weakening of the scattering anomalies induced by defects.

For proper ferroelastics a linear coupling between some components of the dielectric-permeability tensor $\epsilon_{ij}$ and the order parameter is always allowed. In this case it is of interest to find the $\boldsymbol{q}$-dependence ($\boldsymbol{q}$ is the scattering wavevector) of the contribution of the defects to the intensity of light scattering due to fluctuations in those $\epsilon_{ij}$ components which depend linearly on $\eta$. Using eq. (30) one has

$$I_d = VQ_S a_1^2 N \eta_e(\boldsymbol{q}) \eta_e(-\boldsymbol{q}) \,. \tag{43}$$

Here $a_1$ is the constant of a linear coupling between $\epsilon_{ij}$ and $\eta$, and $\eta_e(\boldsymbol{q})$ is the Fourier-transform of the equilibrium value of the order parameter induced by an isolated defect. It is natural that considering the angular dependence of the scattering intensity in a real crystal one must also take into account the angular dependence of the geometric factor $Q_s$).

Yagi et al. (1976, 1977) and Tanaka et al. (1978) have experimentally observed a strong increase in integral intensity of light scattering when approaching the phase-transition temperature in the hydrogen-bonded ferroelastic $KH_3(SeO_3)_2$. The strong divergence of the studied central peak near $T_c$ in the Brillouin-scattering spectra may be explained by the presence of defects: not only point but linear or plane defects as well as by formation of interfaces in the region of coexistence of both phases.

The theoretical study of a static central peak near ferroelastic phase transition has been performed by Chages and Blinc (1979). The authors consider the central peak caused by the influence of dislocations and find the strong temperature dependence giving a good fit to experimental data available. Nevertheless, the results obtained do not correspond to a real situation because, when minimizing thermodynamic potential, Chaves and Blinc neglect the long-range character of elastic fields arising at spatially inhomogeneous distribution of the order parameter. On account of this reason one should carry out the analysis anew.

Recently Sawafuji et al. (1979) have published the results of observation of the Brillouin and Rayleigh scattering spectra near the transition temperature in the paraelectric phase of $KD_2PO_4$. There has been discovered the strong dependence of scattering intensity $I_d$ on the direction of the scattering wavevector. As to the theoretical interpretation given in the reference cited, the strong angular dependence of the intensity is considered to be caused by charged impurities in the presence of strong anisotropy of the dielectric constant along with the piezoelectricity. However we believe that the consideration is not consistent enough and at present it is hard to speak about an agreement between the experimental results and theoretical conclusions because a precise theory has not been developed yet.

It follows from eqs. (38)–(42) that the contribution of defects to the refractive index of a crystal must increase when approaching a phase-transition temperature. The effect can be discovered experimentally by measuring the temperature dependence of the refractive index. However, owing to the temperature dependence of the refractive index of the host matrix, the total refractive index of a crystal remains a monotonic function of temperature. A handling of results of several experiments (Smolensky et al. 1977, Fousek 1978) has shown that the temperature dependence of the refractive index of the crystal can be represented as a sum of two curves, one corresponding to $\epsilon(\eta_\infty)$ and the other having a maximum at the phase-transition temperature. This maximum can, in principle, be connected with either the influence of thermal fluctuations in the order parameter or the presence of defects. In the latter case, using appropriate experimental data and formulae (38)–(42) or their analogues, one can determine directly the dependence $\Delta\epsilon_d(\tau)$ from the correlation

$$\epsilon(\tau) = \Delta\epsilon_d VN + \epsilon(\eta_\infty),$$

and, therefore, predict the character of the anomaly of light scattering from defects of a specific type.

The value of the maximum intensity of light scattering by defects $I_{d,\max}$ (the maximum contribution is due to the $s$ defects) can be estimated from eqs. (37) and (38):

$$I_d = Q_S 4\pi^2 a^2 d^4 (\eta_0 - \eta_\infty)^4 N r_c^2,$$

by putting here, formally, $Nr_c^3 = 1$, though this assumption already exceeds the limits of applicability of the independent-defect approximation. Then at $\tau > 0$ we obtain for the ratio of the intensity of light scattering by the defects to the intensity $I_\rho$ of the noncritical light scattering from thermal fluctuations in the mass density $\rho$ (Levanyuk et al. 1976, for $I_\rho$ see also ch. 1)

$$\frac{I_{\mathrm{d,\,max}}}{I_\rho} \approx \left[\frac{4\pi a\eta_0^2}{\rho(\partial\epsilon/\partial\rho)_T}\right]^2 \frac{d^3}{k_\mathrm{B}T\beta_T}(Nd^3)^{1/3}, \tag{44}$$

where $\beta_T \equiv K_T^{-1}$ is the isothermal compressibility. If we assume $d \sim 10^{-8}$–$10^{-7}$ cm, $\beta_T \sim 10^{11}$–$10^{12}$ erg/cm$^3$, $T_\mathrm{c} \sim 100$ K, $a\eta_0^2/\rho(\partial\epsilon/\partial\rho)_T \sim 0,1$ and put, e.g., $N \sim 10^{18}$ cm$^{-3}$ then we find from eq. (44)

$$I_{\mathrm{d,\,max}}/I_\rho \sim 10^{-2}\text{–}10^4 .$$

This estimate is influenced mostly by the uncertainty in the magnitude of defect core radius *d*, which is the parameter of the theory. One can judge only the order of magnitude of this parameter just as the dimension of a dislocation core in the continuum theory of dislocations. Nevertheless, it is clear that due to the presence of defects in a crystal a great increase in the light-scattering intensity near phase transitions is possible. Apparently, it is defects that cause the light scattering anomaly to be observed in $SrTiO_3$ by Steigmeier et al. (1973). Indeed, according to the results obtained by Hastings et al. (1978), in the $SrTiO_3$ crystal being doped by deuterium, the light scattering intensity near $T_\mathrm{c}$ increases proportionally to the deuterium concentration in the crystal. A rather intense central peak has been also observed in $KH_2PO_4$ (Lagakos and Cummins 1974, Durvasula and Gammon 1977). The peak nearly disappears after annealing (Courtens 1977, 1978) as well as in $KH_3(SeO_3)_2$ mentioned above (Yagi et al. 1977). Strong anomalies of light scattering from static imperfections of a lattice have been investigated at the ferroelectric phase transition in lead germanate (Lockwood et al. 1976, Fleury and Lyons 1976, 1979, Lyons and Fleury 1978). The observed central component of an elastic nature is very narrow (of width less than 2 MHz) and near $T_\mathrm{c}$ its intensity exhibits a power-law divergence close to that given by eq. (38). In the case of a proper uniaxial ferroelectric a theoretical treatment yields $I_\mathrm{d} \sim \ln^2 \tau$, so a disagreement between theory and experiment still exists (see, e.g., Lyons and Fleury (1978)).

The defects of "random local transition temperature" type in symmetrical phase may have influence on thermal fluctuations in various quantities, so their influence upon the integral intensity of elastic scattering of light at $T > T_\mathrm{c}$ is reduced only to renormalization of the transition temperature. At $T < T_\mathrm{c}$ the influence of such defects is quite analogous to that of defects considered on p. 148, but the temperature dependence of the integral intensity is much weaker due to $\eta_0 \sim \eta_\infty$ for this case.

## 4. Central peak in systems with hopping defects

The estimates given in the foregoing section are relevant to the processes of light scattering by fluctuations in the concentration of defects. The width of a central peak in the spectrum of such a scattering is determined by the inverse time of defect diffusion over a distance of the order of the light wavelength. In solids these diffusion times are, as a rule, very long and so this part of the scattering may be considered as a static one.

However, the presence of defects in a crystal can considerably influence the spectrum of the inelastic scattering too. The mechanisms of such an effect are varied and can lead both to a change in the width (and shape) of the spectral maxima and to the appearance of new lines in the spectrum of light scattering.

One of such mechanisms of a central peak formation has been considered by Halperin and Varma (1976). The authors supposed thermal fluctuations in the order parameter to be able to redistribute the defects in the state $s$ among their equilibrium positions in a unit cell with the opposite signs of $\eta_0$ (see fig. 1). If the characteristic time of the redistribution is more than the period of the $\eta$ vibrations, a central peak along with side components will appear in the spectrum of the order-parameter fluctuations. Halperin and Varma (1976) treat randomly distributed defects in the mean-field approximation for the case of a rather high defect concentration when the correlation radius of the order parameter exceeds the mean-defect spacing (i.e., for $Nr_c^3 \gg 1$). Ginzburg et al. (1980) have carried out the analogous treatment for the opposite limiting case (independent-defect approximation) when the interaction between defects may be neglected. For the given mechanism of a dynamic central peak the basic results obtained for these opposite cases differ insignificantly and, moreover, when considering the independent defects we do leave in fact the assumptions formulated in section 2 of the present chapter. For this simple reason we discuss briefly the outline of the results by Ginzburg et al. (1980).

Consider the nonequilibrium crystal state being characterized by a given value of $\eta = \eta_\infty$ far from a defect and assume that in a unit volume of the crystal a certain number of defects $N^+ = \frac{1}{2}(N+n)$ is in the state with the same sign of $\eta_0$ as that of $\eta_\infty$ and the remaining $N^- = \frac{1}{2}(N-n)$ defects are in the state with the opposite sign of $\eta_0$. Then the total free-energy density in this nonequilibrium state can be written as

$$\Phi(\eta_\infty, n) = \Phi_0(\eta_\infty) + \tfrac{1}{2}N[F_d^+(\eta_\infty) + F_d^-(\eta_\infty)] + \tfrac{1}{2}n[F_d^+(\eta_\infty) - F_d^-(\eta_\infty)] + (k_B T/2N)n^2\,, \tag{45}$$

where $F_d^+$ and $F_d^-$ designate the energies of the corresponding states of an isolated defect (see eq. (12)), $\Phi_0(\eta_\infty)$ is the free-energy density for a "pure" crystal and the term $(k_B T/2N)n^2$ describes the entropy of a system of $N^+$ and

$N^-$ noninteracting defects in two different states (see, e.g., Kittel (1958)). Substituting formulae (1) and (12) into eq. (45) one finds at $h = 0$,

$$\Phi(\eta_\infty, n) = \tilde{\Phi}_0 + \tfrac{1}{2}\tilde{A}\eta_\infty^2 + \tfrac{1}{4}B\eta_\infty^4 - f\eta_\infty n + \tfrac{1}{2}gn^2, \tag{46}$$

where

$$\tilde{\Phi}_0 = \Phi_0 + \tfrac{1}{2}Nf|\eta_0|, \quad \tilde{A} = A + Nf/|\eta_0|,$$

$$f = 4\pi Dd(1 + d/r_c)|\eta_0|, \quad g = k_B T/N.$$

It can be seen from this expression that there exists a linear coupling between the order parameter $\eta = \eta_\infty$ and the "dynamic" variable $n$ corresponding to the difference of defect population levels for states with different signs of $\eta_0$. The variable $n$ is quite evident to be of a relaxation nature. The evolution of the spectrum for the general case of the linear coupling between $\eta$ and another relaxative variable $\xi$ is analyzed in detail in ch. 1. A specific feature of the linear coupling between $\eta$ and $n$ is that the corresponding relaxation time $\Omega_{Rn}^{-1}$ (the characteristic time of hopping of the defect between two equilibrium positions) is determined by the processes of thermal activation and depends strongly on the height of the energy barrier $\Delta$ between the states and on temperature: $\Omega_{Rn}^{-1} = \Omega_{at}^{-1}\exp(-\Delta/k_B T)$, where $\Omega_{at}$ is the characteristic phonon frequency. For example, $\Omega_{Rn} \sim 10^{13}\,s^{-1}$ for $\Delta \sim k_B T_c$ and $\Omega_{Rn} \sim 10^3\,s^{-1}$ for $\Delta \sim 40 k_B T_c$. In particular, if the time $\Omega_{Rn}^{-1}$ proves to be greater than the characteristic time of the experiment (e.g., than the time of crystal heating), then the phase transition takes place at a fixed value of the "dynamic" variable $n$, i.e., under the same conditions as in the absence of a coupling between $\eta$ and $n$ (the interaction of $\eta$ with $n$ does not display itself in this case). If, however, during the experiment the defects have time to relax into an equilibrium state but the relaxation rate $\Omega_{Rn}$ is still less than that of the $\eta$ vibrations, then the coupling between $\eta$ and $n$ must be taken into account and the phase-transition temperature remormalized (shifted) due to the coupling is determined from the condition $A - f^2/g = 0$. The influence of the coupling leads to the appearance of a central peak along with the side components in the spectrum of the $\eta$ fluctuations (see ch. 1). The intensity of the central peak is proportional to defect concentration and its width is determined by the relaxation rate $\Omega_{Rn}$.

Writing the coefficient $\tilde{A}$ in the form $\tilde{A} = A_0\tau'$ where $\tau' = (T - T_{c0})/T_{c0}$, $T_{c0}$ is the original (unshifted) transition temperature without an account of the difference of defect population levels for states with different sign of $\eta_0$, and introducing the shift in the transition temperature owing to defect reorientation (the coupling between $n$ and $\eta$) $\Delta T_c = T_c - T_{c0} = (4\pi Dd\eta_0)^2 N/k_B A_0$, we obtain from eq. (46) the expression for the fractional intensity of this "dynamic" central peak at $T > T_c$,

$$\frac{I_{central}}{I_{side}} = \frac{f^2}{gA_0\tau} = \frac{\Delta T_c}{T\tau}\left[1 + \frac{d}{r_{c0}}(\tau')^{1/2}\right]^2, \quad (\tau > 0). \tag{47}$$

Here $r_{c0}$ is the value of the correlation length for the order-parameter far from the phase transition. For estimates one can allow that $\eta_0^2 d^2 \sim D/B$ and $k_B T_c \sim A_0^2 r_{c0}^3 / 2B$, so that $\Delta T_c / T_c \sim 32\pi^2 N r_{c0}^3$. Consequently, the central peak intensity becomes comparable with that of side components already at $\tau \sim 10^{-2}$ if one puts $r_{c0} \sim 5 \times 10^{-8}$ cm and assumes for the value of defect concentration $N \sim 10^{18}$ cm$^{-3}$ which is typical of nominally "pure" crystals.

As to experimental investigation, the superposition of various central peaks caused by all possible mechanisms of their formation is always observed. The main contribution to the intensity of central peaks observed in experiments is as a rule, due to elastic scattering by fluctuations in the defect concentration (discussed in the foregoing section). In all the experiments we know at present one has not succeeded in separating the relatively small contribution of the hopping defects out of the background of this "static" central peak. That is why we are not able to list direct experimental corroborations of the "dynamic" peak existence. One can conclude the presence of a very narrow "dynamic" central peak, which is probably due to the mechanism discussed above, only from indirect data, mostly from the temperature dependence of the side components in the spectrum of light scattering (see, e.g., the survey paper by Müller 1979). The width of the "dynamic" central peak can be so small that it lies under the limits of spectral sensitivity of direct experimental methods of light-scattering investigation. Therefore, it is of interest to study the possibilities of an influence of orientable (hopping) defects on other physical phenomena which, in principle, can be detected experimentally. For example, Vugmeister (1982) has shown in theory that orientable defects of the type discussed by Halperin and Varma (1976) can give rise to considerable broadening of the electron paramagnetic resonance lines. The estimates by the critically increasing width of the EPR line of the axial center $Fe^{3+} - V_0$ in $SrTiO_3$ measured by Reiter et al. (1980) give the peak width $\sim 10$ MHz at $T = 140$ K ($T_c \approx 105$ K). The narrow central peak in $SrTiO_3$ has been discovered in neutron scattering spectra by Riste et al. (1971). Thus, apparently, there appears the opportunity to obtain in an indirect way the information about temperature behavior of the "dynamic" central peak in the spectrum of the $\eta$ fluctuations.

## 5. *Influence of immobile defects on spectral density of scattered light*

We continue the consideration of changes in the spectrum of the first-order light scattering due to the influence of defects in the state $s$. In this section we consider the static (frozen-in) defects only, disregarding the processes of their hopping between one and another equilibrium position in a unit cell. We shall show that static defects near phase transitions can make substantial changes in the spectrum resulting in the appearance of new lines and in temperature anomalies of the shapes of spectral maxima inherent in a perfect crystal. For simplicity we restrict ourselves to consideration of symmetrical phase ($T > T_c$).

In order to facilitate further explanation we recall some results of the "mode-coupling" theory discussed in detail in ch. 1. Let an $\eta$ oscillator be coupled linearly with some other $\xi$ oscillator. In the case when both oscillators are underdamped and the eigen frequency of $\xi$ vibrations, $\Omega_{0\xi}$, is less than that of $\eta$ vibrations, $\Omega_{0\eta}$, the spectrum of the $\eta$ fluctuations contains four maxima at the frequencies $\Omega_0 \approx \pm \Omega_{0\eta}$ and $\Omega_0 \approx \pm \Omega_{0\xi}$. If the $\xi$ oscillator is overdamped and if its relaxation rate $\Omega_{R\xi}$ is less than $\Omega_{0\eta}$, the spectrum of the $\eta$ fluctuations contains three maxima: a central peak at $\Omega_0 = 0$ and two side components at $\Omega_0 \approx \pm \Omega_{0\eta}$.

The static defects in the state $s$ are, in fact, the nuclei of nonsymmetrical phase at $T > T_c$. It will be shown below that in the framework of the "mode-coupling" theory this defect participation leads to ensuring a coupling of the $\eta$ fluctuations with one or another degree of freedom $\xi$.

For example, it may occur due to defects the interaction between the soft mode and the heat-conductive (heat-diffusion) mode (Ginzburg et al. 1979). Indeed, in nonsymmetrical phase the $\eta$ vibrations are followed, as is generally known, by temperature alterations ($T' = T - T_{eq}$) and that gives rise to the "heat-conductive" (Landau–Placzek) central peak in the spectrum of the order-parameter fluctuations. In symmetrical phase of a pure crystal there is no linear coupling between $\eta$ and $T'$ as the corresponding interaction term in the free-energy density has the form $\eta^2 T'$. However, the defect in the state $s$ induces in a host matrix the local distortions corresponding to the order parameter $\eta$, so the linear coupling between $\eta$ and $T'$ in the vicinity of the defect takes place in the symmetrical phase of an imperfect crystal. Using the Fourier transformation one can write the interaction term in the free-energy density in the form

$$\sum_{\boldsymbol{k}} \eta_e(-\boldsymbol{k} - \boldsymbol{q})\eta'(\boldsymbol{q})T'(\boldsymbol{k}) .$$

Thus, the oscillator $\eta'(\boldsymbol{q})$($\boldsymbol{q}$ is the wavevector of the soft phonon, $\boldsymbol{q} \approx 0$ for light scattering) is coupled linearly with a continuous set of $T$ relaxors, $T'(\boldsymbol{k})$. The Fourier component of the equilibrium space distribution of the order parameter given by eq. (11) is

$$\eta_e(\boldsymbol{k}) = \frac{4\pi D d\eta_0}{V} \frac{1}{A + Dk^2} \equiv \frac{4\pi}{V} d\eta_0 r_c^2 \frac{1}{1 + r_c^2 k^2} . \tag{48}$$

One can see that the function $\eta_e(k)$ is practically independent of $k$ for $k \lesssim r_c^{-1}$ and rapidly decreases to zero for $k > r_c^{-1}$. On the other hand, a number of relaxors corresponding to small values of $k$ is small because of the smallness of statistical weight of their states, so it makes sense to introduce the characteristic rate of the $T$ relaxor, $\Omega_T \equiv \kappa / r_c^2 C_\eta$, where $\kappa$ is the heat conductivity and $C_\eta$ is the specific heat of the system at a fixed value of the order parameter. The relaxation rate $\Omega_T$ has the meaning of an inverse time of heat diffusion over a distance $r_c$. If $\Omega_{0\eta} > \Omega_T$ then a central peak of non-Lorentzian shape would be present in the spectrum of the $\eta$ fluctuations whilst the position and intensity of the soft-mode side bands would be determined by the adiabatic generalized susceptibility corresponding to $\eta$, which does not change under the influence of the defects. Indeed, for $\Omega_T < \Omega_{0\eta}$ the "slow" temperature variations have no time to follow the "rapid" $\eta$ vibrations, and therefore the $\eta$ vibrations are not strongly influenced by the defects, as opposed to temperature variations. The total magnitude of the $\eta$ fluctuations is determined by the isothermal susceptibility so that the fractional intensity of the central peak makes up $\Delta C_d / C$, where $\Delta C_d$ is the contribution of the defects to the specific heat $C$ of a crystal. Levanyuk et al. (1979) have shown the ratio $\Delta C_d / C$ to increase like $\tau^{-3/2}$ as $\tau \to 0$ and to be able to reach 0.1 of the order of magnitude near the phase transition (for $N r_c^3 \sim 1$) even for moderate defect concentrations (about $10^{18}\,\mathrm{cm}^{-3}$) so that the central peak must certainly be observable. According to calculations this peak has a rather unusual shape, the contribution of the defects to the scattering intensity close to $\Omega = 0$ being described by the following formula

$$I_{\mathrm{central}} \approx \frac{k_B N d^5 \eta_0^2}{V \kappa \tau^{7/2}} \left(1 - \sqrt{\Omega / 2\Omega_T}\right). \tag{49}$$

In proper ferroelectrics such an unusual lineshape of the central peak is practically the same in spite of a strong influence of long-range electric forces on the order-parameter distribution, although in a uniaxial ferroelectric the temperature dependence of the intensity becomes weaker than that given by eq. (49).

In an analogous way one may consider the defect-induced coupling of the soft-phonon branch with any fully symmetrical phonon branch. If the latter is overdamped and if its relaxation rate is less than the eigen frequency $\Omega_{0\eta}$, a non-Lorentzian dynamic central peak appears in the spectrum of the $\eta$ fluctuations, the intensity of the peak being of the same order of magnitude as that discussed above.

In solid solutions the $\xi$ variable may have the meaning of the concentration of a component. The central peak arising in this case due to coupling of $\eta$ and $\xi$ fluctuations near the defects is completely analogous to that caused by coupling between the soft and thermal diffusion modes, the linewidth of this peak being determined by the inverse time of diffusion of the component to the distance of order $r_c$.

It is possible that some of defect-induced interactions mentioned above give rise to temperature-dependent contributions to dynamic central peaks observed in various crystals, including $SrTiO_3$, $Pb_5Ge_3O_{11}$ and $BaMnF_4$, near the phase transition temperature (see, e.g., Lyons and Fleury (1979)).

Now we discuss mechanisms for the appearance of additional side bands in the light-scattering spectrum in a system with static defects in the state $s$. One more example of a variable which is invariant under operations of the high-symmetry group of a crystal is the mass density $\rho$ or the dilatation (volume deformation) $v = \operatorname{div} \boldsymbol{u} = \boldsymbol{u}_{//} = -\Delta\rho/\rho_e$. The corresponding term of interaction between each oscillator $\eta'(\boldsymbol{q})$ and a set of oscillators $v(\boldsymbol{k})$ is proportional to

$$\int \eta^2 v \, \mathrm{d}V = \sum_{\boldsymbol{k},\boldsymbol{q}} \eta_e(-\boldsymbol{q}-\boldsymbol{k})\eta'(\boldsymbol{q})v(\boldsymbol{k}) \,. \tag{50}$$

Such an interaction has been treated in detail by Levanyuk and Sigov (1980). The $v(\boldsymbol{k})$ oscillators are underdamped, because they correspond to long-wave acoustical phonons. The eigen frequencies of these oscillators are $\Omega_{Sk} = c_S k$ (where $c_S$ is the longitudinal sound velocity), $\Omega_{Sk}$ being substantially less than that of the soft mode. So the defect-induced linear coupling between the given longitudinal acoustical mode $v(\boldsymbol{k})$ and the soft mode $\eta'(\boldsymbol{q})$ gives rise to the appearance of two narrow side maxima at the frequencies $\pm\Omega_{Sk}$ in the $\eta$ fluctuation spectrum. For the whole acoustical branch these maxima are superimposed forming two wide side bands at the frequencies $\Omega_S \approx \pm c_S r_c$. Of the same order is the width of these lines. Note that within the framework of the Landau theory the frequencies $\Omega_{0\eta}$ and $\Omega_S$ have the same temperature dependence, hence we may say that the temperature evolution of additional side components must be in complete agreement with the temperature behaviour of the soft mode. Probably the maximum of such a kind has been observed experimentally by Grigas and Belyatskas (1978) at the frequency of $\sim\Omega_{0\eta}/30$ in SbSI, where the concentration of defects is large enough. The fractional intensity of these side components makes $\Delta K_d/K$ ($\Delta K_d$ is the contribution of the defects to the hydrostatic compression modulus $K$) and increases like $\tau^{-3/2}$ as $\tau \to 0$. It has been already mentioned that relative contributions of the defects to thermodynamic quantities of crystals can reach about 10% (Levanyuk et al. 1979) so the side maximum of the discussed origin seems to be observable by light or neutron scattering. One can draw similar conclusions for proper ferroelectrics too.

Owing to the defects, the soft phonon mode $\eta$ may be coupled linearly with any fully symmetrical phonon mode $\xi$. If $\xi(\boldsymbol{k})$ corresponds to an underdamped optical-phonon branch lying below the soft branch then two additional side bands will appear in the spectrum of the $\eta$ fluctuations. Consider however, this situation to be not too common for crystals in reality, because usually the frequencies of optical-phonon branches much exceed those of soft phonons.

Thus, it can be seen from the given examples that there may be many reasons for the appearance of central peaks and additional side bands in crystals with the static defects inducing a local lowering of the symmetry above a phase transition temperature. The defects in question can give rise to the coupling of the soft mode with various normal vibrations of the crystal lattice.

Last years interest is shown in experimental and theoretical study of the problem of the appearance of those lines in the Raman spectrum near a phase transition which are forbidden by the symmetry selection rules for symmetrical phase and allowed for nonsymmetrical phase only (Barta et al. 1977, Benoit et al. 1978, Ipatova and Klochihin 1978, Yacoby 1978, 1981, Yacoby and Just 1974, Yacoby et al. 1977, Prater et al. 1981 a, b, Barta et al. 1982, Yacoby et al. 1978, Grabinskii and Sigov 1980, Godefroy et al. 1983).

The physical reason for the appearance of such lines is rather obvious. It has been already mentioned above that in symmetrical phase of a crystal the defect in the state $s$ (i.e., the defect with $\eta_0 \neq 0$) can be regarded as a nucleation center of nonsymmetrical phase. Therefore, at $T > T_c$ this defect must give rise to the appearance of spectral lines which are allowed only for a nonsymmetrical phase of a perfect crystal. When giving a theoretical interpretation of the nature of forbidden Raman lines above a transition temperature one has taken into account both the second-order light scattering by fluctuations in the order parameter in a host matrix and the scattering by structural imperfections. Here we do not touch on a number of concrete questions of the microscopic theory of light scattering relevant to the problem of forbidden lines, which are discussed in literature in detail (see, e.g., Loveluck and Sokoloff (1973), Sokoloff (1972), Geisel and Keller (1978)), but present simple qualitative consideration within the framework of the Landau theory of phase transitions. We discuss briefly the results of the theory developed by Yacoby (1978), Ipatova and Klochihin (1978), Grabinskii and Sigov (1980), Godefroy et al. (1983) and afterwards speak of some experimental data available. As before we restrict our consideration to chaotically distributed frozen-in, independent defects.

Recur now to eq. (33). The second sum in this formula determines the intensity of the first-order light scattering by the $\eta$ fluctuations in a crystal with defects in the state $s$. The calculations of such defect scattering have been carried out by Ipatova and Klochihin (1978), but the authors neglect the attenuation of the soft mode which is essential in most cases and can lead to considerable changes in the spectrum (see below). Assume the equation of motion for the order parameter to be similar to that for the harmonic oscillator with a damping. On account of eqs. (10) and (32) it takes the form

$$m\ddot{\eta}' + \gamma\dot{\eta}' + A\eta' - D\nabla^2\eta' = 0\,. \tag{51}$$

Then we have at $k_B T \gg \hbar\Omega$,

$$\langle \eta'(\boldsymbol{k}, \Omega)\eta'(-\boldsymbol{k}, -\Omega)\rangle \approx \frac{k_B T\gamma}{\pi[(A + Dk^2 - m\Omega^2)^2 + \gamma^2\Omega^2]}, \tag{52}$$

and using the solution (48) obtain the intensity of the first-order light scattering

$$I'_{\mathrm{d}}(0, \Omega) \approx \frac{2NQ_{\mathrm{S}}a^2(4\pi Dd)^2\eta_0^2k_{\mathrm{B}}T\gamma}{\pi^3}$$

$$\times \int_0^{k_m} \frac{k^2\,\mathrm{d}k}{(A+Dk^2)^2[(A+Dk^2-m\Omega^2)^2+\gamma^2\Omega^2]}. \tag{53}$$

Here $k_m \sim \pi/d$ is the cutting parameter (we deal with the long-wave fluctuations solely). When integrating we put $k_m \to \infty$ or, more exactly, $k_m \gg r_{\mathrm{c}}^{-1}$ and $k_m \gg [(A-m\Omega^2)^2+\gamma^2\Omega^2]^{1/2}/D$. Hence we get

$$I'_{\mathrm{d}}(0, \Omega) \approx \frac{16Q_{\mathrm{S}}Nk_{\mathrm{B}}TD^{1/2}d^2a^2\eta_0^2}{(m^2\Omega^2+\gamma^2)^2\Omega^2}\left[\frac{\gamma^3}{2A^{1/2}} - \frac{\gamma^2(R+m\Omega^2-A)^{1/2}}{\sqrt{2}\Omega}\right.$$

$$+\frac{\gamma m^2\Omega^2}{2A^{1/2}} + \gamma m\sqrt{2}(R-m\Omega^2+A)^{1/2} - 2\gamma m A^{1/2}$$

$$\left.+\frac{m^2\Omega}{\sqrt{2}}(R+m\Omega^2-A)^{1/2}\right], \tag{54}$$

where $R = [(A-m\Omega^2)^2+\gamma^2\Omega^2]^{1/2}$. The maximum of the intensity $I'_{\mathrm{d}}(0, \Omega)$ appears at $\Omega = 0$ for $\gamma^2 \gtrsim 2.4Am$ (in contrast to a "normal" oscillator which becomes overdamped for $\gamma^2 \geqslant 2Am$). In the absence of damping the maximum in the spectrum of the first-order Raman scattering lies at $\Omega_m = (5A/4m)^{1/2}$,

$$I'_{\mathrm{d}}(0, \Omega_m)|_{\gamma=0} = 256Q_{\mathrm{S}}Nk_{\mathrm{B}}T(da\eta_0)^2m^{1/2}/25\sqrt{5}D^{3/2}. \tag{55}$$

For the integral intensity of the first-order light scattering $I'_{\mathrm{d}}$ one has

$$I'_{\mathrm{d}} = 2Q_{\mathrm{S}}Nk_{\mathrm{B}}T\pi(da\eta_0)^2r_{\mathrm{c}}^3/D \sim \tau^{-3/2}. \tag{56}$$

Taking into account the result for the integral intensity $I''$ of the second-order scattering by the $\eta$ fluctuations in a pure crystal (see ch. 1) and eqs. (38) and (56) one can compare magnitudes of the intensities:

$$\frac{I'_{\mathrm{d}}}{I_{\mathrm{d}}} = \frac{k_{\mathrm{B}}Tr_{\mathrm{c}}}{2\pi Dd^2\eta_0^2}, \qquad \frac{I'_{\mathrm{d}}}{I''} = \frac{8\pi^2d^2\eta_0^2Nr_{\mathrm{c}}^2D}{k_{\mathrm{B}}T}. \tag{57}$$

Just as in sect. 3 of the present chapter we may put here $Nr_{\mathrm{c}}^3 = 1$ in order to estimate the maximum value of the defect contribution. Then, on the same assumptions of the order of magnitude of the parameters as those made when estimating the ratio (44) we have from expressions (57),

$$I'_{\mathrm{d\,max}}/I_{\mathrm{d\,max}} \sim 10^{-1}\text{–}10^3, \quad I'_{\mathrm{d\,max}}/I'' \sim 10^{-2}\text{–}10^2.$$

We see that, in principle, the first-order light scattering by the $\eta$ fluctuations in imperfect crystals can be quite observable in experiments even at $N \sim 10^{18}\,\mathrm{cm}^{-3}$. One can arrive at the analogous conclusions for systems with long-range forces (proper ferroelectrics and proper ferroelastics) as well, although we do not adduce here the results of the calculations because they have a rather complicated form.

Consider now a certain "rigid" (not softening as $T \to T_c$) normal vibration $\zeta$ whose display in the Raman spectrum is forbidden by the selection rules for symmetrical phase and allowed in nonsymmetrical phase (Godefroy et al. 1983). Then, after the analogy of eqs. (31) and (32), the contribution of $\zeta$ fluctuations to an appropriate component of the dielectric permeability tensor is

$$\Delta\epsilon = a\eta\zeta = a\eta_e\zeta + a\eta'\zeta\,, \tag{58}$$

where the second summand describes the second-order light scattering. By close analogy with the second term in eq. (33) the spectral distribution of scattering by the $\zeta$ fluctuations takes the form

$$I'_{d\zeta}(q \approx 0, \Omega) = NQ_S a^2 \sum_k \eta_e(\boldsymbol{k})\eta_e(-\boldsymbol{k})\langle\zeta(\boldsymbol{k}, \Omega)\zeta(-\boldsymbol{k}, -\Omega)\rangle\,. \tag{59}$$

In this formula we have already performed the averaging over positions of randomly distributed uncorrelated point defects. Neglecting a dispersion in the $\zeta$ branch we obtain from eq. (59) the same lineshape as for the first-order scattering by the $\zeta$ vibrations in a perfect crystal. But on account of the dispersion

$$\Omega_\zeta^2(\boldsymbol{k}) = \Omega_{0\zeta}^2 + D_\zeta k^2\,,$$

and regardless of a damping of the $\zeta$ vibrations the spectral intensity becomes

$$I'_{d\zeta}(0, \Omega) = \begin{cases} 0\,, & \text{for} \quad \Omega < \Omega_{0\zeta}\,, \\ \dfrac{Q_S N k_B T (ad\eta_0)^2}{D_\zeta^{3/2}(D_\zeta r_c^{-2} + \Omega^2 - \Omega_{0\zeta}^2)^2}\sqrt{1 - \dfrac{\Omega_{0\zeta}^2}{\Omega^2}}\,, & \\ \quad \text{for} \quad \Omega_{0\zeta} < \Omega < \sqrt{\Omega_{0\zeta}^2 + D_\zeta k_m^2}\,. & \end{cases} \tag{60}$$

The maximum intensity of this spectral feature is located at

$$\Omega_{\max} \approx \Omega_{0\zeta} + D_\zeta r_c^{-2}/2\Omega_{0\zeta}\,, \tag{61}$$

so $\Omega_{\max} \to \Omega_{0\zeta}$ as $T \to T_c$, and the lineshape is becoming highly asymmetrical. The asymmetry is smeared, of course, if we take into account the damping of the $\zeta$ vibrations.

Putting $D_\zeta = 0$ and assuming the value $\langle\zeta^2\rangle$ to be independent of temperature we can see that the integral intensity of this line is proportional to $\tau^{-1/2}$. A marked temperature dependence of such a "noncritical" line is caused by the influence of defects in the state $s$. As to the estimates, the defect contribution to the integral intensity $I'_{d\zeta}$ may be rather large.

The contribution of defects in the state $p$ is much smaller in comparison with that of the $s$ defects and has a weaker temperature dependence ($\sim$ (const-$\tau^{1/2}$)).

We turn to an analysis of experimental data. In the Raman spectra of crystals $Hg_2Cl_2$ and $Hg_2Br_2$ above the phase transition temperature (Barta et al. 1977) the line whose frequency decreases when $T \rightarrow T_c$ has been observed (the soft mode?), though this line appearance is forbidden by the selection rules for symmetrical phase. A neutron-scattering study has also indicated the presence of such a line (Benoit et al. 1978). Ipatova and Klochihin (1978) as well as Barta et al. (1977) related the presence of the $\eta$ line at $T > T_c$ with a local breaking of symmetry due to structural imperfections (i.e., with the first-order scattering by the $\eta$ fluctuations in an inhomogeneous crystal) while Benoit et al. (1978) proposed that it should be connected to the second-order scattering from the $\eta$ fluctuations. A detailed analysis of the experimental data on the basis of a simple phenomenological theory (Grabinskii and Sigov 1980) shows that the interpretation connected with the second-order scattering from the $\eta$ fluctuations is true in the case of $Hg_2Cl_2$. However, in mixed crystals of $Hg_2(Cl_{1-x}Br_x)_2$ even for small concentrations of one of the components, the temperature-dependent line at the soft-mode frequency is observed at $T > T_c$ (Barta et al. 1982). Seemingly, the ion of Br for $x \ll 1$ or the ion of Cl for $1 - x \ll 1$ can be treated as a defect in the state $p$ which induces static distortions of the host matrix at $T > T_c$ and thus gives rise to a violation of the symmetry selection rules, i.e., to a permission of the first-order scattering from the soft-mode vibrations being forbidden in symmetrical phase of a pure crystal.

Studying the Raman spectra of mixed crystals $K_{1-x}Li_xTa_{1-y}Nb_yO_3$ Prater et al. (1981a, b) have observed, in the paraelectric phase, the line corresponding to the softening phonon. This line is visible at any Nb concentration including $x = 0$. Prater et al. (1981a) interpret the line as the spectral maximum of the first-order light scattering from the soft-phonon branch. Such scattering can be induced either by intrinsic disorder or by remanent impurities which differ from Nb and can be present in even the purest available crystals. For example, the original starting materials may contain the atoms of Li and Na, or, these atoms can be introduced during the crystal growth process. The influence of Nb brings, apparently, to a renormalization (shift) of the phase-transition temperature $T_c \rightarrow T_c'$ and can be taken into account by a change in the coefficient $A' = A_0(T - T_c')$. In terms of the continuum theory of lattice distortions induced by defects, which was discussed in sect. 2, one may suppose the ion $Nb^{5+}$ to occupy a symmetrical position with $\eta_0 = 0$ in a unit cell of the perovskite structure. A mixed crystal can, therefore, be considered as an "effective host matrix" containing the off-center ions $Li^+$ and $Na^+$ which act as symmetry-breaking defects with $\eta_0 \neq 0$. So it seems quite possible in such a case to use formulae (54)–(56) for a description of the soft-mode line observed by Prater et al. (1981a, b).

In potassium tantalate doped with the same impurities a temperature dependence of the narrow "noncritical" line corresponding to the "rigid" phonon $\zeta$ has been investigated by Yacoby (1978) and Prater et al. (1981a). The frequency of this line $\Omega_{0\zeta}$ is much higher than the soft-mode frequency $\Omega_{0\eta}$. Yacoby (1978) proposes to connect the anomalous growth of this line intensity with increasing dynamic distortions of the lattice, i.e., with the increase in the intensity of the order-parameter fluctuations $\langle|\eta'|^2\rangle \sim \tau^{-1}$. However, it follows from the calculations (Grabinskii and Sigov 1980) that the character of temperature dependence of the line intensity may be changed considerably if parallel with dynamic distortions ($\eta'$) one takes into account also static ones ($\eta_e$) corresponding to the order parameter. A consistent theoretical study of the problem (Godefroy et al. 1983) yields a weaker dependence $I'_{d\zeta} \sim \tau^{-1/2}$.

It is interesting to note that both peculiar spectral features discussed above (we may designate them as $I'_d$ and $I'_{d\zeta}$) are enhanced by the addition of small concentrations of Li or Na in otherwise pure $KTaO_3$ (Prater et al. 1981a, b, Yacoby 1981). In particular, the intensity $I'_{d\zeta}$ increases linearly with Li or Na concentration (Yacoby 1981), the same does the transition temperature (Prater et al. 1981b).

One more anomalous feature of the soft-mode type has been investigated by Yacoby and Just (1974) in the Raman spectra of paraelectric phase of the doped crystals $K_{1-y}Na_yTaO_3$ and $K_{1-y}Li_yTaO_3$. But these lines markedly differ in frequency, depending on the sort of doping atoms introduced into potassium tantalate. Most likely, these lines can be associated with the first-order scattering from the impurity modes of Li or Na. We would like to recall the fact which had been established by a number of experiments (Yacoby and Just 1974, Yacoby et al. 1977, Yacoby 1981) that the $Li^+$ or $Na^+$ ions occupy off-center positions in a unit cell, even in the cubic (paraelectric) phase, hence the first-order scattering is allowed at $T > T_c$.

We continue now with the problem of the influence of static defects on the width of Raman lines. Levanyuk et al. (1979) have shown that symmetry-breaking defects may lead to important changes in phonon damping constants near a phase transition. In the end the reason is that the region distorted by a defect is of a heightened level of energy dissipation. Since dimensions of the region increase as $T \to T_c$ the defect contribution to the dissipation of corresponding vibrations increases too. It is very likely that just the defects must be responsible for a substantial increase in the width of the soft-mode side bands revealed in numerous experiments (see, e.g., Shapiro and Cummins (1968), Laulicht et al. (1972), Burns and Scott (1970), Fleury (1970), Shigenari et al. (1976), Uwe and Sakudo (1976)). An anomalous widening of the soft-mode components can be explained neither by the mean-field theory nor by theories allowing for the $\eta$ fluctuations. Therefore one cannot find any other possibility of explanation of such a widening except the consideration of the real structure of a crystal. Let us discuss in more detail some mechanisms for the soft-mode damping due to defects.

We discuss again a certain fully symmetrical normal vibration, $\xi$, coupled linearly with the $\eta$ vibration. If $\Omega_{0\xi} \gg \Omega_{0\eta}$ one may neglect the inertia term in the equation of motion for the $\xi$ oscillator when considering the $\eta$ vibrations, i.e., it is possible to treat $\xi$ as a relaxor. Representing the interaction term in the form

$$2r\sum_{\boldsymbol{k}}\eta_{\rm e}(-\boldsymbol{k}-\boldsymbol{q})\eta'(\boldsymbol{q})\xi'(\boldsymbol{k}),$$

and using eq. (48), one can easily estimate the contribution of the $\xi$ branch to the damping constant $\Gamma=\gamma/m$ of the soft mode,

$$\Delta\Gamma\approx\frac{\pi}{8}\frac{(4\pi\eta_0 dr_{\rm c}^2 r)^2 N\Gamma_\xi}{m_\xi m[(\Omega_{0\xi}^2-\Omega^2)^2+\gamma_\xi^2\Omega^2]}\left(r_{\rm c}+\sqrt{\frac{D_\xi}{\Omega_{0\xi}^2-\Omega^2}}\right)^{-3}$$

$$\approx\frac{\pi^3 r^2 N\Gamma_\xi\eta_0^2 d^2 r_{\rm c}^4}{m_\xi m\Omega_{0\xi}^4 r_{\rm c}^3}\sim\tau^{-1/2}. \qquad (62)$$

Here $\gamma$, $\gamma_\xi$ and $m$, $m_\xi$ are damping coefficients and "masses" in the equations of motion for $\eta$ and $\xi$, respectively, $\Gamma_\xi=\gamma_\xi/m_\xi$ the damping constant for the $\xi$ vibrations and the coefficient $D_\xi$ characterises the dispersion in the $\xi$ branch. As to numerical estimates, the ratio $\Delta\Gamma/\Gamma$ may reach up to values $\sim 0, 1$ at the boundary of applicability of used approximation of independent defects ($Nr_{\rm c}^3=1$).

In the defect vicinity under the influence of the $\eta$ vibrations there may be excited oscillations of elastic ("acoustic") deformations along with optical deformations which are not those corresponding to the soft-phonon branch. Since when approaching $T_{\rm c}$ the linear dimensions of the region near an isolated defect where these oscillations are excited increase, the corresponding defect contribution to the soft-mode damping constant increases too.

The damping of the soft-mode vibrations may also be due to time-dependent temperature gradients provoked near defects by the $\eta$ vibrations (Levanyuk et al. 1979, Ginzburg et al. 1980). Let us consider this mechanism more widely. Using the entropy $S$ of a system as a generalized coordinate conjugate to the temperature changes one has for the entropy-balance equation:

$$T\dot{S}=T\left[\left(\frac{\partial S}{\partial T}\right)_\eta\dot{T}'+\left(\frac{\partial S}{\partial\eta}\right)_T\dot{\eta}'\right]=C_\eta\dot{T}'-A_0\eta_{\rm e}(\boldsymbol{r})\dot{\eta}'=\kappa\nabla^2T'. \qquad (63)$$

Add to this equation the equation of motion for the $\eta'$ variable (for the range of applicability of the Landau theory) taking into account temperature changes:

$$m\ddot{\eta}'+\gamma\dot{\eta}'+A\eta'-D\nabla^2\eta'+A_0T_{\rm c}^{-1}\eta_{\rm e}(\boldsymbol{r})T'=0. \qquad (64)$$

Thus we obtain a system of two coupled equations. It is more convenient to write this system in the Fourier-representation. Considering only one Fourier-component $\eta'(\boldsymbol{q},\Omega)$, from eqs. (63) and (64) we obtain:

$$(-{\rm i}\Omega C_\eta+\kappa k^2)T'(\boldsymbol{k},\Omega)+{\rm i}\Omega A_0\eta_{\rm e}(\boldsymbol{k}-\boldsymbol{q})\eta'(\boldsymbol{q},\Omega)=0, \qquad (65)$$

$$[-m\Omega^2+{\rm i}\gamma\Omega+A+Dq^2]\eta'(\boldsymbol{q},\Omega)+A_0T_{\rm c}^{-1}\sum_{\boldsymbol{k}}\eta_{\rm e}(\boldsymbol{q}-\boldsymbol{k})T'(\boldsymbol{k},\Omega)=0. \quad (66)$$

Expressing from eq. (65) the variable $T'(\boldsymbol{k}, \Omega)$ in terms of $\eta'(\boldsymbol{q}, \Omega)$ and substituting in eq. (66) the expression obtained, one finds for the defect contribution to the soft-mode damping coefficient:

$$\gamma_d(q=0, \Omega) = \frac{A_0^2}{T_c} \sum_{\boldsymbol{k}} \frac{|\eta_e(\boldsymbol{k})|^2 \kappa k^2}{\kappa^2 k^4 + C_\eta^2 \Omega^2}. \tag{67}$$

From the viewpoint of studying the temperature dependence of the soft-mode components' linewidth, of most interest is the frequency region $\Omega \gg \Omega_T \equiv \kappa / C_\eta r_c^2$, since it is this region that usually includes the soft-mode frequency. Indeed, although far from the transition point both these frequencies are of the same order of magnitude, the frequency $\Omega_T$ decreases as $\tau \to 0$ according to a stronger power law than $\Omega_0$: $\Omega \sim |\tau|$ whereas $\Omega_0 \sim |\tau|^{1/2}$. For $\Omega \approx \Omega_0 \gg \Omega_T$ we get

$$\gamma_d(q \approx 0, \Omega \approx \Omega_0) = \frac{\pi\sqrt{2 A_0^2 \eta_0^2 d_0^2}}{\kappa T_c} N r_c^3 \left(\frac{\Omega_T}{\Omega_0}\right)^{3/2} \sim |\tau|^{-3/4}. \tag{68}$$

Levanyuk et al. (1979) estimated the defect contribution to the soft-mode damping coefficient for the region of small frequencies, $\Omega \ll \Omega_T$, only. At small frequencies $\gamma_d \sim |\tau|^{-3/2}$, i.e., it diverges much stronger in such a case.

At the boundary of applicability of our approach ($N r_c^3 \sim 1$) the relative contribution of the defects is

$$\frac{\gamma_d}{\gamma} \sim 2\pi\sqrt{2} \frac{\Delta C}{C_\eta} \frac{B \eta_0^2}{\gamma \Omega_{at}} \left(\frac{\kappa}{d^2 C_\eta \Omega_{at}}\right)^{1/2} (N d^3)^{1/2}, \tag{69}$$

where the frequency $\Omega_{at} = (D/md^2)^{1/2}$ has the meaning of a characteristic "atomic" frequency ($\Omega_{at} \sim 10^{13}\,\mathrm{s}^{-1}$). To make estimations one may take into account that at $\eta_0 \sim \eta_{at}$ the value $B\eta_0^2 \sim m\Omega_{at}^2$ and assume also that $\kappa / d^2 C_\eta \sim \Omega_{at}$. Then it yields from eq. (69)

$$\frac{\gamma_d}{\gamma} \sim 2\pi\sqrt{2} \frac{\Delta C}{C_\eta} \frac{\Omega_{at}}{\Gamma} (N d^3)^{1/2}. \tag{70}$$

For displacive phase transitions the width of side components is usually $5 \times 10^{10}$–$10^{12}\,\mathrm{s}^{-1}$ and $\Delta C / C_\eta$ equals 0.1 as to the order of magnitude. Thus we see that defects may cause an appreciable increase in the soft-mode component linewidth already for a defect concentration of $N \sim 10^{18}\,\mathrm{cm}^{-3}$ ($N d^3 \sim 10^{-4}$).

The increase in the dimensions of the region of a heightened level of energy dissipation near a defect may lead also to temperature anomalies of defect contributions to the damping coefficients for other than $\eta$ ("rigid") phonon branches $\xi$. If $\Omega_{0\xi} \gg \Omega_{0\eta}$ one may find the value $\Delta\Gamma_\xi$ from eq. (62). Therefore, we may say that the defects are able to induce a rather significant widening of components corresponding to some "rigid" high-frequency optical and acoustic vibrations. In light scattering experiments this phenomenon has been observed by Aronov et al. (1977). The theory of anomalies of this kind has been considered by Levanyuk et al. (1979) and in more detail by Yermolov et al. (1979), (1981), for the example of the anomaly of absorption coefficients for longitudinal and transversal sound waves.

A widespread class of potentialities of the broadening of phonon lines near phase transitions is not limited to a number of mechanisms of a general type considered above which are based on a coupling between various vibrations in the vicinity of a defect. For example, an inhomogeneous widening of spectral lines due to a variation of local values of the phase transition temperature over a sample may take place. Such a variation may be caused by heterogeneous chemical composition or by inhomogeneities of space distribution of defects, by nonuniform internal stresses, etc. A variation in $T_c$ gives rise to a variation of the soft-mode frequency:

$$\Delta\Omega_{0\eta} \approx \frac{d\Omega_{0\eta}}{dT_c}\Delta T_c = \sqrt{\frac{A_0}{m}\frac{\Delta T_c}{T_c}}|\tau|^{-1/2}, \tag{71}$$

so that

$$\frac{\Delta\Omega_{0\eta}}{\Omega_{0\eta}} = \frac{\Delta T_c}{T_c}|\tau|^{-1/2}.$$

Hence we see, that at $\tau \sim \Delta T_c/T_c$ an inhomogeneous widening of soft-mode components leads to their apparent merging even when one neglects the soft-mode damping.

## 6. Concluding remarks

It is clear even from the short review given in the previous sections that various defects of crystalline structure may play an extremely essential role in the processes of light scattering near phase transitions. At the same time a systematic study of the problem of the influence of defects on physical properties of substances near phase transitions has just begun.

In the present chapter we have not discussed the results of the well elaborated approach (however, as a rule this approach does not take into account specific features of a system near phase transitions) which deals with the microscopic theory of interaction of light quanta with elementary excitations of a nonideal lattice. Such a microscopic approach allows one to link the dynamic properties of a nonideal lattice with the optical characteristics of a real crystal under investigation. This approach is reflected in a whole number of original and review papers as well as in monographs (see, e.g., Loveluck and Sokoloff (1973), Elliott and Taylor (1967), Maradudin (1966), Stoneham (1975), and the literature cited therein).

Along with the phenomenological approach accepted in the chapter, the quasi-microscopic approach which brings into use the renormalization group has been actively developed. The quasi-microscopic approach has to do with the immediate vicinity of the phase stability loss points where thermal fluctuations of the order parameter become rather large, so that the Landau theory proves to be not applicable (this region of temperatures is usually called a critical or scaling region). As to systems containing defects this approach appears to be helpful only with additional assumptions of the "weakness" (extremely small value of $\eta_0$) of randomly distributed defects whose concentration satisfies the condition $Nr_c^3 \gg 1$. Using sophisticated methods of modern theory of phase transitions one has succeeded in an approximate calculation of the asympthotic temperature dependence of various physical quantities of a system with the defects within the scaling region (see, e.g., Ma (1976), Patashinsky and Pokrovsky (1982)). Such dependences are described by the so-called critical indices (critical exponents) which depend on the dimensionality of the system and its symmetry properties. Note, however, that for the majority of structural phase transitions the scaling region represents a very close vicinity of $T_c$ which has not been reliably observed in experiments up to now. For this reason one cannot speak of a comparison of the scaling-theory results with experimental data available. Owing to these circumstances we have not considered methods and results of the approach in the given chapter.

There exists one more branch of the theory whose results are valid within the range of applicability of either the "classical" Landau theory of phase transitions or the generalized Landau theory (Ginzburg and Sobyanin 1976) if a critical region is under consideration. This approach deals with such concentrations of defects and such temperature region near a phase transition which allow one to regard defects as weakly interacting or noninteracting at all, for a first approximation. Then the contributions of defects to various physical quantities appears to be additive. When calculating the contributions one may treat the substance as a continuous medium and introduce defects by means of boundary conditions to the equations of continuous media. In the present chapter we use just the independent-defect approach to describe the light scattering anomalies. The universality and simplicity are the obvious advantages of this approach and they give rise to consider (at least qualitatively) a number of specific features in scattering spectra of real crystals.

In the problem of the role of defects in light scattering of great importance is the type of the order-parameter symmetry. In a whole consideration we restrict ourselves to the case of a discrete symmetry of the order parameter (the so-called Ising-like systems). To a separate class belong the so-called degenerate systems where the energy does not depend on a "phase" of the order parameter; the latter corresponds to a representation of an infinitive group. Such systems are examplified by isotropic ferromagnets of the Heisenberg type, by the superfluid liquid helium and by incommensurate superstructures. All degenerate systems possess anomalously large "compliance" over the temperature range of existence of the appropriate phase: we mean the compliance corresponding to the angle of rotation of magnetic moment or to the phase of a macroscopic wavefunction in the first or in the second cases, respectively, while the third example is relevant to the translation of the supersturcture as a whole. In other words one of the correlation lengths in such systems appears to become infinite within that temperature region where the appropriate degenerate phase exists. In consequence of this the dimension of the space region of the influence of a defect of "random local field" type (which fixes, e.g., a direction of the magnetization in an isotropic magnet or a shift of an incommensurate superstructure with respect to the basic lattice) becomes infinite. We expect, therefore, that experimentally investigated properties of some degenerate systems, e.g., of incommensurate superstructures, can not be understood when the influence of defects is disregarded. A theoretical study of incommensurate phases in the presence of defects is extremely complicated due to the fact that the independent-defect approximation is inapplicable here at all. The existing theoretic results are in these cases of a character of estimates obtained from dimensionality considerations (see, e.g., Ma (1976), Sonin (1980)).

Adduce once more some difficulties connected with the limitations of applicability of the phenomenological theory discussed in this chapter. The defects are assumed to be distributed homogeneously and light scattering from the fluctuations in their concentration is considered. More generally one may say that correlations in distribution of defects are neglected. We also do not take into account the presence of large-scale inhomogeneities (comparable with a sample dimension) which may lead to a phase-transition rounding. These limitations may be essential for the interpretation of experimental data because in many real crystals the defects are inhomogeneously distributed throughout a sample owing to irregular nonisotropic growth of a crystal. It is possible in particular that peculiarities of light scattering in $NH_4Cl$ and in quartz are caused just by macroscopic imperfections in their structure. When being used for interpretation of such experimental results the theory discussed above obviously fails, and must be improved.

The interaction between defects is also neglected here so that for a further development of the theory one must evaluate peculiar features in the spectrum of light scattering within that range of temperatures and concentrations of defects where the mean distance between defects is less than the correlation radius of the order parameter.

Note in addition that a theoretical analysis of light scattering from defects in anisotropic media described by a multi-component order parameter is needed as well.

Passing to experimental problems we would like to emphasize the necessity to perform combined experiments on one and the same specimen with a controlled concentration and correlation of defects of a given type. From these experiments it must result the correspondence between both the intensity and polarization of scattered light from one side and different thermodynamic and kinetic characteristics of the substance from the other. The fundamental problems include undoubtedly the discovery and investigation of the scaling region for structural phase transitions in imperfect crystals.

All above said allows one to conclude that joint efforts of both experimentalists and theoreticians are highly recommended to study the anomalies of light scattering in crystals containing defects. No doubt that they will yield new important results.

## *References*


Andreev, A.F., V.I. Marchenko and A.E. Meierovich, 1977, Pis'ma ZhETF, **26**, 40.

Aronov, A.G., D.N. Mirlin, I.I. Repina and F.F. Chudnovskii, 1977, Fiz. Tverd. Tela, **19**, 193.

Axe, J.D. and G. Shirane, 1973, Phys. Rev. **B 8**, 1965.

Axe, J.D., S.M. Shapiro, G. Shirane and T. Riste, 1974, Neutron Scattering Studies of Soft Mode Dynamics, in: Anharmonic Lattices, Structural Transitions and Melting, ed., T. Riste (Noordhoff, Leiden) p. 23.

Barta, Ch., B.S. Zadohin, A.A. Kaplyanskii and Yu.F. Markov, 1977, Pis'ma ZhETF, **26**, 480.

Barta, Ch., G.F. Dobrzhanskii, M.F. Limonov, Yu.F. Markov and A.S. Sigov, 1982, Pis'ma ZhETF, **34**, 62.

Benoit, J.P., Cao Xuan An, Y. Luspin, J.P. Chapelle and J. Lefebre, 1978, J. Phys. C: Solid State Phys. **11**, L 721.

Blinc, R., 1977, Ferroelectrics, **16**, 33.

Bray, A.J., 1974, J. Stat. Phys. **11**, 29.

Bray, A.J. and M.A. Moore, 1977, J. Phys. A: Math. Gen. Phys. **10**, 1927.

Bulaevskii, L.N., V.V. Kusii and A.A. Sobyanin, 1978, Solid State Commun. **25**, 1053.

Burns, G. and B.A. Scott, 1970, Phys. Rev. Lett. **25**, 1678.

Chaves, A.S., R. Blinc, 1979, Phys. Rev. Lett. **43**, 1037.

Courtens, E., 1977, Phys. Rev. Lett. **39**, 561.

Courtens, E., 1978, Phys. Rev. Lett. **41**, 1171.

De Dominicis, C., 1979, Lect. Notes Phys. **104**, 251.

Durvasula, L.N. and R.W. Gammon, 1977, Phys. Rev. Lett. **38**, 1081.

Elliott, R.J. and D.W. Taylor, 1967, Proc. Roy. Soc. Lond. A **296**, 161.

Eshelby, J.D., 1956, The Continuum Theory of Lattice Defects, in: Solid State Physics, vol. **3** (Academic Press, New York) ch. 2.

Fleury, P.A., 1970, Solid State Commun. **8**, 601.

Fleury, P.A. and K.B. Lyons, 1976, Phys. Rev. Lett. **37**, 1088.

Fleury, P.A. and K.B. Lyons, 1979, Solid State Commun. **32**, 103.

Fousek, J., 1978, private communication.

Geisel, T. and J. Keller, 1978, J. Phys. Chem. Solids, **39**, 1.

Ginzburg, V.L., A.P. Levanyuk, A.S. Sigov and A.A. Sobyanin, 1979, Light Scattering near Structural Phase Transition Points in Pure Crystals and in Crystals Containing Defects, in: Light Scattering in Solids, eds., J.L. Birman, H.Z. Cummins and K.K. Rebane (Plenum Press, New York) p. 331.

Ginzburg, V.L., A.P. Levanyuk and A.A. Sobyanin, 1980, Phys. Re. **57**, 151.

Ginzburg, V.L. and A.A. Sobyanin, 1976, Usp. Fiz. Nauk, **120**, 153.
Godefroy, G., B. Jannot, A.P. Levanyuk and A.S. Sigov, 1983, Izv. AN SSSR, Ser. Fiz. **47**, 688.
Grabinskii, N.V. and A.S. Sigov, 1980, Fiz. Tverd. Tela, **22**, 1308.
Grabinskii, N.V., A.P. Levanyuk and A.S. Sigov, 1982, Fiz. Tverd. Tela, **24**, 1936.
Grigas, I. and R. Belyatskas, 1978, Fiz. Tverd. Tela, **20**, 3675.
Halperin, B.I. and C.M. Varma, 1976, Phys. Rev. B **14**, 4030.
Hastings, J.B., S.M. Shapiro and B.C. Frazer, 1978, Phys. Rev. Lett. **40**, 237.
Höck, K.-H. and H. Thomas, 1977, Z. Phys. B **27**, 267, 314.
Ipatova, I.P. and A.A. Klochihin, 1978, in: Proc. II Conf. Raman Scattering (Moscow State University, Moscow) p. 126.
Kaganov, M.I., 1972, ZhETF, **62**, 1196.
Kaganov, M.I. and A.N. Omelyanchuk, 1971, ZhETF, **61**, 1679.
Kittel, C., 1958, Elementary Statistical Physics (Wiley, New York) § 6.
Kosevich, A.M., 1978, Crystal dislocations and the theory of elasticity, in: Dislocations in Solids, Vol. 1, ed., F.R.N. Nabarro (North-Holland, Amsterdam) ch. 1.
Kristoffel, N.N., 1979a, Optica i Spectroscopiya, **47**, 609.
Kristoffel, N.N., 1979b, Fiz. Tverd. Tela, **21**, 895.
Lagakos, N. and H.Z. Cummins, 1974, Phys. Rev. B **10**, 1063.
Laulicht, I., J. Bagno and G. Shlesinger, 1972, J. Phys. Chem. Solids **33**, 319.
Leibfried, G. and N. Breuer, 1978, Point Defects in Metals I (Springer, Berlin, Heidelberg, New York) ch. 6.
Levanyuk, A.P., 1976, ZhETF, **70**, 1253.
Levanyuk, A.P. and A.S. Sigov, 1979, Izv. AN SSSR, Ser. Fiz. **43**, 1562.
Levanyuk, A.P. and A.S. Sigov, 1980, Fiz. Tverd. Tela, **22**, 1744.
Levanyuk, A.P. and A.S. Sigov, 1980, J. Phys. Soc. Jpn. **49**, Suppl. B, 4.
Levanyuk, A.P., V.V. Osipov and A.A. Sobyanin, 1976, On the influence of impurities on light scattering at phase transitions, in: Theory of Light Scattering in Condensed Matter, eds., B. Bendow, J.L. Birman and V.M. Agranovich (Plenum Press, New York) p. 517.
Levanyuk, A.P., V.V. Osipov and A.S. Sigov, 1978, Ferroelectrics, **18**, 147.
Levanyuk, A.P., V.V. Osipov, A.S. Sigov and A.A. Sobyanin, 1979, ZhETF, **76**, 345.
Levanyuk, A.P., A.S. Sigov and A.A. Sobyanin, 1980, Ferroelectrics, **24**, 61.
Levanyuk, A.P., B.V. Mostchinskii and A.S. Sigov, 1981, Fiz. Tverd. Tela, **23**, 2037.
Lockwood, D.J., J.W. Arthur, W. Taylor and J.T. Hosea, 1976, Solid State Commun. **20**, 703.
Loveluck, J.M. and J.B. Sokoloff, 1973, J. Phys. Chem. Solids, **34**, 869.
Lyons, K.B. and P.A. Fleury, 1977, Solid State Commun. **23**, 477.
Lyons, K.B. and P.A. Fleury, 1978, Phys. Rev. B **17**, 2403.
Lyons, K.B. and P.A. Fleury, 1979, Quasielastic Light Scattering near Structural Phase Transitions, in: Light Scattering in Solids, eds., J.L. Birman, H.Z. Cummins and K.K. Rebane (Plenum Press, New York) p. 357.
Lyubov, B.Ya. and V.S. Solov'ev, 1965, Fiz. Met. Metalloved. **19**, 333.
Ma, Shang-keng, 1976, Modern Theory of Critical Phenomena (Benjamin, London) ch. 10.
Maradudin, A.A., 1966a, Solid State Phys. **18**, 273.
Maradudin, A.A., 1966b, Solid State Phys. **19**, 1.
Mermelstein, M.D. and H.Z. Cummins, 1977, Phys. Rev. B **16**, 2177.
Mills, D.L., 1972, Phys. Rev. B **3**, 3887.
Müller, K.A., N.S. Dalal and W. Berlinger, 1976, Phys. Rev. Lett. **36**, 1504.
Müller, K.A., 1979, Lect. Notes Phys. **104**, 210.
Nagaev, E.L., 1968, ZhETF, **54**, 228.
Patashinsky, A.Z. and V.L. Pokrovsky, 1982, Fluctuatsionnaya Teoriya Fazovikh Perekhodov (Nauka, Moscow).
Prater, R.L., L.L. Chase and L.A. Boatner, 1981a, Phys. Rev. B **23**, 221; 5904.

Prater, R.L., L.L. Chase and L.A. Boatner, 1981b, Solid State Commun. **40**, 697.
Reiter, G.F., W. Berlinger and K.A. Müller, P. Heller, 1980, Phys. Rev. B **21**, 1.
Riste, T., E.J. Samuelsen, K. Otnes and J. Feder, 1971, Solid State Commun. **9**, 1455.
Sawafuji, M., M. Tokunaga and I. Tatsuzaki, 1979, J. Phys. Soc. Jpn. **47**, 1860; 1870.
Schmidt, H. and F. Schwabl, 1977, Phys. Lett. **61 A**, 476.
Schmidt, H. and F. Schwabl, 1978, Z. Phys. B **30**, 197.
Shapiro, S.H. and H.Z. Cummins, 1968, Phys. Rev. Lett. **21**, 1578.
Shigenari, T., Y. Takagi and Y. Wakabayashi, 1976, Solid State Commun. **18**, 1271.
Shirane, G. and J.D. Axe, 1971, Phys. Rev. Lett. **27**, 1803.
Schmidt, V.V., 1966, Pis'ma ZhETF, **3**, 141.
Smolensky, G.A., R.V. Pisarev, P.A. Markovin and B.B. Krichevzov, 1977, Physica, **86–88B**, 1205.
Sobyanin, A.A., 1971, ZhETF, **61**, 433.
Sokoloff, J.B., 1972, Phys. Rev. B **5**, 4962.
Sonin, E.B., 1980, J. Phys. C. Sol. St. Phys. **13**, 3293.
Steigmeier, E.F., H. Auderset and G. Harbeke, 1973, Solid State Commun. **12**, 1077.
Stoneham, A.M., 1975, Theory of Defects in Solids (Clarendon Press, Oxford) ch. 10.
Suhl, H., 1975, Appl. Phys. **8**, 217.
Tanaka, H., T. Yagi and I. Tatsuzaki, 1978, J. Phys. Soc. Jpn, **44**, 2009.
Uwe, H. and T. Sakudo, 1976, Phys. Rev. B **13**, 271.
Vugmeister, B.E., 1982, Pis'ma ZhETF, **36**, 26.
Yacoby, Y., 1978, Z. Phys. B **31**, 275.
Yacoby, Y., 1981, Z. Phys. B **41**, 269.
Yacoby, Y. and S. Just, 1974, Solid State Commun. **15**, 715.
Yacoby, Y., W.B. Holzapfel and D. Bäerle, 1977, Solid State Commun. **23**, 947.
Yacoby, Y., R.A. Cowley, T.J. Hosea, S.J. Lockwood and W. Taylor, 1978, J. Phys. C: Sol. St. Phys. **11**, 5065.
Yagi, T., H. Tanaka and I. Tatsuzaki, 1976, J. Phys. Soc. Jpn, **41**, 717.
Yagi, T., H. Tanaka and I. Tatsuzaki, 1977, Phys. Rev. Lett. **38**, 609.
Yakovlev, I.A., T.S. Velichkina and L.F. Mikheeva, 1956a, Doklady Acad. Nauk SSSR **107**, 675.
Yakovlev, I.A., T.S. Velichkina and L.F. Mikheeva, 1956b, Krystallographiya, **1**, 123.
Yermolov, A.F., A.P. Levanyuk and A.S. Sigov, 1979, Fiz. Tverd. Tela, **21**, 3628.
Yermolov, A.F., A.P. Levanyuk and A.S. Sigov, 1981, Fiz. Tverd. Tela, **23**, 2134.

CHAPTER 3

# Light Scattering from Incommensurate Phases

V.A. GOLOVKO

*Moscow Evening Metallurgical Institute*
*Moscow 111250, USSR*

A.P. LEVANYUK

*A.V. Shubnikov Institute of Crystallography, Academy of Sciences of the USSR*
*Moscow 117333*
*USSR*

*Light Scattering near Phase Transitions*
*Edited by*
*H.Z. Cummins and A.P. Levanyuk*

© *North-Holland Publishing Company, 1983*

# Contents

## 1. Introduction

At the present time a growing attention is given to the study of incommensurate phases, i.e., phases that have a superstructure with a period incommensurate with that of the underlying lattice structure (for a review see, for example, Przystawa (1980)). They differ from ordinary crystals by a lack of translational symmetry although a long-range order remains. This alone implicates that in the case of incommensurate phase, in contrast to an ordinary crystal, the concept of the vibrational spectrum in the form of a small number of phonon branches is strictly speaking inapplicable. A less trivial feature of the incommensurate phase is the specific character of the vibrational spectrum in the long-wavelength excitation region. The vibrational spectrum contains an additional acoustic-like branch corresponding to Goldstone excitations typical of degenerate systems, which the incommensurate phase belongs to. For reasons explained below such an excitation is referred to as a phason. Since light scattering technique is the most convenient tool for investigating the long-wavelength region, it is not surprising that in recent years rather many papers on experimental study of light scattering from incommensurate phases have appeared (see, for example, the review by Petzelt (1981) and ch. 8 of the present book).

Unfortunately, one cannot yet speak of a reliable experimental observation of the phason. In our opinion this was connected not in the last place with an insufficient development of the theory of light scattering by an incommensurate phase and with the absence of its clear presentation. As a result there appeared in the literature very different propositions concerning the phason manifestations in light scattering. Thus, Cowley (1980), Poulet and Pick (1981) stated, in fact, that the phason mode is wholly equivalent to an acoustic mode and must manifest itself in a similar manner in Mandelstam–Brillouin scattering. On the other hand, there are papers (Wada et al. 1977, Dvorak and Petzelt 1978, Golovko and Levanyuk 1980) suggesting that the Goldstone mode is inactive in light scattering. There also exists an opinion (Lyons et al. 1980) that a phason is active in light scattering only if the crystal is not centre symmetric in the high-temperature phase. The absence of a sound theory also led to a controversy in the interpretation of experimental data. In some experiments one sought for the manifestation of a phason in light scattering with a change of polarization (Fleury et al. 1979, Unruh et al. 1979, Takashige and Nakamura 1980, Inoue et al. 1980), in others, in light scattering without a change of polarization (Lyons and Guggenheim 1979, Lyons et al. 1980). Attempts were made to detect the phason contribution both to the central peak (Lyons and Guggenheim 1979, Lyons et al. 1980) and to Mandelstam–Brillouin satellites (Lockwood et al. 1981).

At the same time the phenomenological theory of light scattering in an incommensurate phase, which reflects adequately the fundamental features of this phenomenon, may be constructed by analogy with the theory of light scattering in the case of ordinary (commensurate) phase transitions (ch. 1 of the present book). Such an approach was also applied to incommensurate phases (Dvorak and Petzelt 1978, Golovko and Levanyuk 1980). However in these papers the specific features of the incommensurate phase were not taken into account completely (Golovko and Levanyuk 1981c).

Note that in some papers (see, for example, Janner and Janssen (1980)) in attempts to describe the specific features of the incommensurate phase the so-called superspace groups are introduced. In our opinion such an approach is not obligatory and moreover it complicates the problem. At the same time in the phenomenological approach, as always within the Landau theory, one should take into account only the high-temperature phase symmetry that is an ordinary one, the properties of the incommensurate phase being established straightforwardly.

The present chapter is concerned with the phenomenological theory of light scattering from the incommensurate phase in a defect-free crystal. According to some concepts the temperature range of the incommensurate phase may be a sequence of long-period commensurate phases (the so-called "devil's staircase"). A special section of the chapter deals with light scattering under the existence of the devil's staircase, unexpected features being observed in this case. The theory of light scattering in the incommensurate phase presented here is based on a series of papers by the present authors (Golovko and Levanyuk 1980, 1981b, c), but we tried to refer to corresponding results obtained by other authors. In this chapter we discuss for the sake of illustration only some of the experimental works and no pretense is made of an exhaustive review of the state of experiment in this field, which is the subject of a separate chapter of the book.

## 2. *Incommensurate phases*

### 2.1. *Incommensurate phase transitions*

In this section we consider the notion of the incommensurate phase and discuss its properties necessary for what follows on an example of displacive-type phase transition. To provide a better insight into the nature of the incommensurate phase let us start with an ordinary (commensurate) second-order phase transition. In fig. 1 open and closed circles denote two sorts of atoms in the high-temperature symmetrical (i.e., undistorted or prototype) phase, all the atoms being placed on the same straight line for the sake of simplicity. Arrows indicate displacement of some atoms (closed circles) with respect to other atoms (open circles) below a phase transition point. As is known, the phase transition is associated with the existence in the symmetrical phase of a soft mode, whose frequency becomes zero at the transition point*. As a result, in the crystal the atom displacement wave described by the normal coordinate corresponding to the soft mode is frozen-in. For the phase transition this normal coordinate plays the role of order parameter. Figure 1 illustrates also the dispersion of the soft-phonon branch, i.e., the dependence of the frequency $\omega$ on the wavevector $\boldsymbol{k}$, broken lines indicate the dispersion above the phase transition point and solid lines do the same at the phase transition point.

In fig. 1(a) all the atoms are displaced in a similar manner and there occurs a phase transition which does not modify the crystal period. This means that at the transition point the frequency $\omega$ becomes zero at $\boldsymbol{k} = 0$ (fig. 1(b)). If a softening mode corresponds to the symmetrical phase Brillouin zone boundary, i.e., to the wavevector $\boldsymbol{k} = \boldsymbol{b}/2$, where $\boldsymbol{b}$ is a basic reciprocal lattice vector of the symmetrical phase (fig. 1(d)), the crystal period below the phase transition point becomes equal to $4\pi/b$, i.e., it doubles (fig. 1(c)). If the soft mode has a wavevector $\boldsymbol{k} = m\boldsymbol{b}/n$, where $m$ and $n$ are integers, the phase transition involves $n$-fold increase of the crystal period.

In some cases the soft mode is characterized by a wavevector $\boldsymbol{k} = \boldsymbol{K}_0$ incommensurate with $\boldsymbol{b}$, i.e., such a wavevector that the number $K_0/b$ is

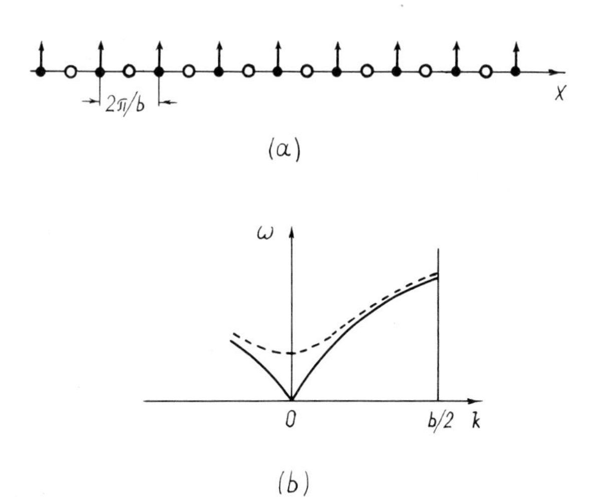


*Strictly speaking, it is not the frequency but a generalized elasticity coefficient that goes to zero as the transition point is approached, this coefficient being proportional to the frequency squared only in the simplest cases (for more details see ch. 1 of the book).

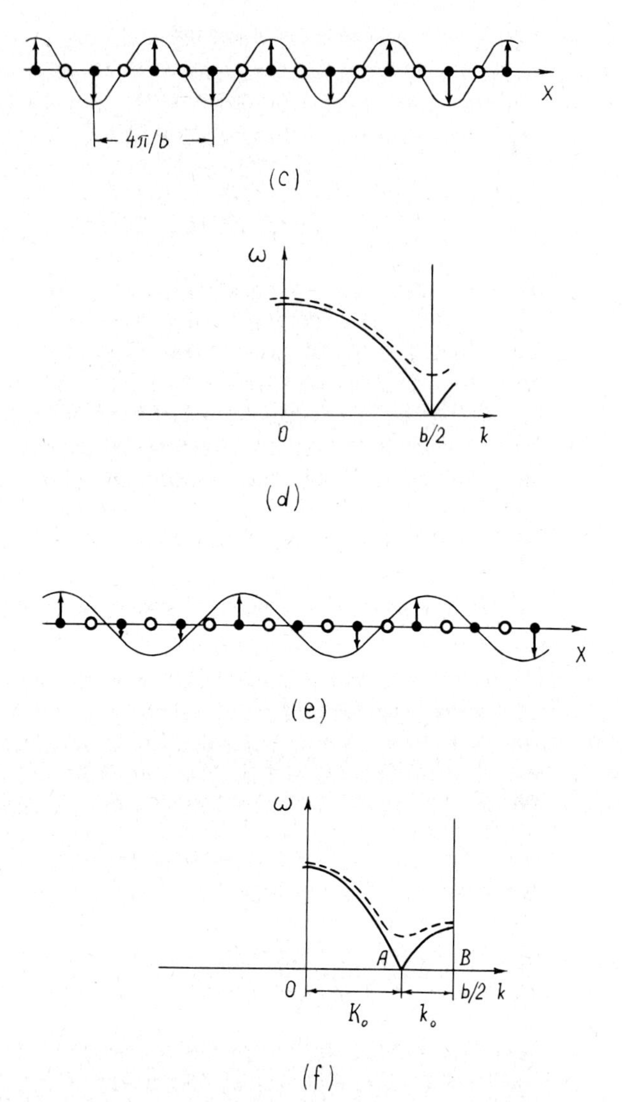


Fig. 1. Atom displacements due to phase transitions and dispersion of the soft-phonon branch in the symmetrical phase ($k = k_x$). Atoms assumed to be immobile are denoted by open circles, arrows stand for atom displacements. (a),(b) Phase transition without change of translational symmetry. (c),(d) Phase transition with cell doubling. (e), (f) Incommensurate phase transition.

irrational (fig. 1(f)). The period of the superstructure which is formed in the crystal below the phase transition point is then not a multiple of that of the underlying lattice structure (fig. 1(e)). On the whole the new phase does not at all possess any periodicity along the coordinate axis in question and is referred to as incommensurate. Incommensurability may, naturally, occur along two or three coordinate axes. Although periodicity, the fundamental feature of the crystalline state, is lost in the incommensurate phase, the latter cannot be assigned to an amorphous state even in the case of a three-dimensional incommensurability since the long-range order is present due to a non-random position of atoms in the space.

For our purposes there is no need to consider microscopic theories which explain the origin of the incommensurate phase. For structural phase transitions these theories are reduced, in fact, to calculating the dependence, $\omega(\boldsymbol{k})$, and establishing the values of $\boldsymbol{k}$ at which the frequence $\omega$ may vanish. This may be exampled by calculations carried out by Haque and Hardy (1980) for $K_2SeO_4$ showing that at the values of atomic interaction constants typical of this crystal one of the eigenfrequencies may vanish when the value of $\boldsymbol{k}$ is incommensurate with $\boldsymbol{b}$. In the case of quasi-one-dimensional conductors the appearance of incommensurate phase is ascribed to Peierls instability (see, for example, Berlinsky (1979)).

Going back to fig. 1, we see that the value of the wavevectors of the superstructure formed below a phase transition point is determined by the position of the minimum on the soft mode dispersion curve in the symmetrical phase. The slope of the dispersion curve to the abscissa at some $\boldsymbol{k} = \boldsymbol{K}_0$ (and in the minimum this slope is zero) is characterized by the term linear in $\Delta\boldsymbol{k} = \boldsymbol{k} - \boldsymbol{K}_0$ in the Taylor expansion of the function $\omega(\boldsymbol{k})$ about the point $\boldsymbol{k} = \boldsymbol{K}_0$. Whether such a linear term is present or absent can be established from symmetry considerations. To this end we note that $[\omega(\boldsymbol{k})]^2/2$ is the coefficient of $|Q(\boldsymbol{k})|^2$ in the potential energy expansion in powers of the normal coordinates $Q(\boldsymbol{k})$, and in the case under consideration it is the thermodynamic potential density that can be taken as potential energy (cf. the equations of motion for the order parameter in ch. 1 of this book). When investigating the behaviour of the function $\omega(\boldsymbol{k})$ in the vicinity of the point $\boldsymbol{k} = \boldsymbol{K}_0$, we may take into account that a wave with the wavevector $\boldsymbol{k} = \boldsymbol{K}_0 + \Delta\boldsymbol{k}$ may be represented as a wave with the wavevector $\boldsymbol{k} = \boldsymbol{K}_0$ but with an amplitude varying in space proportionally to $\exp(\mathrm{i}\Delta\boldsymbol{k}\boldsymbol{r})$. In this case the presence in the function $\omega(\boldsymbol{k})$ of terms linear in $\Delta\boldsymbol{k}$ implies that the thermodynamic potential density which always consists of invariants under symmetry transformations of the high-temperature phase space group (Landau and Lifshitz 1976), when written as a function of the normal coordinates $Q(\boldsymbol{K}_0)$, includes an invariant quadratic in these coordinates and contains linearly their first spatial derivatives. Such an invariant is referred to as a Lifshitz invariant, and we shall discuss it in some more detail, for it plays an important role, as will be seen below, in a theoretical consideration of light scattering by an incommensurate phase. In the case, for example, of normal coordinates $Q_1$ and $Q_2$, which are transformed according to a two-dimensional irreducible representation, the Lifshitz invariant takes the form (Landau and Lifshitz 1976)

$$Q_1 \frac{\partial Q_2}{\partial x_i} - Q_2 \frac{\partial Q_1}{\partial x_i} . \tag{1}$$

From this it is seen, in particular, that for a Lifshitz invariant to exist, the physically irreducible representation, which describes the transformation properties of the normal coordinates with a wavevector $\boldsymbol{k} = \boldsymbol{K}_0$, must be of dimension two or more since one cannot form an expression of the type (1) when having only one quantity $Q$.

If a phonon branch is not degenerate, only one normal coordinate corresponds to the centre or to the boundary of the Brillouin zone, i.e., the irreducible representations related to these points are one-dimensional. The Lifshitz invariant is absent in these cases, which is associated with the well-known fact that a non-degenerate branch has extrema at the centre and at the boundary of the Brillouin zone. For degenerate phonon branches it is multidimensional representations that correspond to the centre or to the boundary of the Brillouin zone, and such representations can admit a Lifshitz invariant. Figure 2(a) gives an example of a phonon branch degenerate at the Brillouin zone boundary. A non-zero slope of the branch at $k = b/2$ (therefore, a Lifshitz invariant exists in this case) and symmetry of the dispersion curves relative to the zone centre ($\omega(\boldsymbol{k}) = \omega(-\boldsymbol{k})$) make it possible to conclude that at the zone boundary two branches intersect whose slopes are of different signs. The lower branch possesses necessarily a minimum inside the zone at some $k = K_0$. Therefore, as temperature lowers, it is not the frequency at $k = b/2$ but that at $k = K_0$ that vanishes and there appears a superstructure with the wavevector $k = K_0$. When $K_0$ is close to $b/2$, the value of $K_0$ is determined first of all by the slope angle of the dispersion curve at $k = b/2$, i.e., by the relevant coefficient of the Lifshitz invariant in the thermodynamic potential. There is no reason for the ratio $K_0/b$ to be rational, i.e., generally speaking, it is irrational,

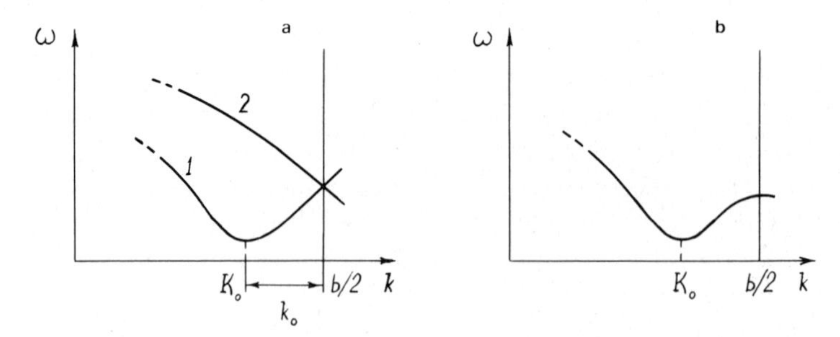


Fig. 2. Possible forms of dispersion curves at temperatures higher than phase transition temperature.

and the superstructure appearing below the phase transition point will be incommensurate. Naturally, for a branch non-degenerate at the centre and at the boundary of the Brillouin zone a minimum inside the zone is also possible (fig. 2(b)).

Irreducible representations that correspond to internal points of the Brillouin zone, excluding the point $\boldsymbol{k} = 0$, always have dimensions not less than two. This is due to the fact that if a vector $\boldsymbol{k} = \boldsymbol{K}_0 \neq 0$ lies inside the zone, the star of the physically irreducible representation contains a vector $-\boldsymbol{K}_0$, non-equivalent to $\boldsymbol{K}_0$ and, therefore, there exist at least two different basis functions of the irreducible representation proportional to $\exp(\pm \mathrm{i}\boldsymbol{K}_0\boldsymbol{r})$ (Landau and Lifshitz 1976). There are no symmetry reason for the dispersion curve slope to be zero inside the zone at $\boldsymbol{k} \neq 0$. This means that physically irreducible representations associated with internal points of the zone at $\boldsymbol{k} \neq 0$ always admit a Lifshitz invariant. A dispersion curve minimum at such points is due to the fact that the coefficient of the Lifshitz invariant in the thermodynamic potential vanishes at these points accidentally but not for symmetry reasons.

Let us consider some properties of incommensurate phases. Since the value of $K_0$ is not fixed by the crystal symmetry, the ratio $K_0/b$ must depend on external parameters, for example on temperature, which is most often the case (see, e.g., Iizumi and Gesi (1980)). If the value of $K_0$ is commensurate with $b$, in the thermodynamic potential additional terms appear (see subsect. 7.1) which may decrease this potential. These terms are particularly large when the ratio $K_0/b$ is a simple rational number like 0, $\frac{1}{2}$, $\frac{1}{3}$, $\frac{1}{4}$. As a result, when the temperature decreases, the ratio $K_0/b$ tends to one of those values (usually to the nearest one), reaches it, and then there occurs a phase transition to a corresponding commensurate phase. Such type of transition is referred to as a lock-in transition. Therefore the incommensurate phase exists, as a rule, only in some temperature regions between $T_i$ (temperature of transition from the symmetrical phase) and $T_c$ (temperature of the lock-in transition). The temperature interval $T_i$–$T_c$ may not be large (for example, about 1,5K in $NaNO_2$, Hoshino and Motegi 1967), but sometimes it reaches high values (about 500 K in $Na_2CO_3$, De Pater and Helmholdt 1979).

Near the temperature $T_i$ the incommensurate phase structure is sufficiently well described by a frozen-in wave with the wavevector $\boldsymbol{K}_0$. As temperature lowers, the role of frozen-in waves with other wavevectors increases, and the incommensurate phase becomes domain-like (Dzialoshinskii 1964b, Levanyuk and Sannikov 1976a, McMillan 1976). Inside the domains the incommensurate phase structure practically coincides with that of the low-temperature commensurate phase, and differences are observed only in the domain walls called in this case discommensurations or solitons. As temperature goes on decreasing, the distance between solitons grows, and some authors believe (Dzialoshinskii 1964b, McMillan 1976, Bak and Timonen 1978) that the lock-in transition to the commensurate phase takes place when this distance becomes infinite, the transition being continuous. However, in experiment the incommensurate–commensurate transition always proves to be of first order, which also has some theoretical grounds (Bruce et al. 1978, Shiba and Ishibashi 1978). There are experiments (Blinc et al. 1981, Milia and Rutar 1981, Rehwald and Vonlanthen 1981) suggesting either that this transition occurs before the incommensurate phase becomes domain-like, or that the domain-like structure exists only in a narrow temperature region near $T_c$, though experimental data showing the opposite are available too (Suits et al. 1980, Chen et al. 1981). The presence or absence of the domain-like structure is possibly associated with the type of crystal. Light scattering in the incommensurate phase with a well developed domain-like structure, although may have its specific features, will not be considered here, since this question, as far as we know, has not yet been give theoretical contemplation.

In theoretical description of incommensurate phase transitions there exist in the literature two equivalent approaches. In one of them (Iizumi et al. 1977, Dvorak and Ishibashi 1978) the order parameter characterizing directly the symmetrical-incommensurate transition is utilized. Such an order parameter, as mentioned above, is the normal coordinate $Q(\boldsymbol{K}_0)$ which corresponds to the wavevector $\boldsymbol{k} = \boldsymbol{K}_0$ of the incommensurate superstructure appearing just below the phase transition temperature $T_i$. The value of $\boldsymbol{K}_0$ is distinguished by the fact that with it the coefficient of the Lifshitz invariant in the thermodynamic potential vanishes at the phase transition point. The other approach (Dzialoshinskii 1964a, Levanyuk and Sannikov 1976a) is used in a unified description of phase transitions from a symmetrical phase to an incommensurate one and from the latter to a low-temperature commensurate phase. In this case one introduces only the order parameter corresponding to the symmetrical-commensurate transition, and the structure of the incommensurate phase is described then as spatially inhomogeneous with a frozen-in wave of the order parameter. This is connected with the fact that a frozen-in wave with the wavevector $\boldsymbol{K}_0$ may be represented as a wave with the wavevector $\boldsymbol{k}_c$, that corresponds to the commensurate phase, but the amplitude of this wave (which the order parameter is proportional to) will be periodically modulated in space, this modulation being characterized by the wavevector $\boldsymbol{k}_0 = \boldsymbol{K}_0 - \boldsymbol{k}_c$. Below the lock-in transition point to the commensurate phase, where $\boldsymbol{K}_0 = \boldsymbol{k}_c$, the order parameter becomes, as is usually the case, spatially homogeneous. From fig. 1(f) it is seen that softening of a mode corresponding to point A may be described proceeding, for example, from point B and with due regard of the fact that at points A and B the wavevectors differ by the quantity $k_0$. Naturally, such a description of the incommensurate phase is meaningful if the value of $k_0$ is small and in the commensurate phase $k_0 = 0$. Note that a somewhat different consideration is required for the case when the irreducible representation corresponding to $\boldsymbol{k} = \boldsymbol{k}_c$ admits the Lifshitz invariant (Dzialoshinskii 1964a, Levanyuk and Sannikov 1976a) and when it does not admit this invariant (Levanyuk and Sannikov 1976b, Ishibashi and Shiba 1978). If $\boldsymbol{k}_c = \boldsymbol{b}/2$, the former case corresponds to fig. 2(a), and the latter to fig. 2(b).

Both the indicated approaches to the description of the incommensurate phase must, naturally, give identical results. In the present chapter we mainly use the first approach since in the study of light scattering by an incommensurate phase it enables one to obtain a number of general results by simpler methods. However in sect. 6 we show how the second approach may be applied to the problem under discussion.

According to what has been said above, the order parameter used in the first approach has at least two components since the vector $\boldsymbol{K}_0$ lies inside the Brillouin zone of the symmetrical phase and the star of the physically irreducible representation contains a vector $-\boldsymbol{K}_0$ non-equivalent to $\boldsymbol{K}_0$. Since in many crystals with incommensurate superstructure (Ishibashi 1980) the order parameter is two-component, this case is the first to be contemplated as a simpler one. The case of the order parameter with more than two components is analyzed in section five of this chapter. For the components of the order parameter one may take two complex conjugate quantities $Q(\boldsymbol{K}_0)$ and $Q(-\boldsymbol{K}_0) = Q^*(\boldsymbol{K}_0)$. Instead of the complex quantities it is convenient to introduce, as is usually done, two real quantities, the amplitude $\rho$ and the phase $\varphi$, according to the relation $Q = \rho e^{i\varphi}$.

Although the symmetry of an incommensurate phase cannot be described in terms of ordinary space groups, the incommensurate phase possesses a certain point symmetry which is a macroscopic characteristic. If the order parameter is two-component, the point symmetry of the incommensurate phase always coincides with that of the high-temperature phase. To prove this fact we note that the point symmetry is directly connected with the presence of non-zero components in the vectors and tensors that characterize the macroscopic properties of the medium under consideration. Vectors and tensors are translational invariants, i.e., they remain unchanged under translations, in particular under those entering the symmetry group of the high-temperature phase. The components of the order parameter are transformed like basis functions of the irreducible representation, whose variation under translations is due to the factors $\exp(\pm i\boldsymbol{K}_0\boldsymbol{r})$. If the vector $\boldsymbol{K}_0$ is incommensurate with any reciprocal lattice vectors of the high-temperature phase, then the translational invariants depending on the order parameter and containing no derivatives are only $QQ^* = |Q|^2$ and the powers of this quantity, which are also invariant under all symmetry transformations of the space group of the high-temperature phase (in what follows we refer to such invariants as perfect). Therefore, after the transition to the incommensurate phase, when $Q$ takes a spontaneous value, no tensorial quantities forbidden by the high-temperature phase symmetry appear since in this phase it is just perfect invariants that differ from zero. As a result we see that the point symmetry remains unchanged with the appearance of the incommensurate phase. The same fact can be proved somewhat differently by using the second of the above-mentioned approaches to the description of the incommensurate phase, when the latter is considered as a spatially inhomogeneous structure (Golovko and Levanyuk 1979). Note that if the order parameter contains more than two components, the point symmetry of the incommensurate phase may differ from that of the high-temperature phase (see subsect. 5.1).

To find the equilibrium value of the order parameter, we write, according to the Landau theory, the thermodynamic potential as a function of the order parameter. The thermodynamic potential may depend exclusively on perfect invariants which, as we have seen, are the quantities $(QQ^*)^n = \rho^{2n}$ alone if we disregard derivative-dependent invariants needless in the case under consideration. Accordingly, the first terms in the expansion of the thermodynamic potential density $\Phi$ in powers of the order parameter have the form

$$\Phi = \tfrac{1}{2}A\rho^2 + + \tfrac{1}{4}B\rho^4 . \tag{2}$$

Here and below we do not write in $\Phi$ the term of zeroth order in $\rho$. As is usually done within the Landau theory, we consider the coefficient $A$ to be a linear function of temperature: $A = A'(T - T_i)$, $A' > 0$; the coefficient $B$ will be considered positive since, according to experiment, phase transitions to an incommensurate phase are, as a rule, of second order.

By minimizing the thermodynamic potential, we find the equilibrium value of the amplitude $\rho_e$:

$$\rho_e^2 = -\frac{A}{B} = \frac{A'(T_i - T)}{B}, \qquad T < T_i , \tag{3}$$

and $\rho_e = 0$ when $T > T_i$. The phase $\varphi$ does not enter eq. (2) and remains arbitrary.

Note that the formulae describing light scattering and obtained below from symmetry considerations hold, not only within the region of applicability of the Landau theory, where expression (3) is valid, but also in the scaling region, wherein the quantity $\rho_e^2$ is proportional to $(T_i - T)^{2\beta}$; here $\beta$ is one of the critical indices (see ch. 1 and the literature cited there). Hereafter we restrict ourselves for the sake of simplicity to the limits of applicability of the Landau theory; an extension of corresponding formulae to the scaling region can be carried out taking into account the fact that the essential temperature dependences are determined in them, as seen below, either by the quantity $\rho_e$ alone or, in some cases, also by other quantities, the temperature dependence of which in the scaling region is discussed in ch. 1.

### 2.2. *Incommensurate phase excitations: phason and amplitudon*

Let us see what happens, after the transition to an incommensurate phase, with the soft mode which existed in the high-temperature symmetrical phase. Since

$Q$ and $Q^*$ correspond to one and the same frequency (phonon branches are symmetric with respect to the Brillouin zone centre: $\omega(\boldsymbol{k}) = \omega(-\boldsymbol{k})$), the corresponding oscillations may be formally considered to be degenerate. This degeneracy is lifted below the phase transition point due to the superstructure formation, which leads to branch splitting, and instead of one, two soft mode frequencies appear. One of the frequencies is associated with shifts with respect to the crystal of the frozen-in wave (presented in fig. 1(e)) as a whole, without alterations of its shape (fig. 3(a)). Such shifts result from variations of the wave phase, and a corresponding excitation is referred to as a phason. The other frequency is connected with the frozen-in wave amplitude oscillations (fig. 3(b)), and a corresponding excitation is called an amplitudon (Overhauser 1971, Lee et al. 1974, Bruce and Cowley 1978).

In the case of fluctuations corresponding to amplitudon, commensurability or incommensurability of the frozen-in wave period with that of underlying lattice does not play an essential role. Accordingly, the properties of an amplitudon must be similar to those of the soft mode below the point of an ordinary (commensurate) second-order phase transition. In particular, the properties of the amplitudon must not vary noticeably at the phase transition from an incommensurate to a commensurate low-temperature phase excluding, for example, a jump in the frequency if this is a first-order transition.

As concerns the phason, its properties differ essentially from those of the soft modes in commensurate phases. Since in an incommensurate phase the period of the frozen-in wave is incommensurate with that of the underlying lattice, i.e., the structure is non-periodic, in an unlimited equilibrium crystal there exists an infinite set of various atom displacements (fig. 1(e)), and in this set there may occur any displacement smaller than the wave amplitude. If the frozen-in wave is shifted without changing its shape, as shown in fig. 3(a), the same infinite set of displacements is observed. Hence, different positions of the frozen-in wave relative to the crystal are energetically equivalent and the phason frequency is equal to zero. Naturally, this is valid only if the fluctuation is homogeneous, i.e., the wavevector $\boldsymbol{q}$ of the fluctuation is equal to zero. If $\boldsymbol{q} \neq 0$, the phason frequency will be non-zero since such a fluctuation induces a change in the

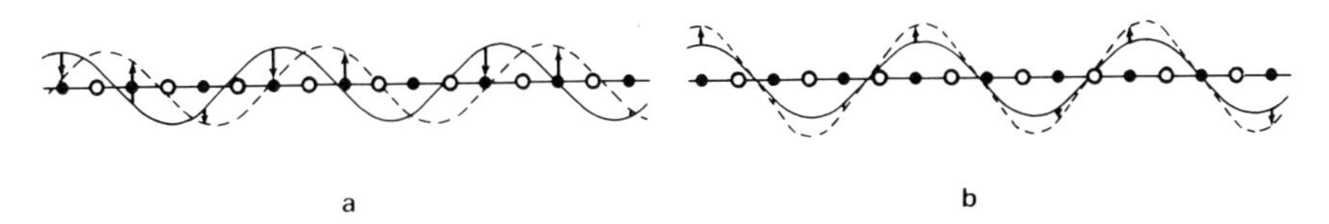


Fig. 3. Incommensurate phase excitations. Solid lines indicate equilibrium positions of atoms corresponding to fig. 1(e), broken lines show positions of the atoms displaced as a result of excitations. Arrows stand for displacements, closed and open circles for atom positions in the symmetrical phase. (a) Displacements corresponding to the phason. (b) Displacements corresponding to the aplitudon.

shape of the frozen-in wave and possibly also in the amplitude. The existence in the incommensurate phase of a mode whose frequency is equal to zero at a zero wavevector $\boldsymbol{q}$ follows also from the fact that the thermodynamic potential (2) is $\varphi$-independent, and below the phase transition point the symmetry is broken and there appears a spontaneous value of $\varphi$. This result is a consequence of the well-known Goldstone theorem, and therefore a phason is called the Goldstone mode (Bruce and Cowley 1978).

Phason-like excitations are also possible in a commensurate phase. So, for example, the atom displacement wave, shown in fig. 1(c), may also oscillate as a whole. However, in a commensurate phase the set of atom displacements in the frozen-in wave is finite; in fig. 1(c), for example, only two types of atom displacements are observed. If the wave presented in fig. 1(c) is shifted as a whole, one will also observe only two types of atom displacements, but different from the first mentioned in magnitude. Hence, in the commensurate phase different positions of the frozen-in wave are not equivalent energetically, and the phason frequency is not equal to zero even at $\boldsymbol{q} = 0$.

The phason branch is sometimes termed a new acoustic branch in an incommensurate phase, for the phason frequency vanishes, like an acoustic phonon frequency, at a zero excitation wavevector $\boldsymbol{q}$. However a phason and an acoustic phonon differ essentially in what concerns their damping. If the wavevector $\boldsymbol{q}$ tends to zero, acoustic vibration does not lead to energy dissipation, because it corresponds to a shift of the whole crystal without variation of relative atom positions. As far as phason is concerned, although different positions of the frozen-in wave in a crystal are energetically equivalent, the passage from one position of the wave to another is connected with the change of relative atom positions, i.e., with dissipative processes. This is quite obvious from fig. 3(a). When passing from one position of the wave to another the atoms are displaced at different distances, i.e., they are displaced relative to one another and, besides, relative to those atoms, which are assumed for simplicity to be immobile at the phase transition. The general character of atom displacements in this case does not differ essentially from that of the atom displacements caused by optical vibrations in the symmetrical phase. Therefore, the damping constant of the phason must be of the same order as that for ordinary optical phonons, i.e., it must have a substantial value at $\boldsymbol{q} = 0$ and a weak dependence on $\boldsymbol{q}$, whereas in the case of acoustic phonons the damping constant vanishes as $\boldsymbol{q}^2$ when $\boldsymbol{q} \to 0$. (Landau and Lifshitz 1959, 1965). Since the phason frequency is small as $\boldsymbol{q} \to 0$, the phason is always overdamped as is distinct from the acoustic phonon, whose quality factor, proportional to the ratio of the frequency to the damping constant, increases infinitely as $\boldsymbol{q} \to 0$.

Let us now confirm by calculations the qualitative results obtained. To consider spatially inhomogeneous fluctuations, we supplement the thermodynamic potential density (2) with the invariants $|\partial Q/\partial x_i|^2$. Passing over to the variables $\rho$ and $\varphi$, we have

$$\Phi = \frac{A}{2}\rho^2 + \frac{B}{4}\rho^4 + \frac{1}{2}\sum_{i=1}^{3} D_i\left[\left(\frac{\partial\rho}{\partial x_i}\right)^2 + \rho^2\left(\frac{\partial\varphi}{\partial x_i}\right)^2\right]. \tag{4}$$

For the thermodynamic potential to be limited from below, we assume $D_i > 0$. According to what has been said in subsect. 2.1, the Lifshitz invariant is not included in (4) since with the order parameter chosen the coefficient of this invariant in $\Phi$ is equal to zero. We introduce also a kinetic energy $\mathscr{K}$ and a dissipation function $\mathscr{R}$ which correspond to the normal coordinate $Q$,

$$\mathscr{K} = \frac{\mu}{2}\frac{\partial Q}{\partial t}\frac{\partial Q^*}{\partial t} = \frac{\mu}{2}\left[\left(\frac{\partial\rho}{\partial t}\right)^2 + \rho^2\left(\frac{\partial\varphi}{\partial t}\right)^2\right], \qquad \mathscr{R} = \frac{\gamma}{2}\frac{\partial Q}{\partial t}\frac{\partial Q^*}{\partial t}. \tag{5}$$

The coefficients $\mu$ and $\gamma$ have the meaning of the effective mass and the friction coefficient, like in the case of $Q$ oscillations in the symmetrical phase. Within the Landau theory the temperature dependence of these coefficients may be neglected, i.e., they may be assumed to be the same both above and below $T_i$.

Let us clarify what estimates can be obtained for the coefficients entering eqs. (4) and (5), for they will be needed in what follows. The quantity $D_i b^2/\mu$ has a dimensionality of the frequency squared (see formulae obtained below). This frequency must coincide, in the order of magnitude, with the characteristic phonon frequencies in the symmetric phase $\Omega_a \sim 10^{13}\,\mathrm{s}^{-1}$. The damping constant $\gamma/\mu$ determines the width of the Raman line (see ch. 1 of this book). Even if one takes into account the existence of extremely narrow and extremely wide lines, the Raman line widths lie in the limit from $10^{-2}\,\Omega_a$ to $\Omega_a$. Hence, the final estimates are as follows

$$D_i b^2/\mu \sim \Omega_a^2, \qquad \Omega_a\mu/\gamma \sim 1 - 100\,. \tag{6}$$

In the case of crystals in which incommensurate phase transitions are observed, the widths of the soft-mode lines in the phase transition temperature region usually lie in the middle of the interval used in estimating $\Omega_a\mu/\gamma$ (Wada et al. 1977, 1978, 1981, Unruh et al. 1979, Sakai and Tatsuzaki 1980, Lockwood et al. 1981).

To obtain the equations of motion, we proceed from the least action principle. When considering the thermodynamic potential density as a potential energy density, we introduce the Lagrange function $\mathscr{L} = \mathscr{K} - \Phi$ and the action integral $\mathscr{S} = \int \mathscr{L}\,\mathrm{d}\boldsymbol{r}\,\mathrm{d}t$. Varying $\mathscr{S}$ in $\rho$ and $\varphi$, we obtain the Lagrange equations which, due to the presence of he dissipation function $\mathscr{R}$, should be supplemented on the right with $-\partial\mathscr{R}/\partial\dot{\rho}$ or $-\partial\mathscr{R}/\partial\dot{\varphi}$, respectively, where the point implies a time derivative (Landau and Lifshitz 1976). Since we are interested in small fluctuations $\Delta\rho = \rho - \rho_e$ and $\Delta\rho = \varphi - \varphi_e$ where $\rho_e$ and $\varphi_e$ are equilibrium values, the equations of motion are linearized in $\Delta\rho$ and $\Delta\varphi$

and finally we are led to the following equations:

$$\mu \frac{\partial^2 \Delta\rho}{\partial t^2} + \gamma \frac{\partial \Delta\rho}{\partial t} - \sum_{i=1}^{3} D_i \frac{\partial^2 \Delta\rho}{\partial x_i^2} + (A + 3B\rho_e^2)\Delta\rho = 0\,, \tag{7}$$

$$\mu \frac{\partial^2 \Delta\varphi}{\partial t^2} + \gamma \frac{\partial \Delta\varphi}{\partial t} - \sum_{i=1}^{3} D_i \frac{\partial^2 \Delta\varphi}{\partial x_i^2} = 0\,. \tag{8}$$

From this it is seen that the amplitude $\rho$ and the phase $\varphi$ vary independently, i.e., an amplitudon and a phason are normal excitations of the incommensurate phase. If solutions of eqs. (7) and (8) are sought for in the form of the waves ($\Delta\rho$, $\Delta\varphi \sim \exp \mathrm{i}(\boldsymbol{qr} - \omega t)$), then for the amplitudon and phason we respectively obtain the dispersion equations for complex frequency,

$$\mu\omega^2 + \mathrm{i}\gamma\omega = 2B\rho_e^2 + D_1 q_x^2 + D_2 q_y^2 + D_3 q_z^2\,, \tag{9}$$

$$\mu\omega^2 + \mathrm{i}\gamma\omega = D_1 q_x^2 + D_2 q_y^2 + D_3 q_z^2\,. \tag{10}$$

In the derivation of eq. (9) eq. (3) has been taken into account. The dispersion (9) of the amplitudon has the same form as in the case of soft modes below the point of ordinary phase transitions and is shown in fig. 4 (curve 2) in the assumption that the value of $\rho_e$ is not very small and the frequency $\omega$ has a real part, which is shown in fig. 4.

The dispersion (10) of the phason at small $q \equiv |\boldsymbol{q}|$ gives only purely imaginary frequencies. The frequencies acquire a real part when $4\mu(D_1 q_x^2 + D_2 q_y^2 + D_3 q_z^2) > \gamma^2$. Using the estimates (6), we find that this is possible if

$$q > q_0 \sim \tfrac{1}{2} b\,(1 - 10^{-2})\,. \tag{11}$$

In the case of light scattering the values of $q$ are of the order of the wavevectors of light waves or smaller, i.e., $q \lesssim 10^{-3} b$. For such $q$ the roots of eq. (10) are purely imaginary, and the root vanishing at $q = 0$ has the form,

$$\omega = -\,\mathrm{i}\omega_0, \qquad \omega_0 = (1/\gamma)(D_1 q_x^2 + D_2 q_y^2 + D_3 q_z^2)\,. \tag{12}$$

The other root gives $\omega_0$ of the order of $\Omega_a$. Note that expression (12) does not involve the mass $\mu$ and therefore it is of the same form also for order-disorder phase transitions, for which $\mu \approx 0$.

Thus, as follows from qualitative considerations, the phason is a purely relaxation mode with the relaxation frequency $\omega_0$. Figure 4 shows the real part of the frequency according to eq. (10) (curves 1), points D and D′ being situated at a distance $q_0$ (eq. (11)) from point A. The broken line stands for phason dispersion if $\gamma = 0$.

A theory indicating that the phason is overdamped was proposed rather long ago (McMillan 1975b), but Boriack and Overhauser (1978) have criticized McMillan's results. In studying charge-density wave systems they have considered phason attenuation associated only with the electron–phason interaction and found that the phason has practically no attenuation if the vector $\boldsymbol{q}$ is

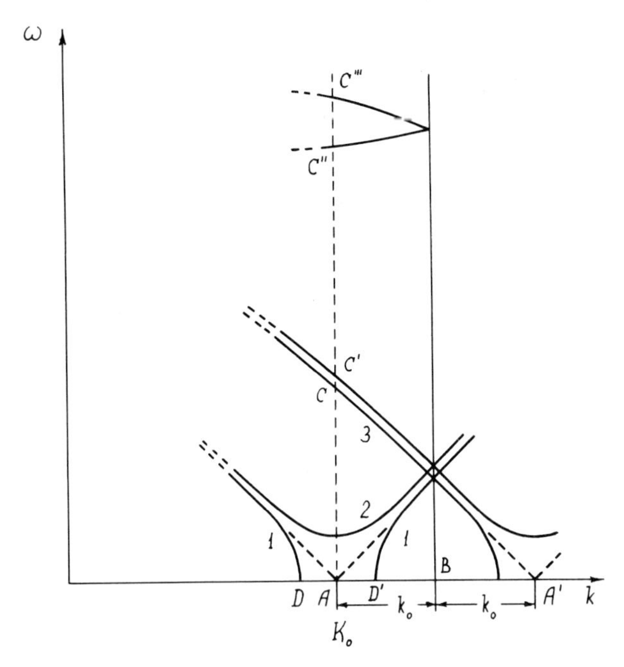


Fig. 4. Dispersion curves (the real part of the frequency) for the incommensurate phase shown in the Brillouin zone of the symmetrical phase. Point B is the zone boundary ($k = b/2$), some of the branches are prolonged into the second zone. Splitting of upper branches is not represented.

perpendicular to the vector $\boldsymbol{K}_0$ and the damping constant of the phason is of the same order as its frequency if the vector $\boldsymbol{q}$ is parallel to the vector $\boldsymbol{K}_0$. The qualitative and quantitative analysis presented above referred, in fact, to dielectrics since we have considered only relative atom displacements. However in the case of charge-density wave systems, when such wave shifts as a whole, there occur both the displacement of electrons with respect to ions and the displacement of ions with respect to one another since spatial modulation of an electron charge induces a corresponding modulation in ion positions. All these processes are associated with energy dissipation and therefore the damping constant of a phason in charge-density wave systems must have a substantial value, like in the case of dielectrics. Boriack and Overhauser have also suggested that the phason frequency must strongly depend upon the direction of the vector $\boldsymbol{q}$ relative to the vector $\boldsymbol{K}_0$. From eq. (10) it is seen that such a strong anisotropy must not be observed in the phason dispersion since there is no reason to assume in a general case the coefficients $D_i$ in the thermodynamic potential density (4) to differ in the order of magnitude.

Concluding the present subsection we note that the terminology in the questions under consideration has not yet become unified. The term phason is sometimes used not only for Goldstone excitations, i.e., excitations corresponding to a close vicinity of point A (fig. 4), but also for higher-lying excitations of branch 1, termed phasons with wavevectors $k_0$, $2k_0$, etc., or inhomogeneous phasons. Such excitations do not possess specific features of the Goldstone phason (in particular, they may be underdamped) and are considered in subsect. 4.2 of the present chapter, along with other incommensurate phase excitations. In this connection we should note that in neutron scattering studies of incommensurate phase excitations in biphenyl (Cailleau et al. 1980) one of the dispersion curves obtained was interpreted as the one corresponding to an underdamped phason. In these experiments the value of $q$ was rather large and apparently exceeded the value $q_0$ in eq. (11), i.e., inhomogeneous phasons were observed, and the curve obtained corresponded to curve 1 (fig. 4) to the left of point D and to the right of point D′. In addition we would say also that if in the thermodynamic potential density (4) higher derivatives are taken into account, which give in eqs. (9) and (10) higher powers of $q_i$, the equations of motion for $\Delta\rho$ and $\Delta\varphi$ are no longer decoupled, i.e., it becomes impossible to divide excitations into oscillations of the phase and those of the amplitude. Hence, at large $q$ the terms phason and amplitudon lose their original meaning.

## 3. *Light scattering by a phason*

Since a phason is a specific excitation of an incommensurate phase, in the question of light scattering from an incommensurate phase the scattering by a phason is of particular interest. Light scattering by a phason has received great interest both in theoretical and experimental works. However, as mentioned in the introduction to the present chapter, there has been controversy in what concerns phason activity, in some papers, even at correct initial premises, not all of the specific features of the phason having been given due regard. For this reason we begin the study of light scattering from an incommensurate phase with a detailed analysis of scattering by the Goldstone phason.

First we present some qualitative considerations. Disregarding damping, which affects spectral distribution of the scattered light, a phason mode is analogous to acoustic modes and therefore the phason may be expected to manifest itself in the light scattering like acoustic phonons if we are dealing with the scattered-light integral intensity. At the same time even in this case one should anticipate some peculiarities concerning scattering by the phason. The point is that acoustic excitations are characterized by a displacement vector, which has three components, whereas the phase $\varphi$ has one component (we are discussing the case of a two-component order parameter). In other words, excitations corresponding to the phason may be considered one-dimensional, but they occur in a three-dimensional crystal. Accordingly, for the scattering by the phason one may expect specific angular dependences which are absent in the scattering by acoustic phonons. Naturally, the phason overdamping affects essentially the scattered-light spectral distribution: the phason must contribute not to the Mandelstam–Brillouin components but to the central peak.

### 3.1. *Integral intensity*

As is known, light scattering is associated with fluctuations of the dielectric permeability tensor $\epsilon_{ij}$ (Landau and Lifshitz 1959, Ginzburg et al. 1980, for a detailed discussion of this question see ch. 1 of the present book). To clarify the manifestations of a phason in the light scattering, one should first of all consider the variations of $\epsilon_{ij}$ associated with those of the order parameter. To this end, one should establish the dependence of the tensor $\epsilon_{ij}$ on the order parameter components, i.e., to find out which combinations of these components that are transformed like the quantities $\epsilon_{ij}$ are translational invariants just as components of any tensor. Recall that within the Landau phase transition theory we speak of establishing the dependence of the tensor $\epsilon_{ij}$ on the normal coordinate $Q$ in the symmetrical phase. This dependence remains, naturally, below the phase transition point, but the normal coordinate $Q$ acquires a nonzero spontaneous value, which is the order parameter.

If the vector $\boldsymbol{K}_0$ characterizing the superstructure is incommensurate with any reciprocal lattice vector of the symmetrical phase, then, as has already been mentioned in subsect. 2.1, the invariants under translations in the symmetrical phase that depend on the order parameter and do not contain its derivatives are only $QQ^* = \rho^2$ and the powers of this quantity. In the study of light scattering near the points of ordinary (commensurate) phase transitions, the invariants containing order parameter spatial derivatives play an insignificant role and are as a rule disregarded (see ch. 1 of the book). Since the quantity $QQ^*$ does not depend on the phase $\varphi$, some authors (Dvorak and Petzelt 1978, Golovko and Levanyuk 1980) concluded that the Goldstone phason is inactive in scattering. However this conclusion is erroneous for, as will be seen below, in the light scattering by an incommensurate phase an essential role is also played by translational invariants containing linearly the first spatial derivatives of the order parameter (we shall refer to them as gradient invariants). Note that owing to gradient invariants the phason also contributes to the dispersion of elastic and dielectric susceptibilities (Golovko and Levanyuk 1981c).

The simplest gradient invariant is $\partial(QQ^*)/\partial x_i = \partial\rho^2/\partial x_i$. Such an invariant is $\varphi$ independent and therefore is of no interest for the question of phason activity. There exist, however, phase-containing gradient invariants, too. An example is the Lifshitz invariant (1), which can be expressed in terms of $\rho$ and $\varphi$, the factor i/2 being introduced for convenience:

$$\frac{\mathrm{i}}{2}\left(Q\frac{\partial Q^*}{\partial x}-Q^*\frac{\partial Q}{\partial x}\right)=\rho^2\frac{\partial\varphi}{\partial x}, \tag{13}$$

where $x$ stands for the axis, along which the incommensurate-superstructure wavevector $\boldsymbol{K}_0$ is directed. In accordance with what has been said in subsect. 2.1, the Lifshitz invariant always exist if the phase transition is incommensurate. From the fact that the Lifshitz invariant (13) is not only a translational invariant but also an invariant under the total symmetry transformation group of the high-temperature phase, i.e., a perfect invariant, it follows that the phase $\varphi$ is transformed like $x$ since $\rho^2$ is a perfect invariant. This result can be explained as follows. Our order parameter has two components $Q(\boldsymbol{K}_0)=\rho\mathrm{e}^{\mathrm{i}\varphi}$ and $Q(-\boldsymbol{K}_0)=Q^*=\rho\mathrm{e}^{-\mathrm{i}\varphi}$. Since the star of the irreducible representation used here contains the vectors $\boldsymbol{K}_0$ and $-\boldsymbol{K}_0$ only, any symmetry transformation of the high-temperature phase either leaves the vector $\boldsymbol{K}_0$ unchanged, or changes its direction to the opposite, this being valid also with respect to the $x$-axis, along which $\boldsymbol{K}_0$ is directed. But from the above expressions for the order parameter components it is seen that the change of the sign in front of $\boldsymbol{K}_0$ and, therefore, in front of $x$ is equivalent to the change of the sign in front of $\varphi$.

Since $\varphi$ is transformed like $x$, gradient invariants, that transform just as components of a tensor of rank two, are also $\rho^2\partial\varphi/\partial y$ and $\rho^2\partial\varphi/\partial z$, their transformation laws being quite clear: they are transformed like $\epsilon_{xy}$ and $\epsilon_{xz}$, respectively. In the case of a two-component order parameter there are no other gradient invariants of the second power in $Q$, except those mentioned above. Naturally, there exist gradient invariants of higher power in $Q$, but they are of little interest since the leading role near a phase transition point is played by quantities of the lowest power in $Q$, translational invariants linear in $Q$ being absent.

For the sake of concreteness we shall be confined to the consideration of the crystals, whose symmetrical phase belongs to the orthorhombic system, as is the case with many crystals with an incommensurate phase (Ishibashi 1980). In crystals of this system an incommensurate phase transition is described by a two-component order parameter if the vector $\boldsymbol{K}_0$ is parallel to one of the crystallographic axes. It is not difficult to extend these considerations to other crystal systems and other $\boldsymbol{K}_0$ directions. First we consider fluctuations of the diagonal components of the tensor $\epsilon_{ij}$, which induce light scattering without change of polarization if the incident light is polarized along one of the crystallographic axes (see ch. 1 of this book). Such incident light polarization is implied hereafter for the sake of simplicity; the results obtained below are easily extended to the case of arbitrary polarization by using formulae of ch. 1. The $\epsilon_{ij}$ diagonal components in the indicated system being perfect invariants, their dependence on the phase $\varphi$ may be associated with the Lifshitz invariant

(13) only, for example,

$$\epsilon_{zz} = \epsilon_{zz}^{(0)} + a\rho^2 \frac{\partial \varphi}{\partial x}, \tag{14}$$

where $\epsilon_{zz}^{(0)}$ denotes the part of the component $\epsilon_{zz}$ which is of no importance now.

Below the phase transition point, when $\rho$ takes an equilibrium value $\rho_e \neq 0$, between the fluctuations of $\epsilon_{zz}$ and of $\varphi$ there appears, according to eq. (14), a linear coupling leading to the first-order light scattering. From eq. (14) we find the part of the $\epsilon_{zz}$ fluctuations associated with the $\varphi$ fluctuations,

$$\Delta\epsilon_{zz} = a\rho_e^2 \frac{\partial \Delta\varphi}{\partial x}. \tag{15}$$

The corresponding scattered-light integral intensity $I(q)$ is proportional to the mean-square fluctuation $\langle |\Delta\epsilon_{zz}(\boldsymbol{q})|^2 \rangle$, where $\Delta\epsilon_{zz}(\boldsymbol{q})$ is the Fourier transform of $\Delta\epsilon_{zz}(\boldsymbol{r})$ and $\boldsymbol{q}$ is the scattering vector, i.e., the difference between the wavevectors of the incident and the scattered light (see ch. 1 of this book). Like in ch. 1, we expand the fluctuations $\Delta\epsilon_{zz}$ and $\Delta\varphi$ into Fourier series, and for mean-square fluctuations we obtain from eq. (15) that

$$\langle |\Delta\epsilon_{zz}(\boldsymbol{q})|^2 \rangle = a^2 \rho_e^4 q_x^2 \langle |\varphi(\boldsymbol{q})|^2 \rangle, \tag{16}$$

where $\varphi(\boldsymbol{q})$ is the Fourier transform of the phase fluctuation $\Delta\varphi$.

To calculate the mean-square fluctuation of the quantity $\varphi(\boldsymbol{q})$, we make use of the procedure presented in detail in ch. 1. Let us represent $\varphi$ as $\varphi = \varphi_e + \Delta\varphi$, where $\varphi_e = \text{const.}$ is an equilibrium value of $\varphi$, expand $\Delta\varphi$ into Fourier series and substitute this in the thermodynamic potential density (4). Note that since phason and amplitudon are normal excitations of an incommensurate phase (see subsect. 2.2), the $\rho$ fluctuations in this case may be disregarded. The coefficient in front of $|\varphi(\boldsymbol{q})|^2$ in the thermodynamic potential makes it possible to find directly the mean-square fluctuation of $\varphi(\boldsymbol{q})$, for which we get the expression

$$\langle |\varphi(\boldsymbol{q})|^2 \rangle = \frac{k_B T}{V \rho_e^2 (D_1 q_x^2 + D_2 q_y^2 + D_3 q_z^2)}, \tag{17}$$

where, like in ch. 1, $V$ is the crystal volume.

Substituting eq. (17) into eq. (16), we find

$$I(q) \propto \langle |\Delta\epsilon_{zz}(q)|^2 \rangle = \frac{a^2 \rho_e^2 q_x^2 k_B T}{V(D_1 q_x^2 + D_2 q_y^2 + D_2 q_z^2)}. \tag{18}$$

Note that the expressions for fluctuations of other diagonal components of the tensor $\epsilon_{ij}$ differ only in the value of the constant $a$, i.e., the numerator will, as before, contain $q_x^2$ since the $\varphi$ dependence of these components is also determined by the Lifshitz invariant (13).

Let us now consider off-diagonal components of the tensor $\epsilon_{ij}$, whose fluctuations lead to light scattering with a change of polarization (see ch. 1 of this book). With regard to what has been said above, their $\varphi$ dependence is determined by the relations

$$\epsilon_{xy} = \epsilon_{xy}^{(0)} + a'\rho^2 \frac{\partial \varphi}{\partial y}, \qquad \epsilon_{xz} = \epsilon_{xz}^{(0)} + a''\rho^2 \frac{\partial \varphi}{\partial z}. \tag{19}$$

The component $\epsilon_{yz}$ does not depend on $\varphi$ if one considers only order parameter fluctuations (see below). Acting as above, we obtain for mean-square fluctuations some formulae of the type (18), but the quantity $\langle |\Delta\epsilon_{xy}|^2 \rangle$ in the numerator contains $q_y^2$ instead of $q_x^2$, and the quantity $\langle |\Delta\epsilon_{xz}|^2 \rangle$ contains $q_z^2$.

The above considerations based on the symmetry arguments only (Golovko and Levanyuk 1981c) are advantageous for their simplicity and generality. Similar results were obtained by Poulet and Pick (1981) by using a more complicated method and considering centrosymmetrical crystals only.

Passing over to the discussion of the expressions obtained, we would like to pay attention to the fact that in contrast to the light scattering near the points of ordinary phase transitions the terms in $\epsilon_{ij}$, that depend on the order parameter derivatives, make a contribution non-vanishing in the limit $\boldsymbol{q} \to 0$. This is due to the fact that the phase fluctuations (17) diverge as $\boldsymbol{q} \to 0$. It is also clear that the powers of the first derivatives or higher derivatives in the dependence of $\epsilon_{ij}$ on the order parameter are of no interest since their contribution vanishes as $\boldsymbol{q} \to 0$ (remember that for the light $q \lesssim 10^{-3} b$, and it is only small $\boldsymbol{q}$ that are of interest). A specific feature of light scattering by phase fluctuations is a strong dependence of the light intensity on the direction of the vector $\boldsymbol{q}$, which remains in the $\boldsymbol{q} \to 0$ limit too. In other words, the $\boldsymbol{q}$ dependence of the scattering intensity is not analytical as $\boldsymbol{q} \to 0$, which shows similarity between the phason and acoustic modes in a solid, the character of the dependence on the $\boldsymbol{q}$ direction in these two cases being, of course, different. Since the numerator of eq. (18) involves $\rho_e^2$, the intensity is weak near the phase transition point and increases with $T_i - T$ as temperature lowers. Note also that the phason contributes to light scattering both without and with the change of polarization. As opposed to Lyons et al. (1980), the phason activity is not associated with the absence or presence of the centre of symmetry in the crystal.

In some cases one may conclude that the constants $a$ in eqs. (14) and (19) are small. If the soft branch is not degenerate at the centre or at the boundary of the symmetrical phase Brillouin zone, as is shown in fig. 2(b), for these points the Lifshitz invariant is absent (see subsect. 2.1) and the constants $a$ in eqs. (14) and (19) must vanish. Therefore, if the extremity of the wavevector $\boldsymbol{K}_0$ of the incommensurate superstructure lies near the above-mentioned points, the constants $a$ and, according to eq. (18), the corresponding intensity of light scattered by the phason must be small. This question is given a more detailed consideration in sect. 6 of the present chapter.

We would like to emphasize that eq. (18) simplifies the character of the dependence of the quantity $\langle|\Delta\epsilon_{zz}(\boldsymbol{q})|^2\rangle$ and the scattered-light intensity (proportional to this quantity) on the direction of the vector $\boldsymbol{q}$. The point is that, as has been mentioned in ch. 1 of this book, when considering light scattering below a phase transition point, along with the dependence of the tensor $\epsilon_{ij}$ on the order parameter one should also take into account its dependence on other quantities, whose fluctuations are linearly coupled with the order parameter fluctuations (such a quantity is, for example, the strain tensor). This account leads to renormalization of the coefficients in eq. (18), it being important that the renormalization is non-trivial, which means that the indicated non-analiticity, as $\boldsymbol{q}\to 0$, may also be present in the renormalized coefficients. As a result, the angular dependence of scattered-light intensity will be quite complicated, but the main features of this dependence are described correctly by eq. (18): at $q_x = 0$ the intensity vanishes.

What has been said can be confirmed by an example of an account taken of the interaction between the phason and elastic deformations. We supplement the thermodynamic potential density (4) with the term $\Phi_1$ that involves mixed invariants comprising the phase $\varphi$ and the components of the strain tensor $u_{ij}$ as well as invariants composed of the $u_{ij}$ components. The form of the mixed invariants is easily found with the use of eqs. (14) and (19), whereas the form of the $u_{ij}$ dependent invariants is known (Landau and Lifshitz 1965). As a result we have

$$\Phi_1 = (r_1 u_{xx} + r_2 u_{yy} + r_3 u_{zz})\rho^2 \frac{\partial\varphi}{\partial x} + 2r_4\rho^2 \frac{\partial\varphi}{\partial y} u_{xy} + 2r_5\rho^2 \frac{\partial\varphi}{\partial z} u_{xz} + \tfrac{1}{2} C_{ijkl} u_{ij} u_{kl} ,$$

where $C_{ijkl}$ is the elasticity modulus tensor, and the repeated indices imply summation from 1 to 3 (in the orthorhombic system under consideration there are nine non-zero quantities $C_{ijkl}$).

Since the strain tensor $u_{ij}$ is expressed in terms of the elastic displacement vector $\boldsymbol{u}$ according to the relation $u_{ij} = \frac{1}{2}(\partial u_i/\partial x_j + \partial u_j/\partial x_i)$, it is the components of the vector $\boldsymbol{u}$ but not the $u_{ij}$ components that should be taken here as independent variables. We are interested here only in $\boldsymbol{u}$ fluctuations induced by $\varphi$ fluctuations. Therefore in accordance with the considerations presented in ch. 1 we look for "equilibrium" values of $\boldsymbol{u}$ corrresponding to a given fixed $\varphi$, i.e., we minimize $\Phi_1$ with respect to $\boldsymbol{u}$. To this end, it is convenient first to expand all the quantities in Fourier series, to substitute them into $\Phi_1$, to integrate over volume, and then to minimize the thermodynamic potential with respect to the Fourier transforms $\boldsymbol{u}(\boldsymbol{q})$. Thus we are led to the following dependence between the Fourier transforms of $\boldsymbol{u}$ and $\varphi$

$$u_x(\boldsymbol{q}) = -\rho_e^2 \frac{\delta_1}{\delta}\varphi(\boldsymbol{q}), \qquad u_y(\boldsymbol{q}) = -q_x q_y \rho_e^2 \frac{\delta_2}{\delta}\varphi(\boldsymbol{q}),$$

$$u_z(\boldsymbol{q}) = -q_x q_z \rho_e^2 \frac{\delta_3}{\delta}\varphi(\boldsymbol{q}) .$$

Here $\delta = \det[C_{iklj}\, q_k q_l]$ is a determinant similar to the one appearing in the theory of elastic waves in a crystal (Landau and Lifshitz 1965) if we put $\omega = 0$. Naturally, $\delta \neq 0$ if $\boldsymbol{q} \neq 0$. The expression for $\delta_1$ is obtained if the quantities $r_1 q_x^2 + r_4 q_y^2 + r_5 q_z^2$, $q_x q_y (r_2 + r_4)$ and $q_x q_z (r_3 + r_5)$ are substituted into the first column of the determinant. The expressions for $\delta_2$ and $\delta_3$ are obtained if those quantities are substituted into the second and third columns, respectively and the appearing factors $q_x q_y$ and $q_x q_z$ are disregarded (they are factored out in the above expressions). The quantities $\delta$ and $\delta_1$ are homogeneous polinomials of third degree in $q_x^2$, $q_y^2$ and $q_z^2$, and the quantities $\delta_2$ and $\delta_3$ are those of second degree.

If the expressions for $\boldsymbol{u}$ are substituted into $\Phi_1$, one obtains terms of the order of $\rho^4 q^2$ which can be neglected as compared with the terms contained in eq. (4). Therefore, the $\varphi$ fluctuations are determined, as before, by eq. (17).

Let us now consider the tensor $\epsilon_{ij}$ taking into account its dependence both on $\varphi$ and on $u_{ij}$ and illustrate it on an example of the components

$$\epsilon_{zz} = a\rho^2 \frac{\partial \varphi}{\partial x} + c_1 u_{xx} + c_2 u_{yy} + c_3 u_{zz}, \qquad \epsilon_{xy} = a'\rho^2 \frac{\partial \varphi}{\partial y} + c_4 u_{xy}\,,$$

$$\epsilon_{yz} = c_5 u_{yz}\,.$$

On finding from this the $\epsilon_{ij}$ fluctuations and expanding them in Fourier series, we substitute there the expressions obtained for $\boldsymbol{u}(\boldsymbol{q})$. In the end we see that the fluctuations of $\epsilon_{zz}$ and $\epsilon_{xy}$ are analogous to the fluctuations that follow from eqs. (15) and (19), but the coefficients are now renormalized, which is denoted by a tilde,

$$\tilde{a} = a - \frac{1}{\delta}(c_1\delta_1 + c_2 q_y^2 \delta_2 + c_3 q_z^2 \delta_3), \qquad \tilde{a}' = a' - \frac{c_4}{2\delta}(\delta_1 + q_x^2 \delta_2).$$

According to what has been said above, the corrections to $a$ and $a'$ do not tend to zero as $\boldsymbol{q} \to 0$, and in this limit there also remains the dependence on the $\boldsymbol{q}$ direction which is rather complicated.

As to the fluctuations of $\epsilon_{yz}$, they are determined by the expression

$$\Delta\epsilon_{yz}(\boldsymbol{q}) = -\frac{\mathrm{i}}{2\delta} q_x q_y q_z c_5 \rho_{\mathrm{e}}^2 (\delta_2 + \delta_3)\varphi(\boldsymbol{q}),$$

the mean-square fluctuations of $\Delta\epsilon_{yz}(\boldsymbol{q})$ neither tending to zero as $\boldsymbol{q} \to 0$. In the above analysis without an account taken of elastic deformations, we have obtained that $\epsilon_{yz}$ does not depend linearly on $\varphi$. Now we see that a linear coupling between the fluctuations of $\epsilon_{yz}$ and those of $\varphi$ does exist. The reason is as follows. Since the renormalized coefficient $a$ are $\boldsymbol{q}$ dependent, they can not only be scalars but also tensors if they are proportional to $q_i q_j$. Naturally, a correction to a scalar can only be a scalar too, which is exampled by expressions for $\tilde{a}$ and $\tilde{a}'$. However if for symmetry reasons one obtains that some coefficient is zero when $\boldsymbol{q}$ dependence is not taken into account, then it is not excluded that this coefficient is non-zero if regarded as a tensor by taking into account the $\boldsymbol{q}$ dependence. In fact, one could have written the relation $\epsilon_{yz} = a'''\rho^2 \partial\varphi/\partial x$ immediately, had $a'''$ been considered as transformed like $q_y q_z$.

The possibility of the indicated $\boldsymbol{q}$ dependence of some coefficients should be implied in what follows, and as a rule is not discussed specially.

### 3.2. *Spectral intensity*

To find the spectral intensity of light scattered by a phason, one should consider the kinetics of phase fluctuations, i.e., to clarify the time dependence of $\Delta\varphi(\boldsymbol{r}, t)$. The spectral intensity can be obtained then by going to space and time Fourier transforms of $\Delta\varphi(\boldsymbol{r}, t)$. We shall follow the method, the principles of which are presented in ch. 1 of this book. Let us supplement the right-hand side of eq. (8) with a random generalized force $h(\boldsymbol{r}, t)$, which induces fluctuations

$$\mu \frac{\partial^2 \Delta\varphi}{\partial t^2} + \gamma \frac{\partial \Delta\varphi}{\partial t} - \sum_{i=1}^{3} D_i \frac{\partial^2 \Delta\varphi}{\partial x_i^2} = h(\boldsymbol{r}, t) . \tag{20}$$

To calculate the Fourier transform, $\varphi(\boldsymbol{q}, \Omega)$, of the fluctuation $\Delta\varphi$, we take one of the harmonics of the random force $h(\boldsymbol{r}, t) = h(\boldsymbol{q}, \Omega) \exp \mathrm{i}(\boldsymbol{qr} - \Omega t)$ and find the solution of eq. (20) in the form

$$\Delta\varphi(\boldsymbol{r}, t) = \varphi(\boldsymbol{q}, \Omega)\, \mathrm{e}^{\mathrm{i}(\boldsymbol{qr} - \Omega t)},$$

$$\varphi(\boldsymbol{q}, \Omega) = \frac{h(\boldsymbol{q}, \Omega)}{D_1 q_x^2 + D_2 q_y^2 + D_3 q_z^2 - \mu\Omega^2 - \mathrm{i}\gamma\Omega} .$$

After being statistically averaged, the quantity $|\varphi(\boldsymbol{q}, \Omega)|^2$ will contain the random force spectral density $\langle |h(\boldsymbol{q}, \Omega)|^2 \rangle$ which is frequency-independent (see ch. 1). Therefore, one can find this density by integrating $\langle |\varphi(\boldsymbol{q}, \Omega)|^2 \rangle$ over $\Omega$ and comparing the result with eq. (17). Finally, for the spectral density of phase fluctuations we obtain

$$\langle |\varphi(\boldsymbol{q}, \Omega)|^2 \rangle = \frac{\gamma k_{\mathrm{B}} T}{\pi V \rho_{\mathrm{e}}^2 [(\mu\Omega^2 - D_1 q_x^2 - D_2 q_y^2 - D_3 q_z^2)^2 + \gamma^2 \Omega^2]} . \tag{21}$$

Using the same estimates as in the derivation of eq. (12), one can easily verify that the quantities $\mu D_i q_i^2$ can be neglected as compared with $\gamma^2$. From eq. (21) now we see that the spectral density $\langle |\varphi(\boldsymbol{q}, \Omega)|^2 \rangle$ decreases monotonically as the frequency $\Omega$ increases, i.e., the $\varphi$ fluctuations are of relaxation character, which is in agreement with the results of subsect. 2.2. In the most interesting region (small $\Omega$) from eq. (21) we have

$$\langle |\varphi(\boldsymbol{q}, \Omega)|^2 \rangle = \frac{\gamma k_{\mathrm{B}} T}{\pi V \rho_{\mathrm{e}}^2 [(D_1 q_x^2 + D_2 q_y^2 + D_3 q_z^2)^2 + \gamma^2 \Omega^2]} . \tag{22}$$

From this it follows that the central-peak width of the fluctuation spectrum is given by

$$\Delta\Omega = (2/\gamma)(D_1 q_x^2 + D_2 q_y^2 + D_3 q_z^2) . \tag{23}$$

Like in the case of diffusive-type fluctuations, this width is proportional to $q^2$.

The scattered-light spectral intensity is proportional to the fluctuation spectral density $\langle|\Delta\epsilon_{ij}(\boldsymbol{q}, \Omega)|^2\rangle$, where $\Omega$ is the change in the light frequency due to scattering (see ch. 1). If we are dealing, for example, with scattering by the fluctuations $\Delta\epsilon_{zz}(\boldsymbol{r}, t)$, then using expression (15) and passing over to the space and time Fourier transform of the fluctuations, we again come to eq. (16), where the $\Omega$ dependence will be also involved. Then a substitution of eq. (22) gives the spectral density of fluctuations

$$\langle|\Delta\epsilon_{zz}(\boldsymbol{q}, \Omega)|^2\rangle = \frac{a^2\rho_e^2 q_x^2 \gamma k_B T}{\pi V[(D_1 q_x^2 + D_2 q_y^2 + D_3 q_z^2)^2 + \gamma^2\Omega^2]} . \tag{24}$$

Let us discuss the width of the phason central peak, whose maximum value $\Delta\Omega_{\max}$ can be estimated by taking for $\boldsymbol{q}$ the wavevector of a light wave. Taking into account that in this case $q \sim 10^{-3}b$ and using estimates (6), from eq. (23) we have

$$\Delta\Omega_{\max} \sim \frac{D_i q^2}{\gamma} = \frac{D_i b^2}{\gamma}\frac{q^2}{b^2} \sim (1 - 100)\frac{q^2}{b^2}\Omega_a \sim (10^{-6} - 10^{-4})\Omega_a . \tag{25}$$

From this it is seen that even the maximum width of the phason peak is rather small, and to observe it in an experiment is no easy problem. The estimate (25) is, naturally, approximate, and for a more accurate evaluation one should know more precisely the values of the constants involved. Note that the equation of motion (7) for the amplitudon comprises the same constants as the equation of motion (8) for the phason. Since scattering by the amplitudon can be observed somewhat easier (see subsect. 4.1), this scattering can be used for specification of the parameters describing scattering by the phason.

Thus, as has been noted above on the basis of qualitative considerations, light scattering by a phason differs essentially from Mandelstam–Brillouin scattering since it is the central peak, but not the satellites that correspond to the phason in the scattering spectrum. Note that this difference between scattering by a phason and Mandelstam–Brillouin scattering was not stressed in the literature even when the phason activity in the light scattering was stated (see, for example, Cowley (1980), Poulet and Pick (1981)). Some attempts have been made (Fleury et al. 1979, Unruh et al. 1979, Takashige and Nakamura 1980) to observe scattering by a phason in an incommensurate phase in the same scattering geometry and in the same frequency band as for the phason in the low-temperature commensurate phase. A negative result of such experiments seems to be quite natural since in a commensurate phase the scattering by a phason is practically determined by the terms in the expansion of $\epsilon_{ij}$ in a power series of the order parameter, which contain no space derivatives and are absent in the case of the incommensurate phase, and this may lead to the phason manifestation in a scattering geometry different from that in the incommensurate phase. Besides, the phason frequency in the commensurate phase does not vanish as $\boldsymbol{q} \to 0$, i.e., such a phason gives in the scattering either satellites or the central peak, whose width does not differ noticeably from $\Omega_a$ and is $\boldsymbol{q}$ independent.

Note that in the consideration of the spectral density of light scattered by a phason, an account taken of the coupling between order parameter fluctuations and fluctuations of other quantities causes no qualitative changes. This is due to the fact that the phason relaxation frequency $\omega_0$ in eq. (12), whose maximum value coincides with the estimate (25), is much lower than the other characteristic crystal frequencies that correspond to a given $\boldsymbol{q}$, i.e., the variations of all other quantities have time to "fit" themselves for the $\varphi(\boldsymbol{q})$ variations. Here one must take into account that even the acoustic frequencies equal to $sq \sim \Omega_a q/b$, where $s \sim \Omega_a/b$ is the sound velocity, are much higher than $\omega_0$ at one and the same $q$ since with due regard for eq. (6) we have $\omega_0/sq \sim (1-100)\, q/b$.

## 4. Light scattering by non-Goldstone excitations

### 4.1. Amplitudon

As mentioned in subsect. 2.2, an amplitudon behaves in much the same way as soft modes below the point of ordinary (commensurate) phase transitions. Therefore, one may expect that in the light scattering the amplitudon manifests itself like an ordinary soft mode; this assumption has been confirmed by theoretical investigations (Dvorak and Petzelt 1978, Golovko and Levanyuk 1980), that, as distinguished from the scattering by a phason, have not led to any controversy.

Just as in the case of ordinary soft modes, in the question of light scattering by an amplitudon an account taken in the dielectric constant tensor $\epsilon_{ij}$ of the terms containing spatial derivatives of the order parameter leads to inessential corrections since the amplitudon frequency (9) does not vanish at $\boldsymbol{q} = 0$. According to what has been said in subsect. 2.1, the only translational invariants depending on the the order parameter and containing none of its derivatives are the quantities $\rho^{2n}$ that are also perfect invariants. In the orthorhombic system under discussion, only the diagonal $\epsilon_{ij}$ components are perfect invariants, and it is exclusively they that may depend on the quantities $\rho^{2n}$. Therefore, an amplitudon contributes only to the light scattering without change of polarization induced by fluctuations of the diagonal $\epsilon_{ij}$ components.

Let us write the dependence of some diagonal component of the tensor $\epsilon_{ij}$ on the amplitude $\rho$:

$$\epsilon_{ii} = \epsilon_{ii}^{(0)} + a_i\rho^2. \tag{26}$$

For the mean-square fluctuation of the Fourier transform $\Delta\epsilon_{ii}(\boldsymbol{q})$ by analogy with eq. (16) we obtain

$$\langle|\Delta\epsilon_{ii}(\boldsymbol{q})|^2\rangle = 4a_i^2\rho_e^2\langle|\rho(\boldsymbol{q})|^2\rangle, \tag{27}$$

where $\rho(\boldsymbol{q})$ is the Fourier transform of the amplitude fluctuation $\Delta\rho$. The mean-square fluctuation $\langle|\rho(\boldsymbol{q})|^2\rangle$ can be calculated on the basis of the thermodynamic potential (4) in the same way as the phase fluctuation (17)

$$\langle|\rho(\boldsymbol{q})|^2\rangle = k_B T/V(2B\rho_e^2 + D_1q_x^2 + D_2q_y^2 + D_3q_z^2). \tag{28}$$

If we exclude a narrow region near the phase transition point, the quantities $D_iq_i^2$ may be neglected as compared with $2B\rho_e^2$. Then after a substitution of eq. (28) into eq. (27), the quantity $\rho_e$ is cancelled, and we see that at the transition point the fluctuation $\langle|\Delta\epsilon_{ii}|^2\rangle$ undergoes a jump:

$$\langle|\Delta\epsilon_{ii}(\boldsymbol{q} \approx 0)|^2\rangle = 2a_i^2k_BT/BV. \tag{29}$$

The integral intensity of light scattered by an amplitudon also undergoes a jump proportional to this quantity, just as is observed in the case of ordinary soft modes (see ch. 1).

The spectral intensity of light scattered by an amplitudon can be found by supplementing eq. (7) with a random generalized force and using the same considerations as in the case of scattering by a phason (subsect. 3.2, see also ch. 1). Finally, in the $\boldsymbol{q}\to 0$ limit for the spectral density $J(\Omega)$ we obtain

$$J(\Omega) = \frac{2B\gamma\rho_e^2 I}{\pi[(2B\rho_e^2 - \mu\Omega^2)^2 + \gamma^2\Omega^2]}, \tag{30}$$

where $I$ is the integral intensity in the $\boldsymbol{q}\to 0$ limit.

The spectral distribution (30) is the same as in the case of ordinary soft modes (see ch. 1). Near the phase transition point an amplitudon gives a central component, which widens in proportion to $T_i - T$ as temperature lowers. Then it splits into two side components, whose width is equal to $\gamma/\mu$ when their splitting is sufficiently large. Since eq. (26) holds also in the commensurate phase, the character of scattering by the amplitudon must change insignificantly below the lock-in transition point: the amplitudon frequency must go through a jump, if this is a first-order transition and $\rho_e$ undergoes a jump. The integral intensity proportional to expression (29) must not change at all in a first approximation.

Light scattering by an amplitudon has been observed in many experiments ((Fleury et al. 1979, Unruh et al. 1979, Inoue et al. 1980, Francke et al. 1980, Siapkas 1980, Takashige and Nakamura 1980, Winterfeldt and Schaack 1980, Wada et al. 1981) if we refer only to some recent papers), the regularities of scattering being, generally, in agreement with those discussed above. In particular, the amplitudon manifested itself not only in the incommensurate but also in the commensurate phase.

### 4.2. Hard and quasi-hard modes

Having considered the activity of the phason and amplitudon, let us now discuss the manifestations of other incommensurate phase excitations in the scattered-light spectrum. As is known, below the point of any phase transition associated with a change in the translational symmetry of a crystal, in the Raman spectrum new lines appear due to the change in the number of atoms in the unit cell. In the case of transition to an incommensurate phase the size of the unit cell along the incommensurate axis becomes equal to the crystal dimension and the cell will comprise a huge amount of atoms. Therefore, it seems that the number of new lines in the spectrum must be practically infinite, i.e., the scattering spectrum will become continuous without any characteristic lines. We shall show, however, that in reality the number of new lines with a noticeable intensity in the incommensurate phase spectrum is comparatively small.

Let us consider first some normal coordinate $Q_1$, which does not belong to the soft branch but pertains to the same point $\boldsymbol{K}_0$ of the Brillouin zone. In fig. 4 this coordinate corresponds to one of the points C placed above point A. The quantity $QQ_1^*$ is also a translational invariant since under multiplication of the basis functions of corresponding irreducible representations, the factors $\exp(\mathrm{i}\boldsymbol{K}_0\boldsymbol{r})$ and $\exp(-\mathrm{i}\boldsymbol{K}_0\boldsymbol{r})$ contained in these functions vanish, the period of these factors being incommensurable with that of the underlying lattice. Being a translational invariant, the quantity $QQ_1^*$ is transformed as a component of a vector or a tensor and, in particular, some components of the tensor $\epsilon_{ij}$ may prove to be transformed just as the quantity $QQ_1^*$ and, accordingly, to be coupled with it. Below the phase transition point there appear a spontaneous non-zero value $Q = Q_e$ and, therefore, the variations of $Q_1$ will cause proportional variations of $\epsilon_{ij}$. This means that in the scattering spectrum there will appear new frequencies that are absent when $T > T_i$, since the indicated point A (fig. 4) is not at the Brillouin zone centre of the symmetrical phase. The intensity of the new lines is proportional to $|Q_e|^2 = \rho_e^2 \sim (T_i - T)$, for the mean-square fluctuation of the $\epsilon_{ij}$ component will be proportional to this quantity.

As distinguished from $QQ^*$, the quantity $QQ_1^*$ may be or may not be a perfect invariant. In the former case the diagonal components of $\epsilon_{ij}$ alone may be coupled with $QQ_1^*$, in the latter it is off-diagonal components forbidden by the high-temperature phase symmetry (remember that we mean the orthorhombic symmetry). For this reason the above-mentioned lines may appear in the scattered-light spectra both with and without the change of polarization.

Consider now a normal coordinate $Q_2$ which corresponds to the wavevector $2\boldsymbol{K}_0 + \boldsymbol{B}$, where $\boldsymbol{B}$ is an arbitrary reciprocal lattice vector of the symmetrical phase that can be chosen, for example, such that the vector $2\boldsymbol{K}_0 + \boldsymbol{B}$ appears to be in the first Brillouin zone. Note that the normal coordinate $Q_2$ may pertain either to a high-lying or to the soft branch, the corresponding excitations being in the latter case an inhomogeneous phason and amplitudon (see the end of subsect. 2.2). Using $Q_2$, one can form a translational invariant $Q^2 Q_2^*$ since the basis functions of corresponding irreducible representations contain the factors $\exp(\mathrm{i}\boldsymbol{K}_0\boldsymbol{r})$ and $\exp(-2\mathrm{i}\boldsymbol{K}_0\boldsymbol{r})$. Reasoning as above, we see that the vibration of $Q_2$ may be active in the Raman spectrum, but the spectral line intensity is proportional to $|Q_\mathrm{e}|^4 = \rho_\mathrm{e}^4$. It is clear that a normal coordinate $Q_n$ corresponding to the wavevector $n\boldsymbol{K}_0 + \boldsymbol{B}$ with an arbitrary integer $n$ may also be active in light scattering, the intensity of the line being proportional to $\rho_\mathrm{e}^{2n}$.

Thus, as has been expected, in the Raman spectrum of an incommensurate phase there appears an infinite number of new lines (no frequencies coincide strictly because of incommensurability between $\boldsymbol{K}_0$ and $\boldsymbol{B}$), but one can observe only a limited number of lines corresponding to small $n$ since, as a rule, the quantity $\rho_\mathrm{e}$ in the entire region of the existence of an incommensurate phase is small and, therefore, the intensity of lines will rapidly decrease as $n$ increases. Knowing the transformational properties of the order parameter and of the normal coordinates $Q_n$, one can clarify in any particular case which component of the tensor $\epsilon_{ij}$ a certain quantity $Q^n Q_n^*$ is coupled with (if such a coupling does exist) and, therefore, establish the scattering geometry in which the spectral line under investigation must be observed. The appearance in an incommensurate phase of new lines with the intensity $\rho_\mathrm{e}^{2n}$ $(n = 2, 3 \ldots.)$ was first considered by Dvorak and Petzelt (1978), but these authors analyzed only the vibrations that correspond to the soft branch (we speak of non-Goldstone excitations), disregarding the possibility of more intense new lines associated with vibrations of $Q_1(\boldsymbol{K}_0)$ pertaining to other branches.

The frequency splitting due to the appearance of a superstructure has not been taken into account in the above considerations. The origin of this splitting in this case can be easily explained as follows. Since there exists a translational invariant $Q(\boldsymbol{K}_0)\ Q_1^*(\boldsymbol{K}_0)$, the thermodynamic potential comprises a perfect invariant $Q^2(Q_1^*)^2$ (plus the complex-conjugate quantity). In an incommensurate phase, where a spontaneous value $Q = Q_\mathrm{e}$ appears, the thermodynamic potential acquires $Q_1$-dependent non-diagonal quadratic terms, i.e., terms that are not reduced to $|Q_1|^2$, which fact just explains frequency splitting at the wavevector $\boldsymbol{k} = \boldsymbol{K}_0$ (see also Poulet and Pick 1981). In fig. 4 two points C and C′ correspond, for example, to split frequencies. Frequency splitting is observed not only at $\boldsymbol{k} = \boldsymbol{K}_0$. A still stronger splitting occurs at $\boldsymbol{k} = \boldsymbol{K}_0/2$ if the translational invariant $Q(\boldsymbol{K}_0)\ [Q_0^*(\boldsymbol{K}_0/2)]^2$ is perfect and therefore is present in the thermodynamic potential. However this splitting is not observed in optical spectra since the point $\boldsymbol{k} = \boldsymbol{K}_0/2$ corresponds to the boundary of the "Brillouin zone" of the superstructure. Note that there also exist weaker splittings at wavevectors $\boldsymbol{k} = 2\boldsymbol{K}_0 + \boldsymbol{B}$, $3\boldsymbol{K}_0 + \boldsymbol{B}$ etc., which can be explained in a similar manner.

Let us now discuss the temperature dependence of the frequencies of newly appeared lines. The frequencies of new lines, which correspond to high-lying (hard) optical branches, in a general case have a weak temperature dependence for the most part due to the change of the vector $\boldsymbol{K}_0$. On the other hand, they may correspond both to the soft branch itself (at $\boldsymbol{k} = 2\boldsymbol{K}_0 + \boldsymbol{B}$, $3\boldsymbol{K}_0 + \boldsymbol{B}, \ldots$, because we disregard here "homogeneous" phason and amplitudon) and to the one genetically related to it, if such a branch does exist (fig. 2(a), branch 2), and then the frequencies can be low and vary noticeably with tempreature. Corresponding modes may be referred to as quasi-hard. Let us discuss in more detail the temperature dependence of the quasi-hard mode frequencies, but at first we notice that since the magnitude of the damping constant in this case is of the same order as for other optical modes, the lines corresponding to low frequencies may, in fact, be not satellites but contribute to the central peak.

First we consider the case when the extremity of the vector $\boldsymbol{K}_0$ lies near the Brillouin zone boundary of the symmetrical phase, and assume for the sake of simplicity that the vector $\boldsymbol{K}_0$ is parallel to a basic reciprocal lattice vector $\boldsymbol{b}$, i.e., $K_0 = b/2 - k_0$, where $k_0$ is small. We start with the situation widespread in the case of incommensurate phase transitions when the softening optical branch is degenerate at the zone boundary, i.e., there exists a branch genetically related to the softening one (branch 2 in fig. 2(a) and branch 3 in fig. 4). As is seen from fig. 4, the frequency $\omega(K_0)$ of branch 3 at $k = K_0$ (at $T < T_i$ we are practically dealing with two close frequencies, points C and C′) is small because $k_0$ is small, the intensity of a corresponding line being proportional to $\rho_e^2$. When $\boldsymbol{K}_0$ takes the indicated position, at low temperatures a commensurate phase is observed as a rule, whose order parameter is a normal coordinate corresponding to the wavevector $\boldsymbol{k} = \boldsymbol{b}/2$. Then in the incommensurate phase the value of $k_0$ decreases as temperature lowers (see subsect. 2.1). If the lock-in transition were continuous, the value of $k_0$, and along with it one of the two frequencies $\omega(K_0)$, would vanish at the phase transition point. However, in all the experiments known to us this transition turns out to be of first order, the value of $k_0$ within the incommensurate phase changing at best only by a small factor; $\omega(K_0)$ must change approximately by the same factor. Since in this case $k_0$ is by one or two orders of magnitude smaller than $b$, the frequencies $\omega(K_0)$ are by one or two orders of magnitude smaller than the characteristic frequencies of a crystal.

Now let us consider the case when the extremity of the vector $\boldsymbol{K}_0$ lies also near the Brillouin zone boundary, but degeneration at the zone boundary is absent (fig. 2(b)). Since in this case there is no branch genetically related with the soft one, new spectral lines with the intensity $\sim \rho_e^2$ correspond only to high-lying hard branches and their frequencies depend weakly on temperature.

As has been mentioned above, the oscillations with normal coordinates $Q_n(n\boldsymbol{K}_0)$ too may manifest themselves in light scattering. In the case of the normal coordinate $Q_2(2K_0)$ the spectral line intensity is proportional to $\rho_e^4$, and the soft-branch frequency $\omega(2K_0)$ is not small and has a weak temperature dependence since the point $2K_0 = b - 2k_0$ turns out to be in the vicinity of the Brillouin zone centre if $b$ is subtracted from $2K_0$. Therefore, the corresponding mode is, in fact, hard but not quasi-hard. The most intense line corresponding to a temperature-dependent low frequency appears due to the fact that the point $3K_0 = 3b/2 - 3k_0$, if placed in the first Brillouin zone, lies near the point $K_0$, but according to what has been said above, the intensity of this line is proportional to $\rho_e^6$ (under more accurate estimations one should take into account, of course, that a low frequency gives a small term in the denominator of the expression for the intensity, the frequency squared, which increases the intensity). Naturally, all the lines just discussed do exist also in the case when the soft branch is degenerate at the Brillouin zone boundary.

If in a low-temperature commensurate phase the period is triple as compared with that in the symmetrical phase, the value of $K_0$ is close to $b/3$, i.e., $K_0 = b/3 - k_0$. Now it is already the frequency $\omega(2K_0)$ that appears to be low, since the reciprocal lattice vector being neglected, the difference between $2K_0$ and $-K_0$ is equal to $-3k_0$. The intensity of a corresponding line is proportional to $\rho_e^4$. This case was considered in detail by Dvorak and Petzelt (1978). Using the above arguments, it is easy to establish the existence of other quasi-hard modes that can manifest themselves in light scattering by an incommensurate phase, but the line intensities will be even weaker than in the cases discussed above.

As concerns formulae for the spectral distribution in newly appeared lines, they can be obtained by using a standard technique; they have a usual form and do not at all reveal specific features of the incommensurate phase since we are dealing with non-Goldstone excitations. This is seen, in particular, on an example of eq. (30), that gives the spectral intensity of light scattered by an amplitudon.

## 5. *Light scattering in the case of a four-component order parameter*

### 5.1. *Specificity of an incommensurate phase in the case of an order parameter with more than two components*

Let us discuss specific features of an incommensurate phase in the case when the order parameter has more than two components. We consider here as an example a four-component order parameter, when the star of a corresponding physically irreducible representation contains four non-equivalent vectors $\boldsymbol{K}_1$, $-\boldsymbol{K}_1$, $\boldsymbol{K}_2$, $-\boldsymbol{K}_2$ incommensurable with the reciprocal lattice periods of the symmetrical phase. For the components of the order parameter one may take corresponding normal coordinates $Q_1(\boldsymbol{K}_1)$, $Q_1^*$, $Q_2(\boldsymbol{K}_2)$, $Q_2^*$.

If the order parameter is two-component, the point symmetry of an incommensurate phase is identical to that of the symmetrical high-temperature phase (see subsect. 2.1). The situation is different if the order parameter has more than two components. In the case of four components, besides the invariant $|Q_1|^2 + |Q_2|^2$, a translational invariant is also, e.g., $|Q_1|^2 - |Q_2|^2$. The latter invariant cannot be perfect since there is only one perfect quadratic invariant, namely $|Q_1|^2 + |Q_2|^2$. Therefore, if a phase with $|Q_{1e}|^2 \neq |Q_{2e}|^2$ corresponds to thermodynamic equilibrium, there appears a spontaneous tensorial quantity $|Q_{1e}|^2 - |Q_{2e}|^2$ forbidden by the high-temperature phase symmetry, i.e., the point symmetry changes. As shown below, such a change in the symmetry may be observed, e.g., in light scattering.

In the case of a four-component order parameter the soft mode in the symmetrical phase is formally four-fold degenerate. Below the phase transition point this degeneration is lifted either partially or completely, i.e., there may appear up to four different modes and, as will be seen below, not all of these modes must necessarily contribute to Raman scattering. In the case of a four-component order parameter there may appear phases with two-dimensional incommensurability and, correspondingly, with two Goldstone modes (McMillan 1975a, Rice 1975) and both of them are shown to be active in light scattering.

When considering incommensurate transitions with a four-component order parameter, one should distinguish two cases (Cowley 1980). In the first one a certain combination of the vectors $\boldsymbol{K}_1$ and $\boldsymbol{K}_2$ with integral coefficients coincides with one of the reciprocal lattice vectors of the symmetrical phase, and, accordingly, incommensurability is possible only in one direction. In the second case no combination of the vectors $\boldsymbol{K}_1$ and $\boldsymbol{K}_2$ with integral coefficients gives a reciprocal lattice vector, which may result in the appearance of a two-dimensional incommensurability. The first case may be exampled by the incommensurate transition in barium manganese fluoride $BaMnF_4$, the second one is realized in barium sodium niobate $Ba_2NaNb_5O_{15}$. Both cases will be discussed on an example of these crystals for the sake of clearness and also for the reason that rather many experiments reported have been devoted to these crystals, particularly to $BaMnF_4$. We confine ourselves to a consideration of light scattering by order parameter fluctuations. It is not difficult to take into account also other fluctuations, just as it was done for a two-component order parameter. A similar analysis can be given also to other crystals in which the incommensurate transition is described by an order parameter containing more than two components.

### 5.2. *Barium manganese fluoride $BaMnF_4$*

The symmetrical phase of $BaMnF_4$ belongs to the space group $C_{2v}^{12}$–$A2_1am$; below the temperature $T_i \approx 250$ K there occurs an incommensurate phase transition, incommensurability being observed along the polar $x$-axis and being characterized by the wavevectors $\boldsymbol{K}_1 = \zeta \boldsymbol{b}_1 + \boldsymbol{b}_2/2$ and $\boldsymbol{K}_2 = \zeta \boldsymbol{b}_1 + \boldsymbol{b}_3/2$, where $\zeta \approx 0.392$ (Cox et al. 1979). The basic vectors $\boldsymbol{b}_1$, $\boldsymbol{b}_2$, $\boldsymbol{b}_3$ of the reciprocal lattice are chosen here as it is usually done in the case of unit cells with centered bases (vector $\boldsymbol{b}_1$ is parallel to the $k_x$-axis, vectors $\boldsymbol{b}_2$ and $\boldsymbol{b}_3$ lie in the $k_y k_z$-plane). The irreducible representation that has vectors $\boldsymbol{K}_1$ and $\boldsymbol{K}_2$ in its star is two-dimensional complex (Kovalev 1961) (in reality there exist two representations that lead in this case to identical results) and the physically irreducible representation is four-dimensional. Analysis (Lyubarsky 1957) of this representation makes it possible to find all the translational invariants of interest; those which do not comprise spatial derivatives are written by Cox et al. (1979) and Dvorak and Fousek (1980).

If we denote $Q_1 = \rho_1 \exp(\mathrm{i}\varphi_1)$ and $Q_2 = \rho_2 \exp(\mathrm{i}\varphi_2)$, the thermodynamic potential density, which may depend only on perfect invariants, has the form

$$\begin{aligned}\Phi &= \frac{A}{2}(\rho_1^2 + \rho_2^2) + \frac{B_1}{4}(\rho_1^4 + \rho_2^4) + \frac{B_2}{2}\rho_1^2\rho_2^2 + \frac{B_3}{2}\rho_1^2\rho_2^2 \cos 2(\varphi_1 - \varphi_2) \\ &+ \frac{1}{2}\sum_{i=1}^{3} D_i \left[\left(\frac{\partial \rho_1}{\partial x_i}\right)^2 + \rho_1^2\left(\frac{\partial \varphi_1}{\partial x_i}\right)^2 + \left(\frac{\partial \rho_2}{\partial x_i}\right)^2 + \rho_2^2\left(\frac{\partial \varphi_2}{\partial x_i}\right)^4\right] \\ &+ \frac{D_4}{2}\left(\frac{\partial \rho_1}{\partial y}\frac{\partial \rho_1}{\partial z} + \rho_1^2 \frac{\partial \varphi_1}{\partial y}\frac{\partial \varphi_1}{\partial z} - \frac{\partial \rho_2}{\partial y}\frac{\partial \rho_2}{\partial z} - \rho_2^2 \frac{\partial \varphi_2}{\partial y}\frac{\partial \varphi_2}{\partial z}\right). \end{aligned} \tag{31}$$

We have not written down here one more perfect invariant quadratic in derivatives of $Q_1$ and $Q_2$ since it can be represented as a sum of exact derivatives, such invariants being of no importance for the thermodynamic potential. Expression (31) involves a term comprising the phases $\varphi_1$ and $\varphi_2$ but not their derivatives. This is connected with the fact that the quantity $2(\boldsymbol{K}_1 - \boldsymbol{K}_2)$ coincides with one of the reciprocal lattice vectors. To limit the thermodynamic potential from below, certain conditions (implied in what follows) should be imposed on the coefficients $B_n$ and $D_n$. If these conditions are not fulfilled, subsequent terms in the thermodynamic potential should be taken into account.

The Lifshitz invariant in this case takes the form

$$\boldsymbol{L} = \rho_1^2 \frac{\partial \varphi_1}{\partial x} + \rho_2^2 \frac{\partial \varphi_2}{\partial x}. \tag{32}$$

The diagonal components of the tensor $\epsilon_{ij}$, which are perfect invariants, are linearly coupled with the combinations of $\rho_n$ and $\varphi_n$ of the type entering eq. (31) and with the Lifshitz invariant (32). The component $\epsilon_{yz}$ is linearly coupled with the combinations

$$\rho_1^2 - \rho_2^2, \qquad \rho_1^2\rho_2^2 \sin 2(\varphi_1 - \varphi_2), \qquad \rho_1^2\frac{\partial\varphi_1}{\partial x} - \rho_2^2\frac{\partial\varphi_2}{\partial x},$$

while the components $\epsilon_{xy}$ and $\epsilon_{xz}$ are coupled respectively with

$$a_1\left(\rho_1^2\frac{\partial\varphi_1}{\partial y} + \rho_2^2\frac{\partial\varphi_2}{\partial y}\right) + a_2\left(\rho_1^2\frac{\partial\varphi_1}{\partial z} - \rho_2^2\frac{\partial\varphi_2}{\partial z}\right), \qquad a_3\left(\rho_1^2\frac{\partial\varphi_1}{\partial z} + \rho_2^2\frac{\partial\varphi_2}{\partial z}\right)$$

$$+ a_4\left(\rho_1^2\frac{\partial\varphi_1}{\partial y} - \rho_2^2\frac{\partial\varphi_2}{\partial y}\right).$$

We do not take into account translational invariants that comprise derivatives of the amplitudes, for they give inessential corrections (see subsect. 4.1).

When $A < 0$ and subject to the coefficients of the thermodynamic potential, there may exist two equilibrium states corresponding to different incommensurate phases (Cox et al. 1979):

$$\text{(1)} \qquad \rho_1^2 = \rho_{1e}^2 \equiv -A/B_1, \qquad \rho_2 = 0 \quad (\text{or } \rho_1 = 0, \quad \rho_2 \neq 0), \qquad \text{if } 0 < B_1 < B_2 - |B_3|; \tag{33}$$

$$\text{(2)} \qquad \rho_1^2 = \rho_2^2 = \rho_e^2 \equiv \frac{-A}{B_1 + B_2 - |B_3|}, \qquad \sin 2(\varphi_1 - \varphi_2) = 0, \qquad \text{if } B_1 > \big|B_2 - |B_3|\big|. \tag{34}$$

In the case (33) the point symmetry of the incommensurate phase is different from that of the symmetrical phase since there appears a non-zero spontaneous value of the component $\epsilon_{yz}$, which is due to freezing in of only one of the vibrations with the wavevector $\boldsymbol{K}_1$ or $\boldsymbol{K}_2$. In the case (34) the point symmetry below the phase transition point remains unchanged (the equilibrium value of $\epsilon_{yz}$ is equal to zero). Note that an account taken in the thermodynamic potential of higher invariants shows that a third state may exist ($0 \neq \rho_1 \neq \rho_2 \neq 0$) (Dvorak and Fousek 1980), which is not analyzed here. Neither do we discuss which of the cases is realized concretely in $BaMnF_4$, but consider for comparison light scattering in the incommensurate phases corresponding to both the possibilities (33) and (34).

In any of these two incommensurate phases there appear four different soft modes (Cox et al. 1979, Lockwood et al. 1981). Corresponding normal coordinates and mode frequencies can be found by introducing a kinetic energy and a dissipation function and applying the methods used in subsect. 2.2. In the case (33) the normal coordinates are $\Delta\rho_1$, $\Delta\varphi_1$ and the two quantities

$$w_1 = Q_2 \exp(-\mathrm{i}\varphi_{1e}) + Q_2^* \exp(\mathrm{i}\varphi_{1e}),$$

$$w_2 = \mathrm{i}[Q_2 \exp(-\mathrm{i}\varphi_{1e}) - Q_2^* \exp(\mathrm{i}\varphi_{1e})],$$

linearly coupled with $Q_2$ and $Q_2^*$ (it is convenient not to introduce $\rho_2$ and $\varphi_2$).

Fluctuations of the phase $\varphi_1$ with the dispersion for complex frequency

$$\mu\omega^2 + \mathrm{i}\gamma\omega = D_1 q_x^2 + D_2 q_y^2 + D_3 q_z^2 + D_4 q_y q_z \tag{35}$$

correspond to the Goldstone mode. The phase mode is overdamped, of course (see subsect. 2.2). The frequencies of vibrations of the amplitude $\rho_1$ and of the quantities $w_1$ and $w_2$ at $\boldsymbol{q} = 0$ are determined respectively from the relations

$$\mu\omega^2 + \mathrm{i}\gamma\omega = 2B_1\rho_{1\mathrm{e}}^2, \qquad \mu\omega^2 + \mathrm{i}\gamma\omega = (B_2 - B_1 \pm B_3)\rho_{1\mathrm{e}}^2 . \tag{36}$$

The fluctuations of $w_1$ and $w_2$ are not observed in first-order light scattering since any translational invariant, including the tensor $\epsilon_{ij}$, is quadratic in these quantities, and their spontaneous value in the incommensurate phase is zero. The diagonal $\epsilon_{ij}$ components and the component $\epsilon_{yz}$ being linearly dependent on the amplitude squared $\rho_1^2$, the amplitudon contributes not only to fluctuations of the diagonal components but also to fluctuations of $\epsilon_{yz}$, i.e., in contrast to the two-component order parameter, the amplitudon manifests itself in light scattering with and without change of polarization, this being referred as before to orthorhombic crystals in the symmetrical phase. At the transition point to the incommensurate phase the intensity of light scattered by the amplitudon undergoes a jump.

Let us now consider the light scattering activity of fluctuations of the phase $\varphi_1$, i.e., of the phason. From the indicated relation between the $\epsilon_{ij}$ components and the order parameter it follows that the mean-square fluctuations of the diagonal components $\Delta\epsilon_{ij}(\boldsymbol{q})$ and of the component $\Delta\epsilon_{yz}(\boldsymbol{q})$ with the wavevector $\boldsymbol{q}$ are proportional to

$$\rho_{1\mathrm{e}}^4 q_x^2 \langle |\varphi_1(\boldsymbol{q})|^2 \rangle , \tag{37}$$

and the mean-square fluctuations of the components $\Delta\epsilon_{xy}(\boldsymbol{q})$ and $\Delta\epsilon_{xz}(\boldsymbol{q})$ are proportional respectively to

$$\begin{aligned} &\rho_{1\mathrm{e}}^4 (a_1 q_y + a_2 q_z)^2 \langle |\varphi_1(\boldsymbol{q})|^2 \rangle , \\ &\rho_{1\mathrm{e}}^4 (a_3 q_y + a_4 q_z)^2 \langle |\varphi_1(\boldsymbol{q})|^2 \rangle . \end{aligned} \tag{38}$$

The mean-square fluctuation $\langle |\varphi_1(\boldsymbol{q})|^2 \rangle$ of the Fourier transform of the phase $\varphi_1$ is calculated according to the procedure described in subsect. 3.1:

$$\langle |\varphi_1(\boldsymbol{q})|^2 \rangle = \frac{\mathrm{k_B} T}{V\rho_{1\mathrm{e}}^2 (D_1 q_x^2 + D_2 q_y^2 + D_3 q_z^2 + D_4 q_y q_z)} \tag{39}$$

Note that under the above-mentioned requirements on the thermodynamic potential coefficients $D_i$ the denominator of (39) is positive if $\boldsymbol{q} \neq 0$.

Expressions (37)–(39) determine the mean-square fluctuations $\langle |\Delta\epsilon_{ij}(\boldsymbol{q})|^2 \rangle$ and proportional to them are intensities of light scattered by the phason. These expressions change as the sign of $q_y$ or $q_z$ changes, i.e., the directional pattern of scattering, in contrast to the case of two-component order parameter, does not possess symmetry of the high-temperature phase. In particular, the quantity $\langle|\Delta\epsilon_{xy}(\boldsymbol{q})|^2\rangle$ vanishes if the vector $\boldsymbol{q}$ is in non-symmetric position with respect to the crystallographic axes $(a_1q_y + a_2q_z = 0)$, the same concerns the quantity $\langle|\Delta\epsilon_{xz}(\boldsymbol{q})|^2\rangle$, but at a different direction of the vector $\boldsymbol{q}(a_3q_y + a_4q_z = 0)$. In the end, this is associated with a lowered point symmetry in the incommensurate phase. Naturally, when a sample is polydomain, this effect is weakened.

In the case (34) the Goldstone mode corresponds to the fluctuations of $\varphi^{(+)} = \varphi_1 + \varphi_2$ with the dispersion for the complex frequency at small $\boldsymbol{q}$,

$$\mu\omega^2 + \mathrm{i}\gamma\omega = D_1q_x^2 + D_2q_y^2 + D_3q_z^2\,, \tag{40}$$

and the three other soft modes are associated with the fluctuations of $\varphi^{(-)} = \varphi_1 - \varphi_2$, $\rho^{(+)} = \rho_1 + \rho_2$ and $\rho^{(-)} = \rho_1 - \rho_2$, the mode frequencies at $\boldsymbol{q} = 0$ being determined respectively from equations

$$\mu\omega^2 + \mathrm{i}\gamma\omega = 4|B_3|\rho_\mathrm{e}^2, \qquad \mu\omega^2 + \mathrm{i}\gamma\omega = 2(B_1 + B_2 - |B_3|)\rho_\mathrm{e}^2\,,$$

$$\mu\omega^2 + \mathrm{i}\gamma\omega = 2(B_1 - B_2 + |B_3|)\rho_\mathrm{e}^2\,. \tag{41}$$

By virtue of eq. (34) the phase fluctuation $\Delta\varphi^{(-)}$ contributes in the $\boldsymbol{q}\to 0$ limit to the $\epsilon_{yz}$ fluctuations only, and in this case,

$$\langle|\Delta\epsilon_{yz}(\boldsymbol{q}\approx 0)|^2\rangle \sim \rho_\mathrm{e}^8\langle|\Delta\varphi^{(-)}(\boldsymbol{q}\approx 0)|^2\rangle \sim \rho_\mathrm{e}^4 \sim (T_\mathrm{i} - T)^2\,,$$

since, as can be shown, $\langle|\Delta\varphi^{(-)}|^2\rangle \sim \rho_\mathrm{e}^{-4}$, as $\boldsymbol{q}\to 0$. Thus, the corresponding scattered-light intensity is low near the phase transition point. The diagonal components of the tensor $\epsilon_{ij}$ are coupled with $\rho_1^2 + \rho_2^2$ and their fluctuations depend on the fluctuations of $\rho^{(+)}$, whereas fluctuations of $\rho^{(-)}$ contribute to the $\epsilon_{yz}$ fluctuations since the component $\epsilon_{yz}$ is coupled with $\rho_1^2 - \rho_2^2$. At the transition to the incommensurate phase, i.e., at $T = T_\mathrm{i}$ the intensity of light scattered by fluctuations of $\rho^{(+)}$ and $\rho^{(-)}$ undergoes a jump, as is usually the case with soft modes.

The Goldstone mode is observed in fluctuations of those components of the tensor $\epsilon_{ij}$ which are transformed like $\rho_1^2\partial\varphi_1/\partial x_i + \rho_2^2\partial\varphi_2/\partial x_i$, i.e., in fluctuations of all the components except $\epsilon_{yz}$ (if elastic deformations are not taken into account, see below). The mean-square fluctuation of the diagonal $\epsilon_{ij}$ components is proportional to

$$\rho_\mathrm{e}^4q_x^2\langle|\varphi^{(+)}(\boldsymbol{q})|^2\rangle\,,$$

and that of the components $\epsilon_{xy}$ and $\epsilon_{xz}$ is proportional to the same quantity with the replacement of $q_x^2$ by $q_y^2$ and $q_z^2$, respectively. The mean-square fluctuation of the Fourier transform $\varphi^{(+)}(\boldsymbol{q})$ has here the form

$$\langle|\varphi^{(+)}(\boldsymbol{q})|^2\rangle = \frac{2k_\mathrm{B}T}{V\rho_\mathrm{e}^2(D_1q_x^2 + D_2q_y^2 + D_3q_z^2)}\,, \tag{42}$$

and as distinguished from eq. (39), it does not change with the change of the signs of $q_y$ or $q_z$. In this case the regularities of light scattering by the Goldstone mode are similar to those which are observed when the order parameter has two components.

Now let us discuss in brief the changes due to an account taken of the interaction between the phason and elastic deformations. Applying the method used in subsect. 3.1, we see that in the case (33) this account comes down only to renormalization of the coefficients connecting $\epsilon_{ij}$ and $\varphi_1$ fluctuations, some of the renormalized coefficients being dependent not only on the value but also on the sign of $q_y$ and $q_z$. In the case (34) there appears coupling between $\Delta\epsilon_{yz}$ and $\Delta\varphi^{(+)}$, which is proportional to $q_x q_y q_z$, while the renormalized coefficients depend on $q_i^2$. In both cases the form of the renormalized coefficients is on the whole similar to that of subsect. 3.1.

As concerns the scattered-light spectral intensity, in the case of non-Goldstone excitations specific features of the incommensurate phase do not practically manifest themselves. For scattering by the Goldstone mode in the case (34) we have the same regularities as for a two-component order parameter (see subsect. 3.2). In the case (33) the central phason peak width is proportional to $(D_1 q_x^2 + D_2 q_y^2 + D_3 q_z^2 + D_4 q_y q_z)/\gamma$ (cf. (35)), i.e., the width as a function of $\boldsymbol{q}$ does not possess the symmetry of the high-temperature phase.

### 5.3. *Barium sodium niobate $Ba_2NaNb_5O_{15}$*

In $Ba_2NaNb_5O_{15}$ the phase transition to the incommensurate phase takes place at a temperature $T_i \approx 300°C$ from a symmetrical phase pertaining to the space group $C_{4v}^2$–P4bm, the incommensurate superstructure being characterized by the wavevectors $\boldsymbol{K}_1 = \zeta(\boldsymbol{b}_1 + \boldsymbol{b}_2) + \boldsymbol{b}_3/2$ and $\boldsymbol{K}_2 = \zeta(-\boldsymbol{b}_1 + \boldsymbol{b}_2) + \boldsymbol{b}_3/2$, where $\zeta \approx 0.279$ (Schneck and Denoyer 1981). To a star containing such wavevectors there correspond two four-dimensional irreducible representations (Kovalev 1961) that yield identical results in the question under consideration. If we denote, as before, $Q_1 = \rho_1 \exp(i\varphi_1)$ and $Q_2 = \rho_2 \exp(i\varphi_2)$, the analysis of these irreducible representations shows that the thermodynamic potential density may be written in the form

$$\begin{aligned}\Phi = {} & \frac{A}{2}(\rho_1^2 + \rho_2^2) + \frac{B_1}{4}(\rho_1^4 + \rho_2^4) + \frac{B_2}{2}\rho_1^2\rho_2^2 \\ & + \frac{D_1}{2}\left[\left(\frac{\partial\rho_1}{\partial x}\right)^2 + \rho_1^2\left(\frac{\partial\varphi_1}{\partial x}\right)^2 + \left(\frac{\partial\rho_2}{\partial x}\right)^2 + \rho_2^2\left(\frac{\partial\varphi_2}{\partial x}\right)^2\right. \\ & \left. + \left(\frac{\partial\rho_1}{\partial y}\right)^2 + \rho_1^2\left(\frac{\partial\varphi_1}{\partial y}\right)^2 + \left(\frac{\partial\rho_2}{\partial y}\right)^2 + \rho_2^2\left(\frac{\partial\varphi_2}{\partial y}\right)^2\right] \\ & + \frac{D_2}{2}\left[\frac{\partial\rho_1}{\partial x}\frac{\partial\rho_1}{\partial y} + \rho_1^2\frac{\partial\varphi_1}{\partial x}\frac{\partial\varphi_1}{\partial y} - \frac{\partial\rho_2}{\partial x}\frac{\partial\rho_2}{\partial y} - \rho_2^2\frac{\partial\varphi_2}{\partial x}\frac{\partial\varphi_2}{\partial y}\right]\end{aligned}$$

$$+\frac{D_3}{2}\left[\left(\frac{\partial\rho_1}{\partial z}\right)^2+\rho_1^2\left(\frac{\partial\varphi_1}{\partial z}\right)^2+\left(\frac{\partial\rho_2}{\partial z}\right)^2+\rho_2^2\left(\frac{\partial\varphi_2}{\partial z}\right)^2\right]. \tag{43}$$

For the same reason as in the case of the thermodynamic potential density (31) we have not written down here one more perfect invariant quadratic in the derivatives. As distinct from (31), expression (43) and other translational invariants involve only derivatives of the phases $\varphi_1$ and $\varphi_2$, but not the phases themselves. This is explained by the fact that no combination of the vectors $\boldsymbol{K}_1$ and $\boldsymbol{K}_2$ with integral coefficients forms a reciprocal lattice vector. We suppose that the coefficients $B_n$ and $D_n$ are restricted to such conditions under which the thermodynamic potential is limited from below.

The Lifshitz invariant in this case has the form

$$\boldsymbol{L}=\rho_1^2\frac{\partial\varphi_1}{\partial x}-\rho_2^2\frac{\partial\varphi_2}{\partial x}+\rho_1^2\frac{\partial\varphi_1}{\partial y}+\rho_2^2\frac{\partial\varphi_2}{\partial y}. \tag{44}$$

In the $\epsilon_{ij}$ components the terms depending on the order parameter, and essential in the problem under consideration, are determined by the expressions

$$\epsilon_{xx}=a_1(\rho_1^2+\rho_2^2)+a_2\left(\rho_1^2\frac{\partial\varphi_1}{\partial x}-\rho_2^2\frac{\partial\varphi_2}{\partial x}\right)+a_3\left(\rho_1^2\frac{\partial\varphi_1}{\partial y}+\rho_2^2\frac{\partial\varphi_2}{\partial y}\right),$$

$$\epsilon_{zz}=a_4(\rho_1^2+\rho_2^2)+a_5\boldsymbol{L},\qquad \epsilon_{xy}=a_6(\rho_1^2-\rho_2^2)$$

$$+a_7\left(\rho_1^2\frac{\partial\varphi_1}{\partial x}+\rho_2^2\frac{\partial\varphi_2}{\partial x}+\rho_1^2\frac{\partial\varphi_1}{\partial y}-\rho_2^2\frac{\partial\varphi_2}{\partial y}\right),$$

$$\epsilon_{xz}=a_8\left(\rho_1^2\frac{\partial\varphi_1}{\partial z}-\rho_2^2\frac{\partial\varphi_2}{\partial z}\right),\qquad \epsilon_{yz}=a_8\left(\rho_1^2\frac{\partial\varphi_1}{\partial z}+\rho_2^2\frac{\partial\varphi_2}{\partial z}\right), \tag{45}$$

the component $\epsilon_{yy}$ being obtained from $\epsilon_{xx}$ by interchanging $a_2$ and $a_3$ because the symmetrical phase belongs to the tetragonal system.

Subject to the values of the coefficients of the thermodynamic potential one of the two incommensurate phases is realized (we are not discussing here the third equilibrium state, the existence of which may be shown when an account is taken of higher invariants):

(1) $\rho_1^2=\rho_{1e}^2=-A/B_1,\ \ \rho_2=0\ \ (\text{or}\ \ \rho_1=0,\ \ \rho_2\neq 0),\qquad \text{if } 0<B_1<B_2;$ (46)

(2) $\rho_1^2=\rho_2^2=\rho_e^2\equiv -A/(B_1+B_2),\qquad \text{if } B_1>|B_2|.$ (47)

In the case (46) one-dimensional incommensurability takes place (a wave with the wavevector $\boldsymbol{K}_1$ or $\boldsymbol{K}_2$ is frozen in), and the point symmetry lowers at the phase transition since there appears a spontaneous value of $\epsilon_{xy}$; in the case (47) the incommensurability is two-dimensional and is characterized by both vectors $\boldsymbol{K}_1$ and $\boldsymbol{K}_2$, the point symmetry being unaffected by the phase transition.

Let us consider soft modes in the incommensurate phase by analogy with what has been done in subsect. 5.2. If the state (46) is realized, there exist two modes corresponding to fluctuations of $\rho_1$ and $\varphi_1$ and one two-fold degenerate mode associated with fluctuations of $Q_2$ and $Q_2^*$. The Goldstone mode (fluctuations of $\varphi_1$) has the following dispersion for complex frequency

$$\mu\omega^2 + \mathrm{i}\gamma\omega = D_1 q_x^2 + D_1 q_y^2 + D_2 q_x q_y + D_3 q_z^2 . \tag{48}$$

The frequencies of the amplitudon and of the two-fold degenerate mode at $\boldsymbol{q} = 0$ are determined respectively by the relations

$$\mu\omega^2 + \mathrm{i}\gamma\omega = 2B_i\rho_{1e}^2, \qquad \mu\omega^2 + \mathrm{i}\gamma\omega = (B_2 - B_1)\rho_{1e}^2 . \tag{49}$$

It is readily seen (cf. subsect. 5.2) that the two-fold degenerate mode is not active in light scattering. The amplitudon makes contribution to fluctuations of the diagonal $\epsilon_{ij}$ components and to fluctuations of the component $\epsilon_{xy}$. The contribution of the phason to $\epsilon_{ij}$ fluctuations is determined by the relations

$$\langle|\Delta\epsilon_{xx}(\boldsymbol{q})|^2\rangle = \rho_{1e}^4(a_2 q_x + a_3 q_y)^2\langle|\varphi_1(\boldsymbol{q})|^2\rangle , \tag{50}$$

$$\langle|\Delta\epsilon_{xz}(\boldsymbol{q})|^2\rangle = \langle|\Delta\epsilon_{yz}(\boldsymbol{q})|^2\rangle = a_8^2\rho_{1e}^4 q_z^2\langle|\varphi_1(\boldsymbol{q})|^2\rangle . \tag{51}$$

The mean-square fluctuations of the remaining components can be obtained from eq. (50): the fluctuations of $\epsilon_{yy}$ are found by interchanging $a_2$ and $a_3$, and the fluctuations of $\epsilon_{zz}$ and $\epsilon_{xy}$ by replacing $a_2$ and $a_3$ in the first case by $a_5$ and in the second case by $a_7$. The mean-square fluctuation of the Fourier transform $\varphi_1(\boldsymbol{q})$ here takes the form

$$\langle|\varphi_1(\boldsymbol{q})|^2\rangle = \frac{k_\mathrm{B}T}{V\rho_{1e}^2(D_1 q_x^2 + D_1 q_y^2 + D_2 q_x q_y + D_3 q_z^2)} . \tag{52}$$

If the state (46) is realized, the regularities of light scattering are on the whole analogous to those discussed in subsect. 5.2 for the state (33), except for the differences in angular dependences. If in the case of $BaMnF_4$ a more complicated dependence on the $\boldsymbol{q}$ direction is inherent to fluctuations of non-diagonal $\epsilon_{ij}$ components, then in this case the dependence on the $\boldsymbol{q}$ direction is inherent, on the contrary, to fluctuations of diagonal components of the tensor $\epsilon_{ij}$.

Let us discuss in some more detail the light scattering in the presence of a two-dimensional incommensurability when the state (47) is realized. Naturally, there exist two Goldstone modes here corresponding to fluctuations of the phases $\varphi_1$ and $\varphi_2$, the dispersions for complex frequencies having the form

$$\mu\omega^2 + \mathrm{i}\gamma\omega = D_1 q_x^2 + D_1 q_y^2 \pm D_2 q_x q_y + D_3 q_z^2 . \tag{53}$$

The two other soft modes are associated with the vibrations of $\rho^{(\pm)} = \rho_1 \pm \rho_2$, at $\boldsymbol{q} = 0$ the mode frequencies being found from the relation

$$\mu\omega^2 + \mathrm{i}\gamma\omega = 2(B_1 \pm B_2)\rho_e^2 . \tag{54}$$

The vibrations of $\rho^{(+)}$ manifest themselves in fluctuations of diagonal $\epsilon_{ij}$

components, the vibrations of $\rho^{(-)}$ make a contribution to fluctuations of the component $\epsilon_{xy}$, the regularities of scattering being here the same as in the case of ordinary soft modes.

To clarify the manifestations of phasons we write, using eqs. (45), relations between the Fourier-transforms of fluctuations of $\epsilon_{ij}$ and those of the phases $\varphi_1$ and $\varphi_2$:

$$\langle|\Delta\epsilon_{xx}(\boldsymbol{q})|^2\rangle = \rho_e^4(a_2q_x + a_3q_y)^2\langle|\varphi_1(\boldsymbol{q})|^2\rangle + \rho_e^4(a_2q_x - a_3q_y)^2\langle\varphi_2(\boldsymbol{q})|^2\rangle\,, \tag{55}$$

$$\langle|\Delta\epsilon_{xz}(\boldsymbol{q})|^2\rangle = \langle|\Delta\epsilon_{yz}(\boldsymbol{q})|^2\rangle = a_8^2\rho_e^4q_z^2[\langle|\varphi_1(\boldsymbol{q})|^2\rangle + \langle|\varphi_2(\boldsymbol{q})|^2\rangle]\,. \tag{56}$$

Fluctuations of other components can be obtained from eq. (55) by making the substitutions mentioned after eqs. (50) and (51). From this it is seen that both the Goldstone modes contribute to fluctuations of all the components of the tensor $\epsilon_{ij}$. Mean-square fluctuations of the Fourier-transforms $\varphi_1(\boldsymbol{q})$ and $\varphi_2(\boldsymbol{q})$ can be found in the usual way; for $\langle|\varphi_1(\boldsymbol{q})|^2\rangle$ we obtain eq. (52) and for $\langle|\varphi_2(\boldsymbol{q})|^2\rangle$ the same but with the sign of $D_2$ reversed. Substituting the expressions obtained into eqs. (55) and (56), we come to

$$\langle|\Delta\epsilon_{xx}(\boldsymbol{q})|^2\rangle = \frac{2k_BT\rho_e^2[(a_2q_x^2 + a_3q_y^2)(D_1q_x^2 + D_1q_y^2 + D_3q_z^2) - 2a_2a_3D_2q_x^2q_y^2}{V[(D_1q_x^2 + D_1q_y^2 + D_3q_z^2)^2 - D_2^2q_x^2q_y^2]}\,,$$

$$\langle|\Delta\epsilon_{xz}(\boldsymbol{q})|^2\rangle = \frac{2a_8^2k_BT\rho_e^2q_z^2(D_1q_x^2 + D_1q_y^2 + D_3q_z^2)}{V[(D_1q_x^2 + D_1q_y^2 + D_3q_z^2)^2 - D_2^2q_x^2q_y^2]}\,. \tag{57}$$

From this it is seen that the angular dependence of the scattered intensity is here rather complicated. In spite of the fact that fluctuations of $\varphi_1(\boldsymbol{q})$ and $\varphi_2(\boldsymbol{q})$ are unsymmetric with respect to the change of the signs of $q_x$ or $q_y$, the contribution of both the Goldstone modes leads to invariance of expressions (57) under the change of signs of $q_x$ or $q_y$, which also explains the complicated form of these expressions.

Let us now consider the spectral intensity of light scattered by the phasons. Acting the same way as in deriving eq. (22), we find the spectral density of the fluctuations

$$\langle|\varphi_1(\boldsymbol{q},\Omega)|^2\rangle = \frac{\gamma k_BT}{\pi V\rho_e^2[(D_1q_x^2 + D_1q_y^2 + D_2q_xq_y + D_3q_z^2)^2 + \gamma^2\Omega^2]}\,. \tag{58}$$

For the fluctuations of $\varphi_2$ we get the same formula but with the sign of $D_2$ reversed. The scattered-light spectral intensity can be obtained from eqs. (55) and (56) by replacing $\langle|\varphi_{1,2}(\boldsymbol{q})|^2\rangle$ by $\langle|\varphi_{1,2}(\boldsymbol{q},\Omega)|^2\rangle$. Thus, in this case in the scattered-light spectrum there exists a central component, which is a superposition of two lines with the widths $2(D_1q_x^2 + D_1q_y^2 \pm D_2q_xq_y + D_3q_z^2)/\gamma$.

We have considered light scattering for two types of incommensurate phase transitions with a four-component order parameter on an example of $BaMnF_4$ and $Ba_2NaNb_5O_{15}$ which pertain in the symmetrical phase to different crystal systems. It is of interest to clarify how different the results would be if the crystals pertained to one and the same space group. We can, for example, consider light scattering by an incommensurate phase for a crystal of the same space group in the symmetrical phase as in $BaMnF_4$ but choosing the vectors $\boldsymbol{K}_1$ and $\boldsymbol{K}_2$ such that none of their combinations with integral coefficients form a reciprocal lattice vector. Calculations show (Golovko and Levanyuk 1981c) that in this case the results obtained above for $Ba_2NaNb_5O_{15}$ remain qualitatively valid too.

## 6. *Incommensurate phase as a spatially modulated structure*

As mentioned in subsect. 2.1, besides the foregoing approach to the description of an incommensurate phase there exists another one based not on the order parameter characterizing the symmetrical–incommensurate transition but on that corresponding to the transition from the symmetrical phase to the commensurate low-temperature phase. Since the latter approach is widely spread in theoretical investigations of the incommensurate phase, it is of interest to discuss the method of light-scattering analysis proceeding from this approach. We consider first the case when the extremity of the wavevector $\boldsymbol{K}_0$, that characterizes the incommensurate superstructure, lies near the boundary of the symmetrical phase Brillouin zone, a corresponding vibration being degenerate at the zone boundary (fig. 2(a)). In the case of such a position of $\boldsymbol{K}_0$ a low-temperatures commensurate phase is usually observed, whose order parameter is a normal coordinate corresponding to the wavevector $\boldsymbol{k} = \boldsymbol{b}/2$. Our description of the incommensurate phase is based just on this order parameter which in the simplest case has, due to degeneracy, two components $\eta_1$ and $\eta_2$ that are assumed to be real. To specify the consideration, we analyze crystals analogous to ammonium fluoroberyllate $(NH_4)_2BeF_4$, wherein phase transitions occur of the type under discussion (Iizumi and Gesi 1977).

The symmetry group of the high-temerature phase of $(NH_4)_2BeF_4$ is $D_{2h}^{16}$–Pnam; in the low-temperature commensurate phase pertaining to the space group $C_{2v}^{2}$–$Pn2_1a$ the period along the $x$-axis is doubled, and along the $y$-axis there appears spontaneous polarization. If instead of $\eta_1$ and $\eta_2$ one introduces the quantities $\rho$ and $\varphi$, whose meaning follows from the relations $\eta_1 = \rho\cos\varphi$, $\eta_2 = \rho\sin\varphi$, the thermodynamic potential density that corresponds to this phase transition has the form (Levanyuk and Sannikov 1976a)

$$\Phi = \frac{A}{2}\rho^2 + \frac{B}{4}\rho^4 + \frac{B_1}{4}\rho^4\cos 4\varphi - C\rho^2\frac{\partial\varphi}{\partial x} + \frac{1}{2}\sum_{i=1}^{3} D_i\left[\left(\frac{\partial\rho}{\partial x_i}\right)^2 + \rho^2\left(\frac{\partial\varphi}{\partial x_i}\right)^2\right]. \tag{59}$$

As distinguished from eq. (4), this expression involves the Lifshitz invariant $\rho^2\partial\varphi/\partial x$, which implies a non-zero slope of dispersion curves at $\boldsymbol{k} = \boldsymbol{b}/2$ (fig.

2(a)). Besides, as a result of a special value of $\boldsymbol{k} = \boldsymbol{b}/2$ there appears an invariant $\rho^4 \cos 4\varphi$.

When $A < A_0$, the thermodynamic potential is minimal if

$$\rho^2 = \rho_e^2 \equiv (A_0 - A)/B, \qquad \varphi = \varphi_e \equiv \boldsymbol{k}_0 \boldsymbol{r} = k_0 x\,, \\ k_0 = C/D_1, \qquad A_0 = C^2/D_1\,. \tag{60}$$

Thus, in equilibrium the order parameter used is spatially inhomogeneous: $\eta_{1e} = \rho_e \cos k_0 x$, $\eta_{2e} = \rho_e \sin k_0 x$, the meaning of the wavevector $k_0$, that characterizes spatial modulation of the order parameter, being clear from fig. 2a.

Simple eqs. (60) for $\rho_e$ and $\varphi_e$ are valid only near the phase transition point, where the quantity $\rho_e^2$ is small. As temperature lowers, the amplitude of the order parameter harmonics increases, and the functions $\rho_e(x)$ and $\varphi_e(x)$ can be found either by numerical (Ishibashi and Dvorak 1978, Shiba and Ishibashi 1978) or by approximate (Golovko 1980) methods. A wide use of a thermodynamic potential of the type (59) in the study of incommensurate phases is connected with the fact that it helps to clarify a large number of properties of the incommensurate phase in the entire region of its existence (Levanyuk and Sannikov 1976a).

The analysis of the incommensurate phase excitations is somewhat more complicated in this case since equations for excitations contain spatially periodic coefficients. However at small $\rho_e^2$, when expressions (60) are valid, these equations turn out to be simple, and for the oscillations of $\rho$ and $\varphi$ we obtain respectively dispersion equations (9) and (10), but now $\rho_e^2$ is determined from eq. (60). Note that the excitation wavevector $\boldsymbol{q}$ is referred, as before, to points A or A′ as the origin (fig. 4) but not to point B that corresponds to the utilized order parameter since the existence of an equilibrium modulation shifts the origin of the vector $\boldsymbol{q}$ from point B by a quantity $\pm k_0$.

When considering light scattering by the incommensurate phase, one should take into account that owing to the order parameter modulation the equilibrium dielectric constant tensor $\epsilon_{ij}$ is in this case periodic in space, and therefore one should modify the formulae that relate the scattered intensity to $\epsilon_{ij}$ fluctuations (the formulae presented in ch. 1 of the book). However to a first approximation these formulae can remain unchanged, for modulation of the equilibrium tensor $\epsilon_{ij}$ has an effect upon light propagation only in higher approximations in powers of the order parameter (Golovko and Levanyuk 1979).

As an example we consider here light scattering by fluctuations of the components $\epsilon_{zz}$ and $\epsilon_{xy}$, induced by order parameter fluctuations. The dependence of these components on the indicated order parameter has the form

$$\epsilon_{zz} = a_1 \rho^2 + a_2 \rho^2 \frac{\partial \varphi}{\partial x}, \qquad \epsilon_{xy} = a_3 \rho^2 \cos 2\varphi + a_4 \rho^2 \frac{\partial \varphi}{\partial y}\,. \tag{61}$$

Note that the component $\epsilon_{xy}$ contains an additional term $a_3\rho^2 \cos 2\varphi$, which was absent from the dependence of $\epsilon_{xy}$ on the order parameter used above. Here we also take into account gradient invariants (see subsect. 3.1) that were omitted in the earlier paper (Golovko and Levanyuk 1980).

Let us first discuss the problem of calculating fluctuations of the amplitude $\rho$ and of the phase $\varphi$. Expanding the fluctuations $\Delta\rho$ and $\Delta\varphi$ in Fourier series and substituting them into the thermodynamic potential density (59) we see that due to the presence of the term $\rho^4 \cos 4\varphi$ the fluctuations $\Delta\rho$ and $\Delta\varphi$ are no longer independent and, besides, fluctuations with different $\boldsymbol{q}$ are related with one another. However such a "mixing" of fluctuations leads to corrections of a higher order of smallness to mean-square fluctuations, and in a first approximation we obtain eqs. (17) and (28), where $\rho_e^2$ is determined from eq. (60).

Taking into account the independence of the fluctuations $\Delta\rho$ and $\Delta\varphi$ in a first approximation, from (61) we find

$$\langle|\Delta\epsilon_{zz}(\boldsymbol{q})|^2\rangle = 4\tilde{a}_1^2\rho_e^2\langle|\rho(\boldsymbol{q})|^2\rangle + a_2^2\rho_e^4 q_x^2\langle|\varphi(\boldsymbol{q})|^2\rangle\,, \tag{62}$$

$$\begin{aligned}\langle|\Delta\epsilon_{xy}(\boldsymbol{q})|^2\rangle = a_3^2\rho_e^2[&\langle|\rho(2\boldsymbol{k}_0-\boldsymbol{q})|^2\rangle + \langle|\rho(2\boldsymbol{k}+\boldsymbol{q})|^2\rangle \\ &+ \rho_e^2\langle|\varphi(2\boldsymbol{k}_0-\boldsymbol{q})|^2\rangle + \rho_e^2\langle|\varphi(2\boldsymbol{k}_0+\boldsymbol{q})|^2\rangle] \\ &+ a_4^2\rho_e^4 q_y^2\langle|\varphi(\boldsymbol{q})|^2\rangle\,,\end{aligned} \tag{63}$$

where $\tilde{a}_1 = a_1 + a_2k_0$. The first term in eq. (62) analogous to eq. (27) is the amplitudon contribution to the $\epsilon_{zz}$ fluctuations; the last terms in eqs. (62) and (63) correspond to the contribution of the Goldstone phason and are analogous to expression (16).

Besides, the $\epsilon_{xy}$ fluctuations are also contributed to by excitations which could be called an amplitudon and a phason with wavevector $2\boldsymbol{k}_0$ (taking into account that $\boldsymbol{q} \approx 0$). However, due to the fact that $K_0$ is close to $b/2$ these excitations belong to another branch, which is readily seen from fig. 4, particularly if for the wavevector point A′ is referred to as the origin since we get at one of the points C or C′ (the Brillouin zone boundary should be passed). So, the excitations in question are vibrations of one of the quantities denoted in subsect. 4.2 by $Q_1$, and in addition we have obtained frequency splitting at $\boldsymbol{k} = \boldsymbol{K}_0$ (points C and C′) also considered in subsect. 4.2. It is quite natural that in the approach used here not only the soft branch but also the one genetically related to it is automatically taken into account since at the Brillouin zone boundary these two branches intersect, and the order parameter used corresponds just to this point. Here the scattering geometry, in which the excitations under discussion are observed, is found immediately, for we know the $\epsilon_{ij}$ component coupled with these excitations. After the substitution of the fluctuations of $\rho$ and $\varphi$, the term with the coefficient $a_3^2$ in eq. (63) takes the form $a_3^2\rho_e^2k_BT/D_1k_0^2V$, the smallness of $\rho_e$ and $q$ being taken into account. The corresponding scattered intensity is proportional to $\rho_e^2$, which agrees with the results of subsect. 4.2, but due to the smallness of $k_0$ the intensity increases rather rapidly as temperature lowers.

Spectral distribution of scattered light can be found by a standard technique (see subsects. 3.2 and 4.1), in a first approximation in $\rho_e^2$ the problem is more simple since in this case all the excitations discussed are independent. The results, naturally, coincide with those obtained above. The estimates show (Golovko and Levanyuk 1980) that the modes corresponding to points C and C′ in fig. 4 can be either overdamped or underdamped.

If the terms of $\epsilon_{ij}$ containing powers of the order parameter higher than in eq. (61) are taken into account, one can establish the Raman activity of other excitations pertaining to the soft branch and to the branch genetically related to it. Since the coefficients of such terms in $\epsilon_{ij}$ are periodic, these excitations correspond to the Brillouin zone points situated at distance $nk_0$ from the zone boundary, the intensities of corresponding lines being proportional to powers of $\rho_e$ higher than two, which is in agreement with the results of subsect. 4.2.

The Raman activity of excitations belonging to high-lying branches can be established like in subsect. 4.2. The $\epsilon_{ij}$ components may involve translational invariants of the type $\eta_1\xi_1 \pm \eta_2\xi_2$ etc., where $\xi_1$ and $\xi_2$ are normal coordinates corresponding to some high-lying branch at $\boldsymbol{k} = \boldsymbol{b}/2$ (in the case of $(NH_4)_2BeF_4$ all the irreducible representations at $\boldsymbol{k} = \boldsymbol{b}/2$ are two-dimensional). When $\eta_1$ and $\eta_2$ acquire spontaneous values, the fluctuations of $\xi_1$ and $\xi_2$ become linearly coupled with the $\epsilon_{ij}$ fluctuations and manifest themselves in light scattering. Since spontaneous values of $\eta_1$ and $\eta_2$ are periodic, the Fourier transform $\Delta\epsilon_{ij}(\boldsymbol{q})$ is coupled with the fluctuations of $\xi_{1,2}(\boldsymbol{q} \pm \boldsymbol{k}_0)$, i.e., those excitations become active that correspond to points C″ and C‴ in fig. 4, which agrees with the results of subsect. 4.2. The activity of other excitations pertaining to high branches may be considered in a similar way.

Now let us pass over to the discussion of the case when the order parameter used in the present section, being, for example, one-component, does not admit a Lifshitz invariant. This is the case when the extremity of the wavevector $\boldsymbol{k}_c$ of the commensurate low-temperature superstructure lies at the Brillouin zone boundary or $\boldsymbol{k}_c = 0$ and the soft branch (in the symmetrical phase) is not degenerate at these points (see subsect. 2.1. and fig. 2b). As an example we can point out sodium nitrite $NaNO_2$ (Hoshino and Motegi 1967) and thiourea $SC(NH_2)_2$ (Shiozaki 1971), in which the value of $K_0$ is small and $k_c = 0$.

To describe such an incommensurate phase transition, we use the method suggested by Ishibashi and Shiba (1978). Let us designate by $\eta$ a one-component order parameter corresponding to the centre or to the boundary of the symmetrical phase Brillouin zone and write the thermodynamic potential density in the form

$$\Phi = \frac{A}{2}\eta^2 + \frac{B}{4}\eta^4 + \frac{C}{2}\left(\frac{\partial^2\eta}{\partial x^2}\right)^2 + \frac{D_1}{2}\left(\frac{\partial\eta}{\partial x}\right)^2 + \frac{D_2}{2}\left(\frac{\partial\eta}{\partial y}\right)^2 + \frac{D_3}{2}\left(\frac{\partial\eta}{\partial z}\right)^2. \quad (64)$$

Suppose that $D_1 < 0$ and $D_2 > 0$, $D_3 > 0$. Since $D_1$ is negative, an additional term with the coefficient $C > 0$ is introduced here, which provides a limitation of the thermodynamic potential from below. A negative value of $D_1$ leads to the result that when $A < A_0$, the spatially modulated order parameter,

$$\eta_e = \rho_e \cos k_0 x, \qquad \rho_e^2 = 4(A_0 - A)/3B, \qquad k_0^2 = -D_1/2C,$$
$$A_0 = D_1^2/4C, \tag{65}$$

corresponds to equilibrium. In these expressions $\rho_e^2$ is assumed to be small. The quantity $k_0$ is equal here to $b/2 - K_0$ if $K_0$ is close to $b/2$, or $k_0 = K_0$ if $K_0$ is small.

The analysis of incommensurate phase excitations, which can be carried out by the methods used above with some modifications, shows (Ishibashi and Takagi 1979) that there exist excitations with the dispersion

$$\mu\omega^2 + i\gamma\omega = 2|D_1|q_x^2 + D_2 q_y^2 + D_3 q_z^2,$$
$$\mu\omega^2 + i\gamma\omega = \tfrac{3}{2} B\rho_e^2 + 2|D_1|q_x^2 + D_2 q_y^2 + D_3 q_z^2.$$

The first excitation corresponds to the phason, the second to the amplitudon.

The consideration of phason manifestations in light scattering requires here an approach somewhat different from that used above since there is no Lifshitz invariant now. In this case the only possible relation between the order parameter and the $\epsilon_{ij}$ components is of the form

$$\epsilon_{zz} = a_1\eta^2 + a_2\left(\frac{\partial\eta}{\partial x}\right)^2, \qquad \epsilon_{xy} = a_3\frac{\partial\eta}{\partial x}\frac{\partial\eta}{\partial y}, \qquad \epsilon_{xz} = a_4\frac{\partial\eta}{\partial x}\frac{\partial\eta}{\partial z}. \tag{66}$$

As an example we find from this the relations between $\eta$ fluctuations and fluctuations of the components $\epsilon_{zz}$ and $\epsilon_{xy}$,

$$\Delta\epsilon_{zz} = 2a_1\eta_e\Delta\eta + 2a_2\frac{\partial\eta_e}{\partial x}\frac{\partial\Delta\eta}{\partial x}, \qquad \Delta\epsilon_{xy} = a_3\frac{\partial\eta_e}{\partial x}\frac{\partial\Delta\eta}{\partial y}. \tag{67}$$

These expressions show that squares of the derivatives may also lead to a linear coupling between $\Delta\epsilon_{ij}$ and $\Delta\eta$ since $\partial\eta_e/\partial x \neq 0$. However this coupling contains a small factor $k_0$ to which the derivative $\partial\eta_e/\partial x$ is proportional. It is also clear that expressions of the type (66) cannot lead to a linear coupling between fluctuations of the component $\epsilon_{yz}$ and those of $\eta$, for $\epsilon_{yz}$ involves derivatives of $\eta$ with respect to $y$ and $z$; it is of no interest if we take into account in $\epsilon_{zz}$ the squares of other derivatives.

Fluctuations of $\eta$ can be found using the thermodynamic potential (64). Omitting intervening calculations, which become somewhat more difficult due to the periodicity of $\eta_e$, we present the final expressions for the mean-square fluctuations,

$$\langle|\Delta\epsilon_{zz}(q)|^2\rangle = \frac{2a_2^2 k_0^2 \rho_e^2 q_x^2 k_B T}{V(2|D_1|q_x^2 + D_2 q_y^2 + D_3 q_z^2)},$$

$$\langle |\Delta\epsilon_{xy}(q)|^2\rangle = \frac{a_3^2 k_0^2 \rho_e^2 k_B T}{2V(2|D_1|q_x^2 + D_2 q_y^2 + D_3 q_z^2)}. \tag{68}$$

Here we have not written down a term which correspond to the amplitudon contribution and has a form similar to eqs. (27)–(28).

Expressions (68) are completely analogous to those which were obtained in subsect. 3.1, but now scattering by the phason is associated with the terms in eq. (66) quadratic in derivatives. Comparing eqs. (68) and (18) we see that in the case under consideration the role of the coefficient $a$ of the Lifshitz invariant in eq. (14) is played by the quantity $a_2 k_0 \sqrt{2}$, which is small if $k_0$ is small. Thus, if the point $\boldsymbol{K}_0$ is near the Brillouin zone points for which the irreducible representation corresponding to the soft branch does not admit a Lifshitz invariant, light scattering by the phason is weak. This is in agreement with the qualitative considerations expressed in subsect. 3.1.

Naturally, terms quadratic in derivatives can be added also in eqs. (61). However an account of these terms is of no interest there, for there are terms that lead to the same effects but without the small factor $k_0$ in the formulae. Note that the incommensurate phase transition, for the consideration of which we have used the thermodynamic potential (64), can be described also in another way (Levanyuk and Sannikov 1976b). It can be shown that in this case, too, one obtains expressions similar to eq. (68).

The examples considered in the present section show that the approach utilized here makes it possible in some cases to obtain more concrete results than the approach exploited in sects 3–5, but at the same time the latter is simpler and more general.

## 7. *Light scattering under the existence of the "devil's staircase"*

### 7.1. *"Devil's staircase"*

In the investigations of incommensurate phases the superstructure wavevector $\boldsymbol{K}_0$ is usually assumed to be a continuous function of temperature (fig. 5a). However, as it was pointed out by Dzialoshinskii (1964b), a commensurability of the superstructure period with the underlying lattice period yields a finite energy gain and therefore the temperature interval where an incommensurate phase exists must, in fact, be a sequence of long-period commensurate phases. In a finite temperature interval the number of such phases can be infinite, the superstructure wavevector (in units of the vector $\boldsymbol{b}$) passing all the rational numbers, i.e., its temperature dependence is the so-called "devil's staircase" of the type shown in fig. 5b (Dzialoshinskii 1964b, Aubry 1978, Bak and Von Boehm 1980, Fisher and Selke 1980, Novaco 1980).

The larger the period of the superstructure, as compared with that of the underlying lattice, the smaller the above-mentioned energy gain. Hence, phases

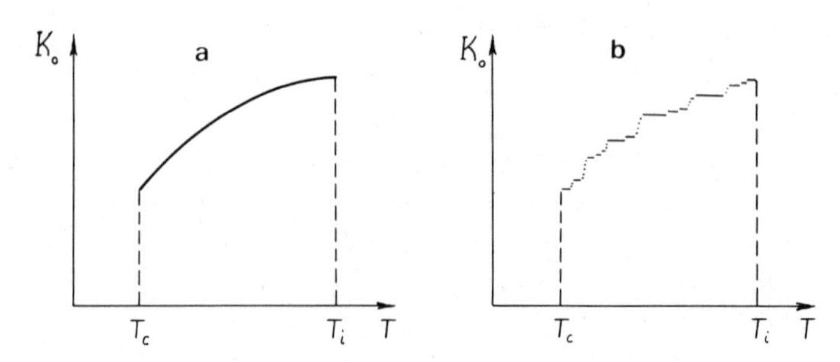


Fig. 5. Temperature dependence of the superstructure wavevector $K_0$: (a) in a truly incommensurate phase; (b) under the existence of the devil's staircase.

with small superstructure periods must exist in a wider temperature range than phases with large periods. Naturally, when speaking of an infinite sequence of phase transitions, one implies an unlimited crystal. In a limited crystal the number of phases must be finite at least because the size of the cell cannot exceed that of the crystal. Moreover, there are grounds to suppose (Bak and Von Boehm 1980, Villain and Gordon 1980) that superstructures with large periods are not realized even in an unlimited crystal. It is also possible that in certain temperature intervals the devil's staircase exists, while in others the superstructure wavevector changes continuously with temperature, i.e., there appears a truly incommensurate phase (Bak and Von Boehm 1980, Bruce 1980). Changes in the superstructure wavevector typical of the devil's staircase were also observed experimentally (Marion et al. 1981); besides, in some crystals long-period (nine-, seven- and five-fold periods as compared with the underlying period) commensurate phases were observed adjacent to incommensurate ones in phase diagrams (Denoyer et al. 1981, Mashiyama et al. 1980).

Using the results of sect. 2, we consider here the devil's staircase on an example of a two-component order parameter and crystals that belong in the symmetrical phase to the orthorhombic system. An essential difference between commensurate phases that form the devil's staircase and a truly incommensurate phase is that in a commensurate phase additional translational invariants are possible. If the period of a commensurate phase is $n$ times larger than that of the underlying lattice, i.e., $\boldsymbol{K}_0 = m\boldsymbol{b}/n$, then such translational invariants of the smallest power in $Q$ are $[Q(\boldsymbol{K}_0)]^n$ and $(Q^*)^n$, instead of which it is more convenient to use their linear combinations $\rho^n \cos n\varphi$ and $\rho^n \sin n\varphi$. Knowing the transformational properties of such invariants, one can determine the point symmetry of the phases that form the devil's staircase, and for a given crystal with a definite soft mode this symmetry depends only on the integers $m$ and $n$ to be even or odd and in certain cases also on the sign of some thermodynamic potential coefficients (Golovko and Levanyuk 1981a). From this it follows in particular that if one moves along the devil's staircase, the symmetry undergoes chaotic changes even if variations of $\boldsymbol{K}_0$ are arbitrarily small. In the above-mentioned paper it is shown how one can easily establish possible point symmetries of the devil's staircase phases, and these symmetries are found for certain crystals with an incommensurate phase.

Let us consider a devil's staircase phase characterized by a wavevector $\boldsymbol{K}_0 = m\boldsymbol{b}/n$. Although the transition to this phase does not occur directly from the high-temperature symmetrical phase, it can nevertheless be described as a distortion of the symmetrical phase in the spirit of the Landau theory, which makes it possible to compare the material constants of these two phases. At given $m$ and $n$ subject to the transformational properties of the order parameter three cases are possible:

(1) the quantities $\rho^n \cos n\varphi$ and $\rho^n \sin n\varphi$ are transformed according to different irreducible representations and none of them is a perfect invariant;

(2) these quantities are transformed identically and they are not perfect invariants;

(3) at least one of these quantities is a perfect invariant.

In the first case among the invariants containing only $\rho$ and $\varphi$, in lowest orders there exist exclusively powers of $\rho^2$, and the first perfect invariant including $\varphi$ can be found by squaring the indicated quantities, which gives an invariant of the form $\rho^{2n} \cos 2n\varphi$. Thus, the thermodynamic potential density in this case takes the form

$$\Phi = \tfrac{1}{2}A\rho^2 + \tfrac{1}{4}B\rho^4 + f(\rho) + E\rho^{2n} \cos 2n\varphi\,, \tag{69}$$

where $f(\rho)$ does not depend on $\varphi$ and contains powers of $\rho$ higher than fourth. The equilibrium value $\rho_e^2 = -A/B$, analogous to eq. (3), can be found by minimization of eq. (69) with respect to $\rho$, the terms of high order being discarded. By minimizing eq. (69) with respect to $\varphi$ we obtain that subject to the sign of $E$ one of the following two phases is realized:

$$\cos 2n\varphi_e = 1, \qquad \sin n\varphi_e = 0 \quad \text{if} \quad E < 0\,, \tag{70}$$

$$\cos 2n\varphi_e = -1, \qquad \cos n\varphi_e = 0 \quad \text{if} \quad E > 0\,. \tag{71}$$

The point symmetry of these phases is determined by the quantity $\rho^n \cos n\varphi$, or $\rho^n \sin n\varphi$, whose spontaneous value is non-zero.

Note that if a higher order invariant $\rho^{4n} \cos 4n\varphi$ is included in the thermodynamic potential, then formally an additional case becomes possible, that corresponds to a third phase with $\sin n\varphi_e \neq 0$ and $\cos n\varphi_e \neq 0$. However in this case $\cos 2n\varphi_e \sim E/\rho_e^{2n}$. Taking into account the smallness of $\rho_e$, we see that at large $n$ the condition under which the third phase exists ($|\cos 2n\varphi_e| \leqslant 1$) is, in fact, the equality $E = 0$. This specific possibility is of no interest and will not be analyzed here.

The oscillations of $\rho$ and $\varphi$ may be considered by analogy with subsect. 2.2.

For the oscillations of $\rho$ we obtain, disregarding high powers of $\rho_e$, the dispersion (9), and for those of $\varphi$, in both cases (70) and (71), the dispersion

$$\mu\omega^2 + i\gamma\omega = \mu\omega_\varphi^2 + D_1 q_x^2 + D_2 q_y^2 + D_3 q_z^2 , \qquad \omega_\varphi^2 = (4/\mu) n^2 |E| \rho_e^{2n-2} . \tag{72}$$

As $n \to \infty$, the quantity $\omega_\varphi$ in eq. (72) vanishes, i.e., we obtain a gapless Goldstone mode typical of a truly incommensurate phase. Note that due to the smallness of $\omega_\varphi$, if $n$ is large, the phase mode is overdamped like in the case of truly incommensurate phase.

In the second of the above-mentioned cases the product of the quantities $\rho^n \cos n\varphi$ and $\rho^n \sin n\varphi$ is a perfect invariant, and in the thermodynamic potential there appears one more $\varphi$-dependent term, which we write as $E'\rho^{2n} \sin 2n\varphi$. Now instead of eqs. (70) and (71) it is the equation $\operatorname{tg} 2n\varphi_e = E'/E$ that corresponds to the extremum of $\Phi$, i.e., only one phase is realized.

In the third case the first term in the thermodynamic potential that depends only on the order parameter and contains $\varphi$ is proportional to $\rho^n$ but not to $\rho^{2n}$. If, for example, it is only $\rho^n \cos n\varphi$ that is a perfect invariant, then supplementing the thermodynamic potential with the term $E_1 \rho^n \cos n\varphi$ from the condition of minimum in $\varphi$ we get $\sin n\varphi_e = 0$. Therefore, the spontaneous value of the quantity $\rho^n \sin n\varphi$, which is not a perfect invariant, is equal to zero, i.e., the phase that has the point symmetry of the high-temperature phase is realized alone. For the frequency of a homogeneous phason we have

$$\omega_\varphi^2 = (n^2/\mu) |E_1| \rho_e^{n-2} , \tag{73}$$

instead of eq. (72). If both the quantities $\rho^n \cos n\varphi$ and $\rho^n \sin n\varphi$ are perfect invariants, the corresponding phase also has the point symmetry of the high-temperature phase.

### 7.2. *Specific features of light scattering under the existence of the devil's staircase*

Let us discuss the reason for which light scattering under the existence of the devil's staircase may differ from light scattering in a truly incommensurate phase. When $\boldsymbol{K}_0$ changes in the occurrence of the devil's staircase, it may turn out that for certain $m$ and $n$ the quantities $\rho^n \cos n\varphi$ or $\rho^n \sin n\varphi$ are transformed as some component of a tensor of rank two, and then in the tensor $\epsilon_{ij}$ a supplementary term appears which is absent in a truly incommensurate phase. If $n$ is large, the spontaneous value of this term is negligibly small by itself, since it is proportional to $\rho_e^n$. However the quantities $\rho^n \cos n\varphi$ and $\rho^n \sin n\varphi$ depend on $\varphi$, and therefore their fluctuations contributed by the $\varphi$ fluctuations are anomalously large at large $n$ since the generalized stiffness corresponding to the $\varphi$ variations and proportional to $\omega_\varphi^2$ in eqs. (72) or (73), if $\boldsymbol{q} \approx 0$, is very small, which may in the end lead to a noticeable contribution to the $\epsilon_{ij}$ fluctuations and will manifest itself in the light scattering. Since the generalized stiffness corresponding to the $\rho$ variations and determined to a first approximation from eq. (9) does not practically depend on $n$, i.e., in a long-period phase it has a usual value, the contribution from the fluctuations of $\rho$ to those of $\rho^n \cos n\varphi$ and $\rho^n \sin n\varphi$ is negligibly small if $n$ is large. From this it is seen that to find the differences between light scattering in the case of a truly incommensurate phase and in the case of the devil's staircase it is sufficient to consider the contribution to the $\epsilon_{ij}$ fluctuations of those of the phase $\varphi$ alone, the fluctuations of normal coordinates unrelated to the soft mode also being of no importance since the corresponding stiffnesses are not small.

What has been said now will be confirmed by calculations. To this end, we first consider fluctuations of off-diagonal $\epsilon_{ij}$ components that lead to light scattering with a change of polarization. Suppose that the transformational properties of the order parameter corresponding to the transition to one of the devil's staircase phases with $\boldsymbol{K}_0 = m\boldsymbol{b}/n$ are such that some off-diagonal component of the tensor $\epsilon_{ij}$, e.g., $\epsilon_{xy}$, is transformed like $\rho^n \cos n\varphi$ or $\rho^n \sin n\varphi$. Taking for definiteness $\rho^n \sin n\varphi$, we obtain the following relation between $\epsilon_{xy}$ and $\varphi$ for the indicated phase

$$\epsilon_{xy} = a_1 \rho^n \sin n\varphi + a_2 \rho^2 \frac{\partial \varphi}{\partial y} \,. \tag{74}$$

Here we have also taken into account the gradient invariant $\rho^2 \partial\varphi/\partial y$ (see subsect. 3.1.), which always exists because the point $\boldsymbol{K}_0$ is neither at the centre nor at the boundary of the symmetrical phase Brillouin zone.

Since in the problem under consideration it is sufficient to take into account the phase fluctuations only, for the mean-square fluctuation of the Fourier transform $\Delta\epsilon_{xy}(\boldsymbol{q})$, that determines the scattered-light integral intensity, by analogy with subsect. 3.1 we have

$$\langle |\Delta\epsilon_{xy}(\boldsymbol{q})|^2 \rangle = (a_1^2 n^2 \rho_e^{2n} \cos^2 n\varphi_e + a_2^2 \rho_e^4 q_y^2) \langle |\varphi(\boldsymbol{q})|^2 \rangle \,. \tag{75}$$

First we assume that the value of $\boldsymbol{K}_0$ is such that the first of the cases discussed in subsect. 7.1. is realized. The mean-square fluctuation of $\varphi(\boldsymbol{q})$ is then found on the basis of the thermodynamic potential density (69) with the addition of terms containing spatial derivatives the same as in eq. (4). Acting like in subsect. 3.1, from (75) we obtain

$$\langle |\Delta\epsilon_{xy}(\boldsymbol{q})|^2 \rangle = \frac{(a_1^2 n^2 \rho_e^{2n-2} \cos^2 n\varphi_e + a_2^2 \rho_e^2 q_y^2) k_B T}{V(\mu\omega_\varphi^2 + D_1 q_x^2 + D_2 q_y^2 + D_3 q_z^2)} \,. \tag{76}$$

In the case of a truly incommensurate phase $\omega_\varphi = 0$ and the first term in eq. (74) is absent ($a_1 = 0$), wherefore eq. (76) takes the form

$$\langle |\Delta\epsilon_{xy}(\boldsymbol{q})|^2 \rangle = \frac{a_2^2 \rho_e^2 q_y^2 k_B T}{V(D_1 q_x^2 + D_2 q_y^2 + D_3 q_z^2)} \,. \tag{77}$$

This expression, naturally, is analogous to eq. (18) with the only difference that here we are dealing with the fluctuations of $\epsilon_{xy}$ and not of $\epsilon_{zz}$.

In the case of the devil's staircase, if $\omega_\varphi$ from eq. (72) is used, eq. (76) at $\boldsymbol{q} = 0$ takes the form

$$\langle |\Delta\epsilon_{xy}|^2 \rangle = a_1^2 \cos^2 n\varphi_e k_B T/4|E|V. \tag{78}$$

It is essential that the last expression does not contain the small quantity $\rho_e^n$ and its dependence on $n$, if $\cos n\varphi_e \neq 0$ (in this case $|\cos n\varphi_e| = 1$ according to eq. (70)), is associated with the quantity $a_1^2/E$ alone. One can argue (Golovko and Levanyuk 1981b) in favour of the fact that the quantity $a_1^2/E$ remains finite at arbitrarily large $n$. Therefore the fluctuation (78) and, accordingly, the scattered-light intensity proportional to this fluctuation does not tend to zero as $n \to \infty$ and differs from eq. (77) by its independence of the $\boldsymbol{q}$ direction and by the fact that it remains finite at $T \to T_i$, too. The first term of eq. (74) is present ($a_1 \neq 0$) only in certain phases which form the devil's staircase, and if $a_1 \neq 0$, the fluctuation (78) is non-zero only when $|\cos n\varphi_e| = 1$; note that in this case the spontaneous value of $\epsilon_{xy}$ is zero according to eq. (74) ($\sin n\varphi_e = 0$). If one implies a concrete crystal, whether the coefficient $a_1$ is zero or not depends on the integers $m$ and $n$ to be even or odd (remember that $\boldsymbol{K}_0 = m\boldsymbol{b}/n$), but not on the values of these integers (Golovko and Levanyuk 1981a). From this it follows that the fluctuation (78) may change from zero to a finite value and vice versa even if a variation of $\boldsymbol{K}_0$ is arbitrarily small. In the end we see that when moving along the devil's staircase, the scattered-light intensity must undergo irregular jumps of the same order of magnitude as at ordinary second-order phase transitions, these jumps being associated with the specificity of scattering by a phason. Note that some susceptibilities also undergo analogous jumps (Golovko and Levanyuk 1981a).

We would like to emphasize that the results obtained are valid only at $|\boldsymbol{q}| \equiv q = 0$. Owing to the smallness of $\rho_e^{2n}$ the $q$-dependent terms in the numerator and denominator of eq. (76) begin, as $q$ increases, to exceed rapidly the first terms, the larger $n$ the faster, and we are led to eq. (77) that determines the scattered intensity in a truly incommensurate phase. The difference between light scattering by a truly incommensurate phase and by a devil's staircase can be found if $D_i q^2 < \mu\omega_\varphi^2$. The quantity $\omega_\varphi^2$ in eq. (72) can be estimated as $\omega_\varphi^2 \sim \Omega_a^2 \rho_e^{2n-2}$, where $\Omega_a \sim 10^{13}\ \mathrm{s}^{-1}$ is the characteristic phonon frequency and $\rho_e$ is expressed in dimensionless units, the value $\rho_e = 1$ corresponding to a complete ordering or displacements at distances of the order of the lattice period. Using eq. (6) we take the inequality $D_i q^2 < \mu\omega_\varphi^2$ in the form $q < b\rho_e^{n-1}$. Due to diffraction effects the quantity $q$ cannot be less than $\sim 1/L$, where $L$ is a crystal dimension. Hence, the limiting value $n = n_{\max}$, at which one may hope to reveal some difference in the light scattering from the truly incommensurate phase and from the devil's staircase, is determined by the relation $\rho_e^{n-1} \sim 1/bL$. Whether or not this relation holds depends radically on the value of $\rho_e$, and the indicated difference can be observed, if $n$ is large, only sufficiently far from the point of incommensurate phase transition when the quantity $\rho_e$ is not small. Assuming $L \sim 1$ cm, we obtain the limiting value $n_{max} \approx 9$ when $\rho_e \sim 0.1$; if $\rho_e \sim \frac{1}{3}$, which can also be expected in some cases, then $n_{max} \approx 17$.

The scattering associated with fluctuations of the component $\epsilon_{xz}$ obeys similar laws, but the numerator of a formula of the type (76) will contain $q_z^2$ instead of $q_y^2$. If one deals with the scattering by the fluctuations of $\epsilon_{yz}$, then according to what has been said in subsect. 3.1, one should assume $a_2 = 0$ in the above expressions. In this case in a truly incommensurate phase, according to eq. (77), scattering by the phason is absent (if elastic deformations are not taken into account), whereas under the existence of the devil's staircase in the $q \to 0$ limit we again obtain an expression of the type (78). We should notice that in the orthorhombic system the transformational properties of the components $\epsilon_{xy}$, $\epsilon_{xz}$ and $\epsilon_{yz}$ are different, and in each phase the quantity $\rho^n \sin n\varphi$ (or $\rho^n \cos n\varphi$) can be coupled with only one of these three components; therefore, when moving along the devil's staircase, the indicated intensity jumps must look differently in different scattering geometries.

In the second case mentioned in subsect. 7.1 the scattered-light intensity associated with fluctuations of some off-diagonal $\epsilon_{ij}$ component in contrast to the first case is not equal to zero at $\boldsymbol{q} = 0$ along with a non-zero spontaneous value of this component if the latter is admitted by the phase symmetry.

More essential differences appear in the third case of subsect. 7.1, when at a certain $\boldsymbol{K}_0 = m\boldsymbol{b}/n$ the transformational properties of the order parameter are such that one of the quantities $\rho^n \cos n\varphi$ or $\rho^n \sin n\varphi$ is a perfect invariant. Suppose that this is the quantity $\rho^n \cos n\varphi$, and $\rho^n \sin n\varphi$ is transformed like $\epsilon_{xy}$, then eqs. (74) and (75) remain valid, but in eq. (75) one should assume $\cos^2 n\varphi_e = 1$, since only the phase with a zero spontaneous value of $\epsilon_{xy}$ can be realized (see subsect. 7.1). Qualitative differences from the previous cases appear here because for the homogeneous-phason frequency $\omega_\varphi$ we have eq. (73) instead of eq. (72). Expression (76) remains valid but with the new value of $\omega_\varphi$. Now at $q = 0$ the scattered-light intensity is proportional to $\rho_e^n$, i.e., it is practically equal to zero if $n$ is large. However since $\omega_\varphi^2 \sim \Omega_a^2 \rho_e^{n-2}$, the intensity differs from the truly incommensurate phase value, determined as before by eq. (77), at much higher values of $q$ than in the previous cases, namely at $q < b\rho_e^{n/2-1}$. Taking, on the contrary, the lowest possible value of $q$ and carrying out estimations as above, we see that one may hope to notice this difference up to the values $n_{max} \approx 18$ if $\rho_e \sim 0.1$ and $n_{max} \approx 34$ if $\rho_e \sim \frac{1}{3}$.

If $q_y \sim b\rho_e^{n/2-1}$, the second term in the numerator of eq. (76) is much larger than the first one, and therefore the scattered intensity will depend on the $\boldsymbol{q}$ direction practically the same way as in the truly incommensurate phase. However, if while moving along the devil's staircase, we get in the phase under discussion and $q$ acquires the above-mentioned value, the intensity jumps down to decrease considerably owing to an increase of the denominator. This effect can be observed only if $a_2 \neq 0$, which is possible in the case in question only when one of the crystallographic axes perpendicular to the incommensurate axis is polar (Golovko and Levanyuk 1981b).

We have so far discussed fluctuations of the off-diagonal components of the tensor $\epsilon_{ij}$. Let us now consider the diagonal-component fluctuations which lead to light scattering without change of polarization. Since in the orthorhombic system under consideration the diagonal components are perfect invariants, they may always be coupled with the invariant $\rho^{2n} \cos 2n\varphi$ (see eq. (69)), as well as with the Lifshitz invariant (13). For fluctuations of the digonal $\epsilon_{ij}$ components we obtain expressions of the type (76), but in the first term of the numerator the factor $\rho_e^{2n}$ is replaced by $\rho_e^{4n}$ and in the second one $q_y^2$ is replaced by $q_x^2$. The regularities of scattering here are analogous to those just discussed (the third case of subsect. 7.1) with the substitution of $2n$ for $n$. If $\rho^n \cos n\varphi$ or $\rho^n \sin n\varphi$ is a perfect invariant, the regularities of scattering are wholly similar to those just discussed, but with the only difference that the second term in the numerator of eq. (76) is always present and the first one is absent (since if, for example, $\rho^n \cos n\varphi$ is invariant, then, according to subsect. 7.1, $\sin n\varphi_e = 0$, but the fluctuations of the quantity $\rho^n \cos n\varphi$ are proportional to $\sin n\varphi_e \Delta\varphi$). Therefore, in the case of crystals, in which phases with the point symmetry of the high-temperature phase are possible on the devil's staircase (and such crystals do exist, see Golovko and Levanyuk (1981a)), one may hope to reveal differences in light scattering as compared with a truly incommensurate phase for much larger $n$ (see the above estimates) than in the case of crystals in which the indicated phases on the devil's staircase are absent. It is important that for the diagonal $\epsilon_{ij}$ components the factor $a_2$ is never equal to zero due to the presence of the Lifshitz invariant.

The spectral density $J(\boldsymbol{q}, \Omega)$ of light scattered by a phason under the existence of the devil's staircase can be found using the same method as in subsect. 3.2 (see also ch. 1 of this book). Taking into account the smallness of $\omega_\varphi^2$ and using the same arguments as in obtaining eq. (22), we see that $J(\boldsymbol{q}, \Omega)$ is essentially non-zero only at small $\Omega$, i.e., in the spectrum of scattering by the phason there exists only a central line, like in a truly incommensurate phase. In the region of small $\Omega$ we obtain

$$J(\boldsymbol{q}, \Omega) = \frac{(\mu\omega_\varphi^2 + D_1 q_x^2 + D_2 q_y^2 + D_3 q_z^2)\gamma I(\boldsymbol{q})}{\pi[(\mu\omega_\varphi^2 + D_1 q_x^2 + D_2 q_y^2 + D_3 q_z^2)^2 + \gamma^2\Omega^2]}, \tag{79}$$

where $I(\boldsymbol{q})$ is the integral intensity.

From this expression it follows that the line width $\Delta\Omega$ is given by

$$\Delta\Omega = (2/\gamma)(\mu\omega_\varphi^2 + D_1 q_x^2 + D_2 q_y^2 + D_3 q_z^2) . \tag{80}$$

In a truly incommensurate phase ($\omega_\varphi = 0$) eq. (80) transforms into eq. (23). In the case when it is quite real to reveal difference between light scattering by a truly incommensurate phase and that by the devil's staircase ($D_i q^2 < \mu\omega_\varphi^2$), the line width (80) becomes practically $q$ independent. For the limiting values of $q^2 \sim \mu\omega_\varphi^2/D_i$, taking into account eq. (6) and assuming $q \sim 1/L$, we come to

$$\Delta\Omega \sim \frac{D_i}{\gamma} q^2 \sim \frac{\mu\Omega_a^2}{\gamma b^2 L^2} \sim \frac{1\text{–}100}{(bL)^2} \Omega_a .$$

Assuming, as before, $L \sim 1$ cm, we obtain $\Delta\Omega \sim 10^{-3}$–$10^{-1}\,\mathrm{s}^{-1}$. If the value of $q$ is by an order of magnitude larger, which decreases the limiting $n_{max}$ (see above) by a few units, the line width can reach $1\,\mathrm{s}^{-1}$. Note that modern technique of light beating spectroscopy permits observation of lines with the width of the order of 1 Hz.

It should be mentioned that a whole set of long-period phases with different $n$ is likely to coexist in the presence of the devil's staircase in real samples (due to inevitable temperature gradients and the influence of crystal defects) particularly if phases with comparatively small $n$, whose temperature interval may be rather large, must not be observed in the temperature region under investigation. Light scattering from such a multiphase system has not yet been theoretically considered.

## 8. *Concluding remarks*

In conclusion we discuss some questions of experimental study of light scattering in incommensurate phases from the viewpoint of the above theoretical considerations. In experiments much attention has been paid to the search of manifestations of the characteristic incommensurate phase excitation, the phason. In our opinion, however, one cannot state that the Goldstone phason has been revealed experimentally although there is a number of papers devoted to light scattering from incommensurate phases, in which phason is reported to be observed. This is partly due to the discrepancy in the terminology, mentioned in subsect. 2.2. So, in one of the reports on phason observation (Inoue et al. 1980) the authors meant, in fact, non-Goldstone low-frequency modes, as may be concluded from the text of the paper.

Lyons and Guggenheim (1979) and Lyons et al. (1980) reported that in the incommensurate phase of $BaMnF_4$ they had observed a central peak with the width proportional to the vector $\boldsymbol{q}$ squared, as is the case with scattering by a Goldstone phason (see subsects 3.2 and 5.2). These experiments were concerned with the study of light scattering without change of polarization, i.e., of the one associated with fluctuations of the diagonal $\epsilon_{ij}$ components. From the expressions obtained in subsect. 5.2 it is seen that in the case of $BaMnF_4$ the contribution of the Goldstone excitation to the mean-square fluctuations of the diagonal $\epsilon_{ij}$ components is proportional to $q_x^2$ since the Lifshitz invariant (32) involves only derivatives with respect to $x$. However, according to the above-mentioned authors the indicated central peak vanished just at the

$x$-directed vector $\boldsymbol{q}$. Besides, the maximum width of this peak made up approximately $10^{10}\,s^{-1}$, which exceeds greatly the estimate (25). Thus, the statement of Lyons and Guggenheim (1979) and Lyons et al. (1980) concerning the observation of scattering by the phason does not at all agree with theory. In a recent paper by Lyons et al. (1982) the central peak in $BaMnF_4$ is reported not to vanish at the vector $\boldsymbol{q}$ parallel to the $x$-axis, however it does not vanish either when the vector $\boldsymbol{q}$ is perpendicular to the $x$-axis, in the latter case the observation being most certain. But according to the theory it must vanish in this case if one deals with scattering by a phason.

Lockwood et al. (1981) also reported the observation of light scattering by a phason in $BaMnF_4$, that was assumed to contribute to the Mandelstam–Brillouin components. However, as has been shown above, the Goldstone phason must contribute to the central component.

From what has been said it is seen already that to observe manifestations of the Goldstone phason is not a simple experimental problem. First, the width of the phason central peak is rather small. Second, this peak must have a comparatively low intensity since it is proportional to $\rho_e^2$ and vanishes at the incommensurate phase transition point when $T = T_i$. Out of the above-mentioned excitations, which must become Raman active below the incommensurate phase transition point, it is only the amplitudon that was confidently observed (see subsect. 4.1) and for the most part only in the crystals in which the incommensurate phase transition is described by a two-component order parameter. As to incommensurate phases characterized by a four-component order parameter, like, for example, that of $BaMnF_4$, there is not yet a full certainty in what concerns the interpretation of the excitations observed in light scattering experiments (Lockwood et al. 1981, Murray et al. 1981). Thus, experimental studies of light scattering from incommensurate phases are practically at the initial stage. As concerns the theory, it has been developed up to now in the application to a defect-free crystal. At the same time the influence of defects upon the incommensurate phase must be especially large since the incommensurate phase is a degenerate system (see ch. 2 of the present book).

## *References*


Aubry, S., 1978, The new concept of transitions by breaking of analyticity in a crystallographic model, in: Solitons and Condensed Matter Physics, eds., A.R. Bishop and T. Schneider (Springer, Berlin) pp. 264–277.

Bak, P. and J. von Boehm, 1980, Phys. Rev. **B 21**, 5297.

Bak, P. and J. Timonen, 1978, J. Phys. **C 11**, 4901.

Berlinsky, A.J., 1979, Rep. Progr. Phys. **42**, 1243.

Blinc, R., V. Rutar, B. Topic, F. Milia, I.P. Aleksandrova, A.S. Chaves and R. Gazzinelli, 1981, Phys. Rev. Lett. **46**, 1406.

Boriack, M.L. and A.W. Overhauser, 1978, Phys. Rev. **B 17**, 4549.

Bruce, D.A., 1980, J. Phys. **C 13**, 4615.

Bruce, A.D. and R.A. Cowley, 1978, J. Phys. **C 11**, 3609.
Bruce, A.D., R.A. Cowley and A.F. Murray, 1978, J. Phys. **C 11**, 3591.
Cailleau, H., F. Moussa, C.M.E. Zeyen and J. Bouillot, 1980, Solid State Commun. **33**, 407.
Chen, C.H., J.M. Gibson and R.M. Fleming, 1981, Phys. Rev. Lett. **47**, 723.
Cowley, R.A., 1980, Adv. Phys. **29**, 1.
Cox, D.E., S.M. Shapiro, R.A. Cowley, M. Eibschütz and H.J. Guggenheim, 1979, Phys. Rev. **B 19**, 5754.
De Pater, C.J. and R.B. Helmholdt, 1979, Phys. Rev. **B 19**, 5735.
Denoyer, F., A.H. Moudden, A. Bellamy, R. Currat, C. Vettier and M. Lambert, 1981, Compt. Rend. Acad. Sci. Ser. 2, **292**, 13.
Dvorak, V. and J. Fousek, 1980, Phys. Status Solidi (a), **61**, 99.
Dvorak, V. and Y. Ishibashi, 1978, J. Phys. Soc. Jpn. **45**, 775.
Dvorak, V. and J. Petzelt, 1978, J. Phys. **C 11**, 4827.
Dzialoshinskii, I.E., 1964a, ZhETF, **46**, 1420 (Sov. Phys. JETP, **19**, 960).
Dzialoshinskii, I.E., 1964b, ZhETF, **47**, 992 (1965, Sov. Phys. JETP, **20**, 665).
Fisher, M.E. and W. Selke, 1980, Phys. Rev. Lett. **44**, 1502.
Fleury, P.A., S. Chiang and K.B. Lyons, 1979, Solid State Commun. **31**, 279.
Francke, E., M. Le Postollec, J.P. Mathieu and H. Poulet, 1980, Solid State Commun. **35**, 183.
Ginzburg, V.L., A.P. Levanyuk and A.A. Sobyanin, 1980, Phys. Rept. **57**, 151.
Golovko, V.A., 1980, Fiz. Tverd. Tela, **22**, 2960 (Sov. Phys. Solid State, **22**, 1729).
Golovko, V.A. and A.P. Levanyuk, 1979, ZhETF, **77**, 1556 (Sov. Phys. JETP, **50**, 780).
Golovko, V.A. and A.P. Levanyuk, 1980, Pis'ma ZhETF, **32**, 104 (Sov. Phys. JETP Lett. **32**, 93).
Golovko, V.A. and A.P. Levanyuk, 1981a, Fiz. Tverd. Tela, **23**, 3170 (Sov. Phys. Solid State, **23**, N 10).
Golovko, V.A. and A.P. Levanyuk, 1981b, Fiz. Tverd. Tela, **23**, 3179 (Sov. Phys. Solid State, **23**, N 10).
Golovko, V.A., and A.P. Levanyuk, 1981c, ZhETF, **81**, 2296 (Sov. Phys. JETP, **54**, 1217).
Haque, M.S. and J.R. Hardy, 1980, Phys. Rev. **B 21**, 245.
Hoshino, S. and H. Motegi, 1967, Jpn. J. Appl. Phys. **6**, 708.
Iizumi, M. and K. Gesi, 1977, Solid State Commun. **22**, 37.
Iizumi, M. and K. Gesi, 1980, J. Phys. Soc. Jpn. **49**, Suppl. B, 72.
Iizumi, M., J.D. Axe, G. Shirane and K. Shimaoka, 1977, Phys. Rev. **B 15**, 4392.
Inoue, K., S. Koiwai and Y. Ishibashi, 1980, J. Phys. Soc. Jpn. **48**, 1785.
Ishibashi, Y., 1980, Ferroelectrics, **24**, 119.
Ishibashi, Y. and V. Dvorak, 1978, J. Phys. Soc. Jpn. **44**, 32.
Ishibashi, Y. and H. Shiba, 1978, J. Phys. Soc. Jpn. **45**, 409.
Ishibashi, Y. and Y. Takagi, 1979, J. Phys. Soc. Jpn. **46**, 143.
Janner, A. and J. Janssen, 1980, Acta Cryst. **A 36**, 399, 408.
Kovalev, O.V., 1961, Irreducible Representation of Space Groups (AN Ukr. SSR, Kiev); English transl. (Gordon and Breach, New York, 1965).
Landau, L.D. and E.M. Lifshitz, 1959, Electrodynamics of Continuous Media (Fizmatgiz, Moscow); English transl. (Pergamon, Oxford, 1960).
Landau, L.D. and E.M. Lifshitz, 1965, Theory of Elasticity (Nauka, Moscow); English transl. (Pergamon, Oxford, 1959).
Landau, L.D. and E.M. Lifshitz, 1976, Statistical Physics (Nauka, Moscow); English transl. (Pergamon, Oxford, 1968).
Lee, P.A., T.M. Rice and P.W. Anderson, 1974, Solid State Commun. **14**, 703.
Levanyuk, A.P. and D.G. Sannikov, 1976a, Fiz. Tverd. Tela, **18**, 423 (Sov. Phys. Solid State, **18**, 245).
Levanyuk, A.P. and D.G. Sannikov, 1976b, Fiz. Tverd. Tela, **18**, 1927 (Sov. Phys. Solid State, **18**, 1122).
Lockwood, D.J., A.F. Murray and N.L. Rowell, 1981, J. Phys. **C 14**, 753.

Lyons, K.B. and H.J. Guggenheim, 1979, Solid State Commun. **31**, 285.
Lyons, K.B., T.J. Negran and H.J. Guggenheim, 1980, J. Phys. **C 13**, L415.
Lyons, K.B., R.N. Bhatt, T.J. Negran and H.J. Guggenheim, 1982, Phys. Rev. B **25**, 1791.
Lyubarskii, G.Ya., 1957, The Application of Group Theory in Physics (Gostekhizdat, Moscow); English trans. (Pergamon, Oxford, 1960).
Marion, G., R. Almairac, J. Lefebvre and M. Ribet, 1981, J. Phys. **C 14**, 3177.
Mashiyama, H., K. Hasebe and S. Tanisaki, 1980, J. Phys. Soc. Jpn. **49**, Suppl. B, 92.
McMillan, W.L., 1975a, Phys. Rev. **B 12**, 1187.
McMillan, W.L., 1975b, Phys. Rev. **B 12**, 1197.
McMillan, W.L., 1976, Phys. Rev. **B 14**, 1496.
Milia, F. and V. Rutar, 1981, Phys. Rev. **B 23**, 6061.
Murray, A.F., G. Brims and S. Sprunt, 1981, Solid State Commun. **39**, 941.
Novaco, A.D., 1980, Phys. Rev. **B 22**, 1645.
Overhauser, A.W., 1971, Phys. Rev. **B 3**, 3173.
Petzelt, J., 1981, Phase Transitions, **2**, 155.
Poulet, H. and R.M. Pick, 1981, J. Phys. **C 14**, 2675.
Przystawa, J., 1980, Modulated structures, in: Physics of Modern Materials, Vol. II (IAEA, Vienna) pp. 213–264.
Rehwald, W. and A. Vonlanthen, 1981, Solid State Commun. **38**, 209.
Rice, T.M., 1975, Solid State Commun. **17**, 1055.
Sakai, A. and I. Tatsuzaki, 1980, J. Phys. Soc. Jpn. **49**, 2287.
Schneck, J. and F. Denoyer, 1981, Phys. Rev. **B 23**, 383.
Shiba, H. and Y. Ishibashi, 1978, J. Phys. Soc. Jpn. **44**, 1592.
Shiozaki, Y., 1971, Ferroelectrics, **2**, 245.
Siapkas, D.I., 1980, Ferroelectrics, **29**, 29.
Suits, B.H., S. Couturie and C.P. Slichter, 1980, Phys. Rev. Lett. **45**, 194.
Takashige, M. and T. Nakamura, 1980, Ferroelectrics, **24**, 143.
Unruh, H.-G., W. Eller and G. Kirf, 1979, Phys. Status Solidi (a), **55**, 173.
Villain, J. and M.B. Gordon, 1980, J. Phys. **C 13**, 3117.
Wada, M., H. Uwe, A. Sawada, Y. Ishibashi, Y. Takagi and T. Sakudo, 1977, J. Phys. Soc. Jpn. **43**, 544.
Wada, M., A. Sawada, Y. Ishibashi and Y. Takagi, 1978, J. Phys. Soc. Jpn. **45**, 1905.
Wada, M., A. Sawada and Y. Ishibashi, 1981, J. Phys. Soc. Jpn. **50**, 531.
Winterfeldt, V. and G. Schaack, 1980, Z. Phys. **B 36**, 303.

CHAPTER 4

# Light Scattering and Phase Transitions in Liquid Crystals

V.A. BELYAKOV

*All-Union Surface and Vacuum Research Centre*
*117965 Moscow*
*USSR*

E.I. KATS

*L.D. Landau Institute for Theoretical Physics*
*Academy of Sciences of the USSR*
*117334 Moscow V-334*
*Vorobjevskoe shosse 2*
*USSR*

*Light Scattering near Phase Transitions*
*Edited by*
*H.Z. Cummins and A.P. Levanyuk*

© *North-Holland Publishing Company, 1983*

# Contents

## *Introduction*

The great role of optical methods and, in particular, of studies of phase transitions in condensed media is widely acknowledged (Ginzburg et al. 1980, Bendow et al. 1976, Birman et al. 1979). Liquid crystals, physical and applied studies of which recently have been intensively developing are no exception in this respect (De Gennes 1974, Pikin 1981, Blinov 1978). Moreover it is possible to assert that the role of optical methods for studies of a liquid-crystal state on the whole and phase transitions in particular is more important than for other condensed media. This is accounted for by the nature of the liquid-crystal state characterized by correlations of orientations of its optically anisotropic molecules on macroscopic distances. In this respect the structure of a liquid-crystal phase as well as its variations and dynamics in the process of a phase transition inevitably and explicitly manifest themselves in the interaction with optical radiation. These phenomena become more versatile than in liquids or solids studied separately. In liquids a strong pretransitional light scattering in the region of second-order phase transitions is caused by density fluctuations or component concentration. In solids it is caused by order-parameter fluctuations related to fluctuations of dielectric properties in a more complicated way than in liquids and, however, restricted in magnitude for energetic reasons peculiar to solids. In liquid crystals a pretransitional scattering may display features of either phase and is not subjected to restrictions valid for solids. It should be mentioned that the conventional method of studying phase transitions, namely, the calometric method in the application to liquid crystals is fairly complicated in realization owing to the closeness of phase transitions in liquid crystals to second-order transitions and to certain hence entailing technical difficulties. This also enhances the importance of optical methods, not to mention the fact that optical properties are the basis of applications of liquid crystals and therefore – a stimulus for studying them by optical methods. The present paper is devoted to light scattering and optics of liquid crystals in the vicinity of phase transition points. Intensive investigations performed recently on the problem of phase transitions make reasonable generalization and summation of the results obtained so far, which is actually the aim of the present review with respect to theoretical studies. The experimental side of the corresponding research is not touched upon and experimental results will be referred to only sometimes for a complete physical picture or comparison of theoretical conclusions with measurement results.

A most impressive progress has lately been made in the field of smectic liquid crystals as well as chiral and particularly chiral smectic crystals. This progress amounts to a discovery and study of various smectic phases, a discovery of ferroelectric smectics (Meyer et al. 1975, Pikin and Indendom 1975), advance in understanding the cholesteric blue phase (Brazovskii and Dmitriev 1975, Stegemeyer and Bergman 1980). There has also been found a new modification of liquid crystals formed by disc-like molecules, the so-called discotics (Chandrasekhar et al. 1977, Kats 1978, Kats and Monastyrskii 1981). A certain success has been achieved in a more detailed study of the most well-known nematic phase; its biaxial modification is discovered (Yu and Saupe 1980) and the influence of biaxial fluctuations in a nematic phase upon its properties is theoretically studied (Pokrovskii and Kats 1977). Investigations of two-dimensional systems and thin liquid-crystal layers are especially noteworthy since the results of these investigations give rise to hope to realize certain model two-dimensional systems proposed theoretically (Halperin and Nelson 1979).

Certain consideration in the review is given to studies of light scattering in the above-mentioned new objects of research which have not so far found the proper coverage in review literature. To make the reader completely understand the role and the place of optical methods in the studies of phase transitions in liquid crystals, the review also includes certain results already sufficiently well reflected by literature, for example, light scattering on uniaxial fluctuations of the director in nematics and cholesterics (Chistyakov 1966, Stephen and Straley 1974, Chandrasekhar 1980).

The material given below is of interest not only for experts in the field of liquid crystals and optics but is also of a general physical interest. This is accounted for by the fact that certain phase transitions in liquid crystals are analogues of phase transitions in other systems, e.g., liquid helium (de Gennes 1974) and it is not excluded that a number of problems of phase-transition physics may be solved by studying phase transitions in liquid crystals by means of optical methods.

Comparing theoretical conclusions with the results of measurements one must bear in mind that in the overwhelming majority of theoretical papers a description of light scattering in the phase transition region is performed with multiple scattering neglected. Therefore the corresponding theoretical conclusions are directly applicable to the description of the experiments where special measures are taken towards suppression of multiple scattering, which may cause certain technical difficulties. In liquid crystals these difficulties are usually very hard to overcome and this is explained by their high scattering ability and liability to even weak environmental influences. Therefore the present review alongside with the traditional description also describes optical quantities associated with multiple scattering and affected by the closeness to phase transitions. As for manifestations of such optical characteristics like birefringence, optical rotation, etc., they are very strong in liquid crystals. This difference of liquid crystals from other condensed media is important and helpful for studying phase transitions.

Before considering light scattering in various types of liquid crystals it is necessary to touch upon certain common characteristics of scattering in the vicinity of second-order transition points (or of transitions close to them). Usually (see, e.g., preceding articles of the collection) the main contribution comes from order-parameter fluctuations. Order-parameter fluctuations are related to fluctuations of the dielectric permeability tensor either linearly (first-order scattering) or quadratically (second-order scattering). The latter situation is more frequent. Liquid crystals are above all characterized by the fact that their order parameter has a sufficiently complex structure. In the simplest case (i.e., of nematic liquid crystals) this is a symmetric two-rank traceless tensor (i.e., five independent quantities). Therefore in liquid crystals, depending on geometry, material parameters and environmental conditions, scattering may be different and determined by fluctuations of various components of the order parameter. It is also of interest that in liquid crystals the relation of the order parameter to the dielectric permeability tensor is in most cases linear, i.e., there is a first-order scattering. There is another difference of scattering in liquid crystals. The fact is that many types of liquid crystals are degenerate systems, i.e., to excite certain types of fluctuations no energy barrier should be overcome. For instance, the ground state of a nematic liquid crystal is degenerate with respect to the orientation of $\boldsymbol{n}$. The order parameter is determined not by the general symmetric two-rank tensor but by the uniaxial tensor only:

$$Q_{\alpha\beta} = s(n_\alpha n_\beta - \tfrac{1}{3}\delta_{\alpha\beta}),$$

i.e., is determined by a unit vector-director $\boldsymbol{n}^2 = 1$ and modulus $s$ (three components but not five like in the general case). This degeneracy leads to an anomalously strong scattering due to transversal fluctuations of $n$. To excite inhomogeneous fluctuations of this type it is required that the energy barrier $\sim q^2$ ($\boldsymbol{q}$ is a wave vector of the scattering) be overcome and therefore the integral scattering cross section be $\sim q^{-2}$. The scattering has the character of the critical opalescence in the overall temperature range of a nematic phase (and is not in any way associated with the closeness to the phase transition point). However in the vicinity of the phase transition point there appear additional mechanisms of scattering due, for example, to fluctuations of the order-parameter modulus or, to fluctuations that break the uniaxiality of a nematic. To a certain degree the above said holds for other types of liquid crystals. A concrete investigation of this range of problems will be performed below.

## 1. Light scattering by nematics

Let us start a study of light scattering in liquid crystals with a theoretically and experimentally most well-known case of nematics. Light scattering by nematics is roughly $10^6$ times stronger than in ordinary isotropic liquids, therefore, they are opaque. It is tempting to assume that this strong scattering is caused by the existence of ordered birefringence regions with dimensions of the order of the visible light wavelength in nematics. This assumption gives rise to the appearance of the "swarm theory" in the physics of nematic liquid crystals which looks very attractive for the interpretation of the results of light scattering measurements. Yet detailed experiments carried out by Chatelain (1948) on light scattering and the theoretical interpretation of them given by De Gennes (1968) show that the reason for the strong light scattering by nematics lies in spontaneous fluctuations of molecular orientations in nematics but not in the existence of "swarms" with fixed boundaries.

The fluctuations of molecular orientations may be considered within the framework of the continual theory of nematics and are described in terms of fluctuations of the director orientations. A physical reason for the anomalously strong scattering by the director fluctuations is the fact that homogeneous transversal fluctuations (i.e., homogeneous rotations of the director in the nematic) are excited without any energy barrier being overcome. Therefore the light scattering corresponding to them has the character of the critical opalescence. This situation is common for a wide range of systems with a continuous symmetry group, i.e., so-called degenerate systems (Patashinskii and Pokrovskii 1975).

To derive quantitative relations it is necessary to find expressions for correlation functions of director fluctuations in a nematic. For this purpose we must find an expression for the elastic energy associated with the fluctuations. Assuming that the average orientation of the director $n_0$ coincides with the $z$-axis, and describing the fluctuations by small non-zero components up to the second order with respect to the introduced small quantities, we get for the distortion energy:

$$F = \tfrac{1}{2}\int\left[K_1\left(\frac{\partial n_x}{\partial x} + \frac{\partial n_y}{\partial y}\right)^2 + K_2\left(\frac{\partial n_x}{\partial y} - \frac{\partial n_y}{\partial x}\right)^2 + K_3\left\{\left(\frac{\partial n_x}{\partial z}\right)^2 + \left(\frac{\partial n_y}{\partial z}\right)^2\right\}\right]. \quad (1)$$

Hence, for fluctuation correlators in a regular way (applying the equipartition theorem) (see Landau and Lifshitz 1976), we obtain the following expressions:

$$\begin{aligned}\langle n_x(\boldsymbol{r}_1)n_x(\boldsymbol{r}_2)\rangle &= \langle n_y(\boldsymbol{r}_1)n_y(\boldsymbol{r}_2)\rangle \\ &= \frac{T}{8\pi}\{[(K_3K_1)^{-1/2}R_1^{-1}] + [(K_3K_2)^{-1/2}R_2^{-1}]\}\,, \end{aligned} \quad (2)$$

where

$$R_\alpha = \sqrt{x^2 + y^2 + z_\alpha^2}\,, \qquad Z_\alpha = (K_2/K_3)^{1/2}z\,.$$

The decay of the correlations according to the law $R^{-1}$ is characteristic, as has already been mentioned, of degenerate systems where the ordered state is characterized by the preferred axis which orientation may be arbitrary and interaction forces are short range.

To find the light scattering associated with these fluctuations, suffice it to express correlations of the dielectric permeability tensor $\delta\epsilon$ in terms of the already derived correlators and employ the well-known expression for the scattering cross section (Landau and Lifshitz 1957),

$$I(\boldsymbol{q}, \omega) = \frac{\omega^4}{32\pi^3} \langle \delta\epsilon_{\alpha\beta} \delta\epsilon_{\gamma\delta} \rangle P_\alpha P_\gamma P'_\beta P'_\delta, \tag{3}$$

where $\langle \delta\epsilon_{\alpha\beta} \delta\epsilon_{\gamma\delta} \rangle$ is the Fourier component of the space–time correlation functions corresponding to the frequency $\omega$ and momentum $\boldsymbol{q}$, $\boldsymbol{P}$, $\boldsymbol{P}'$ are the incident and scattered light polarizations, respectively.

### 1.1. Quasielastic scattering by nematics

Inserting into eq. (3) an expression for the dielectric-permeability tensor in the form $\epsilon_{\alpha\beta} = \epsilon_\perp \delta_{\alpha\beta} + \epsilon_a n_\alpha n_\beta$ where $\epsilon_a = \epsilon_\parallel - \epsilon_\perp$ and $\epsilon_\parallel$ and $\epsilon_\perp$ are components, longitudinal and transversal to the director, and using eq. (2) for the integral light scattering intensity, i.e., quasielastic scattering, we get:

$$I(\boldsymbol{q}) = V\left(\frac{\epsilon_a \omega^2}{4\pi}\right)^2 \sum_{\alpha=1,2} \frac{T(P_\alpha P'_z + P_z P'_\alpha)^2}{K_3 q_\parallel^2 + K_2 q_\perp^2}, \tag{4}$$

where $q_\parallel$ and $q_\perp$ are components of the momentum transfer, longitudinal and transversal with respect to the director, and the $x$- and $y$-axes lie in the plane perpendicular to $\boldsymbol{n}_0$, and the $x$-axis is perpendicular to $\boldsymbol{q}$.

It follows from eq. (4) that at $\boldsymbol{q} \to 0$ the cross section diverges as $q^{-2}$. The ratio of this cross section to the cross section of the scattering by density fluctuations in an isotropic liquid is of the order of $(qa)^{-2}$ where $a$ is a characteristic intermolecular distance. This estimate gives the above-mentioned value of $10^6$ characterizing the scattering intensity by nematics in comparison with ordinary liquids.

The cross-section divergence at $\boldsymbol{q} \to 0$ corresponds to anomalously large orientation fluctuations (with the wave vector $q = 0$) which is physically explained by the fact that such fluctuations do not require an overcome of an energy barrier. Actually in a bulk nematic its energy is independent of the orientation of the director $\boldsymbol{n}_0$ in homogeneous states. Naturally in a finite sample a certain orientation is preferred and a certain energy is required for the excitation of fluctuations, yet, as usual, most energetically profitable in a sample which is sufficiently large are homogeneous fluctuations of the director. The reason leading to an existence of the preferred orientation in a nematic may not be only surface effects but, for instance, an external electric or magnetic field applied to the sample. From what precedes one may expect that the application of a constant, for example, magnetic field must lead to a light scattering decrease due to the suppression of the director orientation fluctuations by the field.

To describe quantitatively the influence of the applied external field, the expression for the free energy (1) should include the "magnetic" term:

$$F_m = \frac{1}{2} \int \chi_a H^2 (n_x^2 + n_y^2)\, dv \, .$$

As a result, the denominator of the expression for the cross section (4) will also involve the term $\chi_a H^2$.

To make the interpretation of this "magnetic" term more transparent, let us study the one-constant approximation of the continual theory. In this case the denominator of eq. (4) can be represented in the form: $K(q^2 + \xi^{-2})$, where $\xi = (K/\chi_a)^{1/2} H^{-1}$ is a magnetic coherence length. It means that the long-wave fluctuations with $q \ll \xi^{-1}$ are suppressed and a contribution to the scattering comes largely from fluctuations of the director orientations with the sizes smaller or of the order of $\xi$. In practice to suppress light scattering it is required that sufficiently strong fields $H > 10^3$ G should be applied. Smaller fields affect only forward scattering.

Another feature making light scattering by nematics different from scattering by density fluctuations in liquids is polarization characteristics of scattering. While in an isotropic liquid scattered light polarization either coincides with incident light polarization, or is maximally close to it (the difference is determined by the transversality condition of incident and scattered waves), in a nematic polarization of a scattered wave, as follows from eq. (4) is sharply different from polarization of an incident wave. In the typical scattering geometry where the director $\boldsymbol{n}_0$ is perpendicular to the scattering plane, as is clear from eq. (4), the polarization plane of linearly polarized light after the scattering turns by 90°.

Let us now discuss the data on nematics provided by light scattering experiments. The scattering cross section (4) involves two terms, each being proportional to the mean-squared Fourier harmonic amplitude of thermal director fluctuations $\langle |n_\alpha(\boldsymbol{q})|^2 \rangle$ with the wave vector $\boldsymbol{q}$ and the corresponding value of the index $\alpha$, i.e., for a definite projection of the director. Studying in these terms the dependence of the cross section on the angle between $\boldsymbol{q}$ and $\boldsymbol{n}_0$, we can find the ratio of the elastic moduli $K_\alpha / K_3$. Studying the dependence of the cross section on $q$ in the magnetic field and assuming that $\chi_a$ is known, we can measure the elastic moduli $K_i$ employing experimental intensity measurements for various $\boldsymbol{q}$. (Martinaud and Durand 1972).

It should be pointed out that the temperature dependence of scattering in nematics as it follows from the theory and is confirmed by the experiment (Chandrasekhar 1980), proves to be weak up to the temperature of a transition into an isotropic liquid. Equation (4) reveals that at a fixed $\boldsymbol{q}$ the cross section is proportional to $\epsilon_a^2/K$. Although separatley $\epsilon_a$ and $K$ are strongly temperature dependent, decreasing near $T_c$, their ratio weakly depends on temperature. So in the self-consistent field theory by Maier and Saupe (1960) $\epsilon_a$ is linear with respect to the order parameter $s$ and $K \sim s^2$, which in our approach does not lead to any temperature dependence of scattering at all. Therefore the temperature dependence of the scattering cross section may manifest itself in the vicinity of the temperature of a transition into an isotropic liquid where this theory is not applicable and experimentally observable. Therefore the region and the character of the temperature dependence of scattering may be used for the establishment of the applicability limits of the self-consistent field theory and of the character of the nematic–isotropic liquid transition.

### *1.2. Scattering in an isotropic phase*

After we have studied the specific features peculiar to scattering in nematics, it is natural to presume that these features should also manifest themselves above, but sufficiently close to, the temperature of the transition of a nematic into an isotropic liquid. The physical basis for it is the closeness of a nematic–isotropic liquid transition to a second-order transition and, as a consequence, the existence of well-developed pretransitional fluctuations in the isotropic phase. It means that in the isotropic phase in the pretransitional region in contrast to ordinary liquids orientation fluctuations, but not density fluctuations, are responsible for light scattering.

Like in the nematic phase, scattering on molecular orientations in the isotropic phase may be described within the framework of the continual theory. As a result, for the light scattering cross section we can use the general expression (4) where however the correlator of the dielectric permeability tensor fluctuations $\epsilon$ should be calculated in a slightly different way from the one conventionally used.

As has already been mentioned, the anisotropy of the tensor $\epsilon$, i.e., $\epsilon_a$ is proportional to the order parameter $s$. Therefore in the isotropic phase where the order parameter is zero, the fluctuations are simply proportional to fluctuations of the tensorial order parameter $Q_{\alpha\beta}$ and the scattering cross section may directly be expressed in terms of the order parameter fluctuation correlators $\langle Q_{\alpha\beta} Q_{\gamma\delta} \rangle$. Then it is clear that depending on the polarization of the incident light and the detected polarization of the scattered light, the scattering cross section is determined by definite components of the order parameter.

Similarly, the order parameter correlators may be expressed in terms of thermal averages of the Fourier components of the order-parameter fluctuations. For this purpose we shall use the following expression:

$$F = \tfrac{1}{2} L_1 (\nabla_\alpha Q_{\beta\gamma})^2 + \tfrac{1}{2} L_2 (\nabla_\alpha Q_{\alpha\gamma})^2 \,, \tag{5}$$

where the constants $L_1$, $L_2$ are related to the moduli of elasticity $K_i$ as $2s^2(L_1 + \frac{1}{2}L_2) = K_1 = K_3$; $2s^2 L_1 = K_2$. Note that expression (5) is an expansion of the elastic energy with an accuracy up to $s^2$. For not small $s$ (immediately below the point of transition into a nematic $s \sim 0.4$–$0.6$) higher invariants should necessarily be taken as well. This will enable us to make the accuracy of the homogeneous and elastic parts of the thermodynamical potential consistent. Higher-order invariants are obtained from all possible convolutions consisting of the components of $\boldsymbol{n}$ and $\nabla_\gamma Q_{\alpha\beta}$ (since $\boldsymbol{n}^2 = 1$, invariants of the degree higher than four are not existent). Yet for most problems discussed in the review the invariants $\sim s^3$ and $s^4$ in the elastic energy are not important.

As a result, using the equipartition theorem for mean-squared fluctuations, we get:

$$\begin{aligned} \langle |Q^+(\boldsymbol{q})|^2 \rangle &= 2T/3A(1 + \xi_1^2 q^2 + \tfrac{2}{3}\xi_2^2 q^2) \\ \langle |Q^-(\boldsymbol{q})|^2 \rangle &= 4\langle |Q_{xy}(\boldsymbol{q})|^2 \rangle = 2T/A(1 + \xi_1^2 q^2), \\ \langle |Q_{xz}(\boldsymbol{q})|^2 \rangle &= \langle |Q_{yz}(\boldsymbol{q})|^2 \rangle = \tfrac{1}{2}T/A(1 + \xi_1^2 q^2 + \tfrac{1}{2}\xi_2^2 q^2), \end{aligned} \tag{6}$$

where $Q^{\pm} = Q_{xx} \pm Q_{yy}$ and the direction $\boldsymbol{q}$ is taken as the $z$-axis. Expression (6) involves now the two coherence lengths $\xi_1^2 = L_1/|A|$ and $\xi_2^2 = L_2/|A|$ ($A$ is a temperature dependent quantity) corresponding to the constants of eq. (5).

In the mean field approximation, for example, each of the coherence lengths is described by the temperature dependence of the form $\xi_0|(T/T^*) - 1|^{-1/2}$ where $\xi_0$ is the microscopic length determined by the radius of intermolecular forces and the temperature $T^*$ is the parameter of the theory which is somewhat smaller than $T_c$. Thus we can consider that the temperature dependence of $A$ is described by the factor $(T - T^*)$.

As a result the quasielastic scattering cross section above $T_c$ is described by eq. (4) where the summation involves expression (6) and polarization factors. To estimate the temperature dependence of the order-parameter correlators and thus the scattering cross section, we can consider that $T_c - T^* \sim 1K$. It means that in the vicinity to $T_c$ the coherence lengths increase by 10 to 20 times, then for a qualitative study we may assume that $\xi_1 \sim \xi_2 \sim \xi$. The typical value of $q$ in expression (6) for light-scattering experiments is $10^5\,\mathrm{cm}^{-1}$ and therefore even near the transition point the angular dependence of the scattering is weak since $\xi q \sim 0, 1$. The light-scattering intensity increases as $T$ approaches $T_c$ according to the law $(T - T^*)^{-1}$ and becomes very large near $T_c$. The above theoretical conclusions are in agreement with a number of experimental measurements (see, e.g., Chandrasekhar (1980)).

### *1.3. Inelastic scattering*

#### *1.3.1. Nematic*

Above we have studied quasielastic scattering only without dwelling upon the spectral distribution of scattered light. Now we shall deal with the information on nematics which can be obtained if the spectral distribution of scattered light is studied.

The relations of the preceding sections formally describe static fluctuations. Actually, fluctuations in nematics, though sufficiently slow, are not static and light frequency variations in the scattering can experimentally be measured at the up-to-date level of methods. In fact the dynamical character of fluctuations in nematics has been discovered in early experiments on the "flicker effects" (see De Gennes (1974)). These time dependent fluctuations lead to frequency modulation of the scattered light which can be observed experimentally. Yet, the corresponding broadening of the line is not large, i.e., is of the order of megaherz or even kiloherz.

The dynamics of fluctuations is described by the equations of liquid-crystal hydrodynamics relating variations of the director orientations with hydrodynamic flows. Therefore the study of the frequency broadening provides data both on the elastic constants and hydrodynamical parameters (friction or viscosity coefficients) the number of which for liquid crystals is higher than for an isotropic liquid, and which are conventionally called Lesli parameters. Attenuation of fluctuations in liquid crystals is of a relaxational character and for each mode may be described by the corresponding relaxation time $\tau_\alpha$ or, which is the same, by the scattering line width $\Gamma_\alpha = 1/\tau_\alpha$. For each mode involved in eq. (4) a Fourier transform of the correlation function may be factorized and represented in the form:

$$I_\alpha(\boldsymbol{q}, \omega) = I_\alpha(q) \frac{2\Gamma_\alpha}{\omega^2 + \Gamma_\alpha^2}, \tag{7}$$

where $I_\alpha(\boldsymbol{q})$ is a Fourier transform of the spatial correlation function of eq. (4). The relaxation time for each mode is inversely proportional to the elastic restoring force $K_\alpha(q) = K_3 q_\parallel^2 + K_\alpha q_\perp^2$ (see eq. (4)) and is proportional to the effective viscosity $\eta_\alpha(\boldsymbol{q})$, i.e., can be represented in the form:

$$\tau_\alpha(\boldsymbol{q}) = \eta_\alpha(\boldsymbol{q})/K_\alpha(\boldsymbol{q}) . \tag{8}$$

Without dwelling upon a detailed discussion of the expressions for the effective viscosity (Martin et al. 1972) which may be obtained at the study of the dynamics of nematics and depend on the $\boldsymbol{q}$ orientation, we shall give their explicit expressions:

$$\begin{aligned} \eta_1(\boldsymbol{q}) &= \gamma_1 - \frac{(q_\perp^2 \alpha_3 - q_z^2 \alpha_2)}{q_\perp^4 \eta_b + q_\perp^2 q_z^2(\alpha_1 + \alpha_3 + \alpha_4 + \alpha_5) + q_z^4 \eta_c}, \\ \eta_2(\boldsymbol{q}) &= \gamma_1 - \frac{\alpha_2^2 q_z^2}{q_\perp^2 \eta_a + q_z^2 \eta_c}, \end{aligned} \tag{9}$$

where $\eta_a$, $\eta_b$, $\eta_c$ are Miesovicz viscosities and $\gamma_1$, $\alpha_1$ are Lesli coefficients.

Thus, as follows from eqs. (7) and (9), for each mode the distribution of the scattered light frequencies is of the Lorenz form. However to extract experimentally the contribution of a definite mode to the scattering it is necessary to use the difference of polarization scattering characteristics of the both modes and to conduct polarization measurements. In the general case the frequency scattering spectrum is a superposition of two Lorenz curves.

As is clear from eq. (9), width of the frequency distribution for each mode is proportional to $\boldsymbol{q}^2$ and involves a certain dependence on the $\boldsymbol{q}$ orientations. It should be pointed out that the expressions for inelastic scattering given here are obtained if the parameter $\mu = K\rho/\eta^2$ ($\rho$ is the density) is small in a nematic. The typical value of this parameter is $\sim 10^{-2}$, therefore in the studied range only slow relaxational modes are excited. In experiments on scattering of a higher-frequency radiation and acoustic experiments other kinds of excitation modes in nematics, namely shear waves (Stephen and Straley 1974) may appear.

*1.3.2. Isotropic phase*

Like in the nematic phase, in the isotropic phase time characteristics of orientation fluctuations reveal themselves in the spectral distribution of the scattered light in a certain temperature interval in the vicinity of the transition point. For sufficiently long-wave fluctuations we may neglect shear modes in the scattering. Then for each mode there is factorization of the form (7) for frequency and momentum dependent factors of the correlation function. The expression for the frequency line width acquires the form:

$$\Gamma_q = v_1 A(1 + \xi^2 q^2),$$

where $\xi$ is the quantity which is a combination of the correlation lengths, $v_1$ is the phenomenological viscosity coefficient and the quantity $A$ is defined in eq. (6).

For shorter-wave fluctuations (Pokrovskii and Kats, 1977) we must take the interaction of orientational and shear modes and in this case the frequency dependence of the correlation function $\langle|Q_{xz}|^2\rangle = \langle Q_{yz}|^2\rangle$ is not of a simple Lorenz form of eq. (7). In any case to extract experimentally the contribution of an individual mode to the scattering it is necessary to suppress the others by means of the polarization properties of the scattering (see eq. (4)).

As has already been mentioned, in the self-consistent field approximation the quantity $A$ as well as $\Gamma$ is linear with respect to $(T - T^*)$. The results of the measurements of the temperature dependence of the line width in the isotropic phase of MBBA (Stinson and Litster 1970) are in agreement with this theoretical prediction. As for the value of the line width, it is of the order of a few megaherz.

Hence, as becomes evident from the discussion, the dominating mechanism of scattering in nematics both below the temperature of a transition into the isotropic phase and above it is the scattering by fluctuations of the director orientations. Experiments on light scattering (including polarization measurements) provide important information on the physics of nematics and the nature of phase transitions in them. So elastic moduli of nematics can be found from the scattering intensity and its angular dependence. The corresponding measurements in the isotropic phase near the transition point provide data on the temperature behavior of elastic moduli and can be interpreted in terms of the temperature dependence of correlation lengths. The latter, as is known, are essentially determined by the nature of phase transitions. Measurements of the frequency line width provide data on ratios of elastic moduli to viscosity coefficients and above the transition point – provide the respective temperature dependence, which may be used for the study of a transition character.

Yet we must point out the complexity of an unambiguous interpretation of the frequency measurements results, which is associated with a large number of insufficiently well studied quantities (viscosity coefficients, elastic moduli, etc.) involved in the respective formulae. On the whole on the example of most well studied modification of liquid crystals, i.e. nematics, we can conclude that light scattering is an extremely informative means for studying liquid crystals. Light scattering in other types of liquid crystal phases has much in common with the case of nematics, each of these phases however, having its own specific properties, and as will be shown below, interesting possibilities for investigation by means of optical methods.

In conclusion, note another remarkable property of liquid crystals. Softness of liquid crystals, i.e., relative weakness of forces responsible for the orientational ordering, leads to a number of characteristic features of light scattering. These features are associated with the fact that small influences which are hardly observable at light scattering in other condensed systems, may give a noticeable contribution to the light scattering cross section in liquid crystals. The contribution of the long-range Van der Waals forces to light scattering in the nematic phase has been calculated by Dzyaloshinskii et al. (1975). Due to the Van der Waals forces formulae (4) for the scattering intensity change. At sufficiently large transferred momenta $q\lambda_0 \gg 1$ ($\lambda_0$ is a characteristic wavelength of the absorption spectrum of a liquid crystal) starting from which there exists dielectric-permeability dispersion) instead of eq. (4) we have for the two intensity components $\alpha = 1, 2$:

$$\begin{aligned} I_1 &= T/[(K_3 - 8L)q_\parallel^2 + (K_1 - 8L)q_\perp^2 + 12M(q_\parallel^2 q_\perp^2/q)], \\ I_2 &= T/[(K_3 - 8L)q_\parallel^2 + (K_2 + 8L)q_\perp^2 + 4Mqq_\perp^2], \end{aligned} \tag{10}$$

where

$$L = \frac{\hbar}{192\pi^2 C} \int_0^\infty \frac{\epsilon_a^2(\mathrm{i}|\omega|)}{\epsilon_\perp^{3/2}(\mathrm{i}|\omega|)}, \qquad M = \frac{\hbar}{2048\pi} \int_0^\infty \frac{\epsilon_a^2(\mathrm{i}|\omega|)}{\epsilon_\perp^2(\mathrm{i}|\omega|)}. \tag{11}$$

In the region of small wave vectors $q\lambda_0 \ll 1$ formulae (4) are preserved. Note also that the coefficients $L$ and $M$ for the Van der Waals contribution to the elastic energy may not be very small in comparison with the elastic Frank moduli, which agrees with many facts, e.g., the value of the modulus $K_2$ is smaller in comparison with the moduli $K_1$ and $K_3$.

### *1.4. Longitudinal scattering by nematics*

A simple analysis shows that the above studied transversal fluctuations do not lead to light scattering at certain conditions. For example, scattering is absent if the vectors of the incident and scattered light polarization belong to the equatorial plane, i.e., the plane perpendicular to the director orientation. Strictly speaking, transversal scattering is small in a small range of angles at which polarization is close to the equatorial plane. In this region light scattering is governed by fluctuations of another type, namely, longitudinal and biaxial. When the wave vectors of the incident and scattered light lie in the equatorial plane, transversal fluctuations do not lead to scattering for polarizations along the director either.

By longitudinal fluctuations we imply fluctuations of the order-parameter modulus. From the general theory of degenerate systems (Patashinskii and Pokrovskii 1975) we know that longitudinal fluctuations are also anomalously large, though they are weaker than transversal.

Biaxial fluctuations are determined in the following way. As has already been mentioned, the order parameter in nematics is a symmetric two-rank traceless tensor. This quantity in the general case is determined by five independent components. Yet in all known nematics the tensor $Q_{\alpha\beta}$ is uniaxial and therefore is determined by only one vector, i.e., by three quantities. Naturally in long-wave fluctuations local properties of a liquid crystal are unaltered, i.e., in each point the tensor $Q_{\alpha\beta}$ is still uniaxial. However with the decrease of the fluctuation wavelength, more and more important become fluctuations breaking the uniaxiality of the tensor $Q_{\alpha\beta}$. Even in the region of wavelengths where these fluctuations are small, they prove to be important for a definite geometry of scattering, for instance, in the equatorial plane at certain polarizations of the incident and scattered light.

#### *1.4.1. Integral light scattering cross section*

In their work Pokrovskii and Kats (1977) have calculated the integral (with respect to frequency) light scattering intensity associated with longitudinal and biaxial fluctuations of the order parameter. Let us give the main conclusions of this work necessary for further analysis. The thermodynamical potential $\Phi$ is in the general case an arbitrary function of two variables $x = \mathrm{tr}\,\boldsymbol{Q}^2$ and $y = \mathrm{tr}\,\boldsymbol{Q}^3$. Since the equilibrium configuration of the order parameter is uniaxial, in the general case we can write down:

$$Q_{\alpha\beta} = s(n_\alpha n_\beta - \tfrac{1}{3}\delta_{\alpha\beta}) + \delta Q^{\perp}_{\alpha\beta}, \tag{12}$$

where the fluctuation $\delta Q^{\perp}_{\alpha\beta}$ obeys the orthogonality conditions:

$$n_\alpha n_\beta \delta Q^{\perp}_{\alpha\beta} = 0\,, \qquad \delta Q^{\perp}_{\alpha\alpha} = 0\,. \tag{13}$$

It is convenient to introduce new variables automatically ensuring the fulfillment of the condition (13):

$$\delta Q^{\perp}_{\alpha\beta} = \theta_1(n_\alpha e_{1\beta} + n_\beta e_{1\alpha}) + \theta_2(n_\alpha e_{2\beta} + n_\beta e_{2\alpha}) + \theta_3(e_{1\alpha}e_{2\beta} + e_{1\beta}e_{2\alpha}) + \theta_4(e_{1\alpha}e_{1\beta} - e_{2\alpha}e_{2\beta})\,, \tag{14}$$

where $\boldsymbol{n}$, $\boldsymbol{e}_1$, $\boldsymbol{e}_2$ is a system of unit orthogonal vectors. It becomes evident that $\delta Q^{\perp}_{\alpha\beta}$ defined by eq. (14), satisfies conditions (13) for arbitrary $\theta_i$. Bearing in mind the conditions of the thermodynamical potential minimum with respect to the variable $Q_{\alpha\beta}$, from eqs. (12) and (14) we get:

$$\delta\Phi = 6\frac{\partial\Phi}{\partial x}(\theta_3^2 + \theta_4^2) \equiv \Delta(\theta_3^2 + \theta_4^2)\,, \tag{15}$$

where $\delta\Phi$ is the variation of the thermodynamical potential due to the fluctuation $\delta Q^{\perp}_{\alpha\beta}$. Formula (15) is valid with the accuracy up to the terms cubic with respect to $\theta_i$. Note that as should be expected, the energy increases only at the appearance of biaxial fluctuations corresponding to the variables $\theta_3$ and $\theta_4$.

Longitudinal fluctuations of the order parameter are fluctuations of the quantity $s$ from eq. (12),

$$\delta Q^{\parallel}_{\alpha\beta} = \delta s(n_\alpha n_\beta - \tfrac{1}{3}\delta_{\alpha\beta})\,. \tag{16}$$

From the general theory of degenerate systems (Patashinskii and Pokrovskii 1975) it is known that in virtue of the principle of the modulus conservation strong transversal fluctuations entail weaker (but also singular) longitudinal fluctuations:

$$2s\delta s = -(\delta Q^{\perp}_{\alpha\beta})^2\,. \tag{17}$$

Apart from these singular fluctuations there are also the so-called classical longitudinal fluctuations $\delta s$. They are determined directly by the expansion of the Landau free energy. Such fluctuations at $T < T_c$ ($T_c$ is the temperature of the transition into the isotropic liquid) are not singular at the fluctuation wave vector $\boldsymbol{q} \to 0$. Classical fluctuations exist in any system experiencing phase transitions but no longer possess any liquid-crystal specific features. We are studying singular longitudinal scattering characteristic of only such degenerate systems like nematics. These singular longitudinal fluctuations are associated with the existence of Goldstone transversal fluctuations (in the given case – fluctuations of the director).

If, to be interested only in long-wave fluctuations, $\theta_3$ and $\theta_4$ should be assumed equal to zero (since to excite biaxial fluctuations it is required that the energy barrier $\sim \Delta$ be overcome). Let us make use of the well-known expression for the elastic energy of a nematic (5). If to pass over to the variables of eq. (14) and chose the vector $\boldsymbol{e}_1$ perpendicular to the plane determined by the vectors $\boldsymbol{n}$ and $\boldsymbol{q}$, and $\boldsymbol{e}_2$ lying in this plane:

$$\boldsymbol{e}_1 = \frac{\boldsymbol{n} \times \boldsymbol{q}}{q \sin \theta}, \qquad \boldsymbol{e}_2 = \boldsymbol{n} \times \boldsymbol{e}_1 , \tag{18}$$

where $\theta$ is the angle between the vectors $\boldsymbol{n}$ and $\boldsymbol{q}$. From eqs. (5) and (14) we obtain:

$$\delta\Phi = L_1 q^2(|\theta_1|^2 + |\theta_2|^2) + \tfrac{1}{2} L_2 q^2[|\theta_1|^2 \cos^2\theta + |\theta_2|^2] . \tag{19}$$

Hence,

$$\begin{aligned} \langle |\theta_1(q)|^2 \rangle &= T/q^2(2L_1 + L_2 \cos^2\theta) , \\ \langle |\theta_2(q)|^2 \rangle &= T/q^2(2L_1 + L_2) . \end{aligned} \tag{20}$$

In their turn the quantities $\langle |\theta_1|^2 \rangle$ and $\langle |\theta_2|^2 \rangle$ determine non-zero averages of the type $\langle \delta Q^{\perp}_{\alpha\beta} \delta Q^{\perp}_{\gamma\delta} \rangle$,

$$\begin{aligned} \langle \delta Q^{\perp}_{\alpha\beta}(q) \delta Q^{\perp}_{\gamma\delta}(-q) \rangle = {} & (n_\alpha e_{1\beta} + n_\beta e_{1\alpha})(n_\gamma e_{1\delta} + n_\delta e_{1\gamma}) \\ & \times \langle |\theta_1|^2 \rangle + (n_\alpha e_{2\beta} + n_\beta e_{2\alpha})(n_\gamma e_{2\delta} + n_\delta e_{2\gamma}) \langle |\theta_2|^2 \rangle . \end{aligned} \tag{21}$$

Hence according to the modulus conservation principle (17) we find

$$\delta s^2 \sim T^2/s^2 L^2 q .$$

Note that classical modulus fluctuations can easily be calculated. For instance, within the framework of the Landau theory

$$\Phi = \tfrac{1}{2} A \operatorname{tr} \boldsymbol{Q}^2 - \tfrac{1}{3} B \operatorname{tr} \boldsymbol{Q}^3 + \tfrac{1}{4} C (\operatorname{tr} \boldsymbol{Q}^2)^2 , \tag{22}$$

we have

$$\langle \delta s^2 \rangle = \tfrac{81}{4} TC/B^2 . \tag{23}$$

A comparison of the two contributions to the longitudinal cross section yields a region of wave vectors where the singular longitudinal fluctuations are dominating:

$$q \lesssim \frac{T}{160 s^2} \frac{B^2}{L^2 C} \sim 10^{-4} a_0^{-1} ,$$

where $a_0$ is the interatomic distance. Actually this value corresponds to the wavelengths $\sim 5000$ Å.

In virtue of the symmetry $\delta\epsilon_{\alpha\beta}$ is related with the order-parameter fluctuation $\Delta Q^{\parallel}_{\alpha\beta}$,

$$\delta\epsilon_{\alpha\beta} = M \delta Q^{\parallel}_{\alpha\beta} , \tag{24}$$

where $M = \partial\epsilon_a/\partial s$, $\epsilon_a$ is the dielectric-permeability anisotropy.

Inserting all the values into the formula for the scattering cross section, we obtain:

$$I(q) - \frac{\omega^4 M^2}{32\pi^3} \langle \delta s^2 \rangle [(pn)(p'n) - \tfrac{1}{3}(pp')] \tag{25}$$

At the deviation of polarization from the equatorial plane a rapidly increasing transversal scattering suppresses all the effects of the longitudinal fluctuations.

Let us estimate the region of angles where the longitudinal fluctuations are important. The contribution of the transversal fluctuations to the scattering cross section has been written above (see eq. (4)). Let us designate the angle of the deviation of polarization from the equatorial plane as $\alpha$. Then eq. (4) shows that the cross section behaves as $\max[\alpha^2, (\theta - \frac{1}{2}\pi)^2]$. Comparing eq. (4) and eq. (25) we find out that the longitudinal scattering is important in the region of angles:

$$\max(\alpha, \theta - \tfrac{1}{2}\pi) \sim (TqM^2/Ks^2\epsilon_a^2)^{1/2} . \tag{26}$$

Assuming in eq. (26) that $M \sim \epsilon_a$, $T \sim 300$ K, $K \sim 10^{-6}$ erg cm$^{-1}$, $s \sim 0.8$, $q \sim 10^5$ cm$^{-1}$, we find $(\alpha, \theta - \frac{1}{2}\pi) \sim 5$–$10°$ which is quite experimentally attainable.

*1.4.2. Dynamics of fluctuations*

Above we have considered the integral (with respect to frequency) intensity. Yet, it is more important to measure not integral but spectral intensity, i.e., the form of the line. This is associated with a higher accuracy of the applied methods as well as with the fact that in the spectral intensity measurements it is easy to distinguish between the effect due to impurities and other defects, i.e., it is possible to use less pure samples. To calculate the spectral intensity we must study the dynamics of the order-parameter fluctuations. (Kamenskii and Kats 1980).

The case of singular longitudinal fluctuations is especially simple. The modulus conservation principle (17) is also valid with the time dependence of transversal fluctuations taken into account. Such local equality takes place in the coordinate space but in the momentum space it is required that the integral of transversal correlators be calculated, which can symbolically be represented in the form of a loop diagram (fig. 1). The lines of the diagram correspond to the correlators of the director fluctuations $\theta_{1,2}$. The value of the transversal correlator is calculated by linearized equations of hydrodynamics and is well known (Stephen and Straley 1974). Unfortunately direct calculations of longitudinal fluctuations can possibly be performed only numerically. However with the accuracy sufficient for the establishment of qualitative dependences, we can use the following approximation. In all so far known nematic crystals Lesli viscosity parameters can roughly be divided into two groups: small parameters of the order of a few centipoise and larger parameters of the order

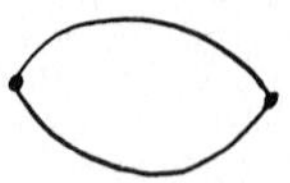

Fig. 1. Relation between longitudinal and transversal correlations.

of a few dozens of centipoise. Then the signs of the large Lesli coefficients are fixed, $\alpha_2 < 0$, $\alpha_4 > 0$, $\alpha_5 > 0$, $\alpha_6 < 0$. Let us neglect small viscosities and elasticity anisotropy. In this approximation (the accuracy of which is $\sim 1\%$),

$$I(q, \omega) = \frac{2T}{\gamma_1}\left(\frac{1}{\omega^2 + \frac{K^2}{\gamma_1^2} q^4}\right)^{-1}, \tag{27}$$

where $\gamma_1$ is the rotation friction coefficient.

It is more convenient to conduct calculations in the coordinate space. After non-complicated but cumbersome calculations,

$$I(r, t) = \frac{T}{4\pi K}\frac{1}{r}\Phi\left(r \Big/ 2\sqrt{\frac{K}{\gamma_1}t}\right), \tag{28}$$

where

$$\Phi(x) = \frac{2}{\sqrt{\pi}}\int_0^x e^{-u^2}\, du$$

with eqs. (17) and (28) yield the longitudinal correlator

$$G(r, t) \equiv \langle \delta s(r, t)\delta s(0, 0)\rangle = \frac{T^2}{64\pi^2 K^2 s^2}\frac{1}{r^2}\Phi^2\left(r \Big/ 2\sqrt{\frac{K}{\gamma_1}t}\right). \tag{29}$$

Formula (29) in principle solves the problem of light scattering by longitudinal fluctuations. Its Fourier transform is

$$G(q, \omega) = \frac{\pi^2}{8s^2}\frac{T^2}{K^2}\left\{\frac{1}{\left(\omega^2 + \frac{K^2}{4\gamma_1^2}q^4\right)^{1/2}\left[\left(\omega^2 + \frac{K^2}{4\gamma_1^2}q^4\right)^{1/2} + \frac{K}{2\gamma_1}q^2\right]^{1/2}}\right\}^{-1}. \tag{30}$$

The main consequence of eq. (30) is a non-Lorenz form of the line. At $\omega \gg q$, $G \sim 1/\omega^{3/2}$ and at $q \gg \omega$, $G \sim 1/q^3$. The light scattering cross section is determined by eq. (25) with the replacement of $\langle|\delta s^2|\rangle$ by $G(q, \omega)$.

### 1.5. *Biaxial scattering*

#### 1.5.1. *Integral scattering cross section*

Let us now study light scattering by biaxial fluctuations. Determination of the integral intensity is factually reduced to the diagonalization of the expression for the thermodynamical potential at $\theta_3 \neq 0$ and $\theta_4 \neq 0$. Using eqs. (14), (22) and (15) we obtain:

$$\delta\Phi = L_1 q^2 \sum_{i=1}^{4} |\theta_i|^2 + \tfrac{1}{2} L_2 q^2 [(\theta_1 \cos\theta + \theta_3 \sin\theta)^2 + \theta_2^2 + \theta_4^2 \sin^2\theta - \theta_2\theta_4 \sin 2\theta] + \tfrac{1}{2}\Delta(\theta_3^2 + \theta_4^2), \tag{31}$$

where $\Delta = 6\,\partial\Phi/\partial x$.

We shall introduce new variables in a standard way:

$$\eta_1 = \theta_1 \cos\beta + \theta_3 \sin\beta, \qquad \eta_2 = \theta_2 \cos\gamma - \theta_4 \sin\gamma \tag{32}$$

$$\eta_3 = -\theta_1 \sin\beta + \theta_3 \cos\beta, \qquad \eta_4 = \theta_2 \sin\gamma + \theta_4 \cos\gamma, \tag{33}$$

where

$$\beta = \tfrac{1}{2}\operatorname{arctg}\frac{\sin 2\theta}{\mu - \cos 2\theta}, \qquad \gamma = \tfrac{1}{2}\operatorname{arctg}\frac{\sin 2\theta}{-\mu + \cos^2\theta}, \qquad \mu = \frac{\Delta}{L_2 q^2}.$$

In terms of the variables $\eta_i$ the form (31) is diagonalized,

$$\delta\Phi = \tfrac{1}{2} L_2 q^2 \sum_{i=1}^{4} Q_i |\eta_i|^2 + L_1 q^2 \sum_{i=1}^{4} |\eta_i|^2, \tag{34}$$

$$\begin{aligned} Q_1 &= \mu \sin^2\beta + \cos^2(\theta + \beta), \\ Q_2 &= \mu \sin^2\gamma + \cos^2\gamma + \sin^2\gamma \sin^2\theta + \cos\gamma \sin\gamma \sin 2\theta, \\ Q_3 &= \mu \cos^2\beta + \sin^2(\theta + \beta), \\ Q_4 &= \mu \cos^2\gamma + \sin^2\gamma + \cos^2\gamma \sin^2\theta - \cos\gamma \sin\gamma \sin 2\theta. \end{aligned} \tag{35}$$

From eq. (34) we have:

$$\langle |\eta_i|^2 \rangle = T/q^2(2L_1 + L_2 Q_i). \tag{36}$$

Let us give also the correlators involving biaxial fluctuations:

$$\begin{aligned} \langle \theta_2 \theta_4^* \rangle &= \langle \theta_4 \theta_2^* \rangle = \sin\gamma \cos\gamma (\langle |\eta_2|^2 \rangle - \langle |\eta_4|^2 \rangle) \\ \langle |\theta_4|^2 \rangle &= \sin^2\gamma \langle |\eta_2|^2 \rangle + \cos^2\gamma \langle |\eta_4|^2 \rangle \\ \langle \theta_1 \theta_3^* \rangle &= \langle \theta_3 \theta_1^* \rangle = \sin\beta \cos\beta (\langle |\eta_1|^2 \rangle - \langle |\eta_3|^2 \rangle) \\ \langle |\theta_3|^2 \rangle &= \sin^2\beta \langle |\eta_1|^2 \rangle + \cos^2\beta \langle |\eta_3|^2 \rangle. \end{aligned} \tag{37}$$

In the case under study the relation of the tensor $\delta\epsilon_{\alpha\beta}$ with the order parameter fluctuation $\delta\theta_{\alpha\beta}^{\perp}$ is more complicated than in the case of uniaxial fluctuations. In the general case we can write down:

$$\delta\epsilon_{\alpha\beta} = M\delta Q_{\alpha\beta}^{\perp} + N(n_\alpha n_\gamma \delta Q_{\beta\gamma}^{\perp} + n_\beta n_\gamma \delta Q_{\alpha\beta}^{\perp}). \tag{38}$$

The general formula for the integral (with respect to frequency) scattering intensity is extremely bulky. (The respective expressions are given in the work of Pokrovskii and Kats (1977)). Here we shall confine ourselves to the scattering in the equatorial plane which corresponds to a conventional setting of the experiment. Let us again take the two above-mentioned types of polarization: in the equatorial plane and along the director. In the first case the general formula is simplified, so the result is:

$$I(q) = (M^2/32\pi^3)\omega^4\langle|\theta_4|^2\rangle\,, \qquad (39)$$

where $\langle|\theta_4|^2\rangle$ is determined by means of eqs. (36) and (37). For the second case the scattering cross section (like in the uniaxial case) equals zero.

*1.5.2. Spectral intensity*

To calculate the spectral intensity due to biaxial fluctuations we must know the dynamics of fluctuations. The system of principle equations of motion is fairly complex and we shall not give it here.

For our purposes suffice it to confine ourselves to equations of hydrodynamics. Therefore in the dissipative function it is sufficient to take the quadratic dependence on the velocity gradients $\partial v_i/\partial x_j$ and the velocity of rotation of the order parameter $Q_{ij}$ with respect to a liquid crystal,

$$\dot{Q}_{ij} - Q_{ij} \times \tfrac{1}{2}\,\mathrm{rot}\,\boldsymbol{v}\,. \qquad (40)$$

It is convenient to introduce the symmetric tensors:

$$\boldsymbol{A}_{ij} = \tfrac{1}{2}(\partial v_i/\partial x_j + \partial v_j/\partial x_i)\,, \qquad (41a)$$

$$\boldsymbol{N}_{ij} = \dot{Q}_{ij} - (e_{inm}Q_{jm} + e_{jnm}Q_{im})\omega_n\,, \qquad (41b)$$

where $\boldsymbol{\omega} = \frac{1}{2}\,\mathrm{rot}\,\boldsymbol{v}$, $e_{inm}$ is a completely antisymmetric tensor. The dissipative function $R$ must be invariantly formed from $Q_{ij}$ and the expressions quadratic with respect to $\boldsymbol{A}_{ij}$ and $\boldsymbol{N}_{ij}$.

The spectrum of the eigenmodes (and consequently, the form of the scattering line) is determined by a set of seven algebraic equations which is reduced to a calculation of the $7 \times 7$ determinant of a general form where most elements are non-zero. Naturally the calculation of such determinant cannot be performed analytically.

However in the studied case of the scattering in the optical frequency range there is a simplifying factor. The thing is that at typical values of the parameters the quantity $\Delta$ of eq. (31) is very large in comparison with the quantity $Kq^2$. For instance, for MBBA at $q \sim 10^5\,\mathrm{cm}^{-1}$ (Poggi et al. 1976, Wong and Shen 1974) $\Delta \sim 4 \times 10^7\,\mathrm{erg/cm^3}$, $Kq^2 \sim 10^4\,\mathrm{erg/cm^3}$. Therefore in all the equations of motion for the parameters $\theta_3$ and $\theta_4$ we can assume that $q = 0$. This cannot be done in the equations for $\theta_1$ and $\theta_2$ involving no large parameter $\Delta$.

Note here that the mode corresponding to biaxial fluctuations is not literally hydrodynamical, i.e., its frequency does not become equal to zero at $q \to 0$. Although the wavelengths corresponding to the parameter $\Delta$ are small ($\sim 300$ Å), they are still larger than the intermolecular distances and therefore the relevant fluctuations can be studied macroscopically.

Thus the scattering cross section for light polarizations in the equatorial plane is determined as follows:

$$I(q, \omega) = \frac{M^2}{32\pi^3}\omega^4 \frac{T}{\nu\omega^2 + \Delta^2}, \tag{42}$$

where $\nu$ is a certain combination of biaxial viscosity coefficients (Kamenskii and Kats 1970). In the general geometry the biaxial scattering cross section is determined also by the correlator $\langle|\theta_3|^2\rangle$ and the cross correlators $\langle\theta_2\theta_4^*\rangle$, and $\langle\theta_1\theta_3^*\rangle$. The relevant formulae can easily be derived.

The main qualitative effects are: (1) at longitudinal scattering the line has a non-Lorenz form; (2) at biaxial scattering the width of the line is practically independent of the transferred momentum.

The both conclusions are essentially different from the results obtained for the scattering due to the director fluctuations where the form of the line is Lorenz and the width of the line is $\sim q^2$.

In conclusion, let us give the dominating types of scattering in certain standard settings of the experiment (Arakelian et al. 1981)

| Phase | Orientation | Angle | | Type of scattering |
|---|---|---|---|---|
| | | $\mathbf{P}; \mathbf{n}$ | $\mathbf{P}; \mathbf{P}'$ | |
| nematic | planar | 0 | 0 | longitudinal |
| nematic | planar | 0 | 90 | transversal |
| nematic | planar | 90 | 0 | longitudinal + biaxial |
| nematic | homeotropic | 90 | 90 | transversal |
| nematic | homeotropic | 0 | 90 | biaxial |
| nematic | homeotropic | 90 | 0 | longitudinal + biaxial |
| isotropic | – | – | 0 | longitudinal + biaxial |
| nematic | – | – | 90 | biaxial |

## 2. *Light scattered by smectics*

### 2.1. *Nematic – smectic-A transition*

#### 2.1.1. *General properties*

Liquid crystals have a great variety of phase diagrams. In this section we shall study light scattering in smectic phases. The simplest out of the smectic phases is smectic-A. The molecules of these liquid crystals are of a prolate form and have the orientation order, the mass centers of these molecules are positioned on equidistant planes (one-dimensional translational order). Then the average molecular orientation (the director $\boldsymbol{n}$) is orthogonal to these planes. The orthogonality and equidistance conditions impose essential restrictions on possible fluctuations of the order parameter of smectics-A. Actually, if alongside with the components of $\boldsymbol{n}$ we introduce one more variable $u_z(x, y)$ describing a displacement of a smectic layer in the direction perpendicular to it ($z$-axis), it becomes evident that the deformations of $\boldsymbol{n}$ are determined by the following relations:

$$\delta n_x = -\partial u_z/\partial x\,, \qquad \delta n_y = -\partial u_z/\partial y\,. \tag{43}$$

The equidistance condition imposes the requirement:

$$\operatorname{rot} \boldsymbol{n} = 0\,. \tag{44}$$

For small fluctuations of $\delta\boldsymbol{n}$, hence it follows that $\delta\boldsymbol{n} = 0$, with the exception of the case $q_z = 0$. Thus only fluctuations with wave vectors parallel to smectic layers are permissible.

Keeping in mind the above remarks, we can write the free energy of the deformation of smectics-A up to the quadratic terms with respect to $\delta\boldsymbol{n}$, $u_z \equiv u$ and the density variation $\rho$ (Stephen and Straley 1974):

$$F = \tfrac{1}{2}\mu\left[\left(\delta n_x + \frac{\partial u}{\partial x}\right)^2 + \left(\delta n_y + \frac{\partial u}{\partial y}\right)^2\right] + \tfrac{1}{2}A(\delta\rho)^2 + \tfrac{1}{2}B\left(\frac{\partial u}{\partial z}\right)^2 + C\delta\rho\frac{\partial u}{\partial z} + \tfrac{1}{2}K_1(\operatorname{div}\delta n)^2\,. \tag{45}$$

In this expression $\mu$ is the elastic modulus describing deviations of molecules from the normal to the layers (elsewhere we shall regard molecules as strictly orthogonal to the layers, and accordingly, $\mu = \infty$), $A$, $B$ and $C$ are compressibilities. It is more convenient to rewrite this expression with the orthogonality and equidistance conditions:

$$F = \tfrac{1}{2}A(\delta\rho)^2 + \tfrac{1}{2}B\left(\frac{\partial u}{\partial z}\right)^2 + C\delta\rho\frac{\partial u}{\partial z} + \tfrac{1}{2}K_1\left[\frac{\partial^2 u}{\partial x^2} + \frac{\partial^2 u}{\partial y^2}\right]^2. \tag{46}$$

In Fourier components we get:

$$F = \tfrac{1}{2}\sum_q \left[A|\rho_q|^2 + (Bq_z^2 + K_1 q_\perp^4)|u_q|^2 + \mathrm{i}Cq_z u_q\rho_{-q}\right]. \tag{47}$$

Hence it is easy to obtain correlators necessary to calculate the scattering intensity,

$$\langle|\rho_q|^2\rangle = \frac{T}{A}\frac{Bq_z^2 + K_1 q_\perp^4}{(B - C^2/A)q_z^2 + K_1 q_\perp^4}\,, \qquad \langle|u_q|^2\rangle = \frac{T}{(B - C^2/A)q_z^2 + K_1 q_\perp^4}\,,$$

$$\langle u_q\rho_{-q}\rangle = \frac{\mathrm{i}Cq_z}{A}\frac{T}{(B - C^2/A)q_z^2 + K_1 q_\perp^4}\,. \tag{48}$$

Note that from the above-given formulae follows the logarithmic divergence of the displacement correlators with the distance

$$\langle u^2 \rangle \sim \frac{T}{[K_1(B - C^2/A)]^{1/2}} \ln \frac{R}{a},$$

where $R$ is a size of the system, $a$ is the intermolecular distance. This is in agreement with the well-known Landau–Peierls theorem (see Landau and Lifshitz 1976) on the impossibility of the one-dimensional translational order in three-dimensional systems.

Light scattering is determined by the dielectric permeability fluctuations. We can relate easily these fluctuations with fluctuations of the introduced quantities $\delta\rho$ and $u$:

$$\delta\epsilon_{xx} = a_\perp \delta\rho + b_\perp \frac{\partial u}{\partial z}, \qquad \delta\epsilon_{xz} = -\epsilon_a \frac{\partial u}{\partial x},$$
$$\delta\epsilon_{zz} = a_\parallel \delta\rho + b_\parallel \frac{\partial u}{\partial z}, \qquad \delta\epsilon_{yz} = -\epsilon_a \frac{\partial u}{\partial y}. \tag{49}$$

Here $\epsilon_a = \epsilon_\parallel - \epsilon_\perp$ is the anisotropy of the dielectric permeability, $a$ and $b$ are certain coefficients.

Most intensive is the scattering at such polarizations where the fluctuations $\delta\epsilon_{xz}$ and $\delta\epsilon_{yz}$ are relevant. The formulae for correlators yield:

$$\langle |\delta\epsilon_{xz}(q)|^2 \rangle = \epsilon_a^2 T q_x^2 / (\tilde{B} q_z^2 + K_1 q_x^4). \tag{50}$$

At $q_z = 0$ we obtain the usual nematic scattering,

$$\langle |\delta\epsilon_{xz}(q)|^2 \rangle = \epsilon_a^2 T / K_1 q_x^2. \tag{51}$$

### 2.1.2. Longitudinal scattering

A specific Landau–Peierls form of the correlator $\langle |u_q|^2 \rangle \sim [\tilde{B} q_z^2 + K_1 q_x^4]^{-1}$ leads to a possibility of observing longitudinal fluctuations of the order parameter in smectics-A. This fact has first been drawn attention to by Liuksiutov (1978). The thing is that the new variable $u$ describes degeneration of the system with respect to translations in the $z$ direction. This degeneration gives rise to an anomalous form of the correlator $\langle |u_q|^2 \rangle$ which is more convenient if it is written as:

$$G_0 = \langle |u_q|^2 \rangle = \frac{T}{K_1} \frac{1}{q_1^2 q_z^2 + q_\perp^4}, \tag{52}$$

where $q_1 \sim q_0 \sim \pi/d$ ($d$ is the distance between smectic layers). The quantity $u$ plays the role of the transversal coordinate (phase). It is convenient to redefine the transversal coordinate,

$$\varphi = u q_0. \tag{53}$$

The role of the longitudinal coordinate (order-parameter modulus) $m$ is played by the amplitude of smectic density modulation,

$$\rho = \rho_0 + m\cos(q_0 z + q_0 u)\,. \tag{54}$$

In accordance with the modulus conservation principle, like in nematics, it is easy to obtain:

$$\delta m = -\tfrac{1}{2} m_0 (\delta\varphi)^2\,. \tag{55}$$

And then the longitudinal correlator is

$$G_1 = \langle \delta m(q)\delta m(-q)\rangle = \frac{T^2 m_0^2 q_0^4}{s\pi K_1^2 q_1}\frac{1}{q_1^2 q_z^2 + q_\perp^4}\,. \tag{56}$$

Thus the longitudinal correlator has the same singular form as the transversal (unlike in nematics, where the longitudinal correlator has a weaker singularity than the transversal correlator). Yet, due to the small factor, longitudinal fluctuations are still weaker than transversal:

$$\frac{G_1}{m_0^2 G_0 q_0^2} \sim \frac{1}{8\pi}\frac{T q_0^2}{K_1 q_1} \sim 0.05$$

($T \sim 350$ K, $K \sim 5\times 10^{-7}$ erg/cm).

To calculate the light scattering cross section it is necessary to relate the quantities $\delta\epsilon$ and $\delta m$:

$$\delta\epsilon = \frac{\partial\epsilon}{\partial\rho}\delta m\cos(q_0 z + q_0 u) + \frac{1}{4}\frac{\partial^2\epsilon}{\partial\rho^2}(\delta m)^2[1 + \cos 2(q_0 z + q_0 u)]\,.$$

Since in the optical range $q_0 \gg \omega/c$ ($c$ is the velocity of light) the contribution of the first term in this expansion is exponentially small due to the oscillating term (thus this is the so-called second order scattering):

$$\langle \delta\epsilon(q)\delta\epsilon(-q)\rangle = \frac{1}{8}\left(\frac{\partial^2\epsilon}{\partial\rho^2}\right)^2 \int \frac{d^3p}{(2\pi)^3} G_1(p+q)G_1(p)$$
$$\approx \frac{1}{2^{11}\pi^3}\left(\frac{\partial^2\epsilon}{\partial\rho^2}\right)^2 \frac{T^4 m_0^4 q_0^8}{K_1^4 q_1^3}\frac{1}{q_1^2 q_z^2 + q_\perp^4}\,. \tag{57}$$

To observe the scattering on the background of a stronger transversal scattering, it is necessary that $q_\perp = 0$ since then $\langle |\delta\epsilon_{xz}|^2\rangle$ from eq. (50) is zero. However there is another contribution to the scattering due to density fluctuations:

$$\langle \delta\rho(q)\delta\rho(-q)\rangle = T/A\,.$$

Estimating $A \sim s^2\rho^{-1}$ ($s$ is the velocity of sound) for the relations of intensities we get:

$$\frac{I_m}{I_\rho} \approx \frac{1}{2^{11}\pi^3}\frac{T^3 \rho q_0^8}{K_1^2 s^2 q_1}\frac{1}{q_z^2}\,. \tag{58}$$

Assuming $s \sim 1.5 \times 10^5$ cm/s, $\rho \sim 1$ g/cm$^3$, $q_0 - q_1 \sim 2 \times 10^7$ cm$^{-1}$, we find that the longitudinal scattering dominates at $q_z^2 \sim 3 \times 10^8$ cm$^{-2}$, which at the wavelengths $\lambda \sim 6000$ Å corresponds to the scattering angles $\lesssim 10°$. The quadratic frequency dependence $\sim \omega^2$ (but not the fourth power as usual) is characteristic of this longitudinal scattering.

### *2.1.3. Nematic – smectic-A transition*

Actually everything that has been said above concerns smectics-A, independently of their closeness to phase transition points. Usually smectics-A become solid at a temperature decrease (i.e., a real solid crystal is formed) and at a temperature increase are melted into the nematic phase. The first of these transitions is a first-order phase transition with a high latent heat and with a high jump of the order parameter, etc. Therefore at this transition there is practically no fluctuational region and we shall not discuss it any longer.

The transition into a nematic is a first-order transition but close to a second-order transition (in certain systems it may be a genuine second-order transition) (McMillan 1971). Therefore in the vicinity of such a phase transition we achieve a region of developed order-parameter fluctuations which can be investigated by light scattering. Let us discuss this transition in more detail, in accordance with the results of de Gennes (1969), Brochard (1973).

Instead of the modulus $m$ and the phase $q_0 u$ of the smectic order parameter it is convenient to introduce a complex order parameter $\psi$. Then instead of eq. (54) we have

$$\rho(r) = \langle \rho \rangle \left\{ 1 + \frac{1}{\sqrt{2}} [\psi \exp(\mathrm{i} q_0 z) + \mathrm{c.c.}] \right\}. \tag{59}$$

In the vicinity of the phase transition the free energy can be expanded in powers of $\psi$ and its gradients. It is also necessary to take into consideration the invariance of the free energy at the simultaneous rotation of the director and smectic layers. This invariance is similar to the gauge symmetry in the superconductivity theory and leads to the expansion of the free energy in the Ginzburg–Landau form:

$$F = F_0 + A|\psi|^2 + B|\psi|^4 + (\nabla + \mathrm{i} q_0 \delta \boldsymbol{n})\psi^* \frac{1}{2M} (\nabla - \mathrm{i} q_0 \delta \boldsymbol{n})\psi + F_{\mathrm{el}}, \tag{60}$$

where $A = a(T - T_{\mathrm{c}})$, and $a$, $B$ are constants, $T_{\mathrm{c}}$ is the transition temperature, $M$ is the "effective mass" tensor, $F_{\mathrm{el}}$ is the elastic Frank energy.

Performing Fourier transformation instead of eq. (47) we arrive at

$$\begin{aligned} F = {} & |\psi_q|^2 \left( A + \frac{q_z^2}{2M_\parallel} + \frac{q_\perp^2}{2M_\perp} \right) + \tfrac{1}{2}|\gamma_q|^2 \left( \frac{\psi_0^2 q_0^2}{M_\parallel} \pm \frac{\psi_0^2 q_0^2}{M_\perp} \frac{q_\perp^2}{q_z^2} \right) \\ & + \tfrac{1}{2}|n_{1q}|^2 \left( \frac{\psi_0^2 q_0^2}{M_\perp} + K_1 q_\perp^2 + \tilde{K}_3 q_z^2 \right) - \frac{q_\perp}{q_z} \gamma_{-q} n_{1q} \frac{\psi_0^2 q_0^2}{M_\perp} \\ & + \tfrac{1}{2}|n_{2q}|^2 \left( \frac{\psi_0^2 q_0^2}{M_\perp} + \tilde{K}_2 q_\perp^2 + \tilde{K} q_z^2 \right). \end{aligned} \tag{61}$$

In this formula to simplify the notation we have introduced the following designations: $\gamma = \partial u/\partial z$ and $M_\parallel$, $M_\perp$ are longitudinal and transversal components of the effective mass, respectively, $n_1$ and $n_2$ are the projections of $\delta\boldsymbol{n}$ on the unit vectors $\boldsymbol{e}_2$ and $\boldsymbol{e}_1$ ($\boldsymbol{e}_2 \perp \boldsymbol{n}_0$ and $\boldsymbol{q}$ but $\boldsymbol{e}_1 \perp \boldsymbol{n}_0$ and $\boldsymbol{e}_2$) $\psi_0$ is the equilibrium value of $\psi$ (at $T < T_c$ we have $\psi_0^2 = -A/B$, but at $T > T_c$: $\psi_0 = 0$) $\tilde{K}_2$ and $\tilde{K}_3$ are renormalized (see below) elastic moduli. From eq. (62) we can in a standard way find all the correlators. We shall not give the respective formulae in order not to overload the exposition (see, e.g., Brochard (1973)). We shall discuss only one fact associated with the appearance of the renormalized elastic moduli $\tilde{K}_2$ and $\tilde{K}_3$. The point is that in the smectic-A phase deformations with $\operatorname{rot}\boldsymbol{n} \neq 0$ are impossible due to the layer equidistance condition. Formally this requirement can be obeyed if in the Frank expansion it is assumed that $\tilde{K}_2 = \tilde{K}_3 = \infty$. The moduli $\tilde{K}_2$ and $\tilde{K}_3$ in the nematic phase are already finite. Yet as the point of a transition into the smectic-A phase is being approached, order-parameter fluctuations affect these moduli and increase them. Physically this increase entails from the fact that fluctuationally there appear clusters of the smectic phase (the cluster size is of the order of the coherence length $\xi$). These clusters cannot be deformed with $\operatorname{rot}\boldsymbol{n} \neq 0$, which leads to an effective increase of the moduli $\tilde{K}_2$ and $\tilde{K}_3$. The exact formulae first derived by de Gennes (1969) confirm this simple assertion,

$$\tilde{K}_2 = K_2 + \frac{1}{24\pi}\frac{Tq_0^2}{M_\perp}\sqrt{M_\parallel/A}\,,$$
$$\tilde{K}_3 = K_3 + \frac{1}{24\pi}\frac{Tq_0^2}{\sqrt{M_\parallel A}}\,. \tag{62}$$

## 2.2. Smectic-C

### 2.2.1. Fluctuating parameters

The above said concerns the solid–smectic-A and nematic–isotropic liquid diagram. However there are often more complicated phase diagrams with the smectic-C phase. Let us first discuss properties of scattering in the smectic-C phase (Saupe 1969).

An undistorted structure of the smectic-C phase is formed by equidistant planes, yet, the molecules are being tilted with respect to the normal. Therefore alongside with the unit vector-director $\boldsymbol{n}$ it is convenient to introduce a vector of the normal $\boldsymbol{\kappa}$. In the undistorted structure $\boldsymbol{\kappa} = \boldsymbol{\kappa}_0$ is parallel to the $z$-axis and $\boldsymbol{n} = \boldsymbol{n}_0$ lies in the $xz$-plane. Thus the following small fluctuations are possible:

$$\kappa_x, \kappa_y, \delta n_x = n_x - n_{0x}, n_y, \delta n_z = n_z - n_{0z}\,.$$

From the condition $\boldsymbol{n}^2 = 1$ follows:

$$\boldsymbol{n}_0\delta\boldsymbol{n} = \delta n_x \sin\theta_0 + \delta n_z \cos\theta_0 = 0\,, \tag{63}$$

where $\theta_0$ is the tilt angle (the angle between $\boldsymbol{n}$ and $\boldsymbol{\kappa}$): $n\kappa = \cos\theta_0$. Hence,

$$\delta n_x = \kappa_x \cos\theta_0 , \quad \delta n_z = -\kappa_x \sin\theta_0 . \tag{64}$$

The equidistance condition, i.e., rot $\boldsymbol{\kappa} = 0$:

$$\frac{\partial \kappa_x}{\partial z} = \frac{\partial \kappa_y}{\partial z} = 0 , \quad \frac{\partial \kappa_x}{\partial y} = \frac{\partial \kappa_y}{\partial x} = 0 . \tag{65}$$

Finally we get six independent variables describing the deformation:

$$a_1 = \frac{\partial n_y}{\partial y}, \quad a_2 = \frac{\partial n_y}{\partial z}, \quad a_3 = \frac{\partial n_y}{\partial x}, \quad a_y = \frac{\partial \kappa_x}{\partial x}, \quad a_5 = \frac{\partial \kappa_y}{\partial y},$$
$$a_6 = \frac{1}{2}\left(\frac{\partial \kappa_x}{\partial y} + \frac{\partial \kappa_y}{\partial x}\right).$$

Hence it follows that for non-polar optically inactive smectic-C phases the elastic energy contains nine elastic constants

$$F = \tfrac{1}{2}\alpha_{11}a_1^2 + \tfrac{1}{2}\alpha_{22}a_2^2 + \tfrac{1}{2}\alpha_{33}a_3^2 + \alpha_{23}a_2a_3 + \tfrac{1}{2}\alpha_{44}a_4^2 + \tfrac{1}{2}\alpha_{55}a_5^2 + \alpha_{45}a_4a_5 + \alpha_{14}a_1a_4 + \alpha_{15}a_1a_5 . \tag{66}$$

If $q_z \neq 0$, from the equidistance condition it follows that $\kappa_{xq} = \kappa_{yq} = 0$. Then the main contribution to the light scattering comes from the fluctuations $n_y$:

$$\langle |n_{yq}|^2 \rangle = T/(\alpha_{11}q_y^2 + \alpha_{22}q_z^2 + \alpha_{33}q_x^2 + 2\alpha_{23}q_xq_z) . \tag{67}$$

If $q_z = 0$, fluctuations $\kappa_{xq} \neq 0$ and $\kappa_{yq} \neq 0$ become permissible. These fluctuations are similar to the ones in smectics-A.

The light scattering is determined by the dielectric permeability fluctuations. If to neglect weak biaxiality of the smectic-C phase, we get:

$$\begin{aligned} &\delta\epsilon_{xx} = 2\epsilon_a \sin\theta_0 \cos\theta_0 \kappa_x, && \delta\epsilon_{yy} = 0 , \\ &\delta\epsilon_{zz} = -2\epsilon_a \sin\theta_0 \cos\theta_0 \kappa_x, && \delta\epsilon_{xy} = \epsilon_a \sin\theta_0 n_y , \\ &\delta\epsilon_{xz} = \epsilon_a \cos 2\theta_0 \kappa_x, && \delta\epsilon_{yz} = \epsilon_a \cos\theta_0 n_y . \end{aligned} \tag{68}$$

Note that at the incident and scattered light polarizations, where the components of $\delta\epsilon_{xy}$ and $\delta\epsilon_{yz}$ are important, the fluctuations are of a purely nematic character, i.e., $\sim 1/q^2$. This explains in particular higher opaqueness of smectics-C in comparison with smectics-A, where nematic fluctuations occur only at $q_z = 0$.

### 2.2.2. Smectic-A–smectic-C transition

In phase diagrams of many liquid crystals there is a point of transition from smectic-A into smectic-C. Let us discuss properties of the light scattering in the vicinity of this point.

De Gennes (1974) proposed within the Landau theory a model describing this transition. As the order parameter we shall introduce the complex quantity:

$$\chi = \chi_x + \mathrm{i}\chi_y = \sin\theta \exp(\mathrm{i}\varphi)\,, \tag{69}$$

where $\theta$ is the tilt angle and $\varphi$ is the azimuthal angle. In the A-phase the angle $\theta = 0$. If the smectic-C–smectic-A transition is close to a second-order phase transition, in the vicinity of this transition $\chi$ is small and the free energy can be expanded in series:

$$F = F_\mathrm{A} + a|\chi|^2 + \tfrac{1}{2}b|\chi|^4\,, \tag{70}$$

where $F_\mathrm{A}$ is the free energy of the A-phase, $a$, $b$ are the expansion parameters which can in principle be temperature dependent. Additional gradient terms describing the dependence of the order parameter on the coordinates should be added to this formula. This energy is analogous to the elastic energy of nematics. In our case we can formally assume that $\boldsymbol{n} = \chi_x, \chi_y, 1$. Then,

$$F_\mathrm{el} = \tfrac{1}{2}K_1\left(\frac{\partial\chi_x}{\partial x} + \frac{\partial\chi_y}{\partial y}\right)^2 + \tfrac{1}{2}K_2\left(\frac{\partial\chi_x}{\partial y} - \frac{\partial\chi_y}{\partial x}\right)^2 + \tfrac{1}{2}K_3\left[\left(\frac{\partial\chi_x}{\partial z}\right)^2 + \left(\frac{\partial\chi_y}{\partial z}\right)^2\right]. \tag{71}$$

From the order parameter symmetry it follows that in the model under study the phase transition is similar to the $\lambda$ transition in $^4$He. Hence, it follows that $|\chi| \sim (T_\mathrm{c} - T)^\beta$ in the C-phase, and $\langle|\chi_q|^2\rangle \sim (T - T_\mathrm{c})^{-\gamma}$ in the A-phase, where $\beta = \frac{1}{3}$ and $\gamma = 1.3$.

Experimentally the index $\gamma$ can be determined by light scattering. The components of $\delta\epsilon_{xz}$ and $\delta\epsilon_{yz}$, as is clear from the above formulae, are proportional to $\chi_q$, and the scattering intensity, as the transition point is being approached, increases according to the law $(T - T_\mathrm{c})^{-\gamma}$. Certainly a real smectic-C–smectic-A transition may have a more complicated character, for example, it may be accompanied by variations of the interlayer spacing, etc. This problem has not been completely solved so far.

### 2.3. Lifshitz point

Chen and Lubensky (1976) have proposed a phenomenological model describing all the three nematic, smectic-A and smectic-C phases. They have obtained the lines of the nematic–smectic-A and nematic–smectic-C phase transitions. In the phase diagram there is also a triple point – a point of intersection of both lines of the phase transitions. The order parameter fluctuations lead to renormalization of the elastic moduli. For the nematic–smectic-C transition all the three elastic moduli: $\tilde{K}_1$, $\tilde{K}_2$, $\tilde{K}_3 \sim \zeta^2$ diverge (where $\zeta$ is the correlation length). At the nematic–smectic-A transition only $\tilde{K}_2, \tilde{K}_3 \sim \zeta$ diverge (the exact formulae have been given above). Finally in the vicinity of the triple point, $\tilde{K}_1, \tilde{K}_2 \sim \ln \zeta$, where $\tilde{K}_3 \sim \zeta$. Light scattering in each phase is determined by the formulae of this section given above. In the vicinity of the transition lines it is necessary to take into account renormalization of the elastic modulus.

Note that in the simplified model of the nematic–smectic-C phase transition proposed by de Gennes (1974) the elastic modulus diverge according to another law $\tilde{K}_1, \tilde{K}_2, \tilde{K}_3 \sim \zeta^{3/2}$ and in the model proposed by Chu and McMillan (1977) $\tilde{K}_1, \tilde{K}_2, \tilde{K}_3 \sim \zeta$. Which of the models adequately describes real systems is not clear so far. Experimental data pertaining to this problem are not unambiguous (Chandrasekhar 1980). The light scattering studied in this section provide additional possibilities for investigating this problem. It is possible that depending on the molecular structure and other characteristics of the substance, the phase transition between smectics-A and -C and a nematic is described by different models.

When molecules forming the smectic-C phase are optically active, the ground state of the liquid crystal may be inhomogeneous. Then there arises a chiral C* phase. The introduced order parameter describes possible transitions between the smectic-A, -C and -C* phases. Such a general investigation of the problem has first been carried out by Michelson (1977). Within the framework of the Landau theory the free energy can be expanded in series over $\chi$ and its derivatives. By means of the symmetry, let us write down the free energy

$$F = \int d^3r \left[ A_1|\chi|^2 + iA_2\left(\chi\frac{\partial\chi^*}{\partial z} - \chi^*\frac{\partial\chi}{\partial z}\right) + A_3\left|\frac{\partial\chi}{\partial z}\right|^2 + A_4|\chi|^4 \right]. \tag{72}$$

Here $A_1 = a(T - T^*)$, $a$, $A_2$, $A_3$, $A_4$, $T^*$ are constants. The second term, i.e., the so-called Lifshitz invariant, describes a possibility of the existence of the C* phase characterized by the non-zero quantity $\partial\chi/\partial z$.

To determine the transition temperature $T_c$ and the type of ordering below $T_c$, let us study the harmonic part of the free energy. In terms of Fourier components:

$$F^{(2)} = \sum_q \alpha_q|\chi_q|^2 , \tag{73}$$

where

$$\alpha_q = A_1 + 2A_2q + A_3q^2 . \tag{74}$$

The minimum of eq. (73) occurs at $q = q_c = -A_2/A_3$, then $\alpha_{q_c} = A_1 - A_2^3/A_3$. The ordering below $T_c$ is determined by the order parameter $\chi(z) = \chi_{q_c}\exp(iq_cz)$ describing the director precession $\boldsymbol{n}$ with respect to the $z$-axis with the wave vector $q_c$. The transition temperature is determined by the condition $\alpha_{q_c} = 0$. At $A_2 = 0$ we deal with a transition directly from the smectic-A into smectic-C phase and the parameter $T^*$ is the transition temperature. In the presence of a magnetic field the term $-\frac{1}{2}\chi_a(\boldsymbol{Hn})^2$ ($\chi_a$ is anisotropy of the diamagnetic susceptibility) is usually added to the free energy. Diagonalization of the complete free energy yields the following spectrum of soft modes:

$$\omega_q^{\pm} = A_1 + A_3 q^2 - \tfrac{1}{4}\chi_a H^2 \pm \tfrac{1}{4}(\chi_a^2 H^4 + 64 A_2^2 q^2)^{1/2} . \tag{75}$$

The softening is now dependent on the magnitude of the magnetic field. If $H < H_L$ ($H_L$ designates the field for the Lifshitz point), where $H_L = (8A_2^2/A_3|\chi_a|)^{1/2}$, then $\omega_{\min}$ is achieved at $q_0 = q_c(1 - H^4/H_L^4)^2$. If $H > H_L$, then $\omega_{\min}$ occurs at $q = 0$. The transition temperature $T_\lambda$ from the smectic-A phase is:

$$T_\lambda = \begin{cases} T^* + \Delta T\left(1 + \dfrac{\chi_a H^2}{|\chi_a| H_L^2}\right)^2, & H \leqslant H_L , \\ T^* + 2\left(1 + \dfrac{\chi_a}{|\chi_a|}\right)\Delta T \dfrac{H^2}{H_L^2}, & H \geqslant H_L , \end{cases} \tag{76}$$

where

$$\Delta T = T_c - T^* .$$

At $H < H_L$ we observe a transition from the A into the C* phase and at $H > H_L$, a second-order phase transition from the A phase into the usual smectic-C phase. The point $H = H_L$ and $T = T_L = T_\lambda(H_L)$ is the triple point where the disordered (smectic-A), homogeneously ordered (smectic-C) and helicoidal (smectic-C*) phases coexist. This point is sometimes called the critical Lifshitz point (Hornreich et al. 1975). The light scattering near the Lifshitz point can be described like above (for nematic and smectic phases) by the formulae of this section.

## 3. *Scattering by cholesterics*

### 3.1. *Scattering by ordered phase*

In the preceding sections the influence of the closeness to a phase transition upon scattering in nematics and smectics has been studied. In this section we shall deal with inelastic light scattering close to the point of a transition into the cholesteric phase which like in nematics is accompanied by variations of light frequency, direction of its propagation and its polarization properties. However, the helical ordering of molecules in a cholesteric affects light scattering in them.

Due to the closeness of phase transitions in liquid crystals to second-order transitions, like in other varieties of liquid crystals, in cholesterics near the transition points there must be observed a strong light scattering. Here it should be pointed out that the theory of inelastic scattering develops with the multiple scattering neglected, therefore to check experimentally the validity of its predictions it is required that certain restrictions on the experimental setting conditions should be obeyed. Therefore experiments on inelastic scattering prove to be more complicated than on "optical" scattering since optical characteristics are in principle due to multiple scattering. In particular, when one uses thin layers where multiple scattering is not important, to interpret the results of scattering it is necessary to get rid of the influence on surfaces on the pretransitional behaviour of a cholesteric which cannot be assumed to be weak.

To describe inelastic scattering, and in particular its angular, frequency and polarization characteristics we must know the excitation spectrum or modes for a cholesteric in the vicinity of the phase-transition point. As is known, (De Gennes 1974, Brazovsky and Dmitriev 1975) excitation modes in cholesterics have a fairly complex structure and they are sufficiently numerous, namely, there are five different modes. They allow for a simple physical interpretation at special directions of the wave vector of the respective modes and in particular if the wave vector $\boldsymbol{q}$ is directed along the cholesteric axis. Therefore keeping in mind the momentum conservation law in the scattering $\boldsymbol{K}' - \boldsymbol{K} = \boldsymbol{q}$ (where $\boldsymbol{q}$ is the scattering wave vector) we can easily understand that scattering experiments allow for a simple physical interpretation if the scattering wave vector $\boldsymbol{q}$ is parallel to the cholesteric axis. As for the scattered light frequency variations, we can neglect them since they are smaller than the frequency of light. It enables us to assume that $\boldsymbol{K}' = \boldsymbol{K}$ and regard light scattering by cholesterics as quasielastic, and for the description of the scattering cross section to employ eq. (3) inserting into it fluctuational additions to the dielectric permeability tensor. Then the scattering cross section on fluctuations is of the form:

$$I(\boldsymbol{q}, \omega) = \left(\frac{\omega^2}{4\pi}\right)^2 \int \mathrm{d}\boldsymbol{r}_{12} < \boldsymbol{P}'^* \delta\epsilon(r_1)\boldsymbol{P} \cdot \boldsymbol{P}'\delta\epsilon(r_2)\boldsymbol{P}^* \rangle \,, \tag{77}$$

where only quadratic quantities over fluctuational additions are retained since the thermodynamic averaging procedure leads to the fact that all the terms linear with respect to $\delta\epsilon$ vanish.

### *3.2. Fluctuations and angular distribution of scattering for ordered phases*

Thus, as follows from eq. (77), the fluctuational scattering cross section is determined by the correlation function of the dielectric permeability fluctuations of a cholesteric. The two reasons can cause fluctuations of $\epsilon$: (a) fluctuations of $\epsilon$ due to small local variations of temperature, density, etc., and (b) fluctuations of $\epsilon$ due to the director orientation. The second reason largely affects fluctuations of dielectric properties since small fluctuations of the director orientation in the vicinity of the phase transition point are associated with much smaller energy variations than, for example, fluctuations of temperature and density. This fact is well known for nematics (De Gennes 1974), and naturally the situation in cholesterics become similar since locally cholesterics and nematics are identical.

Thus, relating fluctuations of $\boldsymbol{\epsilon}$ only with fluctuations of the director orientation $\boldsymbol{n}$ and using the expression,

$$\epsilon_{\alpha\beta} = \epsilon_{\perp}\delta_{\alpha\beta} + (\epsilon_{\parallel} - \epsilon_{\perp})n_{\alpha}n_{\beta}\,, \tag{78}$$

it is easy to find the fluctuational addition to $\boldsymbol{\epsilon}$.

Inserting the director of the form $\boldsymbol{n} = \boldsymbol{n}_0 + \delta\boldsymbol{n}$, where $\boldsymbol{n}_0$ is the equilibrium local value of the director and $\delta\boldsymbol{n}$ is a small fluctuational addition perpendicular to $\boldsymbol{n}_0$, for the quantity $\boldsymbol{P}'^*\delta\boldsymbol{\epsilon}\boldsymbol{P}$ we shall find:

$$\boldsymbol{P}'^*\delta\boldsymbol{\epsilon}\boldsymbol{P} = \epsilon_{\mathrm{a}}[(\boldsymbol{P}'^*\delta\boldsymbol{n})(\boldsymbol{n}_0\boldsymbol{P}) + (\boldsymbol{P}'^*\boldsymbol{n}_0)(\delta\boldsymbol{n}\boldsymbol{P})]\,. \tag{79}$$

Then it is convenient to expand fluctuations of the director $\delta\boldsymbol{n}$ over eigenmodes of a cholesteric (Brazovsky and Dmitriev 1975, Fan et al. 1970). For this purpose we can use the Fourier transforms:

$$\boldsymbol{n}(\boldsymbol{q}) = \int \boldsymbol{n}(\boldsymbol{r})\,\mathrm{e}^{\mathrm{i}\boldsymbol{q}\boldsymbol{r}}\,\mathrm{d}\boldsymbol{r}\,. \tag{80}$$

It turns out that for each $\boldsymbol{q}$ there are two orientational modes strongly interacting with light. For $\boldsymbol{q}$ parallel to the cholesteric axis, these modes are especially simple and allow for a clear physical interpretation. One of them corresponds to pure twist and can be represented in the form:

$$\begin{aligned} n_x &= \cos(q_0 z + u) \simeq n_x^0 - u \sin q_0 z\,, \\ n_y &= \sin(q_0 z + u) \approx n_y^0 + u \cos q_0 z\,, \\ n_z &= 0\,, \end{aligned} \tag{81}$$

where $u = u_0\,\mathrm{e}^{\mathrm{i}qz}$ and as usual the cholesteric axis ($q_0 = 2\pi/p$, $p$ is the pitch) is chosen as the $z$-axis.

This type of deformation gives a contribution only to off-diagonal components $\epsilon_{xy}$ of the tensor $\boldsymbol{\epsilon}$

$$\begin{aligned} \delta\epsilon_{xy} &= \epsilon_{\mathrm{a}}(n_x^0\delta n_y + n_y^0\delta n_x) \\ &= \epsilon_{\mathrm{a}}\cos(2q_0 z)u_0\,\mathrm{e}^{\mathrm{i}qz}\,. \end{aligned} \tag{82}$$

From eqs. (77) and (82) it follows that for the mode under study the maximum of scattering is achieved if

$$\boldsymbol{K}' - \boldsymbol{K} = \boldsymbol{q} \pm 2\boldsymbol{q}_0\,. \tag{83}$$

The scattering intensity is determined by the mean squared amplitude $\langle|u_0(q)|^2\rangle$ which expression has the form:

$$\langle u_0(q)|^2\rangle = T/K_2q^2\,. \tag{84}$$

Expression (84) diverges at $q \to 0$ and, as follows from eq. (83) the maximum of fluctuational inelastic scattering corresponds to the region of Bragg peaks in elastic scattering. Let us consider now the scattering on the second mode strongly interacting with light. It can be called an "umbrella mode" since the director goes out of the plane perpendicular to $z$ and for $\boldsymbol{q}$ along the cholesteric axis it is governed by the relations:

$$n_x = \cos q_0 z \cos v \sim n_x^0 ,$$

$$n_y = \sin q_0 z \cos v \sim n_y^0 , \tag{85}$$

$$n_z = \sin v \sim v , \tag{86}$$

where $v = v_0 \, e^{iqz}$. This mode contributes to the components of $\epsilon$ identically equal to zero in an undistorted structure of cholesterics.

It is clear from eqs. (77), (85) and (86) that now the maximum of scattering corresponds to the condition:

$$\boldsymbol{K}' - \boldsymbol{K} = \boldsymbol{q} \pm \boldsymbol{q}_0 , \tag{87}$$

and the mean squared amplitude $\langle |v_0(q)|^2 \rangle$ is equal to

$$\langle |v_0(q)|^2 \rangle = T/(K_3 q_0^2 + K_1 q^2) , \tag{88}$$

and does not diverge at $q \to 0$, though it achieves the maximum. The position of the maximum corresponds to a Bragg peak with a period twice exceeding the one of an undistorted cholesteric.

The difference in angular distributions of quasielastic scattering on the twist and "umbrella" modes is quite clear. Actually the twist mode retains the symmetry of an undistorted cholesteric. Therefore the angular position of the quasielastic scattering maxima is determined by the symmetry of an undistorted structure and coincides with the position of Bragg peaks in elastic scattering. The "umbrella" mode breaks the static symmetry of a cholesteric and in quasielastic scattering there springs up a Bragg peak due to fluctuations associated with a difference of the dynamical symmetry of a cholesteric from its static symmetry. The study of fluctuations with $\boldsymbol{q}$ non-parallel to the optical axis complicates the picture of scattering, in particular, the conditions for scattering maxima is now the relation $\boldsymbol{K}' - \boldsymbol{K} = \pm n\boldsymbol{q}_0$ where $n$ is an integer and the scattering intensity, however, rapidly decreases with the increase of $n$ (Veshunov 1979). The temperature dependence of quasielastic scattering below the point of transition into the isotropic phase is completely analogous to the corresponding dependence for nematics (de Gennes 1974) which is experimentally confirmed by Chatelain (1948).

### *3.3. Scattering in isotropic phase*

Let us now study light scattering in an isotropic liquid in the vicinity of a transition into the cholesteric phase. Like below the transition point (with respect to temperature) the main source of scattering are fluctuations of molecular orientations. Above the transition point these fluctuations are determined by the order parameter fluctuations $s$. As we have seen above, the order-parameter fluctuations increase as the temperature is approaching the transition temperature from above. The light scattering intensity with the momentum transfer $\boldsymbol{q}$ is proportional to the Fourier transform of the correlation function of the dielectric permeability fluctuations (77):

$$\delta\epsilon_{\alpha\beta} = MQ_{\alpha\beta}\,,$$

where $M$ is the proportionality constant. Therefore now the expression for the scattering cross section is

$$I(\boldsymbol{q},\omega) = \left(\frac{\omega^2}{4\pi}\right)^2 M^2 \int \mathrm{d}\boldsymbol{r}\, \exp[\mathrm{i}(\boldsymbol{K}-\boldsymbol{K}')\boldsymbol{r}] \times \langle Q_{\alpha\beta}(\boldsymbol{r})Q_{\gamma\delta}(0)\rangle P^*_\alpha P_\gamma P'_\beta P'^*_\delta\,. \tag{89}$$

As is clear from eq. (89), the order-parameter components involved in the expression for the cross section are dependent on the polarization of the incident and detected light. Thus the light scattering intensity is determined by the Fourier transform of the correlation function of the order-parameter fluctuations. Employing the explicit form of the expression for the free energy of the cholesteric phase (De Gennes 1974) and the equipartition theorem for the correlators of eq. (89) we get:

$$\begin{aligned}
\langle|Q_{xx}+Q_{yy}|^2\rangle &= 2T/3A(1+\zeta_1^2q^2+\tfrac{2}{3}\zeta_2^2q^2)\,,\\
\langle|Q_{xx}-Q_{yy}|^2\rangle = 4\langle|Q_{xy}|^2\rangle &= \frac{2T(1+\zeta_1^2q^2)}{A[(1+\zeta_1^2q^2)^2-4\zeta_1^4q_0^2q^2]}\,,\\
\langle|Q_{xy}|^2\rangle = \langle|Q_{yz}|^2\rangle &= \frac{\tfrac{1}{2}T(1+\zeta_1^2q^2+\tfrac{1}{2}\zeta_2^2q^2)}{A[(1+\zeta_1^2q^2+\tfrac{1}{2}\zeta_2^2q^2)^2-\zeta_1^4q_0^2q^2]}
\end{aligned} \tag{90}$$

(Stephen and Straley 1974, Zeldovich and Tabiryan 1978). In this formula $\zeta_1$ and $\zeta_2$ are coherence lengths determined by the ratio of the elastic constants. In the one-constant approximation, $\zeta_1 = \zeta_2$. In the mean-field approximation these coherence lengths are proportional to $A^{-1/2} \sim \zeta_0(T/T^* - 1)^{-1/2}$ where $\zeta_0$ is the microscopic length determined by the radius of the intermolecular forces.

It is evident from eq. (90) that the scattering intensity as the transition point $T_c$ is being approached, increases as $(T - T^*)^{-1}$ since the whole expression $\sim A^{-1}$. From the expression for $\langle|Q_{xy}|^2\rangle$ it is clear that at fixed $\zeta_1$ the scattering maximum corresponds to $q = 0$ if $\zeta_1 q_0$ is smaller than $\frac{1}{2}$ and the scattering cross section behaves like in nematics. If $\zeta_1 q_0 > \frac{1}{2}$, the scattering maximum corresponds to the finite angle determined by the expression $\zeta_1^2 q^2 = 2\zeta_1 q_0^{-1}$. In this case in scattering there appears a wide peak resembling a Bragg peak in the ordered phase.

As has been pointed out by de Gennes (1971), the temperature $T^{**}$ for which $\zeta_1 q_0 = 1$ corresponds to absolute instability of the isotropic phase with respect to cholesteric fluctuations. The transition then is a first-order transition if $\zeta_1(T_c)q_0 < 1$ and in this case $T_c > T^{**}$. Apparently in most cases $\zeta_1(T_c)q_0 < \frac{1}{2}$ and the helical fluctuations in the isotropic phase are rather weak. The case $T_c < T^{**}$ apparently corresponds to the existence of the so-called blue phase between the isotropic and cholesteric phases.

As for the angular dependence of the scattering, then at $\zeta_1 q_0 < \frac{1}{2}$ it should be weak since $\zeta q \sim 0, 1$ like for the scattering in the isotropic phase of a nematic (Stinson and Litster 1970).

### 3.4. *Inelastic scattering*

Like in nematics attenuation of fluctuations in cholesterics is of a relaxational character. For each mode the relaxation is described by its own time $\tau_\alpha$ (width of the scattering line $\Gamma_\alpha$).

Let us give these times for an ordered phase. So for the twist mode at $\boldsymbol{q}$ directed along the cholesteric axis, the respective relaxation time is

$$1/\tau_t = K_2 q^2/\gamma_1 \,. \tag{91}$$

For the "umbrella" mode, also at $\boldsymbol{q}$ along the cholesteric axis, the relaxation time is described by the expression:

$$\tau_u^{-1} = \frac{(\alpha_3 + \alpha_4 + \alpha_5)(K_3 q_0^2 + K_1 q^2)}{\gamma_1(\alpha_4 + \alpha_5) - \gamma_2 \alpha_2}, \tag{92}$$

where $\gamma$ and $\alpha_i$ are Lesli coefficients.

In accordance with the expressions for relaxation times, the scattered light spectral distribution by each mode is of the form:

$$I = \frac{2/\tau_\alpha}{\omega^2 - \tau_\alpha^{-2}}, \tag{93}$$

where for the "umbrella" mode and twist mode the expressions for $\tau_\alpha$ are given by eqs. (91) and (92).

Above the transition point, i.e., in the isotropic phase, the scattering line width depends on the relaxation of the order parameter $s$, then for long-wave fluctuations the relaxation rate in the vicinity of the transition point decreases since $1/\tau_s \sim (T - T^*)$. In this case the scattered light spectral distribution is determined by eq. (93) if to insert into it the given above value of $\tau_s$.

For shorter-wave fluctuations the relaxation of the order parameter is described by more complicated expressions since in this case there is a coupling of orientational and shear modes. As a result, the correlation function of the order parameter and the respective angular and frequency characteristics of the

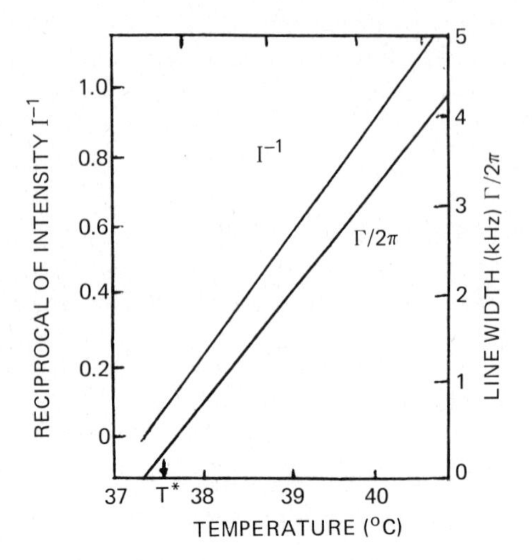


Fig. 2. Temperature dependence of the scattering line width $\Gamma$ on the inverse scattering intensity $I^{-1}$ (upper curve) obtained in accordance with measurements in the isotropic phase of CEEEC cholesteryl 2- (2-Ethoxyethoxy) ethyl carbonate.

scattering are described by more complicated expressions:

$$I_{xz}(q, \omega) = \frac{2\Gamma(4\rho^2\omega^2 + \eta_0\eta' q^4)}{(2\rho\omega^2 - \Gamma\eta_0 q^2) + \omega^2(2\rho\Gamma + \eta' q^2)^2}, \qquad \Gamma = \frac{1}{\tau_s}, \tag{94}$$

where $\eta'$ is the shear viscosity, $\eta_0$ is the effective shear viscosity for flows in capillars, $\rho$ is the density. In the limit of small $q$, $\eta_0 q^2 \ll 2\rho\Gamma$, expression (94) gives a frequency distribution of the form (93) with $\tau_s$. For short-wave fluctuations $\eta' q^2 \gg 2\rho\Gamma$ the distribution is of the same form but with the relaxation time $\tau_s' = (\eta'/\eta_0)\tau_s$. In the general case of arbitrary wavelengths of fluctuations the scattered light frequency distribution is not described by a simple Lorenz curve and relation (94) should be applied.

The width of the Reileigh scattering line in the isotropic phase of a cholesteric CEEEC has been measured by Yang (1972) and it proved to be of the order of 1 kHz (see fig. 2), which is much smaller than the appropriate value for a nematic MBBA (Stinson and Litster 1970). As for the temperature dependence of the line width, it is analogous to the case of MBBA and is sufficiently well described by the linear dependence on $T_c - T^*$, which is in agreement with the theoretical dependence obtained in the self-consistent field approximation. Angular and frequency characteristics of the scattering have been experimentally studied for an ordered phase (Duke and Dupre 1977) and a good agreement with the theory has been obtained.

### *3.5. On the observation of fluctuation scattering*

Summing up the study of quasielastic light scattering in cholesterics close to the phase-transition point, let us stress again that, on the one hand, these measurements give a lot of information on liquid crystals but are rather complicated for experiment and interpretation, on the other hand. Actually, the pretransitional light scattering is most strong on fluctuations of director orientations. Then the basic contribution to the scattering comes either from long-wave fluctuations or from fluctuations with the wave vector close to the pitch. Measurements of the scattering intensity (see eqs. (80) and (84)) provide information on elastic constants of a cholesteric. As for spectral scattering characteristics, they also contain important physical information on parameters of liquid crystals. The scattering line width is determined by slow relaxational modes of a cholesteric. Therefore measurements of the line width give information on the relation between elastic constants and viscosity coefficients (see eqs. (91) and (94)). Since in liquid crystals in the general case there are eight viscosity coefficients (and for incompressible liquid crystals – five), here we meet with difficulties in the interpretation of even authentic results of the measurements. In addition to the enumerated parameters of a cholesteric temperature measurements provide information on the order parameter and its temperature behaviour and measurements in the isotropic phase enable us to obtain temperature dependences of correlation lengths, i.e., information most valuable from the viewpoint of the study of phase transitions and the establishing of their nature. The difficulties encountered at the interpretation of measurement results are the reason for the fact that the amount of experimental works devoted to light scattering in cholesterics in the vicinity of phase-transition points is fairly limited and there is a wide field for experimental research.

It should also be noted that the application of external electric and magnetic fields on a cholesteric, like in nematics (Stephen and Straley 1974), leads to the suppression of orientational director fluctuations and hence to the decrease of the quasielastic scattering intensity. However, it should be borne in mind that the magnetic field, for example, suppresses longer-wave fluctuations more than the magnetic coherence length $\zeta(H)=(K/\chi_a)^{1/2}H^{-1}$. It means that with the exception of forward scattering, practically only sufficiently strong fields ($H > 10^3$ G) may affect light scattering. The effect of fluctuation suppression in cholesterics has been experimentally observed by Belyaev et al. (1979).

We have studied the influence on light scattering in cholesterics of only so-called uniaxial fluctuations, i.e., fluctuations of the director most strongly manifesting themselves in scattering. In principle, apart from uniaxial fluctuations consistent with the symmetry of an ordered phase, there are biaxial fluctuations and fluctuations of the order parameter modulus which should manifest themselves in light scattering by cholesterics similarly to the case of nematics (Pokrovsky and Kats 1977) (see also sect. 1).

## 4. Optics of cholesterics near phase transitions

One usually investigates optical properties of cholesterics without assuming that a cholesteric is in the vicinity of a phase-transition point (Belyakov et al. 1979). In the present section we shall assume that cholesterics are in the immediate vicinity to the point of transition into another phase, either above or below $T_c$. It turns out that with the formulated conditions satisfied, optical characteristics alongside with quasielastic scattering display specific features determined by the nature of phase transitions.

### 4.1. Selective reflection near the phase-transition point

As is known (Belyakov et al. 1979), important parameters determining the optics of cholesterics are the dielectric anisotropy $\delta = \epsilon_a/\epsilon_\perp$ and cholesteric pitch $p$. Near the phase-transition point both quantities experience changes since they are related to the order parameter describing the phase transition. Note that the value of the dielectric anisotropy $\delta$ is related to the order parameter $s$, and at the point of the cholesteric–isotropic-liquid transition jumps into zero. Therefore optical characteristics determined by the quantity $\delta$ provide data on the order parameter and thus on the phase transition. Similarly pretransitional changes of the pitch affect optical pitch-dependent parameters and thus the corresponding optical parameters also contain information on the phase transition and its particularities. However, in this case the relation between the order parameter and the pitch is not so direct as for anisotropy $\delta$. On the whole it should be asserted that in liquid crystals in general and in cholesterics in particular there is a rare possibility of investigating the phase transition by optical characteristics. Note that similar optical characteristics (birefringence) are used for studying phase transitions in nematics (Zwetkoff 1942, Chatelain 1955, Saupe and Maier 1961) and transitions between phases of smectic liquid crystals (Lockhart 1979, Lim et al. 1979). Therefore although below we shall deal with optical characteristics of a cholesteric, the results concerning the relation of the birefringence to the order parameter can be rewritten for other kinds of liquid crystals.

Since in cholesterics like in nematics the problem of the order-parameter behaviour in a liquid crystal phase, as the temperature approaches the temperature of the transition into the isotropic liquid, is in need of further investigation, both theoretical and experimental (De Gennes 1974), the well-developed optical methods for studying cholesterics prove to be especially useful.

Experimental data (e.g., frequency width of the selective reflection band) determine the temperature dependence of $\delta$. A next step in determining the order parameter is the establishment of the relation between $\delta$ and the order parameter. In the simplest assumption that molecular polarizability in a cholesteric coincides with polarization of a free molecule, this relation, like in a nematic, is determined as,

$$\epsilon - 1 + 4\pi N\boldsymbol{\alpha}\,, \tag{95}$$

where $N$ is the number of molecules per a unit volume and $\boldsymbol{\alpha}$ is the polarizability tensor.

However, the dependence of the macroscopic refraction indices on the molecular polarizability should involve corrections for the local (internal) field. Assuming that local dielectric properties of a cholesteric are identical with those of a nematic, we can, following Saupe and Maier (1961), relate the observed dielectric anisotropy with the polarizability tensor as follows:

$$\begin{aligned} \frac{\epsilon_1 - 1}{\epsilon_1 + 2} &= \frac{4\pi}{3}\left[\frac{N}{A_1 + \alpha_{\parallel}^{-1}}\right], \\ \frac{\epsilon_2 - 1}{\epsilon_2 + 2} &= \frac{4\pi}{3}\left[\frac{N}{A_2 + \alpha_{\perp}^{-1}}\right], \end{aligned} \tag{96}$$

where $\alpha_{\parallel}, \alpha_{\perp}$ are the principle values of the polarizability tensor along the director and perpendicular to it, respectively, $\epsilon_1, \epsilon_2$ are the principle values of the tensor $\boldsymbol{\epsilon}$, $A_1$, $A_2$ are parameters dependent on the structure of a liquid crystal and related as $A_1 + 2A_2 = 0$.

The polarizability tensor is naturally determined as

$$\alpha_{ij} = \alpha^{(0)}\delta_{ij} + \alpha_a Q_{ij}\,, \tag{97}$$

where $\alpha^{(0)}$ is the average and $\alpha_a$ the polarizability anisotropy of an individual uniaxial molecule.

Applying eq. (97) for the quantities $\alpha_{\parallel}, \alpha_{\perp}$, we get

$$\begin{aligned} \alpha_{\parallel} &= \alpha^{(0)} + \tfrac{2}{3}\alpha_a s\,, \\ \alpha_{\perp} &= \alpha^{(0)} - \tfrac{1}{3}\alpha_a s\,. \end{aligned} \tag{98}$$

Thus if to assume that molecular characteristics $\alpha^{(0)}$ and $\alpha_a$ are known or measured (for example, in the crystallic phase), by means of eqs. (96) and (98) using the data of optical measurements (frequency width of the selective reflection band) we can find the order parameter $s$ and its temperature dependence. As for the parameters $A_1$ and $A_2$ of eq. (96), their values can be found experimentally if to assume that the average molecular polarizability in a liquid-crystal phase and in a liquid phase coincide. The results of the calculation of the temperature dependence of the order parameter from the selective reflection band width are presented in the work by Sukhenko (1977).

### 4.2. *Optical rotation and circular dichroism*

Another approach to the problem of the order-parameter behaviour is the study of the temperature dependence of the optical rotation (see fig. 3) as well

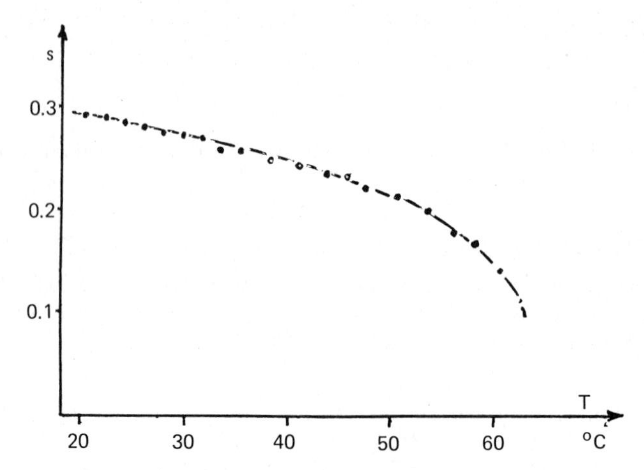


Fig. 3. Temperature dependence of the order parameter (in arbitrary units since the factor is not defined) obtained in accordance with measurements of the optical-rotation dispersion in a cholesteric–nematic mixture.

as of the circular dichroism, the latter being in the region of the absorption line of a cholesteric (or of a dye in a cholesteric). Let us in detail dwell upon the relation of the order parameter to the circular dichroism. The absorption anisotropy and thus the value of the dichroism in a cholesteric depend on the order parameter (Belyakov et al. 1979). Therefore the temperature dependence of the circular dichroism enables us to derive the temperature dependence of the order parameter $s$. Then for this purpose the case when the absorption line is practically outside the selective reflection band is most convenient.

Let us clarify the idea of the method, assuming that important is only one absorption line with the oscillation strength $f$ and the polarization direction at an angle $\varphi$ with the long axis of the molecule. In which case the components of the oscillator strength parallel to the director $f_\parallel$ and perpendicular to $f_\perp$ are expressed as:

$$\begin{aligned} f_\parallel &= \tfrac{1}{3}f + \tfrac{2}{3}f(1 - \tfrac{3}{2}\sin^2\varphi)s\,, \\ f_\perp &= \tfrac{1}{3}f - \tfrac{1}{3}f(1 - \tfrac{3}{2}\sin^2\varphi)s\,, \end{aligned} \tag{99}$$

and eq. (99) yields the following expression for the order parameter

$$s = \frac{f_\parallel - f_\perp}{f(1 - \tfrac{3}{2}\sin^2\varphi)}\,. \tag{100}$$

Yet the experimentally found values of $f_\parallel^{\text{ex}}$ and $f_\perp^{\text{ex}}$ like the above-studied expression relating the observed dielectric anisotropy with the molecular polarizability contain corrections for the internal field. Saupe and Maier (1961) have shown (see also Neugebauer (1950)) that for nematics the expressions for

$f_\parallel$ and $f_\perp$ are related with the experimentally observed quantities as

$$f_\parallel = \frac{3\epsilon'_\perp f_\parallel^{\text{ex}}}{\epsilon'_1 + 2 - 2a(\epsilon'_1 - 1)}, \qquad f_\perp = \frac{3\epsilon'_2 f_\perp^{\text{ex}}}{\epsilon'_2 + 2 + 2a(\epsilon'_2 - 1)}, \tag{101}$$

where $\epsilon'_1, \epsilon'_2$ are the principle values of the tensor $\boldsymbol{\epsilon}$, the contribution of the absorption line being substracted;

$$a = 3A_1/4\pi N.$$

The equivalence of the local ordering in a cholesteric to the nematic ordering enables us to employ eqs. (99) and (101) for the investigation of the order parameter and its temperature dependence near the phase-transition point in cholesterics.

Despite the fact that the optical methods of studying the order parameter are principally simple, systematic experimental investigations of the order parameter in the phase-transition region and in particular of the circular dichroism in the cholesteric phase performed by means of these methods are rather few (Sukhenko 1977). In all probability this is caused by difficulties in preparing perfect samples. This becomes understandable if we take into consideration that imperfection of a sample, say, variations of the cholesteric axis orientation within the sample, lead in the observed quantities to effects imitating the order-parameter variations. Therefore the data of the order-parameter measurements if they are obtained by means of the relations pertaining to perfect cholesterics, contain errors depending on the degree of perfection. Yet, it should be borne in mind that the respective errors have a regular character and could be removed after an appropriate processing of experimental results. Therefore we think that the optical method of studying phase transitions in cholesterics has not yet exhausted its potentialities.

### 4.3. *Refractometric determination of the order parameter*

Another possibility of optical determination of the order parameter in cholesterics lies in application of the classical refractometric method, i.e., measurements of birefringence (Averyanov and Shabanov 1979). The observed birefringence is $\Delta n = n_\parallel - n_\perp$ and the anisotropy of the effective molecular polarizability is $\Delta\gamma = [\gamma_{33} - \frac{1}{2}(\gamma_{11} + \gamma_{22})]$, where $\gamma_{ii}$ are the principle values of the molecular polarizability tensor related as:

$$s\Delta\gamma = \frac{5\mu(n_\parallel + n_\perp)\Delta n}{2\pi N_{\text{A}}\rho(\epsilon_\perp + 2)}, \tag{102}$$

where $\mu$ is the molecular weight, $N_{\text{A}}$ is the Avogadro number, $\rho$ is the density. Equation (102) contains two unknown quantities $s$ and the molecular-polarizability anisotropy. To determine the value of the order parameter we can apply the following extrapolation procedure proposed by Haller et al. (1973)

for nematics. This method is grounded on the fact that the experimentally measured dependence of $\ln(s\Delta\gamma)$ on $\ln[(T_c - T)/T_c]$, where $T_c$ is the temperature of a transition into the isotropic phase, is linear with a good accuracy. Extrapolating the experimentally found linear dependence up to the intersection with the coordinate axis and assuming that in this crossing point corresponding to $T = 0$ the order parameter is $s = 1$, we shall determine $\Delta\gamma$. The slope of the same line is determined by the factor $\eta$ in the dependence of $s = [(T_c - T)/T_c]^\eta$ on the temperature (for the found value of $\Delta\gamma$). In fig. 4 is given the temperature dependence of the order parameter for cholesteril pelargonate. The obtained values of $s$, as is clear from fig. 4, are in a satisfactory agreement with the values found by another method.

### 4.4. *Cholesteric pitch near transition into the smectic phase*

Optical observation of a pretransitional increase of the cholesteric pitch near the cholesteric–smectic transition (de Gennes 1974) deserves a special mention. This increase is due to the existence of small fluctuations with local smectic structure immediately above the transition point. These fluctuations have first been observed by de Vries (1970) in the nematic–smectic transition region and called cibotactic groups. The existence of cibotactic groups in a cholesteric, as has been theoretically shown by De Gennes (1972), causes an increase of the twist modulus $K_2$ and the bend modulus $K_3$, and as a consequence, an increase of the pitch near the point of transition into the smectic phase. Approximately this increase is described as

$$p = p_0(K_{20}/K_{20} + \delta K_2)^{-1}, \tag{103}$$

where $p_0$, $K_{20}$ are values of the pitch and the twist modulus in the absence of

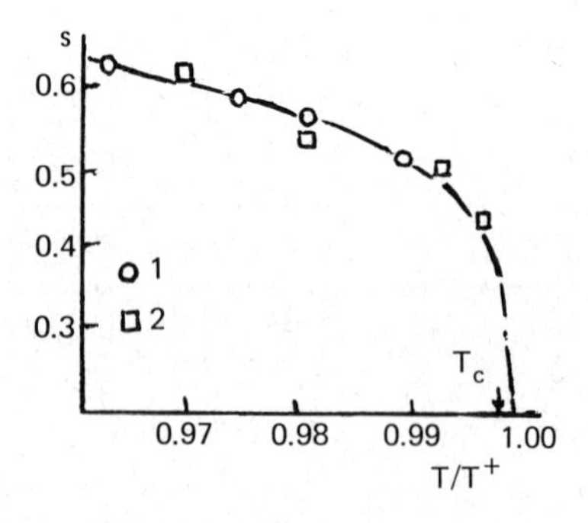


Fig. 4. Temperature variations of the orientational order parameter in the cholesteryl pelargonate obtained by the refractometric method (Averyanov and Shabanov 1979). Solid curve – $s = (1 - T/T^*)^{0.142}$, (1) according to formulae (102), (2) by circular optical dichroism.

cibotactic groups, respectively, and $\delta K_2$ is an addition to the twist modulus due to these groups. Since the addition to the twist modulus $\delta K_2$ is linearly dependent on the correlation length of the smectic order, the study of the pitch behaviour near the transition into the smectic phase provides a simple approach to the pretransitional behaviour of cibotactic groups. Yet, no systematic experimental study of this problem has so far been carried out and there are only very few publications on this subject (Pindak et al. 1974). The performed measurements show that the pitch increase $p \sim (T - T_c)^{-0.67}$, i.e., pitch behaviour, is closer to "helium analogy" (De Gennes 1974) than to calculations in the self-consistent field approximation (the critical exponents are $-\frac{2}{3}$ and $-0.5$, respectively). However, as has been pointed out by Huang et al. (1974), the index of this dependence is very sensitive to impurity concentration, therefore it is desirable that further research on this problem be carried out.

Moreover it should be stressed (see Wiegman and Filyev (1975)), that such a simple formula (103) holds for sufficiently large values of the pitch $p_0 \sim 10^5$ Å (in fact it corresponds to the mixtures of cholesteric and nematic liquid crystals). In pure cholesterics $p_0 \sim 10^3$–$10^4$ Å the pitch increase cannot be described only as an increase of the twist modulus $K_2$. Then for the quantity $p$ we get a very complicated expression.

### 4.5. *Pretransitional optical rotation in isotropic phases*

A most characteristic feature of phase transitions in liquid crystals is their closeness to second-order phase transitions. It means that the corresponding phase transitions are first-order transitions, however, since the latent heat of the transition is very small, therefore, in the process of a transition we can observe features typical of second-order transitions. In the first turn it concerns pretransitional fluctuational phenomena characteristic of second-order transitions.

As for the phase isotropic liquid–cholesteric transition, these general considerations mean that in the isotropic liquid near the phase transition point regions must develop fluctuationally with a short-range cholesteric order. Then the lifetime and sizes of these fluctuations increase as the point of the transition into the cholesteric phase is being approached (De Gennes 1974). This order can be compared with the appearance and disappearance in the isotropic phase of arbitrarily positioned and oriented "pieces" of the cholesteric spiral, their correlation being absent at distances exceeding correlation lengths.

Naturally there arises a question how the described fluctuational effects display themselves in optical characteristics and in particular in optical rotation in the isotropic phase near the transition temperature. Keeping in mind an anomalously high optical rotation in the cholesteric phase, in the pretransitional region of the isotropic phase one should expect a strong optical rotation exceeding the ordinary molecular rotation power. For cholesterics we can almost always neglect the optical molecular rotation power, but in the isotropic phase this cannot be done. At least it is clear that such neglect is not applicable far from the temperature of the phase transition point.

Let us estimate the order of magnitude of the optical rotation in the isotropic phase following the works by Cheng and Meyer (1972), Kats (1973), Cheng and Meyer (1974), Dolganov et al. (1980). Since optical rotation is determined by a difference of refraction indices $\Delta n_+$ and $\Delta n_-$ for the right-hand and left-hand circular polarizations, the problem is reduced to the calculation of corrections to these quantities due to cholesteric fluctuations in the isotropic phase. Since the sizes of the cholesteric ordering regions are small, i.e., do not exceed 200 Å (Stinson and Litster (1975), Chu et al. (1972), Stinson and Litster (1973)), it is clear that in each individual region the influence of the cholesteric ordering is weak. In particular, multiple scattering effects are weak and the interaction of a light beam with a fluctuation can be described by a simple kinematic approximation. These properties of the fluctuations do not imply that the "cholesteric effect" is weak in the whole sample since a large number of fluctuating regions may compensate a small effect in an individual region. On the whole the observed effect should be characterized by averaged characteristics of the fluctuations, their sizes in particular, i.e., by the value of the correlation lengths $\zeta$. Using the relation between the dielectric permeability $\epsilon$ and the forward scattering amplitude $\epsilon = 1 + 4\pi\rho f_0/K^2$, where $\rho$ is the density and $f_0$ is the forward scattering amplitude of an individual element of the medium, for the difference of the refractive indices of circular polarizations from the refraction index of the isotropic liquid we get:

$$\Delta n_\pm = 2\pi\rho\Delta f_0^\pm/K^2\,, \tag{104}$$

where $\Delta f_0^\pm$ is the forward scattering amplitude at fluctuations of the right-hand and left-hand circular polarizations. The form of the fluctuational correction to the dielectric-permeability tensor coincides with the form of the cholesteric-phase tensor (Belyakov et al. 1979) therefore the quantity $\Delta f_0^\pm$ and its symmetric properties are such that the forward scattering amplitude becomes zero. Thus the quantity $\Delta f_0^\pm$ is non-zero only with the multiple scattering taken into account and can be found in the second order of the perturbation theory expansion over fluctuational additions to the tensor $\epsilon$. Performing Fourier transformation and solving the Maxwell equation for the isotropic liquid with $\epsilon$ containing the fluctuational addition, in the second order of the perturbation theory we find (Kats 1976):

$$\Delta n_\pm = \mp\frac{\delta^2 K^2}{2}\int\frac{\boldsymbol{P}_\pm\delta\epsilon(\boldsymbol{K}-\boldsymbol{q})\delta\epsilon(\boldsymbol{q}-\boldsymbol{K})\boldsymbol{P}_\pm}{K^2-q^2}\,\mathrm{d}^3q\,, \tag{105}$$

where $\delta\epsilon(\boldsymbol{q})$ is the Fourier transform of the fluctuational addition to $\bar{\epsilon}$ and $K^2 = \omega^2\bar{\epsilon}c^{-2}$ where $\bar{\epsilon}$ is the dielectric permeability of the liquid. To obtain the observed quantity in eq. (105) we should perform thermodynamic averaging, then the expression for the optical rotation power is of the form:

$$\varphi = \varphi_0 + \delta^2 K^2 \int \frac{\langle \boldsymbol{P}_+ \delta\epsilon(\boldsymbol{\kappa})\delta\epsilon(-\boldsymbol{\kappa})\boldsymbol{P}_+ \rangle}{\boldsymbol{\kappa}(2\boldsymbol{K} - \boldsymbol{\kappa})} \mathrm{d}^3 K, \tag{106}$$

where we have introduced the already mentioned usual molecular rotation power $\varphi_0$. The fluctuational part of the rotation is determined in eq. (106) by the correlation function which may be expressed in terms of the order parameter correlator *s*, if to keep in mind that the local value of the fluctuational addition is proportional to the order parameter $\delta\epsilon_{\alpha\beta} = \mu Q_{\alpha\beta}$. Thus the calculation of the rotation power of the liquid near the point of a transition into the cholesteric phase is reduced to the calculation of the order-parameter correlation function. Explicit theoretical expressions for the corresponding correlators in eq. (106) are dependent on the approximation used for the description and will be discussed in what follows. Yet independently of the concrete form of these correlators the rotation angle in liquids in contrast to cholesterics is linearly dependent on the thickness of the sample. For example, applying the results obtained by de Gennes (1969) by means of eq. (106) we can express the optical rotation in terms of the temperature dependence of the correlation lengths:

$$\varphi = \varphi_0 + \tfrac{1}{8}\delta^2 T K^2 q_0/\sqrt{\tau}\,, \tag{107}$$

where

$$\tau = (T - T)^*/T^*\,.$$

From eq. (107) it follows that the optical rotation power diverges as $(T - T)^{-1/2}$ as the transition point is being approached. The direction of the rotation is determined by the sign of the cholesteric spiral (sign of $q_0$) and does not depend on the ratio of the light wavelength to the pitch $p$ and in contrast to the cholesteric phase, does not change its sign at $\lambda = 2p$. The relation (107) is derived according to the perturbation theory. A more detailed study leads to an expression of the same type with a few somewhat different coefficients (Filyev 1978):

$$\varphi = \varphi_0 + \frac{TK^2 q_0 \zeta_1}{24\pi^2 \bar{\epsilon} b(1 + \frac{1}{2}c)^{3/2}} \left[ \operatorname{arctg} \zeta q_{\max} - \frac{\zeta q_{\max}}{1 + (\zeta q_{\max})^2} \right], \tag{108}$$

where $b$, $c$ are the coefficients of the free-energy expansion and $q_{\max} \sim 1/a$ is the cut-off parameter in the integral of the (106) type ($a$ is the intermolecular distance), $\zeta_1$ and $\zeta_2$ are correlation lengths.

It should be pointed out that expressions (107) and (108) are valid in the region not too close to the transition temperature. The De Gennes approach, our study is based upon, is in need of certain improvement in a small temperature range near the transition where the correlation length $\zeta$ exceeds the pitch and in the approximation under study the isotropic phase is absolutely

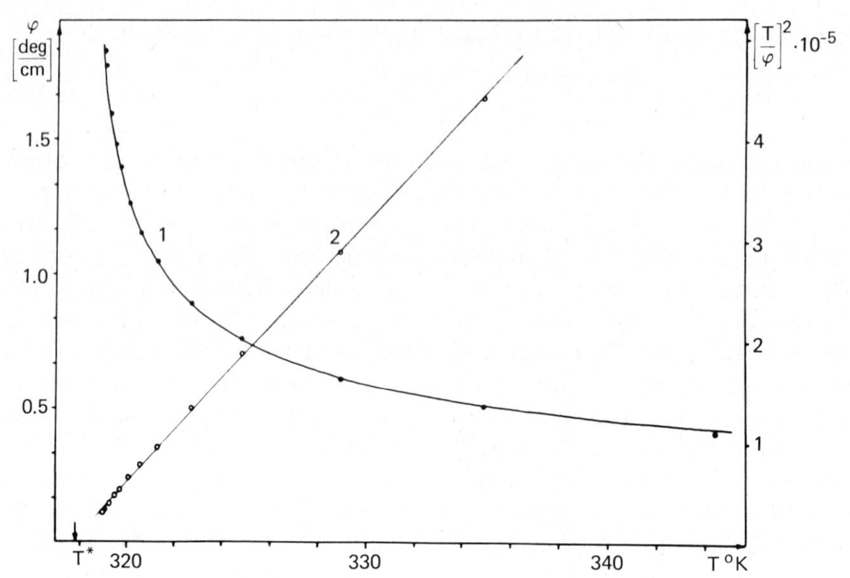


Fig. 5. Temperature dependence of the optical rotation in MBBA with the mixture of x–5 in the pretransitional region (curve (1)); curve (2) $(T/\varphi)^2$ (Cheng and Meyer 1974).

unstable. If the corresponding temperature range is really achieved, then in this region expressions (107) and (108) are inapplicable and another approach to the study of the pretransitional behaviour of the liquid and its optical properties in this region is required. Apparently this situation is in fact realized in some cases and this is testified by the existence of the "blue phase" in a narrow temperature range between the cholesteric and isotropic phases.

The experimental measurements of the pretransitional optical rotation in the isotropic phase (see fig. 5) are well described by eqs. (107) and (108) and the optical rotation divergence is $(T - T^*)^{-1/2}$ as the transition point is being approached. It means that according to eqs. (107) and (108) the observed property is associated with the temperature behaviour of the correlation length $\zeta$ which in the self-consistent theory has the same divergence. The universal character of this dependence is demonstrated by the results of the work by Cheng and Meyer (1974) who have shown that the temperature behaviour of the pretransitional optical rotation for different substances is described by the same temperature dependence.

### *4.6. Blue phase*

As far as in the beginning of this century Lehman discovered that in a narrow temperature range (of the order of 1°) near the point of the isotropic liquid–cholesteric transition certain substances have an intermediate phase

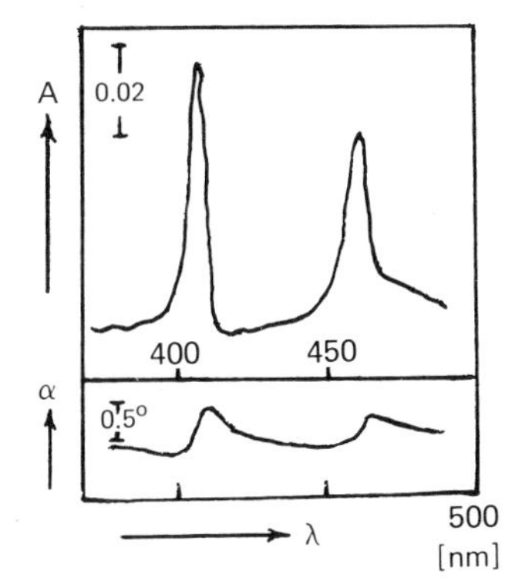


Fig. 6. Selective reflection and optical rotation dispersion CN at a temperature of 91.35°C corresponding to the point of transition between two varieties of the "blue phase" (Bergman et al. 1979) (CN is cholesteryl nonanoate).

different from the isotropic liquid and cholesteric phases (Lehman 1906). Later the results obtained by Lehman have been confirmed by other scholars, in particular, by Gray (1956) who called this intermediate state a "blue phase". Though discovered so long ago, the "blue phase" is still a mysterious phenomenon. Its existence has for long been doubted, and only recently essential progress has been made in its understanding, the most effective methods of its investigation being optical methods.

Let us first describe experimentally observable properties of the "blue phase" (see fig. 6). So, this phase is observed in certain cholesteril derivatives (e.g., cholesteril oleil carbonate (COC) and cholesteril oleate (CO)) and exists in a narrow temperature range between the isotropic liquid and usual cholesteric phase. Then it is easier to observe the "blue phase" at a temperature decrease. It has been experimentally established that for the "blue phase":

(1) There exists a selective light scattering in the visible range of the spectrum. Therefore this phase has a bluish colouring which the name of the phase originated from. In many cases brightly coloured platelets become noticeable.

(2) There is a strong optical rotation (Bergman and Stegemeyer 1979a).

(3) There is no birefringence, i.e., it is optically isotropic (Saupe 1969).

(4) The phase transition from the isotropic liquid to the "blue phase" is thermodynamically analogous to the isotropic liquid–cholesteric transition (Armitage et al. 1977) which is testified by specific heat anomalies being of the same order of magnitude for either case.

(5) The transition from the cholesteric into the "blue phase" (observed at a temperature increase) is characterized by a low latent heat (Armitage et al. 1977).

(6) The NMR spectra for the "blue phase" show that it is stable and have

a long-range orientational order, the spectra being similar to the spectra of the isotropic liquid and different from the cholesteric phase spectra (Collings and McColl 1978).

(7) The dynamics of pretransitional fluctuations in the isotropic phase of liquid crystals is different for substances experiencing a transition into the "blue phase" and substances experiencing a transition directly into the cholesteric phase.

The above said proves that the "blue phase" is a new phase of liquid crystals but is not a special cholesteric texture (Gray 1956).

The nature of molecular ordering in the "blue phase" is not finally established. Yet advances in the theoretical and experimental research give rise to hope that the puzzle of the "blue phase" will soon be solved. Independently of the fact which of the "blue phase" models is correct, it is an example of an extraordinarily interesting and pretty complex structural state of condensed matter. Therefore the study of the "blue phase" is of general interest and the significance of the obtained results exceeds the problem of liquid crystals only.

A possible "blue phase" structure has been proposed by Saupe (1969). Basing upon the observed optical isotropy, he made an assumption that the "blue phase" structure is the bcc lattice of point-like defects of the director orientation. Outside these defect points the "blue phase" structure is similar to the cholesteric structure. However, Saupe has not given sufficient theoretical foundation for the proposed structure.

Studying the isotropic liquid–cholesteric transition in the framework of the Landau phase transition theory Brazovsky and Dmitriev (1975), Brazovsky and Filyev (1978) have shown that structures similar to that of Saupe can be thermodynamically stable. They concluded that for some values of the parameters of the Landau free energy expansion there exists an intermediate phase between the isotropic liquid and cholesteric phase. They pointed to hexagonal and cubic phases as possible intermediate phases, the priority as a possible "blue phase" structure being given to the hexagonal phase.

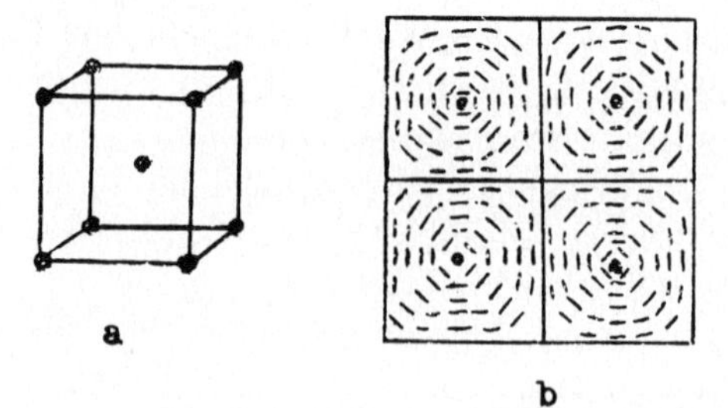


Fig. 7. Structure of the elementary cell of the "blue phase", proposed by Saupe (1969): (a) elementary cell; (b) field of the director on the face of the elementary cell.

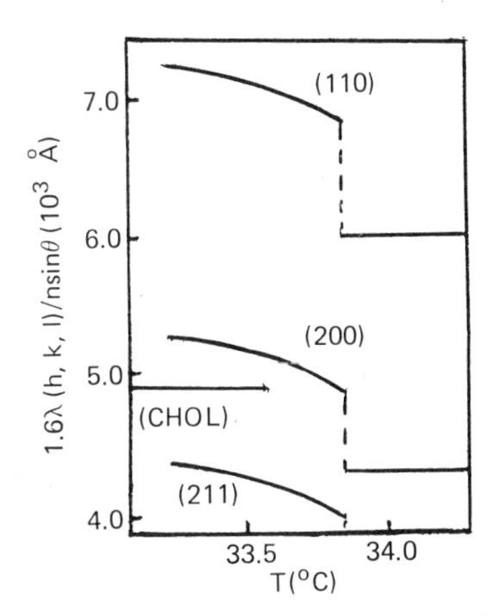


Fig. 8. Temperature dependence of the selective scattering wavelength in a cholesteric, "blue phases" of BPI and BPII in the 50–50% mixture of CB15 and E9.

In experiments by Meiboom and Summon (1980) on cholesteril nonanoate and the mixture of this compound with cholesteril chloride two intermediate phases have been observed: one is stable in the temperature range 0.2° below the transition temperature into the isotropic liquid, the other is preceding the cholesteric phase and stable in the temperature range 0.5°. The structure of the first high temperature "blue phase" has not been unambiguously established but either the bcc or primitive cubic structure corresponds to the second low temperature "blue phase". The conclusions of this work are grounded on optical measurements.

The optical measurements by Johnson et al. (1980) have also confirmed the conclusions of Meiboom and Summon on the existence of two "blue phases" and cubic symmetry of the low temperature "blue phase". Besides, they have asserted that the translational symmetry of the both "blue phases" is the same, i.e., they have either body-centered or primitive cubic lattice. Johnson et al. have made an interesting conclusion based upon the observed temperature dependence of linear sizes of unit cells of both the phases on the character of the transition between them (see fig. 8). For the low-temperature phase the sizes of unit cells decrease with a temperature increase. When the edge of a unit cell coincides with the cholesteric pitch there occurs a phase transition into the high-temperature phase where the sizes of unit cells are temperature independent in the overall temperature range of its existence. The transition between the low- and high-temperature phases may be either continuous or jump-like, depending on the composition of the mixtures.

The experimental determination of the "blue phase" structure agrees with the results of the theoretical paper by Hornreich and Shtrikman (1980). In the framework of the Landau theory they have come to the conclusion that the existence of phases with cubic structure is possible. Since this structure between the isotropic and cholesteric phases is formed by "defects" in the order parameter field, the decrease of lattice parameters with a temperature increase (observed by Johnson et al. (1980)) is especially noteworthy. Such temperature behaviour of lattice parameters (the contraction of the lattice but not its expansion at a temperature increase) agrees with the behaviour of the "defects" lattice known for usual crystals (Alexander and McTague 1978). It turns out that the "defects" lattice in contrast to the usual crystal lattice, must experience temperature contraction but not expansion at a temperature increase.

The studies of the "blue phases" although so far incomplete (Stegemeyer and Bergman 1980, Meiboom et al. 1981, Hornreich and Shtrikman 1981, Flack and Gooker 1981) show that the "blue phase (or "blue phases") is a very interesting object for phase transition physics. Furthermore, these investigations again demonstrate the importance of optical methods for phase transition in liquid crystals (Belyakov et al. 1982).

## 5. Two-dimensional liquid crystals

### 5.1. Smectic-B films

Optical studies of liquid crystals enable us in principle to study properties of two-dimensional systems. The point is that the order in certain smectic phases (e.g., smectic-B) is a superposition of two-dimensional and three-dimensional periodic distributions of density and orientation. As is well known, a real two-dimensional long-range order in the density and orientation distribution is not possible. Therefore the presence of the long-range order in layered smectic liquid crystals automatically implies a three dimensionality of interactions in the system (and consequently, of the ordering). Nevertheless the existence of layers in the system implies that the interaction inside a layer is much larger than the interaction between layers. Therefore the character of the scattering differs from that of the scattering in "ordinary" weakly anisotropic systems. This difference is especially evident at the scattering in films of a liquid crystal. Sufficiently thick films behave similar to a bulk three-dimensional liquid crystal. Yet, the more the thickness (actually the ratio of the thickness to the transversal size) decreases, the more specific two-dimensional fluctuations become dominating. It means, for example, that the amplitude of the coherent scattering on an infinite sample is zero. As for the amplitude of the incoherent scattering, it has maxima (peaks) at certain characteristic momentum transfers, close to certain vectors of the reciprocal lattice. The number of such peaks and the character of the singularity is essentially temperature dependent. On the other hand, in a three-dimensional crystal the coherent scattering cross section is non-zero and in a perfect infinite crystal has a $\delta$-like character. The incoherent scattering has peaks of a Lorenz form near the Bragg vectors of the reciprocal lattice. In their work Feigelman and Pokrovsky (1981) considered diffraction on multilayered films of a smectic-B liquid crystal. In fact, they have dealt with the X-ray wavelength range. Although the investigation of the X-ray diffraction is beyond the framework of our survey, we shall discuss the results of this work since it contains a very interesting physical conclusion important for understanding the nature of ordering in smectic liquid crystals.

*5.1.1. Elastic energy*

So let us study a film of a smectic-B liquid crystal with a sufficiently large number of layers (Feigelman and Pokrovsky 1981). This system may be described as a continuous elastic medium, isotropic in the $xy$-plane, which corresponds to the hexagonal symmetry of molecular ordering in a smectic layer. The elastic energy of such medium is described by the Hamiltonian:

$$H_{\mathrm{el}} = \tfrac{1}{2}\int\Bigg[(D_1 + D_2)\left(\frac{\partial u_\alpha}{\partial x_\alpha}\right)^2 + D_2\left(\frac{\partial u_\alpha}{\partial x_\beta}\right)^2 + D_3\left(\frac{\partial v}{\partial z}\right)^2 + D_4\left(\frac{\partial v}{\partial x_\alpha}\right)^2 + D_4\left(\frac{\partial u_\alpha}{\partial z}\right)^2 + D_5\frac{\partial u_\alpha}{\partial x_\alpha}\frac{\partial v}{\partial z}\Bigg]\mathrm{d}^3x\,, \tag{109}$$

here $u_\alpha$ is the displacement vector in the plane of a smectic layer, $u_z \equiv v$ and $D_1$, $D_2$, $D_3$, $D_4$, $D_5$ are elastic moduli. The existence of layers in the system means that the elastic moduli $D_4$ and $D_5$ are small in comparison with other elastic constants. The elastic energy in Fourier components is:

$$H_{\mathrm{el}} = \tfrac{1}{2}d\int\sum_{\kappa}[(D_1 + D_2)|K_\alpha u_\alpha|^2 + D_2K^2|u_\alpha|^2 + D_3\kappa^2|v|^2 + D_4K^2|v|^2 + D_4\kappa^2|u_\alpha|^2 + D_5K_\alpha\kappa u_\alpha v^*]\,\mathrm{d}^2K\,. \tag{110}$$

The finite thickness of the film leads to quantization of the $z$ component of the momentum:

$$\kappa d = \pm 2\pi n/N\,, \qquad n = 0, 1 \ldots [N/2]$$

($d$ is the interplane spacing).

From the expression for the elastic energy using the equipartition theorem it is easy to calculate the correlation displacement functions:

$$\langle u_\alpha(K,\kappa)u_\beta(-K,-\kappa)\rangle = \frac{T}{d}\left[\frac{\delta_{\alpha\beta} - n_\alpha n_\beta}{D_2K^2 + D_4\kappa^2} + \frac{n_\alpha n_\beta}{(D_1 + 2D_2)K^2 + D_4\kappa^2}\right], \tag{111}$$

$$\langle v(K,\kappa)v(-K,-\kappa)\rangle = \frac{T}{d}\frac{1}{D_4K^2 + D_3\kappa^2}, \tag{112}$$

$$\langle u_\alpha(K,\kappa)v(-K,-\kappa)\rangle = \frac{T}{d}\frac{K_\alpha\kappa D_5}{[(D_1 + 2D_2)K^2 + D_4\kappa^2]}\frac{1}{[D_4K^2 + D_3\kappa^2]}. \tag{113}$$

These formulae are valid for sufficiently small values of $K$ and $\kappa$. Further we shall regard $\kappa$ as always small but the values of $\kappa$ in some cases are of the order $1/d$. In these cases the quantity $\kappa^2$ in the dominator of the correlation functions should be replaced by its analogue for the discrete system of layers:

$$(2/d^2)(1 - \cos \kappa d) .$$

Besides, in the denominator of the correlation functions we neglect the quantities proportional to $D_5$, since their contribution is small in comparison with other terms.

Since $D_4$ is small, it is necessary to write the subsequent term of the expansion over $\sim cK^4$. Then the coefficient c has the order of magnitude $D_2 a^2$, since it is physically related to the bend of the layer but not to the interaction between different layers ($a$ is a characteristic molecular size).

*5.1.2. Coherent scattering*

Let us now calculate the structural factor:

$$S(q) = \left\langle \sum_{a,a'} \exp(\mathrm{i}\boldsymbol{q}\boldsymbol{z}_a - \mathrm{i}\boldsymbol{q}\boldsymbol{z}'_{a}) \right\rangle , \tag{114}$$

where $\boldsymbol{a}$ and $\boldsymbol{a}'$ denote the lattice points, $\boldsymbol{z}_a = \boldsymbol{a} + \boldsymbol{u}_a$. Thus:

$$S(q) = \left| \sum_{a} \exp(\mathrm{i}\boldsymbol{q}\boldsymbol{a}) \langle \exp(\mathrm{i}\boldsymbol{q}\boldsymbol{u}_a) \rangle \right|^2 + \sum_{a,a'} \exp[\mathrm{i}\boldsymbol{q}(\boldsymbol{a} - \boldsymbol{a}')] \langle \exp[\mathrm{i}\boldsymbol{q}(\boldsymbol{u}_a - \boldsymbol{u}_{a'})] \rangle . \tag{115}$$

The first term of this formula corresponds to the coherent, and the second to the incoherent scattering. Let us start with the calculation of the coherent structural factor:

$$S_c(q) = \left| \sum_{a} \mathrm{e}^{\mathrm{i}\boldsymbol{q}\boldsymbol{a}} \right|^2 \langle \mathrm{e}^{\mathrm{i}\boldsymbol{q}\boldsymbol{u}} \rangle^2 \equiv \left| \sum_{a} \mathrm{e}^{\mathrm{i}\boldsymbol{q}\boldsymbol{a}} \right|^2 \mathrm{e}^{-2W} .$$

We shall be interested in the values of $q$ equal to the reciprocal lattice vectors:

$$\boldsymbol{q} = m\boldsymbol{g}_1 + n\boldsymbol{g}_2 + l\boldsymbol{g}_3 ,$$

where $\boldsymbol{g}_1$ and $\boldsymbol{g}_2$ are basis vectors of the hexagonal plane lattice, $\boldsymbol{g}_3$ is the vector perpendicular to the plane, $m$, $n$, $l$, are integers.

Let us first study the case when the vector $\boldsymbol{q}$ lies in the plane $l = 0$. Then the Debye–Waller factor $W$ is

$$2W = q_\alpha q_\beta \frac{1}{N} \sum_{\kappa} \int \frac{\mathrm{d}^2 K}{(2\pi)^2} \langle u_\alpha(K, \kappa) u_\beta(-K, -\kappa) \rangle . \tag{116}$$

Let us pick out the term with $\kappa = 0$ from the sum over $\kappa$ in the preceding expression and replace the remaining sum with the integral over $\kappa$,

$$\langle u_\alpha u_\beta \rangle = \frac{1}{N} \int \frac{d^2K}{(2\pi)^2} \langle u_\alpha(K,0) u_\beta(-K,0) \rangle + d \int \frac{d^2K \, d\kappa}{(2\pi)^3} \langle u_\alpha(K,\kappa) u_\beta(-K,-\kappa) \rangle . \tag{117}$$

The first term in the first part of this formula corresponds to the two-dimensional fluctuations of the crystallic lattice. With the logarithmic accuracy, the first term is,

$$\langle u_\alpha u_\beta \rangle^{(2)} = \delta_{\alpha\beta} \frac{1}{Nd} \frac{T}{4\pi} \frac{D_1 + 3D_2}{D_2(D_1 + 2D_2)} \ln \frac{R}{a} \tag{118}$$

($R$ is the transversal size of the system).
At a fixed number of layers and $R \to \infty$ the Debye–Waller factor tends to zero according to the law:

$$e^{-2W} \approx (a/R)^{\eta_q/N}$$

where

$$\eta_q = \frac{Tg_1^2}{4\pi d} \frac{D_1 + 3D_2}{D_2(D_1 + 2D_2)} (m^2 + n^2 + mn) . \tag{119}$$

The contribution to the quantity $\langle u_\alpha u_\beta \rangle$ from the three-dimensional fluctuations equals:

$$\langle u_\alpha u_\beta \rangle^{(3)} = \frac{T\delta_{\alpha\beta}}{8\pi d} \left( \frac{1}{D_2} \ln \frac{D_2}{D_4} + \frac{1}{D_1 + 2D_2} \ln \frac{D_1 + 2D_2}{D_4} \right) . \tag{120}$$

At $N \to \infty$ we obtain the Debye–Waller factor of the three-dimensional system

$$e^{-2W} = (D_4/D_2)^{\eta_q/2}$$

In the general case when the reciprocal lattice vector $g$ has also the $z$ component, the term of the following form is added to the quantity,

$$l^2 g_3^2 \langle v^2 \rangle = \frac{1}{N} \zeta_q \ln \frac{R}{a^*} + \omega_q \ln \frac{a^*}{d} \tag{121}$$

where

$$\zeta_q = \frac{Tg_3^2 l^2}{2\pi D_4 d}, \qquad a^* = \sqrt{\frac{c}{D_4}}, \qquad \omega_q = \frac{Tg_3^2 l^2}{2\pi \sqrt{cD_3}} .$$

The contribution of the cross terms $\langle u_\alpha v^* \rangle$ to the quantity $W$ is zero in virtue of the symmetry.

Finally we have:

$$e^{-2W} = \left( \frac{D_4}{D_2} \right)^{\eta_q/2} \left( \frac{D_4 a^2}{c} \right)^{\omega_q/2} \left( \frac{a}{R} \right)^{\eta_q/N} \left( \frac{a}{R} \sqrt{\frac{c}{D_4}} \right)^{\zeta_q/N} . \tag{122}$$

At $l = 0$ the dimensional effects in the Debye–Waller factor can be observed at a number of layers equal to

$$N \lesssim N_1 \approx \frac{\ln(R^2/a^2)}{\ln(D_2/D_4)}, \quad \text{i.e.,} \quad N_1 \sim 10\text{–}15\,. \tag{123}$$

At $l \neq 0$ similar estimates yield

$$N \lesssim N_2 = \left\{\frac{\sqrt{cD_3}}{D_4 d}\right\} \frac{\ln(R^2 D_4/c)}{\ln(c/D_4 a^2)}, \quad \text{i.e.,} \quad N_2 \sim 300\,. \tag{124}$$

At sufficiently large $q$ the coherent scattering cross section in the maximum becomes comparable with the average incoherent scattering cross section. Therefore one can observe only Bragg peaks for which the incoherent scattering is small in comparison with the coherent scattering.

### *5.1.3. Incoherent scattering*

Let us now consider the incoherent scattering. Let $\boldsymbol{q}$ be close to a certain vector of the reciprocal lattice $\boldsymbol{b}$. The small difference $\boldsymbol{q} - \boldsymbol{b}$ will be as usual denoted as $\boldsymbol{q}$, and the components of the vector lying in the plane of the layer and normal with respect to it as $\boldsymbol{K}$ and $\kappa$, respectively. In this notation the incoherent structural factor is:

$$S_i \sim \int \mathrm{d}^3 r \, \exp(\mathrm{i}\boldsymbol{q}\boldsymbol{z}) \langle \exp[\mathrm{i}\boldsymbol{b}(\boldsymbol{u}(\boldsymbol{r}) - \boldsymbol{u}(0))] \rangle\,. \tag{125}$$

At large distances the harmonic approximation holds:

$$\langle \exp[\mathrm{i}\boldsymbol{b}(\boldsymbol{u}(\boldsymbol{r}) - \boldsymbol{u}(0))] \rangle = \exp[-b_i b_j G_{ij}(r)] - \exp[-b_i b_j G_{ij}(\infty)]\,, \tag{126}$$

where $i, j = 1, 2, 3$,

$$G_{ij} = \tfrac{1}{2}\langle (u_i(\boldsymbol{r}) - u_i(0))(u_j(\boldsymbol{r}) - u_j(0)) \rangle$$
$$= \int \frac{\mathrm{d}^3 q}{(2\pi)^3} \langle u_i(\boldsymbol{q}) u_j(-\boldsymbol{q}) \rangle (1 - \cos \boldsymbol{q}\boldsymbol{z})\,.$$

We shall not give here the complicated calculations of the quantity $G_{ij}$. The main idea of the calculations is the same as in the determination of the coherent scattering cross section. Namely, we shall consider separately the contribution of the two-dimensional fluctuations corresponding to $\kappa = 0$ and the remaining sum (the three-dimensional contribution) is replaced by the integral. The correlator $\langle u_i u_j \rangle$ at small distances $r \ll a\sqrt{D_2/D_4}$ has the simplest interpretation. At such distances fluctuations of parallel displacements in different planes are independent and therefore in each plane the correlators are like the ones in a two-dimensional crystal.

At large distances for $b_3 \neq 0$, the formulae become very cumbersome. We shall point out only physical consequences. A weak interplane interaction leads to two effects. In the first one – the background of the incoherent scattering increases, in the second – the incoherent scattering peaks (at $b_3 \neq 0$) have a non-Lorenz form.

### 5.2. *Possible types of quasi-two-dimensional phases*

In fact, what is said above concerns smectic-B liquid crystals regarded as genuine three-dimensional crystals. More exotic situations are also possible. The point is that apart from the order parameter with respect to the displacements (it will be denoted as $\psi_d$) we can introduce the order parameter $\psi_0$ describing the orientation of bonds. The presence of the long-range order in two-dimensional systems does not contradict the well-known Bogoliubov inequalities and therefore in principle may take place. In different types of smectic phases a combination becomes possible of various types of ordering in the fields $\psi_d$ and $\psi_0$ with an ordinary orientational order (orientation of long molecular axes).

Such analysis had first been conducted by Halperin and Nelson (1978). They found that seven various phases are possible.

(a) The correlators $\langle \psi_0^*(r)\psi_0(0)\rangle$ and $\langle \psi_d^*(r)\psi_d(0)\rangle$ decay exponentially; this behaviour corresponds to smectic-A liquid crystals.

(b) $\psi_0$ has a quasi-long-range order (i.e., algebraic decay of the correlators) and $\psi_d$ has a short-range order; this corresponds to the so-called hexatic phase.

(c) $\psi_0$ has a genuine long-range order, and $\psi_d$ a short-range order.

(d) $\psi_d$ and $\psi_0$ have a genuine long-range order; this corresponds to solids.

(e) $\psi_0$ and $\psi_d$ have a quasi-long-range order (smectic-C).

(f) $\psi_0$ and $\psi_d$ have an independent quasi-long-range order.

(g) A long-range order in the field $\psi_0$ and a quasi-long-range order in the field $\psi_d$.

It should be pointed out that the phase with a long-range or quasi-long-range order in the field $\psi_d$ but with a short-range order in the field $\psi_0$ is impossible. The non-zero value of $\psi_d$ in a certain region produces the effective field linearly interacting with $\psi_0$. The character of the scattering in each of these phases may be easily established by the given formulae and, namely, the two-dimensional contribution to the structural factor corresponds to the case of the quasi-long-range order and the three-dimensional contribution to a genuine long-range order. In the case of a short-range order only, more important become fluctuations of the director orientations investigated in another section of the review.

### 5.3. *Role of spontaneous polarization*

An interesting possibility of stabilizing a genuine two-dimensional order had been considered by Pelcovits and Halperin (1979). The thing is that there exist ferroelectric smectic-C liquid crystals. A dipole–dipole interaction may lead to stabilization of the two-dimensional order. The free energy of this system is

$$F = \int \mathrm{d}^2 r \left\{ \tfrac{1}{2} K_s (\mathrm{div}\, \boldsymbol{n})^2 + \tfrac{1}{2} K_b (\mathrm{rot}\, \boldsymbol{n})^2 - \boldsymbol{P}_0 \boldsymbol{E} + \int \mathrm{d}^2 r' (\nabla \boldsymbol{P}_0)(\nabla \boldsymbol{P}_0')/2|\boldsymbol{r} - \boldsymbol{r}'| \right\}. \tag{127}$$

Here $\boldsymbol{n}$ is the director tilted by the angle $\theta$ to the plane, $\boldsymbol{P}_0 \sim [\boldsymbol{n} \times \boldsymbol{z}]$ is a ferroelectric polarization, $\boldsymbol{E}$ is the external field, $K_s$ and $K_b$ are "two-dimensional" elastic constants.

If to neglect surface effects and to assume that the elastic behaviour of the smectic-C phase is similar to the nematic behaviour, it is easy to express the modules $K_s$ and $K_b$ in terms of the usual Frank constants:

$$\begin{aligned} K_s &= dK_1 \sin^2\theta\,, \\ K_b &= d[K_2 \sin^2\theta \cos^2\theta + K_3 \sin^4\theta]\,, \end{aligned} \tag{128}$$

$d$ is the thickness of the film.
Similarly

$$P_0 = \mathrm{d}P \tag{129}$$

where $P$ is polarization of a unit volume.

In an ordered system (the director along the $x$-axis) there are two fluctuational modes corresponding to the splay and bend ($K_s$ and $K_b$). For amplitudes of these modes with the wave vectors $q_y$ and $q_x$ we have:

$$\begin{aligned} I_1 &= T/(K_s q_y^2 + \boldsymbol{P}_0 \boldsymbol{E})\,, \\ I_2 &= T/(K_b q_x^2 + 2\pi P_0^2 |q_x| + P_0 E)\,. \end{aligned} \tag{130}$$

These formulae determine the integral scattering intensity. In the simplest approximation (with only one viscosity coefficient of $\eta_s$ and $\eta_b$, respectively) the line widths are determined by similar expressions:

$$\begin{aligned} \Gamma_s &= (K_s q_y^2 + \boldsymbol{P}_0 \boldsymbol{E})/\eta_s\,, \\ \Gamma_b &= (K_b q_x^2 + 2\pi P_0^2 |q_x| + \boldsymbol{P}_0 \boldsymbol{E})/\eta_b\,. \end{aligned} \tag{131}$$

Variation of these quantities near the phase transition into the smectic-C phase depend on the behaviour of the spontaneous polarization $P_0$. At $q_y = 0$ the scattering intensity by splay fluctuations increases $\sim P_0^{-1}$. Similarly at the transition into the smectic-A phase the scattering intensity (and the line width) is determined by the behaviour of the tilt angle $\theta$.

### 5.4. *Two-dimensional nematics*

In conclusion let us point to a purely hypothetical possibility of the two-dimensional nematic phase first investigated by De Gennes (1974). As is known

(Landau and Lifshitz 1976), the orientational nematic order in two-dimensional systems is not possible. Yet since the fluctuations increase sufficiently slowly (logarithmically), we may assume that the existence of the orientational order in two-dimensional nematic films of finite sizes is possible. The study performed by De Gennes does not take into account the role of the finite size of the film, the influence of the substrate, etc. Actually these facts lead qualitatively to the same effects as the increase of a number of the smectic-B phase layers studied above. And in such systems fluctuations and consequently light scattering are determined by the competition of two-dimensional and three-dimensional contributions. De Gennes studied a purely two-dimensional situation. Then the light scattering cross section is:

$$I(q) \sim a^x/q^{2-x}, \tag{132}$$

where $a$ is a characteristic molecular length, $x = 2T/\pi K$ is the two-dimensional exponent of the correlation function, $K$ is the elastic modulus (for simplicity we use the one-constant approximation). Since $x > 0$, the divergence $I(q)$ at small $q$ is weaker than in the three-dimensional case. We shall not analyse the appropriate formulae in detail.

## 6. Conclusion

The content of this review demonstrates, on the one hand, significant progress in the study and explanation of the physical nature of phase transitions in liquid crystals and, in particular, in the most well studied modifications of liquid crystals – nematics, and, on the other hand, shows that in liquid crystals a large number of interesting physical states is realized. The study of the nature and phase transitions between these states is still in an initial stage of its development. Therefore it is natural that the present review does not reflect completely all the versatile problems associated with phase transitions in liquid crystals. Yet we hope that the review gives a sufficiently complete idea of possibilities of optical methods for the study of phase transitions in liquid crystals and of their importance for liquid crystals in comparison with other systems.

In conclusion we shall briefly mention certain new trends of research the advances of which will no doubt be associated with optical methods.

Let us start with mesophases formed by disc-like molecules and biaxial nematics for which practically no experimental data are available.

The case of discotic liquid crystals is investigated relatively better. In these systems the following phase diagram is possible: solids–discotic liquid crystals–nematics–isotropic liquid. Light scattering in nematics (and naturally in the isotropic phase) does not differ from the case of usual nematics considered in sect. 1. A nematic formed by disc-like molecules differs from the usual nematic only by the fact that the role of the director in it is attributed to the normal with respect to the plane of the dominating molecular orientation. This, in its turn, leads to the different signs (in comparison with the case of usual nematics) of anisotropy of a lot of parameters important for the process of scattering ($\epsilon_a$, $\gamma_1/\gamma_2$, etc.).

Processes of scattering would look more exotic in the discotic phase (a system of liquid columns forming a two-dimensional lattice). Fluctuating parameters in the given case are displacements of liquid columns ($\boldsymbol{u}_\alpha$; $\alpha = 1.2$) and the director orientation $\boldsymbol{n}$. If the director orientation is rigidly connected with the system of liquid columns ($\boldsymbol{n} \parallel \boldsymbol{u}_1 \times \boldsymbol{u}_2$) there remains only scattering by fluctuations of liquid column displacements. In the opposite case there is also an orientational contribution. The respective formulae for the intensity may be easily derived by means of the explicit generalization of the expressions obtained for smectics and nematics. Unfortunately, however, phase transitions between the discotic phase and nematics or solids are first-order phase transitions with a high latent heat. Therefore the region of developed fluctuations and intensive light scattering is absent from them.

This remark also concerns biaxial nematics (which have been discovered in systems of disc-like molecules). In principle depending on the form of the free energy of this liquid crystal various systems of fluctuating parameters and, accordingly, various scattering mechanisms become possible. If intermolecular interactions fix two anisotropy directions in a biaxial nematic, then in light scattering most evidently become three soft modes associated with variations of the orientation of these axes. In principle, the case when only the tensorial order parameter $Q_{\alpha\beta}$ is fixed, is possible. Then all the five modes of fluctuations are soft and will lead to intensive light scattering. This situation has not been studied at all.

A large freedom for various combinations of fluctuating parameters can be found in lyotropic liquid crystals (attracting the attention of scholars at present). Here there is another hydrodynamical variable – concentration. Therefore the combination of scattering by concentration fluctuations (typical of mixtures) with scattering by orientational and translational degrees of freedom (typical of liquid crystals of different kinds) leads to the availability of rich possibilities for them in scattering.

And finally, let us mention possibilities of studying phase transitions in liquid crystals by methods of non-linear optics which, in particular, enable us to obtain data inaccessible for linear optics, i.e., higher moments of the function of the orientational molecular distribution. The application of methods of non-linear optics to the study of liquid crystals (Arakelian et al. 1980) looks very perspective and deserves a special attention in the field of the phase transition physics.

## References


Alexander, S. and J. McTague, 1978, Phys. Rev. Lett. **41**, 702.
Arakelian, S.M., G.A. Liachov and U.S. Chilangazian, 1980, Usp. Fiz. Nauk, **131**, 3.
Arakelian, S.M., L.E. Arushanian and U.S. Chilangarian, 1981, Zh. Eksp. Teor. Fiz. **80**, 1186.
Armitage, D. and F.A. Price, 1977, J. Chem. Phys. **66**, 3414.
Averyanov, E.M. and V.F. Shabanov, 1979, Krystallografiya, **21**, 184.
Belyaev, S.V., L.M. Blinov and V.A. Kizel, 1979, Pis'ma ZhETF, **29**, 17.
Belyakov, V.A. and A.S. Sonin, 1982, Optics of Cholesteric Liquid Crystals (Nauka, Moscow).
Belyakov, V.A., V.E. Dmitrienko and V.P. Orlov, 1979, Usp. Fiz. Nauk, **127**, 221.
Belyakov, V.A., V.E. Dmitrienko and S.M. Osadchii, 1982, Zh. Eksp. Teor. Fiz. **83**, 585.
Bendow, B., J.L. Birman and V.M. Agranovich, 1976, The Theory of Light Scattering in Solids, in: Proc. 1st USSR–USA Symp. (Plenum Press, New York).
Bergman, H. and H. Stegemeyer, 1979a, Z. Naturforsch. **34a**, 253.
Birman, J.L., H.Z. Cummins and K.K. Rabane, eds., 1979, Light Scattering in Solids, Proc. 2nd USSR–USA Symp. (Plenum Press, New York).
Blinov, L.M., 1978, Electro- and Magneto-Optics of Liquid Crystals (Nauka, Moscow).
Brazovsky, S.A. and S.G. Dmitriev, 1975, Zh. Eksp. Teor. Fiz. **69**, 79.
Brazovsky, S.A. and V.M. Filyov, 1978, Zh. Eksp. Teor. Fiz. **75**, 1140.
Brochard, F., 1973, J. de Phys. **34**, 411.
Chandrasekhar, S., 1980, Liquid Crystals (Mir, Moscow).
Chandrasekhar, S., B.K. Sudashiva and K.A. Surech, 1977, Pramana, **9**, 471.
Chatelain, P., 1948, Acta Crystallogr. **1**, 315.
Chatelain, P., 1955, Bull. Soc. Pr. Mineral. Crystallogr. **78**, 262.
Chen, J.R. and T.C. Lubensky, 1976, Phys. Rev. **A 14**, 1202.
Cheng, J. and R.B. Meyer, 1972, Phys. Rev. Lett. **29**, 1240.
Cheng, J. and R.B. Meyer, 1974, Phys. Rev. **A 9**, 2744.
Chistyakov, I.G., 1966, Usp. Fiz. Nauk, **89**, 563.
Chu, B., C.S. Bak and F.L. Lin, 1972, Phys. Rev. Lett. **28**, 1111.
Chu, R.C. and W.C. McMillan, 1977, Phys. Rev. **A 15**, 1181.
Collings, P.J. and J.R. McColl, 1978, J. Chem. Phys. **69**, 3371.
De Gennes, P.G., 1968, Compt. Rend. **266**, 15.
De Gennes, P.G., 1969a, J. de Phys. **30C**, 65.
De Gennes, P.G., 1969b, Phys. Lett. **A30**, 454.
De Gennes, P.G., 1971, Mol. Cryst. Liq. Cryst. **12**, 193.
De Gennes, P.G., 1972, Solid State Commun. **10**, 753.
De Gennes, P.G., 1974, The Physics of Liquid Crystals (Oxford U.P., London).
De Vries, A., 1970, Mol. Cryst. and Liq. Cryst. **11**, 361.
Dolganov, V.K., S.P. Krylova and V.M. Filyov, 1980, Zh. Eksp. Teor. Fiz. **78**, 2343.
Duke, R.W. and D.B Dupre, 1977, Mol. Cryst. Liq. Cryst. **43**, 33.
Dzyaloshinsky, I.E., S.G. Dmitriev and E.I. Kats, 1975, Zh. Eksp. Teor. Fiz. **68**, 2335.
Fan, C., L. Kramer and M.J. Stephen, 1970, Phys. Rev. **A 2**, 2482.
Feigelman, M. and V.L. Pokrovsky, 1981, J. de Phys. **42**, 125.
Filyov, V.M., 1978, Pis'ma ZhETF, **27**, 625.
Flack, J.H. and P.P. Crooker, 1981, Phys. Lett. **A 82**, 247.
Ginzburg, V.L., A.P. Levanyuk and A.A. Sobyanin, 1980, Phys. Rep. **57**, 153.
Gray, G.W., 1953, J. Chem. Soc. **37**, 33.
Haller, J., H.A. Huggins and H.R. Lielienthal, 1973, J. Phys. Chem. **77**, 950.
Halperin, B.I. and D.R. Nelson, 1978, Phys. Rev. Lett. **41**, 121.
Halperin, B.I. and D.R. Nelson, 1979, in: Light Scattering in Solids, 1980, ed., J.L. Birman (Plenum Press, New York).

Hornreich, R.M. and S. Shtrikman, 1980, J. de Phys. **41**, 335.
Hornreich, R.M. and S. Shtrikman, 1981, Phys. Lett. **A 84**, 20.
Hornreich, R.M., M. Luban and S. Shtrikman, 1975, Phys. Rev. Lett. **38**, 1678.
Huang, C., R. Pindak and J.T. Ho, 1974, Phys. Lett. **A 47**, 263.
Johnson, D.L., J.H. Flack and P.P. Crooker, 1980, Phys. Rev. Lett. **45**, 641.
Kamensky, V.G. and E.I. Kats, 1980, Zh. Eksp. Teor. Fiz. **78**, 1606.
Kats, E.I., 1973, Zh. Eksp. Teor. Fiz. **65**, 2487.
Kats, E.I., 1976, Zh. Eksp. Teor. Fiz. **70**, 1394.
Kats, E.I., 1978, Zh. Eksp. Teor. Fiz. **75**, 1819.
Kats, E.I. and M.I. Monastyrsky, 1981, Pis'ma ZhETF, **34**, 543.
Keyes, P.H. and C.C. Yang, 1979, J. de Phys. **40C**, 376.
Landau, L.D. and E.M. Lifshits, 1957, Electrodynamics of Continuous Media (Gostechizdat, Moscow).
Landau, L.D. and E.M. Lifshits, 1976, Statistical Physics, part 1 (Nauka, Moscow).
Lehman, O., 1906, Z. Phys. Chem., **56**, 750.
Lim, K.C. and J.T. Ho, 1979, Phys. Rev. Lett. **43**, 1167.
Liuksiutov, I.P., 1978, Zh. Eksp. Teor. Fiz. **75**, 760.
Lockhart, T.E., 1979, Phys. Rev. **A 20**, 1619.
Lubensky, T.C. and J.H. Chen, 1976, Phys. Rev. **B 17**, 366.
Mahler, D.S., A.H. Keyes and W.B. Daniels, 1976, Phys. Rev. Lett. **36**, 491.
Maier, W. and A. Saupe, 1960, Z. Naturforsch. **A 15**, 287.
Martin, P.C., O. Parodi and P.J. Pershan, 1972, Phys. Rev. **A 6**, 2401.
Martinaud, J.L. and G. Durand, 1972, Solid State Commun. **10**, 815.
McMillan, W.L., 1971, Phys. Rev. **A 4**, 1238.
Meiboom, S. and M. Sammon, 1980, Phys. Rev. Lett. **44**, 882.
Meiboom, S., J.P. Sethna, P.W. Anderson and P.W. Brinkman, 1981, Phys. Rev. Lett. **46**, 1216.
Meyer, R.B., L. Liebert, L. Strzelecki and P. Keller, 1975, J. de Phys. Lett. **36**, 69.
Michelson, A., 1977, Phys. Rev. Lett. **39**, 464.
Neugebauer, H.E.J., 1950, Can. J. Phys. **18**, 292.
Patashinsky, A.Z. and V.L. Pokrovsky, 1975, Fluctuation Theory of Phase Transitions (Nauka, Moscow).
Pelcovits, R.A. and B.I. Halperin, 1979, Phys. Rev. **B 19**, 4614.
Pikin, S.A., 1981, Structural Transformation in Liquid Crystals (Nauka, Moscow).
Pikin, S.A. and V.L. Indenbom, 1978, Usp. Fiz. Nauk, **125**, 251.
Pindak, R.S., C.C. Huang and J.T. Ho, 1974, Phys. Rev. Lett. **32**, 43.
Poggi, Y., P. Atten and R. Aleonard, 1976, Phys. Rev. **A 14**, 466.
Pokrovsky, V.L. and E.I. Kats, 1977, Zh. Eksp. Teor. Fiz. **73**, 774.
Saupe, A. and W. Maier, 1961, Z. Naturforsch. **16**, 816.
Saupe, A., 1969, Mol. Cryst. Liq. Cryst. **17**, 59.
Stegemeyer, H. and K. Bergman, 1980, in: Liquid Crystals of One- and Two-Dimensional Order, eds., W. Helfrich and G. Heppke (Springer, Berlin, Heidelberg, New York).
Stephen, M.J. and J.P. Straley, 1974, Rev. Mod. Phys. **46**, 617.
Stinson, T.W. and J.D. Litster, 1970, Phys. Rev. Lett. **25**, 503.
Stinson, T.W. and J.D. Litster, 1973, Phys. Rev. Lett. **30**, 688.
Stinson, T.W. and J.D. Litster, 1975, Phys. Rev. Lett. **35**, 503.
Suchenko, E.P., 1977, Thesis, Moscow.
Veshunov, M.S., 1979, Zh. Eksp. Teor. Fiz. **76**, 1515.
Wiegmann, P.B. and V.M. Filyov, 1975, Zh. Eksp. Teor. Fiz. **69**, 1466.
Wong, G. and Y.R. Shen, 1974, Phys. Rev. **A 10**, 1277.
Yang, C.C., 1972, Phys. Rev. Lett. **28**, 955.
Yu, L.J. and A. Saupe, 1980, Phys. Rev. Lett. **45**, 1000.
Zel'dovich, Ya.B. and N.V. Tabiryan, 1978, Preprint N°201, FIAN.
Zwetkoff, W., 1942, Acta Physicochim., USSR, **16**, 132.

# PART II

## *Experiment*

H.Z. CUMMINS
P.A. FLEURY
W.I. GOLDBURG
M.V. KLEIN
J.D. LITSTER
K.B. LYONS
J.F. SCOTT
O.A. SHUSTIN
I.A. YAKOVLEV

CHAPTER 5

# Raman Spectroscopy of Structural Phase Transitions

J.F. SCOTT

*Department of Physics, University of Colorado*
*Boulder, Colorado 80309*
*USA*

This work was supported in part by NSF grants DMR 78–02552, DMR 80–25238 and AFOSR Grant NP77–3105A.

*Light Scattering near Phase Transitions*
*Edited by*
*H.Z. Cummins and A.P. Levanyuk*

© *North-Holland Publishing Company, 1983*

# Contents

## 1. Introduction

The idea of phase transitions and spontaneous symmetry breaking has been widely used in physics since the 1930s (Landau 1937). Its application has been significant in both elementary particle physics, where spontaneous symmetry breaking is described by the creation of Higgs mesons, and in solid state physics, where it is described (Cochran*, 1960, 1961) by the collapse of the energy of an optical phonon, i.e., "soft mode". The diagram in fig. 1 is now familiar in a variety of contexts. The free energy of a system is plotted versus a general displacement coordinate $x$. Below a certain critical temperature $T_0$ the lowest free-energy state is obtained at nonzero values of $x$. The transition may be either continuous (second-order) of discontinuous (first-order) depending upon the sign of the quartic coefficient $B$ in the free energy. In many cases it is useful to consider two-dimension displacements $\boldsymbol{x}$, such that below $T_0$ fig. 1 takes on a structure analogous to a "Mexican hat."

Free energies used in these applications are often written as,

$$F(x, T) = A_0(T - T_0)x^2 + Bx^4 + Cx^6 , \tag{1.1}$$

where the linear approximation $A = A_0(T - T_0)$ for the quadratic coefficient results from the mean-field assumption that all particles interact equally in the system; this is equivalent (Stanley 1971) to the assumption of interaction forces of infinite range, and is expected to be a good approximation for systems in which Coulomb forces are dominant. The coefficient $B$ is an explicit function of other variables, such as pressure $P$, applied electric field $E$ (for a ferroelectric), or stress $S$ (for a ferroelastic). If $B(P)$ changes sign, a point in the $(T, P)$ phase space exists at $(T_c, P_c)$ where the transition changes from second order to first order. This is called a *tricritical* point, because in the three-dimension $T$, $P$, $E$ space of a ferroelectric (or $T$, $P$, $S$ of a ferroelastic) there are *three* lines of second-order phase transition boundaries which intersect at $(T_c, P_c)$. This is diagrammed in fig. 1(c) (Scott 1975b). Such situations are well known in ferroelectric $KH_2PO_4$ and SbSI, and have been studied in detail by light scattering for SbSI by Peercy (1976).

These free energy descriptions have been applied successfully in some steady-state but nonequilibrium systems, particularly lasers (Haken 1975, Scott et al. 1975, Scott 1975b). They account for all aspects of optical bistability

*It is interesting to note that both Higgs and Cochran are at the University of Edinburgh.

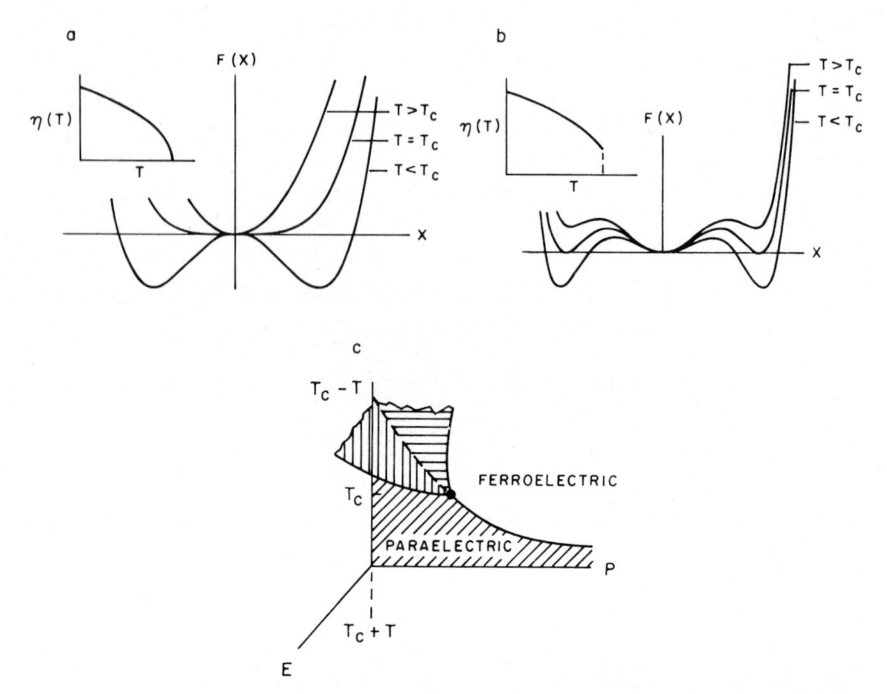


Fig. 1. Free energy for a system undergoing a phase transition plotted versus displacement parameter $x$ for various temperatures. The value of $x$ which minimizes the free energy at a given temperature $T$ is called the expectation value of the order parameter and is usually denoted $\eta(T)$. Figure 1(a) shows the dependence of free energy upon $x$ and $T$ for a continuous transition (in which quartic coefficient $B$ in eq. (1.1) is positive); the insert plots $\eta(T)$ versus $T$ for this case. Figure 1(b) shows the dependence of free energy upon $x$ and $T$ for a discontinuous transition (in which the quartic coefficient $B$ in eq. (1.1) is negative); the insert plots $\eta(T)$ versus $T$ for this case. Figure 1(c) shows second-order transitions as solid lines; first-order, as dashed lines. The black dot is the tricritical point.

observed subsequently by Gibbs et al. (1976) in "optical transistors", effects originally predicted (McCall 1974) by more complicated microscopic theory. A curious parallel in the formalisms of both the laser and elementary particle descriptions is the occurrence of logarithmic terms in the free energies (Mahanthappa and Sher 1980).

Many treatises have reviewed the application of these theoretical ideas to solid state structural phase transitions, along with summaries of experimental studies, such as Shirane (1974). Probably the most comprehensive of these are the text by Lines and Glass (1977) and the reviews by Cowley (1980) and Bruce (1980) and Bruce and Cowley (1980). A simple and readable review for magnetic systems is found in Stanley (1971). Both mean field and self-consistent phonon field descriptions (Gillis and Koehler 1971, 1972a, b, 1974, Koehler and

Gillis 1973, Nettleton 1969, 1971, Pytte and Feder 1969) have been applied to ferroelectric and other displacive phase transitions. In both cases the critical exponents describing temperature dependences of phonon frequencies and other parameters agree well with experiment. In particular, Lines (1974) has shown that the proportionality of the temperature dependences of the order parameter and soft-mode frequency is maintained in the self-consistent phonon approximation even when neither obeys mean-field exponential behavior (Steigmeier and Auderset 1973).

In this review we will emphasize Raman studies of phase transitions which are continuous or very nearly so. For a theoretical discussion of nearly continuous transitions the reader is referred to Larkin and Pikin (1969). Until 1971 all such experimental data appeared to satisfy mean-field predictions. However, from 1971 to the present, extensive studies, particularly those of Müller and his coworkers on $SrTiO_3$ (Müller 1979), have permitted the evaluation of non-mean-field exponents for structural transitions. Definitions of these exponents are given in sect. 3. Raman spectroscopy has generally lacked the precision required to evaluate such exponents with accuracy, and most of the unambiguous non-mean-field results to date are from EPR studies (Aharony 1980, Müller and Berliner 1975, Aharony et al. 1977, Aharony and Bruce 1979). However, see the liquid crystal results of Birgeneau and Litster and their colleagues at MIT (Davidov et al. 1979, Litster et al. 1979) which were obtained via light scattering.

## 2. *Diatomic lattices*

SnTe represents the simplest crystal structure which undergoes a displacive phase transition. PbTe was the first compound to be examined by Cochran (1964) from the soft-mode point of view, and the low-frequency TO phonon in both pure PbTe (at $\sim 30\,\mathrm{cm}^{-1}$) and in SnTe (Pawley et al. 1966, Gillis 1969) was examined in early neutron studies. Since 1974 an extensive amount of Raman spectroscopy has been done on PbTe, SnTe, and GeTe, and on mixed crystals of form $Pb_{1-x}Ge_xTe$ and $Pb_{1-x}Sn_xTe$. This work has been primarily stimulated by the discovery of Curie temperatures which vary inversely and monotonically with carrier concentration in nominally pure SnTe (Murase et al. 1978, Sugai et al. 1977).

All of the members of this family have the $O_h$ rocksalt structure at high temperatures and a $C_{3v}$ rhombohedral structure at low temperatures; although pure PbTe remains cubic down to zero degrees kelvin, germanium doping even at the 1% level is sufficient to produce the $O_h \rightarrow C_{3v}$ transition at finite temperatures. Raman spectroscopy of the soft modes in pure GeTe was first reported by Steigmeier and Harbeke (1970) and has subsequently been extended to SnTe and to mixed crystals by Sugai et al. (1977a), Sugai et al.

(1977b) and Shimada et al. (1977). The results of these authors are compatible with X-ray investigations (Bierly et al. 1963, Goldak et al. 1966, Zhukova and Zaslavskii 1967, Brebick 1971, Hohnke et al. 1972, Muldawer 1973, Bocchi and Chezzi 1975, and Valassiades and Economou 1975). The basic conclusion of all of these authors is that $Pb_{1-x}Ge_xTe$ and SnTe compounds undergo the simple distortion diagrammed in fig. 2. There has been a variety of other experimental techniques employed which complement the Raman soft-mode measurements graphed in fig. 3. Most important are the neutron scattering results (Cochran et al. 1966, Cowley et al. 1969, Lefkowitz et al. 1970, Alperin et al. 1972, Dolling and Buyers 1973, and Iizumi et al. 1975). In addition, infrared measurements of the $q=0$ soft TO phonon are given by Kinch and Buss (1972) and Grosse (1977), and microwave measurements of the phase transition have been reported by several authors (Sawada et al. 1965, Takano et al. 1974, Kawamura 1977, and Nishi et al. 1979).

At present there are two fundamentally different and competitive interpretations of the microscopic dynamics of the phase transitions in this family of crystals. The first is the anharmonic hypothesis. As put forth by Cochran (1960, 1961) and Cowley (1963, 1965), this well-known hypothesis assumes that anharmonic corrections to the soft TO phonon frequency (at $q=0$ in this case) can be expressed as,

$$\omega_{\mathrm{TO}}^2 = \omega_0^2 + \frac{\hbar}{4N} \sum_j \frac{\Phi[\omega_0, \omega_0, \omega_j(q), \omega_j(-q)]}{\omega_j} [2n(\omega_j + 1)] , \tag{2.1}$$

where $\omega_0$ is the bare, harmonic TO frequency; $\Phi[\omega_0, \omega_0, \omega_j(q), \omega_j(-q)]$ is a fourth-order anharmonic coefficient which couples $\omega_0(q=0)$ with pairs of phonons on branch $j$ at wave vectors $\pm q$; and where $N$ is the number of unit cells and $n$ is the Bose population factor. If $\Phi$ is approximated as a constant over the Brillouin zone and the density of states is represented by a Debye distribution, eq. (2.1) becomes:

$$\omega_{\mathrm{TO}}^2 = \omega_0^2 + \frac{9\hbar\Phi}{2\omega_{\mathrm{D}}^3} \int_0^{\omega_{\mathrm{D}}} \omega \coth\left(\frac{\hbar\omega}{2kT}\right) \mathrm{d}\omega , \tag{2.2}$$

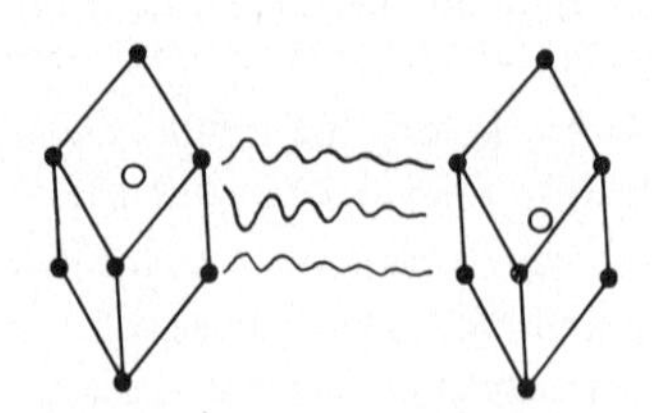

Fig. 2. Eigenvector for structural distortion in SnTe and its isomorphs. This soft mode induces a transition from $O_h$ to $C_{3v}$ point symmetry.

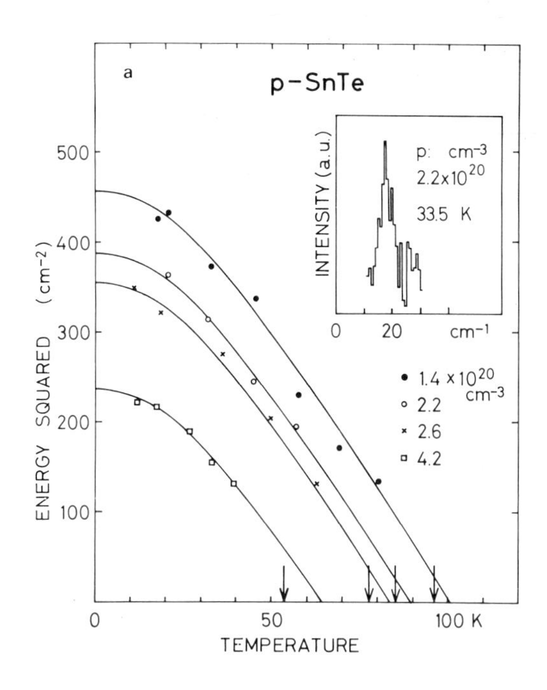


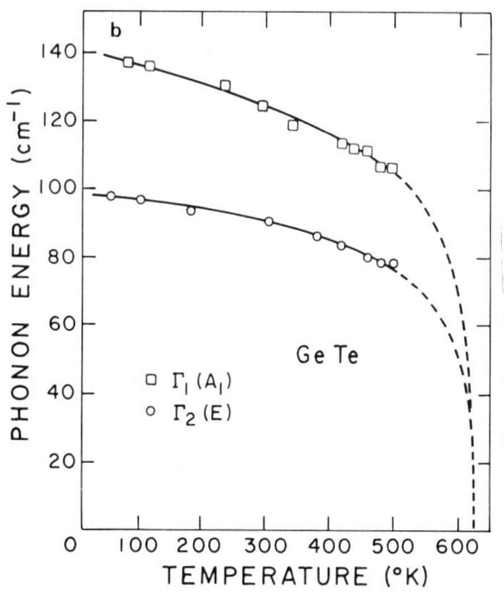


Fig. 3. Soft-mode frequency dependences upon temperature for SnTe and GeTe family. After Sugai et al. (1977), Shimada et al. (1977), and Steigmeier and Harbeke (1970).

where $\omega_D$ is the Debye frequency and $k$ is the Boltzmann constant. Gillis has shown (1969) that this theory gives good agreement with $\omega(T)$ data from neutron scattering experiments on SnTe having carrier concentration of $\sim 8 \times 10^{20}\,cm^{-3}$.

A very different approach to the soft-mode problem has been developed in a long series of papers by Kristoffel and Konsin (1967, 1968, 1971, 1973, 1976, Konsin 1976, 1978, Konsin and Kristoffel, 1971, 1973). These authors have developed a purely vibronic theory of soft modes, originally with application to $BaTiO_3$ but lately for the SnTe family; Kawamura et al. (1974, 1975) have followed Kristoffel and Konsin in this application to SnTe and have fitted the variation of $T_c$ with carrier concentration to a purely phenomenological vibronic model in which a single parameter (a deformation potential of about 10 eV) suffices to fit the $T_c$ data. Despite a great abundance of independent data taken near $T_c$ in this family of crystals, including electrical conductivities (Novikova et al. 1975, Kobayashi et al. 1976, Katayama 1976, and Murase et al. 1976), elastic coefficient behavior (Seddon et al. 1975, 1976, and Rehwald and Lang 1975), thermal measurements of expansion and specific heat (Shelimova et al. 1965, Smith et al. 1976, Novikova and Shelimova 1967, Hatta and Kobayashi 1977, and Hatta and Rehwald 1977), and Mossbauer spectroscopy (Keune 1974, Fano and Ortalli 1974), it has not been possible to obtain clear evidence ruling out one of the two hypotheses. The difficulty is that any change in the carrier concentration (due either to doping or to stoichiometric deficiencies) is always accompanied by important lattice changes. If the carrier concentrations could be adjusted in a purely electronic way, the hypotheses could be tested. Since p–n junctions of $Pb_{1-x}Ge_xTe$ have been fabricated, it is possible that steady-state, nonequilibrium experiments with injected carriers could be performed, but so far only relatively routine capacitance measurements on such p–n junctions have been reported (Bate et al. 1970, Antcliffe et al. 1973).

From a purely spectroscopic viewpoint the Raman studies of SnTe and $Pb_{1-x}Ge_xTe$ are relatively complete. Early reports by Brillson and Burstein (1971) and Brillson et al. (1974) gave some erroneous identifications, particularly for lines at 126 and 144 $cm^{-1}$, which were thought to be unscreened LO phonons in surface depletion layers, but which were subsequently assigned as free Te and Te-O vibrations in nonstoichiometric surface regions (Cape et al. 1977). Phonon data reported for $Pb_{0.95}Ge_{0.05}Te$ (Sugai et al. 1979) are especially complete, including anisotropy, and are shown below in figs. 4 and 5.

The vibronic theory of Kristoffel and Konsin is reminiscent of the theories of Labbe and Friedel (1966a, b), Pytte (1970b), Anderson and Blount (1965), and Sham (1971) for the A-15 structures $Nb_3Sn$ and $V_3Si$. All of these authors proposed basically vibronic mechanisms for the structural phase transitions in these materials. Their work suggested that partially filled, narrow d-bands played a key role in the elastic coefficient anomalies and structural phase

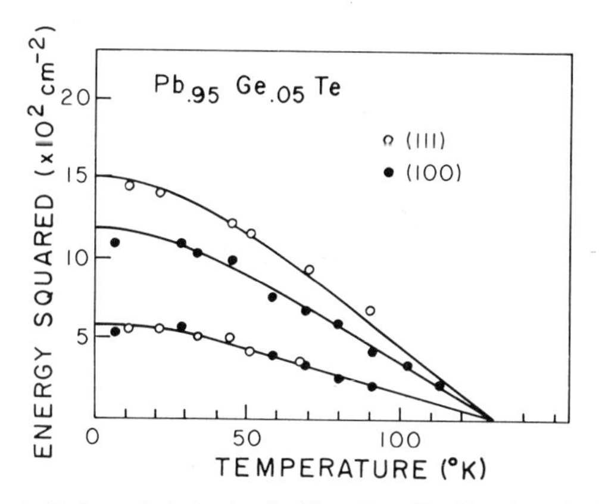


Fig. 4. Soft-mode behavior in $Pb_{0.95}Ge_{0.05}Te$ (Sugai et al. 1979).

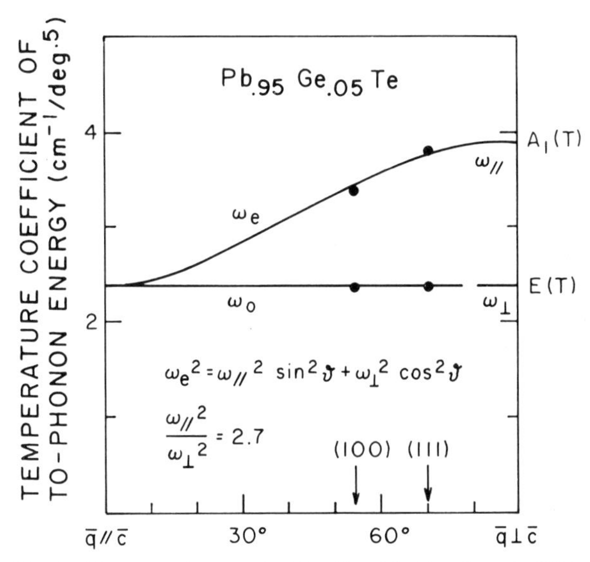


Fig. 5. Phonon anisotropy in $Pb_{0.95}Ge_{0.05}Te$ (Sugai et al. 1979).

transitions. The difference between $Nb_3Sn$ or $V_3Si$ and the SnTe, $Pb_{1-x}Ge_xTe$ family, however, is that the latter exhibit distinct softenings of their $q = 0$ transverse optical phonons.

There is no question that electron–phonon interactions are at least partially responsible for the phase transitions in A-15 family high-temperature superconductors; this was demonstrated experimentally in 1972 by Maita and Bucher, who observed a 0.3 K decrease in the transition temperature in the presence of a 9 T magnetic field. More recently a very small shift in $T_c$ was also observed in $Pb_{0.99}Ge_{0.01}Te$ by Murase et al. (1976), which supports the vibronic hypothesis in that material. However, in the ferroelectric insulators exemplified by $BaTiO_3$ and $SrTiO_3$, evidence is overwhelmingly for a purely anharmonic interpretation. In $SrTiO_3$ it has been possible to separate the vibronic and anharmonic effects on the phase transition. Deis et al. (1969) measured $T_0$ as a function of carrier concentration for both self-doped (oxygen deficient) and impurity doped samples. At $n = 10^{20}\,cm^{-3}$ the oxygen deficient samples had $T_0$ decrease approximately 25 K (to 80 K), whereas the $n = 10^{20}\,cm^{-3}$ impurity doped sample has the same $T_0$ as insulating specimens. This shows that vibronic contributions to the transition dynamics are negligible, and anharmonic effects dominant; the same conclusion was reached independently more recently by Comes et al. (1981), who found absolutely no effect of large magnetic fields upon $T_0$. Despite these results it is still claimed elsewhere that the vibronic mechanism is dominant for titanates: Bersucker et al. (1976, 1978ab), Konsin and Ord (1980)*. Even in the case of the SnTe family, where the majority opinion now seems to favor a vibronic interpretation (see especially Volkov and Pankratov (1978)), we find that a completely anharmonic interpretation of the transition dynamics still appears compatible with the existing data (Bussman-Holder et al. 1980, Bilz et al. 1982). Unfortunately, no simple experiments exist, to this author's knowledge, on systems such as $Pb_{1-x}Ge_xTe$ in which the changes in carrier concentration are not also accompanied by local alterations of stoichiometry; and the latter of course have a strong effect upon anharmonicity.

Several other aspects of importance have had attention focussed upon them by the work of Konsin and Kristoffel. For example, they propose explicitly that crystals which have direct bandgaps should, on average, exhibit phase transitions which do not alter the size of the unit cell (e.g., proper ferroelectric transitions), whereas those with indirect gaps should have statistically more cell-doubling, antiferroelectric transitions. I know of no data for or against this conjecture. This line of argument obviously applies to incommensurate systems. In metallic incommensurates the electron–phonon interactions near $q_F$ have been described as gian Kohn anomalies and yield a very physical picture of the phase-transition dynamics. No such physical picture is generated by the data on incommensurate insulators, where the algebraic formalism of Lifshitz invariants (gradient terms in the free energies) arises from Dzyaloshinskii's work on spiral magnets (1959) and has no obvious connection to electron–phonon interactions. An important gap for us to bridge in the next few years is the connection between band structure and the critical wave-vector location

*The important shift in $T_c(H)$ reported by Ismailzade et al. (1981a,b) must await independent confirmation since it appears incompatible with NMR measurements, usually also done around 10kG, which show no $H$ dependence of $T_c$. Also, repetition of the experiment of Ismailzade on Rochelle salt by A. Levstik (private communication) showed no effect.

in the Brillouin zone for incommensurate insulators. Unfortunately, all such crystals known thus far have complicated formulas and are not amenable to band-structure calculations. The best discussion to date of the anharmonic versus vibronic hypotheses for the SnTe family is given by Natori (1976) and by Kawamura (1979). (See, however, Motizuki et al. (1981) Haque and Hardy (1980), and Bilz et al. (1982).)

## 3. *Molecular crystals*

During the past five years Raman studies of second-order phase transitions have been extended from ionic crystals to organic, molecular crystals. Two prototype systems have been emphasized: The first is chloranil (p-tetrachlorobenzoquinone), which undergoes a now-classic soft mode, displacive phase transition with underdamped soft-mode frequency decreasing to $\omega \sim 2\,\mathrm{cm}^{-1}$ at $T_0$. The second system is s-triazine, which has a continuous order–disorder transition from an ordered $C_{2h}$ structure to a disordered $D_{3d}^6(R\bar{3}c)$ point-group symmetry. Since the $D_{3d}$ high-temperature phase has doubly degenerate vibrations which necessarily split into resolvable doublets as temperature is reduced below $T_0$, this frequency splitting provides a convenient measurement of the temperature dependence of the order parameter.

### *3.1. Chloranil (p-tetrachlorobenzoquinone)*

Figure 6 (Terauchi et al. 1975) shows the structure of chloranil as well as the soft-mode eigenvector, which is that of a rigid, out-of-phase rotation of chlorinated benzene rings. Most of the information we have concerning this $C_{2h}(P2_1/a) \rightarrow P\bar{1}(C_i^1)$ transition has come from the work by Chihara and colleagues (Chihara and Masukane 1973, Chihara and Nakamura 1973, Chihara et al. 1971, 1973, Terauchi et al. 1975). Earlier structural work was done by Chu et al. (1962), and Richardson (1963), however; and the Raman soft-mode results shown in fig. 7a were obtained by Hanson (1975); the soft-mode frequency is proportional to the square root of the X-ray superlattice intensity (Terauchi et al. 1975) shown in fig. 7b. This is expected, since in mean-field theory both are proportional to the order parameter $\phi(T)$. See also Yamada et al. (1974). Both the rigid, out-of-phase rotation and the underdamped soft-mode behavior of chloranil are strikingly reminiscent of strontium titanate (Fleury et al. 1968). In chloranil the soft mode lies at $(0,\,0,\,1)\pi/c$ of the high-temperature phase. $SrTiO_3$ continues to serve as the paradigm for this kind of antiferro-distortive transition, and in addition to the related lanthanide aluminates studied a decade ago (Scott 1969), we find the same kind of transition showing up in more recent work on $Cs_2LiFe(CN)_6$ and $Cs_2LiCo(CN)_6$ cyanides. In the latter structures the tight covalent CN "molecules" form face centers of octahedra – as the oxygen ions do in $SrTiO_3$ – and undergo nearly rigid, out of phase rotations similar to those in strontium titanate. This shows dramatically that the basic cause of such phase transitions is a geometric, steric instability which has little to do with subtle details of band structure or bonding

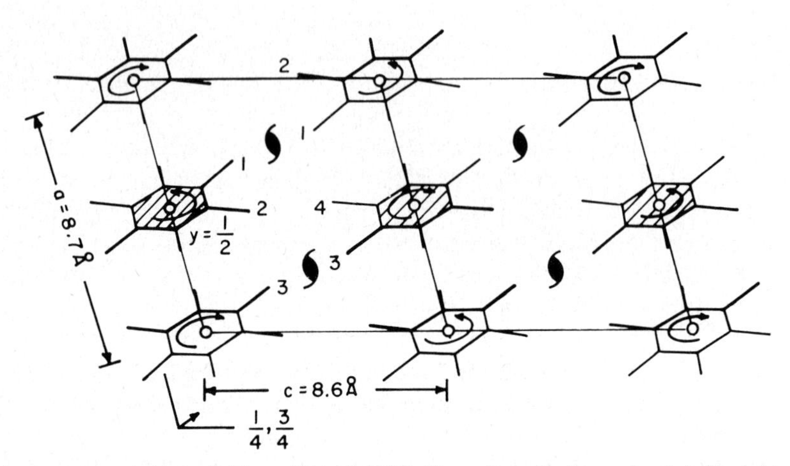


Fig. 6. Structure of chloranil (Terauchi et al. 1975). The soft-mode eigenvector is indicated by the arrows.

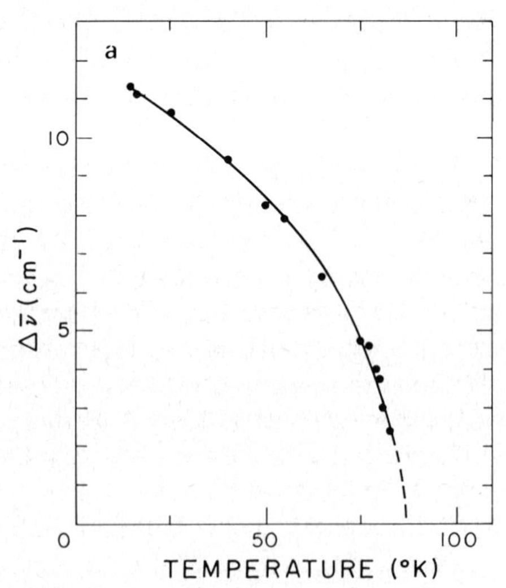


Fig. 7a. Soft-mode behavior near the displacive, second-order transition in chloranil (Hanson 1975).

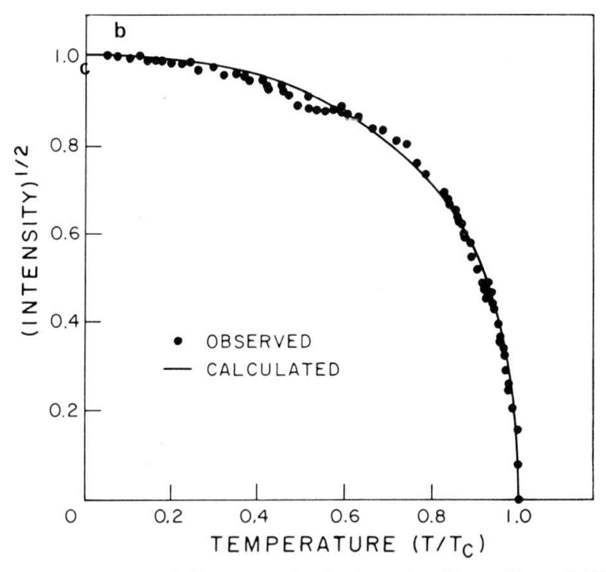


Fig. 7b. Square root of X-ray superlattice intensity (Terauchi et al. 1975).

(Konsin 1978). The data for $Cs_2LiM(CN)_6$ shown below in fig. 8 are from Ryan and Swanson (1976); see also Swanson and Jones (1974) and Jones et al. (1974). The transition studied at 310 K is $D_{4h}(P4/mnc) \rightarrow C_{2h}(P2_1/c)$. At 348 K the $O_h(Fm3m)$ phase is reached via a first-order transition.

### 3.2. *s-Triazine* ($C_3N_3H_3$)

Chloranil, to this author's knowledge, is the only purely molecular crystal with an underdamped soft mode characterizing its phase-transition dynamics. A continuous phase transition more typical of organic crystals is that in s-triazine, $C_3N_3H_3$. This material has a transition at $200 \pm 5$K in the fully deuterated form. Its structure is shown in fig. 9. Extensive Raman studies were made independently by Daunt et al. (1975) and Elliott and Iqbal (1975). Some of their results are shown in fig. 10. In fig. 11 it is shown (Scott 1975) that the E′-symmetry mode splittings $\Delta\omega(T)$ scale with both the X-ray data (Coppens and Sabine 1968) and the NQR results (Zussman 1974). The spectra in the high-temperature phase agree with the predictions of $D_{3d}$ from X-ray studies (Siegel and Williams 1954, Wheatley 1955); the data below $T_0$ indicate a $C_{2h}$ point group symmetry.

We note that no evidence of direct scattering from order-parameter fluctuations in s-triazine have been reported (central mode scattering). Since the order parameter for this $C_{2h} \rightarrow D_{3d}$ transition is not dipolar, infrared and/or dielectric measurements are not apt to be useful. Large single crystals are

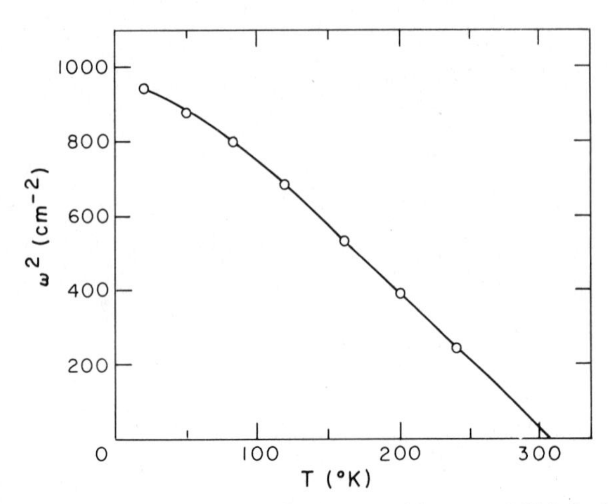


Fig. 8. Soft-mode behavior in $Cs_2LiM(CN)_6$, after Ryan and Swanson (1976). In these systems the phase-transition dynamics resemble those of $SrTiO_3$. The cyanogen ions are tightly bound and sit at the face centers of octahedra; the transition involves the out-of-phase rotation of CN octahedra in adjacent cells. This is the same as in $SrTiO_3$, with the CN ions playing the role of the oxygen ions in strontium titanate.

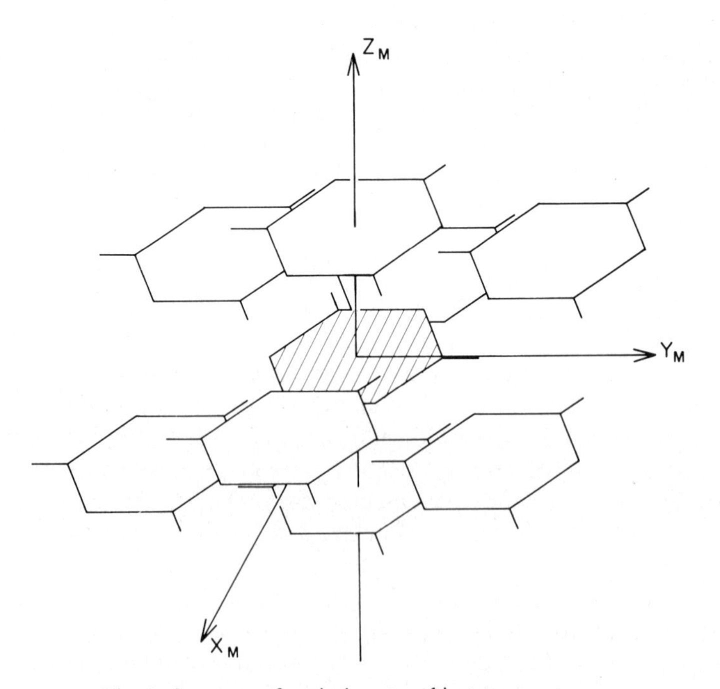


Fig. 9. Structure of s-triazine at ambient temperatures.

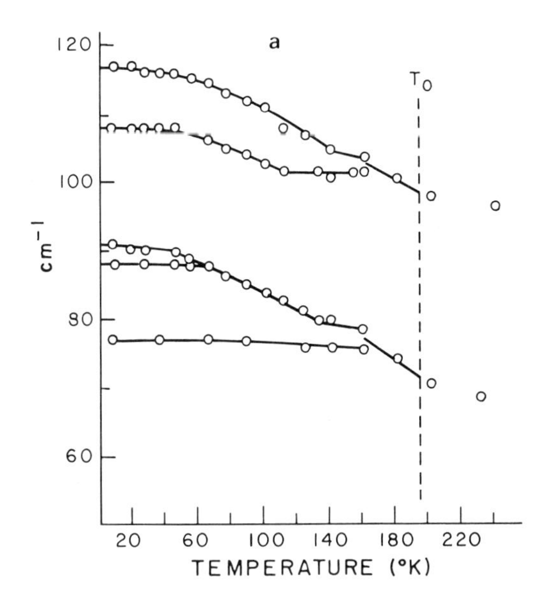


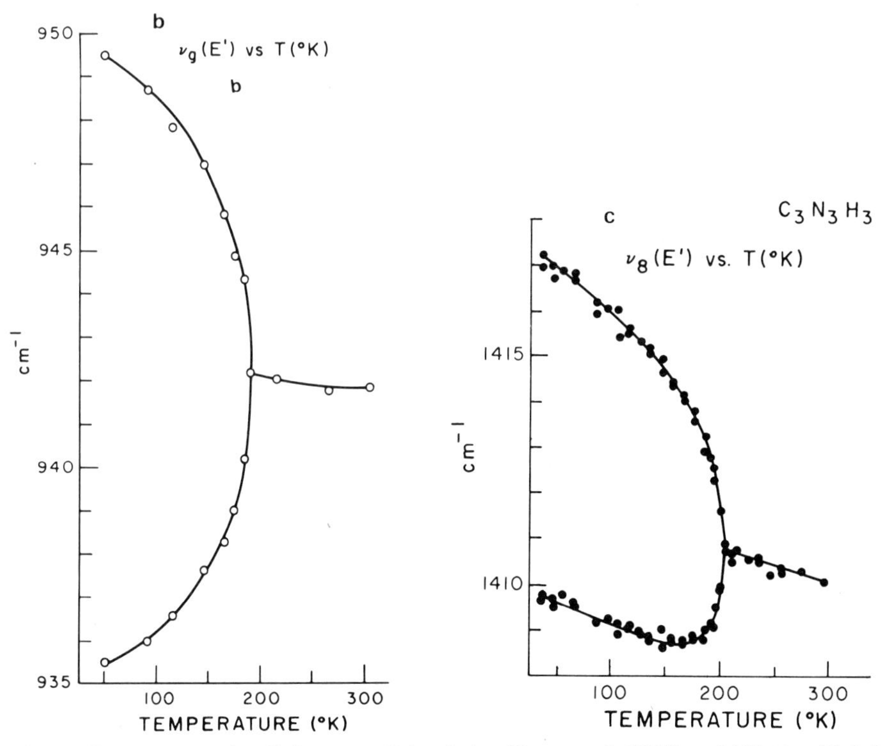


Fig. 10. Degenerate mode splittings near $T_0$ in triazine (Daunt et al. (1975), and Elliott and Iqbal (1975)).

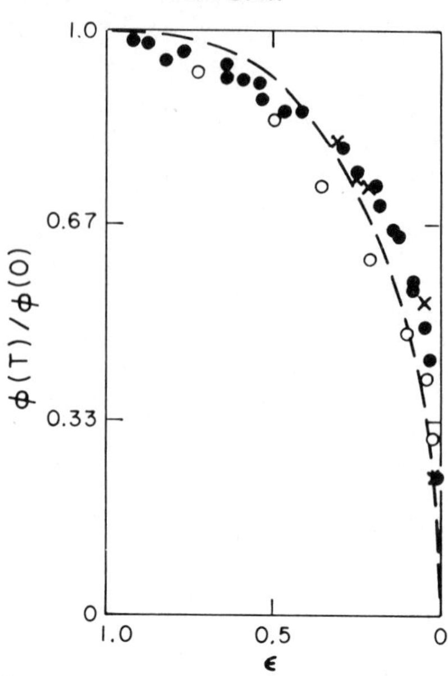


Fig. 11. E′-symmetry mode splittings versus $T$, illustrating the scaling with X-ray displacement data (Scott 1975). Solid curve is mean field; open circles $\Delta\nu_9$ (deuterated); solid circles $\Delta\nu_8$ (undeuterated); and (x) are X-ray data for the angle between a* and (Hoh̄h).

available, however, and thus Rayleigh/Brillouin measurements are suggested.

Very recently some attempts have been made to analyze all of the diverse experimental data on the triazine transition (Rae 1978, 1979, Mason and Rae, 1968, Smith and Rae 1978a,b, Raich and Bernstein 1980). The latter analysis uses the fact that the transition is very slightly first-order with a latent heat of 75 J/mol (Briels and Miltenburg 1979) and fits the available data to a first-order, mean-field description. Such a slightly first-order transition is compatible with the Lifshitz criterion, since the number of symmetry elements is reduced by a factor of three in the $D_{3d}\rightarrow C_{2h}$ transition. Cowley's criterion (1976) for an elastic instability in a $D_{3d}$ point group symmetry crystal is that $(c_{11}-c_{12})c_{44}-2c_{14}^2=0$. Heilman et al. (1979) show via inelastic neutron scattering that this situation is obtained by $c_{44}$ approaching zero in triazine; this implies that the order parameter is proportional to the $e_5$ strain. Note that $c_{11}-c_{12}$ also softens, however. Folk et al. (1976, 1979) have included triazine as among a very small class of ferroelastics with two-dimensional order parameters; however, Raich and Bernstein show, to the contrary, that the order parameter is restricted to a one-dimensional subsection of momentum space. In this case no logarithmic corrections to mean-field behavior are expected.

Bernstein and Lal (1980) have extended these triazine Raman studies to trioxane ($C_3O_3H_6$) which has an analogous phase transition [$C_{3v}(R3c) \xrightarrow{?} C_{2h}$] at

63K*. The transition appears to be discontinuous. When we recall that the $D_{3d}$ symmetry of triazine is only a statistical average for a disordered array of rings with local $C_{3v}$ symmetry, it appears that the transitions in trioxane and triazine are probably isomorphic. It is not clear whether the trioxane phase transition at 63 K and ambient pressure is the same phase transition as that reported at high pressure by Brasch et al. (1970).

In summarizing this section, it is clear that molecular crystals display continuous phase transitions of both displacive and order–disorder character and that both are amenable to Raman spectroscopy and analysis. The next decade should extend such studies of organics and involve more chemists and solid state physicists in collaboration.

### *3.3. Tris-sarcosine calcium chloride*

Tris-sarcosine calcium chloride, $(CH_3NHCH_2COOH)_3 \cdot CaCl_2$, is a ferroelectric with a Curie temperature of approximately 127 K (Makita 1965). Its structure shown in fig. 12 consists of three sarcosine

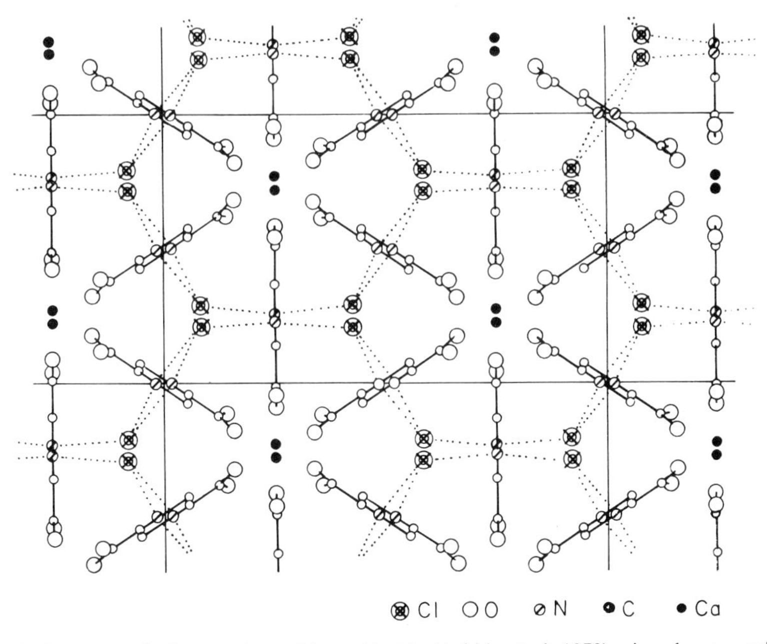


Fig. 12. Structure of tris-sarcosine calcium chloride (Ashida et al. 1972), view down *a*-axis. The N–H–Cl hydrogen bonds are dotted.

*Unfortunately an earlier Raman study of s-trioxane went down only to 100 K and erroneously assumed "that no (phase transition) would occur down to 30K"; Thomas (1977).

```
    H       H    O
    |       |   //
H — C — N — C — C
    |   |   |    \
    H   H   H     OH
```

molecules contained within an ionic $CaCl_2$ lattice. The sarcosine nitrogen ions are hydrogen bonded to chlorines (Sawada et al. 1977).

At room temperature, tris-sarcosine calcium chloride (abbreviated TSCC) crystallizes in the Pnma ($D_{2h}^{16}$) space-group symmetry with four formula groups per primitive cell. Below $T_c$ the structure (Pepinsky and Makita 1962, Makita 1965) is thought to be $Pn2_1a$ space group ($C_{2v}^{9}$) with the $b$-axis becoming the ferroelectric axis (Ashida et al. 1972). The dielectric behavior near $T_c$ is shown in fig. 13. The reported value for the Curie–Weiss constant, $(4.7 \pm 0.2)$K, is small and comparable with those of other hydrogen-bonded systems. Note that the dielectric anomaly shown in fig. 13 persists from $10^4$Hz to $10^{10}$Hz but does exhibit some dispersion over this range of frequencies. Makita (1965) originally reported a $(0.4 \pm 0.1)$K discrepancy between $T_c$, the Curie temperature extrapolated from dielectric data, and $T_0$, the observed transition temperature.

Specific-heat measurements show a lambda-shaped anomaly at about 127K from which a transition energy of $\Delta Q = 47 \pm 5$ cal/mole and an entropy change of $\Delta S = 0.41$ cal/mol degree can be deduced. These values are compatible with other glycine ferroelectrics, which vary from about 0.1 to 1.1 cal/moledegree (Triebwasser 1957, Hoshino et al. 1957, Mitani 1964). However, the Curie constant of 4.7K is almost two orders of magnitude smaller than those of the other glycine ferroelectrics, glycine-$AgNO_3$, -$H_2BeF_4$, or -$H_2SO_4$.

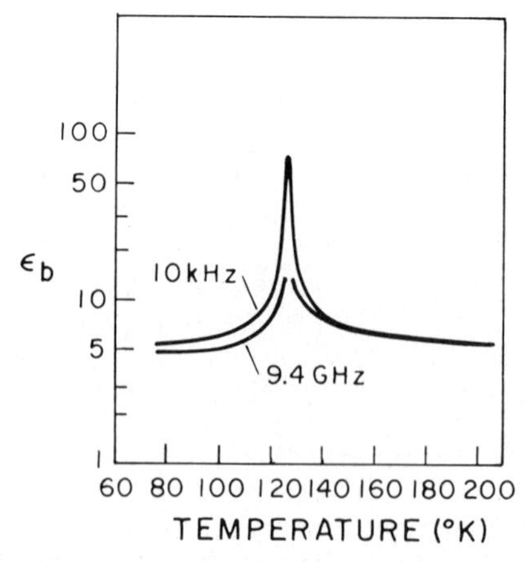


Fig. 13. Dielectric data near $T_c$ in tris-sarcosine calcium chloride (Makita 1965), $\epsilon_b$ is plotted at 10 kHz and 9.4 GHz, showing dispersion.

Sorge and Straube (1977, 1979) conclude from dielectric data that the transition is perfectly continuous. They shared the view of all earlier authors (Windsch et al. 1975, Windsch 1976, Levstik et al. 1976) that the transition is of the order–disorder variety, a conclusion we shall show is false, or at best, oversimplified. Sawada et al. (1977) have shown that the $D_{2h}$ phase is ferroelastic, with a hypothetical higher-temperature hexagonal phase. Using these ferroelastic properties Sorge and Straube applied uniaxial stress to obtain single-domain samples with $\epsilon_b$ at $T_c$ of 2000, in agreement with Ivanov and Arndt (1978). More recent measurements by Windsch yield ~ 5000. The single-domain measurements yielded Curie constants from 20 to 58K, an order of magnitude larger than Makita's results, and comparable to those in glycine ferroelectrics.

Accurate measurements of the critical exponents $\beta$, $\gamma$, and $\delta$ were reported by Levstik et al. (1976). These exponents are defined below:

$$\lim_{T\to T_c+} \epsilon(T) = C(T - T_c)^{-\gamma}, \tag{3.1}$$

$$\lim_{T\to T_c-} \epsilon(T) = C'(T_c - T)^{-\gamma'}, \tag{3.2}$$

$$\lim_{T\to T_c-} P_s(T) = B(T_c - T)^{\beta}, \tag{3.3}$$

$$\lim_{E\to 0} P(E) = DE^{1/\delta}. \tag{3.4}$$

Some care must be used in applying eqs. (3.1) to (3.4) to the fitting of experimental data, however. For example, Levstik et al. (1976) show that impurity effects cause rounding within ~ 0.1 K of $T_c$ for eqs. (3.1) to (3.3), and similar effects distort the predictions of eq. (3.4) for very small fields. The net result is that eqs. (3.1) to (3.3) should be employed for temperatures typically more than 0.1 K from $T_c$ but less than 15–20 K away. Similarly, eq. (3.4) does not truly apply as $E\to 0$, but for $E \gg 25$ V/cm but less than, say 10kV/cm.

For a ferroelectric, the soft-mode frequency $\omega(T)$ will generally *not* be proportional to the order parameter $P_S(T) \propto (T_c - T)^{\beta}$; but under less restrictive assumptions than mean field, i.e., that the Lyddane–Sachs–Teller relation remains valid, $\omega_{TO}(T)$ will vary as $(T_c - T)^{\gamma'/2}$:

$$\omega_{TO}^2(T) \propto \frac{1}{\epsilon(T)} \propto (T_c - T)^{\gamma'}. \tag{3.5}$$

See Fleury (1972 a,b) for further discussion of this point.

In general, the soft-mode frequency will be expected to scale not as the order parameter $\phi(T)$ for any phase transition, with its critical exponent $\beta$, but as the square root of the susceptibility $\chi(T)$, with its associated exponent $\gamma$. Even this statement naively neglects central modes, however.

For TSCC Levstik et al. find $\gamma = \gamma' = 1.00 \pm 0.02$; $\beta = 0.49 \pm 0.01$; and $\delta = 3.1 \pm 0.1$. All of these agree within their uncertainties with the mean-field values $\gamma = \gamma' = 1$, $\beta = \frac{1}{2}$, $\delta = 3$. (This is expected for uniaxial ferroelectrics (Craig 1966, Stanley 1971).) They are in complete disagreement with the values $\beta \sim \frac{1}{3}$ reported by Windsch (1975, 1979) deduced from EPR of Mn-doped TSCC.

We note, however, an extremely curious aspect of the data published by Levstik et al.: whereas the ratio $C'/C$ of the Curie constant in eqs. (3.1) and (3.2) is required to be 2:1 in mean-field systems (Stanley 1971) and is observed experimentally to be approximately 2:1 in other ferroelectrics, such as TGS (Triebwasser 1958), the data on tris-sarcosine calcium chloride yield a ratio 1:2 – exactly the reverse of the theory! This reversal seems not to have been noticed by Levstik et al., and could be due to strain coupling (see sect. 4) or to an improper ferroelectric transition. Improper transitions, if second order, have $\lambda$-shaped dielectric anomalies with increases as $T \to T_c+$ much more rapid than those as $T \to T_c-$, in agreement with the TSCC data. However, Bartuch and Windsch (1972), Lippe et al. (1976), Volkel et al. (1975), and Windsch (1976) and Windsch et al. (1975) show via EPR that the application of an electric field along $\hat{b}$ is equivalent to lowering $T$ below $T_c$. This shows that there is *not* a cell doubling transition at $T_c$, at least as far as the Ca sites are concerned, and thus that the transition is probably proper (it is possible to have improper transitions which do not alter the primitive cell size).

The paradox posed by the discrepancy between these mean-field exponents of Levstik et al. and the $\beta = \frac{1}{3}$ values of Windsch and of Sorge and Straube is easily resolved: Sorge and Straube obtain $\beta = \frac{1}{2}$ for $T_c - T$ between 0.15K and 3.2K but claim $\beta = \frac{1}{3}$ for temperatures more than 3.2K from $T_c$. Their problem is that they do not recognize that $\beta$ can be obtained only from data near $T_c$, and that mean-field theories do *not* predict a $(T_c - T)^{1/2}$ dependence of $P_s$ down to $T = 0$ (this would in fact violate the laws of thermodynamics, which require $\partial P_s/\partial T|_{T=0} = 0$). The mean-field theory predicts, approximately,

$$P_s(T) = P_s(0) = \tanh[T_c P_s(T)/P_s(0)T]\,, \tag{3.6}$$

which varies as $(T_c - T)^{1/2}$ near $T_c$ and is flat as $T \to 0$. This may be verified by expanding $\tanh x \approx x - \frac{1}{3}x^3$. Attempts to fit $\beta$ as $P_s \sim (T_c - T)^\beta$ over a wide range of low temperatures will always yield a physically meaningless value of $\beta$ which is less than $\frac{1}{2}$. This should *not* be interpreted as evidence for non-mean-field dynamics, but only as the direct result of the thermodynamic requirement $\partial P/\partial T|_{T=0} = 0$.

A more serious paradox for TSCC is presented by the Raman studies made independently in 1979 by Feldkamp et al. (1980) and Smolensky's group in Leningrad (Smolensky et al. 1979). As shown in fig. 14 these groups find an underdamped soft mode in the ferroelectric phase! That in itself contrasts with the hypothesis, based upon earlier thermal work, of an order–disorder transition.

The puzzle deepens in complexity when one observes that there is a negligible LO/TO splitting of the soft mode. Smolensky's group used a back-scattering geometry to prove this, measuring A(LO) via $b(aa)\bar{b}$ geometry and A(TO) via $a(bb)c$. Feldkamp et al. employed the right-angle scattering geometries originally discussed by Scott and Porto (1967), with faces cut at 45° to the crystallographic axes: A(TO) via $[b+c(aa)b-c]$, and A(LO) via $[b-c(aa)-b-c]$. Here the notation (Damen et al. 1966) $x(yy')x'$ designates incident photon propagating along $x$ with polarization along $y$; scattered photon propagating along $x'$ with polarization along $y'$. The results of these two groups are nearly the same.

The negligible LO/TO splitting, together with the Lyddane–Sachs–Teller (1941) relationship,

$$\frac{\epsilon_0(T)}{\epsilon_\infty}=\left[\frac{\omega_{\mathrm{LO}}(T)}{\omega_{\mathrm{TO}}(T)}\right]^2, \tag{3.7}$$

shows that the soft mode predicts no dielectric divergence until very near $T_c$.

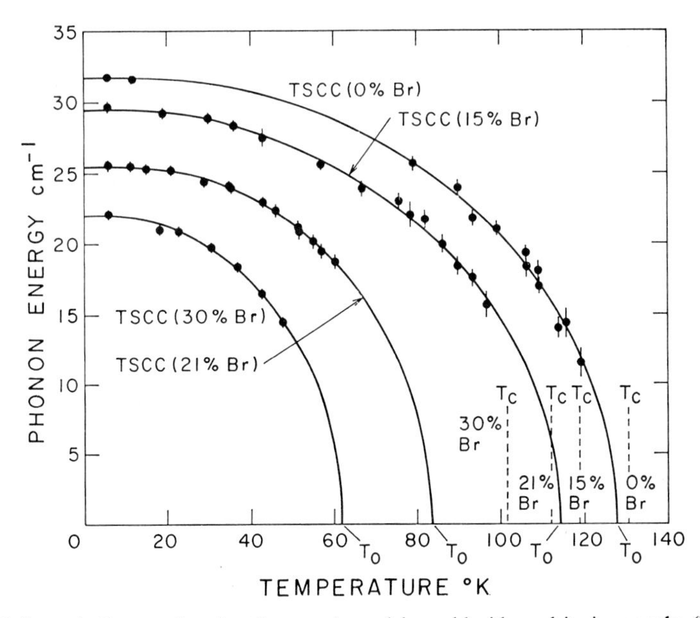


Fig. 14. Soft-mode Raman data in tris-sarcosine calcium chloride and its isomorphs (Feldkamp et al. 1980 and Smolensky et al. 1979). The solid curves are least-squares fits to eq. (3.6), but for Brillouin function with spin 3, rather than spin $\frac{1}{2}$ (spin $\frac{1}{2}$ Brillouin function is tanh).

Note that in the ferroelectric phase there must be some LO/TO splitting by symmetry; soft modes must be totally symmetric in the low-symmetry phase, and all totally symmetric representations are split into nondegenerate TO/LO pairs in all ferroelectric point groups.

In an attempt to explain their data, Smolensky et al. proposed that the phase transition in TSCC is improper. However, Windsch (1979) does not detect any doubling of the primitive cell in TSCC in his EPR studies, which would be unlikely for an improper transition. No evidence for cell doubling has been provided yet by nuclear magnetic resonance, either (Blinc et al. 1970, 1975, Kadaba et al. 1975). However, these authors do establish a large change in the bond length of the N–H...Cl hydrogen bond as $T$ nears $T_c$. Thus, the dynamics of the phase transition are seen to be rather displacive and not simple proton tunnelling in a temperature-independent double well. All of the Raman studies confirm that the unit cell remains unchanged in size at $T_c$.

In very recent work (Kozlov et al. 1983) the most puzzling aspects of the ferroelectric phase transition in TSCC have been explained. Far infrared spectroscopy in the 2 to 15 $cm^{-1}$ region, using a backward-wave oscillator as source, has permitted measurement of the soft mode in the paraelectric $D_{2h}$ phase, as well as determination of oscillator strengths and LO/TO splittings in both ferroelectric and paraelectric phases. The results are shown in fig. 15. The principal conclusion is that the transition in TSCC is proper and displacive; there is no doubling of the primitive unit cell. What makes the TSCC transition unique among displacive ferroelectrics is the extremely small oscillator strength of the soft mode. Expressed in terms of Cochran's original shell model description, the TO and LO frequencies at $q = 0$, given by sums and differences of short-range and long-range terms

$$\mu\omega_{\mathrm{LO}}^2 = R_0' + \frac{8\pi(Ze)^2}{9\epsilon_\infty V}, \tag{3.8}$$

$$\mu\omega_{\mathrm{TO}}^2 = R_0' - \frac{4\pi(Ze)^2}{3V}, \tag{3.9}$$

are both "soft"—that is, have frequencies which decrease substantially as the transition temperature is reached. This implies that the transition arises from an intrinsic instability in the short-range force constant, not (as proposed in Cochran's original theory) from some subtle and accidental cancellation of long-range and short-range terms in eq. (3.9). In this sense the small Coulombic term in eqs. (3.8) and (3.9) is quite negligible, except that it is responsible for the dielectric anomaly very near $T_c$, and the phase transition in TSCC is more like that in non-ferroelectrics, such as quartz, than in orthodox ferroelectrics.

Figure 16 shows data very near $T_c$, where the overdamped soft mode has a quasiharmonic frequency of about 2 $cm^{-1}$ and the LO component has a frequency estimated from the "zero" of the dielectric constant, with algebraic allowance for the role of damping, of about 4 $cm^{-1}$. The Lyddane-Sachs-Teller relation predicts a dielectric constant of about 20 at this temperature, in reasonable agreement with direct measurement at low frequencies (Sorge and Straube 1979). The Curie constant evaluated from the far infrared data is $25 \pm 2$ and agrees reasonably well with that measured at lower frequencies (MHz region), which vary from 5 to 58, depending upon specimen and domain properties. In summary, TSCC is now viewed as a proper, displacive ferroelastic with an instability in a short-range force constant and a pathologically small oscillator strength.

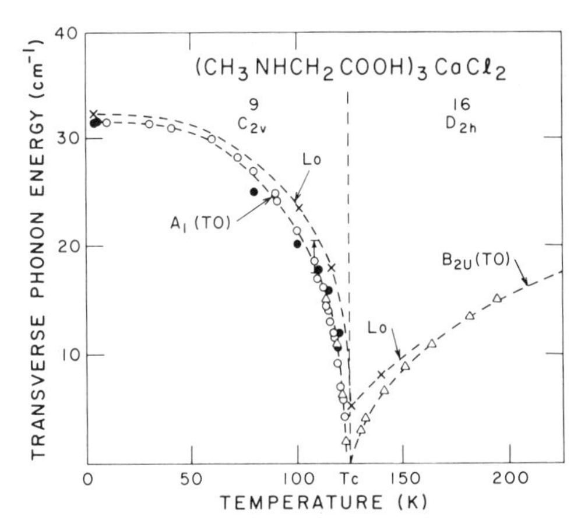


Fig. 15. Soft-mode frequencies for TO and LO components in TSCC; the data are a compilation of Raman and infrared in the ferroelectric phase, and infrared only in the paraelectric phase.

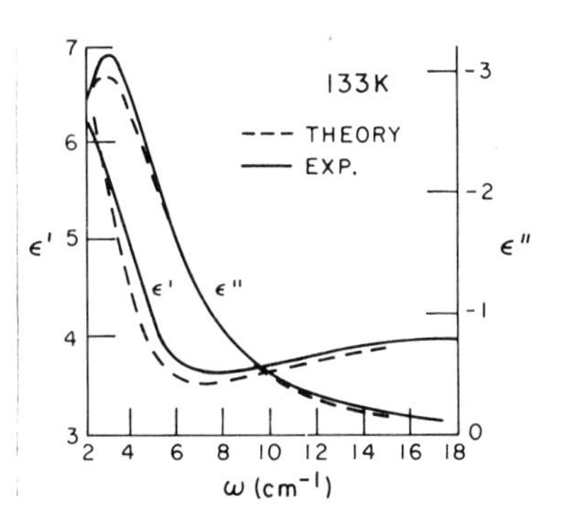


Fig. 16. Real and imaginary parts of the dielectric constant for TSCC for temperatures very near $T_c$. The soft mode becomes underdamped for frequencies above approximately 10 $cm^{-1}$ in each phase.

## 4. Ferroelastics

### 4.1. Ferroelastic $NdP_5O_{14}$

Neodymium pentaphosphate and its lanthanum, praesodymium and terbium isomorphs are the ferroelastics which have been most studied since 1974. $NdP_5O_{14}$ has been of considerable device interest as a solid state laser; it has a pumping threshold less than 1 mW and emits c.w. at 1 μm, an optimum wavelength for fiber optics (Danielmeyer and Weber 1972, Tofield et al. 1975). $LaP_5O_{14}$, $PrP_5O_{14}$, $NdP_5O_{14}$ and $TbP_5O_{14}$ all crystallize in a $C_{2h}$ structure at ambient temperatures, usually called "monoclinic I". $TbP_5O_{14}$ also exists in a different monoclinic structure (monoclinic II) which characterizes $ErP_5O_{14}$, $GdP_5O_{14}$ and other pentaphosphates of lanthanides heavier than Tb (Bagieu et al. 1973). At high temperatures (140–190°C) the monoclinic I crystals transform continuously into a $D_{2h}$ point-group symmetry phase (orthorhombic I). A highly schematic phase diagram is given in fig. 17. The monoclinic I to monoclinic II phase transition has also been observed for some pentaphosphates. It is highly first order (necessarily) but has been reported in $YP_5O_{14}$ at pressures of $10^5$ bar (Hong and Pierce 1974). Little is known about the other phase boundaries.

The continuous $C_{2h}$–$D_{2h}$ phase transition ($P2_1/c \rightarrow Pcmn$) in the lanthanide pentaphosphates is ferroelastic, i.e., exhibits hysteresis in its stress–strain relations (Aizu 1971). The macroscopic order parameter for this phase transition is the monoclinic distortion angle $\Phi'$, or more precisely, $\Phi = (90° - \Phi')$, see Hong (1974). The quantity $\Phi$ approaches zero as $T \rightarrow T_0$ from below and satisfies mean-field theory; it has been measured by a technique involving classical optics. $\Phi$ is of order 0.6° of arc at low temperatures. From a more microscopic point of view, the order parameter is a small, nearly rigid rotation of $PO_4$ tetrahedra. The structure is shown in fig. 18. The atomic displacement required to transform the $C_{2h}$ structure to $D_{2h}$ is less than 0.1 Å at 20°C. This displacement corresponds to an optical phonon of $B_{2g}$ symmetry in the $D_{2h}$ phase and $A_g$ symmetry in the $C_{2h}$ phase. The temperature dependence of this mode was first determined by Fox et al. (1976) for $NdP_5O_{14}$ and $LaP_5O_{14}$ and subsequently (Fox et al. 1978) for $TbP_5O_{14}$, using very thin platelets. Their data are shown in fig. 19. Subsequent Raman measurements by Unger (1979) and by Errandonea and Sapriel (1979) on large bulk samples yielded cleaner polarization data and showed that the upper of the two soft branches in fig. 19 is of $B_{3g}$ symmetry and not $B_{2g}$. The reason for its softening near $T_c$ is a slight mystery, since it is of $B_g$ symmetry below $T_c$ and does not transform as do the ionic displacements at $T_c$. This is discussed in considerable detail by Errandonea (1980).

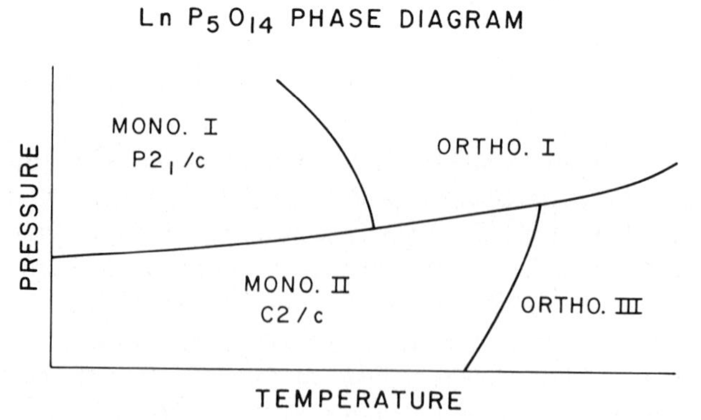


Fig. 17. Phase diagram (schematic) for lanthanide pentaphosphates (Scott 1978).

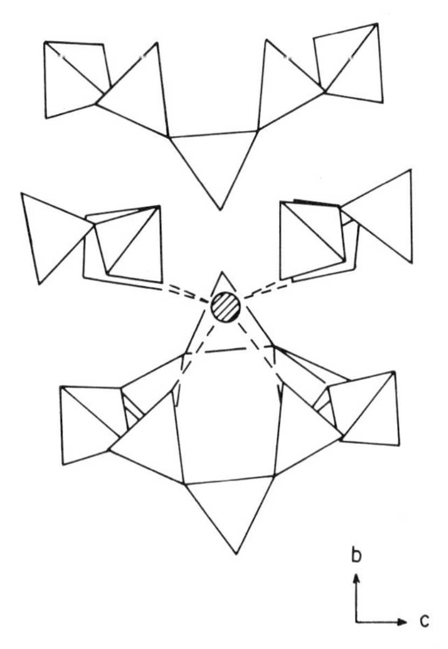


Fig. 18. $LnP_5O_{14}$ structure (Tofield et al. 1975).

The acoustic-mode instabilities in these pentaphosphates were first measured independently by Peercy (1976), using $La_{0.5}Nd_{0.5}P_5O_{14}$ and by Toledano et al. (1976) in pure $LaP_5O_{14}$. Their data are shown in fig. 20. The solid curves are fits (Scott 1978) to a simple theory originally developed by Miller and Axe (1967). It assumes:

$$c_{55}(T) = c_{55}(0) - F^2/\omega^2_{TO}(T) . \tag{4.1}$$

Here $F$ is an opto-acoustic coupling parameter and $\omega_{TO}(T)$ is the $A_g/B_{2g}$ soft-optic-mode frequency graphed in fig. 19. Since $c_{55}(0)$ far away from $T_0$ is an independent measurement, eq. (4.1) involves only a single fitting parameter, $F$.

Both the acoustic and optic mode data are mean field:

$$\omega(T) = A(T - T_c)^{\gamma/2} , \tag{4.2}$$

$$\omega(T) = A'(T_c - T)^{\gamma'/2} , \tag{4.3}$$

yield $\gamma = \gamma' = 1.00 \pm 0.10$. Similar mean-field data are found for $\Phi(T)$, as shown in fig. 21.

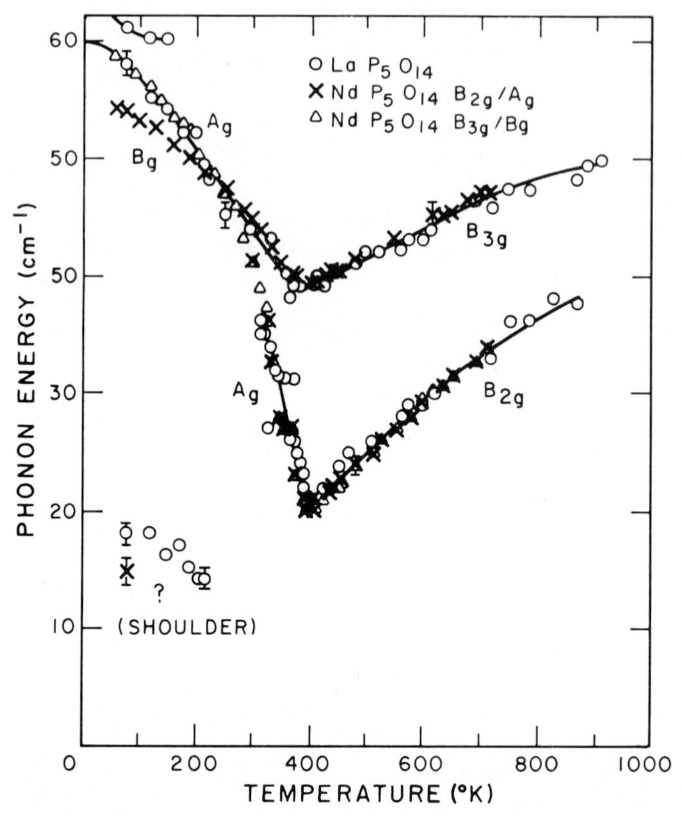


Fig. 19. Soft-mode dependences upon temperature for $LaP_5O_{14}$ (Fox et al. 1976, Unger 1979, Sapriel and Errandonea 1979).

*4.1.1. Fox model for $LnP_5O_{14}$ (Fox* 1979):

The complexity (eighty atoms per primitive cell) and bonding in the lanthanide pentaphosphates does not make them suitable for conventional lattice dynamical models. However, Fox (1979) has developed a simple model which describes many of their characteristics. His model is based upon the fact that the $LnP_5O_{14}$ structure consists of long $(P_5O_{14})^{-3}$ ribbons parallel to the $a$-axis (Albrand et al. 1974). Each ribbon consists of two chains of corner-sharing $PO_4$ tetrahedra, cross linked at every second tetrahedron by a bridging $PO_3$ group. Fox has represented these ribbons in fig. 22 by featureless rods. Each pair of rods (ribbon) may be described by a single coordinate, for example the angle $\beta$ the cross links make with the rod. The problem may be further simplified by making the mean-field assumption, that is, considering the dynamics of one pair of rods in an effective potential determined by the rest of the crystal.

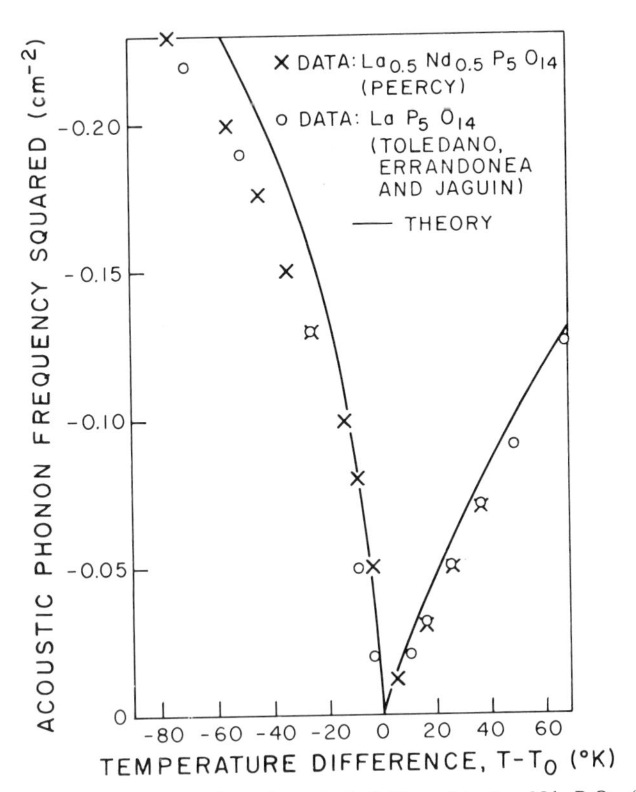


Fig. 20. Acoustic-mode temperature dependence in $LaP_5O_{14}$ and in $La_{0.5}Nd_{0.5}P_5O_{14}$ (Peercy 1976, Toledano et al. 1976). The solid line is according to theory due to Scott (1978) and Miller and Axe (1967).

Fox solves the equations of motion from his assumed model Hamiltonian and by calculating the turning points of the motion obtains information on both the high-temperature orthorhombic phase and the low-temperature monoclinic phase. Some of his results are shown in fig. 23. In this figure the ordinate is given by $z = \tan(\beta/2)$, where $\beta$ is the angle the cross links make in fig. 22. The abscissa is a measure of $\alpha$, a lattice parameter assumed to vary linearly with temperature. Above a critical temperature the expectation value of $z$ is unity, corresponding to $\beta = 90°$ of arc (the orthorhombic phase). Below that temperature two different regions of stability are observed (the cross-hatched regions in the figure); the transition from one of these regions to the other corresponds to ferroelastic switching. It can be seen from this figure that the average value of the monoclinic distortion angle increases as temperature drops below $T_0$, since the separation (in $z$) of the center of the shaded regions monotonically decreases; but in addition, this figure shows explicitly the

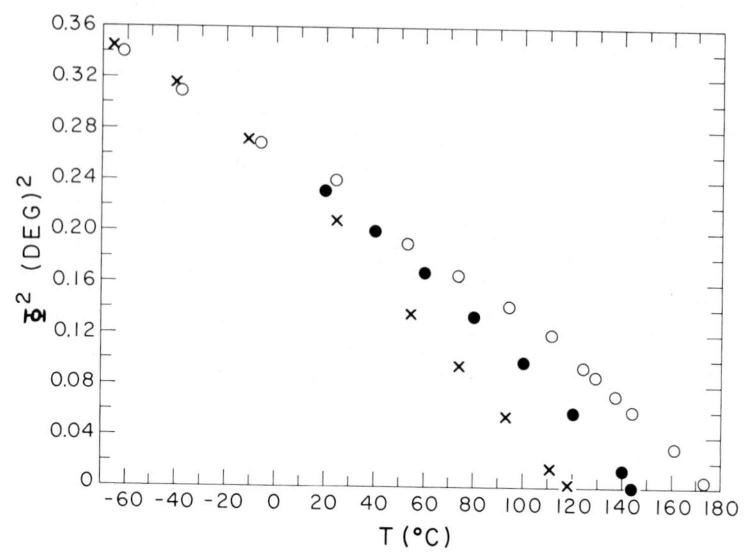


Fig. 21. Monoclinic distortion angle versus temperature in $LaP_5O_{14}$, showing mean-field behavior(Weber et al. 1975, Budin et al. 1975).

dependence of the turning points of the motion upon $\alpha$ (i.e., upon temperature), information not usually available from such mean-field model calculations.

*4.1.2. Errandonea's phenomenological model* (*Errandonea* 1980)
Errandonea (1980) constructs a Landau free-energy for $LnP_5O_{14}$ as:

$$F = (F_Q + F_e + F_c)\,, \tag{4.4}$$

with:

$$F_Q = (\alpha/2)Q^2 + (B/4)Q^4\,, \tag{4.5}$$

$$F_e = (\tfrac{1}{2}) \sum_{i,j=1,3} c^0_{ij} e_i e_j + \tfrac{1}{2} \sum_{k=4,6} c^0_{kk} e^2_k\,, \tag{4.6}$$

and

$$F_c = G e_5 Q + \sum_{i=1,3} \delta_i e_i Q^2\,. \tag{4.7}$$

This kind of phenomenological free-energy approach has been traditionally successful for ferroelectric and ferroelastic transitions. See particularly the work of Dvorak (1970, 1971a,b, 1972a,b, 1974), Dvorak and Petzelt (1971, 1978), and of Levanyuk and Sannikov (1968, 1970a,b, 1971, 1974, and 1976) and the earlier work of Ginzburg (1945, 1949a,b, 1960). Most recently this method has been applied to incommensurates by Ishibashi and colleages (1978).

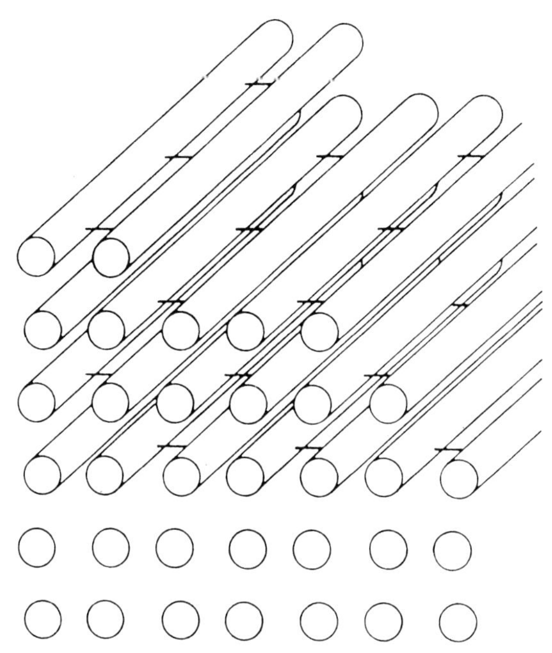

Fig. 22. Simplified structure of $LnP_5O_{14}$ pentaphosphates, in which the phosphate ribbons are represented by rigid rods (Fox 1979).

$F_Q$ contains the terms relative to the "bare" normal coordinate $Q$ associated with the soft optic mode of $B_{2g}$ symmetry. It is assumed that $\alpha = A(T - T_0)$. $F_e$ is the elastic energy of the crystal. $F_c$ contains the coupling terms between $Q$ and the components of the strain tensor.

To account for a second-order transition Errandonea assumes that $A$ and $B > 0$ and that the coefficients $A$, $B$, $c_{ij}^0$, $c_{kk}^0$, $G$, $\delta_i$ vary slowly with temperature. He did not take into account the thermal expansion of the prototype phase which would introduce linear terms in $F$. These are unimportant if one restricts the comparison of the theoretical predictions to the change in thermal expansion observed at the transition. Also, he kept only the lowest-order coupling term between $Q$ and the strain tensor, and neglected some of the coupling terms of fourth degree.

With these assumptions, we can derive from the free-energy expansion the temperature dependence of the various quantities of interest. The equilibrium values of $Q$ and $e_i$ in a mechanically free crystal are determined by minimizing the free energy $F$. The frequency of the soft optic mode and the elastic stiffness tensor components are related to the second derivatives of $F$. The resulting expressions are listed in table 1.

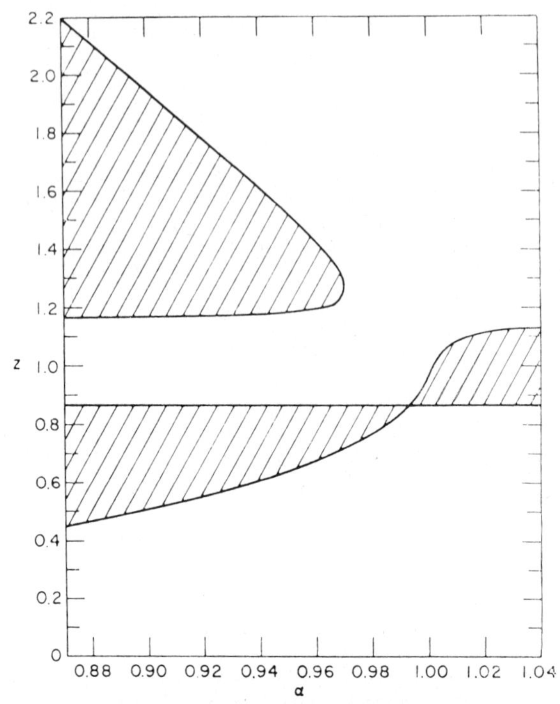


Fig. 23. Stable phases of $LnP_5O_{14}$ as a function of parameter $\alpha$. $\alpha$ is assumed to vary linearly with lattice temperature (Fox 1979).

As shown by this table, all the quantities are expected to be continuous at the transition, which occurs at $T_c = T_0 + (G^2/Ac_{55}^0)$. The coupling between $Q$ and $e_5$ produces an upward shift of the transition temperature in a mechanically free crystal, $T_0$ being the temperature where $\omega_0$ would vanish in the absence of coupling. The spontaneous quantities $e_5$ and $Q$ vary as $\sim (T_c - T)^{1/2}$, whereas the elastic strains $e_1$, $e_2$, $e_3$ vary as $(T_c - T)$ because of the quadratic coupling with $Q$.

This coupling has another important effect: whereas both $c_{55}$ and $\omega_Q^2$ are predicted to increase linearly on either side of $T_c$, the ratio of the slopes below and above the transition is $2B/B'$ instead of 2, which would be the case (Stanley 1971) in the absence of coupling. Here $B' = B - 2\Sigma\, s_{ij}^0\delta_i\delta_j$, where $s_{ij}^0$ are compliances. Farther from $T_c$, $c_{55}$ increases more slowly and reaches a saturation value equal to $c_{55}^0$. This saturation is predicted without need of introducing terms of degree higher than four in the free energy; it results from the coupling between $e_5$ and $Q$. $c_{55}$ and $\omega_Q$ are related in both phases by the Miller and Axe formula (eq. (4.1)).

Table 1
Temperature dependences of optic and acoustic parameters in $LnP_5O_{14}$ (Errandonea 1980). $s^0_{ij}$ is the inverse of $c^0_{ij}$; $g = Ac^0_{55}/G^2$; $g' = 2Bg/B'$; $t = T_c - T$; $t' = T - T_c$; and $B' = 2\Sigma s^0_{ij}\delta_i\delta_j$.

| | Orthorhombic phase | Monoclinic phase |
|---|---|---|
| $Q$ | 0 | $\pm(At'/B')^{1/2}$ |
| $e_5$ | 0 | $\mp(G/c^0_{55})(At'/B')^{1/2}$ |
| $e_i$ | 0 | $-\left(\sum_{j=1}^{3} s^0_{ij}\delta_j\right)(At'/B')$ |
| $e_4$ | 0 | 0 |
| $e_6$ | 0 | 0 |
| $m\omega^2_Q$ | $a\left(t+\frac{1}{g}\right)$ | $A\left(\frac{2B}{B'}t'+\frac{1}{g}\right)$ |
| $c_{55}$ | $c^0_{55}\frac{gt}{1+gt}$ | $c^0_{55}\frac{g't'}{1+g't'}$ |
| $c_{i5}$ | 0 | $\mp\delta_i\left(\frac{2c^0_{55}}{B}\right)^{1/2}\left(\frac{g't'}{1+g't'}\right)^{1/2}$ |
| $c_{ij}$ | $c^0_{ij}$ | $c^0_{ij}-\frac{2\delta_i\delta_j}{B}\frac{g't'}{1+g't'}$ |
| $c_{44}$ | $c^0_{44}$ | $c^0_{44}$ |
| $c_{66}$ | $c^0_{66}$ | $c^0_{44}$ |
| $c_{46}$ | 0 | 0 |

In the framework of the present phenomenological theory, the temperature dependence of the 13 elastic constants depends on only four "reduced" parameters of the expansion namely: $(\delta_i/B^{1/2})$ $(i = 1, 2,$ and $3)$ and $(G^2/A)$ in addition to the $c^0_{ij}$ parameters. These variations are therefore related to each other. In particular, the spontaneous quantities $c_{i5}/\delta_i$ and $\Delta(c_{ij})/\delta_i\delta_j$ with $i, j = 1$ to 3 (where $\Delta c_{ij}$ is the difference between $c_{ij}$ values in the monoclinic phase and in the orthorhombic phase extrapolated below $T_c$) are independent of $i$ and $j$.

Experimental results summarized in table 2 are in accord with the predictions above.

On the basis of these values in table 2 the predicted temperature dependence of the elastic constants are plotted in fig. 24. These variations correspond to isothermal conditions. The difference between isothermal and adiabatic values is small and does not affect the results: $(c^S_{11} - c^T_{11})/c^T_{11} \approx 2 \times 10^{-3}$; $[(c^S_{55} - c^T_{55})/$

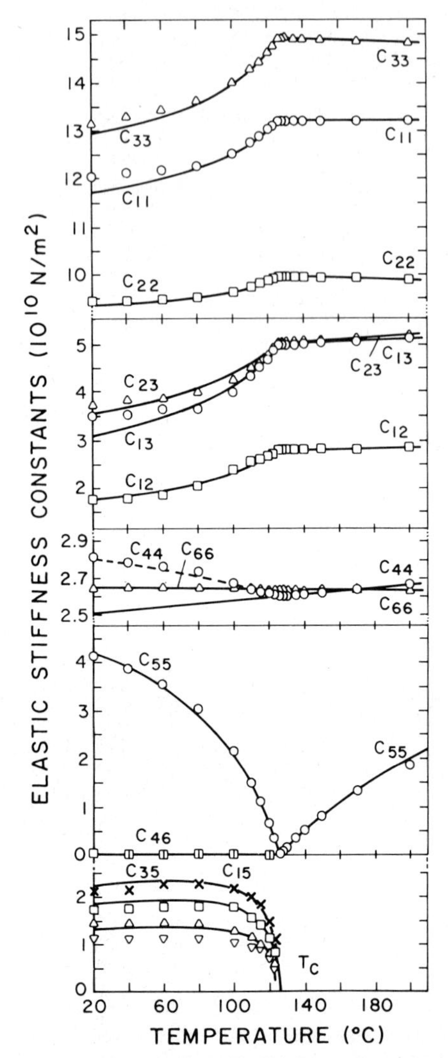


Fig. 24. Calculated and observed acoustic-mode frequencies in $LaP_5O_{14}$ (Errandonea 1980).

$c_{55}^T \leqslant 5 \times 10^{-3}$]. Therefore the experimental data deduced from Brillouin scattering can be directly compared to the calculated ones given in fig. 24.

The thermal expansion coefficients ($\Delta e_i/\Delta T$) coincide with the experimental values near $T_c$. In addition, a large shift of the transition temperature, $(T_c - T_0) = 170$ K is predicted from the calculated value of the coupling coefficient $GA^{-1/2}$ which matches the value ($161 \pm 11$K) deduced from the Raman spectra (Unger 1979, Sapriel and Errandonea 1979).

At low temperatures various quantities are more slowly temperature dependent than the model predicts. This saturation effect is not unexpected and

Table 2

Values of the coefficients of the free-energy expansion for $LaP_5O_{14}$. $c_{ij}$ coefficients are slowly temperature dependent and are given at $T_c$ ($c^0_{55}$ is constant in the whole temperature range). From Errandonea (1980) (cgs units).

| Coefficient | Numerical value | Coefficient | Numerical value |
|---|---|---|---|
| $B/A^2$ | $4.48 \times 10^{-3}$ | $c^0_{33}$ | $14.94 \times 10^{10}$ |
| $G/A$ | $3.36 \times 10^6$ | $c^0_{44}$ | $2.59 \times 10^{10}$ |
| $\delta_1/A$ | $7.14 \times 10^3$ | $c^0_{55}$ | $6.62 \times 10^{10}$ |
| $\delta_2/A$ | $5.01 \times 10^3$ | $c^0_{66}$ | $2.63 \times 10^{10}$ |
| $\delta_3/A$ | $8.62 \times 10^3$ | $c^0_{12}$ | $2.79 \times 10^{10}$ |
| $c^0_{22}$ | $13.21 \times 10^{10}$ | $c^0_{13}$ | $5.01 \times 10^{10}$ |
| $c^0_{22}$ | $9.96 \times 10^{10}$ | $c^0_{23}$ | $5.03 \times 10^{10}$ |

could be accounted for by the introduction of high-order terms in the free-energy expansion. As an example, an additional $\epsilon Q^6/6$ term in eq. (4.5) leads to:

$$e_5^2 = \frac{G^2}{(c^0_{55})^2}\frac{B'}{2\epsilon}\left[\left(1 + \frac{4A\epsilon(T_c - T)}{B'^2}\right)^{1/2} - 1\right], \tag{4.8}$$

which gives an excellent fit of the accurate gamma-ray diffractometry measurements.

The experimental data obtained for the transitions of the isostructural rare-earth pentaphosphates coincide with the data of $LaP_5O_{14}$ when we take into account the shift of the transition temperature. This is the case of the soft optic modes behavior in $NdP_5O_{14}$ and $TbP_5O_{14}$, the acoustic mode in $La_{0.5}Nd_{0.5}P_5O_{14}$ the optical shear measurement in $La_{0.25}Nd_{0.75}P_5O_{14}$, and the thermal expansion data on $NdP_5O_{14}$.

The jump of specific heat derived from the free energy (4.4–4.7) is:

$$\Delta c = T_c A^2/2B' . \tag{4.9}$$

Equation (4.9) yields 4.7 cal $g^{-1}K^{-1}$ for $LaP_5O_{14}$ and 4.9 cal $g^{-1}K^{-1}$ for $NdP_5O_{14}$ ($T_c = 143°C$) and $PrP_5O_{14}$ ($T_c = 139°C$) in good agreement with the experimental value $5.3 \pm 0.4$ cal $g^{-1}K^{-1}$ deduced from the work of Loiacono et al. (1978) for the two latter materials.

Finally, the phenomenological model allows an evaluation of the pressure dependence of the soft optic mode. From Clapeyron's relation we have:

$$(\mathrm{d}T_c/\mathrm{d}p) = 2 \sum_{i,j=1,3} s^0_{ij}(\delta_j/a) = 18.9°/\mathrm{kbar}\,, \tag{4.10}$$

where $s^0_{ij}$ is the elastic compliance tensor for $T > T_c$. The signs of the jumps of the expansion coefficients at $T_c$ lead to a positive value of $\mathrm{d}T_c/\mathrm{d}p$. As noticed by Samara (1977) the rare-earth pentaphosphates are unusual: most displacive ferroelectric crystals whose transition is induced by a Brillouin zone center soft optic mode have $\mathrm{d}T_c/\mathrm{d}p < 0$. The pressure dependence of the soft-mode frequency $\omega_Q$ is given by:

$$\mathrm{d}(\log \omega_Q)/\mathrm{d}p = -(1/2\omega_Q)(\mathrm{d}\omega_Q^2/\mathrm{d}T)(\mathrm{d}T_c/\mathrm{d}p)\,. \tag{4.11}$$

Errandonea (1980) finds $\mathrm{d}(\log \omega_Q)/\mathrm{d}p \approx 11.64\%$ per kbar at $(T_c - 49\mathrm{K})$, whereas measurements of Peercy et al. (1976) in $TbP_5O_{14}$ give 11.08% (see also Assaumi et al. (1980)).

In summary, these experimental data clearly show that the phenomenological model used to explain the properties of $LaP_5O_{14}$ can be extended to all the ferroelastic $LnP_5O_{14}$ compounds with close values of all the parameters of eq. (4.4) except $T_0$, in all these materials. However, two features of the considered transitions are not accounted for by the present model. In the first place, the experimental ratio of the slopes of $\omega_Q^2$ on either side of $T_c$ is $4.80 \pm 0.25$ and is significantly different from those for $c_{55}$ (3.90) whereas both numbers are expected to be equal (though not equal to the 2.0 predicted in the absence of coupling). However, this equality relies on the existence of only one bilinear coupling term $Ge_5Q$ in the expansion. The introduction of higher-degree coupling terms in the expansion such as $Ge_5^2Q^2$, leading to a renormalized temperature dependent $G$ coefficient, will yield different slope ratios for $\omega_Q^2$ and $c_{55}$. Considering the strength of the linear coupling, which is disclosed by the large shift $(T_c - T_0)$, we can understand that these higher-degree coupling terms can contribute significantly to the results.

Secondly, the $c_{44}$ elastic constant is predicted to be unaffected by the transition, whereas a slight anomaly has been observed at $T_c$. It can be understood if we remember the existence of a second soft optic mode of the same $B_{3g}$ symmetry as the strain $e_4$. Thus, this mode which was not included in the free-energy expansion can linearly couple with $e_4$ and as the $B_{2g}$ soft optic mode couples with $e_5$. Here again, the softening of the $B_{3g}$ mode on both sides of $T_c$ induces the anomaly of the $c_{44}$ temperature dependence since we have:

$$c_{44} = c_{44}^0 - \frac{\gamma'^2/m'}{\omega'^2}\,, \tag{4.12}$$

in both phases by analogy with eq. (4.1), where $\gamma'$ is the coupling coefficient between $e_4$ and the $B_{3g}$ mode of mass $m'$ and frequency $\omega'$. Thus, we can fit the temperature dependence of $c_{44}$ with $\omega'$ being determined by previous Raman scattering measurements, $c_{44}^0 = 3.15 \times 10^8$ and $G'^2/m' = 815 \times 10^{12}$, which is nearly three times smaller than the coupling coefficient between the $B_{2g}$ symmetry optic and acoustic modes.

Figure 25 shows one last anomaly in this $LnP_5O_{14}$ family: on both $LaP_5O_{14}$ and $TbP_5O_{14}$ an asymmetry is manifest as a "shoulder" at about $16\,\mathrm{cm}^{-1}$ on the soft mode near $T_0$. This is not understood.

In summary, $LaP_5O_{14}$ present an especially simple case of opto-acoustic interactions. The optic mode is Raman-active and underdamped in both phases right up to $T_0$. The frequency decrease of the acoustic mode can be measured to 99.7%. All data indicate a mean-field system. The absence of overdamped excitations near $T_0$ make this an ideal material for an inelastic neutron or

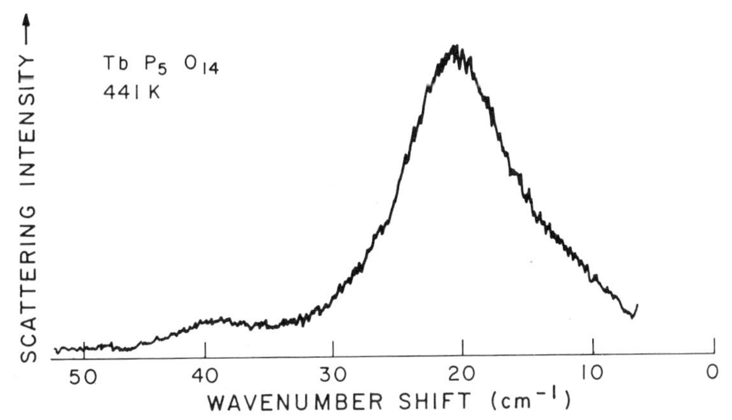


Fig. 25. Raman spectra in $TbP_5O_{14}$ very near $T_0$, showing unexplained "shoulder" on the soft mode at about $16\,cm^{-1}$ (Fox et al. 1978).

Rayleigh-scattering study of central modes, since the spectrum below $19\,cm^{-1}$ should be very clean. Since the material is ferroelastic, domains can be eliminated by very small applied stress. The measurement of central-mode scattering is thought to be due to domain-wall scattering in both KDP (Durvasula and Gammon 1977) and in $Pb_5Ge_3O_{11}$, lead germanate (Lockwood et al. 1976, Fleury and Lyons 1976, Cowley et al. 1976) and has occasionally been linked with critical (non-mean-field) phenomena (Müller 1979). Consequently the search for central modes in mono-domain samples of $LnP_5O_{14}$ with demonstrable mean-field dynamics should be illuminating.

One might ask why the $B_{3g}$ soft mode does not continue to soften in the monoclinic phase, producing a second phase transition at a lower temperature. The answer is that, microscopically, the eigenvectors for both $B_{2g}$ and $B_{3g}$ soft modes are (Scott 1978, Nakashima 1980) rigid rotations of $PO_4$ rotation: for $B_{2g}$ in the *ac*-plane; for the $B_{3g}$ mode in the *bc*-plane. When the $B_{2g}$-induced ferroelastic phase transition is reached, the monoclinic distortion stabilizes the $PO_4$ rotational instabilities in *both ac*- and *bc*-planes. This a purely steric effect which renormalizes the $B_{3g}(B_g)$ mode frequency below $T_0$.

### 4.2. $BiVO_4$

An analogous ferroelastic system of recent Raman interest is $BiVO_4$ (Pinczuk et al. 1977).

Bismuth vanadate has the simple scheelite structure shown in fig. 26. This $C_{4h}$ structure characterizes $CaWO_4$, $CaMoO_4$ and a large family of tungstates and molybdates with relatively large divalent metal ions (Ba, Ca, Sr). Smaller metal ions (e.g., Zn) produce a distortion to a related $C_{2h}$ monoclinic structure.

The scheelites have been studied rather completely by Raman techniques (Porto and Scott 1967, Scott 1968a,b) and were in fact chosen in the text by Hayes and Loudon (1978) to illustrate Raman analysis techniques for centric

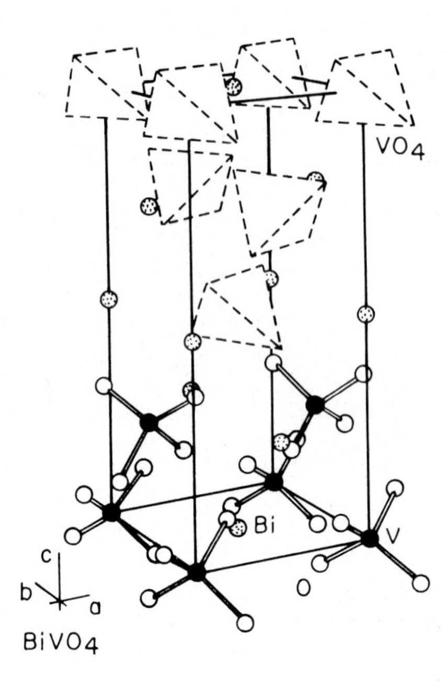


Fig. 26. Scheelite structure of bismuth vanadate at high temperatures.

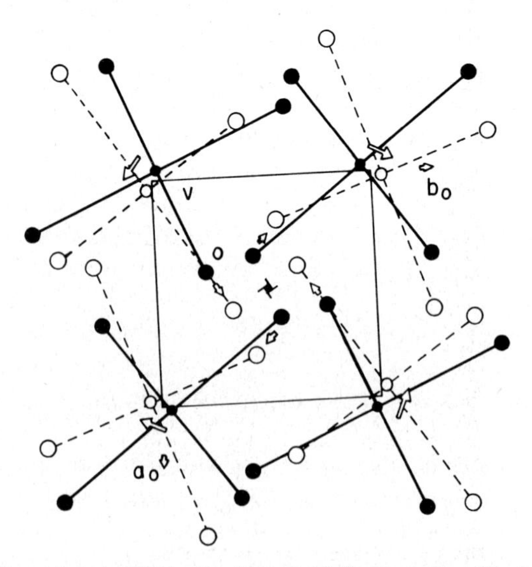


Fig. 27. Domain configurations in $BiVO_4$ (David et al. 1979). The soft-mode eigenvector is the set of translations which rotate the solid lines into the dashed lines.

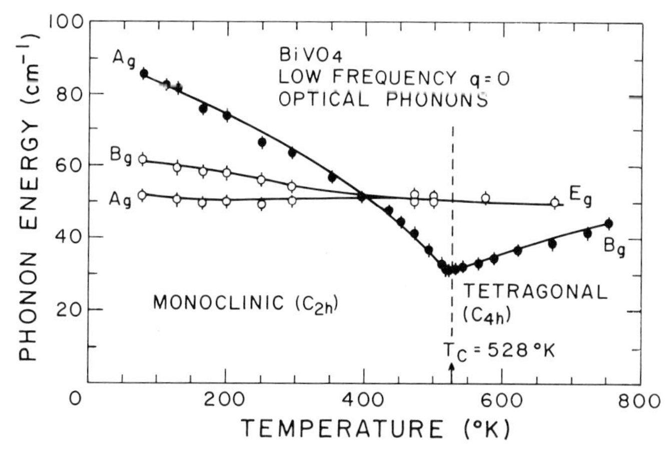


Fig. 28. Soft optic phonon frequency versus temperature in $BiVO_4$ (Pinczuk et al. 1977).

solids. They exhibit excellent examples of both factor group splittings and Davydov splittings. Figure 27 illustrates the eigenvector for the $B_g$ symmetry soft mode producing the $C_{4h}\rightarrow C_{2h}$ ferroelastic phase transition (David et al. 1979). This phase transition was first analyzed by Bierlein and Sleight (1975) and is characterized by an underdamped soft mode (Pinczuk et al., 1977), as shown in fig. 28. Note that, as with the $LnP_5O_{14}$ ferroelastics, strong optoacoustic interactions keep the soft-mode frequency finite at $T_c$. The theory of Miller and Axe (1967) provides a quantitative explanation of these data, as in the case of La, Nd, Pr, and $TbP_5O_{14}$. See Gu et al. (1981) for more recent details.

The key structural element in both scheelites (including $BiVO_4$) and the pentaphosphates is the covalent $MO_4$ ion (phosphate, tungstate, molybdate or vanadate); this ion is tightly bound so that only bending or rotational distortions are possible.

### *4.3. Boracites*

The boracites, particularly nickel iodine boracite ($Ni_3B_7O_{13}I$), have received considerable attention in the scientific literature of the last decade, by virtue of their ferroelectric and ferromagnetic properties, although their apparent potential as magnetoelectric devices (Ascher et al. 1966, Ascher 1970, Schmidt 1967, 1970) has never been realized. Because of their small size and opacity, Raman spectroscopic studies of their phase transitions were successful only recently. Most of this work is by Murray and Lockwood (1978a,b). Typical data are shown in fig. 29.

A point of concern has been the behavior of the dielectric constant near $T_c$. As reviewed by Levanyuk and Sannikov (1974), the *continuous decrease* in $\epsilon(T)$

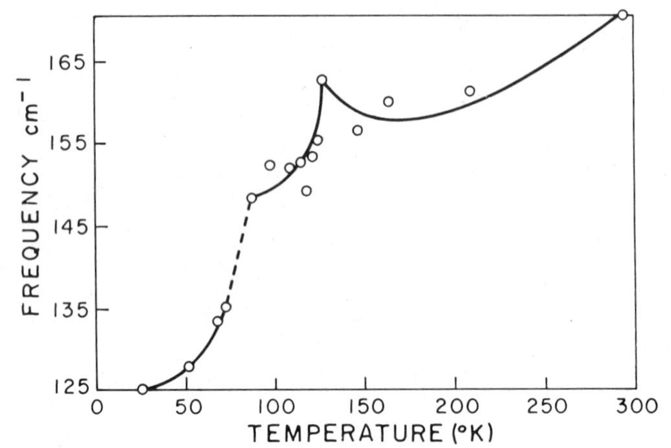


Fig. 29. Raman frequency of anomalous mode in nickel iodine boracite (Murray and Lockwood 1978b).

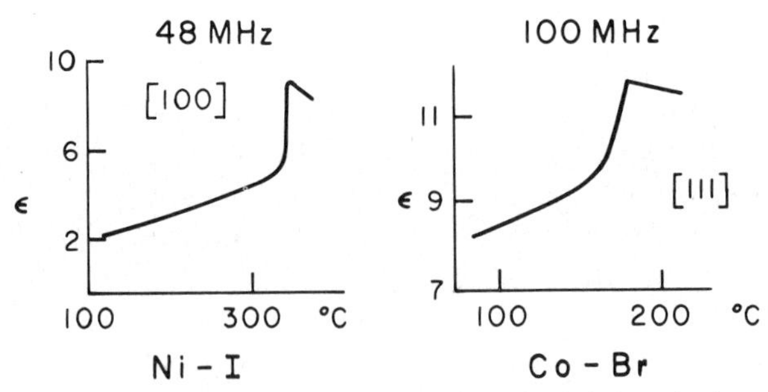


Fig. 30. Dielectric behavior in nickel iodine boracite (left) and cobalt bromine boracite (right) near $T_0$ (Levanyuk and Sannikov 1974).

as $T$ drops below $T_c$ is contrary to all published models, which generally predicted a discontinuous increase (see fig. 30). Following the suggestion of Gufan and Sakhnenko (1972), Lavrencic (1979, unpublished) pointed out to the author that these characteristics can all be explained by free energies with key term of form,

$$H' = \phi^2 P^2 , \tag{4.13}$$

where $\phi$ is the order parameter and $P$ is the spontaneous polarization. Fox et al. (1980) have used such a term to explain the dielectric behavior in magnetoelectric $BaMnF_4$ near $T_N$, and it would appear that such an analysis is appropriate in the boracites as well (see, Shaulov et al. (1981)). Note that Fox et al. (1980) provide a counter example to the assertion of Levanyuk and Sannikov (1974) that the order parameter for an improper ferroelectric must be two dimensional.

Other light scattering studies of continuous or nearly continuous ferroelastic transitions include $Hg_2Cl_2$, $Hg_2Br_2$, $Hg_2I_2$ (Kaplyanskii 1976) and $Pb_3(PO_4)_2$ (Chapelle et al. 1976).

### 4.4. *Pure ferroelastics*

It has been proposed at various times that there exist "purely elastic" phase transitions (or "pure" ferroelastics) in which the true order parameter is some soft $c_{ij}$ elastic coefficient intrinsically unstable and not driven by a soft optic phonon or low-lying electronic state. This hypothesis was first considered by Anderson and Blount (1964) in the context of discussions of the A–15 structure high-temperature superconductors $V_3Si$ and $Nb_3Sn$. In that case they were able to show group theoretically that the order parameter could *not* be acoustoelastic in $Nb_3Sn$, $V_3Si$, or any crystal with cubic symmetry in its high-temperature phase.

Whereas Anderson and Blount argue that a transition from cubic to tetragonal structures with strain as the order parameter cannot be second order (except with probability zero) within the Landau theory, such transitions are listed at special points in the $T$–$p$ phase diagrams summarized by Janovek et al. (1976). However, this is what is meant by the parenthetical comment of Anderson and Blount. The possibility exists, however, that such transitions would appear quasi-continuous under many experimental studies. (The author is indebted to H.Z. Cummins for this observation.) It is useful to compare the conclusions of Anderson and Blount with those of Aubry and Pick (1971) and Cowley (1976) regarding acoustic phonon instabilities. Anderson and Blount show that for cubic–tetragonal transitions, which preserve the unit primitive cell size, the transitions must be first order if the order parameter transforms like a strain, i.e., acoustic phonon. Aubry and Pick add that any symmetry-breaking transition (including cubic–tetragonal at $q = 0$), if second order, has a soft acoustic phonon. The two statements are logically compatible. Cowley (1976), on the contrary, says that "in the hexagonal or trigonal classes . . . there is in general no acoustic wave whose velocity is zero when the elastic constant stability condition given in table 1 is zero." This is simply wrong and conflicts with the general proof given by Aubry and Pick (and similar work written by Toupin and Thomas in 1969 as an unpublished IBM report). Much of Cowley's Table 1 is wrong and has unfortunately misled experimentalists; e.g., the B-symmetry entry for $4/m$ says there is no "soft"-shear wave; actually the soft shear exists and propagates at an angle* given by $\tan 2\phi = c_{16}/(c_{12} - c_{11})$. We conclude that, except for Cowley's non-symmetry-breaking "type-zero" transitions, there is always at second-order structural phase transitions a shear wave whose velocity goes to zero; Aubry and Pick are viewed as completely correct (except for an error in their table, where $D_{2h} \rightarrow C_{4h}$ should read $D_{2h} \rightarrow C_{2h}$) and not in disagreement with Anderson and Blount.

*For $BiVO_4$ Cho et al. (1982) calculated $\tan\theta = c_{66}[c_{16} \pm (c_{16}^2 + c_{66}^2)^{1/2}]^{-1}$

The idea of a purely elastic instability resurfaced with the study of the phase transition at low temperatures in $PrAlO_3$. In this case Fleury et al. (1973) made the erroneous claim that $PrAlO_3$ exhibited a unique kind of purely elastic phase transition with oxygen octahedra distorting along [110]. Unfortunately, as discussed by Scott (1974), it was already known from the work of Thomas and Müller (1970) that this was impossible; for whereas the perovskite order parameter is three-fold degenerate and may (group theoretically) produce distortions along [100] (strontium titanate), [111] (lanthanum aluminate), or [110], only the [100] and [111] distortions minimize the free energy. Thus, the [110] distortion in $PrAlO_3$ cannot be a stable phase unless it is driven by some other instability. Harley (1977) has shown that this is in fact a cooperative Jahn–Teller effect driven by the lowest $Pr^{3+}$ electronic state and its temperature dependence.

In paratellurite ($TeO_2$) Peercy and Fritz (1974) made a similar claim of uniqueness for the peculiar phase transition in this material. An elastic instability was found with no apparent driving force. However subsequent studies by Uwe and Tokumoto (1979) have shown that the pressure dependence $(\partial/\partial P)(c_{11}-c_{12})$ of the pertinent elastic coefficient in $TeO_2$ is not unusual; rather, it is the extremely low value of $c_{11}-c_{12}$ at ambient pressure which is anomalously low. Uwe and Tokumoto explain this fact via a two-dimensional model of $TeO_2$ in which $c_{11}-c_{12}$ would be exactly zero. Their mathematical formalism is analogous to that of Shirane and Axe (1971), involving a bilinear term $duQ$ in the free energy, where $u$ is strain and $Q$ is an optical phonon coordinate. As shown by Shirane and Axe, this kind of interaction can introduce an elastic phase transition without significant optical phonon softening (see, Scott (1974)). Since $TeO_2$ is tetragonal in its high-symmetry phase, the Anderson–Blount criterion against purely elastic instabilities does not apply. Nevertheless it would be interesting to analyze the electronic states of the Te ion to see if they play any role in the anomalously low ambient value of $c_{11}-c_{12}$. Perhaps $TeO_2$ is similar to $V_3Si$.

The most recent candidate for a purely elastic phase transition is lithium ammonium tartrate (LAT), whose acoustic instability was reported by Nakamura et al. (1979). However, in this case it has been suggested (Abe 1980) that the driving force and true order parameter is ammonium-ion ordering.

## *5. Order–disorder systems*

Until recently the order–disorder systems studied most extensively by light scattering were the amonium halides, particularly $NH_4Cl$ and $NH_4Br$ (Lazay et al. 1969, Wang 1971, Bartis 1973). The spectroscopy of these materials and their phase-transition dynamics were very well examined by Raman techniques and by inelastic neutron scattering (Yamada et al. 1974). Although, as pointed out by Scott (1974), there is some confusion in the work of Wang and Fleury

(1969) concerning critical-point assignments for several of the phonons active in the disordered phases. This is clarified by Geisel et al. (1973).

### 5.1. $CaF_2$ family ionic conductors and AgI

During the past five years most Raman work on order–disorder systems has focussed upon "super-ionic conductors", that is, solid state electrolytes. The two principal materials examined were AgI, because of its simplicity, and the $CaF_2$ family, for similar reasons. AgI Raman spectra have been reported by several authors, including Bottger and Damsgard (1972), Burns et al. (1976a,b, 1977a,b), Delaney and Ushioda (1976), and Fontana et al. (1978). Infrared studies of AgI are reported by Bruesch et al. (1975, 1978), among others. These studies show order–disorder behavior with an abrupt, discontinuous transition. In the low-temperature phase there is a characteristic low-frequency phonon at $17\,cm^{-1}$ which has its intensity disappear completely over a temperature span of less than 1 K near $T_0$. The two-phonon spectra of AgI have been analyzed by Fukushi et al. (1978). Their assignments have been questioned by Vardeny and Brafman (1979). On the basis of polariton data the latter authors suggest that all observed scattering features in AgI are first order and thus that the true symmetry is lower than commonly accepted. This hypothesis rests upon the assumption that second-order Raman features do not produce polariton effects; in fact, it is well known that any feature which is manifest in the infrared absorption spectrum of a crystal will also produce polariton dispersion, whether first order or second order. Thus, we regard the conclusion of Vardeny and Brafman as a non-sequitur.

### 5.2. $Ag_{26}I_{18}W_4O_{16}$

A variety of light scattering mechanisms in superionic conductors have been described, including static effects (local distortions of the symmetry of scattering centers, strain fields, etc.) and truly dynamic effects (hopping and diffusion of ions on time scales commensurate with that of the scattered light frequency shifts). Readers are referred to Dieterich et al. (1978), Geisel (1977), Gurevich and Khorkats (1977) Halperin and Varma (1976), Hayes et al. (1977), Chase (1976), Huberman (1974), Klein (1976), and Rice et al. (1974). In both AgI and $RbAg_4I_5$, it has been possible to relate the Raman data in a quantitative way to transport properties such as AC electrical conductivity (Junod et al. 1971) or diffusion coefficients (Funke 1976). There has been little quantitative success with microscopic models, however. The microscopic analysis of Raman data has been much more successful in the $CaF_2$ family (Harley et al. 1975, Hayes et al. 1977, Shand et al. 1976). In these systems fluorine (or chlorine) vacancies play a key role and theories have been developed from the dilute vacancy concentration limit. The primary experimental limitation on these systems are the very high temperatures required, which produce anharmonic thermal broadening of all features in addition to the conducting effects. One advantage however, of systems such as $SrCl_2$ or $PbF_2$, is the slow but structured evolution of their spectra with temperature changes, in contrast to AgI or $RbAg_4I_5$ (Burns et al. 1976a,b, 1977a,b, Delaney and Ushioda 1976). A system which combines the low $T_0$ of AgI with the nearly continuous phase-transition characteristics of the fluorides is $Ag_{13}I_9W_2O_8$ (usually written as $Ag_{26}I_{18}W_4O_{16}$ in order to display explicitly its characteristic $W_4O_{16}^{-8}$ ion). This material is used as the electrolyte in commercial batteries (Ikeda 1977, Takahashi 1976) and its ionic conductivity and chemical stability rival those of $RbAg_4I_5$. It was discovered by Takahashi et al. (1973), who analyzed the $Ag_xI_y(WO_4)_z$ phase diagram and found ionic conduction in all of the stable phases including iodine-free $Ag_2WO_4$. The latter crystal evidences an apparent phase transition characterized by a change in slope of electrical resistivity versus temperature; however, this was subsequently shown to be an irrversible chemical degradation involving oxygen loss at elevated temperatures and not a true phase transition (Turkovic et al. 1977). Large single crystals of $Ag_{26}I_{18}W_4O_{16}$ have been reported by Skarstad and Geller (1975), and the structural determination via X-ray crystallography was made by Chan and Geller (1977). Raman spectra for this material are shown in fig. 31. Together with the dielectric data of fig. 32 they illustrate three phase transitions: a ferroelectric transition at 198 K, and transitions to a fully ionic inductor at about 247 K and 278 K. All three transitions are order–disorder, and each involves the nearly continuous overdamping of a silver ion vibration. Thus, the spectra in fig. 31 show true sublattice melting, and the line widths of the overdamped spectra can be related to the silver ion diffusion coefficient. Assuming a jump length of 3 Å, a $10^{-4}\,cm^2s^{-1}$ diffusion coefficient is inferred at $\sim 310$ K, in good agreement with typical values for AgI (Funke 1976). Specific heat anomalies at these three transitions are shown in fig. 33.

Most striking in the $Ag_{26}I_{18}W_4O_{16}$ conductivity data shown in fig. 32 is the effect of low-power laser illumination. A power density of $\sim 50\,mW/cm^2$ at 514.5 nm is sufficient to reduce the conductivity an order of magnitude over a wide range of temperature. This is interpreted as photoexcitation of electron–hole pairs and concomitant neutralization of silver ions, viz.

$$\gamma + Ag^+ \rightarrow e^- + h^+ + Ag^+ \rightarrow h^+ + Ag^\circ . \tag{5.1}$$

Since the hole mobility is negligible, this results in a net decrease of one carrier ($Ag^+$ ion) per incident photon. It also suggests the possibility of a new class of electric devices: "optically induced ferroelectrics". Usually ionic conductors and ferroelectrics are regarded as mutually exclusive classes of crystals, since in one case an applied electric field induces switching whereas in the other case the same field produces a current flow. $Ag_{26}I_{18}W_4O_{16}$ is a rather unique ferroelectric superionic conductor. Below $T_0 = 198$ K it has all of the usual characteristics of a ferroelectric, but its ionic conductivity is still too high to readily permit demonstration of switching; application of sufficient light intensities should reduce ionic conductivity to the point where switching can be demonstrated.

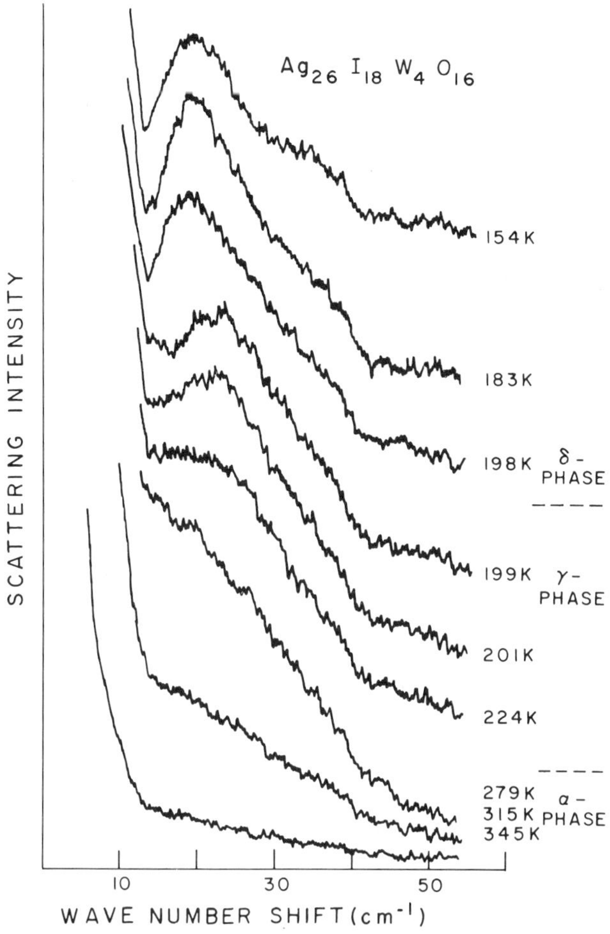


Fig. 31. Raman spectra of low-frequency modes in $Ag_{26}I_{18}W_4O_{16}$ at various temperatures showing the evolution of silver-ion motion from underdamped vibrations to diffusion (Habbal et al. 1978).

This would produce a three-position device: + or − under illumination, or zero (conducting) in the dark. Such experiments are in progress.

An additional interesting aspect of the work on $Ag_{26}I_{18}W_4O_{16}$ is the observation that ionic AC conductivity varies linearly with frequency for small frequencies (Scott et al. 1980). This is in qualitative contrast to the quadratic dependence predicted by Lines (1979a,b), but in agreement with experimental results on AgI (Armstrong and Taylor 1975, Hodge et al. 1976) as well as with low-dimensional theories (Habbal 1979, Bottger et al. 1979, Bernasconi et al. 1979). Data are shown in fig. 34. The implication is that the $Ag^+$ ion conduction paths are spatially three dimensional, but topologically one dimen-

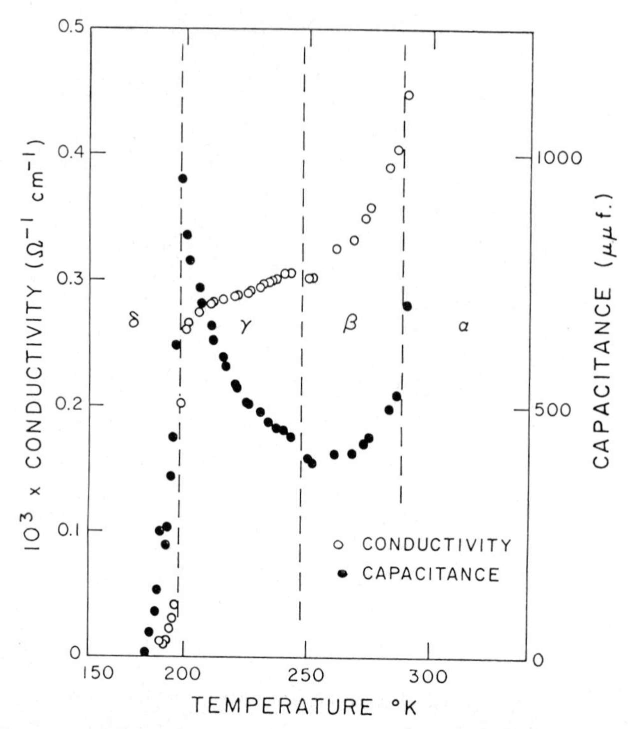


Fig. 32. Dielectric and conductivity data versus temperature for $Ag_{26}I_{18}W_4O_{16}$ (Habbal et al. 1978).

sional, i.e., the $Ag^+$ diffusion is highly anisotropic at each site. This interpretation is compatible with all existing data but conflicts with the general philosophy expressed by Geller and Skarstad (1974).

A variety of other order–disorder systems are also subject to recent Raman experiments, including continuous or nearly continuous transitions in $KH_3(SeO_3)_2$: Yagi et al. (1976, 1977a, 1977b), Vaks and Zein (1974), Tanaka and Tatsuzaki (1979), Tanaka et al. (1978a,b), Peercy (1970), and in $LiNH_4C_4H_4O_6{\cdot}H_2O$ (Udagawa et al. 1978). The latter crystal is a proper ferroelectric with $T_c = 98$ K and no soft mode (Sawada et al. 1977), Gorbatyi et al. (1972), and Grande et al. (1978). Symmetry assignments are also given for the selenates by Dvorak (1972), and spontaneous polarization measurements by Ivanov et al. (1970). Although the bulk of the volume of publications on this interesting family has come from Shuvalov's group in Moscow (Shuvalov et al. 1970, 1972, 1967) or from Japan, where interest has centered on its dynamic (?) central mode, Raman studies on the KTS: $KH_3(SeO_3)_2$ family really began with Peercy's work (1970), which is summarized below in fig. 35. See Makita et al. (1976, 1977), Kasohara (1978).

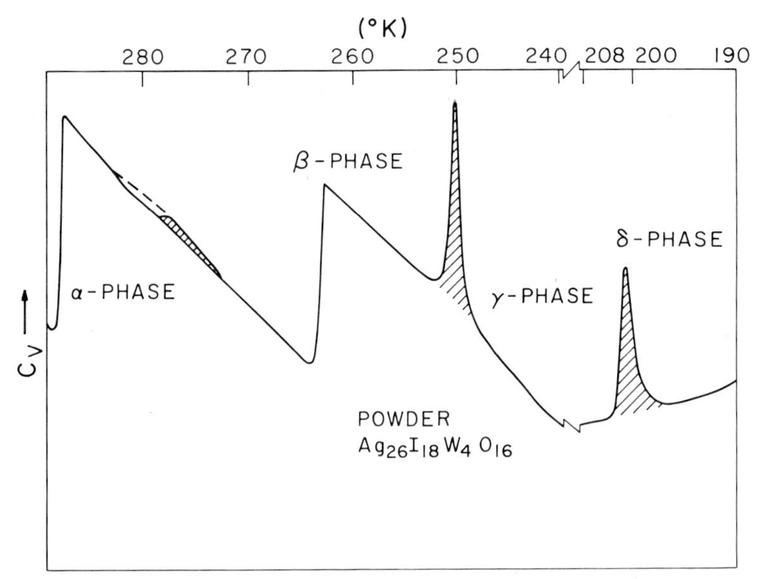


Fig. 33. Specific heat anomalies in $Ag_{26}I_{18}W_4O_{16}$, showing two first-order transitions at about 199 and 247 K and a second-order transition at 278 K. The transition at 278 K is the transition to a "superionic" conducting phase, not that at 247 K, as erroneously concluded by Geller et al. (1980).

### *5.3. $SrTiO_3$*

Strontium titanate has intentionally been ignored in the review, despite, or perhaps because of, the fact that more has been published on its phase transitions over the past decade than on those of any other crystal (see the papers by Müller et al., Aharony, and Aharony and Bruce). Müller has rightly termed $SrTiO_3$ the "drosophila" of solid state physics, after the fruit fly whose myriad transitions educated biologists. However, before leaving this brief summary of order–disorder systems, I wish to call attention to the unambiguous analysis by Bruce et al. (1979), which shows that very near $T_0$, the dynamics of $SrTiO_3$ are order–disorder (and the central mode is dynamic and intrinsic). Ironically, this text-book example of a continuous displacive phase transition has its order parameter ($TiO_6$ octahedron rotation angle) decrease not to zero, but to 0.22° of arc, at which point order–disorder transitions of clusters (from + 0.22° to − 0.22°) occur. The Raman data (e.g., Steigmeier and Anderset (1973)) are compatible with this result, showing optic-mode softening to only $\sim 13\,cm^{-1}$, but they do not prove it, since similar effects can be produced through other mechanisms.

Within the structures isomorphic with $SrTiO_3$, special attention has been received by $RbCaF_3$, whose phase transition should be second order according

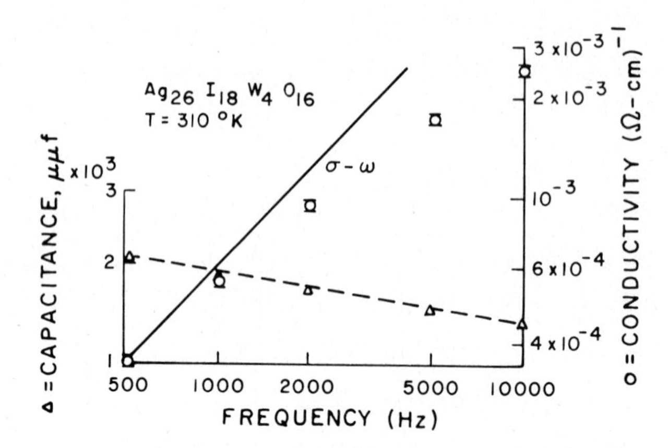


Fig. 34. Frequency dependence of the AC conductivity in $Ag_{26}I_{18}W_4O_{16}$ (Scott et al. 1980). The linear dependence observed conflicts with the quadratic dependence upon frequency predicted by Lines (1979ab).

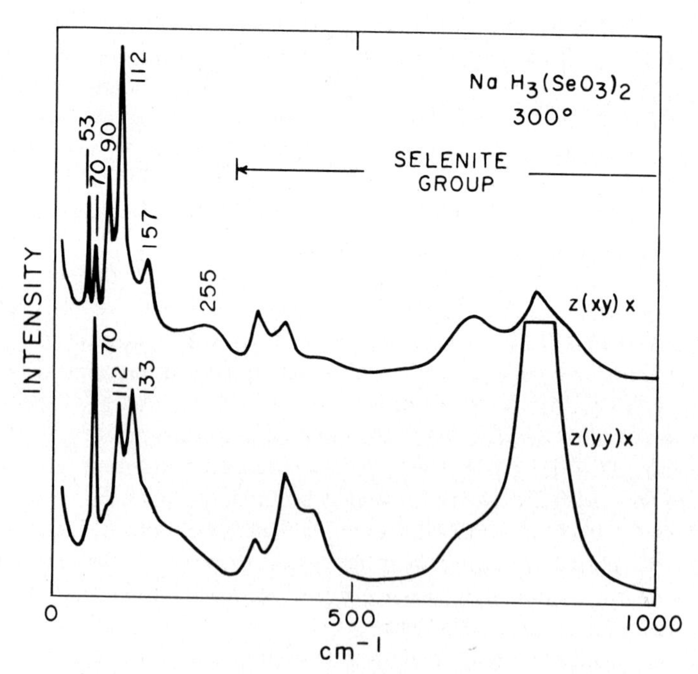


Fig. 35. Raman spectra of $NaH_3(SeO_3)_2$ (from Peercy (1970)).

to Landau theory, but is apparently rendered discontinuous by fluctuations (Aharony 1980, Ridou et al. 1977, Hirotsu et al. 1979).

## 6. Incommensurate transitions

Incommensurate magnetic structures have been understood for two decades. These spiral magnets were first analyzed by Dzyaloshinskii (1964). Using the same symmetry criteria, in particular the existence of "Lifshitz invariants" in the free energies of form, for example for $K_2SeO_4$, $H' \sim \rho^2(\mathrm{d}\theta/\mathrm{d}x)$ where $\rho$ and $\theta$ describe a two-dimensional order parameter ($p = \rho \sin\theta$; $q = \rho \cos\theta$) and $x$ is a spatial coordinate, Levanyuk and Sannilov (1976) have extended this approach to structurally incommensurate insulators, with application to such systems as potassium selenate, ammonium fluoberyllate, and sodium nitrate.

A qualitatively different approach to the understanding of incommensurate phase transitions has been developed by Overhauser (1968, 1971, 1978) for metals. He proposed that, due to the spin of the conduction electrons, the time-averaged translational symmetry of the metallic crystal might be lower than that which would arise from ions and electrons in the absence of spin. Such incommensurate metals are analyzed by a charge density wave description. Figure 36 shows schematically an incommensurate distortion of a lattice. The displacements shown are for a wave vector of approximately $\boldsymbol{k} = \pi/a\,(0.4, \frac{1}{2}, \frac{1}{2})$. Note that these are static displacements. Small oscillations about this static configuration occur and are termed "amplitudons" if they occur perpendicular to $\boldsymbol{k}$ (i.e., fluctuations in the amplitude of the wave) and "phasons" if they occur along $\boldsymbol{k}$ (fluctuations in the phase of the wave).

A primary problem at present is that there exists in the published literature no simple connection between the theories of incommensurate insulators (Dzyaloshinskii–Levanyuk–Sannikov) with its Lifshitz invariants and the theories of incommensurate metals (Overhauser–McMillan) with their charge density waves. The most physical of all published explanations for incommensurate lattices is that of Emery and Axe (1978) for $Hg_3AsF_6$. This steric argument shows that one-dimensional columns of Hg ions do not fit into an integer number of $AsF_6$ cells (Hastings et al. 1977, Pouget et al. 1978).

### 6.1. Metallics

Overhauser (1968, 1971) first proposed that because electrons have spin as well as charge, the lowest energy state in crystals might well be one in which the time-averaged charge distribution does not have the lattice periodicity. The strong Coulomb imbalance would be compensated by a modulation in the *ion* positions, relative to a periodic lattice.

Raman experiments on incommensurate metals, particularly the $TaS_2$ family,

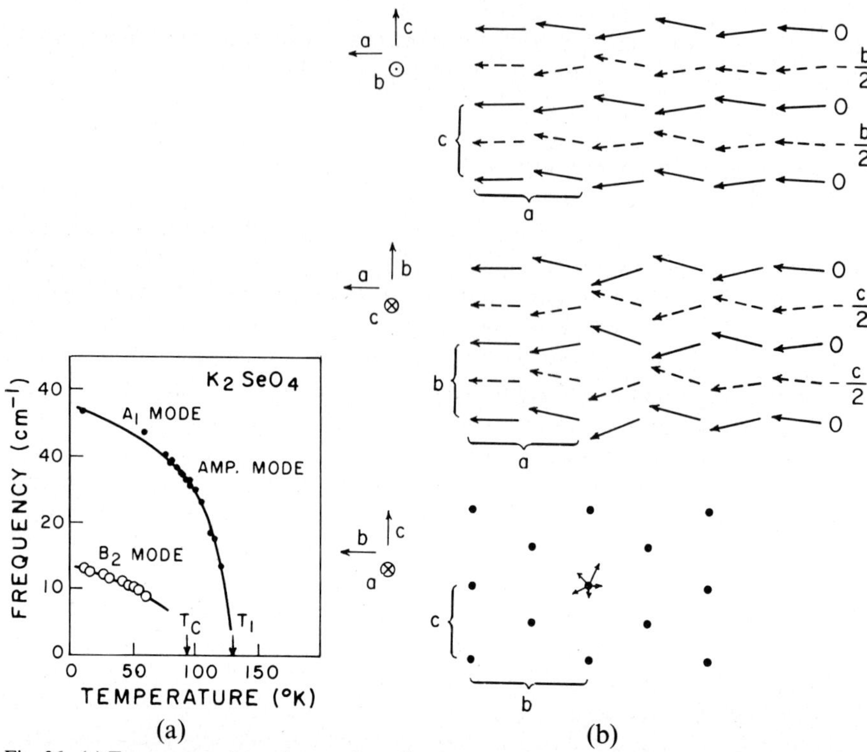


Fig. 36. (a) Temperature dependences of amplitude and phase modes in $K_2SeO_4$ (Caville et al. 1976 and Wada et al. 1977). Note that the amplitude mode persists through the incommensurate phase as an underdamped propagating mode, whereas the phason has a finite frequency only below the lock-in transition temperature. (b) Schematic diagram of the $BaMnF_4$ incommensurate lattice.

are summarized in this book in the chapter by Miles Klein. Review articles on this subject may be found in several recent journals (Di Salvo and Rice 1979, Wilson et al. 1975, Overhauser 1978).

### 6.2. *Insulators; theory*

Ishibashi and Dvorak (1978) present a particularly lucid phenomenological description of incommensurate phase transitions in insulators. They assume, following Levanyuk and Sannikov (1976), that incommensurate phase transitions are a special case of improper phase transitions and therefore necessarily have two-dimensional order parameters. If we call the components of the order parameter $r$ and $s$, then the "Lifshitz invariant" is of form $r(\partial s/\partial x_i) - s(\partial r/\partial x_i)$, where $x_i$ is any vector component. As pointed out elsewhere in this review, the role of Lifshitz invariants is well known in magnetism (Dzyaloshinskii 1964). When it is possible to construct a free energy (scalar) containing a Lifshitz invariant at a particular wave vector $q_0$, the occurrence of a second-order phase transition at $q_0$ induced by the basis functions $r$ and $s$ of the irreducible representation is forbidden. The transition must therefore occur at a (slightly) different wave vector $q$.

Ishibashi and Dvorak construct a free-energy density suitable for point-group symmetry 422, viz.:

$$f(x)=\tfrac{1}{2}\alpha(r^2+s^2)+\tfrac{1}{4}\beta(r^4+s^4)+\tfrac{1}{2}\gamma r^2s^2 + \delta\left(r\frac{\mathrm{d}s}{\mathrm{d}x}-s\frac{\mathrm{d}r}{\mathrm{d}x}\right)+\frac{\kappa}{2}\left[\left(\frac{\mathrm{d}r}{\mathrm{d}x}\right)^2+\left(\frac{\mathrm{d}s}{\mathrm{d}x}\right)^2\right], \tag{6.1}$$

$\alpha$ is assumed to be mean field: $\alpha=\alpha_0(T-T_0)$; and the total free energy is given by the integral of $f(x)$ over all space. Using variational calculus, one constructs two Euler equations for the variables $r$ and $s$, which are to be solved for suitable boundary conditions:

$$\kappa\frac{\mathrm{d}^2r}{\mathrm{d}x^2}-2\delta\frac{\mathrm{d}s}{\mathrm{d}x}-(\alpha r+\beta r^3+\gamma rs^2)=0, \tag{6.2}$$

$$\kappa\frac{\mathrm{d}^2s}{\mathrm{d}x^2}+2\delta\frac{\mathrm{d}r}{\mathrm{d}x}-(\alpha s+\beta s^3+\gamma r^2s)=0. \tag{6.3}$$

There are two simple cases of interest. First, if $\mathrm{d}r/\mathrm{d}x=\mathrm{d}s/\mathrm{d}x=0$ for all $x$, the eqs. (6.2) and (6.3) become:

$$\alpha r+\beta r^3+\gamma rs^2=0, \tag{6.4}$$

$$\alpha s+\beta s^3+\gamma r^2s=0, \tag{6.5}$$

which have solutions $r=s=0$ for $\alpha>0$ and $T>T_0$ and

$$r^2=s^2=-\alpha/(\beta+\gamma), \tag{6.6}$$

for $T<T_0$. Solution (6.6) is commensurate and has free energy

$$F_c=-2L\frac{\alpha^2}{2(\beta+\gamma)}, \tag{6.7}$$

where $2L$ is the length of the crystal. The phase with $r=s=0$ permits eqs. (6.2) and (6.3) to be rewritten as,

$$\kappa\frac{\mathrm{d}^2r}{\mathrm{d}x^2}-2\delta\frac{\mathrm{d}r}{\mathrm{d}x}-\alpha r\cong 0, \tag{6.8}$$

$$\kappa\frac{\mathrm{d}^2s}{\mathrm{d}x^2}+2\delta\frac{\mathrm{d}s}{\mathrm{d}x}-\alpha s\cong 0, \tag{6.9}$$

where we have dropped all terms higher than linear in $r$ and $s$ to examine the stability of this phase to small variations about the $r=s=0$ values. The instability condition for this phase is that one of the principal minors of the determinant for eqs. (6.8) and (6.9) be zero for real $q$ which minimizes it; and from the assumption $r \cong t_0 \exp[i(q-x)]s \cong s_0 \exp[i(q-x)]$ this yields an instability at a transition temperature such that

$$\alpha\kappa - \delta^2 = 0 . \qquad (6.10)$$

Thus, a transition occurs at a temperature $T_1 = T_0 + \delta^2/\alpha_0 q$, where $T_0$ is the transition temperature which would occur in the absence of gradient terms in the free energy. The wave number of the modulation which sets in at $T_I$ is $q_0 = |\delta|/\kappa$, which is, in general, not an integer fraction of the reciprocal lattice constant.

Ishibashi and Dvorak show that for an improper phase transition, such as in $BaMnF_4$, the dynamics can be continuous, with electric susceptibility near $T_I$ varying as,

$$\chi(T) = \chi_0 + \chi_0^2 \xi \frac{T_I - T}{(3\beta + \gamma')(T_I - T_0) + \gamma'(T_I - T)} . \qquad (6.11)$$

Here $\xi$ is a coefficient in the free energy for a polarization term

$$F' = -P(x)E + (1/2\chi_0)P^2(x) + \xi r(x)s(x)P(x) , \qquad (6.12)$$

and

$$\gamma' = \gamma - \chi_0 \xi^2 , \qquad (6.13)$$

with $\gamma$ from eq. (6.1).

### 6.3. *Insulators; experiment*

Raman studies of transition dynamics and soft modes in materials later understood to display incommensurate phase transitions were first done on the ferroelectric insulators $K_2SeO_4$ and $BaMnF_4$ by Fawcett et al. (1974, 1975), Caville et al. (1976), and Ryan and Scott (1974). The $K_2SeO_4$ soft-mode data of Caville et al. are shown in fig. 36 below. In the jargon currently used for such incommensurate transitions, this mode is the "amplitudon". The "phason" was not reported until later (Wada et al. 1977a,b); their data are shown in fig. 36(a) as the lower-frequency mode of $B_2$ symmetry.

There has been some controversy concerning the Raman activity and intensity permitted for phasons in the incommensurate phases of crystals. This topic is beyond the scope of the present chapter, but readers are referred to the chapter in this book by Fleury and Lyons, as well as to the earlier paper by Dvorak and Petzelt (1978). Generally it is considered that phason scattering in the incommensurate phases of crystals can occur only through coupling to the soft optic mode or to acoustic phonons; the form of coupling has not been unambiguously established for any experimental data yet. See Inoue et al. (1980) for an example of recent data.

The first experimental analysis of phasons as overdamped, characterized by Debye-like diffusion in the incommensurate phase, was that of Bechtle et al. (1978). A subsequent analysis made the same assumptions of phason diffusion and linear coupling to acoustic phonons in the incommensurate phase of thiourea (An et al. 1979). These analyses, stimulated in part by the early theoretical model of Bhatt and McMillan (1975), focus on the dispersion observed for acoustic phonons near the incommensurate phase transitions. Complementary information on such dispersion is also available from dielectric studies, of course; in $BaMnF_4$ early dielectric studies extended into the GHz regime (where dispersion due to phasons is expected), see Samara and Richards (1976), whereas in $K_2SeO_4$ the initial reports were limited to much lower frequencies (Aiki et al. 1969, 1970). The present author has suggested elsewhere that dispersion due to phasons might be expected in the frequency region 1 to 30 GHz. This appears to be true in $Ba_2NaNb_5O_{15}$ (Young and Scott 1981), as well as in $BaMnF_4$ (Bechtle et al. 1978), and in $RbH_3(SeO_3)_2$ (Tsukui et al. 1980). Similar relaxation times for phasons in $K_2SeO_4$ are not incompatible with the data (Petzelt et al. 1979, Inoue et al. 1980), but the situation is somewhat less clear in that material.

*6.3.1. $BaMnF_4$*

Barium manganese tetrofluoride and potassium selenate remain the two incommensurate crystals most studied by light scattering. $BaMnF_4$ has been studied by Raman, Brillouin, and Rayleigh spectroscopy by at least four different groups; its structure is shown in fig. 37a (Keve et al. 1969). Much of this work has already been summarized by the author in a lengthy review (Scott

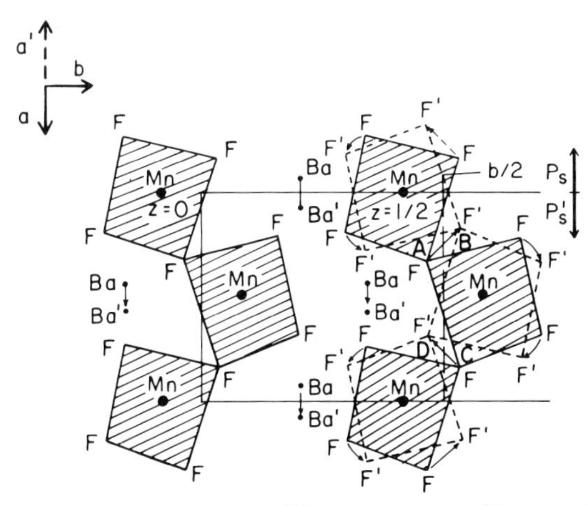


Fig. 37a. $BaMnF_4$ structure at ambient temperatures (Keve et al. 1969).

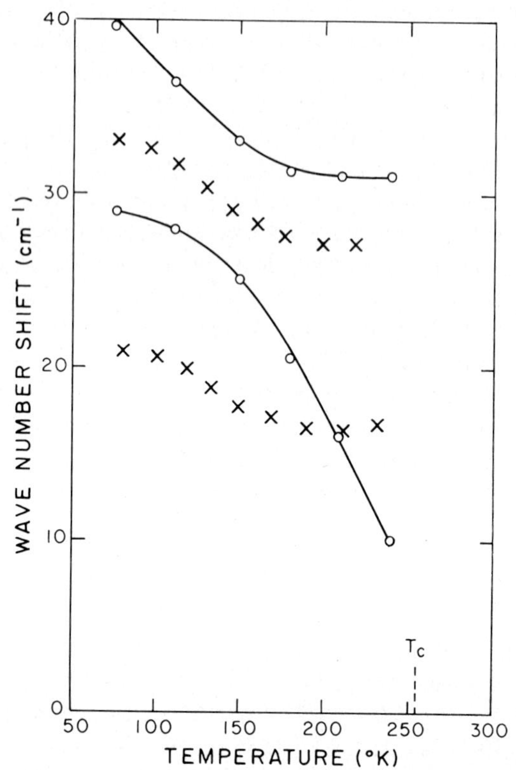


Fig. 37b. Frequencies versus temperature for the two amplitude modes in the incommensurate phase of $BaMnF_4$ (Ryan and Scott 1974).

1979) however, several research reports have been made since that review was prepared, and in addition, several aspects of the early data and interpretation need clarification in view of recent theories. Furthermore, some unpublished data on $BaMnF_{4-x}$ are included which show shifts in the incommensurate transition temperature.

The original Raman data of Ryan and Scott (1974) are summarized in fig. 37b. More recent data by Ryan at Oxford, by Feldkamp and Douglas at Colorado, and by Murray and Lockwood at Edinburgh are lower at most temperatures, as shown. The more recent data are shown as ×'s in fig. 37b. The $C_{2v} \to C_2$ phase transition involves a nearly rigid rotation of $MnF_6$ octahedra (Fritz 1975, Scott 1975, 1976).

A second aspect of the original data which requires explicit correction is that the lower mode (at $\sim 25\,cm^{-1}$) in fig. 37b was originally assigned as a non-totally symmetric vibration which coupled to the A(TO) soft mode due to the oblique propagation direction employed. It is now known that both modes have the same A(TO) character. As discussed further below, the theory of Cox et al. (1979) predicts two amplitude modes and two phase modes for the $q = 0$ excitations of $BaMnF_4$ below $T_c$, and it is likely that the two optical modes shown in fig. 37b are the *two* predicted amplitudons.

Inelastic neutron scattering studies (Shapiro et al. 1976, Cox et al. 1979) of $BaMnF_4$ over the past several years have reported that this material is incommensurate at all temperatures below $T_0 = 247$ K. There are four respects in which the incommensurate structure is unique and puzzling:

First, the incommensurate translation vector $q_0 = (0.392, \frac{1}{2}, \frac{1}{2})$ reported is not nearly a reciprocal integer of the form $1/(n_a - \delta)$; $1/n_b$; $1/n_c$. Second, the vector $q_0$ is almost totally independent of temperature from $T_I = 247$ K down to $T = 0$. Neither of these facts is typical of incommensurates (Ishibashi 1980). Third, a possible explanation of the temperature independence of $q_0$, according to Ishibashi and Dvorak's theory (1978), is that the order parameter is perfectly isotropic; yet Lyons et al. (1980) have measured its anisotropy, which is found to be large. Fourth, the exact temperature of the phase transition (247 K) measured via neutron scattering (Shapiro et al. 1976) agrees with that from dielectric measurements (Samara and Richards 1976), sound velocity anomalies (Fritz 1975a, Bechtle et al. 1978), but disagrees substantially from that (254–255 K) determined from the original ultrasonic attenuation experiments (Spencer et al. 1970) or Raman measurements (Ryan and Scott 1974). These differences cannot arise from sample heating or sample differences, because they have been observed via simultaneous measurements on the same specimen (Lyons et al. 1982).

For these reasons new measurements have been made on $BaMnF_4$ (Scott et al. 1982) via specific heat, X-ray, electron and neutron diffraction techniques. The $C_p$ results are shown in fig. 38. Surprisingly, two transitions are observed at 247.1 K and 255.3 K. The large anomaly at 255 K can best be fitted to a power-law divergence with an exponent $\alpha = 0.5$. Such a large exponent differs, of course, from the 0.1 value expected from most statistical mechanical theories. Similar values are found experimentally for KDP (Reese 1969) and $K_2SeO_4$ (Lopez-Echarri 1980). They may arise from at least three different sources: (1) extrinsic causes: the defect theory of Levanyuk et al. (1979) yields such values; (2) the presence of tricritical points nearby in phase space: $\alpha$ at tricritical points is of order 0.5 (mean field) to 0.65 (Shang and Salamon 1980, Bastie et al. 1980); (3) the values may arise simply from fluctuations neglected in mean field (Levanyuk and Sobyanin 1970). The presence of large values of $\alpha$ then is not in itself sufficient to draw unambiguous conclusions.

The smaller specific heat anomaly at 247 K is similar in size and shape to that observed in $K_2SeO_4$ or $(NH_4)_2BeF_4$ (Aiki et al. 1970, Strukov et al. 1973). This similarity suggests that perhaps some samples of $BaMnF_4$ have a small temperature region of incommensurability ranging from 247 to 255 K. Similar conclusions were suggested much earlier by Levstik et al. (1975), who found

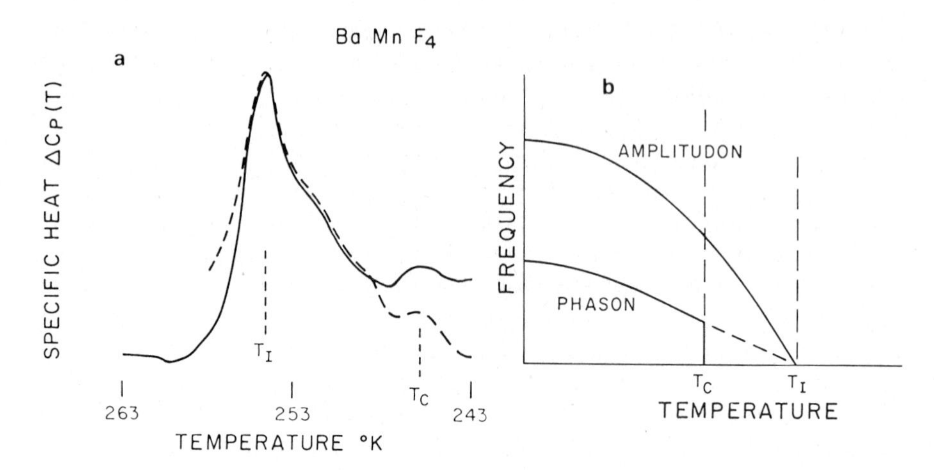


Fig. 38. (a) $\Delta C_p(T)$ in $BaMnF_4$ for heating (solid curve) and cooling (dashed curve). The data were taken on a Perkin–Elmer DSCII; absolute background value of $c_p$ is not accurately determined. For experimental details, see Greer (1981). (b) Proposed phason and amplitudon behavior in $BaMnF_4$. Acoustic phonon–phason interaction should be strong near $T_c$, as observed.

a pair of phase transitions near 250 K separated by 3.95 K. The most convincing evidence for two transitions in $BaMnF_4$ at 247 and 255 K comes from piezoelectric resonance data, which display sharp dips at each temperature (Scott et al. 1982); however, no data show directly that the phase between 247 and 255 K is incommensurate.

More serious sample-dependent aspects of the $BaMnF_4$ problem have surfaced recently. Hidaka et al. (1982) have found that their samples do not have translation vectors involved in the phase transition which lie at $(0.392, \frac{1}{2}, \frac{1}{2})$, but rather at $(0.399, 0, \frac{1}{2})$. This is important for two reasons: first, the 0.399 value is, within experimental uncertainty, 0.4 and would correspond to a commensurate lattice with $Z = 20$ formula units per primitive cell (compared with $Z = 2$ in the ambient phase); second, the new data show no doubling along the $b$-axis – i.e., $(0.4, 0, \frac{1}{2})$ compared with $(0.4, \frac{1}{2}, \frac{1}{2})$ (see fig. 39). Such extreme sample variations may arise from fluorine vacancies and resulting stacking faults. Indeed, Hidaka et al. have found that some $BaMnF_4$ specimens exist in a polytype of $D_2^3$ space group, in which the $b$-axis stacking of fluorine octahedra is antiparallel, rather than parallel (as in the usual $C_{2V}^{12}$ phase). Such stacking faults, which are observed to exhibit only short-range order, may account for the forbidden Bragg scattering reported by Almairac et al. (1981), as well.

### 6.3.2. *Phasons in $BaMnF_4$*

On the basis of this new understanding of the $BaMnF_4$ phase-transition dynamics, we can reinterpret the dynamic central mode data at $T \simeq 247$ K

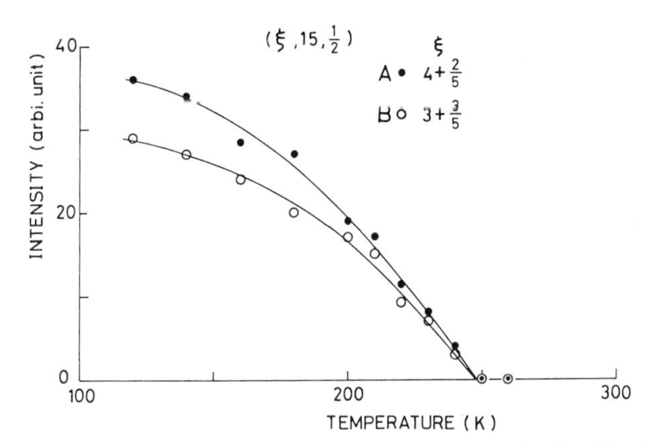


Fig. 39. Inelastic neutron scattering intensities versus temperature, showing (5*a*, *b*, 2*c*) structure below $T_c$ (data taken at I.L.L., Grenoble).

reported by Bechtle et al. (1978), Lyons and Guggenheim (1979), and Lyons et al. (1980).

In the original Brillouin study (Bechtle et al. 1978), the existence of a diffusive mode was inferred, with a temperature dependent Debye relaxation time satisfying the theory of Bhatt and McMillan (1975). The spectral distribution junction was calculated as:

$$S(a) = \bar{n}(\omega, T)\,\mathrm{Im}\,G_{ij}(\omega)P_iP_j\,, \tag{6.14}$$

where

$$G_{ij}^{-1}(\omega) = \begin{bmatrix} \omega_0^2 - \omega^2 + \mathrm{i}\Gamma\omega & A & B \\ A & \omega_A^2 - \omega^2 + \mathrm{i}\gamma\omega & \alpha \\ B & \alpha & C(1 + \mathrm{i}\omega\tau) \end{bmatrix}, \tag{6.15}$$

is the response function describing the optic mode at $\omega_0$, acoustic mode at $\omega_A$, and diffusion with time $\tau(T)$. This formalism was applied to the analysis of Rayleigh–Brillouin data in $BaMnF_4$ with two important approximations: first, $A$ and $B \ll \alpha$; and second, the $q^2$ term in $\tau(T)$, viz.,

$$\frac{1}{\tau(T)} = \frac{1}{\tau_1}\left|\frac{T_c - T}{T_c}\right| + \frac{1}{\tau_2} + Fq^2\,, \tag{6.16}$$

from Bhatt and McMillan, was assumed small and neglected. The first approximation yields the wrong width and polarizability for the central-mode scattering; if only $\alpha$ is included of the off-diagonal matrix elements in eq. (6.15) a dynamic central mode width of 0.73 GHz is calculated, (Bechtle et al. 1978) in contrast to the larger value of $\sim$ 3 GHz measured experimentally (Lyons and Guggenheim 1979, Lyons et al. 1982). And predominantly depolarized scattering is predicted, in contrast to the polarized $\alpha_{aa}$ measured. Both results show that $A$ is greater than $\alpha$ in eq. (6.15).

The second approximation, neglecting the third term in eq. (6.16), was shown false by Lyons and Guggenheim. The $q^2$ dependence measured there is of course *required* for any diffusive process, such as the phason diffusion proposed by Bechtle et al., but was thought small compared with $1/\tau_2$ in eq. (6.16). Lyons and Guggenheim (1979) show that these two terms are comparable in magnitude.

Thus, the work of Lyons et al. (1982) confirms in detail the analysis of Bechtle et al. (1978) based upon phason diffusion. (Note that Lyons et al. show that this cannot be due to entropy diffusion since the quantitative Rayleigh linewidths have order-or-magnitude disagreement with that calculated from thermal conductivity).

We believe that this diffusion at $T \approx T_c$ is due to the diffusion of discommensurations. We note that in this region very near the lock-in transition $T_c$ phasons are expected to be very soliton-like, and have been characterized by a single relaxation time of the same magnitude (2.0 ns) at $T \approx T_c + 0.5$ K in the incommensurate phase of $RbH_3(SeO_3)_2$ from the analysis of dielectric dispersion by Tsukui et al. (1980). This suggests to us that phasons may be diffusive in the incommensurate phases of all insulators, a hypothesis in accord with both infrared and Raman data on $K_2SeO_4$ (Petzelt et al. 1979, Inoue et al. 1980). The hypothesis that some $BaMnF_4$ specimens are incommensurate only between 247 and 255 K is yet unproved. It is appealing, however, in that it would provide a simple explanation of the central-mode scattering reported by Lyons et al. (1982) in this material; their central-mode scattering has maximum intensity at 247 K and disappears at exactly 255 K. However, there are other possible interpretations of such data, including the possibility of scattering from relaxing, mobile defects such as fluorine vacancies. Such mechanisms seem likely to explain the anomalously large ultrasonic attenuation exponents, which are too great for existing theories (Rehwald 1973). It would be pleasant to be able to invoke the same mechanism for both the ultrasonic attenuation anomalies and the central-mode scattering in $BaMnF_4$.

### 6.4. *Other incommensurate insulators*

Other ferroelectric incommensurate crystals are listed in table 3. $NaNO_2$ has been studied by Raman spectroscopy, but with little information obtained regarding its incommensurate structure. Thiourea has been examined extensively via inelastic neutron scattering (Denoyer et al. 1980) and has $q_0 = b^*/9$; and the amplitude mode has been studied by Raman techniques (Wada et al. 1977, 1978) and is shown in fig. 40. $Ba_2NaNb_5O_{15}$ ("bananas") has been analyzed by Brillouin and neutron techniques (Schneck 1979,

Table 3
Examples of some IC transitions in ferroelectrics.

| Crystal | Initial group | Critical Point | References |
|---|---|---|---|
| $K_2SeO_4$ | $D_{2h}^{16}$ | $\Sigma = (\frac{1}{3} - \delta, 0, 0)$ | a |
| $K_2ZnCl_4$ | $D_{2h}^{16}$ | $\Lambda = (0, 0, \frac{1}{3} - \delta)$ | b |
| $Rb_2ZnCl_4$ | $D_{2h}^{16}$ | $\Lambda = (0, 0, \frac{1}{3} - \delta)$ | b |
| $(NH_4)_2BeF_4$ | $D_{2h}^{16}$ | $\Sigma = (\frac{1}{2} - \delta, 0, 0)$ | a |
| $NaNO_2$ | $D_{2h}^{25}$ | $\Sigma = (\delta, 0, 0)$ | c |
| $SC(NH_2)_2$ | $D_{2h}^{16}$ | $\Lambda = (0, \frac{1}{9} - \delta, 0)$ | c |
| ARS | $D_2^3$ | $X = (\frac{1}{2}, 0, 0)$ | a |
| $RbLiSO_4$ | orthorhombic | ? | d |
| $BaMnF_4$ | $C_{2v}^{12}$ | ? | e |

Notes: $\delta$ represents a small, temperature dependent value which vanishes at IC-commensurate transitions.

References

a. Y. Ishibashi (1978) Ferroelectrics **20**, 103.
b. K. Gesi and M. Iizumi, 1979, J. Phys. Soc. Japan **46**, 697.
c. Y. Ishibashi and H. Shiba, 1978, J. Phys. Soc. Jpn. **45**, 409.
d. Y. Shiroshi, A. Nakata and S. Sawada, 1976, J. Phys. Soc. Jpn. **40**, 911.
e. J.F. Scott, F. Habbal and M. Hidaka, 1982, Phys. Rev. **B 25**, 1805.

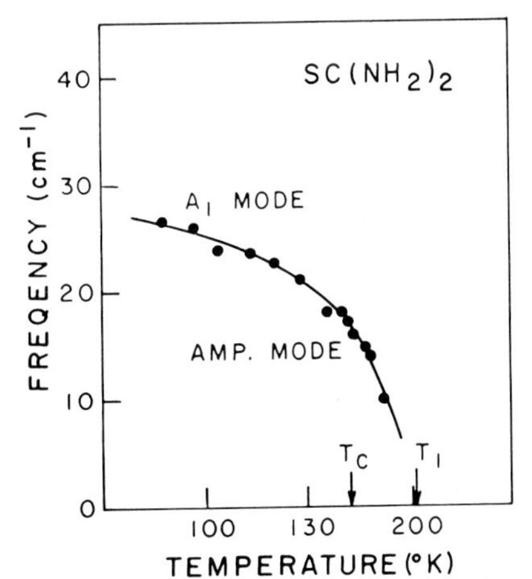


Fig. 40. Amplitude mode frequency versus temperature near the lock-in transition temperature in thiourea, $SC(NH_2)_2$. (Wada et al. 1978).

Schneck et al. 1979a,b); and a $(NH_4)_2BeF_4$ Brillouin study was executed by Kudo and Hikita (1978) but only for the less interesting longitudinal modes. For related studies of ammonium fluoberyllate see Makita and Yamauchi (1974) and Strukov et al. (1973). Some details of the incommensurate structure of ammonium Rochelle Salt have also been provided by Sawada and Takagi (1971, 1972), but light-scattering studies of this system have not been enlightening.

In concluding this section we draw attention to the work of Blinc and Zeks (1978), who have succeeded in treating smectic liquid crystals by a formalism analogous to that used for incommensurate crystals, complete with phason-like excitations.

## 7. Summary

In 1974 when I published an earlier review on Raman spectroscopy of phase transitions (Scott 1974) there were a few well-understood prototype systems, including SnTe, $SrTiO_3$, and $KH_2PO_4$. Now there are a hundred, including order–disorder organics (triazine) and displacive organics (chloranil). In 1974 virtually all systems exhibited mean-field behavior, although naive analysis sometimes yielded curious exponents. Now we have seen the realization of bicritical and tricritical points and their associated exponents, and even the three-states Potts model, which was viewed even by its creator as unphysical! Liquid crystals have yielded accurate, non-mean-field exponents (e.g., $\gamma = 1.26$) from light scattering. In this sense, structural phase transitions in 1974 were about where magnetic transitions were in 1962, just as Heller and Benedek measured the first $\beta = \frac{1}{3}$ values.

In 1980 incommensurate systems are popular, both metallic and ferroelectric. Viewed as a special category of improper phase transitions, these systems still suffer from the historically different approaches of metal physics and spiral magnets with its resulting dichotomy.

Even without the considerations of dynamic central modes (fortunately assigned to other authors as a separate chapter of this book), we find the introduction of true critical (fluctuation-dominated) phenomena, organic and liquid crystals, and incommensurate structures evidence that Raman spectroscopy of structural phase transition will have at least another fruitful decade.

## References

Abe, R., 1980, Proc. Conf. Jpn. Phys. Soc. Fukui, Oct. 1, 1980.
Aharony, A., 1980, Ferroelectrics, **24**, 313.
Aharony, A. and A.D. Bruce, 1979, Phys. Rev. Lett. **42**, 462.
Aharony, A., K.A. Muller and W. Berlinger, 1977, Phys. Rev. Lett. **38**, 33.

Aiki, K., K. Hukuda and O. Matumura, 1969, J. Phys. Soc. Jpn. **26**, 1064.
Aiki, K., K. Hukuda, H. Koga and T. Kobayashi, 1970, J. Phys. Soc. Jpn. **28**, 389.
Aizu, K., 1971, J. Phys. Soc. Jpn. **31**, 802.
Almairac, R., M. Regis, J. Nouet and C. Filippini, 1981, J. Physique Lett.
Alperin, H.A., S.J. Pickart, J.J. Rhyne and V.J. Minkiewicz, 1972, Phys. Lett. **40 A**, 295.
Anderson, P.W. and E.I. Blount, 1965, Phys. Rev. Lett. **14**, 217.
Antcliffe, G.A., R.T. Bate and D.D. Buss, 1973, Solid State Commun. **13**, 1003.
Armstrong, R.D. and K. Taylor, 1975, J. Electroanal. Chem. **63**, 9.
Ascher, E., 1970, J. Phys. Soc. Jpn., Suppl. **28**, 7.
Ascher, E., H. Rieder, H. Schmid and H. Stossel, 1966, J. Appl. Phys. **37**, 1404.
Ashida, T., S. Bando and M. Kakudo, 1972, Acta. Cryst. **B 28**, 1560.
Aubry, S. and R. Pick, 1971, J. Physique **32**, 657.
Bagieu, M., I. Tordjman, A. Durif and C. Bossi, 1973, Cryst. Struct. Commun. **3**, 387.
Bartis, F.J., 1973, Phys. Lett. **A43**, 61.
Bartuch, H. and W. Windsch, 1972, Phys. Status Solidi, (a) **14**, K51.
Bastie, P., G. Dolino and M. Vallade, Ferroelectrics, **25**, 431.
Bate, T.R., D.L. Carter and J.S. Wrobel, 1970, Phys. Rev. Lett. **25**, 159.
Bechtle, D.W. and J.F. Scott, 1977, J. Phys. **C10**, 1209.
Bechtle, D.W., J.F. Scott and D.J. Lockwood, 1978, Phys. Rev. **B 18**, 6213.
Bernasconi, J., W.R. Schneider and W. Wyss, 1980, Z. Phys. **B 37**, 175.
Bernstein, E.R. and B.B. Lal, 1980, Molec. Cryst. Liquid Cryst. **58**, 407.
Bersucker, I.B. and B.G. Vekhter, 1976, Ferroelectrics, **13**, 373.
Bersucker, I.B. and B.G. Vekhter, 1978a, Ferroelectrics, **19**, 137.
Bersucker, I.B. and B.G. Vekhter, 1978b, Ferroelectrics, **20**, 163.
Bhatt, R.N. and W.L. McMillan, 1975, Phys. Rev. **B 12**, 2042.
Bierlein, J.D. and A.W. Sleight, 1975, Solid State Commun. **16**, 69.
Bierly, J.N., L. Muldawer and O. Beckman, 1963, Acta Metall. **11**, 447.
Bilz, H., 1980, Proc. Int. Conf. Semicond. Kyoto, Sept. 1980.
Bilz, H., H. Büttner, A. Bussmann-Holder, W. Kress and U. Schröder, 1982, Phys. Rev. Lett. **48**, 264.
Blinc, R. and B. Zeks, 1978, Phys. Rev. **A 18**, 740.
Blinc, R., M. Jamsek-Vilfan, C. Lahajnar and G. Hajdukovic, 1970, J. Chem. Phys. **52**, 6407.
Blinc, R., M. Mali, R. Osredkar and J. Selizer, 1975, J. Chem. Phys. **63**, 35.
Bocchi, C. and C. Chezzi, 1975, Phys. Status Solidi, **671**, 461.
Bottger, G. L. and C.V. Damsgard, 1972, J. Chem. Phys. **53**, 1215.
Brasch, J.W., A.J. Melveger, E.R. Lippincott and S.D. Hamann, 1970, Appl. Spectrosc. **24**, 184.
Brebick, R.F., 1971, J. Chem. Phys. Solids, **32**, 551.
Briels, W.J. and J.C. van Miltenberg, 1979, J. Chem. Phys. **70**, 1064.
Brillson, L.J. and E. Burstein, 1971, Phys. Rev. Lett. **27**, 808.
Brillson, L.J., E. Burstein and L. Muldawer, 1974, Phys. Rev. **139**, 1547.
Bruce, A.D., 1980, Adv. Phys. **29**, 111.
Bruce, A.D. and R.A. Cowley, 1978, J. Phys. **C 11**, 3609.
Bruce, A.D. and R.A. Cowley, 1980, Adv. Phys. **29**, 219.
Bruce, A.D., K.A. Muller and W. Berlinger, 1979, Phys. Rev. Lett. **42**, 185.
Bruesch, P., S. Strassler and H.R. Zeller, 1975, Phys. Status Solidi (a) **31**, 217.
Bruesch, P., H.U. Beyler and W. Buhrer, 1978, Proc. Int. Conf. Lattice Dynamics, ed. M., Balkanski.
Budin, J.P., A. Milato-Roufos, Duc Chinh Nguyen and G. Le Roux, 1975, J. Appl. Phys. **46**, 2867i.
Burns, G., F.H. Dacol and M.W. Shafer, 1976a, Solid State Commun. **19**, 287.
Burns, G., F.H. Dacol and M.W. Shafer, 1976b, Solid State Commun. **19**, 291.
Burns, G, F.H. Dacol and M.W. Shafer, 1977a, Solid State Commun. **24**, 753.
Burns, G., F.H. Dacol and M.W. Shafer, 1977b, Phys. Rev. **B 16**, 1416.

Bussmann-Holder, A., H. Bilz and W. Kress, 1980, Proc. 15th Int. Conf. Semicond., Kyoto, Sept. 1, 1980, paper 2pB-7, p. 81.
Cape, J.A., L.G. Hale and W.E. Tennant, 1977, Surf. Sci. **62**, 639.
Caville, C., V. Fawcett and D.A. Long, 1976, Proc. 5th Int. Conf. Raman Spectroscopy, eds. E.D. Schmid, J. Brandmuller, W. Kiefer, B. Schrader, and H.W. Schrotter (Hans-Ferdinand-Schulz, Freiberg) p. 626.
Chan, L.Y.Y. and S. Geller, 1977, J. Solid State Chem. **21**, 331.
Chapelle, J.P., C.X. An and J.P. Benoit, 1976, Solid State Commun. **19**, 573.
Chase, L.L., 1976, Superionic Conductors, eds., G.D. Mahan and W.L. Roth (Plenum, New York) p. 299).
Chihara, H. and K. Masukane, 1973, J. Chem. Phys. **59**, 5397.
Chihara, H. and N. Nakamura, 1973, J. Chem. Phys. **59**, 5392.
Chihara, H., N. Nakamura and M. Tachiki, 1971, J. Chem. Phys. **54**, 3640.
Chihara, H., N. Nakamura and M. Tachiki, 1973, J. Chem. Phys. **59**, 5387.
Cho, M., T. Yagi, A. Sawada and Y. Ishibashi, 1982, J. Phys. Soc. Jpn. **51**, 2914.
Chu, S.S.C., G.A. Jeffrey and T. Sakurai, 1962, Acta Crystallogr. **15**, 661.
Cochran, W., 1960, Adv. Phys. **9**, 387.
Cochran, W., 1961, Adv. Phys. **10**, 401.
Cochran, W., 1964, Phys. Lett. **13**, 193.
Cochran, W., R.A. Cowley, G. Dolling and M.M. Elcombe, 1966, Proc. Roy. Soc. (London) **A 293**, 433.
Comes, R., S.M. Shapiro, B.C. Frazer and G. Shirane, 1981, Phys. Rev. **B 24**, 1559.
Coppens, P., and Sabine T.M., 1968, Mol. Cryst. **3**, 507.
Cowley, E.R., J.K. Darby and G.S. Pawley, 1969, J. Phys **C 2,** 1916.
Cowley, R.A., 1963, Adv. Phys. **12**, 421.
Cowley, R.A., 1965, Phil. Mag. **11**, 673.
Cowley, R.A., 1976, Phys. Rev. **B 13**, 4877.
Cowley, R.A., 1980, Adv. Phys. **29**, 1.
Cowley, R.A., J.D. Axe and M. Iijumi, 1976, Phys. Rev. Lett. **36**, 806.
Cox, D.E., S.M. Shapiro, R.A. Cowley, M. Eibschutz and H.J. Guggenheim, 1979, Phys. Rev. **B 19**, 5754.
Craig, P.P., 1966, Phys. Lett. **20**, 140.
Greer, A., F. Habbal, J.F. Scott and T. Takahashi, 1980, J. Chem. Phys. **73**, 5833.
Damen, T.C., S.P.S. Porto and B. Tell, 1966, Phys. Rev. **142**, 570.
Danielmeyer, H.G. and H.P. Weber, 1972, I.E.E.E. J. Quant-Electron. **QE 8**, 805.
Daunt, S.J., H.F. Shurvell and L. Pazdernik, 1975, J. Raman Spect. **4**, 205.
David, W.I.F., A.M. Glazer and A.W. Hewat, 1979, Phase Transitions, **1**, 155.
Davidov, D., C.R. Safinya, M. Kaplan, S.S. Dana, R. Schaetzing, R.J. Birgeneau and J.D. Litster, 1979, Phys. Rev. **B19**, 1657.
Deis, D.W., J.K. Hulm and C.K. Jones, 1969, Bull. Am. Phys. Soc. **14**, 61.
Delaney, M.J. and S. Ushioda, 1976, Solid State Commun. **19**, 297.
Denoyer, F., A.M. Moudden and M. Lambert, 1980, Ferroelectrics, **24**, 43.
Di Salvo, F.J., Jr. and T.M. Rice, 1979, Physics Today, p. 32.
Dieterich, W., T. Geisel and I. Peschel, 1978, Z. Phys. **B 29**, 5.
Dolling, G. and W.J.L. Buyers, J. Nonmetals, **1**, 159.
Durvasula, L.N. and R.W. Gammon, 1977, Phys. Rev. Lett. **38**, 1081.
Dvorak, V., 1970, J. Phys. Soc. Jpn. Suppl. **28**, 252.
Dvorak, V., 1971a, Phys. Status Solidi (b) **45**, 147.
Dvorak, V., 1971b, Phys. Status Solidi (b) **46**, 763.
Dvorak, V., 1972a, Phys. Status Solidi, (b) **51**, K129.
Dvorak, V., 1972b, Phys. Status Solidi, (b) **52**, 93.
Dvorak, V., 1974, Ferroelectrics, **7**, 1.

Dvorak, V. and J. Petzelt, 1971, Phys. Lett **A 35**, 209.
Dvorak, V. and J. Petzelt, 1978, J. Phys. **C 11**, 4827.
Dzyaloshinskii, I.E., 1964, Zh. Eksp. Teor. Fiz. **46**, 1420 (translation: 1964, Sov. Phys. JETP, **19**, 960).
Elliott, G.R. and Z. Iqbal, 1975, J. Chem. Phys. **63**, 1914.
Emery, V.J. and J.D. Axe, 1978, Phys. Rev. Lett. **40**, 5107.
Errandonea, G., 1980, Phys. Rev. **B 21**, 5221.
Errandonea, G. and J. Sapriel, 1979, Solid State Commun.
Fano, V. and I. Ortalli, 1974, J. Chem. Phys. **61**, 5017.
Fawcett, Y., R.J.B. Hall, D.A. Long and V.N. Sankaranarayanan, 1974, J. Raman Spectrosc. **2**, 629.
Fawcett, Y., R.J.B. Hall, D.A. Long and V.N. Sankaranarayanan, 1975, J. Raman Spectrosc. **3**, 229.
Feldkamp, G.E., K. Douglas, B.B. Lavrencic and J.F. Scott, 1980, Bull. Am. Phys. Soc. **25**, 171.
Fleury, P.A., 1972a, Comm. Solid State Phys. IV, 149.
Fleury, P.A., 1972b, Comm. Solid State Phys. VIII, 67.
Fleury, P.A. and K.B. Lyons, 1976, Phys. Rev. Lett. **37**, 1088.
Fleury, P.A., J.F. Scott and J.M. Worlock, 1968, Phys. Rev. Lett. **21**, 16.
Fleury, P.A., P.D. Lazay and L.G. Van Uitert, 1974, Phys. Rev. Lett. **33**, 492.
Folk, R., H. Iro and F. Schwabl, 1976, Z. Phys. **B 25**, 69.
Folk, R., H. Iro and F. Schwabl, 1979, Phys. Rev. **B 20**, 1229.
Fontana, A., G. Mariotto, M. Montagna, V. Capp, V. Capozzi, E. Cazzanelli and M.P. Fontana, 1978, Solid State Commun. **28**, 35.
Fox, D.L., 1979, Ph.D. Thesis, Univ. Colorado (University Microfilms, Ann Arbor, Mich.).
Fox, D.L., J.F. Scott and P.M. Bridenbaugh, 1976, Solid State Commun. **18**, 111.
Fox, D.L., J.F. Scott, P.M. Bridenbaugh and J.W. Pierce, 1978, J. Raman Spectrosc. **7**, 41.
Fox, D.L., D.R. Tilley, J.F. Scott and H.J. Guggenheim, 1980, Phys. Rev. **B 21**, 2926.
Fritz, I.J., 1975, Phys. Rev. Lett. **35**, 1511.
Fritz, I.J., 1975a, Phys. Lett. **51A**, 219.
Fukushi, K., M. Nippus and R. Claus, 1978, Phys. Status Solidi, **686**, 257.
Funke, K., 1976, Prog. Solid State Chem. **11**, 345.
Geisel, T., 1977, Solid State Commun. **24**, 155.
Geller, S. and P.M. Skarstad, 1974, Phys. Rev. Lett. **33**, 1384.
Geller, S., S.A. Wilber, G.F. Ruse, J.R. Akridge and A. Turkovic, 1980, Phys. Rev. **B 21**, 2506.
Gesi, K. and M. Iizumi, 1979, J. Phys. Soc. Jpn. **46**, 697.
Gibbs, H.M., S.L. McCall and T.N.C. Venkatesan, 1976, Phys. Rev. Lett. **36**, 1135.
Gillis, N., 1969, Phys. Rev. Lett. **22**, 1251.
Gillis, N.S. and T.R. Koehler, 1971, Phys. Rev. **B 4**, 3971.
Gillis, N.S. and T.R. Koehler, 1972a, Phys. Rev. **B 5**, 1925.
Gillis, N.S. and T.R. Koehler, 1972b, Phys. Rev. Lett. **29**, 369.
Gillis, N.S. and T.R. Koehler, 1974, Phys. Rev. **B 9**, 3806.
Ginzburg, V.L., 1945, Zh. Eksp. Teor. Fiz. **15**, 739.
Ginzburg, V.L., 1949a, Zh. Eksp. Teor. Fiz. **19**, 35.
Ginzburg, V.L., 1949b, Usp. Fiz. Nauk **38**, 490.
Ginzburg, V.L., 1960, Fiz. Tverd. Tela, **2**, 2031 (translation: 1961, Sov. Phys. Solid State, **2**, 1824).
Goldak, J., Barrett C.S., Innes D. and Youdelis W., 1966, J. Chem. Phys. **44**, 3323.
Gorbatyi, I.V., Ponomareo V.I. and Kheiker D.M., 1972, Kristallografiya, **16**, (translation: 1972, Sov. Phys. Crystallogr. **16**, 781).
Greer, A.L., 1982, Thermochimia Acta (in press).
Grande, S., Mecke H.D. and Shuvalov L.A., 1978, Phys. Status Solidi, (a) **46**, 547.
Grosse, P., 1977, Proc. Third Int. Conf. Phys. Narrow-Gap Semiconductors (Warsaw) 1977.
Gu, Benyuan, M. Copic and H.Z. Cummins, 1981, Phys. Rev. **B 24**, 4098.

Gufan, Yu.M. and V.P. Sakhnenko, 1972, Zh. Eksp. Teor. Fiz. **63**, 1909 (translation: 1973, Sov. Phys. JETP, **36**, 1009).
Gufan, Yu.M. and V.P. Sakhnenko, 1972b, Fiz. Tverd. Tela, **14**, 1915 (translation: 1973, Sov. Phys. Solid State, **14**, 1660).
Gurevich, Y.Y. and Y.I. Kharkats, 1977, Electrochem. Acta **22**, 735.
Habbal, F., 1979, J. Phys. **C12**, L789.
Habbal, F., J.A. Zvirgzds and J.F. Scott, 1978, J. Chem. Phys. **69**, 4984.
Haken, H., 1975, Rev. Mod. Phys. **47**, 67.
Halperin, B.I. and C.M. Varma, 1976, Phys. Rev. **B 14**, 4030.
Hanson, D.M., 1975, J. Chem. Phys. **63**, 5046.
Haque, M.S. and J.R. Hardy, 1980, Phys. Rev. **B 21**, 245.
Harley, R.T., 1977, J. Phys. **C10**, L205.
Harley, R.T., W. Hayes, A.J. Rushworth and J.F. Ryan, 1975, J. Phys. **C 8**, 530.
Hastings, J.M., J.P. Pouget, G. Shirane, A.J. Heeger, N.D. Miro and A.C. MacDiarmid, 1977, Phys. Rev. Lett. **39**, 1484.
Hatta, I. and K.L.I. Kobayashi, 1977, Solid State Commun. **22**, 775.
Hatta, I. and W. Rehwald, 1977, J. Phys. **C 10**, 2075.
Hayes W. and R. Loudon, 1978, Scattering of Light by Crystals (Oxford University Press) Oxford.
Hayes, W., A.J. Rushworth, J.F. Ryan, R.J. Elliott and W.G. Kleppmann, 1977, J. Phys. **C 10**, 111.
Heilman, I.U., W.D. Ellenson and J. Echert, 1979, J. Phys. **C 12**, L185.
Heller, P. and G.B. Benedek, 1962, Phys. Rev. Lett. **8**, 428.
Hidaka, M., K. Inoue, S. Yamashita and J.F. Scott, 1982 (unpublished).
Hirotsu, S., K. Toyota and K. Hamano, 1979, J. Phys. Soc. Jpn. **46**, 1389.
Hodge, I.M., M.D. Ingram and A.R. West, 1976, J. Am. Chem. Soc. **59**, 360.
Hohnke, D.K., H. Hooloway and S. Kaiser, 1972, J. Phys. Chem. Solids, **33**, 2053.
Hong, H.Y.-P., 1974, Acta Cryst. **B 30**, 468.
Hong, H.Y.-P. and J.W. Pierce, 1974, Mat. Res. Bull. **9**, 179.
Hoshino, S., T. Mitsui, F. Jona and R. Pepinsky, 1957, Phys. Rev. **107**, 1255.
Huberman, B.A., 1974, Phys. Rev. Lett. **32**, 1000.
Iizumi, M., Y. Hamaguschi, D.F. Komatsubara and Y. Kato, 1975, J. Phys. Soc. Jpn. **38**, 443.
Ikeda, H., 1977, Rechargeable Batteries in Japan, Eds., Y. Miyake and A. Kozawa (JEC Press, Cleveland) p. 441.
Inoue, K., S. Koiwai and Y. Ishibashi, 1980, J. Phys. Soc. Jpn. **48**, 1785.
Ishibashi,Y., 1978, Ferroelectrics, **20**, 103.
Ishibashi, Y. and V. Dvorak, 1978, J. Phys. Soc. Jpn. **44**, 32.
Ishibashi, I. and M. Shiba, 1978, J. Phys. Soc. Jpn. **45**, 409.
Ismailzade, I.H., R.M. Ismailov and I.S. Rez, 1981, Ferroelectrics, **34**, 117.
Ismailzade, I.H., N.A. Eyubova, R.M. Ismailov, A.I. Alekberov, A.M. Habibov and O.A. Samedov, 1981, Ferroelectrics, **34**, 149.
Ivanov, N.R. and H. Arndt, 1978, Vortrag 6 Fruhjahrsschule Ferroelektrigitat, Rosslau.
Ivanov, N.R., I.T. Tukhtasunov and L.A. Shuvalov, 1970, Kristallografiya, **15**, 752 (translation: 1971, Sov. Phys. Crystallogr. **15**, 647).
Janovec, V., V. Dvorak and J. Petzelt, 1976, Czech. J. Phys. **B 25**, 1362.
Jones, L.M., B.I. Swanson and G.J. Kubas, 1974, J. Chem. Phys. **61**, 4650.
Junod, P., B. Kilochor and J. Wallschleger, 1971, Helv. Phys. Acta, **44**, 563.
Kadaba, P.K., J. Pirnat and Z. Trontelj, 1975, Chem. Phys. Lett. **32**, 382.
Kaplyanskii, A.A., 1976, The Theory of Light Scattering in Solids, eds., V.M. Agranovich and J.L. Birman (Nauka, Moscow) p. 29.
Kasohara, M., 1978, J. Phys. Soc. Jpn. **44**, 537.
Katayama, S., 1976, Solid State Commun. **19**, 381.
Katayama, S. and H. Kawamura, 1977, Solid State Commun. **21**, 521.

Kawamura, H., 1979, Comm. on Solid State Phys. **9**, 55.
Kawamura, H., S. Katayama, S. Takano and S. Hotta, 1974, Solid State Commun. **14**, 259.
Kawamura, H., K. Murase, S. Nishikawa, S. Nishi and S. Katayama, 1975, Solid State Commun. **17**, 341.
Keune, W., 1974, Phys. Rev. **B 10**, 5057.
Keve, E.T., S.C. Abrahams and J.L. Bernstein, 1971, J. Chem. Phys. **54**, 3185.
Kinch, M.A. and D.D. Buss, 1972, Solid State Commun. **11**, 319.
Klein, M.V., 1976, Light Scattering in Solids, eds., M. Balkanski, R.C.C. Leite and S.P.S. Porto (Flammarion, Paris) p. 503.
Kobayashi, K.L.I., Y. Kato, Y. Katayama and K.F. Komatsubara, 1976, Phys. Rev. Lett. **37**, 772.
Koehler, T.R. and N.S. Gillis, 1973, Phys. Rev. **B 7**, 4980.
Konsin, P., 1976, Phys. Status Solidi, (b) **76**, 487.
Konsin, P., 1978, Phys. Status Solidi (b) **86**, 57.
Konsin, P.I. and N.N. Kristoffel, 1971, Izv. Akad. Nauk. Estonian SSR, Ser. Fiz – Mat. **20**, 37.
Konsin, P.I. and N.N. Kristoffel, 1973, Izv. Akad. Nauk Estonian SSR, Ser. Fiz. – Mat. **22**, 1973.
Konsin, P.I. and T. Ord, 1980, Phys. Status Solidi (b), **97**, 609.
Kozlov, G.V., A.A. Volkov, J.F. Scott, G.E. Feldkamp and J. Petzelt, 1983, Phys. Rev. **B** (in press).
Kristoffel, N. and P. Konsin, 1967, Phys. Status Solidi, **21**, k39.
Kristoffel, N. and P. Konsin, 1968, Phys. Status Solidi, **28**, 731.
Kristoffel, N.N. and P.I. Konsin, 1971, Fiz. Tverd. Tela, **13**, 2513 (translation: 1972, Sov. Phys. Solid State, **13**, 2113).
Kristoffel, N. and P. Konsin, 1973, Ferroelectrics, **6**, 3.
Kristoffel, N.N. and P.I. Konsin, 1976, Izv. Akad. Nauk Estonian SSR, Ser. Fiz.-Mat. **25**, 23.
Kudo, S. and T. Hikita, 1978, J. Phys. Soc. Jpn. **45**, 1775.
Labbe, J. and J. Friedel, 1966a, J. Phys. Radium, **27**, 153.
Labbe, J. and J. Friedel, 1966b, J. Phys. Radium **27**, 303.
Landau, L.D., 1937, Phys. Z. Sowjetunion, **11**, 26.
Larkin, A.I. and S.A. Pikin, 1969, Zh. Eksp. Teor. Fiz. **56**, 1664 (translation: 1969, Sov. Phys. JETP, **29**, 891).
Lazay, P.D., J.H. Lunacek, N.A. Clark and G.B. Benedek, 1969, Light Scattering Spectra of Solids, ed. G.B. Wright (Springer Verlag, New York) p. 593.
Leftowitz, I., M. Shields, G. Dolling, W.J.L. Buyers and R.A. Cowley, 1970, J. Phys. Soc. Jpn. **28**, Suppl. p. 249.
Levanyuk, A.P. and D.G. Sannikov, 1968, Zh. Eksp. Teor, Fiz. **55**, 256 (translation: 1969, Sov. Phys. JETP, **28**, 134).
Levanyuk, A.P. and D.G. Sannikov, 1970a, Fiz. Tverd. Tela, **12**, 2997 (translation: 1971, Sov. Phys. Solid State, **12**, 2418).
Levanyuk, A.P. and D.G. Sannikov, 1970b, Zh. Eksp. Teor. Fiz. **11**, 68. (translation: 1970, Sov. Phys. JETP Lett. **11**, 43).
Levanyuk, A.P. and D.G. Sannikov, 1971, Zh. Eksp. Teor. **60**, 1109 (translation: 1971, Sov. Phys. JETP, **33**, 600).
Levanyuk, A.P. and D.G. Sannikov, 1974a, Usp. Fiz. Nauk **112**, 561 (translation: 1974, Sov. Phys. Usp. **17**, 199).
Levanyuk, A.P. and D.G. Sannikov, 1976, Fiz. Tverd. Tela, **18**, 423 (translation: 1976, Sov. Phys. Solid State, **18**, 245).
Levanyuk, A.P. and A.A. Sobyanin, 1970, JEPT Lett. **11,** 371.
Levanyuk, A.P., A.S. Sigov and A.A. Sobyanin, 1980, Ferroelectrics, **24**, 61.
Levstik, A., R. Blinc, P. Kadaba, S. Cizikov, I. Levstik and C. Filipic, 1976, Ferroelectrics, **14**, 703.
Lines, M.E., 1974, Phys. Rev. **B 9**, 950.
Lines, M.E., 1979a, Phys. Rev. **B 19**, 1183.
Lines, M.E., 1979b, Phys. Rev. **B 19**, 1189.

Lines, M.E. and A.M. Glass, 1977, Principles and Applications of Ferroelectrics and Related Materials (Clarendon Press, Oxford).
Lippe, R., W. Windsch, G. Volkel and W. Schluga, 1976, Solid State Commun. **19**, 587.
Litster, J.D., J. Als-Nielsen, R.J. Birgeneau, S.S. Dana, D. Davidov, F. Garcia-Golding, M. Kaplan, C.R. Safinya and R. Schaetzing, 1979, J. Physique **C 3**, 339.
Lockwood, D.J., J.W. Arthur, W. Taylor and T.J. Hosea, 1976, Solid State Commun. **20**, 703.
Loiacono, G.M., M. Delfino and W.A. Smith, 1978, Appl. Phys. Lett. **32**, 595.
Lopez-Echarri A., M.J. Tello and P. Gili, 1980, Solid State Commun. **36**, 1021.
Lyons, K.B. and H.J. Guggenheim, 1979, Solid State Commun. **31**, 285.
Lyons, K.B., T.J. Negran and H.J. Guggenheim, 1980, J. Phys. **C 13**, L415.
Lyons, K.B., T.J. Negran and R.N. Bhatt, 1982, Phys. Rev. **B 25**, 1791.
Mahanthappa, K.T. and M. Sher, 1980, Phys. Rev. **D 22**, 1711.
Makita, Y., 1965, J. Phys. Soc. Jpn. **20**, 2073.
Makita, Y. and Y. Yamauchi, 1974, J. Phys. Soc. Jpn. **37**, 1470.
Makita, Y., T. Yagi and I. Tatsuzaki, 1976, Phys. Lett. **A 55**, 437.
Makita, Y., F. Sakurai, T. Osaka and I. Tatsuzaki, 1977, J. Phys. Soc. Jpn. **42**, 518.
Mason, R. and A.I.M. Rae, 1968, Proc. Roy. Soc. (London) **A 301**, 501.
McCall, S.L., 1974, Phys. Rev. **A 9**, 1515.
Miller, P.B. and J.D. Axe, 1967, Phys. Rev. **163**, 924.
Mitani, S., 1964, J. Phys. Soc. Jpn. **19**, 481.
Motizuki, K., N. Suzuki, Y. Yoshida and Y. Takoka, 1981, Solid State Commun. **40**, 995.
Muller, K.A., 1979, Dynamical Critical Phenomena and Related Topics, ed. C.P. Enz (Springer Verlag, Berlin) (Vol. 104 of Lecture Notes in Physics) p. 210.
Muller, K.A. and W. Berlinger, 1975, Phys. Rev. Lett. **35**, 1547.
Murase, K., S. Sugai, S. Takaoka and S. Katayama, 1976, Proc. 13th Int. Conf. Phys. Semi Conductors, Rome, ed. F.G. Fumi (North-Holland, Amsterdam, 1977) p. 305.
Murase, K., S. Sugai, T. Higuchi, S. Takaoka, T. Fukunaga, and H. Kawamura, 1978, Proc. 14th Int. Conf. Phys. Semicond., Edinburgh.
Murray, A.F. and D.J. Lockwood, 1978a, J. Phys. **C 11**, 2349.
Murray, A.F. and D.J. Lockwood, 1978b, J. Phys. **C 11**, 4651.
Nakamura, T. and M. Maeda, 1979, J. Phys. Soc. Jpn. **47**, 869.
Nakashima, S., 1980, Proc. Conf. Japanese Phys. Soc., Fukui, Oct. 1, 1980.
Natori, A., 1976, J. Phys. Soc. Jpn. **41**, 782.
Nettleton, R.E., 1969, Z. Phys. **220**, 401.
Nettleton, R.E., 1971, Ferroelectrics, **2**, 77.
Nishi, S., H. Kawamura and K. Murase, 1980, Phys. Status Solidi, **6**.
Novikova, S.I. and L.E. Shelimova, 1967, Fiz. Tverd. Tela, **9**, 1336, (translation: 1967, Sov. Phys. Solid State, **9**, 1046).
Novikova, S.I., L.E. Shelimova, E.S. Avilov and M.A. Korzhuev, 1975, Fiz. Tverd. Tela, **17**, 2379 (translation: 1976, Sov. Phys. Solid State, **17**, 1570).
Oron, M., A. Zussman and E. Rapoport, 1978, J. Chem. Phys. **68**, 794.
Overhauser, A.W., 1968, Phys. Rev. **167**, 691.
Overhauser, A.W., 1971, Phys. Rev. **B 3**, 3173.
Overhauser, A.W., 1978, Adv. Phys. **27**, 343.
Pawley, G.S., W. Cochran, R.A. Cowley and G. Dolling, 1966, Phys. Rev. Lett. **17**, 753.
Peercy, P.S., 1970, Optics Commun. **2**, 270.
Peercy, P.S., 1976, Proc. 5th Int. Conf. Raman Spectroscopy, eds., E.D. Schmid, J. Brandmuller, W. Kiefer, B. Schrader and H.W. Schrotter (Hans Ferdinand Schulz Verlag, Freiburg) p. 571.
Peercy, P.S. and I.J. Fritz, 1974, Phys. Rev. Lett. **32**, 466.
Peercy, P.S., J.F. Scott and P.M. Bridenbaugh, 1976, Bull. Am. Phys. Soc. **21**, 337.
Pepinsky, R. and Y. Makita, 1962, Bull. Am. Phys. Soc. **7**, 241.
Petzelt, J., G.V. Kozlov, A.A. Volkov and Y. Ishibashi, 1979, Z. Phys. **B 33**, 369.

Pinczuk, A., G. Burns and F.H. Dacol, 1977, Solid State Commun. **24**, 163.
Porto, S.P.S. and J.F. Scott, 1967, Phys. Rev. **157**, 716.
Pouget, J.P., G. Shirane, J.M. Hastings, A.J. Heeger, N.D. Miro and A.G. MacDiarmid, 1978, Phys. Rev. **B 18**, 3645.
Pytte, E., 1970, Phys. Rev. Lett. **25**, 1176.
Pytte, E. and J. Feder, 1969, Phys. Rev. **187**, 1077.
Quilichini, M., J.F. Ryan, J.F. Scott and H.J. Guggenheim, 1975, Solid State Commun. **16**, 471.
Rae, A.I.M., 1978, J. Phys. **C 11**, 1779.
Rae, A.I.M., 1979, J. Chem. Phys. **70**, 639.
Raich, J.C. and E.R. Bernstein, 1980, J. Chem. Phys. **73**, 1955.
Reese, W., 1969, Solid State Commun. **7**, 969.
Rehwald, W., 1973, Adv. Phys. **22**, 721.
Rehwald, W. and G.K. Lang, 1975, J. Phys. **C 8**, 3287.
Rice, M.J., S. Strassler and G.A. Toombs, 1974, Phys. Rev. Lett. **32**, 596.
Richardson, C.B., 1963, J. Chem. Phys. **38**, 510.
Ridou, C., M. Rousseau and A. Freund, 1977, J. Physique Lett. **38**, L359.
Robertson, J., 1979, J. Phys. **C 12**, 4767.
Ryan, J.F. and J.F. Scott, 1974, Solid State Commun. **14**, 5.
Ryan, R.R. and B.I. Swanson, 1976, Phys. Rev. **B 13**, 5320.
Samara, G.A., 1977, Commun. Solid State Phys. VIII, 13.
Samara, G.A. and P.M. Richards, 1976, Phys. Rev. **B 14**, 5073.
Samara, G.A. and J.F. Scott, 1977, Solid State Commun. **21**, 167.
Sawada, A. and Y. Takagi, 1971, J. Phys. Soc. Jpn. **31**, 952.
Sawada, A. and Y. Takagi, 1972, J. Phys. Soc. Jpn. **33**, 1071.
Sawada, Y., E. Burstein, D.L. Cater and L. Testardi, 1965, Plasma Effects in Solids, ed., J. Bok (Dunod, Paris) p. 71.
Sawada, A., Y. Makita and Y. Takagi, 1977a, J. Phys. Soc. Jpn. **42**, 1918.
Sawada, A., M. Udagawa and T. Nakamura, 1977b, Phys. Rev. Lett. **39**, 829.
Sawada, Sh., Y. Shiroishi, A. Yamamota, M. Takashize and M. Matusuo, 1977c, J. Phys. Soc. Jpn. **43**, 2089.
Schmid, H., 1967, Rost kristallov (Nauka, Moscow) Vol. 7 (translation: 1969, Growth of Crystals (Consultants Bureau, New York) p. 25).
Schmid, H., 1970, Phys. Status Solidi, **37**, 209.
Schneck, J., 1979, unpublished.
Schneck, J. et al., 1979 (private communication).
Scott, J.F., 1968a, J. Chem. Phys. **48**, 874.
Scott, J.F., 1968b, J. Chem. Phys. **49**, 98.
Scott, J.F., 1969, Phys. Rev. **183**, 823.
Scott, J.F., 1974, Rev. Mod. Phys. **46**, 83.
Scott, J.F., 1975a, Vibrational Spectra and Structure, ed., J.R. Durig (Elsevier, Amsterdam) p. 67.
Scott, J.F., 1975b, Optics Commun. **15**, 343.
Scott, J.F., 1976, Molecular Spectroscopy of Dense Phases, eds., M. Grosmann, S.G. Elkomoss and J. Ringeissen (Elsevier Scientific Publishing Co., Amsterdam) p. 203.
Scott, J.F., 1978, Ferroelectrics, **20**, 69.
Scott, J.F., 1979, Rep. Prog. Phys. **12**, 1055.
Scott, J.F. and S.P.S. Porto, 1967, Phys. Rev. **161**, 903.
Scott, J.F., M. Sargent and C.D. Cantrell, 1975, Opt. Commun. **15**, 13.
Scott, J.F., F. Habbal and J.A. Zvirgzds, 1980, J. Chem. Phys. **72**, 2760.
Scott, J.F., F. Habbal and M. Hidaka, 1982, Phys. Rev. **B. 25**, 1805.
Seddon, T., J. Farley and G.A. Saunders, 1975, Solid State Commun. **17**, 55.
Seddon, T., S.C. Gupta and G.A. Saunders, 1976, Phys. Lett. **56A**, 45.

Sham, L.J., 1971, Phys. Rev. Lett. **27**, 1725.
Shand, M., R.C. Hanson, C.E. Derrington and M. O'Keefe, 1976, Solid State Commun. **18**, 769.
Shannon, R.D. and C.T. Prewitt, 1969, Acta Cryst. **B 25**, 925.
Shapiro, S.M., R.A. Cowley, D.E. Cox, M. Eibschutz and H.J. Guggenheim, 1976, Proc. Conf. Neutron Scattering, ed., R.M. Moon (Nat. Tech. Info. Service, Springfield) p. 399.
Shaulov, A., M.E. Rosar and W.A. Smith, 1981, Ferroelectrics, **36**, 467.
Shelimova, L.E., N.Kh. Abrikosov and V.V. Zhdanova, 1965, Russ. J. Inorg. Chem. **10**, 650.
Sheng, H.T. and M.B. Salamon, 1980, Phys. Rev. **B 22**, 4401.
Shimada, T., K.L.I. Kobayaski, Y. Katayama and K.F. Komatsubara, 1977, Phys. Rev. Lett. **39**, 143.
Shirane, G., 1974, Rev. Mod. Phys. **46**, 438.
Shirane, G. and J.D. Axe, 1971a, Phys. Rev. Lett. **27**, 1803.
Shirane, G. and J.D. Axe, 1971b, Phys. Rev. **B 4**, 2957.
Shiroshi, Y., A. Nakata and S. Sawada, 1976, J. Phys. Soc. Jpn. **40**, 911.
Shuvalov, L.A., N.R. Ivanov and T.K. Sitnik, 1967, Kristallografija, **12** (translation: 1967, Sov. Phys. Crystallogr. **12**, 315).
Shuvalov, L.A., N.R. Ivanov, L.F. Kirpichnikova and N.V. Gordeyeva, 1970, Phys. Lett. **A 33**, 490.
Shuvalov, L.A., N.R. Ivanov, L.F. Kirpichnikova and N.M. Shchazina, 1972a, Kristallografija, **17**, 966 (translation: 1973, Sov. Phys. Crystallogr. **17**, 851).
Shuvalov, L.A., A.M. Shirokov, N.R. Ivanov, A.I. Baranov, L.F. Kirpichnikova and N.M. Shchazina, 1972b, J. de Phys. Suppl. **53**, C2–165.
Sigel, L.A. and E.F. Williams, 1954, J. Chem. Phys. **22**, 1147.
Skarstad, P.M. and S. Geller, 1975, Mater. Res. Bull. **10**, 791.
Smith, J.H. and A.I.M. Rae, 1978a, J. Phys. **C 11**, 1767.
Smith, J.H. and A.I.M. Rae, 1978b, J. Phys. **C 11**, 1771.
Smith, T.F., J.A. Birch and J.G. Collins, 1976, J. Phys. **C 9**, 4375.
Smolensky, G.A., I.G. Siny, H. Arndt, S.L. Prozorova, E.G. Kuzmirov, V.D. Mikvabya and N.N. Kolpakova, 1979, Izvest. Akad. SSSR, **43**, 1664.
Sorge, G. and U. Straube, 1978, Ferroelectrics, **21**, 533.
Sorge, G. and U. Straube, 1979, Phys. Status Solidi (a) **51**, 117.
Spencer, E.G., H.J. Guggenheim and G.J. Kominiak, 1970, Appl. Phys. Lett. **17**, 300.
Stanley, H.E., 1971, Introduction to Phase Transitions and Critical Phenomena (Oxford University Press, London).
Steigmeier, E. and H. Auderset, 1973, Solid State Commun. **12**, 565.
Steigmeier, E.F. and G. Harbeke, 1970, Solid State Commun. **8**, 1275.
Steigmeier, E.F., H. Auderset and G. Harbeke, 1974, Anharmonic Lattices, Structural Transitions and Melting, ed., T. Riste (Noordhoff, Leiden) p. 153.
Strukov, B.A., 1964, Fiz. Tverd. Tela, **6**, 2862 (1965, Sov. Phys. Solid State, **6**, 2278).
Strukov, B.A., T.L. Skomorkhova, V.A. Koptski, A.A. Boiko and A.N. Izrailenko, 1973, Kristallografija, **18**, 143 (translation: 1973, Sov. Phys. Crystalogr. **18**, 86).
Strukov, B.A., S.A. Taraskin, K.A. Minaeva and V.A. Fedorikhin, 1980, Ferroelectrics, **25**, 399.
Sugai, S., K. Murase and M. Kawamura, 1977a, Solid State Commun. **23**, 127.
Sugai, S., K. Murase, S. Katayama, S. Takaoka, S. Nishi and H. Kawamura, 1977b, Solid State Commun. **24**, 407.
Sugai, S., K. Murase, T. Tsuchihira and H. Kawamura, 1979, J. Phys. Soc. Jpn. **47**, 539.
Swanson, B.I. and L.H. Jones, 1974, Inorg. Chem. **13**, 313.
Takahashi, T., 1976, Superionic Conductors, eds., G.D. Mahan and W.L. Roth (Plenum Press, New York) p. 379.
Takahashi, T., S. Ikeda and O. Yamamoto, 1973, J. Electro Chem. Soc. **120**, 647.
Takano, S., S. Hotta, H. Kawamura, Y. Kato, K.L.I. Kobayashi and K.T. Komatsubara 1974, J. Phys. Soc. Jpn. **37**, 1007.

Tanaka, H. and I. Tatsuzaki, 1979, J. Phys. Soc. Jpn. **47**, 878.
Tanaka, H. and T. Yagi and I. Tatsuzaki, 1978a, J. Phys. Soc. Jpn. **44**, 1257.
Tanaka, H., T. Yagi and I. Tatsuzaki, 1978b, J. Phys. Soc. Jpn. **44**, 2009.
Terauchi, H., T. Sakai and H. Chihara, 1975, J. Chem. Phys. **62**, 3832.
Thomas, D.M., 1977, J. Raman Spectrosc. **6**, 169.
Tofield, B.C., P.M. Bridenbaugh and H.P. Weber, 1975, Mater. Res. Bull. **10**, 1091.
Toledano, J.C., G. Errandonea and J.P. Jaguin, 1976, Solid State Commun. **20**, 905.
Triebwasser, S., 1957, Bull Am. Phys. Soc. **2**, 127.
Triebwasser, S., 1958, IBM J. Res. Develop. **3**, 212.
Tsukui, M., M. Sumita and Y. Makita, 1980, J. Phys. Soc. Jpn. **49**, 427.
Turkovic, A., D.L. Fox, J.F. Scott, S. Geller and G.F. Ruse, 1977, Mater. Res. Bull. **12**, 189.
Udagawa, M., K. Kohn and T. Nakamura, 1978, J. Phys. Soc. Jpn. **44**, 1873.
Unger, W.K., 1979, Solid State Commun. **29**, 601.
Uwe, H., H. Tokumoto, M. Udagawa, Y. Tominaga and K. Kohn, 1979, Phys. Rev. **B 19**, 3700.
Vaks, V.G. and N.E. Zein, 1974, Ferroelectrics, **6**, 265.
Valassiades, O. and N.A. Economou, 1975, Phys. Status Solidi (a) **30**, 187.
Van Treeck, E. and W. Windsch, 1978, Krist. u. Tech. **13**, 513.
Vardeny, Z. and O. Brafman, 1979, Solid State Commun. **32**, 859.
Volkel, G., W. Brunner and W. Winsch, 1975, Solid State Commun. **17**, 345.
Volkov, B. and O. Pankratov, 1978, Sov. Phys. JETP, **48**, 687.
Wada, M., A. Sawada, Y. Ishibashi, Y. Takagi and T. Sakudo, 1977a, J. Phys. Soc. Jpn. **42**, 1229.
Wada, M., H. Uwe, A. Sawada, Y. Ishibashi, Y. Takagi and T. Sakudo, 1977b, J. Phys. Soc. Jpn. **43**, 544.
Wada, M., A. Sawada, Y. Ishibashi and Y. Takagi, 1978, H. Phys. Soc. Jpn. **45**, 1905.
Wang, C.H., 1971, Phys. Rev. Lett. **26**, 1226.
Wang, C.H. and P.A. Fleury, 1969, Light Scattering Spectra of Solids, ed., G.B. Wright (Springer Verlag, New York) p. 651.
Weber, H.P., B.C. Tofield and P.F. Liao, 1975, Phys. Rev. **B 11**, 1152.
Wheatley, P.J., 1955, Acta Crystallogr. **8**, 224.
Wilson, J.A., F.J. de Salvo and S. Mahajon, 1975, Adv. Phys. **24**, 117.
Windsch, W., 1976, Ferroelectrics, **12**, 1.
Windsch, W., 1979, 4th Eur. Conf. Ferroelectricity Portoroz Yugoslavia, Abstracts, p. 172.
Windsch, W., R. Lipjse and G. Volkel, 1975, Solid State Commun. **17**, 1375.
Yamada, H., M. Saheki, S. Fukushima and T. Nagasao, 1974, Spectrochim. Acta **30 A**, 295.
Yagi, T., H. Tanaka and I. Tatsuzaki, 1976, J. Phys. Soc. Jpn. **41**, 717.
Yagi, T., H. Tanaka and I. Tatsuzaki, 1977a, Phys. Rev. Lett. **38**, 609.
Yagi, T., I. Tatsuzaki and H. Tanaka, 1977b, Proc. 6th Int. Conf. Internal Friction and Ultrasonic Attenuation in Solids (Univ. Tokyo Press) p. 209.
Zhukova, T.B. and A.I. Zaslavskii, 1967, Kristallografiya, **12**, 37 (translation: 1967, Sov. Phys. Crystallogr. **12**, 28).
Zussman, A., 1974, Phys. Lett. **47 A**, 195.
Zussman, A. and M. Oron, 1977, J. Chem. Phys. **66**, 743.

## *Additional recent references not cited in the text*

### *Molecular crystal soft modes and/or phase transitions*

*Diamantane*: ($C_{14}H_{20}$)
Andrews, S.R., R.T. Harley and T.E. Jenkins, 1982, J. Phys. **C 15**, L243.
Jenkins, T.E. and A.R. Bates, 1979, J. Phys. **C 12**, 1003.

Jenkins, T.E., A.R. Bates and E.H.M. Evans, 1978, J. Phys. **C 11**, L83.
*Triamantane*:
Jenkins, T.E. and P. O'Brien, 1981, Phys. Status Solidi **A 67**, K161.
*Chloranil*:
Ecolivet, C., 1981, Solid State Commun. **40**, 503.
Girard, A., Y. Delugeard, C. Ecolivet and H. Cailleau, 1982, J. Phys. **C 15**, 2127.
*s-Triazine*:
Rae, A.I.M., 1982a, J. Phys. **C 15**, 1883.
Rae, A.I.M., 1982b, J. Phys. **C 15**, L287.
Raich, J.C. and E.R. Bernstein, 1982, J. Phys. **C 15**, L283.
*tris-Sarcosine Calcium Chloride*:
Reichelt, H., W. Windsch and A. Sienkiewicz, 1981, Ferroelectrics, **34**, 195.

## *Inorganic crystal soft modes and/or phase transitions*

$Hg_2Br_2$:
Lemanov, V.V., S.Kh. Esayan and J.P. Chapelle, 1981, Fiz. Tverd. Tela, **23**, 262 (Sov. Phys. Solid State **23**, 146).
$BiVO_4$:
Cho, M., T. Yagi, T. Fujii, A. Sawada and Y. Ishibashi, 1982, J. Phys. Soc. Jpn. **51**, 2914.
$Rb_2ZnCl_4$:
Pezeril, M. and J.C. Fayet, 1982, J. Phys. (Paris) **43**, L267.
*Boracites*
Shaulov, A., W.A. Smith and H. Schmid, 1981, Ferroelectrics, **34**, 219.
$Pb_{1-x}Ge_xTe$:
Littlewood, P.B., 1979, J. Phys. **C 12**, 4441.
Littlewood, P.B., 1980a, J. Phys. **C 13**, 4855.
Littlewood, P.B., 1980b, J. Phys. **C 13**, 4875.
Suski, T., M. Baj and K. Murase, 1982, J. Phys. **C 15**, L377.

CHAPTER 6

# Brillouin Scattering Studies of Phase Transitions in Crystals

HERMAN Z. CUMMINS

*Department of Physics*
*City College of The City University of New York*
*New York 10031*
*USA*

*Light Scattering near Phase Transitions*
*Edited by*
*H.Z. Cummins and A.P. Levanyuk*

© *North-Holland Publishing Company, 1983*

# Contents

## 1. Introduction

Crystals undergoing phase transitions almost always exhibit anomalous elastic properties in the transition region. Consequently, during the past fifteen years, Brillouin scattering spectroscopy which probes long-wavelength acoustic modes has been increasingly applied to the study of solid state phase transitions, supplementing the closely related techniques of acoustic resonance and ultrasonic propagation. The results of these experiments, in conjunction with Raman scattering, inelastic neutron scattering and X-ray diffraction measurements have contributed significantly to recent progress in our understanding of both structural and dynamic aspects of phase transitions.

In this chapter I present a survey of Brillouin scattering investigations of phase transitions in crystals. Fluids and liquid crystals, which will be discussed in ch. 9 by Goldburg and ch. 10 by Litster in this volume, are not included. Incommensurate transitions are included in the survey, but will not be discussed in detail since they will be covered fully in ch. 8 by Klein. Similarly I will omit the subject of central peaks which, though closely associated with Brillouin scattering, is covered separately in ch. 7 by Fleury and Lyons.

The central topic of this review will be the changes in Brillouin shifts and linewidths observed in the light scattering spectra of crystals near phase transitions. In most cases the transitions occur between two crystalline phases. However, under the rubric of order–disorder transitions, I will also include transitions between crystalline phases and mesophases where the disordered mesophase has either sublattice rotational freedom (plastic crystals) or translational freedom (superionic conductors).

The crystals covered in the present survey are listed in table 1. They are arranged alphabetically except that groups of isomorphic crystals with similar phase transitions are grouped together under the most common example (e.g., gadolinium molybdate and terbium molybdate). Although the 73 crystals listed constitute a small subset of all known crystal phase transitions, they represent all Brillouin scattering studies reported through June 1981 revealed by a bibliographic search and by inquiries sent to workers in the field. Nevertheless there may well be some omissions. The letters A, B, C, D at the right indicate the category of the transition, to be described below. P indicates that only preliminary studies are available.

Previous review articles and conference proceedings have already covered many of the phase transitions to be described here, usually in the context of

Table 1
Crystal phase transitions investigated by Brillouin scattering spectroscopy

| Crystal | |
|---|---|
| $C_{10}H_{16}$: adamantane | P |
| $AlPO_4$: aluminum phosphate (berlinite) | B2 |
| $(NH_4)_2Cd_2(SO_4)_3$: di-ammonium di-cadmium sulfate | P |
| $Tl_2Cd_2(SO_4)_3$: di-thallium di-cadmium sulfate | P |
| $NH_4Cl$: ammonium chloride | D1 |
| $NH_4Br$: ammonium bromide | D1 |
| $(NH_4)_2BeF_4$: ammonium fluoberyllate | P |
| $NH_4HSO_4$: ammonium hydrogen sulfate (ammonium bisulfate) | D1 |
| $RbHSO_4$: rubidium hydrogen sulfate | D1 |
| $(NH_4)_2SO_4$: ammonium sulfate | B2 |
| $C_6H_5NH_3Br$: aniline hydrobromide | D1 |
| $Sb_5O_7I$: penta-antimony heptaoxide iodide | C1 |
| $BaF_2$: barium fluoride | D3 |
| $SrCl_2$: strontium chloride | D3 |
| $BaMnF_4$: barium manganese tetrafluoride | C2 |
| $Ba_2NaNb_5O_{15}$: barium sodium niobate | C2 |
| $BaTiO_3$: barium titanate | B1 |
| $(C_6H_5\text{-}CO)_2$: benzil | C1 |
| $(C_6H_5)_2$: biphenyl | P |
| $(C_6H_5)\text{-}(C_6H_4)\text{-}(C_6H_5)$: para-terphenyl | P |
| $BiVO_4$: bismuth vanadate | A |
| $Ca_2Pb(C_2H_5CO_2)_6$: di-calcium lead propionate | P |
| $CCl_4$: carbon tetrachloride | D2 |
| $CBr_4$: carbon tetrabromide | D2 |
| $CMe_4$: neopentane | D2 |
| $CsPbCl_3$: cesium lead chloride | P |
| $C_6Cl_4O_2$: chloranil | C1 |
| $DyVO_4$: dysprosium vanadate | A |
| $TbVO_4$: terbium vanadate | A |
| $Gd_2(MoO_4)_3$: gadolinium molybdate (GMO) | C1 |
| $Tb_2(MoO_4)_3$: terbium molybdate (TMO) | C1 |
| $LaP_5O_{14}$: lanthanum pentaphosphate | A |
| $Nd_{0.5}La_{0.5}P_5O_{14}$: neodymium lanthanum pentaphosphate | A |
| $Pb_5Ge_3O_{11}$: lead germanate | B2 |
| $PbMgNb_2O_9$: lead magnesium niobate | P |
| $PbHPO_4$: lead monohydrogen phosphate (lead di-orthophosphate) | D1 |
| $Pb_3(PO_4)_2$: lead phosphate | C1 |
| $LiNH_4C_4H_4O_6 \cdot H_2O$: lithium ammonium tartrate (LAT) | A |
| $Hg_2Cl_2$: mercury chloride (calomel) | C1 |
| $Hg_2Br_2$: mercury bromide | C1 |
| $TeO_2$: paratellurite | A |
| $(CH_3)_3 \cdot C \cdot COOH$: pivalic acid | D2 |
| KCN: potassium cyanide | D1 |
| NaCN: sodium cyanide | D1 |
| RbCN: rubidium cyanide | D1 |
| $KH_2PO_4$: potassium dihydrogen phosphate (KDP) | A |
| $KD_2PO_4$: potassium dideuterium phosphate (DKDP) | A |
| $KH_2AsO_4$: potassium dihydrogen arsenate (KDA) | A |
| $RbH_2PO_4$: rubidium dihydrogen phosphate (RbDP) | A |

Table 1 (*contd.*)

| | |
|---|---|
| $CsH_2AsO_4$: cesium dihydrogen arsenate (CDA) | A |
| $CsD_2AsO_4$: cesium dideuterium arsenate (DCDA) | A |
| $KMnF_3$: potassium managanese fluoride | C1 |
| $K_2Hg(CN)_4$: potassium mercury cyanide | P |
| $K_2SeO_4$: potassium selenate | C2 |
| $Rb_2ZnCl_4$: rubidium tetrachlorozincate | C2 |
| $K_2SnCl_6$: potassium tin chloride | C1 |
| $KH_3(SeO_3)_2$: potassium trihydrogen selenite (KTS) | A |
| $KD_3(SeO_3)_2$: potassium trideuterium selenite (DKTS) | A |
| $PrAlO_3$: praseodymium aluminate | A |
| $RbCaF_3$: rubidium calcium fluoride | C1 |
| $TlCdF_3$: thallium cadmium fluoride | C1 |
| $RbAg_4I_5$: rubidium silver iodide | D3 |
| $NaNO_2$: sodium nitrite | C2 |
| $SiO_2$: silicon dioxide (quartz) | B2 |
| $NaKC_4H_4O_6{\cdot}4H_2O$: sodium potassium tartrate (Rochelle salt) | D1 |
| $NaH_3(SeO_3)_2$: sodium trihydrogen selenite | C1 |
| $C_4O_2(OH)_2$: squaric acid | C1 |
| $SrTiO_3$: strontium titanate | C1 |
| $(CH_2CN)_2$: succinonitrile | D2 |
| $SC(NH_2)_2$: thiourea | C2 |
| $(NH_2CH_2COOH)_3{\cdot}H_2SO_4$: triglycine sulfate (TGS) | D1 |
| $(NH_2CH_2COOH)_3{\cdot}H_2SeO_4$: triglycine selenate (TGSe) | D1 |
| $(C_3NO_2H_7)_3CaCl_2$: tris-sarcosine calcium chloride (TSCC) | C1 |

more general discussions in which Brillouin scattering is not a central topic, or as part of a detailed analysis of particular aspects of phase transitions. Fleury (1971) reviewed the theory of coupling between soft modes and acoustic modes and discussed several examples. Rehwald (1973) and Luthi and Rehwald (1981) also reviewed the theory and surveyed a large number of ultrasonic studies. Scott (1974) discussed several examples of elastic anomalies in a major review of soft mode spectroscopy. Additional examples were discussed in review articles by Worlock (1971), Fleury (1976, 1980), Fleury and Lyons (1981) and in a recent comprehensive article on light scattering near phase transitions in solids by Ginzburg, Levanyuk and Sobyanin (Ginzburg et al. 1980). Many ferroelastic phase transitions have been reviewed by Toledano (1974), Sapriel (1975) and Toledano and Toledano (1980). (Also see the recent book by Lines and Glass (1980).)

Crystal phase transitions can be divided into two major categories according to the nature of the observed elastic anomaly. One category (A) includes all crystals exhibiting major anomalies in which an elastic constant falls to (or very nearly to) zero as the transition is approached from either side. For these *proper ferroelastic transitions* strain either is the order parameter or has the same symmetry as the order parameter and couples to it linearly. The second category includes all other transitions marked by (usually) weak elastic anomalies, often seen predominantly on one side of the transition, resulting from anharmonic interaction of strain with the order parameter. These transitions can be further subdivided according to the nature of the fundamental soft mode which can be: (B) a zone center optic mode (optic-ferrodistortive) whose symmetry does not correspond to any strain (e.g., quartz). Note that the term optic-ferrodistortive is being used here to indicate the appearance at the transition of a homogeneous order parameter produced by internal atomic motion in each unit cell, i.e., the condensation of a $\Gamma$-point optic mode. If the condensed optic mode is polar, then the order parameter is the electric polarization and the transition is proper ferroelectric. (C) A cell-multiplying zone-boundary mode (anti-ferrodistortive as in strontium titanate or the improper ferroelastic ferroelectric gadolinium molybdate) or a mode at a general point in the Brillouin zone (incommensurate transition). Finally, (D) the underlying transition may be an order–disorder transition (e.g., ammonium chloride).

In the following four sections we will take up each type of transition, first discussing the general aspects and then reviewing individual examples. The section in which each of the transitions in table 1 is discussed is indicated by the letter appearing to its right:

(A) *Proper ferroelastic transitions.* Homogeneous distortion of the unit cell occurs at $T_0$ accompanied by major elastic anomalies. The transition may be ferroelastic and also involve atomic displacements within the unit cell corresponding to a soft zone-center optic mode. If it does, then close to the transition, the soft optic and acoustic modes are hybridized. The order parameter is either strain or the amplitude of the condensed optic soft mode which are proportional to each other in the low-symmetry ferroelastic phase. Both the acoustic and soft optic modes are Raman active in both phases. Ferroelasticity may also involve coupling of strain to other degrees of freedom such as the electronic energy levels in the cooperative Jahn–Teller transitions of the rare-earth vanadates.

(B) *Proper optic ferroic transitions.* There is a zone-center soft optic mode which condenses at $T_0$, lowering the symmetry, but linear coupling to strain is symmetry forbidden. Improper spontaneous strain can occur in the ferroic phase through higher-order coupling, with temperature dependence proportional to the square or higher power of the order parameter. The soft mode is Raman active in the ferroic phase but not in the prototype (high-symmetry) phase. If the improper spontaneous strain is itself symmetry breaking, the transition is an improper ferroelastic one (e.g., the proper ferroelectric/improper ferroelastic transition in $BaTiO_3$ at 393 K) ($B_1$). Otherwise it is non-ferroelastic ($B_2$).

(C) *Cell-multiplying transitions.* There is a soft lattice mode which is not at the center of the Brillouin zone (e.g., the R-point instability in the 106 K cell doubling transition of strontium titanate). This category includes antiferroelectric as well as incommensurate phase transitions. Proper ferroelasticity (or ferroelectricity) is automatically excluded by translational symmetry since the order parameter is not homogeneous. (Transitions which do *not* change the size of the unit cell are, by contrast, called cell preserving or equi-translational). The transition may be an improper ferroelastic ($C_1$) or ferroelectric one, or incommensurate ($C_2$).

(D) *Order–disorder transitions.* The high-symmetry prototype phase is a mesophase since no true translational symmetry exists. The prototype symmetry applies to a unit cell averaged over positions and orientations of the disordered constituents. This category includes orientational order–disorder transitions (e.g., $NH_4Cl$) ($D_1$), translational sublattice disorder (super-ionic conductors) ($D_3$) and rotational sublattice melting (plastic crystals) ($D_2$). (It should be noted that there is often no clear boundary between order–disorder and displacive transitions, but it is conventional to classify specific transitions as being one or the other.) Finally, several crystal phase transitions for which only preliminary results (P) are currently available will be discussed in sect. 6 along with conclusions based on the entire survey.

### *1.1. Experimental techniques*

Although Brillouin scattering spectra are sometimes analyzed with grating spectrometers, the usual instrument of choice is the Fabry–Perot interferometer because of its high resolution and excellent light gathering power. Both plane and confocal interferometers have been employed, with scanning accomplished either by pressure variation within an enclosure containing a fixed etalon, or by piezoelectric scanning of one of the Fabry–Perot mirrors.

The resolution available in Brillouin scattering experiments depends on the type of interferometer used. For a plane Fabry–Perot with 1 $cm^{-1}$ free spectral range and a finesse of 60, the resolution is 1/60th $cm^{-1}$ or 500 MHz. For a 20 cm confocal Fabry–Perot, with 1/80th $cm^{-1}$ free spectral range and finesse of 40, the resolution is 10 MHz. Although Brillouin linewidth and *relative* Brillouin shift measurements in the range of a few MHz are possible, *absolute* measurements in this range are much harder due to uncertainties in the plate separation as well as non-linearities in the scan. Recently, Sussner and Vacher (1979) have developed a scheme with which absolute Brillouin shifts can be measured with an accuracy approaching that of ultrasonic experiments, $\sim$0.5 M/S. In their apparatus a small reference portion of the beam traverses an electrooptic modulator and then enters the Fabry–Perot collinearly with the signal. By tuning of the modulation frequency, the side bands on the reference signal are brought to the center of the Brillouin components, and the Brillouin shift is then read directly as a frequency from the r.f. oscillator which produces the modulation.

Pressure scanned plane Fabry–Perots have been widely utilized in the investigation of crystals of superior optical quality where parasitic scattering is not a problem. With crystals of poor optical quality, the piezoelectric-scanned multi-pass Fabry–Perot system introduced by John Sandercock has become the preferred instrument because its high contrast allows weak Brillouin components to be observed in the presence of strong elastic scattering.

It is also possible to suppress the elastic scattering with the molecular iodine reabsorption technique which allows weak-inelastic or quasielastic components to be studied without interference by the strong elastic line. This technique, which has been particularly useful in the study of central peaks, is described in ch. 7 by Fleury and Lyons in this volume.

Because piezoelectric scanning is inherently less stable than pressure scanning (where the mirror spacing is maintained by a quartz or invar spacer between the mirrors) some form of active stabilization is required for piezoelectrically scanned multi-pass Fabry–Perots. Fabry–Perot systems with complete stabilization systems can be purchased commercially. Alternatively, a mini-computer used for data acquisition can also provide the necessary stabilization.

An extensive review of the theory and practice of modern Brillouin scattering spectroscopy with applications in many areas of condensed-matter physics (including phase transitions) has recently been completed by Dil (1982). The interested reader should consult this review for additional discussion and references.

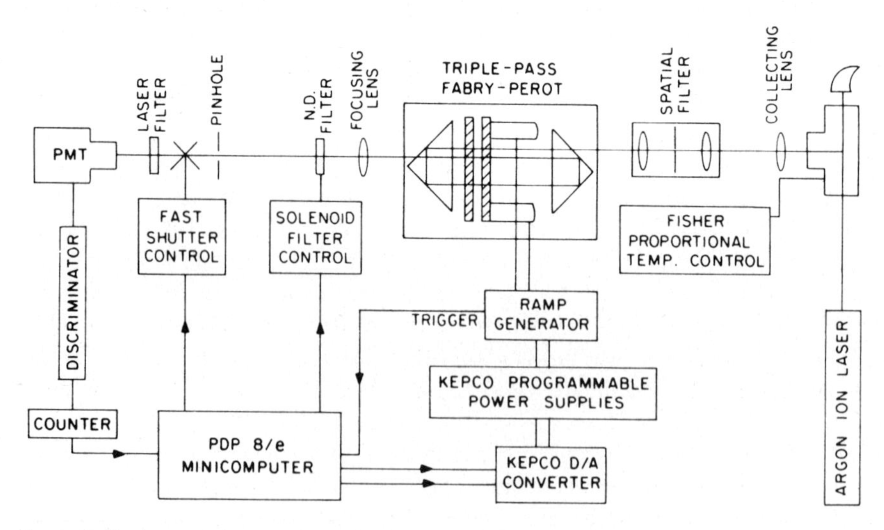


Fig. 1. Brillouin scattering spectrometer consisting of an ion laser, computer controlled triple pass Fabry–Perot interferometer and digital data collection system (Yao et al. 1981).

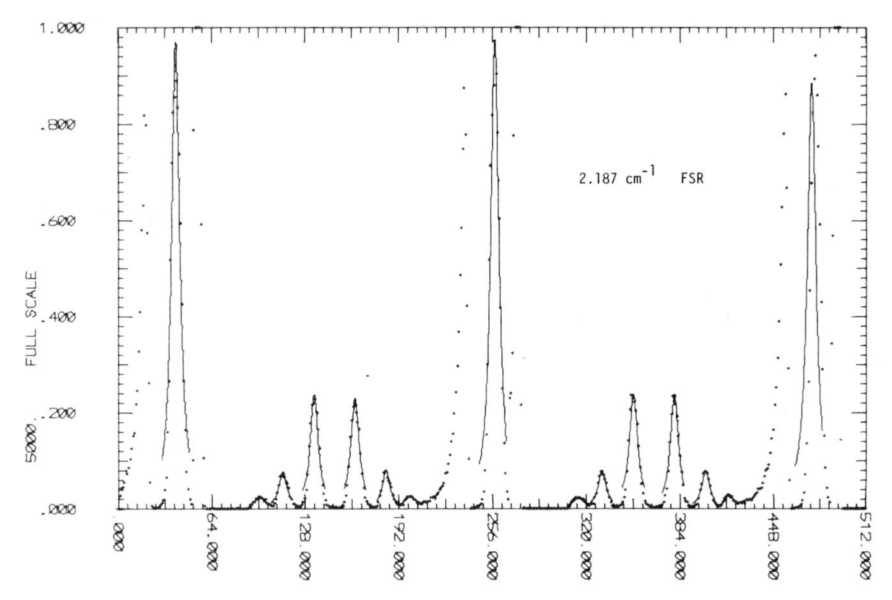


Fig. 2. Brillouin spectrum of crystalline quartz obtained with the spectrometer of fig. 1. The 512 points are the data. The solid lines are computer-fit Lorentzians from which the Brillouin shifts and linewidths are determined (provided by B.Y. Gu).

Figure 1 shows a triple pass piezoelectric Fabry–Perot interferometer controlled by a PDP-8E minicomputer, currently in use in the author's laboratory (Yao et al. 1981). The Fabry–Perot scan is driven by an associated linear ramp generator (Tropel, Fairport, N.Y.) which also provides a trigger signal to the minicomputer for multi-scaling data read in. During the scan, a fast shutter closes automatically whenever the count rate exceeds a preset maximum, and remains closed during a preset number of channels in order to protect the photomultiplier from excessive light.

After a preset number of scans, the computer jumps to an alignment routine. With the shutter open and a neutral density filter inserted in the optical path, the bias voltages to two of the three piezoelectric stacks are alternately stepped via two programmable power supplies until the peak height of the central Rayleigh line is maximized. The position of the central Rayleigh line is then recorded and used to reset a delay in the triggered multi-scaling routine to compensate for interferometer and laser drifts. Before returning to the regular data acquisition mode, one scan is added to the memory with the ND filter in place in order to provide reduced Rayleigh peaks in the spectrum for subsequent computer analysis of the data.

A spectrum of crystalline quartz obtained with the instrument illustrated in fig. 1 is shown in fig. 2. In each order there are one longitudinal and two transverse acoustic modes. The dots are the 512 data points; the solid lines are computer-fit Lorentzians to each component from which the Brillouin shifts and linewidths are found by a non-linear least-squares fitting program.

Recently, Sandercock (1980) has introduced a new Fabry–Perot interferometer which I believe will supercede many existing instruments. This system combines the high contrast and finesse of the multi-pass interferometer with the extended free spectral range of a tandem instrument (two incommensurate etalons in series). A recent article by Sandercock (1980) describes the operation of the instrument and discusses applications to the study of surface excitations.

### *1.2. Analysis of Brillouin spectra*

In many cases Brillouin components are narrow and symmetric (as in the quartz spectrum of fig. 2) and the Brillouin shifts can be simply determined from the positions of the line centers. The linewidths can be found by first fitting the Rayleigh line to a parametrized instrument function (often taken to be Gaussian) and then convolving the instrument function with a Lorentzian in which the width $\Delta\omega$ is adjusted to give a best fit to the data.

If the linewidth is not small compared to the Brillouin shift, this procedure is inadequate since the Brillouin components become asymmetric and markedly non-Lorentzian. It then becomes preferable to determine both the Brillouin shift and linewidth by convolving the instrument function with the theoretical spectrum:

$$I(\omega) = R\begin{bmatrix} n(\omega)+1 \\ n(\omega) \end{bmatrix} \mathrm{Im}[P^2 G(\omega)]\,, \tag{1.1}$$

where $R$ is a normalization constant, $[n(\omega)+1]$ and $[n(\omega)]$ are the Bose factors for the Stokes and anti-Stokes spectra, $P$ is an optical coupling constant and $G$ is a Green's function (or response function) for the acoustic modes. $G(\omega)$ is usually taken to have the damped-oscillator form:

$$G(\omega) = (\omega_0^2 - \omega^2 - \mathrm{i}\omega\gamma)^{-1}\,. \tag{1.2}$$

Fitting spectra to eq. (1.1) is usually adequate, even in the case of strong damping, as long as no mode coupling exists.

In cases where interaction between the acoustic modes and other modes is significant, eq. (1.1) can be generalized to

$$I(\omega) = R\begin{bmatrix} n(\omega)+1 \\ n(\omega) \end{bmatrix} \mathrm{Im} \sum P_i P_j G_{ij}(\omega)\,. \tag{1.3}$$

The $G_{ij}(\omega)$ in eq. (1.3) satisfy the matrix equation

$$\begin{vmatrix} G_1^{-1} & \Delta_{12}^2 - \mathrm{i}\omega\Gamma_{12} \ldots\ldots & \Delta_{1n}^2 - \mathrm{i}\omega\Gamma_{1n} \\ \Delta_{12}^2 - \mathrm{i}\omega\Gamma_{12} & G_2^{-1} \ldots\ldots & \Delta_{2n}^2 - \mathrm{i}\omega\Gamma_{2n} \\ \vdots & & \\ \Delta_{1n}^2 - \mathrm{i}\omega\Gamma_{1n} \ldots\ldots\ldots\ldots & & \Delta_{nn}^2 - \mathrm{i}\omega\Gamma_{nn} \end{vmatrix} \times \begin{vmatrix} G_{11} & G_{12} \ldots G_{1n} \\ G_{12} & G_{22} \ldots G_{2n} \\ \vdots & \\ G_{1n} \ldots\ldots & G_{nn} \end{vmatrix} = \begin{vmatrix} 1\,0 \ldots 0 \\ 0\,1 \ldots \\ \vdots \\ 0\,0 \ldots 1 \end{vmatrix}. \tag{1.4}$$

Here $G_j$ is the uncoupled response function of the $j$th mode, and $\Delta_{ij}^2 - \mathrm{i}\omega\Gamma_{ij}$ is the (complex) mode interaction between modes $i$ and $j$ to lowest order in $\omega$.

Coupled mode analyses of this type have been discussed extensively in the literature and have been widely applied to the analysis of Brillouin and Raman spectra. The method is particularly useful for analyzing Brillouin spectra in crystals with linearly coupled soft optic and coustic modes. (cf. Lagakos and Cummins (1974) and ch. 7 of this volume).

## *2. Proper ferroelastic transitions (A)*

This category actually subsumes three closely related classes of phase transitions. (i) Pure ferroelastics where strain is the only order parameter and the soft acoustic mode is the only strongly temperature dependent excitation. (ii) Linearly coupled ferroelastic transitions where the order parameter is either a strain or the eigenvector of a condensed optic mode which are proportional to each other in the low-symmetry phase (there are two soft modes – a soft optic mode and a soft acoustic mode which are mixed near the transition – with the soft optic mode exhibiting the primary instability). (iii) Ferroelastics which differ from (ii) only in that it is the soft acoustic mode which exhibits the primary instability. Assignment within these three classes is frequently uncertain since phase transitions initially thought to be purely ferroelastic have often later been found to have a soft optic mode. In crystals such as $DyVO_4$ no soft optic mode has been observed above $T_c$, but the acoustic instability arises from coupling of strain to low lying electronic levels of the Dy ions. Similarly, the distinction between classes (ii) and (iii) which both involve linear coupling between an optic mode and an acoustic mode is often controversial. I will therefore not attempt to separate these three classes but will treat them under a single category, discussing each example separately. From the point of view of symmetry the last distinction is irrelevant in any case since the linearly coupled soft optic and acoustic modes must belong to the same representation of the prototype crystal point group. Dynamically, however, the distinction is important.

We note that the term *pure ferroelastic* has been used by some authors to denote the appearance of ferroelasticity without the simultaneous appearance of either ferroelectricity or ferromagnetism. Thus, for example, Kaplyanskii et al. (1979) describe the improper ferroelastics $Hg_2Cl_2$ and $Hg_2Br_2$ as pure ferroelastics because they have no dielectric anomaly, despite the presence of a zone-boundary soft mode which doubles the volume of the unit cell in the low-temperature phase. Although the soft mode is a zone-boundary acoustic mode in the paraelastic phase, it becomes a zone-center optic mode in the ferroelastic phase where it can be observed in the Raman spectrum.

We will avoid this usage here and reserve the term pure ferroelastic for those transitions in which strain is the only order parameter, and the acoustic soft mode is the only soft mode in both the paraelastic and ferroelastic phases. A phase with a condensed soft optic mode whose condensation lowers the crystal symmetry will be considered as ferroic, even if the optic mode is not polar so that the order parameter is not associated with macroscopic polarization. (The term opto-ferroelastic has been suggested for such cases by R.E. Newnham.) Our generalization of the term ferroic is related to a generalized classification scheme for ferroics proposed by Paquet and Jerphagnon (1980) who also review various other definitions of ferroicity.

Within the framework of the Landau theory of phase transitions, homogeneous strain can produce structural phases transitions only between particular crystal point groups. Specifically, for a transition between a high-symmetry (prototype) phase belonging to point group $G_0$ and a low-symmetry phase belonging to point group G which is a subgroup of $G_0$, a strain $\epsilon$ can be the order parameter of a second-order (continuous) transition only if the symmetrized cube of its representation does not contain the identity representation, and if it transforms according to an irreducible representation of $G_0$ which subduces the identity representation in G (Birman 1966). The subduction principle also provides the formal proof of Worlock's theorem, that "the soft mode is always totally symmetric and always Raman active in the ordered phase" (Worlock 1971, Birman 1973).

A systematic effort to identify those structural phase transitions characterized by homogeneous strain which are consistent with Landau theory was reported by Boccara (1968). Those elastic strains which can serve as order parameters for each crystal class were deduced from the quadratic invariants of the strain tensors, and the existence of cubic invariants was used to identify those transitions which are necessarily first order. Boccara showed that not all subgroups of $G_0$ can be induced by homogeneous elastic strain; if $G_0$ is centric, for example, then only centric subgroups are allowed. Boccara's results were extended by Aubry and Pick (1971) who also found the eigenvectors and wavevectors of the soft acoustic modes associated with each elastic transition, and by several other authors who investigated the critical dynamics of elastic phase transitions which will be discussed further below. A complete compilation of *all* cell-preserving phase transitions was given by Janovec et al. (1975). Their table 1 gives all possible group–subgroup transitions which can be induced by irreducible representations of each prototype group $G_0$ and includes several cases of elastic transitions not given by Boccara or by Aubry and Pick.

The term ferroelasticity was introduced by Aizu (1969, 1970) to designate crystals possessing two or more energetically equivalent states between which they can be switched by application of mechanical stress. When ferroelasticity appears at a phase transition, it is designated a ferroelastic phase transition. The theory of ferroelasticity and the relationship of ferroelasticity to ferroelectricity and ferromagnetism have been discussed extensively by Toledano (1974, 1978) and by Toledano and Toledano (1976, 1977, 1980).

Aizu identified 94 species of ferroelastic crystals and developed a classification scheme which gives for each species both the prototype point group and the ferroelastic point group, separated by the letter F. Although Aizu's classification scheme is convenient for designating ferroelastic crystal structures, it is not entirely appropriate for classifying ferroelastic phase transitions. For example, for the orthorhombic prototype point group mmm($D_{2h}$), Aizu gives two ferroelastic species: mmmF2/m and mmmF$\bar{1}$.

Referring to tables 1 and 2 of Janovec et al. (1975), the transition mmm→2/m can be induced by a shear strain of symmetry $B_{1g}$, $B_{2g}$ or $B_{3g}$, but the transition mmm→$\bar{1}$ cannot be induced by a single strain: it would either require two successive transitions, mmm→2/m followed by 2/m→$\bar{1}$ (induced by a shear strain of symmetry $B_g$), or else a single transition induced by the reducible representation $B_{1g} + B_{2g}$ implying a simultaneous instability against two soft modes (V. Dvorak – private communication). Although such double-instability transitions might occur in principle, they will not be included in the present classification which will include only transitions driven by a single irreducible representation of $G_0$, as required by Landau theory for second-order phase transitions.

In table 2 the 84 possible proper ferroelastic phase transitions are listed along with a number of properties which will be useful in the discussion of particular examples. The 32 prototype point groups $G_0$ listed in order of ascending symmetry are given in column 2. All allowed transitions $G_0$→G for which a strain can be the order parameter, taken from table 1 of Janovec et al. are given in column 3 in order of decreasing symmetry of G.

The symbols in column 4 which indicate the order of the transition have the following meaning. "2" signifies that a line of second-order transitions in the *T–p* plane is allowed by symmetry. A dot (·) signifies that the high- and low-symmetry phases can meet *continuously* at most at a point in the *T–p* plane because two coefficients in the free energy expansion must vanish simultaneously (e.g., entry 19 or 24) or else, if there is a cubic invariant, this is the point where its coefficient passes through zero (e.g., entries 32 and 33). "1" signifies that no continuous transition is possible because, for example, there are two cubic invariants (e.g., entry 28). "$\bar{1}$" signifies that there is a Lifshitz invariant so that the prototype and ferroic phases are separated by an incommensurate phase (V. Dvorak – private communication).

The strain which is the order parameter, and the irreducible representation according to which it transforms in $G_0$, are given in columns 6 and 5 respectively. If the transition is ferroelastic and also involves a soft optic mode, the soft optic mode must also belong to the representation shown in column 5. If the transition can also be ferroelectric (i.e., if the symmetry of the strain inducing the transition is the same as that of a polar optic mode) the orientation

Table 2
Symmetry properties and soft acoustic modes of proper ferroelastic phase transitions (based on Janovec et al. (1975), Aubry and Pick (1971) and Cowley (1976))

| No. | Prototype group $G_0$ | Transition $G_0 \to G$ | Order | Order parameter | | Ferro-elect. | Soft acoustic modes | | | |
|---|---|---|---|---|---|---|---|---|---|---|
| | | | | Rep. | Strain | | Class | $\boldsymbol{q}$ | | Eigenvector |
| | Triclinic | | | | | | | | | |
| | $1(C_1)$ | none | | | | | | | | |
| | $\bar{1}(C_1)$ | none | | | | | | | | |
| | Monoclinic | | | | | | | | | |
| 1 | $2(C_2)$ | $2 \to 1$ | 2 | B | $\epsilon_4, \epsilon_5$ | $P_x, P_y$ | I | $\Vert[001]$ | | $\Vert[a, b, 0]$ |
| 2 | $m(C_2)$ | $m \to 1$ | 2 | A″ | $\epsilon_4, \epsilon_5$ | $P_z$ | I | $\Vert[a, b, 0]$ | | $\Vert[001]$ |
| 3 | $2/m(C_{2h})$ | $2/m \to \bar{1}$ | 2 | $B_g$ | $\epsilon_4, \epsilon_5$ | – | I | $\Vert[001]$ | ↔ | $\Vert[a, b, 0]$ |
| | Orthorhombic | | | | | | | | | |
| 4 | $222(D_2)$ | $222 \to 2_z$ | 2 | $B_1$ | $\epsilon_6$ | $P_z$ | I | $\Vert[010]$ | ↔ | $\Vert[100]$ |
| – | | $222 \to 2_y$ | 2 | $B_2$ | $\epsilon_5$ | $P_y$ | I | $\Vert[001]$ | ↔ | $\Vert[100]$ |
| – | | $222 \to 2_x$ | 2 | $B_3$ | $\epsilon_4$ | $P_x$ | I | $\Vert[001]$ | ↔ | $\Vert[010]$ |
| 5 | $mm2(C_{2v})$ | $mm2 \to 2$ | 2 | $A_2$ | $\epsilon_6$ | – | I | $\Vert[010]$ | ↔ | $\Vert[100]$ |
| 6 | | $mm2 \to m_y$ | 2 | $B_1$ | $\epsilon_5$ | $P_x$ | I | $\Vert[001]$ | ↔ | $\Vert[100]$ |
| – | | $mm2 \to m_x$ | 2 | $B_2$ | $\epsilon_4$ | $P_y$ | I | $\Vert[001]$ | ↔ | $\Vert[010]$ |

| | | | | | | | | | | |
|---|---|---|---|---|---|---|---|---|---|---|
| 7 | mmm($D_{2h}$) | mmm→2/$m_z$ | 2 | $B_{1g}$ | $\epsilon_6$ | – | I | ∥[010] | ↔ | ∥[100] |
| – | | mmm→2/$m_y$ | 2 | $B_{2g}$ | $\epsilon_5$ | – | I | ∥[001] | ↔ | ∥[100] |
| – | | mmm→2/$m_x$ | 2 | $B_{3g}$ | $\epsilon_4$ | – | I | ∥[001] | ↔ | ∥[010] |
| | Tetragonal | | | | | | | | | |
| 8 | 4($C_4$) | 4→2 | 2 | B | $\epsilon_6, \epsilon_1 - \epsilon_2$ | – | I | ∥[$a, b$, 0] | ↔ | ⊥$q$ and $z$ |
| 9 | | 4→1 | $\tilde{1}$ | E | $\epsilon_4, \epsilon_5$ | $P_x, P_y$ | II | ⊥[001] | | ∥[001] |
| 10 | $\bar{4}$($S_4$) | $\bar{4}$→2 | 2 | B | $\epsilon_6, \epsilon_1 - \epsilon_2$ | $P_z$ | I | ∥[$a, b$, 0] | ↔ | ⊥$q$ and $z$ |
| 11 | | $\bar{4}$→1 | 2 | E | $\epsilon_4, \epsilon_5$ | $P_x, P_y$ | II | ⊥[001] | ↔ | ∥[001] |
| 12 | 4/m($C_{4h}$) | 4/m→2/m | 2 | $B_g$ | $\epsilon_6, \epsilon_1 - \epsilon_2$ | – | I | ∥[$a, b$, 0] | ↔ | ⊥$q$ and $z$ |
| 13 | | 4/m→$\bar{1}$ | 2 | $E_g$ | $e_4, \epsilon_5$ | – | II | ⊥[001] | ↔ | ∥[001] |
| 14 | 422($D_4$) | 422→222 | 2 | $B_1$ | $\epsilon_1 - \epsilon_2$ | – | I | ∥[110] | ↔ | ∥[1$\bar{1}$0] |
| – | | 422→222 | 2 | $B_2$ | $\epsilon_6$ | – | I | ∥[100] | ↔ | ∥[010] |
| 15 | | 422→$2_x$ | $\tilde{1}$ | E | $\epsilon_4$ | $P_x$ | II | ⊥[001] | | ∥[001] |
| – | | 422→$2_{xy}$ | $\tilde{1}$ | E | $\epsilon_4 - \epsilon_5$ | $P_x = P_y$ | II | ⊥[001] | | ∥[001] |
| 16 | | 422→1 | $\tilde{1}$ | E | $\epsilon_4, \epsilon_5$ | $P_x, P_y$ | II | ⊥[001] | | ∥[001] |
| 17 | 4mm($C_{4v}$) | 4mm→mm2 | 2 | $B_1$ | $\epsilon_1 - \epsilon_2$ | – | I | ∥[110] | ↔ | ∥[1$\bar{1}$0] |
| – | | 4mm→mm2 | 2 | $B_2$ | $\epsilon_6$ | – | I | ∥[100] | ↔ | ∥[010] |
| 18 | | 4mm→$m_x$ | 2 | E | $\epsilon_4$ | $P_y$ | II | ⊥[001] | | ∥[001] |
| – | | 4mm→$m_{xy}$ | 2 | E | $\epsilon_4 - \epsilon_5$ | $P_x = -P_y$ | II | ⊥[001] | | ∥[001] |

Table 2 (*cont.*)

| No. | Prototype group $G_0$ | Transition $G_0 \to G$ | Order | Order parameter: Rep. | Order parameter: Strain | Ferro-elec. | Soft acoustic modes: Class | $\boldsymbol{q}$ | | Eigenvector |
|---|---|---|---|---|---|---|---|---|---|---|
| 19 | | $4mm \to 1$ | · | E | $\epsilon_4, \epsilon_5$ | $P_x, P_y$ | II | $\perp[001]$ | | $\parallel[001]$ |
| 20 | $\bar{4}2m(D_{2d})$ | $\bar{4}2m \to mm2$ | 2 | $B_2$ | $\epsilon_6$ | $P_z$ | I | $\parallel[100]$ | $\leftrightarrow$ | $\parallel[010]$ |
| 21 | | $\bar{4}2m \to 222$ | 2 | $B_1$ | $\epsilon_1 - \epsilon_2$ | – | I | $\parallel[110]$ | $\leftrightarrow$ | $\parallel[1\bar{1}0]$ |
| 22 | | $\bar{4}2m \to m_{xy}$ | 2 | E | $\epsilon_4 - \epsilon_5$ | $P_x = -P_y$ | II | $\perp[001]$ | | $\parallel[001]$ |
| 23 | | $\bar{4}2m \to 2_x$ | 2 | E | $\epsilon_4$ | $P_x$ | II | $\perp[001]$ | | $\parallel[001]$ |
| 24 | | $\bar{4}2m \to 1$ | · | E | $\epsilon_4, \epsilon_5$ | $P_x, P_y$ | II | $\perp[001]$ | | $\parallel[001]$ |
| 25 | $4/mmm(D_{4h})$ | $4/mmm \to mmm$ | 2 | $B_{1g}$ | $\epsilon_1 - \epsilon_2$ | – | I | $\parallel[110]$ | $\leftrightarrow$ | $\parallel[1\bar{1}0]$ |
| – | | $4/mmm \to mmm$ | 2 | $B_{2g}$ | $\epsilon_6$ | – | I | $\parallel[100]$ | $\leftrightarrow$ | $\parallel[010]$ |
| 26 | | $4/mmm \to 2/m_x$ | 2 | $E_g$ | $\epsilon_4$ | – | II | $\perp[001]$ | $\leftrightarrow$ | $\parallel[001]$ |
| – | | $4/mmm \to 2/m_{xy}$ | 2 | $E_g$ | $\epsilon_4 - \epsilon_5$ | – | II | $\perp[001]$ | $\leftrightarrow$ | $\parallel[001]$ |
| 27 | | $4/mmm \to \bar{1}$ | · | $E_g$ | $\epsilon_4, \epsilon_5$ | – | II | $\perp[001]$ | $\leftrightarrow$ | $\parallel[001]$ |
| | Trigonal | | | | | | | | | |
| 28 | $3(C_3)$ | $3 \to 1$ | $\tilde{1}$ | E | $\epsilon_1 - \epsilon_2, \epsilon_4, \epsilon_5, \epsilon_6$ | $P_x, P_y$ | I | | | |
| 29 | $\bar{3}(C_{3i})$ | $\bar{3} \to \bar{1}$ | 1 | $E_g$ | $\epsilon_1 - \epsilon_2, \epsilon_4, \epsilon_5, \epsilon_6$ | – | I | | | |
| 30 | $32(D_3)$ | $32 \to 2_x$ | $\tilde{1}$ | E | $\epsilon_1 - \epsilon_2, \epsilon_4$ | $P_x$ | I | | | |
| 31 | | $32 \to 1$ | $\tilde{1}$ | E | $\epsilon_1 - \epsilon_2, \epsilon_4, \epsilon_5, \epsilon_6$ | $P_x, P_y$ | I | | | |

| | | | | | | | | | | |
|---|---|---|---|---|---|---|---|---|---|---|
| 32 | 3m($C_{3v}$) | 3m→$m_x$ | · | E | $\epsilon_1-\epsilon_2, \epsilon_4$ | $P_y$ | I | | | |
| 33 | | 3m→1 | · | E | $\epsilon_1-\epsilon_2, \epsilon_4, \epsilon_5, \epsilon_6$ | $P_x, P_y$ | I | | | |
| 34 | $\bar{3}$m($D_{3d}$) | $\bar{3}$m→2/$m_x$ | · | $E_g$ | $\epsilon_4, \epsilon_1-\epsilon_2$ | – | I | | | |
| 35 | | $\bar{3}$m→$\bar{1}$ | · | $E_g$ | $\epsilon_1-\epsilon_2, \epsilon_4, \epsilon_5, \epsilon_6$ | – | I | | | |
| | <u>Hexagonal</u> | | | | | | | | | |
| 36 | 6($C_6$) | 6→2 | $\tilde{1}$ | $E_1$ | $\epsilon_1-\epsilon_2, \epsilon_6$ | – | II | ⊥[001] | | ⊥[001] and ***q*** |
| 37 | | 6→1 | $\tilde{1}$ | $E_2$ | $\epsilon_4, \epsilon_5$ | $P_x, P_y$ | II | ⊥[001] | | ‖[001] |
| 38 | $\bar{6}$($C_{3h}$) | $\bar{6}$→m | 1 | E′ | $\epsilon_1-\epsilon_2, \epsilon_6$ | $P_x, P_y$ | II | ⊥[001] | | ⊥[001] and ***q*** |
| 39 | | $\bar{6}$→1 | 2 | E″ | $\epsilon_4, \epsilon_5$ | $P_x, P_y, P_z$ | II | ⊥[001] | ↔ | ‖[001] |
| 40 | 6/m($C_{6h}$) | 6/m→2/m | 1 | $E_{1g}$ | $\epsilon_1-\epsilon_2, \epsilon_6$ | – | II | ⊥[001] | | ⊥[001] and ***q*** |
| 41 | | 6/m→$\bar{1}$ | $\tilde{2}$ | $E_{2g}$ | $\epsilon_4, \epsilon_5$ | – | II | ⊥[001] | ↔ | ‖[001] |
| 42 | 622($D_6$) | 622→222 | $\tilde{1}$ | $E_2$ | $\epsilon_1-\epsilon_2$ | – | II | ⊥[001] | | ⊥[001] and ***q*** |
| 43 | | 622→$2_x$ | $\tilde{1}$ | $E_1$ | $\epsilon_4$ | $P_x$ | II | ⊥[001] | | ‖[001] |
| – | | 622→$2_y$ | $\tilde{1}$ | $E_1$ | $\epsilon_5$ | $P_y$ | II | ⊥[001] | | ‖[001] |
| 44 | | 622→$2_z$ | $\tilde{1}$ | $E_2$ | $\epsilon_1-\epsilon_2, \epsilon_6$ | $P_z$ | II | ⊥[001] | | ⊥[001] and ***q*** |
| 45 | | 622→1 | $\tilde{1}$ | $E_1$ | $\epsilon_4, \epsilon_5$ | $P_x, P_y$ | II | ⊥[001] | | ‖[001] |
| 46 | 6mm($C_{6v}$) | 6mm→mm2 | · | $E_2$ | $\epsilon_1-\epsilon_2$ | – | II | ⊥[001] | | ⊥[001] and ***q*** |
| 47 | | 6mm→$m_x$ | 2 | $E_1$ | $\epsilon_4$ | $P_y$ | II | ⊥[001] | | ‖[001] |
| – | | 6mm→$m_y$ | 2 | $E_1$ | $\epsilon_5$ | $P_x$ | II | ⊥[001] | | ‖[001] |
| 48 | | 6mm→$2_z$ | · | $E_2$ | $\epsilon_1-\epsilon_2, \epsilon_6$ | – | II | ⊥[001] | | ⊥[001] and ***q*** |
| 49 | | 6mm→1 | · | $E_1$ | $\epsilon_4, \epsilon_5$ | $P_x, P_y$ | II | ⊥[001] | | ‖[001] |

Table 2 (*cont.*)

| No. | Prototype group $G_0$ | Transition $G_0 \to G$ | Order | Order parameter | | Ferro-elec. | Soft acoustic modes | | | |
|---|---|---|---|---|---|---|---|---|---|---|
| | | | | Rep. | Strain | | Class | $\boldsymbol{q}$ | | Eigenvector |
| 50 | $\bar{6}m2(D_{3h})$ | $\bar{6}m2 \to mm2$ | · | E′ | $\epsilon_1 - \epsilon_2$ | $P_y$ | II | $\perp[001]$ | | $\perp[001]$ and $\boldsymbol{q}$ |
| 51 | | $\bar{6}m2 \to m_z$ | · | E′ | $\epsilon_1 - \epsilon_2, \epsilon_6$ | $P_x, P_y$ | II | $\perp[001]$ | | $\perp[001]$ and $\boldsymbol{q}$ |
| 52 | | $\bar{6}m2 \to m_x$ | 2 | E″ | $\epsilon_4$ | $P_y, P_z$ | II | $\perp[001]$ | $\leftrightarrow$ | $\parallel[001]$ |
| 53 | | $\bar{6}m2 \to 2_y$ | 2 | E″ | $\epsilon_5$ | $P_y$ | II | $\perp[001]$ | $\leftrightarrow$ | $\parallel[001]$ |
| 54 | | $\bar{6}m2 \to 1$ | · | E″ | $\epsilon_4, \epsilon_5$ | $P_x, P_y, P_z$ | II | $\perp[001]$ | $\leftrightarrow$ | $\parallel[001]$ |
| 55 | $6/mmm(D_{6h})$ | $6/mmm \to mmm$ | · | $E_{2g}$ | $\epsilon_1 - \epsilon_2$ | – | II | $\perp[001]$ | | $\perp[001]$ and $\boldsymbol{q}$ |
| 56 | | $6/mmm \to 2/m_z$ | · | $E_{2g}$ | $\epsilon_1 - \epsilon_2, \epsilon_6$ | – | II | $\perp[001]$ | | $\perp[001]$ and $\boldsymbol{q}$ |
| 57 | | $6/mmm \to 2/m_x$ | 2 | $E_{1g}$ | $\epsilon_4$ | – | II | $\perp[001]$ | $\leftrightarrow$ | $\parallel[001]$ |
| – | | $6/mmm \to 2/m_y$ | 2 | $E_{1g}$ | $\epsilon_5$ | – | II | $\perp[001]$ | $\leftrightarrow$ | $\parallel[001]$ |
| 58 | | $6/mmm \to \tilde{1}$ | · | $E_{1g}$ | $\epsilon_4, \epsilon_5$ | – | II | $\perp[001]$ | | $\parallel[001]$ |
| | <u>Cubic</u> | | | | | | | | | |
| 59 | 23(T) | $23 \to 3$ | $\tilde{1}$ | T | $\epsilon_4 + \epsilon_5 + \epsilon_6$ | $P_x = P_y = P_z$ | II | $\perp[100]$ | | $\parallel[100]$ |
| 60 | | $23 \to 222$ | 1 | E | $\epsilon^*$ | – | I | $\parallel[110]$ | | $\parallel[1\bar{1}0]$ |
| 61 | | $23 \to 2_z$ | $\tilde{1}$ | T | $\epsilon_6$ | $P_z$ | II | $\perp[100]$ | | $\parallel[100]$ |
| 62 | | $23 \to 1$ | $\tilde{1}$ | T | $\epsilon_4, \epsilon_5, \epsilon_6$ | $P_x, P_y, P_z$ | II | $\perp[100]$ | | $\parallel[100]$ |
| 63 | $m3(T_h)$ | $m3 \to \bar{3}$ | · | $T_g$ | $\epsilon_4 + \epsilon_5 + \epsilon_6$ | – | II | $\perp[100]$ | | $\parallel[100]$ |
| 64 | | $m3 \to mmm$ | 1 | $E_g$ | $\epsilon^*$ | – | I | $\parallel[110]$ | | $\parallel[1\bar{1}0]$ |
| 65 | | $m3 \to 2/m_z$ | · | $T_g$ | $\epsilon_6$ | – | II | $\perp[100]$ | | $\parallel[100]$ |
| 66 | | $m3 \to \bar{1}$ | 1 | $T_g$ | $\epsilon_4, \epsilon_5, \epsilon_6$ | – | II | $\perp[100]$ | | $\parallel[100]$ |

| | | | | | | | | | |
|---|---|---|---|---|---|---|---|---|---|
| 67 | 432(O) | 432→32 | $\tilde{1}$ | $T_2$ | $\epsilon_4+\epsilon_5+\epsilon_6$ | – | II | ⊥[100] | ‖[100] |
| 68 | | 432→422 | · | E | $\epsilon_1+\epsilon_2-2\epsilon_3$ | – | I | ‖[110] | ‖[1$\bar{1}$0] |
| 69 | | 432→222 | $\tilde{1}$ | $T_2$ | $\epsilon_6$ | – | II | ⊥[100] | ‖[100] |
| 70 | | 432→222 | · | E | $\epsilon^*$ | – | I | ‖[110] | ‖[1$\bar{1}$0] |
| 71 | | 432→2 | $\tilde{1}$ | $T_2$ | $\epsilon_4-\epsilon_5, \epsilon_6$ | – | II | ⊥[100] | ‖[100] |
| 72 | | 432→1 | $\tilde{1}$ | $T_2$ | $\epsilon_4, \epsilon_5, \epsilon_6$ | – | II | ⊥[100] | ‖[100] |
| 73 | $\bar{4}$3m($T_d$) | $\bar{4}$3m→3m | · | $T_2$ | $\epsilon_4+\epsilon_5+\epsilon_6$ | $P_x=P_y=P_z$ | II | ⊥[110] | ‖[1$\bar{1}$0] |
| 74 | | $\bar{4}$3m→$\bar{4}$2m | · | E | $\epsilon_1+\epsilon_2-2\epsilon_3$ | – | I | ‖[110] | ‖[1$\bar{1}$0] |
| 75 | | $\bar{4}$3m→mm2 | · | $T_2$ | $\epsilon_6$ | $P_z$ | II | ⊥[100] | ‖[100] |
| 76 | | $\bar{4}$3m→222 | · | E | $\epsilon^*$ | – | I | ‖[110] | ‖[1$\bar{1}$0] |
| 77 | | $\bar{4}$3m→m | 1 | $T_2$ | $\epsilon_4-\epsilon_5, \epsilon_6$ | $P_x=-P_y, P_z$ | II | ⊥[100] | ‖[100] |
| 78 | | $\bar{4}$3m→1 | 1 | $T_2$ | $\epsilon_4, \epsilon_5, \epsilon_6$ | $P_x, P_y, P_z$ | II | ⊥[100] | ‖[100] |
| 79 | m3m($O_h$) | m3m→$\bar{3}$m | · | $T_{2g}$ | $\epsilon_4+\epsilon_5+\epsilon_6$ | – | II | ⊥[100] | ‖[100] |
| 80 | | m3m→4/mmm | · | $E_g$ | $\epsilon_1+\epsilon_2-2\epsilon_3$ | – | I | ‖[110] | ‖[1$\bar{1}$0] |
| 81 | | m3m→mmm | · | $E_g$ | $\epsilon^*$ | – | I | ‖[110] | ‖[1$\bar{1}$0] |
| 82 | | m3m→mmm | · | $T_{2g}$ | $\epsilon_6$ | – | II | ⊥[100] | ‖[100] |
| 83 | | m3m→2/m | 1 | $T_{2g}$ | $\epsilon_4-\epsilon_5, \epsilon_6$ | – | II | ⊥[100] | ‖[100] |
| 84 | | m3m→$\bar{1}$ | 1 | $T_{2g}$ | $\epsilon_4, \epsilon_5, \epsilon_6$ | – | II | ⊥[100] | ‖[100] |

$\epsilon^* = \epsilon_1, \epsilon_2, \epsilon_3 = \frac{1}{4}(\epsilon_1+\epsilon_2)$

of the resulting ferroelectric polarization is given in column 7. In numbering the 84 transitions shown (column 1), groups of transitions arising from a prototype structure $G_0$ to different low-symmetry structures which are crystallography equivalent (related by rotations of the axes) are given a single number. (For orientations in other cases see Janovec et al. (1975).)

The last three columns of table 2 give information related to the soft acoustic mode, based on the work of Aubry and Pick (1971), Cowley (1976), and Folk et al. (1976a, b). Column 8 gives a Roman numeral designating the dimensionality class. For any elastic instability, i.e., a decrease to zero of an eigenvalue of the $6 \times 6$ elastic constant tensor, the dimensionality class specifies the directions of propagation for which the velocity of one of the three acoustic modes approach zero as $T \to T_0$ (Cowley 1976). Class 0 implies that no such direction exists so that there are no soft acoustic modes. Class I implies that soft acoustic modes exist for special directions of propagation, while class II implies that soft acoustic modes exist for propagation directions lying in planes perpendicular to particular directions. The wavevectors ($\boldsymbol{q}$) and eigenvectors ($\boldsymbol{u}$) of the soft acoustic modes are given in columns 9 and 10. The symbol $\leftrightarrow$ signifies that $\boldsymbol{q}$ and $\boldsymbol{u}$ can be interchanged.

Every transition listed in table 2 is seen to correspond to either class I or II, in agreement with the proof by Aubry and Pick (1971 that "a (symmetry changing) second-order elastic transition always carries with it an acoustic phonon the speed of which is zero"*. An interesting example of class 0 transitions occurs in a ferroelectric in an electric field. Since the symmetry of the prototype phase is already broken by the field, the transition occurs with no further change of symmetry, hence with no microscopic critical fluctuations (Courtens et al. 1979, Courtens and Gammon 1981).

The primary significance of the dimensionality class lies in the effect of the limited dimensionality of the critical fluctuations on the critical behavior. Khmel'nitskii (1974) noted that the sound velocity at the critical point may not vanish at all – as in isomorphous transitions where there is no soft acoustic mode – and Landau theory is then exact (class 0). Or it may vanish only for certain directions of $\boldsymbol{q}$. This anisotropy reduces the fluctuation corrections to thermodynamic properties making the critical singularities weaker. He examined transitions originating from a cubic prototype phase using the self-consistent field method to evaluate integrals and concluded that for one-dimensional soft modes (restricted to particular directions) the integrals converge giving no infinite singularities (class I) while for two-dimensional soft modes (restricted to planes) there would be logarithmic divergences (class II) (also see Larkin and Khmel'nitskii (1969)).

*Some details of the proof are incorrect, but they do not invalidate the results (R. Pick, private communication).

The renormalization group method was applied to the analysis of this problem concurrently by Cowley (1976), Folk et al. (1976a, 1976b, 1977, 1979) and Schwabl (1980a, 1980b)

Cowley (1976) considered all elastic transitions for each of the prototype crystal classes and assigned each to one of the classes 0, 1 or II*. Renormalization group theory was then used to evaluate the relevant integrals for those transitions in which second-order transitions are not forbidden by the presence of a cubic invariant.

Cowley found that the borderline between classical and non-classical behavior (marginal dimensionality) which is at $d_c = 4$ for isotropic short-range interactions occurs at $d_c = 2$ for class I transitions and at $d_c = 3$ for class II transitions. Thus Landau theory should be completely valid for class I, while for class II classical exponents should still apply but with logarithmic corrections. The susceptibility $\chi$, for example, should be given for class II transitions by:

$$\chi \propto (T - T_0)^{-1} |\ln(T - T_0)|^{1/3} .$$

The analysis of Folk et al. which was developed concurrently with that of Cowley uses a somewhat different extension of the three-dimensional elastic Hamiltonian to the lower dimensionality of the "soft subspace". For class I transitions they find the critical dimensionality $d_c = 2.5$ rather than 2 found by Cowley. Their predictions for critical behavior agree with Cowley and Khmel'nitskii (1974): class I transitions exhibit classical Landau behavior; class II systems exhibit classical behavior with logarithmic corrections, with the exponent of the $\ln(T - T_0)$ term either $\frac{1}{3}$ or $\frac{4}{9}$.

I have not included any information on the number and orientation of ferroelastic domain walls in table 2, nor have I tabulated the combination of elastic constants which determine the velocity of the acoustic soft mode. A detailed discussion and tabulation of ferroelastic domain walls has been given by Sapriel (1975). The soft mode elastic constants will be discussed in the following section (sect. 2.1). Finally, we note that the inclusion of first-order (discontinuous) transitions in this table constitutes an *ad hoc* extension of Landau theory which is strictly applicable only to continuous transitions; nevertheless, first-order transitions are often marked by soft modes with considerable temperature dependence, particularly if the transition is only weakly first order. Some additional complications may arise in applying Landau theory to first-order transitions however, as we shall see in the case of the improper ferroelastic benzil in sect. 4.1.

*Although table 1 of Cowley (1976) gives 14 examples of class 0 transitions, only the non-symmetry changing examples are actually class 0. The remaining four (two trigonal E, one tetragonal B and one monoclinic B), although originally listed by Cowley as class 0 are actually class I (R.A. Cowley, private communication).

### 2.1. Critical elastic constants and soft acoustic modes

In many proper ferroelastic transitions the prototype lattice becomes unstable against a single strain or a simple combination of strains. The connection between the critical elastic constants $C_c$ and the soft acoustic mode velocity $v_c$ is then straightforward. Thus, for example, in the transition $\bar{4}2m \rightarrow mm2$ (# 20) which occurs in potassium dihydrogen phosphate (KDP), the only strain with the appropriate $B_2$ symmetry is $\epsilon_6$. The crystal becomes unstable against homogeneous $xy$ shear strain when $C_c = C_{66}$ reaches zero. The velocity $v_c$ of the soft acoustic mode, with $\boldsymbol{q} \parallel [100]$ and displacement $\boldsymbol{u} \parallel [010]$ (or vice versa), is determined by $\rho v_c^2 = C_{66}$.

A more subtle situation occurs when more than one strain is involved. In bismuth vanadate, for example, the $4/m \rightarrow 2/m$ transition (# 12) is driven by a soft $B_g$ mode. Since both $\epsilon_6$ and $\epsilon_1 - \epsilon_2$ have $B_g$ symmetry, the spontaneous strain which appears in the ferroelastic phase will be a linear combination of $\epsilon_6$ and $\epsilon_1 - \epsilon_2$.

The stability condition on the elastic free energy in the tetragonal phase

$$F_{\epsilon} = \tfrac{1}{2} \sum_{ij} C_{ij} \epsilon_i \epsilon_j \tag{2.1}$$

is that all the eigenvalues of the 6 by 6 tensor $[C_{ij}]$ be positive. In the subspace of $\epsilon_1$, $\epsilon_2$ and $\epsilon_6$, these eigenvalues are given by the characteristic equation

$$\begin{vmatrix} C_{11} - \lambda & C_{12} & C_{16} \\ C_{12} & C_{11} - \lambda & -C_{16} \\ C_{16} & -C_{16} & C_{66} - \lambda \end{vmatrix} = 0 .$$

Since $\epsilon_1 + \epsilon_2$ transforms as $A_g$ while $\epsilon_1 - \epsilon_2$ and $\epsilon_6$ transform as $B_g$, partial diagonalization is achieved by transforming to $(1/\sqrt{2})(\epsilon_1 \pm \epsilon_2)$, $\epsilon_6$:

$$\begin{vmatrix} (C_{11} + C_{12} - \lambda) & 0 & 0 \\ 0 & (C_{11} - C_{12} - \lambda) & C_{16}\sqrt{2} \\ 0 & C_{16}\sqrt{2} & C_{66} - \lambda \end{vmatrix} = 0 .$$

The $2 \times 2$ submatrix of $B_g$ strains has as its smaller eigenvalue

$$C_c = \tfrac{1}{2}(C_{11} - C_{12} + C_{66} - [(C_{11} - C_{12} - C_{66})^2 + 8C_{16}^2]^{1/2}) . \tag{2.2}$$

This is the critical elastic constant derived by Boccara (1968) which goes to zero at $T_c$ as the crystal becomes unstable against homogeneous deformation. Boccara lists critical elastic constants for most of the transitions in table 2. Some authors have assumed that $C_c$ also governs the acoustic soft mode.

Recently, Copic has analyzed the Brillouin scattering tensors for bismuth vanadate (Gu et al. 1981) and has shown that the soft-mode velocity $v_c$ is *not* determined by $C_c$ (eq. (2.2)). For acoustic modes with $\boldsymbol{q}$ in the $xy$-plane, the minimum velocity occurs when $\theta$, the angle between $\boldsymbol{q}$ and the $x$-axis, is given by

$$\tan 4\theta = 4C_{16}/(C_{11} - C_{12} - 2C_{66}) . \tag{2.3}$$

For propagation in this direction, the velocity of the lowest-frequency mode, which is a pure transverse mode in the $xy$-plane, is

$$\rho v_c^2 = \tfrac{1}{4}(C_{11} - C_{12} + 2C_{66} - [(C_{11} - C_{12} - 2C_{66})^2 + 16C_{16}^2]^{1/2}) . \tag{2.4}$$

Although $\rho v_c^2$ and $C_c$ vanish simultaneously at $T_c$, their values for $T > T_c$ are in general different.

The origin of this apparent inconsistency lies in the 6-component strain convention. The elastic free energy, written in terms of the symmetrized strains $\epsilon_{ij}$, is (cf., Landau and Lifshitz (1970))

$$F_\epsilon = \tfrac{1}{2}\sum_{ijkl} C_{ijkl}\epsilon_{ij}\epsilon_{kl} . \tag{2.5}$$

Computing the eigenvalues of $[C_{ijkl}]$ in the four dimensional subspace of $\epsilon_{11}$, $\epsilon_{22}$, $\epsilon_{12}$ and $\epsilon_{21}$ one finds as the lowest eigenvalue

$$C_c' = \tfrac{1}{2}(C_{11} - C_{12} + 2C_{66} - [(C_{11} - C_{12} - 2C_{66})^2 + 16C_{16}^2]^{1/2}) . \tag{2.6}$$

The difference between $C_c$ and $C_c'$ arises in contracting from nine strains to six, in which $\epsilon_{11}\to\epsilon_1$, but $\epsilon_{12}\to\frac{1}{2}\epsilon_6$ in order to preserve the simple form of the elastic free energy (eq. (2.1)). Although the curvatures of the two energy paraboloids (eqs. (2.1) and (2.5)) along the respective principal directions of minimum curvature vanish simultaneously, the two curvatures will, in general, differ because of the scale factor of $\frac{1}{2}$ in some of the strains. A relatively simple procedure for finding $C_c'$ starts with the eigenvalue equations for the stress,

$$\sigma_{ij} = \sum_{ijkl} C_{ijkl}\epsilon_{kl} = \lambda\epsilon_{ij} , \tag{2.7}$$

which can readily be solved in the subspace of those strains $\epsilon_{ij}$ which couple to the order parameter. The smallest resulting $\lambda$ (which is also the minimum eigenvalue of the $9 \times 9$ $[C]$ tensor) is $C_c'$. Although this procedure does not give the direction of soft mode propagation, this can usually be found empirically from the orientation of the domain walls near $T_0$ which are perpendicular to $\boldsymbol{q}$. Note, however, that far from $T_0$ the direction of $\boldsymbol{q}_c$ will change with temperature as indicated in eq. (2.3).

The remaining factor of 2 difference between $C_c'$ and $\rho v_c^2$ in eqs. (2.4) and (2.6) results from the fact that while the elastic constants are defined in terms of *symmetrized* strain, the acoustic modes are composed of *pure* strain which, for shear waves, is half symmetric and half antisymmetric. Since the antisymmetric part of pure shear strain is (in the $\boldsymbol{q}\to 0$ limit) a pure rotation which produces no stress, the stress per unit pure shear strain (which determines $\rho v_c^2$) is only one half as large as the stress per unit symmetrized shear strain which determines $C_c'$. The factor of two does not occur in the case of longitudinal modes where the strain associated with the acoustic wave is completely symmetric.

The reader may notice that the distinction between $\rho v_c^2$ and $C_c$ seems not to occur in simpler cases. In most of the examples cited in standard tables (cf., Cummins and Schoen (1972)) the values of $\rho v^2$ for shear modes correspond to eigenvalues of the $6 \times 6$ elastic constant tensor. Thus, for example, in potassium dihydrogen phosphate (KDP), the $xy$ shear mode which is the soft $B_2$ acoustic mode in this $\bar{4}2m \rightarrow mm2$ ferroelastic transition has $\rho v_c^2 = C_{66}$, and inspection of the $\bar{4}2m$ elastic tensor shows that $C_{66}$ is indeed an eigenvalue.

A fortuitous cancellation of two factors of two is responsible. If the $9 \times 9$ tensor of elastic constants is found from eq. (2.7), the appropriate eigenvalue turns out to be $C_c' = 2C_{66}$. The acoustic mode velocity is $\rho v_c^2 = C_{66}$, which is again $\frac{1}{2}$ of $C_c'$. However in this case $C_c = \frac{1}{2}C_c'$ also $\rho v_c^2 = C_c$.

We note that no complete tabulation of elastic constants governing acoustic soft modes yet exists, and that caution should be exercised in associating the critical elastic constants $C_c$ tabulated by Boccara with $\rho v_c^2$ when mixed strains are involved.

### 2.2. *Proper ferroelastics* (*A*)

*Potassium dihydrogen phosphate* (*KDP*); *Potassium trihydrogen selenite* (*KTS*); *Lanthanum pentaphosphate*; *Praseodymium aluminate*; *Dysprosium vanadate and terbium vanadate*; *Lithium ammonium tartrate monohydrate* (*LAT*); *Paratellurite*; *Bismuth vanadate*

*$KH_2PO_4$: potassium dihydrogen phosphate* (*KDP*) *and isomorphs* ($\bar{4}2m \rightarrow mm2$) (# 20)

We begin our discussion of specific transitions with KDP which has been a workhorse of soft mode spectroscopy. Since KDP is the most widely studied example of a proper ferroelectric which is simultaneously a proper ferroelastic, it will serve as a convenient point of departure for the discussion of other linearly coupled proper ferroelastic transitions.

KDP and its isomorphs have been investigated extensively since the discovery of its ferroelectric transition in 1935 by Busch and Scherrer. Much of the work prior to 1962 was reviewed by Jona and Shirane (1962). Subsequently, a large number of additional publications have appeared, primarily concerned with the soft mode dynamics of these crystals.

I will not undertake a comprehensive review of KDP but will summarize the principle features of the currently accepted interpretation of dynamical phenomena which underlie the onset of spontaneous polarization and shear at $T_0$.

The prototype $\bar{4}2m$ tetragonal phase of KDP (space group $D_{2d}^{12}$) becomes unstable at 122 K against $B_2$ distortions and undergoes a transition to a mm2 polar orthorhombic phase (space group $C_{2v}^{19}$). Since both $z$ and $xy$ transform according to the $B_2$ irreducible representation of $\bar{4}2m$, a polar optic soft mode ($P_z$) and transverse acoustic soft mode ($\epsilon_6$) which are linearly coupled both condense at the transition. The linear coupling coefficient in this case is the piezoelectric constant $a_{36}$.

The microscopic dynamics of this transition are somewhat complicated. The essential instability is generally believed to be a collective excitation of tunneling protons which is linearly coupled to a $B_2$ optic lattice vibration. The lower hybrid branch of this mixed optical phonon-tunneling proton mode, in the absence of piezoelectric coupling to the $xy$ shear strain, obeys the Cochran (mean-field) behavior

$$\omega_1^2 = 42 \times (T - 117.3\ \mathrm{K})\ (\mathrm{cm}^{-1})^2 \tag{2.8}$$

(cf. Lagakos and Cummins (1974)). At atmospheric pressure this soft optic mode is overdamped. However, it becomes underdamped at hydrostatic pressure $\geqslant 6$ kbar (Peercy 1973).

Linear (piezoelectric) coupling of the soft optic mode to the transverse acoustic mode causes optic–acoustic hybridization forcing the elastic constant (nearly) to zero at $T_0 = 122$ K. The difference between the transition temperature $T_0$ and the Curie temperature $T_c$ is $\sim 0.1$ K since the transition is slightly first order. The difference between $T_c \cong 122$ K and the extrapolated zero of the uncoupled soft mode frequency at $T_c^x = 117.3$ K corresponds to the difference between the clamped and free Curie temperatures.

A purely thermodynamic argument originally proposed by Mason in 1946 shows that the elastic constant $C_{66}^E$ should drop towards zero near $T_0$ while the uncoupled elastic constant $C_{66}^P$ remains essentially temperature independent. Miller and Axe (1967) showed that in the presence of linear coupling, lattice dynamics predicts that the anomalous elastic constant should have the form $C(T) = C^0 - \Delta C/(T - T_c^x)$. An equivalent result follows from the elementary theory of coupled harmonic oscillators for the optic and acoustic modes:

$$\omega_2^2 = \omega_a^2 - (q^2 a_{36}^2/\rho m^* \omega_1^2)\,, \tag{2.9}$$

where $\omega_2$ and $\omega_a$ are the coupled and uncoupled acoustic mode frequencies ($\omega_a^2 = q^2 C_{66}^P/\rho$), $a_{36}$ is the piezoelectric coupling constant, and $m^*$ and $\omega_1$ are the effective mass and frequency of the optic soft mode (Brody and Cummins 1968).

Equation (2.9) is reasonably valid only in the limit $\omega_a \to 0$. Coupled-mode spectra should, in general, be analyzed with a fluctuation–dissipation approach which embodies the full frequency dependence of the spectrum rather than extracting the renormalized elastic constant (cf. Reese et al. (1973)). This approach was discussed in sect. 1.2.

The Brillouin scattering spectrum of KDP was investigated by Brody and Cummins (1968, 1974), and Lagakos and Cummins (1974). Figure 3 shows the change in position and lineshape of the Brillouin components as the temperature decreases from $T_0 + 7.2$ K to $T_0 + 0.13$ K. The solid lines show the fit obtained with coupled mode response functions. The strong central peak seen in fig. 3 was subsequently shown by Courtens (1978) to arise from annealable defects. The zero-field elastic constant is shown in fig. 4; the temperature and electric field dependence of $C_{66}^E$ is shown in fig. 5.

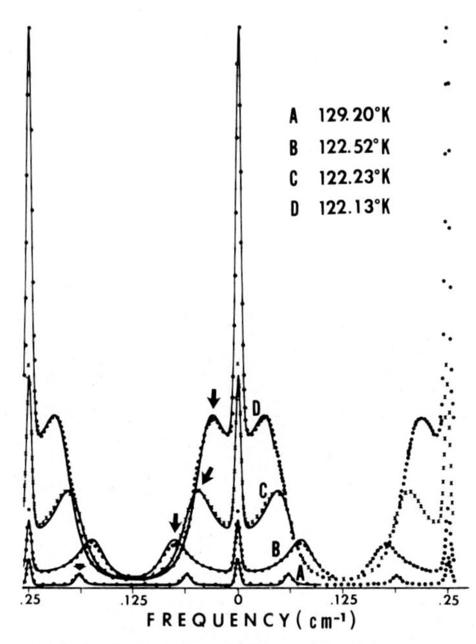


Fig. 3. Brillouin spectra of KDP at T = 129.29, 122.52, 122.23 and 122.13 K with $x + z$ $(x - z, y)$ $x - z$ scattering geometry. The arrows indicate the positions of the anti-Stokes Brillouin peaks of the order centered at zero. Each spectrum consists of 620 data points of which approximately $\frac{1}{4}$ are plotted (from Lagakos and Cummins (1974)).

The Brillouin components in fig. 3 are seen to decrease in frequency and to broaden strongly as the transition is approached. If these spectra are analyzed in terms of a single response function (eq. (1.1) and (1.2)), then the frequency $\omega_0$, damping constant $\gamma$ and coupling constant $P$ are all found to depend strongly on temperature. However, since the soft optic and acoustic mode in KDP are linearly coupled through the piezoelectric constant $a_{36}$, it is more appropriate to use the two-coupled-mode equation (1.3). The spectra were

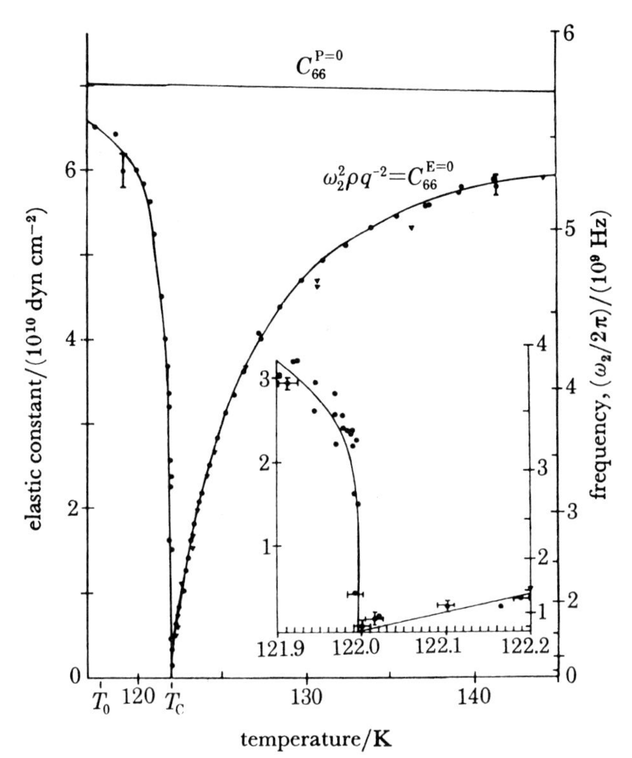


Fig. 4. Observed Brillouin shift ($\omega_2/2\pi$) and resulting elastic constant $C_{66}^{E=0}$ of the soft transverse acoustic mode in KDP (from Brody and Cummins (1968)).

found to fit eq. (1.3) with the soft optic mode frequency following the classic soft mode equation $\omega_1^2 \propto (T - T_c^x)$ where $T_c^x$ is the clamped Curie temperature. All other parameters could be taken as temperature independent. (The solid lines in fig. 3 were actually obtained from a three-oscillator fit, but a simple two-oscillator fit gives equally good fits for the Brillouin spectra, ignoring the central peak). Thus the temperature dependence of the optical soft mode frequency accounts for the observed temperature dependence of both the Brillouin shift and linewidth.

Courtens et al. (1979) and Courtens and Gammon (1981) have also investigated the Brillouin spectrum of KDP in an electric field. In this case, since the electric field already lowers the symmetry from tetragonal to orthorhombic, the phase transition observed with decreasing temperature occurs without further change of symmetry. Although there is a macroscopic elastic instability, there is *no* acoustic soft mode in this case (dimensionality class 0).

Various isomorphs of KDP have also been investigated by Brillouin scatter-

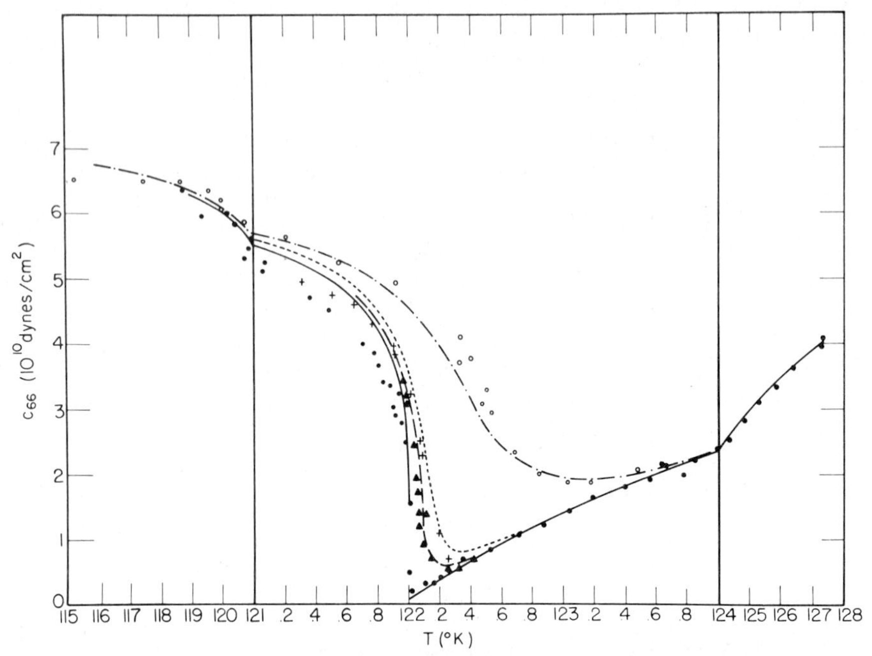


Fig. 5. Temperature and electric field dependence of the elastic constant $C_{66}^{E,S}$ deduced from Brillouin shift measurements. (●) $E = 0$; (▲) $E = 500$ V/cm; (+) $E = 787$ V/cm; (○) $E = 3937$ V/cm. The lines are predictions of the SUS theory with $\epsilon_0/k = 61.78$ K, $\beta/k = 14.7$ K, $\epsilon_1/k = 475.0$ K. ____ ($E = 0$); --- ($E = 500$ V/cm); ----- ($E = 787$ V/cm); –·–·– ($E = 3937$ V/cm) (from Brody and Cummins (1974)).

ing with generally similar results. These include $KD_2PO_4$ (potassium dideuterium phosphate – DKDP) (Reese et al. 1973, Tanaka and Tatsuzaki 1981); $KH_2AsO_4$ (potassium dihydrogen arsenate – KDA), (Durvasula and Gammon 1976, 1977); $RbH_2PO_4$ (rubidium dihydrogen phosphate – RDP), (Hauret et al. 1971); $CsH_2AsO_4$ (cesium dihydrogen arsenate – CsDA) and $CsD_2AsO_4$ (cesium dideuterium arsenate – DCsDA) (Azoulay et al. 1977, Lagakos and Cummins 1975). In the deuterated compounds the soft optic and acoustic modes fall in the same frequency range making full coupled mode analyses particularly important. Tanaka and Tatsuzaki (1981) studied Brillouin scattering in deuterated KDP at different scattering angles. They observed dispersion of the soft acoustic mode frequency as well as a dynamic central peak in the paraelectric phase.

Reese et al. (1973) performed two different light scattering experiments on $KD_2PO_4$. In one, with scattering geometry $x\,(y, x)\,y$, the soft $B_{2g}$ optic mode was seen but not the soft acoustic mode since as shown in table 2, $q$ must be along $x$ or $y$ for the soft acoustic mode. The second geometry

$[x+z\,(-x+z, y)\,-x+z]$ allowed coupling to both the soft optic and acoustic modes. Soft optic mode parameters deduced from the first geometry were used as input parameters for coupled mode analysis of the second geometry.

Finally, we note that both static and dynamic central peaks occur in KDP and they have been studied extensively. Discussion of the central peaks can be found in ch. 7 by Fleury and Lyons in this volume.

*$KH_3(SeO_3)_2$: Potassium trihydrogen selenite (KTS) ($mmm \rightarrow 2/m$) (# 7)*
Potassium trihydrogen selenite and its deuterated isomorph $KD_3(SeO_3)_2$ undergo second-order ferroelastic phase transitions at 211.8 K and 300.2 K respectively from a high-temperature orthorhombic structure (space group $D_{2h}^{14}$) to a low-temperature monoclinic structure (space group $C_{2h}^{5}$). The $B_{3g}$ transverse acoustic mode with $\boldsymbol{q} \parallel [010]$ and $\boldsymbol{u} \parallel [001]$, governed by the elastic constant $C_{44}$, has been shown to approach zero as $T \rightarrow T_c$ in Brillouin scattering experiments by Yagi et al. (1976a) who also found an annealable central peak in KTS (Yagi et al. 1977, Tanaka et al. 1978). Figure 6 from Yagi et al. (1977) illustrates the "history dependence" of the central peak as well as the temperature dependence of the acoustic modes. Later experiments by Tanaka and Tatsuzaki (1979) with partially deuterated KTS showed that as the deuterium concentration is increased the central peak intensity increases in direct proportion to the increase in the transition temperature.

Copic et al. (1981), in a Brillouin scattering study which included applied uniaxial stress, showed that the shear elastic constant in the unstressed crystal is remarkably classical, obeying the Curie–Weiss behavior to within 0.01 K of the transition as shown in fig. 7. (The classical behavior is consistent with the consequences of the "dimensionality class I" property of the elastic soft mode.) Elastic anomalies near the transition have also been studied by ultrasonic techniques (cf. Ivanov et al. (1975), Makita et al. (1977)) with similar results.

The microscopic theory of the KTS phase transition was discussed by Blinc et al. (1980). The large isotope effect indicates that the hydrogen bonds which link the $SeO_3$ groups play a significant role in the transition, also possibly including some coupling to lattice ion displacements in analogy to KDP. However, in contrast to KDP no soft optic mode has been observed and Blinc et al. have analyzed the KTS transition using a Hamiltonian with bilinear coupling between the soft pseudospin mode and the $B_{3g}$ transverse acoustic mode only.

*$LaP_5O_{14}$: Lanthanum pentaphosphate and isomorphs ($mmm \rightarrow 2/m$) (# 7)*
The rare-earth pentaphosphates $ReP_5O_{14}$ (with Re representing any rare earth from La to Tb or mixtures of these elements) undergo second-order proper ferroelastic transitions from orthorhombic ($Pcmn = D_{2h}^{16}$) to monoclinic ($P2_1/c = C_{2h}^{5}$) symmetry driven by a soft $B_{2g}$ zone-center optic mode which is

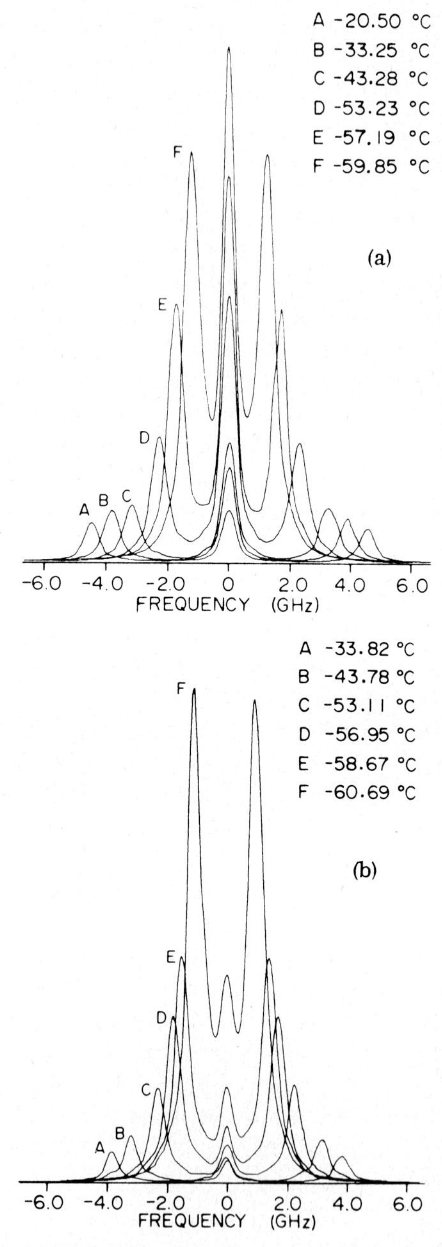


Fig. 6. The central peak of $KH_3(SeO_3)_2$. Each spectrum is plotted on-line on a chart of the XY plotter after a simple dynamic average on a computer. (a) The Brillouin spectra associated with the soft acoustic mode $x_1$ observed with the *as-grown* sample. (b) The spectra observed after keeping the sample below $-30°C$ for about 24 h (from Yagi et al. (1977)).

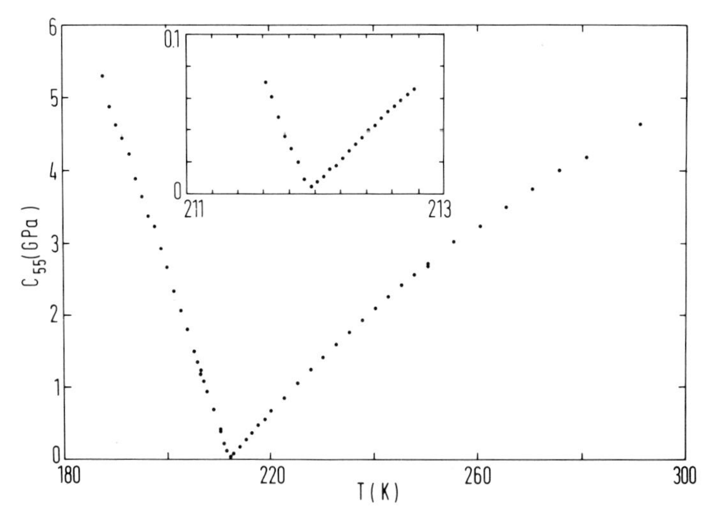


Fig. 7. The $C_{55}$ shear elastic constant in a free $KH_3(SeO_3)_2$ crystal (from Copic et al. (1981)).

linearly coupled to $\epsilon_5$ shear strain. The transition is similar to that in $KH_3(SeO_3)_2$, and is also expected to follow the classical Landau theory predictions appropriate to class I transitions in its critical behavior.

Brillouin scattering measurements on $LaP_5O_{14}$ demonstrating the complete softening of $C_{55}$ at $T_c = 398$ K were reported by Toledano et al. (1976b), and by Errandonea and Bastie (1978) who also studied the temperature dependence of the spontaneous strain $\epsilon_5$. Both the Raman and Brillouin spectra of the mixed crystal $Nd_{0.5}La_{0.5}P_5O_{14}$ have been studied by Peercy (1976, 1978) who also investigated the pressure dependence of the soft optic mode.

Recently, Errandonea (1980) has presented new Brillouin scattering results on $LaP_5O_{14}$ along with a complete review of the literature on this transition. He has emphasized the importance of self-consistent analysis of the critical behavior, and has presented a comprehensive tabulation of temperature dependent properties of this crystal. The spontaneous strain $\epsilon_5$ was found to accurately follow a $(T_c - T)^\beta$ behavior over the range $0.02 \leq T_c - T \leq 10$ K with $\beta = 0.500 \pm 0.007$. He has also recently reported a central peak in this crystal (Errandonea 1981). For an extensive discussion of the phase transitions in these crystals including a review of the theoretical models, the reader should consult sect. 4.1 of ch. 5 by Scott in this volume.

*$PrAlO_3$: Praseodymium aluminate*

Praseodymium aluminate is a cubic perovskite above 1643 K where it undergoes a cubic-trigonal transition driven by a soft $R_{25}$ zone-corner optic mode. A first-order trigonal–orthorhombic transition occurs at 205 K, and a second-order orthorhombic–monoclinic transition at 151 K. The 151 K transition can be described as a cooperative Jahn–Teller transition similar to that in $DyVo_4$.

In 1974, Brillouin scattering experiments revealed a new second-order transition at 118.5 K driven by a soft transverse acoustic mode (Fleury et al. 1974). Since previous studies by numerous other techniques had revealed no anomalies in the properties of this crystal at 118.5 K while the Brillouin scattering indicated a nearly complete acoustic softening, Fleury et al. identified this transition as purely ferroelastic, i.e., as a transition in which strain is the sole order parameter.

An alternative explanation was suggested by Harley (1977) based on the temperature dependence of the lowest-frequency Raman active mode. Harley argued that the acoustic anomaly at the 118.5 K transition results from linear coupling of the transverse acoustic mode to temperature dependent optic modes, and that the temperature dependence of the optic modes is itself due to the changing populations of the $Pr^{3+}$ electronic energy levels. The order parameter would then consist of a linear combination of rotations of $AlO_6$ groups about (010) with $(\epsilon_4 + \epsilon_5)$ shear strain. Harley's explanation has also been contested, however, and the ultimate source of the 118.5 K transition remains an open question. Other aspects of this transition are discussed in sect. 4.4 of ch. 5 by Scott in this volume.

*$DyVO_4$: Dysprosium vanadate*; *$TbVO_4$: Terbium vanadate* ($4/mmm \rightarrow mmm$) (# 25)

These rare-earth vanadates belong to a class of crystals which undergo "pure" ferroelastic phase transitions from the point of view of lattice dynamics. The microscopic origin of the acoustic anomaly is the cooperative Jahn–Teller effect which couples elastic strain to the degenerate electronic energy levels of the rare-earth ions. The theory of this transition has been treated extensively in the literature, initially by Kanamori (1960) and subsequently, with the inclusion of both static strain and acoustic phonons, in an extensive series of papers by Elliott and coworkers (cf. Elliott et al. (1972), Sandercock et al. (1972)) and by Pytte (1971). The lifting of electronic degeneracy by elastic distortion is cooperative since the ions interact through the common strain field.

The existence of acoustic instabilities at these transitions was predicted in 1971 by Elliott et al. (1971) and Pytte (1971) and was first observed experimentally in $DyVO_4$ in 1972 in an ultrasonic resonance experiment (Melcher and Scott 1972). Ultrasonic techniques have also been applied to the study of other crystals exhibiting cooperative Jahn–Teller transitions including $TmVO_4$ (Melcher et al. 1973) and nickel zinc chromite (Kino et al. 1972).

Brillouin scattering has been utilized in the study of both $DyVO_4$ and $TbVO_4$ (Sandercock et al. 1972, Harley et al. 1980). Both of these crystals have the tetragonal $D_{4h}^{19}$ (zircon) structure in the high-temperature phase. However the structure of the orthorhombic phase of $DyVO_4$ is $D_{2h}^{28}$ and that of $TbVO_4$ is $D_{2h}^{24}$.

As shown in table 2 ( # 25) the 4/mmm→mmm second-order transition can be driven either by $\epsilon_1 - \epsilon_2$ strain (symmetry $B_{1g}$) or $\epsilon_6$ (symmetry $B_{2g}$). The different electronic ground states of the $Dy^{3+}$ and $Tb^{3+}$ ions result in coupling to the $B_{1g}$ and $B_{2g}$ strains, respectively. Consequently, the soft acoustic mode in $TbVO_4$ is a transverse mode in the $xy$-plane with $\boldsymbol{q}$ along [100] or [010], with soft elastic constant $C_{66}$; the soft acoustic mode in $DyVO_4$ is also a transverse mode in the $xy$-plane with $\boldsymbol{q}$ along [110] or [$1\bar{1}0$], with soft elastic constant $\frac{1}{2}(C_{11} - C_{12})$.

Brillouin scattering experiments in these crystals are complicated by the symmetry of the soft modes which do not scatter light in the backscattering configuration employed by Sandercock et al. (1972) and by the extreme anisotropy of the electron–phonon coupling. Furthermore, scattering from the electronic excitations is important, so that the observed spectra must be treated in a full coupled mode analysis.

Harley et al. (1980) have studied the Brillouin spectrum of $TbVO_4$ in two geometries. In $[y(xy)x]$ geometry the almost "bare" electronic response is measured. In the second geometry (VH with $\boldsymbol{q} \parallel [100]$) both the electronic and acoustic modes are observed (see fig. 8). Harley et al. have analyzed this data with a two-coupled oscillator model and obtained excellent fits to the data which was obtained with an iodine cell that eliminates elastic scattering. For a discussion of the central peak in $TbVO_4$, see ch. 7 by Fleury and Lyons in this volume.

Elliott has shown that the formalism for the cooperative Jahn–Teller transition in $TbVO_4$ is isomorphic with that for the pseudo-spin induced transition in KDP. Interestingly, the two geometries utilized by Harley et al. for $TbVO_4$ (fig. 8) correspond to the two geometries used by Reese et al. (1973) to study the coupled optic and acoustic modes in $KD_2PO_4$. In both cases the soft acoustic mode is $C_{66}$ while the excitation to which it couples is of $B_{2g}$ symmetry.

*$LiNH_4C_4H_4O_6 \cdot H_2O$: Lithium ammonium tartrate monohydrate (LAT)*(222→2) ( # 4)

Lithium ammonium tartrate (LAT), which is structurally and chemically closely related to Rochelle salt, was found to be ferroelectric in 1951. Its structural and dielectric properties have been studied extensively (Jona and Shirane 1962). The orthorhombic–monoclinic ferroelectric transition ($D_2^3 \rightarrow C_2^2$) occurs at $T_c = 98$ K. Below $T_c$, LAT develops spontaneous polarization along the $b$-axis and exhibits Curie–Weiss susceptibility with an unusually small Curie constant ($C \sim 2$ K). For this transition the strain $\epsilon_5$ has the same $B_2$ symmetry as the order parameter ($P_y$), indicating an acoustic soft mode associated with the transition. Indeed, a large anomaly in the $C_{55}$ elastic constant does occur.

Sawada et al. (1977) and Udagawa et al. (1978a, b) reported Brillouin scattering studies of LAT in the orthorhombic phase which show that both $C_{55}^E$ and $C_{55}^P$ approach zero, at $T_c$ and $T_0 \cong T_c - 4$ K, respectively. They also

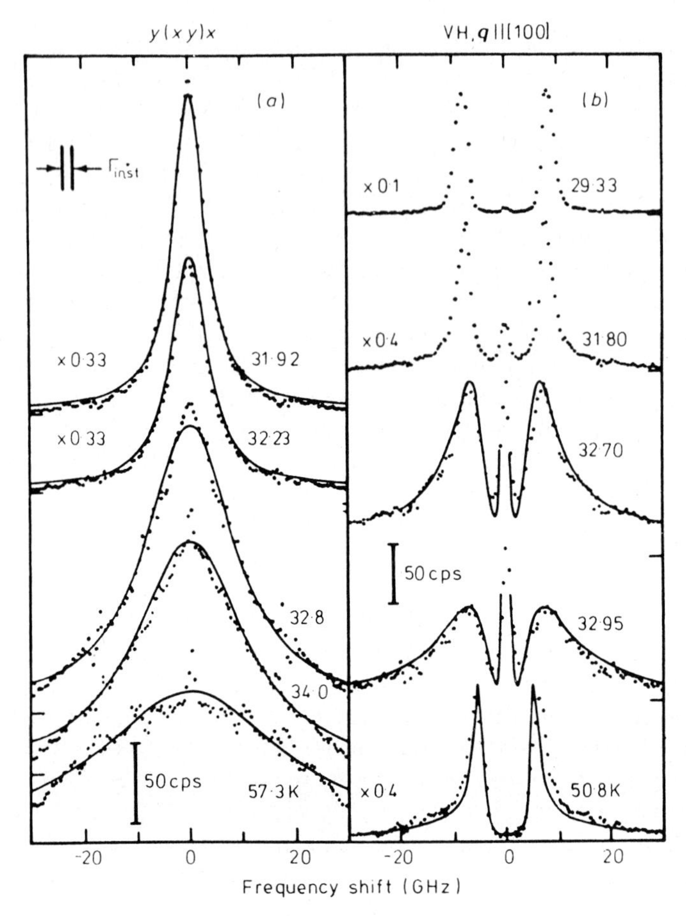


Fig. 8. Brillouin scattering spectra of $TbVO_4$ in (a) uncoupled, and (b) coupled geometries (from Harley et al. (1980)).

performed dielectric-susceptibility measurements which indicated that while the free susceptibility $\chi_2^X$ diverges at $T_0$, the clamped susceptibility $\chi_2^x$ measured at 2 MHz is essentially temperature independent.

The essential implication of these results follows from a simple thermodynamic analysis of the linearly coupled polarization and strain, or of the equivalent linearly coupled equations of motion (Sawada et al. 1977, Brody and Cummins 1968). If the primary instability is in the (polar) optical mode, then the free dielectric susceptibility diverges at $T_c$ and the clamped dielectric constant diverges at $T_0$; the shorted elastic constant $C^E$ approaches zero at $T_0$ while $C^P$ is temperature independent. This is the situation in the KDP-type crystals, for example, where the elastic instability can be considered as a secondary effect of the dielectric instability, induced by the piezoelectric coupling.

The results of Sawada et al. for LAT suggest, on the contrary, that the primary instability is *elastic*, with the dielectric instability appearing as a secondary effect of the elastic instability, again induced by the piezoelectric coupling. As mentioned at the beginning of sect. 2, this distinction (class ii vs class iii) between the two classes of linearly coupled ferroelastic transitions is irrelevant from the point of view of symmetry since both the optic and acoustic modes condense together at the transition. However, the dynamics as well as the relevant response functions are clearly different in the two classes. It appears that LAT is the only case reported so far as belonging to class (iii).

Raman scattering studies of LAT have not revealed an optic soft mode, although an anomalous low-lying response was found (Udagawa et al. 1979). It has been proposed that LAT is an order–disorder ferroelectric in which the ordering occurs due to coupling to the soft acoustic mode which is the primary instability (A. Sawada – private communication).

*$TeO_2$: Paratellurite* (422→222) ( # 14)
In 1974, Peercy and Fritz discovered that paratellurite undergoes a phase transition at room temperature under hydrostatic pressure of 9.0 kbar. Their Brillouin scattering experiments revealed an acoustic soft mode with $\boldsymbol{q}$ along [110] and $\boldsymbol{u}$ along [1$\bar{1}$0] whose frequency approaches zero at the transition. The soft elastic constant $C = \frac{1}{2}(C_{11} - C_{12})$ also decreases slightly with decreasing temperature, but the transition cannot be induced by temperature at atmospheric pressure.

The low-pressure tetragonal and high-pressure orthorhombic structures are $D_4^4$ and $D_2^2$, respectively. Since optic modes with the symmetry of the order parameter ($B_1$) are Raman active in the high-symmetry phase, the temperature dependence of the Raman spectra should prove if the acoustic instability arises from linear coupling of $(\epsilon_1 - \epsilon_2)$ strain to an optic soft mode. Peercy and Fritz (1974) and Peercy et al. (1975) performed extensive Raman and Brillouin scattering, ultrasonic and dielectric studies of paratellurite. Their results indicate that there is no soft optic mode, and that the transition is class (i), or pure ferroelastic. It is apparently second order, and the temperature dependence of the soft mode elastic constant is consistent with mean-field theory, as expected for dimensionality class (i) transitions (Fritz and Peercy 1975).

Peercy et al. (1975) note the resemblance of the strain-induced transition in paratellurite to the 118.5 K transition in praseodymium aluminate which we discussed earlier. However, the $PrAlO_3$ transition may be class (ii) with the acoustic anomaly resulting from linear coupling to a soft optic mode, as proposed by Harley (1977). In the case of paratellurite, however, the possibility of an optic soft mode seems to be ruled out by the Raman results of Peercy et al. All the $B_1$ optic modes were observed, and their frequencies were all found to increase with increasing pressure throughout the tetragonal phase.

A discussion of this transition based on a microscopic model of the coupling of strains to internal displacements was given by Uwe and Tokumoto (1979) who also investigated the third-order elastic constants of $TeO_2$.

*$BiVO_4$: Bismuth vanadate* ($4/m \rightarrow 2/m$) (# 12)
Since Bierlein and Sleight (1975) discovered the ferroelastic phase transition in bismuth vanadate at $T_0 = 528$ K, extensive X-ray and neutron diffraction studies have supported their interpretation of the tetragonal 4/m to monoclinic 2/m transition as a proper ferroelastic transition with paraelastic and ferroelastic space groups $C_{4h}^6(I4_1/a)$ and $C_{2h}^6(I2/b)$ respectively. Furthermore, Raman scattering experiments by Pinczuk et al. (1977, 1978, 1979) revealed a temperature dependent low-frequency zone-center optic mode with $B_g$ symmetry in the 4/m paraelastic phase and $A_g$ symmetry in the 2/m ferroelastic phase, consistent with the requirement of group theory for the soft mode driving this transition (see table 2). It has therefore been generally accepted that the transition is driven by a $B_g$ zone-center soft optic mode which in turn forces the frequency of a soft acoustic mode to which it is linearly coupled to zero at $T_0$, triggering the transition.

There are, however, two possible problems with this interpretation:

(1) The amount of softening of the "soft" optic mode is rather small, and its frequency extrapolates to zero at more than 200 K below $T_0$. Furthermore, the Raman spectrum of $LaNbO_4$ (which is isostructural to $BiVO_4$ and exhibits a similar ferroelastic transition) shows no temperature dependence for the frequency of the equivalent $B_g$ mode above $T_0$ (Wada et al. 1979).

(2) Dudnik et al. have concluded that the $BiVO_4$ transition is cell doubling ($I4_1/c \rightarrow P2_1/c$) which would make it an *improper* ferroelastic transition (Dudnik et al. 1979). We note that the paraelastic phase has the body-centered tetragonal scheelite structure in all analyses; the difference in structure assignments occurs in the ferroelastic phase where the unit cell volume would double if the $VO_4$ group at the center of the unit cell is slightly rotated or translated relative to (and therefore no longer equivalent to) the $VO_4$ groups at the cell corners, a distinction which is very difficult to establish by structural analysis alone.

Recently, Gu et al. (1981) reported the observation by Brillouin scattering of a soft acoustic mode in $BiVO_4$, and of the temperature dependent splitting of a pair of $B_g$ optic modes in the Raman spectrum in the ferroelastic phase which arise from a doubly degenerate $E_g$ mode in the paraelastic phase. These observations demonstrate the correctness of the proper ferroelastic interpretation of this transition, in agreement with X-ray and neutron diffraction results except for those of Dudnik et al. In their experiments, right angle Brillouin scattering in the *xy*-plane was excited by a krypton ion laser at 6471 Å and analyzed with a piezoelectrically scanned Fabry–Perot interferometer. A

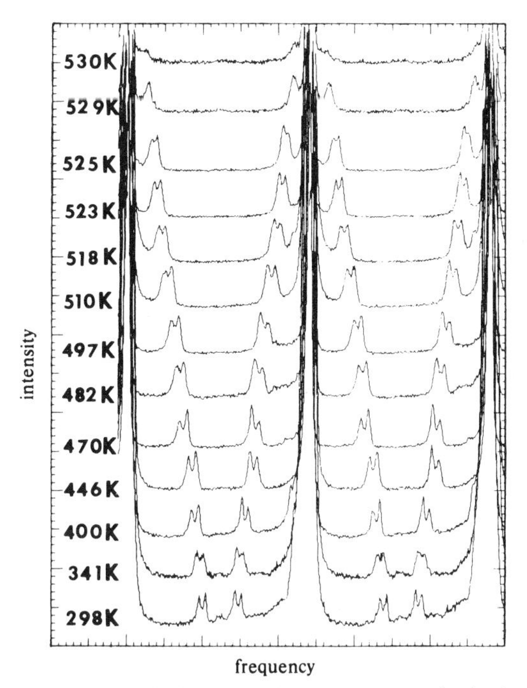


Fig. 9. Brillouin scattering spectra of $BiVO_4$ at various temperatures in the ferroelastic phase with VV polarization. The free spectral range is $0.92\ cm^{-1}$. The splitting of the Brillouin components is associated with the presence of domains (Gu et al. 1981).

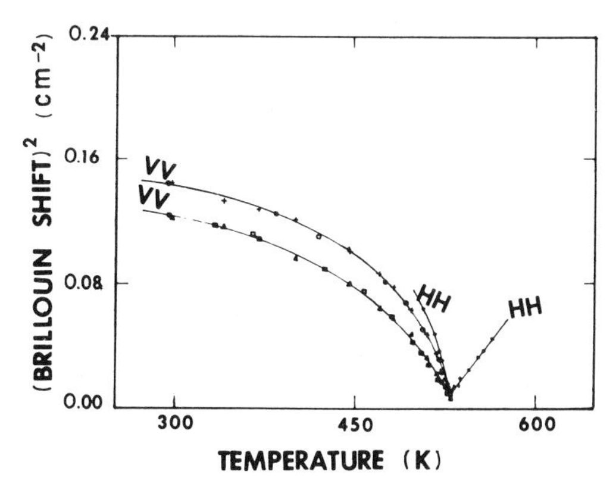


Fig. 10. Observed Brillouin shifts for the soft acoustic mode in $BiVO_4$ as a function of temperature. Different symbols represent different experimental runs (Gu et al. 1981).

typical set of Brillouin spectra in VV polarization for $T < T_0$ is shown in fig. 9.

With the scattering vector $\boldsymbol{q}$ perpendicular to one set of domain walls which are oriented at approximately 30° to the pseudo-tetragonal axes, the Brillouin shift of the in-plane transverse mode decreased from $0.4\,\mathrm{cm}^{-1}$ at room temperature to about $0.1\,\mathrm{cm}^{-1}$ at $T_0$ and then increased as $(T - T_0)^{1/2}$ up to 570 K, the maximum temperature investigated (see fig. 10).

In the ferroelastic phase the Brillouin components observed in VV polarization are split, presumably due to the relative rotation by several degrees of the optical indicatrices in the two types of domains due to the spontaneous piezo-optic effect. The transmitted beam emerged from the sample as two slightly divergent beams for $T < T_0$, but as a single beam for $T > T_0$ when the domain walls are absent.

Standard analysis of the Brillouin scattering tensors predicts Brillouin activity for this mode only for HH polarization in the paraelastic phase, but for both HH and VV polarizations in the ferroelastic phase, consistent with the observations. The analysis also predicts that the angle $\phi$ between the direction of propagation of the soft acoustic mode and the tetragonal axes is given by $\tan(4\phi) = 4C_{16}/(C_{11} - C_{12} - 2C_{66})$, and that the velocity $v$ of the soft acoustic mode is determined by

$$\rho v^2 = \tfrac{1}{4}(C_{11} - C_{12} + 2C_{66} - [(C_{11} - C_{12} - 2C_{66})^2 + 16C_{16}^2]^{1/2}) .$$

We note that the stability condition based on the vanishing of an eigenvalue of the 6 × 6 elastic constant tensor predicts that the transition occurs when the critical constant $C_c$ reaches zero, where

$$C_c = \tfrac{1}{2}(C_{11} - C_{12} + C_{66} - [(C_{11} - C_{12} - C_{66})^2 + 8C_{16}^2]^{1/2})$$

(Boccara 1968). Although $\rho v^2$ of eq. (2.4) and $C_c$ of eq. (2.2) vanish simultaneously, they are not generally equal, as discussed in sect. 2.1.

The failure of the observed soft acoustic mode frequency to reach zero at the transition may be due either to slight misalignment of the sample or to the transition being slightly first order. But its temperature dependence is entirely inconsistent with a zone-boundary instability driven improper ferroelastic transition where the acoustic anomaly would be weak and extremely asymmetric. The distinction is immediately apparent in fig. 14 which shows the results of Gu et al. for $BiVO_4$ and the results of Rehwald (1978) for squaric acid, a crystal which also undergoes a 4/m to 2/m transition, but with a doubling of the volume of the unit cell.

## 3. *Proper optic ferroic transitions* (*B*)

This category includes all transitions driven by a zone-center soft optic mode in the prototype phase $G_0$ which transforms according to an irreducible representation of $G_0$ different from that of any strain, so that proper ferroelasticity is symmetry forbidden. Spontaneous strain may nevertheless occur through higher-order coupling of strains to the order parameter. The possible improper spontaneous strains associated with such transitions are tabulated in Janovec et al. (1975).

If the improper strain would *by itself* break the prototype symmetry, then the transition is an improper ferroelastic one. Barium titanate, for example, undergoes a transition at 393 K from a cubic m3m prototype phase to a tetragonal 4mm phase. This is a proper ferroelectric transition driven by a zone center $T_{1u}$ soft optic mode. Indirect coupling to strains induces improper spontaneous strains $\epsilon_1 = \epsilon_2 = -2\epsilon_3$. This strain would, by itself, break the cubic symmetry also, resulting in the tetragonal point group 4/mmm (transition # 80). The transition is thus a proper ferroelectric/improper ferroelastic one (Janovec et al. (1975) table 5). Note that the ferroic point group (4mm) is of lower symmetry than that which would arise from the strain alone (4/mmm). Transitions of this type are discussed in sect. 3.1.

If, however, the improper strain is not by itself symmetry breaking, then the transition is *not* ferroelastic. Thus for example, in the case of quartz (622→32) the allowed spontaneous strains are ($\epsilon_1 = \epsilon_2, \epsilon_3$) which would not by themselves modify $G_0$. (These transitions are grouped together in sect. 3.2.) The classification of such higher-order transitions based on the rank of the lowest-rank polar tensor characterizing the different orientation states of the ferroic crystal has been discussed extensively by Newnham and Cross (1974) and by Toledano and Toledano (1977).

In those transitions which are improper ferroelastics, the transition results in a low-symmetry space group which is always of lower symmetry than that which would have been induced by the improper strain alone (J.C. Toledano, private communication). Similarly, in the case of improper ferroelectric transitions, the spontaneous order parameter reduces the symmetry further than would the appearance of spontaneous polarization alone; the transition which induces improper ferroelectricity is not to the maximal polar subgroup of the space group of the prototype phase (Levanyuk and Sannikov 1974).

### *3.1. Improper ferroelastics ($B_1$); Barium Titanate (m3m→4mm)*

The ferroelectric transition in barium titanate has been studied exhaustively since its discovery in 1945. Jona and Shirane (1962) reviewed the structural, dielectric, optical, thermal, piezoelectric and elastic properties; Scott (1974) reviewed the X-ray, neutron and Raman scattering experiments.

Barium titanate undergoes three distinct phase transitions. At 393 K the prototype cubic phase (m3m) undergoes a transition to a ferroelectric tetragonal phase (4mm) with the onset of spontaneous polarization ***P*** in one of six equivalent [100] directions. At 278 K a second transition occurs to a ferroelectric orthorhombic phase (mm) with ***P*** along a [110] direction; at 183 K, a final transition takes place to a rhombohedral phase (3m) with ***P*** along a [111] direction.

The m3m→4mm transition at 393 K is driven by a soft $T_{1u}$ (or $F_{1u}$) polar optic mode. The temperature dependence of the soft mode in the cubic prototype phase has recently been determined by infrared reflectivity measurements by Luspin et al. (1980a,b) Although the transition is first order, a continuous transition is allowed by symmetry. Second-order coupling of the soft mode to strains $U_1$ ($\epsilon_1 = \epsilon_2 = -2\epsilon_3$) and $\delta U_c$ ($\epsilon_1 = \epsilon_2 = \epsilon_3$) is allowed by symmetry (Janovec et al. 1975). $\delta U_c$ corresponds to thermal expansion in the cubic phase and is not symmetry changing, but $U_1$ would produce tetragonal distortion. The transition is thus proper ferroelectric and improper ferroelastic.

The importance of coupling between the soft optic mode and the acoustic modes, as well as the extremely anisotropic dispersion of the soft optic mode was revealed in the inelastic neutron scattering studies of Harada et al. (1971). This coupling gives rise to remarkable features in the Raman–Brillouin spectrum in the tetragonal phase where the broken symmetry converts the third order coupling of the cubic phase ($\epsilon P^2$) into an effective bilinear coupling. The temperature dependence of the Brillouin shifts and linewidths, as well as a very large increase in the Brillouin scattering intensity below $T_0$ were reported by Fleury and Lazay (1971) and Lazay and Fleury (1971). The pressure dependence of the coupled acoustic and optic mode spectrum was investigated by Peercy and Samara (1972). Recently, Tominaga et al. (1980) have investigated the dispersion in the velocity of the $C_{44}$ TA mode propagating in the [101] direction.

### *3.2. Non-ferroelastics ($B_2$); Quartz; Aluminum phosphate; Ammonium Sulfate; Lead germanate*

*$SiO_2$: Silicon dioxide (quartz) (622→32)*

Quartz undergoes a structural phase transition at 846 K from a low-temperature trigonal α phase (space group $D_3^4$ or $D_3^6$) to a high-temperature hexagonal β phase (space group $D_6^4$ or $D_6^6$). The transition which is slightly first order is driven by a soft zone-center optic mode with $B_1$ symmetry in the high-temperature β phase whose frequency, as determined by inelastic neutron scattering, exhibits the "Cochran law" behavior $\omega_0^2 \propto (T - T_c)$ with $T_c \cong 836$ K (Axe and Shirane 1970). Raman scattering investigations of the α phase and inelastic neutron scattering investigations of both α and β phases were reviewed by Scott (1974).

Since the $B_1$ soft optic mode transforms neither like a polar vector nor like a second-rank tensor, the transition can be neither proper ferroelectric nor proper ferroelastic. The allowed spontaneous improper strains in the $\alpha$ phase, arising from higher-order coupling to the order parameter, are shown in table 1 of Janovec et al. (1975) to be $\epsilon_1 = \epsilon_2$, $\epsilon_3$. Since these strains would not by themselves break the hexagonal symmetry, the transition is *not* an improper ferroelastic one either.

Following the work of Newnham and Cross (1974), Toledano and Toledano (1977) showed that for this transition (which they classify as ferroelastoelectric and ferrobielastic) the spontaneous tensorial quantities allowed by symmetry are the piezoelectric modulus $d_{11}$ and the elastic constant $C_{14}$. The coupling to $C_{14}$ arises from terms in the free energy linear in the order parameter and quadratic in the strain which cause $C_{14}$, which is zero in the $\beta$ phase, to be proportional to the order parameter in the $\alpha$ phase (Rehwald 1973).

The elastic constants of quartz have been investigated by ultrasonic experiments (cf. Hochli and Scott (1971)) which were reviewed by Rehwald (1973). Inelastic neutron scattering experiments by Axe and Shirane (1970) revealed a downward bending of $C_{11}$, $C_{33}$, $C_{12}$ and $C_{13}$ above the transition which was analyzed as anharmonic renormalization of the acoustic modes through virtual excitation of pairs of optical phonons of $+\boldsymbol{q}$ and $-\boldsymbol{q}$ on the soft mode branch.

Brillouin scattering studies of the quartz $\alpha$–$\beta$ transition were reported by Shapiro and Cummins (1968), Shapiro (1968) and Pelous and Vacher (1976). These experiments are in substantial agreement with the ultrasonic and neutron scattering studies. They reveal a gradual downward trend in $C_{11}$ and $C_{33}$ as the crystal is heated towards the transition, a discontinuous upward step at $T_0$ and gradual leveling off in the $\beta$ phase. Hysteresis of about 1° was also observed (Shapiro and Cummins 1968) consistent with the transition being slightly first order.

Figure 11 from Shapiro (1968) shows the temperature dependence of four elastic constants of quartz determined by Brillouin scattering along with results obtained by ultrasonic methods.

*$AlPO_4$: Aluminium phosphate (berlinite)* (622→32)

Aluminum phosphate is isomorphous to quartz ($SiO_2$) and undergoes an $\alpha$–$\beta$ phase transition closely resembling that in quartz. Ecolivet and Poignant (1981) have measured the Brillouin spectrum of aluminum phosphate from room temperature to 873 K. They have extracted the elastic constant $C_{33}$ and plotted it against temperature.

When $C_{33}$ is divided by its value at room temperature and this ratio is plotted against the ratio $T/T_c$, they find that the results for aluminum phosphate and quartz superimpose almost perfectly.

*$(NH_4)_2SO_4$: Ammonium sulfate* (*mmm*→*mm*2)

Ammonium sulfate (AS) undergoes a ferroelectric phase transition at 223.6 K between two orthorhombic structures. The prototype $D_{2h}^{16}$ and ferroelectric $C_{2v}^{9}$

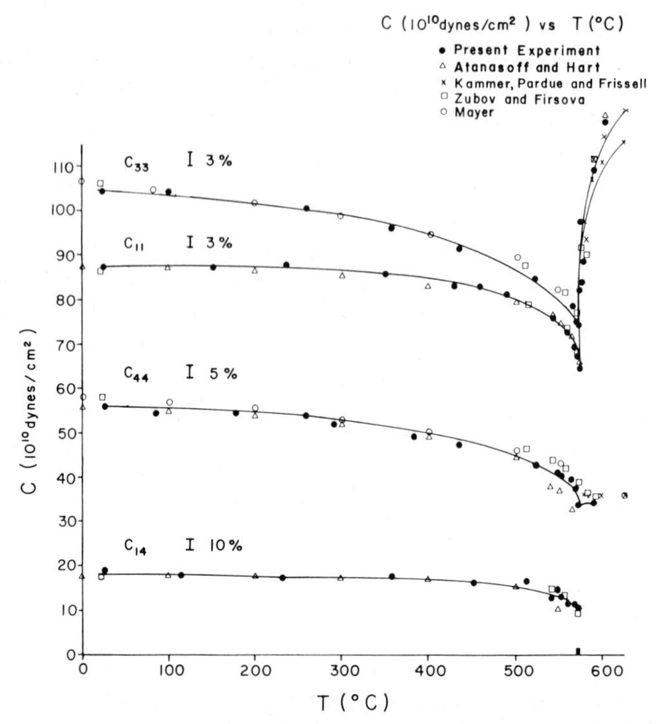


Fig. 11. $C_{33}$, $C_{11}$, $C_{44}$ and $C_{14}$ ($10^{10}$ dynes/cm²) vs $T$ (°C) from Brillouin scattering measurements. Comparison is made with the elastic constants measured by ultrasonic techniques (from Shapiro (1968)).

differ in the loss of a mirror plane perpendicular to the $C$-axis which becomes the ferroelectric axis. The soft mode which would connect these phases (mmm→mm2) and produce $Z$ polarization is $B_{1u}$ with indirect coupling permitted to the elastic strains $\epsilon_1$, $\epsilon_2$ and $\epsilon_3$ (Janovec et al. 1975).

Although the ferroelectric transition in AS has been studied extensively since its discovery in 1956, the mechanism of the transition remains unclear. Ferroelectric AS is often described as a ferrielectric structure in which the polarization is attributed to two crystallographically inequivalent and oppositely polarized sublattices. An underlying soft mode of $B_{1u}$ symmetry is assumed to be anharmonically coupled to the polarization which then appears as an indirect consequence of the fundamental transition (cf. Yoshihara et al. (1978), Luspin (1976)). However, the soft $B_{1u}$ mode has not been found spectroscopically whereas temperature dependent $A_g$ features have been seen, leading to the suggestion that both $A_g$ and $B_{1u}$ optic modes participate in the transition (Unruh et al. 1978a).

AS is also ferroelastic in both phases, and the ferroelasticity has been attributed to an instability at the M point of a hypothetical hexagonal $D_{6h}^4$ phase which could exist above the melting point. In this sense AS is similar to tris-sarcosine calcium chloride (TSCC) which will be discussed in sect. 4.1 (cf. Smolenskii et al. (1979)). TSCC is currently thought to undergo both an order–disorder transition and an improper displacive transition at nearly the same temperature. However, the evidence for the displacive transition in TSCC being improper is not conclusive, and the transitions in these two ferroelectric crystals may be very similar.

The acoustic properties of AS in the transition region are difficult to determine because AS tends to develop cracks in the ferroelectric phase. The first Brillouin scattering investigations of AS were therefore restricted to the prototype phase (Luspin et al. 1974, Luspin 1976).

Yoshihara et al. (1978) managed to overcome the strong elastic scattering due to cracks by the use of a differential interferometer technique and found a number of step-type anomalies in sound velocities near the transition. For longitudinal sound waves propagating along the $X$-axis they observed a particularly strong anomaly; the sound velocity gradually decreased as $T$ was lowered from room temperature to $T_0$ (by about 14% in all) and then decreased abruptly by about 70% just below $T_0$. An analysis of the data based on a model free energy including anharmonic coupling was also presented, but gave only moderate agreement with the experiment.

Unruh et al. (1978a, 1978b) also performed Brillouin scattering experiments on AS, overcoming the cracking problem by the use of thin plates of the crystal ($\sim 0.2$ mm thick) and a triple pass Fabry–Perot system. They also found a major highly asymmetric anomaly in $C_{11}$ and another in $C_{12}$ with only minor anomalies in the other elastic constants (both pure and rubidium doped AS were studied).

Because the elastic anomalies in ammonium sulphate (and the other crystals covered in this section as well) arise through indirect coupling to the order parameter, the shapes of the acoustic anomalies are closely related to those in the cell-multiplying improper ferroelastic transitions covered in sect. 4.1. In his review, Rehwald (1973) (fig. 9) compares the acoustic anomalies in ammonium sulphate and gadolinium molybdate, and they show a striking resemblance. However, Luspin (1976) noted that for $C_{55}$ there is a curious decrease with decreasing temperature in the prototype phase which is not observed for $C_{44}$ or $C_{66}$.

*$Pb_5Ge_3O_{11}$: Lead germanate* ($\bar{6} \rightarrow 3$)

Lead germanate undergoes a second-order hexagonal to trigonal ferroelectric transition ($C_{3h}^1 \rightarrow C_3^1$) at 451 K. The $A_{1g}$ optic soft mode (in the ferroelectric phase) saturates $\sim 6$ K below the transition, and both static and dynamic central peaks have been found (see the discussion in ch. 7 by Fleury and Lyons in this volume).

Linear coupling of strains to $P_z$ in the paraelectric phase is forbidden. The allowed improper strain ($u_{xx} = u_{yy}, u_{zz}$) is not symmetry breaking, so the transition is neither proper nor improper ferroelastic. Nevertheless, small anomalies in both the frequency and linewidth of the longitudinal acoustic modes do occur in the vicinity of the transition due to higher-order interactions as revealed in the Brillouin scattering experiments of Fleury and Lyons (1978).

## 4. *Cell-multiplying transitions*

In the first two classes of phase transitions discussed in the preceding chapters, the prototype crystal structure becomes unstable against one of its zone-center ($\Gamma$-point) normal modes whose condensation at the transition breaks the prototype symmetry. The order parameter $\eta$ in these crystals is a homogeneous quantity – i.e., it has the same value in every unit cell in both phases. (In the high-symmetry phase its mean value $\langle\eta\rangle$ is zero, but its instantaneous value is homogeneous.) Such transitions do not change the size of the unit cell.

Cell-multiplying transitions occur when the instability is at a point other than the $\Gamma$ point. Typically, a zone-edge or zone-corner instability causes the size of the unit cell to double (or quadruple) at the transition. Transitions of this type are often called antiferrodistortive. In the case of incommensurate transitions, the instability occurs at a general point in the Brillouin zone. The condensed phase exhibits a modulation of the underlying structure with a wavelength which is not an integral multiple of a direct lattice vector.

Since the order parameter in the high-symmetry phase of all cell-multiplying transitions is inhomogeneous ($\boldsymbol{q} \neq 0$), linear coupling to homogeneous strain is forbidden and proper ferroelasticity cannot occur. However, higher-order coupling of strains to the order parameter may give rise to improper ferroelasticity. Transitions in this category will be considered in sect. 4.1. Incommensurate transitions will be reviewed in sect. 4.2.

### *4.1. Improper ferroelastics* ($C_1$)

*Gadolinium molybdate (GMO) and terbium molybdate (TMO); Lead phosphate; Strontium titanate; Potassium manganese fluoride; Rubidium calcium fluoride and thallium cadmium fluoride; Potassium tin chloride; Sodium trihydrogen selenite; Squaric acid; Mercury chloride (calomel) and mercury bromide; Chloranil; Pentaantimony heptaoxide iodide; Benzil; Tris-sarcosine calcium chloride (TSCC)*

*$Gd_2(MoO_4)_3$: Gadolinium molybdate (GMO) ($\bar{4}2m \rightarrow mm2$) and $Tb_2(MoO_4)_3$: Terbium Molybdate (TMO)*

The rare-earth molybdate GMO and its Sm, Eu, Tb and Dy isomorphs are improper ferroelectric ferroelastics. Since the discovery of this transition in

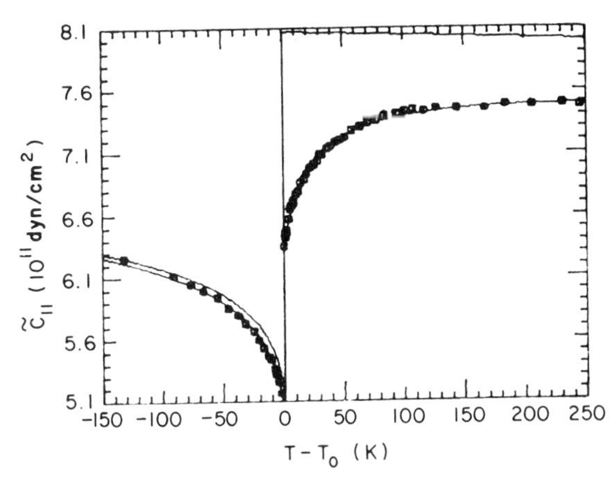


Fig. 12. TMO longitudinal elastic constant deduced from Brillouin scattering experiments. Upper curve: above $T_0$, uncoupled elastic constant; below $T_0$, effect of induced bilinear coupling included. Lower curve (through data points): bilinear coupling plus cubic coupling to fluctuations in the order parameter (Yao et al. 1981).

1966 by Borchardt and Bierstedt, extensive experimental and theoretical research has gradually revealed the structural and dynamical subtleties of the coupled mode system. GMO has also served as a model system for the development of a general understanding of improper ferroelectrics and ferroelastics.

The GMO transition occurs at 432 K between a tetragonal prototype structure $D_{2d}^3$ and an orthorhombic ferroic structure $C_{2v}^8$ with a doubling of the unit cell. The transition is driven by a doubly degenerate optic soft mode in the tetragonal phase at the M point of the Brillouin zone. The soft mode was investigated in terbium molybdate (TMO) in both phases by neutron scattering which established the correctness of the M-point instability model. In the orthorhombic phase, the zone-corner soft mode becomes a pair of zone-center modes which are Raman active. Raman scattering studies of GMO in the orthorhombic phase first revealed the presence of a temperature dependent mode near 47 cm$^{-1}$ at room temperature. There is a small dielectric anomaly near the transition which disappears entirely for the clamped crystal.

GMO has an elastic anomaly which is now recognized as one signature of improper ferroelasticity. First observed in acoustic resonance measurements, it has been investigated extensively in ultrasonic and Brillouin scattering experiments. Brillouin scattering in GMO has been studied by Itoh and Nakamura (1973, 1974), Busch (1974), Busch et al. (1974), Luspin and Hauret (1974, 1976a,b) and Luspin (1976). A combined Raman and Brillouin study of TMO was recently completed by Fleury et al. (1982).

Figure 12, from a recent Brillouin scattering study of TMO by Yao et al.

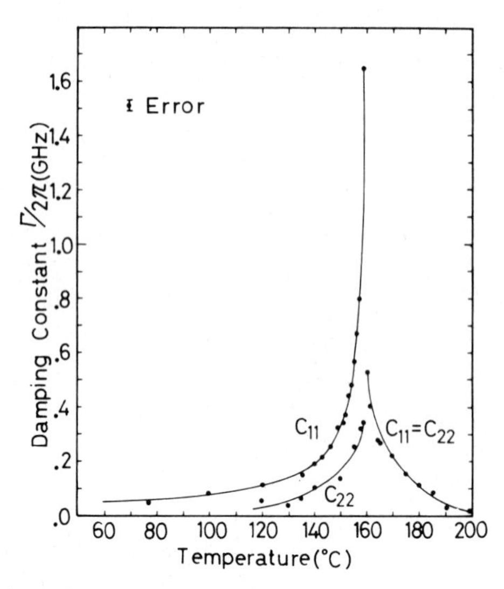


Fig. 13. Temperature dependence of the damping constants for the $C_{11}$ and $C_{22}$ modes in GMO (from Itoh and Nakamura (1974)).

(1981), illustrates the characteristic shape of the acoustic anomaly associated with improper-ferroelastic transitions. (This reference also includes a summary of the experimental literature on this transition.)

The origin of the asymmetric elastic anomaly lies in the non-linear coupling of the strain ($\epsilon$) to the order parameter ($\eta$) which is schematically represented as $\frac{1}{2}K\epsilon\eta^2$ (the correct form of the coupling involves a number of terms connecting the two components of $\eta$ with various strains). In the low-temperature (ferroic) phase, $\langle\eta\rangle$ has a non-zero value which produces a "morphic" bilinear coupling term (cf., Cummins (1979)). This term vanishes for $T > T_0$ so that linearized equations of motion (or equilibrium thermodynamics) predicts a discontinuous upward step at the transition and no temperature dependence above $T_0$ as shown by the upper curve in fig. 12. The additional curvature in both phases comes from anharmonic coupling of strain to pairs of soft modes at $\pm \boldsymbol{Q}$ in the Brillouin zone. This mode–mode coupling effect, when combined with the effective bilinear coupling below $T_0$, explains the distinctive shape of the acoustic anomalies observed in improper ferroelastics.

The temperature dependence of the Brillouin *linewidth* in GMO and TMO has also been investigated by Itoh and Nakamura (1974), Luspin and Hauret (1974, 1976a,b) and Fleury et al. (1981). Figure 13 from Itoh and Nakamura (1974) shows the damping constants for longitudinal modes propagating in the $x$ and $y$ directions. The damping constant $\Gamma$ (and Brillouin shift $\omega_0$) were found from the spectra by convolving the Gaussian instrument function with the spectral intensity

$$S(\omega) \propto \Gamma/[(\omega-\omega_0)^2 + 4\Gamma^2\omega_0^2],$$

and adjusting $\Gamma$ and $\omega_0$ for the best fit. Luspin and Hauret (1976a,b) have shown how the temperature dependence of both the Brillouin shift and linewidths in the high temperature phase can be explained by anharmonic coupling of strain to the square of the fluctuating order parameter.

*$Pb_3(PO_4)_2$: Lead phosphate* ($\bar{3}m \rightarrow 2/m$)
Lead phosphate exhibits an improper ferroelastic transition at 453 K from a rhombohedral prototype structure ($R\bar{3}m = D^5_{3d}$) to a monoclinic ferroelastic structure ($C2/c = C^6_{2h}$). The transition is first order, driven by an L-point instability which doubles the volume of the unit cell (Torres 1975). The soft $A_g$ optic mode in the ferroelastic phase has been studied by Raman scattering (Benoit and Chapelle 1974b, Smirnov et al. 1979); the zone-boundary soft mode in the high-temperature phase was studied by Joffrin et al. (1977) by inelastic neutron scattering. Dilatometry measurements (Toledano et al. 1975) show that at room temperature the spontaneous shear in the (100) plane reaches $22 \times 10^{-3}$.

Brillouin scattering experiments on lead phosphate were reported by An et al. (1975) and An and Chapelle (1976). The anomaly in $C_{11}$ (Chapelle et al. 1976) is similar to the anomaly in TMO shown in fig. 12.

*$SrTiO_3$: Strontium titanate* ($m3m \rightarrow 4/mmm$)
Strontium titanate undergoes a cubic to tetragonal transition ($O^1_h \rightarrow D^{18}_{4h}$) at 106 K driven by an instability at the Brillouin zone corner (111) against a $\Gamma_{25}$ soft optic mode. This R-point transition has been studied extensively by many techniques (cf., Scott (1974)).

Brillouin scattering experiments on $SrTiO_3$ were first reported by Kaiser and Zurek (1966) *before* the cell-doubling transition mechanism was understood. Their measured Brillouin shifts for transverse phonons traveling in the [100] direction resemble fig. 12, exhibiting the weak and strongly asymmetric anomaly typical of cell-doubling transitions. Numerous ultrasonic studies of $SrTiO_3$ have also been performed; stress measurements by Rehwald (1977) revealed an apparent bicritical point at this transition. For a recent review of the ultrasonic experiments plus a discussion of the theory of the elastic critical behavior in $SrTiO_3$, see Hochli and Bruce (1980).

*$KMnF_3$: Potassium manganese fluoride* ($m3m \rightarrow 4/mmm$)
$KMnF_3$ is a cubic perovskite which undergoes an $O^1_h \rightarrow D^{18}_{4h}$ cell doubling transition at $T_1 = 187.6$ K similar to the transition which occurs in strontium titanate at 106 K. This transition, which is slightly first order, is followed by a second cell-doubling transition at $T_2 = 91.5$ K, and an antiferromagnetic ordering transition at $T_n = 88$K.

The 187.6 K transition which has been studied extensively by many techniques, also exhibits a central peak similar to the original central peak in $SrTiO_3$.

Brillouin scattering studies of $KMnF_3$ have been reported by Fomin and Popkov (1976) and Eremenko et al. (1976). Weak asymmetric anomalies were observed at both $T_1$ and $T_2$.

*$RbCaF_3$: Rubidium calcium fluoride and $TlCdF_3$: Thallium cadmium fluoride* ($m3m \rightarrow 4/mmm$)
These fluoroperovskites, which are isomorphic to strontium titanate, undergo R-point instability driven cell doubling transitions at 193 K and 191 K, respectively ($O_h^1 \rightarrow D_{4h}^{18}$). As in $SrTiO_3$, the order parameter is the rotation of $CaF_6$ or $CdF_6$ octahedra around the (100) axis, but with a crystallographic distortion about ten times larger than in strontium titanate.

Berger et al. (1978a, 1978b) reported Brillouin scattering studies of $RbCaF_3$ and $TlCdF_3$, performed with a double pass pressure scanned interferometer. Their results for longitudinal acoustic waves propagating along [100] show the typical assymetric anomaly for both velocity and attenuation.

*$K_2SnCl_6$: Potassium tin chloride* ($m3m \rightarrow mmm$ *or* $4/mmm$)
Henkel et al. (1980) have performed Brillouin scattering experiments on the improper ferroelastic potassium tin chloride through the transition at 262 K, and have used their results to determine the symmetry of the zone boundary optic soft mode.

The hexahalometallates generally crystallize in the cubic antifluorite structure (Fm3m or $O_h^5$). Some (e.g., $(NH_4)_2SnCl_6$ and $(NH_4)_2SiF_6$) remain cubic at all accessible temperatures. $K_2ReCl_6$ undergoes a zone-center transition at 109 K. The improper transition in $K_2SnCl_6$ is driven by a soft optical phonon at the X-point, but the low-temperature structure is uncertain with some measurements indicating a tetragonal structure (P4/mnc or $D_{4h}^6$) and others an orthorhombic structure.

Henkel et al. consider two possible X-point soft modes: an $A_{2g}$ mode consisting of rotations of the $SnCl_6$ octahedra about the $z$-axis, and an $E_g$ mode consisting of rotations about the $x$- and $y$-axes. They show that the $E_g$ soft mode would lead to elastic anomalies in both $(C_{11} - C_{12})$ and $C_{44}$, while the $A_{2g}$ soft mode would lead to an elastic anomaly in $(C_{11} - C_{12})$ but not in $C_{44}$. Since the Brillouin scattering experiments reveal a strong anomaly in $(C_{11} - C_{12})$ and no anomaly in $C_{44}$, they conclude that the soft mode is $A_{2g}$, and that the possible space groups in the low-symmetry phase compatible with Landau theory are $D_{4h}^6$, $D_{4h}^{12}$, $D_{2h}^2$ or $T_h^2$. Thus, the low-symmetry point group is $D_{2h}$ (mmm) if the structure is orthorhombic and $D_{4h}$ (4/mmm) if it is tetragonal.

In analyzing the acoustic anomalies near the transition, Henkel et al. analyze the effect of terms in the free energy of the form $\langle QQ \rangle \epsilon$ where $\epsilon$ is an external strain and $Q$ is the soft mode amplitude. Those terms for which $\langle QQ \rangle$ varies linearly with $\epsilon$ give contributions quadratic in $\epsilon$ which renormalize the bare elastic constants. The elastic anomalies are thus seen to result from coupling of the elastic strain field associated with the acoustic modes to critical fluctuations of the soft mode coordinates.

We note that although the details of the analysis differ, this approach is physically identical to that of Levanyuk which was employed by Yao et al. (1981) in the analysis of the acoustic anomalies in TMO discussed earlier in this section.

*$NaH_3(SeO_3)_2$: Sodium trihydrogen selenite* ($2/m \to 1$)
The alkali trihydrogen selenites undergo a number of different phase transitions. The Li, Na and Rb compounds are ferroelectric while the K and Cs ones are not. The Li ferroelectric transition is proper while the Rb transition is improper.

Sodium trihydrogen selenite crystallizes at room temperature in the $\alpha$-phase with space group $P2_1/b$ or $C_{2h}^5$. At 196.4 K it undergoes a nearly second-order monoclinic to triclinic ferroelectric transition into the $\beta$-phase with space group P1 or $C_1^1$ with a doubling of the unit cell in the $z$ direction.

Lavrencic et al. (1976a, 1976b) have investigated the Brillouin spectrum of sodium trihydrogen selenite. They observed step discontinuities and downward bending characteristic of the $\epsilon\eta^2$ coupling which occurs, inter alia, in improper ferroelastic transitions, and showed that the observed elastic anomalies conform to the expected anharmonic interactions of acoustic modes with a soft $Z_1$ mode at the Z point $(0, 0, \frac{1}{2})$ of the Brillouin zone.

*$C_4O_2(OH)_2$: Squaric acid ($H_2SA$)* ($4/m \to 2/m$)
In 1974, Semingsen and Feder (1974) discovered that squaric acid undergoes a second-order anti-ferrodistortive tetragonal to monoclinic phase transition at 370.8 K. This transition has been studied extensively by many techniques, and was reviewed by Rehwald and Vonlanthen (1978). The temperature dependence of the elastic constants has been investigated by ultrasonics (Rehwald and Vonlanthem 1978, Rehwald 1978b) and by Brillouin scattering (Kruger et al., 1980).

The crystal structure of squaric acid consists of planes of square $C_4O_4$ groups with centers on a square grid oriented perpendicular to the unique axis, linked by hydrogen bonds to the four neighboring $C_4O_4$ groups. The bonding forces between planes are weak, resulting in good cleavage. Successive layers are shifted by $(\frac{1}{2}, \frac{1}{2}, 0)$ so that each $C_4O_4$ molecule is located both above and below an interstice of its neighboring planes. Each square carbon ring is rotated by about 24° relative to the crystal axes.

The transition is driven by ordering of the hydrogen bonds within each plane, as in the ferroelectric lead monohydrogen phosphate (LHP) discussed in sect. 3.2. However, the ferroelectric ordering in squaric acid is reversed in successive planes producing a "two-dimensional ferroelectric with the layers antiferroelectrically stacked" (J. Feder – private communication).

Above $T_c$ the hydrogen bonds are disordered and the unit cell is body-centered tetragonal ($C_{4h}^5$). Below $T_c$ the ordering removes the equivalence of the corner and center $C_4O_4$ groups and the monoclinic unit cell ($C_{2h}^2$) is doubled in volume relative to the tetragonal cell. Since the order parameter is inhomogeneous ($\boldsymbol{Q} = (001)$), linear coupling to strain is forbidden. The important coupling, which is quadratic in the order parameter, affects $(\epsilon_1 - \epsilon_2)$ and $\epsilon_6$ (Rehwald and Vonlanthen 1978).

Quasilongitudinal and quasitransverse waves propagating in (and polarized in) the (110) plane, whose velocities depend on $C_{11}$, $C_{66}$ and $C_{16}$, exhibit strongly asymmetric anomalies.

It is interesting to compare the transition in squaric acid with that in bismuth vanadate (see sect. 2.2). Both undergo a $4/m \rightarrow 2/m$ tetragonal to monoclinic transition in which the strains $(\epsilon_1 - \epsilon_2)$ and $\epsilon_6$ couple to the order parameter. The tetragonal (scheelite) structure of bismuth vanadate is closely related to the squaric acid structure with $VO_4$ tetrahedra at the corners and center of the body-centered tetragonal unit cell, rotated relative to the crystal axes as are the $C_4O_4$ groups in squaric acid. (It is this rotation which makes the scheelite structure less symmetric than the 4/mmm [$D_{4h}^{19}$] zircon structure to which it is otherwise identical.)

In bismuth vanadate, the optic soft mode is at the zone center and the coupling is linear. Consequently the transition is proper ferroelastic with a sound velocity falling to zero; the anomaly is strong and nearly symmetric, typical of a proper ferroelastic. In squaric acid, since the coupling is non-linear, the anomaly is weak and asymmetric, typical of an improper ferroelastic.

In fig. 14 the observed anomalies in $BiVO_4$ and squaric acid are shown, illustrating the dramatic distinction between the acoustic anomalies in proper and improper ferroelastics.

*$Hg_2Cl_2$: Mercury chloride (calomel) (4/mmm→mmm) and $Hg_2Br_2$: mercury bromide*

The mercury halides $Hg_2Cl_2$, $Hg_2Br_2$ and $Hg_2I_2$ were first synthesized as single crystals in 1970, and have since been studied extensively. Much of the experimental work on these crystals was reviewed in 1979 by Kaplyanskii et al. Above $T_0$ (185 K for $Hg_2Cl_2$, 143 K for $Hg_2Br_2$) the crystal structure is body-centered tetragonal $D_{4h}^{17}$ with one molecule in the primitive cell. The molecules form linear chains with weak inter-chain coupling, leading to very large anisotropy in optical and elastic properties. Condensation of a zone-boundary acoustic mode at one of the two X points induces a cell doubling transition to a low-temperature $D_{2h}^{17}$ orthorhombic structure (Benoit et al. 1978, 1980). Non-linear coupling to strain leads to improper ferroelasticity with domain walls oriented parallel to (100) and (010).

Several new lines appear in the Raman spectrum in the low-symmetry phase

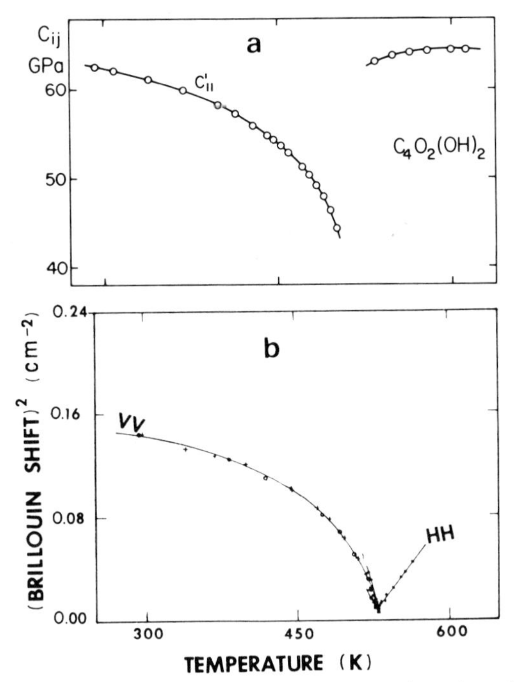


Fig. 14. (a) Elastic anomaly for longitudinal waves propagating along (100) in squaric acid (Rehwald 1978b). (b) Elastic anomaly for transverse waves propagating perpendicular to domain walls in bismuth vanadate (Gu et al. 1981).

due to the doubling of the unit cell. The soft mode, which is Raman inactive above $T_c$, appears with increasing intensity as $T$ is descreased below $T_0$. Its frequency is 15 cm$^{-1}$ ($Hg_2Cl_2$) or 12 cm$^{-1}$ ($Hg_2Br_2$) at 0 K, and can be followed down to 2 cm$^{-1}$ close to the transition where it is still underdamped.

The acoustic modes in $Hg_2Cl_2$ were investigated by Brillouin scattering by An et al. (1977). As in all improper ferroelastics, complete softening of acoustic modes is not observed. Rather, a highly anisotropic anomaly appears which can be analyzed as a jump discontinuity on which downward curvature is superimposed as a result of non-linear coupling of acoustic modes to critical fluctuations of the order parameter. The anomaly is particularly pronounced for $C_{66}$.

Brillouin and Raman scattering studies of $Hg_2Br_2$ have been performed by Daimon et al. (1978, 1981). These authors have analyzed the temperature dependence of the $C_{66}$ acoustic mode for $T < T_0$ through third-order coupling to the order parameter which, in the ferroelastic phase, becomes an induced bilinear coupling due to the non-zero expectation value of the order parameter. This analysis allows some conclusions to be drawn about the nature of the critical exponents $\gamma$ and $\beta$ which govern the temperature dependence of the soft optic mode and the order parameter, respectively.

*$C_6Cl_4O_2$: Chloranil* ($2/m \rightarrow 2/m$)
Chloranil undergoes a second-order structural phase transition near 90 K in which a staggered rotation of the molecules causes a doubling of the unit cell along the $C$-axis. The transition occurs without loss of symmetry between two monoclinic $C_{2h}^5$ structures ($P2_1/a \rightarrow P2_1/n$). Ecolivet (1981b) has performed Brillouin scattering experiments on chloranil and observed step-like discontinuities in the velocity of longitudinal acoustic modes, characteristic of cell multiplying transitions.

Raman and neutron scattering studies of chloranil have also been reported, including a recent temperature and pressure dependent Raman study by Girard et al. (1982). These authors have been able to observe the soft optic mode in the low-temperature phase with both pressure and temperature down to $\sim 2\,cm^{-1}$ where it is still underdamped, a result reminiscent of mercury chloride.

*$Sb_5O_7I$: Pentaantimony heptaoxide iodide* ($6/m \rightarrow 2/m$)
The $\alpha$ – or 2MC – polytype of this crystal undergoes a slightly first-order transition at $T_c = 481$ K from a high-temperature hexagonal phase ($C_{6h}^2$) to a ferroelastic non-ferroelectric monoclinic phase ($C_{2h}^5$) with a doubling of the unit cell along the monoclinic $C$-axis. The spontaneous strain is proportional to the square of the order parameter which is a characteristic of improper ferroelastics.

Rehwald et al. (1980b) investigated the elastic constants of $Sb_5O_7I$ by ultrasonic pulse echo and Brillouin scattering experiments. Their results exhibit the typical assymetric acoustic anomalies consisting of step discontinuities superimposed on curvature due to critical fluctuations.

*$(C_6H_5\text{-}CO)_2$: Benzil* ($32 \rightarrow 2$)
Benzil is a molecular crystal whose room temperature trigonal structure ($P3_121$ or $D_3^4$) and morphology are similar to quartz. At 83.5 K a first-order transition occurs to a monoclinic phase (C2 or $C_2^3$) which is both ferroelectric and ferroelastic. Raman scattering experiments by Sapriel et al. (1979) revealed a soft transverse optic mode of E symmetry in the paraelastic phase whose frequency decreased from $15.5\,cm^{-1}$ at room temperature to $8\,cm^{-1}$ at the transition where it splits into an A mode and a B mode in the monoclinic phase. The temperature dependence of the soft (trigonal) E and (monoclinic) A modes were shown to be consistent with a zone-center transition (# 30 in table 2). Since the order parameter introduces a cubic invariant, the transition is necessarily first order, as observed. Since strains $\epsilon_1 - \epsilon_2$ and $\epsilon_4$ have the same E symmetry, a soft acoustic mode should also occur, and

is predicted to be a transverse mode propagating along the $x$ direction of the trigonal phase.

Vacher et al. (1981) performed Brillouin scattering experiments on benzil using a triple pass Fabry Perot in series with a spherical Fabry Perot. They determined the temperature dependence of the elastic constants through the transition region, and found that the strongest temperature dependence was observed for a transverse acoustic mode propagating in the $x$ direction, associated with the elastic constant $C_{44}$. Thus, the Raman and Brillouin experiments would appear to support a straightforward application of Landau theory to this first order transition.

Structural analysis by X-ray diffraction, however, showed that the transition is actually a cell multiplying one associated with an M-point instability; the primitive cell is increased fourfold in the ferroelastic phase.

An analysis of this curious transition which reconciles the apparent contradictions was given by Toledano (1979). This analysis, which we briefly summarize here, illustrates the caution which must be used in extending standard Landau theory to first-order transitions. Toledano's discussion of the benzil transition begins with a review of the experimental background and a reinterpretation of the X-ray data which resolves an ambiguity in the structural assignment of the ferroelastic phase. He then develops a detailed application to benzil of a consequence of anharmonic coupling in first-order transitions which had previously been noted by Holakovsky.

Suppose that the primary instability in a phase transition is against a zone-center optic mode whose eigenvector $\eta$ is the order parameter. In a second-order transition, the coefficient $a$ of the quadratic term $\frac{1}{2}a\eta^2$ in the free energy expansion passes through zero at $T_c$, and the spontaneous symmetry breaking in the ferroic phase is governed by the irreducible representation of $\eta$ in the prototype space group.

Anharmonic coupling to another degree of freedom $\xi$ of the form $-\delta\eta\xi^2$ can occur if $\xi^2$ contains the representation of $\eta$ in its reduction ($\xi$ must be multicomponent). The leading (quadratic) contribution of $\xi$ to the free energy, $\frac{1}{2}b\xi^2$, will be renormalized by this coupling term to $(\frac{1}{2}b-\delta\eta)\xi^2$. As $T$ is decreased below $T_c$, $\langle\eta\rangle$ increases until this renormalized coefficient reaches zero, triggering a second phase transition.

In the case of first-order phase transitions, however, $\langle\eta\rangle$ jumps to a finite value immediately at the transition temperature $T_0$, and the renormalized coefficient of $\xi^2$ can become negative at once. In this case *both* $\eta$ and $\xi$ will take on non-zero values at $T_0$ ("triggered transition"), and the space group of the low-symmetry phase is the intersection of the symmetries determined separately by the irreducible representations of $\eta$ and $\xi$.

For benzil, Toledano identified the primary order parameter $\eta$ with the two-dimensional soft E model at the $\Gamma$ point ($\boldsymbol{Q}=0$) and the secondary order parameter $\xi$ with a three-dimensional mode at the M point ($\boldsymbol{Q}=0, \frac{1}{2}, 0$). A careful analysis of the experimental data indicated that this model adequately explains the transition in benzil.

The simultaneous appearance of two order parameters of different symmetry at $T_0$ is not allowed in the Landau theory of second-order phase transitions. However, as this example illustrates, first-order transitions may violate this restriction while otherwise obeying the Landau phenomenology completely.

*$(C_3NO_2H_7)_3$ $CaCl_2$: Tris-sarcosine calcium chloride (TSCC) (mmm→mm2)*
TSCC has long been known to undergo a ferroelectric phase transition from a non-polar orthorhombic (Pnma or $D_{2h}^{16}$) phase to a polar ($Pn2_1a$ or $C_{2v}^{9}$) orthorhombic phase near 127 K, similar to that in ammonium sulfate (cf. Smolenskii et al. (1979)). The crystal is also ferroelastic in both phases with room-temperature domain structure characteristic of an orthorhombic structure derived from a hypothetical high-temperature $D_{6h}^{3}$ hexagonal structure (which is not actually observable) by an M-point cell-doubling transition.

The crystal structure and the experimental work including dielectric and Raman scattering measurements are discussed in sect. 3.3 of ch. 5 by Scott in this volume. As he notes, the dynamics of the phase transition(s) in TSCC are presently subject to considerable controversy.

There is strong evidence for both order–disorder and displacive transitions occurring, and somewhat weaker evidence that the displacive transition is improper. As Scott notes, there is no direct evidence for cell doubling at the transition. The Leningrad group (Smolenskii et al. 1979, Prokhorova et al. 1980) concluded that the displacive transition is improper largely because of the small L–T splitting of the soft optic mode. However, a small L–T splitting could occur even for a proper ferroelectric transition if the origin of the dielectric anomaly is in the order–disorder transition rather than in the (weakly) polar soft mode. If the transition is proper (cell preserving), then the soft optic mode would have $B_g$ symmetry in the paraelectric phase and would not be Raman active, consistent with the observation that a soft optic mode is seen in the Raman spectrum in the ferroelectric phase only. Furthermore, since no strain has $B_g$ symmetry in the mmm point group, no major elastic anomalies would occur either.

Smolenskii et al. (1979) and Prokhorova et al. (1980) performed Brillouin as well as Raman scattering experiments on TSCC. The temperature dependence of longitudinal acoustic phonons propagating both along and perpendicular to the polar axis exhibited weak anomalies in both phases, as would be expected for either a proper or improper transition. They assert that the Brillouin data also supports the improper phase transition interpretation, but do not present sufficient information to make a convincing case.

### 4.2. *Incommensurate transitions* ($C_2$)

*Potassium selenate and rubidium tetrachlorozincate; Sodium nitrite; Thiourea; Barium sodium niobate (BSN); Barium manganese tetrafluoride*

The subject of incommensurate phase transitions is covered in this volume in ch. 8 by Klein and in sect. 6 of ch. 5 by Scott. I will therefore limit this section to a brief survey of Brillouin scattering experiments on incommensurates.

Since the wavevector of the soft mode at an incommensurate transition is finite, bilinear coupling to homogeneous strain is forbidden. However, coupling terms of the form $\frac{1}{2}K\epsilon Q^2$ (where $\epsilon$ is a strain and $Q$ is the order parameter) which are responsible, for example, for the elastic anomalies in improper ferroelastics discussed in the preceding section also occur in incommensurates. Therefore, weak and highly asymmetric acoustic anomalies are also frequently observed at incommensurate transitions.

*$K_2SeO_4$: Potassium selenate* ($6/mmm \rightarrow mmm \rightarrow IC \rightarrow mm2$) *and $Rb_2ZnCl_4$: Rubidium tetrachlorozincate*

Potassium selenate exhibits three distinct phase transitions: at $T_1 = 745$ K from the prototype hexagonal phase I ($D_{6h}^4 = P6_3/mmc$) to an orthorhombic phase II ($D_{2h}^{16} = Pnam$); at $T_2 = 127$ K to an incommensurate (IC) phase III in which the modulating wavevector $\boldsymbol{q}_0$ increases with decreasing temperature and changes discontinuously at $T_3$; at $T_3 = 93$ K (the lock-in transition) to a commensurate orthorhombic phase IV ($C_{2v}^9 = Pna2_1$) with $\boldsymbol{q}_0 = a^*/3$ which is also ferroelectric (Rehwald et al. 1980).

Brillouin scattering studies of all three transitions have been reported by Yagi et al. (Yagi et al. 1979, Cho and Yagi 1980, 1981); Brillouin and Raman scattering measurements in the range of $T_2$ and $T_3$ were performed by Fleury et al. (1979); Brillouin and ultrasonic measurements in the range 300 to 30 K (including $T_2$ and $T_3$) were reported by Rehwald et al. (1980); Brillouin shifts and linewidths on both sides of the incommensurate transition were studied by Hauret and Benoit (1981).

The hexagonal to orthorhombic transition at $T_1$ results from a zone-boundary instability and could produce an asymmetric acoustic anomaly of the type encountered, for example, in improper ferroelastic transitions discussed in sect. 4.1. However, Cho and Yagi found no evidence of that type of anomaly. Instead, they found that for the longitudinal mode propagating in the [100] direction there is a rapid decrease in the Brillouin shift with increasing temperature below $T_1$ and no temperature dependence above $T_1$ (see fig. 15).

They considered the influence of two types of coupling terms in the free energy. The first type $\gamma\epsilon Q^2$ couples strain ($\epsilon$) to the square of the soft mode amplitude $Q$, and would lead to the typical step discontinuity and downward bending. Since no anomalies of this type are seen, they concluded that all the

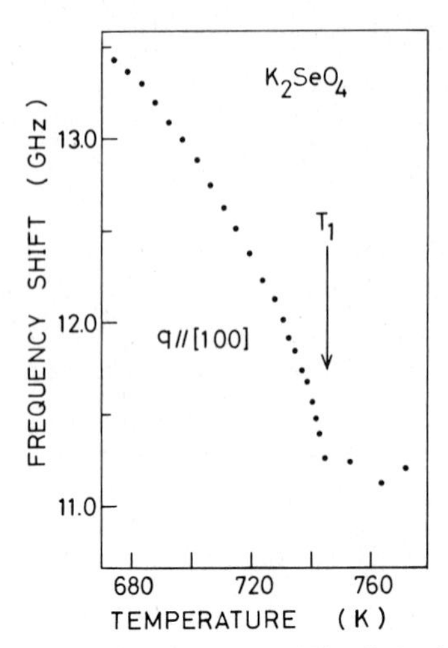


Fig. 15. Temperature dependence of the Brillouin shift of the longitudinal mode in $K_2SeO_4$ propagating along the direction [100] in the orthorhombic axes in the range of the high-temperature hexagonal to orthorhombic transition at $T_1 = 745$ K (from Cho and Yagi (1980)).

$\gamma$ coefficients are negligible. The second type of coupling, $\delta e^2Q^2$ results in the elastic constant $C_{ii}$ having the form

$$C_{ii} = C_{ii}^0 \quad (T > T_1),$$

$$C_{ii} = C_{ii}^0 + 2\delta_i\langle Q\rangle^2 \quad (T < T_1),$$

where $C_{ii}^0$ is the elastic constant at $Q = 0$. Cho and Yagi (1981) showed that the data of fig. 15 is very well described by this equation. The plot of $\ln(C_{ii} - C_{ii}^0)$ against $\ln[(T - T_1)/T_1]$ produced an excellent straight line fit with slope $2\beta = 0.69 \pm 0.02$.

At the incommensurate transition $T_2$, elastic anomalies typical of cubic interactions $K\epsilon Q^2$ were reported by Cho and Yagi (1981), Hauret and Benoit (1982) and Rehwald et al. (1980). Rehwald et al. investigated the complete set of elastic constants and found that near the incommensurate transition anomalies occur in several elastic constants, while at the lock-in transition $T_3$ there is only one anomaly, in $C_{55}$.

Figure 16 (from Rehwald et al. 1980) shows the anomalies in $C_{33}$ at $T_2$ and in $C_{55}$ at $T_3$. (Note the resemblance of the $C_{33}$ anomaly in fig. 16a to the acoustic anomaly at the improper ferroelastic transition in TMO in fig. 12). A thorough analysis of all the elastic anomalies in potassium selenate within the Landau theory formulation is given in the paper of Rehwald et al. In the experiments of Hauret and Benoit (1981), high resolution measurements with a double-pass Fabry–Perot and high-quality untwinned crystals allowed both the Brillouin shifts and linewidths to be accurately determined. Figure 17 shows their results

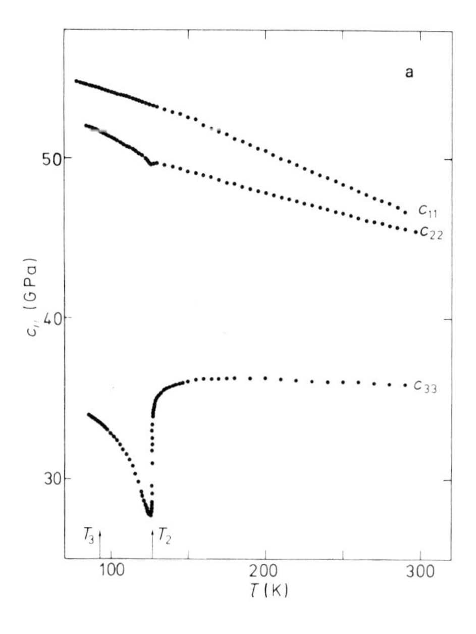


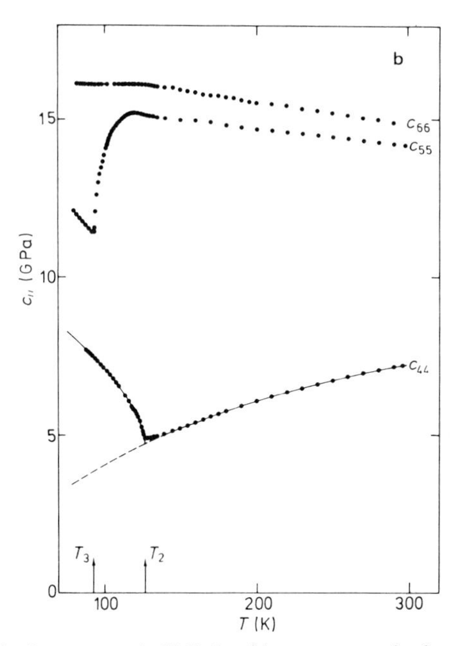


Fig. 16. Variation of elastic constants in $K_2SeO_4$ with temperature in the range of the incommensuarate ($T_2$) and lock-in ($T_3$) transitions from ultrasonic measurements. (a) $C_{11}$, $C_{22}$ and $C_{33}$. (b) $C_{44}$, $C_{55}$ and $C_{66}$ (from Rehwald et al. (1980a)).

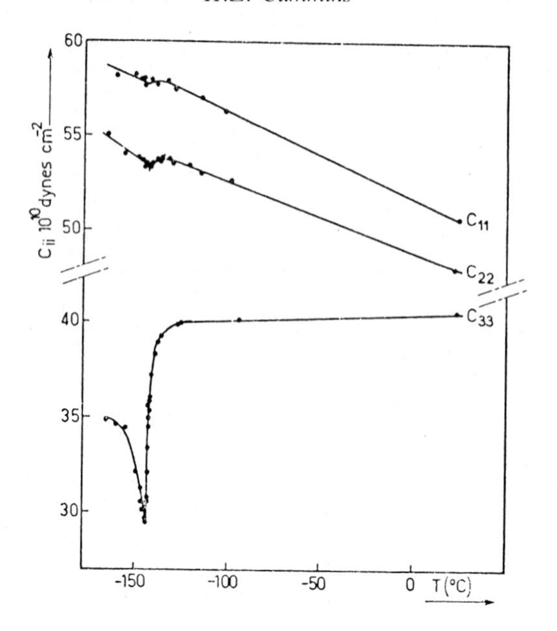


Fig. 17. Variation of the elastic constants $C_{11}$, $C_{22}$ and $C_{33}$ of $K_2SeO_4$ with temperature from Brillouin scattering experiments (from Hauret and Benoit (1981)).

for $C_{11}$, $C_{22}$ and $C_{33}$ which agree well with the ultrasonic results in fig. 16. In fig. 18, $C_{33}$ and the linewidth $\Gamma_3$ found in their experiments are shown.

Rubidium tetrachlorozincate ($Rb_2ZnCl_4$) has the same orthorhombic $D_{2h}^{16}$ structure as $K_2SeO_4$ and undergoes a similar incommensurate transition at $T_i = 301$ K followed by a commensurate transition at $T_c = 192$ K to a ferroelectric phase.

Ultrasonic experiment by Hirotsu et al. (1979) revealed a typical step discontinuity with rounding in the velocity of the longitudinal $C_{33}$ mode at the incommensurate transition, resulting from coupling of strain to the square of the order parameter. Much weaker anomalies were observed for the $C_{11}$ and $C_{22}$ modes. (The results of Hirotsu et al. closely resemble the $K_2SeO_4$ data of fig. 16). Recent Brillouin scattering studies by Yamanaka et al. (1981) showed a gradual decrease in the velocity of the $C_{33}$ mode with decreasing temperature below $T_i$ rather than the sharp dip observed by ultrasonic measurements. This result suggests that in $Rb_2ZnCl_4$ the relaxation time for order-parameter fluctuations near $T_i$ is on the order of $10^{-11}$ s.

*$NaNO_2$: Sodium nitrite* ($mmm \rightarrow IC \rightarrow mm2$)

Sodium nitrite has long been known to undergo a ferroelectric phase transition near 435 K associated with an order–disorder transition in the orientation of the $NO_2$ ions. Scott reviewed the neutron scattering and dielectric experiments

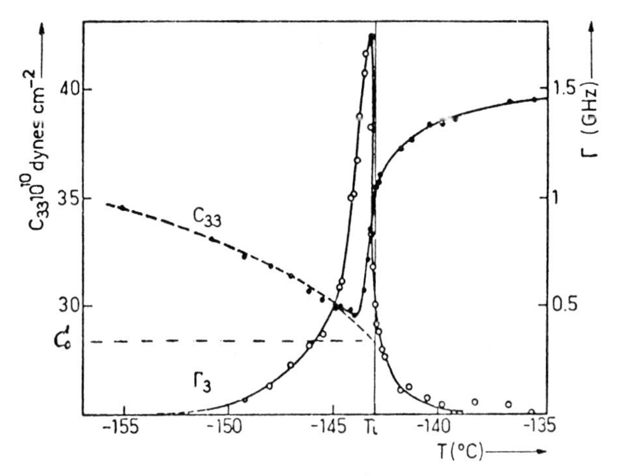


Fig. 18. Detailed variation around $T_2$ of $C_{11}$ and of the FWHM $\Gamma_1$ of the deconvoluted Brillouin peak corresponding to the acoustic wave propagating along (001). The dashed curve represents the best fit to the power law $(T_2 - T)^{\beta}$ with $\beta = 0.7$, extrapolated value $C_{33}(T_2) = 28.3 \pm 0.3 \times 10^{10}$ dynes cm$^{-2}$ (from Hauret and Benoit (1981)).

on $NaNO_2$ briefly in 1974; much of the literature on this material was reviewed by Lines and Glass (1977).

It has been recognized that there are actually two closely spaced transitions in sodium nitrite: at $T_1 = 437.7$ K from a non-polar orthorhombic phase ($D_{2h}^{25}$ = Immm) to an incommensurate (IC) phase, and at $T_2 = 436.3$ K from the incommensurate phase to a polar orthorhombic phase ($C_{2v}^{20}$ = Im2m) which is ferroelectric.

As an aside, we remark here that there are two different classes of incommensurate phase transitions. In some crystals the existence of a Lifshitz invariant makes a homogeneous second-order transition impossible, and a transition to an incommensurate structure must occur (Levanyuk and Sannikov 1976a). $NaNO_2$ and $SC(NH_2)_2$ are examples of the second class of incommensurate phase transitions in which there is no Lifshitz invariant. The incommensurate transition is not required by symmetry, but rather arises as a consequence of the details of the relevant coupling mechanisms.

Phenomenological theories of the $NaNO_2$ transition were given by Levanyuk and Sannikov (1976b), by Ishibashi and Shiba (1978) and by Heine and McConnell (1981). In the Levanyuk–Sannikov and Heine–McConnell theories the main mode whose instability at the lock-in transition causes a commensurate transition is linearly coupled to a second subsidiary mode at some point in the Brillouin zone. The incommensurate transition occurs when the frequency of the low-frequency member of the coupled mode pair reaches zero. Michel (1981) and Ehrhardt and Michel (1981) have developed a microscopic theory of the $NaNO_2$ transition which shows that it is driven by coupling between rotation of the $NO_2$ molecules and shear strain with non-zero wavevector. It is the wavevector dependence of this interaction which determines the modulation wavelength of the incommensurate phase.

Brillouin scattering experiments on sodium nitrite were reported by Shimizu et al. (1974) and Yagi et al. (1980). Although a discontinuous change in the Brillouin shift was observed near $T_2$ by Yagi et al. no dip-like anomalies were seen. Such an anomaly was seen in $C_{22}$ in ultrasonic studies in the MHz region, leading Yagi et al. to conclude that hypersonic dispersion was occurring. (Dispersion in the elastic properties near the incommensurate transition in barium sodium niobate to be discussed later in this section, has recently been observed by Scott, and in thiourea by An et al.)

Since dielectric measurements on $NaNO_2$ indicate that the Debye relaxation time of $NO_2$ orientational fluctuations is (Scott 1974):

$$\tau(T) \sim (1.6 \times 10^{-8})/(T - 160°\text{C})\ \text{s},$$

the Brillouin measurements of Yagi et al. which were performed at $\sim$ 200 GHz would pass from the low- to the high-frequency regime more than 100° above $T_1$, in agreement with the conclusions of Yagi et al. This association of orientational order–disorder transitions with ultrasonic dispersion is frequently encountered in crystals with no intervening incommensurate phase which will be discussed in sect. 5.1.

*$SC(NH_2)_2$: Thiourea ($mmm \to IC \to mm2$)*

Thiourea crystallizes in a centrosymmetric orthorhombic room temperature structure ($D_{2h}^{16}$). At $T_1 = 202$K there is a transition to an incommensurate phase, and at $T_2 = 169$ K to an orthorhombic structure which is ferroelectric ($C_{2v}^2$). There apparently are three different incommensurate phases between $T_1$ and $T_2$, although inelastic neutron scattering measurements indicate a monotonic increase in satellite intensity and decrease in modulation wavevector in this range (Moudden et al. 1978).

Brillouin scattering studies of thiourea have been reported by Benoit and Chapelle (1974), by Hauret and Cao (1977) and by An et al. (1979). Benoit and Chapelle (1974) found anomalies in $C_{33}$ at both $T_1$ and $T_2$.

The results of An et al. (1979) reveal several interesting weak anomalies in the $xy$ shear elastic constant $C_{66}$. At the orthorhombic to incommensurate transition ($T_1$) there is a change of slope; near the incommensurate to ferroelectric transition ($T_2$) there are two discontinuous increases within $\sim$ 3°, presumably associated with two of the incommensurate transitions.

The results of An et al. for $C_{33}$ in the region of $T_1$ are quite striking, revealing strong dispersion in both the Brillouin shift and linewidth (see fig. 19). The data of fig. 19 were fit by An et al. to a relaxational mode coupling linearly to the elastic wave, with Debye relaxation time $\tau$ given by

$$\tau = (6.2 \times 10^{-12})/(200.9 - T)\ \text{s}.$$

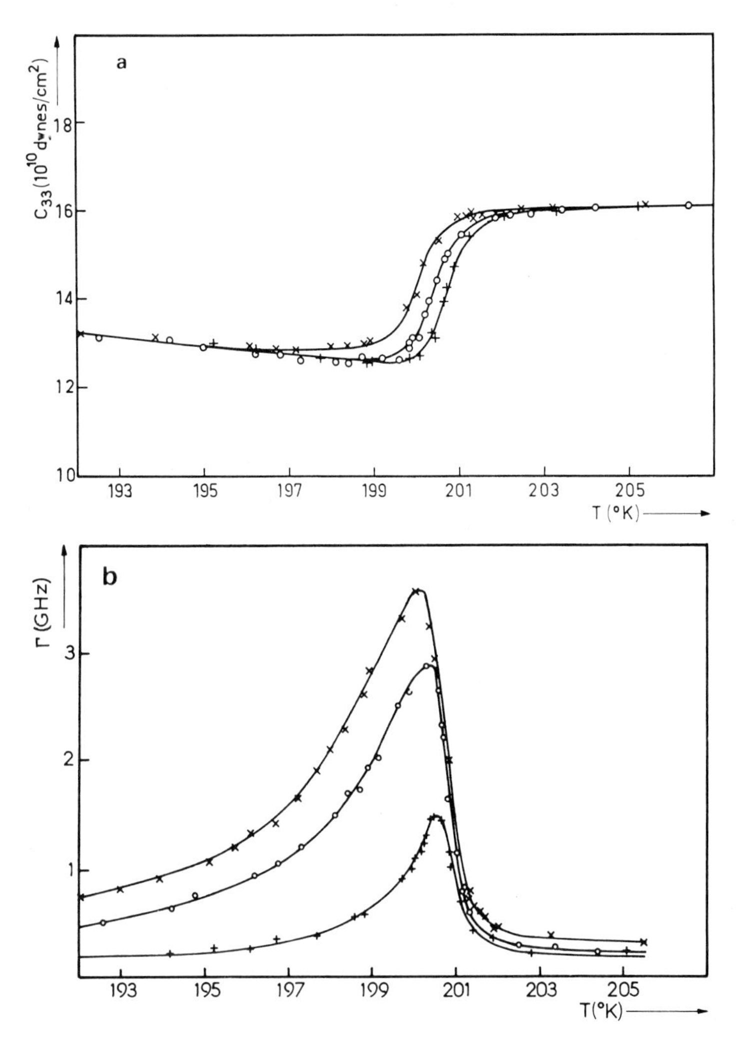


Fig. 19. Brillouin scattering in thiourea. (a) $C_{33}$ as a function of temperature for several scattering angles $\theta$. (b) Half-height width $\Gamma$ of the Brillouin line associated to $C_{33}$ for several scattering angles $\theta$. The lines were deconvoluted assuming a Lorentzian shape. × ($\theta = 135°$), ○ ($\theta = 90°$), + ($\theta = 45°$) (from An et al. (1979) and An (1982) – unpublished revision of (a)).

An et al. tentatively identified the relaxational mode as a phason, the phase degree of freedom associated with the modulation wave present in the incommensurate phase*. It now seems likely, however, that the relaxing degree of freedom which couples to the $C_{33}$ acoustic mode is actually the amplitudon rather than the phason (J.P. Chapelle – private communication).

*Scott has reanalyzed the data of An et al. and finds $\tau = 1 \times 10^{-10}$ s at $T_1$ (J.F. Scott – private communication).

*$Ba_2NaNb_5O_{15}$: Barium sodium niobate* (4/*mmm*→4*mm*→*IC*→?*mm*2→4*mm*)
Barium sodium niobate (BSN or "bananas") belongs to a family of ferroelectrics having the tetragonal tungsten bronze structure. BSN is a technologically important material because of its high electro-optic and non-linear optical coefficients; BSN and lithium niobate are widely used as frequency doublers for the important 1.06 $\mu$m neodymium laser line (cf., Lines and Glass (1977)).

At 858 K, BSN undergoes a ferroelectric transition between two tetragonal phases from the prototype centrosymmetric 4/mmm ($D_{4h}^5$ = P4/mbm) structure to the ferroelectric 4mm ($C_{4v}^2$ = P4bm) structure. This transition, which is driven by an $A_{2u}$ soft mode, is neither proper nor improper ferroelastic (Janovec et al. 1975). A second diffuse transition at 578 K to an orthorhombic phase produces spontaneous $\epsilon_6$ strain. At 110 K, a third transition occurs in which the ferroelectric ferroelastic orthorhombic phase reverts to 4 mm tetragonal symmetry (Schneck et al. 1977).

Toledano et al. (1976a) inferred that 578 K transition is an improper ferroelastic one from the tetragonal $C_{4v}^2$ (P4bm) structure to an orthorhombic $C_{2v}^{12}$ ($Ccm2_1$) structure driven by a soft zone boundary mode at the $Z$ point [$k = (0, 0, \pi/c)$] largely based on the observation in X-ray diffraction measurements of superstructure in the (001) direction. The diffuse nature of the transition which seemed to cover a span of about 30° had also been noted, and was attributed to spatial inhomogeneity.

Recent neutron and X-ray diffraction measurements have shown that the diffuse 578 K transition actually corresponds to an incommensurate phase (Schneck and Denoyer 1981). The first transition at $T_1$ = 578 K corresponds to the onset of an incommensurate orthorhombic structure. This transition is second order. The second transition at $T_2$ = 548 K is first order, and results in a change in the period of incommensurate modulation (which may, however be an incomplete lock-in transition due to internal strain or inhomogeneity).

Brillouin scattering in BSN has been reported by Toledano and Busch (1975), Toledano et al. (1976a) and Young and Scott (1981). Toledano et al. found that the longitudinal elastic constant $C_{11}$ softens continuously as $T$ is decreased in the ferroelectric phase from 800 K to 578 K and that both $C_{11}$ and $C_{22}$ increase throughout the temperature range of the orthorhombic phases, and that the difference $C_{11} - C_{12}$ due to orthorhombic distortion increases rapidly through the range of the incommensurate phase and then slowly decreases below the incomplete lock-in transition at 548 K. Young and Scott (1981) found that the magnitude of the dip in $C_{11}$ at the incommensurate transition depends strongly on scattering angle, indicating interaction of the sound wave with a degree of freedom characterized by relaxation time $\tau$. Their analysis gave

$$\frac{1}{\tau} = \frac{1}{\tau_1}\left(\frac{T_\mathrm{I} - T}{T_\mathrm{I}}\right) + \frac{1}{\tau_2},$$

where

$$2\pi\tau_1 = 1.4 \times 10^{-11}\,\mathrm{s}, \qquad 2\pi\tau_2 = 1.1 \times 10^{-10}\,\mathrm{s}.$$

*$BaMnF_4$: Barium manganese tetrafluoride* ($mm2 \rightarrow IC \rightarrow 2$)

This crystal, which is also discussed in chs. 5, 7 and 8 by Scott, by Fleury and Lyons and by Klein in this volume, manifests an extraordinary range of phenomena associated with its sequence of phase transitions.

From 250 K to its melting temperature $BaMnF_4$ has a polar orthorhombic structure ($C_{2v}^{12} = A2_1m$) and is ferroelectric – technically pyroelectric, since polarization switching has not been observed. An extrapolated divergence in the dielectric susceptibility and decreasing optical soft mode frequency imply a transition from mm2 to a hypothetical non-polar orthorhombic mmm phase at a Curie temperature of 1100 K, about 100° above the melting temperature. At high temperatures it is also an anisotropic ionic conductor.

At $\sim$ 250 K an incommensurate orthorhombic to monoclinic transition occurs, marked by anomalies in the dielectric and elastic constants and by the appearance of new temperature dependent optical modes in the Raman spectrum.

At low temperatures $BaMnF_4$ exhibits two-dimensional antiferromagnetic ordering. The magnetic susceptibility has a maximum near 50 K and decreases to a minimum at 25 K, the Neel temperature $T_N$, marking the onset of three dimensional long-range antiferromagnetic order. Below $T_N$, $BaMnF_4$ also exhibits weak ferromagnetism induced by the ferroelectric polarization through a linear magnetoelectric effect. Much of the extensive literature concerning these phenomena has been reviewed by Scott (1979).

The incommensurate transition has been the subject of much recent research, and is currently somewhat controversial. Scott et al. (1981) have concluded that at $T_2 = 247.1$ K there is a transition to a commensurate phase, and that the incommensurate phase exists only between $T_1 = 255$ K and $T_2$ where the structure again becomes commensurate. For additional discussions of the theory of this transition as well as of the Raman and central peak spectra, the reader should consult chs. 5 and 7 of Scott and of Fleury and Lyons.

Several Brillouin scattering studies of $BaMnF_4$ have been reported in the past few years. Bechtle et al. (1978) in the first Brillouin scattering measurement on this transition observed no anomaly in right angle scattering, but at small angles they found a strongly frequency dependent anomaly in the velocity of transverse modes in the plane perpendicular to the polar axis, approaching the 25% velocity drop previously observed in ultrasonic measurements by Fritz (1975). Bechtle et al. interpreted their results as a linear coupling between the transverse acoustic mode and a relaxing mode, and identified the relaxing mode as the phason associated with the incommensurate modulation. (Their analysis was carried out for three interacting degrees of freedom – the acoustic phonon, the phason and the amplitudon – but the interaction of the acoustic phonon with the amplitudon, which is the soft TO mode, was not found to be significant for the fit). Their analysis also leads to the prediction of a central peak due to the modification of the acoustic phonon response function by coupling to the phason. A central peak was observed, and the temperature dependence of its intensity and linewidth were found to be in reasonable agreement with the predictions of the theory. The temperature dependent linewidth was later shown to be spurious when improved data analysis techniques were employed (Lockwood et al. 1981).

Lyons and Guggenheim (1979) also studied the Brillouin components and the central peak in $BaMnF_4$, and found dispersion in the LA mode propagating along $\boldsymbol{c}$. Their analysis of the polarized central peak differed markedly from that of Bechtle et al. for the depolarized one. Lyons and Guggenheim tentatively concluded that, as in KDP, the central peak arises from coupling of the soft optic mode to entropy fluctuations, while the optic mode in turn couples to the much more strongly scattering acoustic mode. (This is also a three coupled mode theory which involves the acoustic, amplitudon and thermal diffusion modes, but *not* the phason).

A subsequent measurement of the thermal diffusivity (Negran 1981) demonstrated that the thermal diffusivity is too small by a factor of $\sim 25$ to explain the observed central peak linewidth. Lyons et al. (1980) extended their light scattering observations to investigate dependence on the direction of $\boldsymbol{q}$. They observed a strong anisotropy of the central peak linewidth within the *ac* plane in accord with a previous theoretical model which should apply if the observed spectral profile represents the phasons. They concluded that the central peak in $BaMnF_4$ observed below $T_I$ represents scattering from an overdamped phason mode which coupled to the light either directly or indirectly through another excitation (Lyons et al. (1982) have recently shown that the intensity of the central peak can be quantitatively explained by coupling of the phason to an LA phonon; this is also consistent with the LA phonon asymmetry observed by Lockwood et al. (1981) (K.B. Lyons, private communication)).

Murray and Lockwood (1980) and Lockwood et al. (1981) studied the Brillouin and Raman spectra of $BaMnF_4$, and interpreted their results differently from the other groups. They find that the narrow central peak is static, while a weak and rather broad feature seen in the Brillouin spectrum is identified as the underdamped Goldstone phason. Two additional low-frequency components observed in the Raman spectra are tentatively identified as two amplitudons, or as one amplitudon and one non-Goldstone phason.

## 5. *Order–disorder transitions* ($D$)

Order-disorder transitions differ from distortive transitions covered in the previous sections of this review in that there is no soft optic mode, and the high-symmetry phase lacks translational symmetry owing to the randomness of either orientation or position of some constituent of each unit cell. The distinction between distortive and order–disorder is not always obvious, however, owing to the effects of coupling between different degrees of freedom. Thus, for example, the primary driving force in the 122 K ferroelectric phase transition in $KH_2PO_4$ (KDP) is an ordering of protons in the hydrogen bonds between adjacent phosphate groups. However, bilinear coupling between tunneling protons and a polar optic lattice mode generates a hybrid soft mode, and this ferroelectric soft mode in turn couples bilinearly to $xy$ shear strain so that proper ferroelectricity and proper ferroelasticity occur simultaneously. KDP was therefore included in sect. 2.2 as a proper ferroelastic. In potassium cyanide, bilinear coupling between the CN ions which order orientationally at $T_0$ and shear strain causes the elastic constant $C_{44}$ to decrease nearly to zero. In this case, however, there is no optic soft mode and we include KCN in the present section.

In analyzing Brillouin spectra of crystals undergoing order–disorder transitions, the frequency of the hypersonic waves involved can not usually be neglected in comparison with the inverse relaxation time of the order parameter, $\tau_0^{-1}$. Consequently, the static (thermodynamic) limit is usually not valid; significant dispersion between ultrasonic and Brillouin results may occur, and a dynamic analysis is necessary.

The appropriate methods of analysis have been reviewed by various authors (cf., Cummins (1979)). A particularly thorough discussion can be found in Rehwald (1973).

For crystals in which the lowest-order coupling of strain $\epsilon$ to the order parameter $\eta$ allowed by symmetry has the form $-\frac{1}{2}K\epsilon\eta^2$, thermodynamics predicts that the static elastic constant $C$ associated with the strain $\epsilon$ will have its "bare" value $C_0$ above the transition temperature $T_0$, and will drop discontinuously at $T_0$, below which

$$C(T < T_0) = C_0 - K^2/(2\beta + 4\gamma\eta_0^2)\,, \tag{5.1}$$

where $\eta_0$ is the equilibrium value of the order-parameter $\eta$, and $\beta$ and $\gamma$ are the coefficients of the fourth- and sixth-power terms in $\eta$ in the Landau expansion of the free energy. This is the static limit which will govern the velocity of acoustic waves at sufficiently low-frequency $\omega$ so that $\omega\tau_0 \ll 1$.

When the Landau–Khalatnikov equation $\tau_0^{-1} \propto \Gamma|T - T_0|$ for $\tau_0(T)$ where $\Gamma$ is the kinetic coefficient governing the relaxation of $\eta$ is used to compute the complex elastic constant $\tilde{C}(\omega)$, the result is (O'Brien and Litovitz 1964)

$$\tilde{C}(\omega) = C_0 - \frac{K^2}{2\beta + 4\gamma\eta_0^2 + \mathrm{i}\omega/\Gamma\eta_0^2}\,. \tag{5.2}$$

The real part of the complex elastic constant $\tilde{C}(\omega)$ in eq. (5.2), which determines the Brillouin shift, is

$$C_R(\omega) = C_0 - \Delta C/(1 + \omega^2\tau_0^2), \tag{5.3}$$

where $\Delta C = K^2/(2\beta + 4\gamma\eta_0^2)$.

Equation (5.3) predicts that the downward step in $C$ will occur when $\omega\tau_0 = 1$ rather than at the transition temperature $T_0$. The same calculation shows that the Brillouin linewidth (like the ultrasonic attenuation) is maximum at $\omega\tau = 1$, in the middle of the downward step in $C$. Both effects are clearly seen in the thiourea results in fig. 19.

### 5.1. Orientational disorder (D1)

*Potassium cyanide, sodium cyanide and rubidium cyanide; Triglycine sulfate and triglycine selenate; Ammonium chloride and ammonium bromide; Rochelle salt; Ammonium hydrogen sulfate and Rubidium hydrogen sulfate; Lead monohydrogen phosphate; Aniline hydrobromide*

*KCN: Potassium cyanide (m3m→mmm); NaCN: sodium cyanide (m3m→mmm) and RbCN: rubidium cyanide (m3m→m)*

The alkali cyanides have the cubic rocksalt structure ($O_h^5$ = Fm3m) at room temperature where the CN anions are orientationally disordered (phase I). According to Rehwald et al. (1977) it is generally accepted that the cyanide ions in KCN do not rotate freely, and are probably orientated randomly in one of the eight (111) directions, although Fontaine et al. (1976) refer to this as a plastic phase. At $T_0 \sim 168$ K, KCN exhibits a slightly first order transition to an orthorhombic phase II ($D_{2h}^{25}$ = Immm) having quadrupole order, with the CN ions pointing preferentially along one [110] direction, but with residual head-to-tail disorder. The dipole (head-to-tail) order is established at $T_1 = 83$ K where the space group becomes $D_{2h}^{12}$ = Pmmm (phase III).

Brillouin scattering experiments in the cubic phase of KCN by Krasser et al. (1976) revealed a marked softening of $C_{44}$ on approaching $T_0$ which closely followed previous ultrasonic measurements. The close resemblance was a surprising result since the rotational correlation time of the CN ions which was thought to be responsible for the softening of $C_{44}$ had been found from NMR measurement to be $\sim 10^{-10}$ s, so that no softening at hypersonic (Brillouin) frequencies was expected. In a later Brillouin scattering study by Boissier et al. (1978), the frequencies and linewidths were analyzed on the basis of linear coupling between strains $\epsilon_4$, $\epsilon_5$ and $\epsilon_6$, and pseudospin variables representing the CN orientation. Good agreement with experiment was found with $\tau \sim 10^{-12}$ s. Brillouin scattering measurements by Hochheimer et al. (1977, 1978) showed that although $C_{44}$ falls very nearly to zero at $T_0$, its temperature dependence results entirely from interaction with the ordering of the $CN^-$ ions, and that the transition is not driven by the soft acoustic mode. Other high pressure studies by Dultz and his coworkers have revealed several additional phases of KCN (Dultz and Krause 1978, Dultz et al. 1979, Stock and Dultz 1979).

Rehwald et al. (1977) performed both Brillouin scattering and ultrasonic measurements on KCN as well as on mixed KCN–KCl crystals. They succeeded in extending both measurements into the partially ordered and fully ordered phases, and observed small anomalies in $C_{44}$ and $C_{11}$ near the second transition at ~ 81 K (bilinear coupling is forbidden at this transition which is cell multiplying). Their analysis of the anomalies at $T_1$ showed that in the absence of coupling to strains, orientational ordering of the CN ions would occur at − 230 K. The upward shift of the transition temperature by 300 K is qualitatively similar to but generally larger than the shifts occurring in linearly coupled transitions involving soft optic and acoustic modes. In KDP, for example, the equivalent shift is 5 K.

Satija and Wang (1978) observed Brillouin scattering from mixed KCN–KBr crystals and found that the admixture of KBr shifted the stability limit of the cubic phase to lower temperatures, as Rehwald et al. had found for KCN–KCl. Wang and Satija (1977) performed Brillouin scattering experiments on KCN and NaCN and found that the temperature dependence of the elastic constants is qualitatively similar in the two crystals (also see Satija and Wang (1977)). The two transitions in NaCN, at $T_0 = 288$ K and $T_1 = 172$ K involve quadrupole and dipole ordering of CN ions along a (110) direction, as in KCN.

Rubidium cyanide differs from the potassium and sodium compounds in that the CN ions order along a [111] direction rather than [110]. At $T_0 = 110$ K the quadrupole ordering transition results in the monoclinic space group $C_s^4$. Krasser et al. (1979) performed Brillouin scattering experiments in the cubic phase of RbCN. Pronounced softening of $C_{44}$ was observed as $T_0$ was approached from above. Krasser et al. analyzed their data in terms of a microscopic theory of rotation–translation interaction due to Michel and Naudts (1977) rather than with the phenomenological linear strain–quasispin coupling model employed by other investigators. This model has recently been further extended by DeRaedt et al. (1981).

*$(NH_2CH_2COOH)_3 \cdot H_2SO_4$: Triglycine sulfate (TGS); $(NH_2CH_2COOH)_3 \cdot H_2SeO_4$: Triglycine selenate (TGSe)* $(2/m \rightarrow 2)$

Triglycine sulfate is a technologically important ferroelectric material in which the pyroelectric effect (variation of electric polarization with temperature) is used to detect infrared radiation. The tendency of TGS to form domains on cooling after being heated above its transition temperature (322 K) has been overcome by doping with L-alanine which biases the hysteresis loop and assures complete poling when the crystal is cooled. Deuterated TGS and isomorphous triglycine fluoberyllate are also used in pyroelectric vidicon tubes for infrared imaging (Putley 1977).

The high-temperature paraelectric phase of TGS is centrosymmetric monoclinic ($C_{2h}^2 = P2_1/m$). At the second-order transition, the mirror plane disappears and spontaneous polarization appears parallel to the monoclinic $b$-axis. The low-temperature phase has space group $C_2^2 = P2_1$. Despite extensive investigations by many techniques, the microscopic mechanism of the transition is still unclear, with some combination of an order–disorder transition of protons and an orientational ordering of glycine groups seeming probable (Blinc 1976).

The order parameter $P_z$ driving this transition transforms according to the representation $A_u$ in the prototype 2/m phase, and bilinear coupling to strain is forbidden. Quadratic coupling is allowed to $\epsilon_{xx}$, $\epsilon_{yy}$, $\epsilon_{zz}$ and $\epsilon_{xy}$ to which the discussion preceding eq. (5.3) therefore applies.

Following the initial ultrasonic investigation of TGS by O'Brien and Litovitz (1964), several groups have applied Brillouin scattering to this transition. Gammon and Cummins (1966) and Gammon (1967) investigated the temperature dependence of longitudinal and transverse acoustic modes at different scattering angles. The rounded step in $\omega$ below $T_0$ predicted by eq. (5.3) which they observed is shown in fig. 20. Analysis of these results gave $\tau = A(T_0 - T)^{-1}$ s, with $A = (2.9 \pm 0.3) \times 10^{-11}$ sK. A later Brillouin experiment by Luspin and Hauret (1977) gave $A = 3.7 \times 10^{-11}$, in good agreement with ultrasonic measurements by Minaeva et al. (1969). (The smaller value of $A$ found by O'Brien and Litovitz (1964) was later found to be in error, so that no $\tau$ dispersion is involved.) Additional Brillouin studies were reported by Yagi et al. (1976b) who have also recently discovered a central peak in TGS (Miyakawa and Yagi 1980). Yagi et al. found $A = (4.3 \pm 0.3) \times 10^{-11}$ for TGS.

Triglycine selenate (TGSe) is isomorphous to TGS, with $T_0 = 295$ K. Brillouin scattering studies of TGSe have been reported by Yagi et al. (1974, 1976b) who found $A = (2.9 \pm 0.3) \times 10^{-11}$, identical to Gammon's (1967) result for TGS.

*$NH_4Cl$: Ammonium chloride ($m3m \rightarrow \bar{4}3m$); $NH_4Br$: Ammonium bromide ($m3m \rightarrow 4/mmm$)*

Ammonium chloride undergoes an order–disorder cubic–cubic phase transition at $T_0 = 242$ K (at atmospheric pressure). In the high-symmetry phase ($O_h^1 = Pm3m$) the tetrahedral ammonium ions are randomly distributed in two sterically different orientations within the CsCl-type unit cell which gives the structure, on the average, a center of inversion. In the ordered phase below 242 K, ($T_d^1 = P\bar{4}3m$) the center of inversion is lost. Although slightly first order at atmospheric pressure, the transition becomes second order under hydrostatic pressure exceeding 1.5 kbar, and has a tricritical point at $T = 256$ K, $P = 1.5$ kbar. This transition in $NH_4Cl$ has been investigated exhaustively by a large variety of experimental methods. Extensive reviews can be found, inter alia, in Rehwald (1973), Scott (1974) and Michel (1976).

Brillouin scattering studies of $NH_4Cl$ were reported by Lazay et al. (1969),

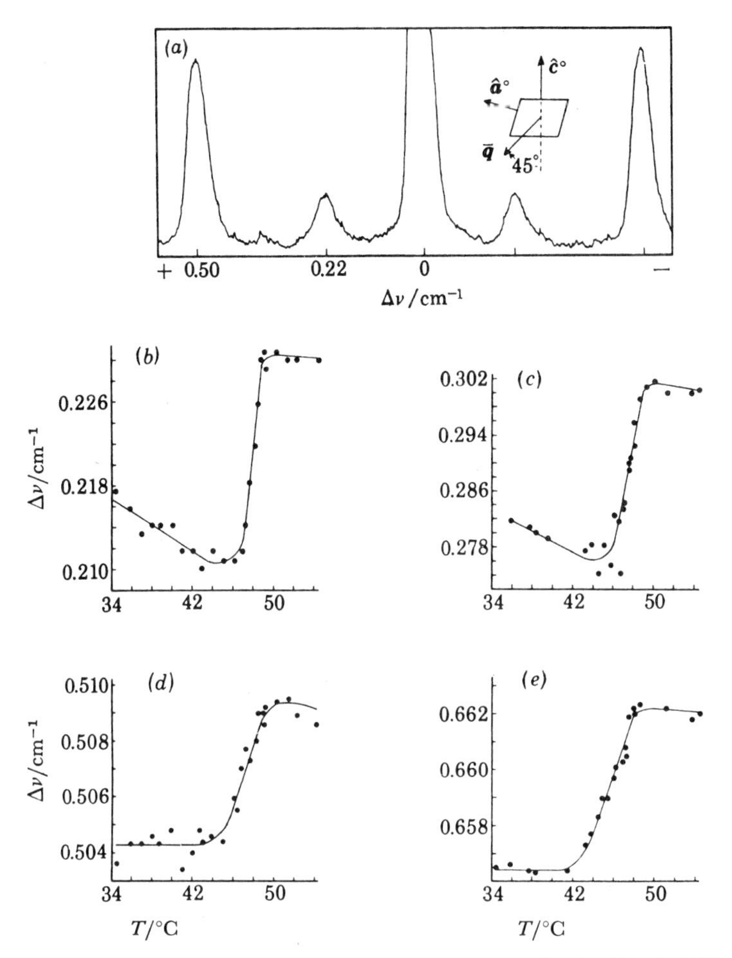


Fig. 20. Brillouin spectrum and temperature dependence of the Brillouin shifts in TGS. (a) (010) plane, $\theta = 90°$; (b) transverse, $\theta = 90°$; (c) transverse, $\theta = 135°$; (d) longitudinal, $\theta = 90°$; (e) longitudinal, $\theta = 135°$ (from Gammon (1967)).

by Gross et al. (1979) and by Hikita et al. (1981). Extensive ultrasonic studies by Garland and coworkers (reviewed by Rehwald (1973)) revealed a strong dip in the sound velocity near $T_0$. The Brillouin measurements of Lazay et al. showed a much weaker dip, and this dispersion in conjunction with the measured linewidth was used to deduce a relaxation time for the order-parameter fluctuations.

The measurements of Hikita et al. (1981) were performed with both hydrostatic pressure and temperature as variables. Disappearance of the step discontinuity in sound velocities for pressures around 2kbar is evident in their data. Curiously, the dip in velocity at $T_0$ observed by Lazay et al. for the longitudinal mode with $\boldsymbol{q} \parallel (110)$ was *not* observed by Hikita et al.

Ammonium bromide is isomorphous to $NH_4Cl$ in the disordered phase, but its transition at $T_0 = 234.5$ K produces antiparallel ordering of adjacent $NH_4$ tetrahedra and a doubling of the unit cell in the low-temperature tetragonal structure ($D_{4h}^7 = P/nmm$) (cf., Garland and Yarnell (1966)). Brillouin scattering in $NH_4Br$ has been reported by Rosasco et al. (1971) above $T_0$ and by Gross and Gerlich (1978) both above and below $T_0$. Gross and Gerlich reported a dip in the sound velocity for longitudinal waves with $\boldsymbol{q} \parallel (100)$ corresponding to an anomaly in the elastic constant $C_{11}$.

*$NaKC_4H_4O_6 \cdot 4H_2O$: Sodium potassium tartrate (Rochelle salt)* (222→2→222)
Rochelle salt is the oldest known ferroelectric crystal. It was studied by the Curies and by Pockels in the 19th century; its dielectric Curie point was discovered in 1921 by Valasek (Jona and Shirane 1962). It has the unusual (but not unique) property of being ferroelectric in a limited temperature range bounded by two Curie points. Above $T_1 = 297$ K, Rochelle salt is orthorhombic (space group $D_2^4 = P2_12_12_1$) and piezoelectric. Between $T_1$ and $T_2 = 255$ K, it is monoclinic with the polar axis along the orthorhombic [100]-axis. Below $T_1$ it is again orthorhombic. Both transitions are second order. The manifestation of relaxation in the high-frequency dielectric properties, and the softening of the elastic constant $C_{44}$ due to piezoelectric coupling to the polarization have been known for many years and were reviewed by Jona and Shirane (1962).

Brillouin scattering in Rochelle salt was investigated by Sailer and Unruh (1975, 1976) and Unruh et al. (1978a). The pure $C_{44}$ transverse mode which would show the maximum anomaly was not visible in the spectrum, however, and they studied a quasilongitudinal phonon propagating along [011] instead. The velocity of this mode is determined by a mixed elastic constant which includes $C_{22}$, $C_{33}$ and $C_{23}$ as well as $C_{44}$. Scattering by polarization fluctuations was also looked for but not seen.

The temperature dependence of the quasilongitudinal mode studied was analyzed in terms of bilinear coupling to relaxing polarization flucutations. No independent determination of the relaxation time $\tau$ was reported, but the temperature dependence of the Brillouin shift and linewidth were shown to agree well with theoretical expressions using $\tau$ values taken from previous dielectric relaxation measurements.

*$NH_4HSO_4$: Ammonium hydrogen sulfate (ammonium bisulfate)* ($2/m \to m$);
*$RbHSO_4$: Rubidium hydrogen sulfate*
Ammonium hydrogen sulfate (AHS) belongs to a large family of ferroelectric sulfates whose dielectric properties have been investigated extensively (cf. Jona and Shirane (1962)). AHS is reminiscent of Rochelle salt in that it is ferroelectric in a limited range of temperatures bounded on both sides by

transitions to paraelectric phases. At the high-temperature transition at 270 K, the prototype centric monoclinic $C_{2h}^5$ structure transforms into another monoclinic phase $C_s^2$ which is ferroelectric. At 154 K a second transition occurs to a triclinic $C_1^1$ structure which is again paraelectric. The 154 K transition is strongly first order, but the 270 K transition appears to be second order from the nature of the dielectric anomaly in $\epsilon_c$.

At the 270 K transition (2/m→m), linear coupling of strain to the $B_u$ order parameter is automatically excluded since the prototype phase is centric. Quadratic coupling is allowed to the strains $\epsilon_1$, $\epsilon_2$, $\epsilon_3$ and $\epsilon_6$ all of which may show anomalies at $T_c$.

Hikita and Ikeda (1977) reported a Brillouin scattering study of the longitudinal $b$-axis mode and the quasilongitudinal $a^*$ and $c$-axis modes through the temperature range 118 K to 313 K. The longitudinal mode propagating along $b(y)$ showed a markedly rounded step discontinuity and an increase in linewidth near the 270 K transition. Only a small anomaly appeared in the $a^*(x)$-axis QL mode, while no anomaly was seen in the $c(z)$-axis QL mode.

Rubidium hydrogen sulfate is isomorphous with ammonium hydrogen sulfate. It exhibits an order–disorder transition from the non-piezoelectric paraelectric monoclinic structure ($P2_1/c = C_{2h}^5$) to the piezoelectric ferroelectric monoclinic structure ($Pc = C_s^2$) at 263 K. Unlike $NH_4HSO_4$, however, there is no lower Curie point.

Tsujimi et al. (1981) have recently performed Brillouin scattering experiments on $RbHSO_4$. The dispersion and linewidth of the Brillouin components corresponding to longitudinal modes propagating along [010] were analyzed to find the relaxation time of polarization fluctuations, which gave

$$\tau = (1.9 \pm 0.3) \times 10^{-11}/(T_c - T)\,\mathrm{s}\,.$$

*$PbHPO_4$: Lead monohydrogen phosphate (LHP) (2/m→m)*

Lead monohydrogen phosphate, lead monohydrogen arsenate and their deuterated isomorphs are a recently discovered family of hydrogen bonded ferroelectric crystals. LHP undergoes a second-order proper ferroelectric transition at 310 K from a monoclinic paraelectric phase ($C_{2h}^4$) to a monoclinic ferroelectric phase ($C_s^2$).

The transition is driven by collective ordering of the protons which reside in double-minimum potentials, as in the KDP ferroelectrics. In contrast to KDP, however, the spontaneous polarization which lies in the plane of the hydrogen bonds is caused directly by proton ordering and does not involve additional interaction with lattice modes. Furthermore, since the paraelectric phase is centrosymmetric, bilinear (piezoelectric) coupling of strain to polarization is forbidden.

Brillouin and Raman scattering experiments in LHP have been performed by Lavrencic and Petzelt (1977) and by Lavrencic et al. (1978). The Brillouin shift of the $C_{22}$ longitudinal acoustic mode exhibits a temperature dependence characteristic of order–disorder ferroelectrics such as TGS and is satisfactorily fit by the Landau Khalanikov theory. Lavrencic et al. deduced from their Brillouin data that the polarization mode is relaxational with $\tau = \tau_0/(T_c - T)$ where $\tau_0 = 9.33 \times 10^{-11}$ sK.

*$C_6H_5NH_3Br$: Aniline hydrobromide* ($mmm \rightarrow 2/m$) (# 7)
Aniline hydrobromide undergoes a zone-center proper ferroelastic transition at 300 K from a high-temperature orthorhombic structure ($D_{2h}^{10}$ or Pnaa) to a low-temperature monoclinic structure ($C_{2h}^{5}$ or $P12_1/al$) with the appearance of spontaneous $yz$ shear strain ($\epsilon_5$). Since the transition is not ferroelectric, no dielectric anomaly occurs. Three degrees of freedom are thought to be involved in this transition: orientational ordering of $NH_3^+$ ions, the acoustic mode whose eigenvector is $\epsilon_5$ strain, and a $B_{2g}$ transverse optical mode.

Ultrasonic, Brillouin and Raman scattering investigations by Sawada et al. (1980, 1981) showed that the $C_{55}$ acoustic mode frequency in the high-temperature paraelastic phase approaches zero as $T \rightarrow T_0$, with no observable dispersion between ultrasonic and Brillouin results. However, no soft optic mode was observed in the Raman spectrum in either phase.

On the basis of the temperature dependence of the soft acoustic mode (which does not follow the $\omega^2 \propto (T - T_0)$ law expected of the primary instability) and the absence of a soft optic mode, Sawada et al. conclude that the transition is primarily driven by the ordering of the $NH_3^+$ ions which couple linearly to the $C_{55}$ acoustic mode. On the basis of the lack of dispersion between the ultrasonic and Brillouin results they conclude that just above the transition the relaxation time for fluctuations in the ionic order must be less than $3 \times 10^{-10}$ s.

### *5.2. Plastic crystals (D2)*

*Succinonitrile; Pivalic acid; Carbon tetrachloride, carbon tetrabromide and neopentane*

The plastic crystalline phase is a special case of orientational disorder in which the orientationally disordered constituent is able to rotate. Plastic crystals are usually cubic, have low melting entropy and are easily deformed plastically. The rate of molecular reorientation changes only slightly on melting indicating that the plastic crystalline phase is a mesophase in which rotational degrees of freedom are already largely melted out. This phase usually lies between the liquid phase and a true crystalline phase. For a review of the plastic crystalline phase, see Timmermans (1961).

The separation of order–disorder transitions (D1) and plastic crystals (D2) is rather arbitrary. In the ammonium halides, for example, reorientation of the

$NH_4$ ions occurs on a fairly rapid time scale indicating that the rotational motion is only weakly hindered. Nevertheless, these compounds are not usually classified as plastic crystals.

Because the elastic constants are small, the velocity of sound (and the Brillouin shifts) in plastic crystals are closer to liquids than to hard crystals. Typically $v \sim 1.5 \times 10^5$ cm/s compared to $6 \times 10^5$ in quartz.

*$(CH_2CN)_2$: Succinonitrile* ($m3m$)

Succinonitrile undergoes a first-order transition at $T_0 = 233$ K from a monoclinic crystalline phase to a body-centered cubic plastic crystalline phase. It melts at $T_M = 331$ K.

Brillouin scattering experiments on succinonitrile were reported by Bird et al. (1971), Boyer et al. (1971) and Bischofberger and Courtens (1976). The rapid reorientation of anisotropic groups gives rise to a broad depolarized component in the spectrum whose width is governed by the orientational relaxation time. Bird et al. found that the relaxation time $\tau$ is $\sim 3 \times 10^{-10}$ s near $T_0$ and decreases logarithmically with $1/T$ to less than $10^{-10}$ s. near $T_M$, in reasonable agreement with the results of NMR and dielectric measurements. The activation energy deduced was 3.8 kcal/mole. They also observed the temperature dependence of the Brillouin shift. Boyer et al. measured both the velocity and linewidth of longitudinal Brillouin components and deduced relaxation times from them comparable to those found by Bird et al.

Bischofberger and Courtens presented a more elaborate analysis based on coupling between orientation and translation. Their results, at 318 K, indicated that there are two different orientational relaxation frequencies, 42 GHz and 24 GHz, associated with rotation and isomerization, respectively. A noteworthy feature in their spectra is the interference between shear waves and reorientation which gives the shear component a characteristic resonance shape.

*$(CH_3)_3C\ COOH$: Pivalic acid* ($m3m$)

Pivalic acid is a face-centered cubic plastic crystal from 280 K to the melting point $T_M = 309.5$ K. Reorientation of the molecules is hindered by hydrogen bonds and is much slower than the typical $\tau \sim 10^{-11}$ s in most other plastic crystals or in organic liquids.

Bird et al. (1973) observed Brillouin scattering in pivalic acid and deduced the correlation time from the width of the narrow depolarized component. Their $\tau \sim 10^{-8}$ s showed characteristic Arrhenius temperature dependence ($\ln\tau \propto 1/T$) from which the activation energy was found to be $\sim 60$ kJ/mole.

*$CCl_4$: Carbon tetrachloride; $CBr_4$: Carbon tetrabromide; $CMe_4$: Neopentane*

Carbon tetrachloride freezes at $T_M = 250$ K into one of two plastic phases, $I_a$ which is face-centered cubic and $I_b$ which is rhombohedral. At $T_3 \sim 225$ K a transition to a monoclinic solid phase II occurs.

Levy-Mannheim et al. (1974) and Djabourov et al. (1977) reported Brillouin scattering measurement of $CCl_4$ in both plastic phases and in the liquid. They observed the longitudinal Brillouin components and found significant pre-transitional softening as the melting point was approached from below. At $T_M$, the velocity decreases and the linewidth increases discontinuously.

Tekippe and Abels (1977) reported Brillouin scattering studies of $CBr_4$ which has a single plastic crystalline phase between $T_M = 365$ K and $T_1 = 320$ K. The longitudinal mode and two transverse modes were observed in the plastic phase. In contrast to the $CCl_4$ results, however, no softening of the longitudinal modes was observed in the vicinity of the plastic–liquid transition.

A preliminary Brillouin scattering experiment was reported by Wergin et al. (1977) on neopentane which is structurally similar to $CCl_4$. The measurements were in the liquid phase close to the plastic–liquid transition at 256.5 K.

### *5.3. Superionic conductors (D3)*
### *Rubidium silver iodide; Barium fluoride and strontium chloride*

Superionic conductors (or solid electrolytes) are ionic crystals in which one subset of ions can move through the background lattice relatively freely, resulting in conductivities comparable to fluid electrolytes, several orders of magnitude higher than in ordinary insulating crystals near their melting points. The mobile ions are in a sense fluidlike so that the transitions to the superionic state is often called sublattice melting. However, the motion of the mobile ions is generally believed to proceed by thermally activated hopping between different possible sites within the unit cell and not by collision limited diffusion.

Superionic conductors have come to prominence in the last decade as electrolytes for high-efficiency primary and secondary batteries. The phenomenon, however, was observed and described almost 150 years ago by Faraday (for a review of the history and properties of superionic conductors, see O'Keefe (1976) and other papers in the same volume).

A considerable number of light scattering studies of superionic conductors have appeared during the last several years. Most of them have been concerned either with the Raman spectra or with the low-frequency quasielastic components associated with hopping motions of the mobile ions. Much of this work is reviewed in chs. 7 and 5 by Fleury and Lyons and by Scott in the present volume.

Relatively few publications have dealt with the Brillouin spectra of these crystals, although velocity dispersion similar to that observed in orientational order–disorder transitions should also occur here.

#### *$RbAg_4I_5$: Rubidium silver iodide*

The room temperature ionic conductivity of rubidium silver iodide, 0.25 $(\Omega\,\mathrm{cm})^{-1}$ is the highest value measured in a solid. Field et al. (1978) have investigated the Rayleigh–Brillouin spectrum of this crystal. The principal results of their work concern two quasielastic features associated with hopping motion of the ions.

Longitudinal Brillouin components with $\boldsymbol{q}$ along [100] were observed, but the transverse modes were not seen. The linewidth of the L modes was less than instrumental, and the measured Brillouin shift resulted in sound velocities in agreement with ultrasonic values indicating that there is no dispersion, presumably because $\tau$ is too short at the 90°C temperature of the observation to cause dispersion between ultrasonic and hypersonic frequencies.

Field et al. note that critical slowing down of the relaxation mode should occur near the 208 K transition temperature. However no spectra could be obtained below room temperature owing to the instability $RbAg_4I_5$ against formation of AgI and $Rb_2AgI_3$ below 300 K.

*$BaF_2$: Barium fluoride*; *$SrCl_2$: Strontium chloride*

Many crystals with the fluorite structure ($O_h^5$) exhibit a transition from insulator to superionic conductor which is spread out over a temperature range of a hundred degrees or more. This contrasts with the transitions in the silver ion materials which are first order and very sharp. The first fluorite superionic conductor, lead fluoride, was discovered by Faraday (O'Keefe 1976).

Catlow et al. (1978) have studied Brillouin scattering in a series of fluorites: $PbF_2$, $BaF_2$, $CaF_2$, $SrF_2$ and $SrCl_2$, all of which show qualitatively similar behavior. They find that for barium fluoride, $C_{11}$ and $(C_{11} - C_{12})$ show rapid

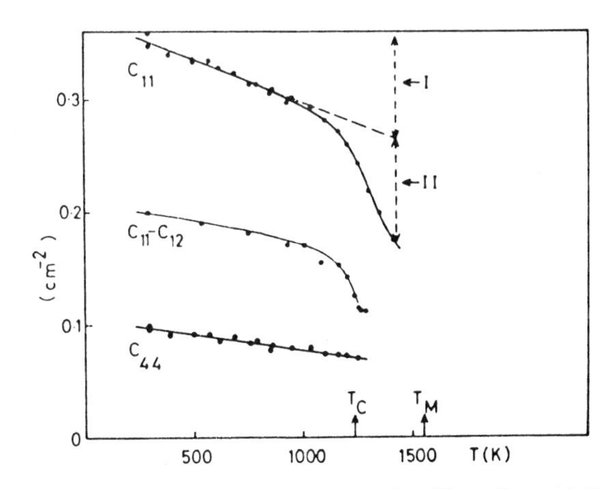


Fig. 21. Temperature dependence of elastic constants $C_{11}$, $(C_{11} - C_{12})$ and $C_{44}$ of $BaF_2$. Contributions of anharmonicity (I) and defects (II) to changes in $C_{11}$ are indicated (from Catlow et al. (1978)).

decreases as the crystal is warmed through the temperature range of the transition (see fig. 21). They explain this decrease as the result of the rapid growth of the defect concentration above $T_c$. They estimate that about 25% of the fluorines form interstitial defects in this range.

Brillouin scattering measurements on strontium chloride were reported by An (1977) who also observed a strong decrease in $C_{11}$ through the temperature range of the diffuse transition. An pointed out that if the temperature dependence of the observed Brillouin shift arose from the dynamical properties of the relaxing ion modes (as found in many of the orientational order–disorder transitions), then the linewidth would have to be much greater than what is observed. Thus, the dependence must result from temperature dependence of the static elastic constants.

## 6. Preliminary results (P)

*Adamantane; Di-ammonium di-cadmium sulfate and di-thallium di-cadmium sulfate; Ammonium fluoberyllate (AFB); Cesium lead chloride; Di-calcium lead propionate; Potassium mercury cyanide; Lead magnesium niobate; Biphenyl and para-terphenyl*

In this final section we briefly review several Brillouin scattering studies of phase transitions in crystals which have been published in a preliminary or incomplete form. The probable assignment among categories covered in the previous sections is indicated when possible in the heading of each example.

$C_{10}H_{16}$: *Adamantane* ($\bar{4}3m$) [*D*1]. Adamantane undergoes an order–disorder transition at 208.6 K. Damien and Deprez (1976) performed Brillouin scattering experiments at 295.7 K at various scattering angles. Analysis of the dispersion and linewidths observed resulted in a relaxation time $\tau = 0.9 \times 10^{-10}$ s, somewhat longer than the value deduced from NMR experiments.

An analysis of the dispersion of the Brillouin shift in adamantane (as well as in ammonium bromide) based on the Mori formalism has been described by Naudts (1981).

$(NH_4)_2Cd_2(SO_4)_3$: *Di-ammonium di-cadmium sulfate* (23) [*A*]; $Tl_2Cd_2(SO_4)_3$: *Di-thallium di-cadmium sulfate*. These crystals belong to a large family of compounds classed as langbeinites after the mineral $K_2Mg_2(SO_4)_3$. The ammonium cadmium salt is cubic at room temperature (space group $P2_13 = T^4$) and becomes ferroelectric below $T_0 = 95$ K (Jona and Shirane 1962).

Sailer et al. (1975) reported Brillouin scattering studies of the ammonium-cadmium and thallium–cadmium compounds. Attempts to verify a predicted elastic anomaly in $C_{11} - C_{12}$ were unsuccessful because of strong scattering from domain walls which masked the weak Brillouin components.

*$(NH_4)_2BeF_4$: Ammonium fluoberyllate (AFB) (mmm → ?IC → mm2)* [?C2]. Ammonium fluoberyllate (AFB – also known as ammonium fluoroberyllate) has an orthorhombic paraelectric structure (Pnam = $D_{2h}^{16}$) at room temperature. At $T_1 = 182$ K, a transition occurs to an intermediate phase of uncertain structure, presumed to be incommensurate. At $T_2 = 176$ K, a second transition occurs to a ferroelectric orthorhombic structure ($Pn2_1a = C_{2v}^{9}$) with a doubling of the unit cell along the *a*-axis relative to the high-temperature phase.

Brillouin scattering studies of AFB were published by Aleksandrov et al. (1976) and by Kudo and Hikita (1978). Weak anomalies in several elastic constants near $T_1$ and $T_2$ were reported by both groups. Aleksandrov et al. also performed ultrasonic measurements which gave similar but more pronounced anomalies indicating a relaxational character for the ordering process.

*$CsPbCl_3$: Cesium lead chloride.* Cesium lead chloride is a perovskite which undergoes a typical perovskite cubic to tetragonal transition at $T_1 = 320$ K. Several additional transitions have been detected at lower temperatures. No dielectric anomalies have been observed at any of the transitions, and X-ray diffraction measurements indicate that one of the transitions (at 310 K) is cell multiplying. Acoustic resonance experiments by Hirotsu (1971) revealed strong elastic anomalies near the 310 K transition. Hirotsu also examined the birefringence, specific heat and Raman spectrum of this crystal.

Gammon and Durvasula (1973) studied the Brillouin scattering spectrum of cesium lead chloride with a triple pass Fabry–Perot, but observed no elastic anomalies at any temperature, in marked contrast to the acoustic resonance results (R. Gammon – private communication, 1980).

*$Ca_2Pb\ (C_2H_5CO_2)_6$: Di-calcium lead propionate* [*D*1]. Although isomorphous with di-calcium strontium propionate which becomes ferroelectric below 281 K, the lead compound is not ferroelectric. It has two phase transitions, at $T_1 = 331$ K and at $T_2 = 191$ K.

Yagi and Tatsuzaki (1974) performed Brillouin scattering experiments on di-calcium lead propionate and found weak anomalies in several elastic constants near the upper transition temperature $T_1$. Since much stronger anomalies are seen in ultrasonic measurements, the transition dynamics appear to be an order–disorder transition of propionate ions with relatively long relaxation times.

*$K_2Hg(CN)_4$: Potassium mercury cyanide.* This crystal has the cubic spinel structure (Fd3m = $O_h^7$) at room temperature. At $T_1 = 110$ K a transition occurs to a low-temperature phase of unknown structure.

Krasser and Haussuhl (1976) examined the Raman and Brillouin spectra in the range 300 K–110 K. A decrease in all elastic constants was observed in agreement with previous ultrasonic experiments. The Raman experiments were extended down to 4 K and revealed a splitting of the $F_{2g}$ modes in the low-temperature phase corresponding to a trigonal distortion of the $Hg(CN)_4$ tetrahedra. Brillouin spectra could not be obtained in the low-temperature phase due to strong elastic scattering.

*$Pb_3MgNb_2O_9$: Lead magnesium niobate.* Smolenskii et al. (1971) reported Brillouin scattering studies of this crystal which exhibits a diffuse ferroelectric phase transition over a large temperature range centered near 273 K. The high-temperature paraelectric phase is cubic ($O_h$). The symmetry of the low-temperature polar phase may be tetragonal, but has not been definitely established. Broad pronounced dips were observed in both LA and TA modes extending over a range of $\sim 300°$.

*$(C_6H_5)_2$: Biphenyl* [*C*2]; *$(C_6H_5)$–$(C_6H_4)$–$(C_6H_5)$: para-terphenyl* ($2/m \rightarrow \bar{1}$) [*C*1]. Recently Ecolivet and his coworkers at the University of Rennes have undertaken a Brillouin scattering investigation of polyphenyls, a family of homologous monoclinic molecular crystals which exhibit unusual low-temperature phase transitions (Ecolivet and Sanquer 1980, Ecolivet 1981a). The basic molecular structural unit is a linear chain of benzene rings (2 in biphenyl, 3 in para-terphenyl, 4 in para-quaterphenyl).

Para-terphenyl exhibits an antiferrodistortive transition at $T_0 = 178$ K between a high-temperature monoclinic phase (space group $P2_1/a$) and a low-temperature triclinic phase (space group $P_{\bar{1}}$). The underlying cell multiplying dynamics, driven by the conformation of the chain molecules may be order–disorder rather than displacive. In either case it is improper ferroelastic. Weak asymmetric anomalies were observed in the acoustic mode frequencies near $T_0$.

Biphenyl undergoes two transitions – at $T_1 = 41$ K and at $T_2 = 17$ K. The $T_1$ transition is apparently displacive to an incommensurate phase, and the $T_2$ transition is a partial "lock-in". Ecolivet observed anomalies in the Brillouin shifts and linewidths in biphenyl near $T_1$.

## Conclusions

The large variety of phase transitions which have been investigated by Brillouin scattering already provides clear evidence of the utility and versatility of the Brillouin scattering technique. Several related points became increasingly apparent during the preparation of this review. First, because of the complex mode interactions occurring in many cases, it is often difficult to assign a particular transition to a definite category, or to determine the identity of the primary instability driving the transition. The interplay of order–disorder and displacive degrees of freedom, the relation between proper and pure ferroelasticity, the origin of triggered transitions and secondary order parameters all seem to be areas in need of some additional clarification.

Brillouin scattering has been particularly fruitful in combination with ultrasonic measurements (or small angle scattering) in probing the dynamics of relaxational modes giving rise to hypersonic dispersion. The occurrence of such dispersion near incommensurate transitions seems particularly intriguing. Also, the distinction between proper and improper ferroelastic transitions, as in bismuth vanadate vs. squaric acid, for example, is particularly suitable to analysis by Brillouin scattering.

Finally, there has been a marked increase in the sophistication of data analysis during the 15 years since the first studies appeared. While earlier studies were generally limited to the determination of Brillouin shifts and linewidths, later studies have revealed that detailed analysis of the entire spectrum often yields additional information about subtle mode interaction phenomena. The richness of the phenomena involved should continue to provide challenging opportunities for new experiments for many years.

## Acknowledgments

I am greatly indebted for invaluable assistance in the preparation of this manuscript to many colleagues who have sent me reprints and preprints of (and references to) relevant publications, and for advice related to the preparation of table 2 and the classification of various transitions covered by this review. Although a complete list of all those whom I wish to thank is impractical, I acknowledge the special help of J.C. Toledano, R.A. Cowley, J.F. Scott, V. Dvorak, R.M. Pick, J.L. Birman, F. Schwabl and M. Copic.

This work was supported, in part, by the National Science Foundation under grant DMR-8020835.

## References

Agyei, A.K. and J.L. Birman, 1977, Phys. Status Solidi, B **82**, 565.

Aizu, K., 1969, J. Phys. Soc. Jpn. **27**, 387.

Aizu, K., 1970, Phys. Rev. **B 2**, 754.

Aleksandrov, K.S., I.P. Aleksandrova, L.I. Zherebtsova, A.I. Kruglik, A.I. Krupnyi, S.V. Mel'nikova, V.E. Shneider and L.A. Shuvalov, 1975, Izv. Akad. Nauk SSSR, Ser. Fiz. **39**, 943.

Aleksandrov, K.S., A.T. Anistratov, A.I. Krupnyi, V.G. Martynov, Yu.A. Popkov and V.I. Fomin, 1976, Sov. Phys. Crystallogr. **21**, 296.

An, C.X., 1977, Phys. Status Solidi, **A 43**, K69.

An, C.X. and J.P. Chapelle, 1976, Ferroelectrics, **13**, 329.
An, C.X., G. Hauret and J.P. Chapelle, 1975, C.R. Acad. Sci. Ser. B **280**, 543.
An, C.X., G. Hauret and J.P. Chapelle, 1977, Solid State Commun. **24**, 443.
An, C.X., J.P. Benoit, G. Hauret and J.P. Chapelle, 1979, Solid State Commun. **31**, 581.
Anderson, P.W. and E.I. Blount, 1965, Phys. Rev. Lett. **14**, 217.
Aubry, S. and R. Pick, 1971, J. Phys. **32**, 657.
Axe, J.D. and G. Shirane, 1970, Phys. Rev. **B 1**, 342.
Azoulay, J., D. Gerlich, E. Wiener-Avnear and I. Pelah, 1977, Phys. Status Solidi, **B 81**, 295.
Barta, C., J.P. Chapelle, G. Hauret, C.X. An, A. Fouskava and C. Konak, 1976, Phys. Status Solidi, **A 34**, K51.
Bechtle, D.W., J.F. Scott and D.J. Lockwood, 1978, Phys. Rev. **B 18**, 6213.
Benoit, J.P. and J.P. Chapelle, 1974a, Solid State Commun. **14**, 883.
Benoit, J.P. and J.P. Chapelle, 1974b, Solid State Commun. **15**, 531.
Benoit, J.P., C.X. An, Y. Luspin, J.P. Chapelle and J. Lefebvre, 1978, J. Phys. **C 11**, L721.
Benoit, J.P., G. Hauret, Y. Luspin and C.X. An, 1980, Ferroelectrics, **25**, 569.
Berger, J., G. Hauret and M. Rousseau, 1978a, Solid State Commun. **25**, 569.
Berger, J., F. Plicque, M. Rousseau and A. Zarembowitch, 1978b, Ultrasonic and Brillouin scattering investigation of the structural phase transition of antiferrodistortive crystals, in: Proc. FASE 78, Second Congress of Federation of Acoustical Societies of Europe (Polish Academy of Sciences, Warsaw).
Bierlein, J.O. and A.N. Sleight, 1975, Solid State Commun. **16**, 69.
Bird, M.J., D.A. Jackson and H.T.A. Pentecost, 1971, Light scattering studies of molecular motion in the plastic crystalline phase of succinonitrile, in: Light Scattering in Solids, ed., M. Balkanski (Flammarion Sciences, Paris) p. 493.
Bird, M.J., D.A. Jackson and J.G. Powles, 1973, Mol. Phys. **25**, 1051.
Birman, J.L., 1966, Phys. Rev. Lett. **17**, 1216.
Birman, J.L., 1973, Phys. Lett. **A 45**, 196.
Bischofberger, J. and E. Courtens, 1976, Observation and interpretation of a new mode-coupling in a plastic crystal: Succinotrile, in: Light Scattering in Solids, eds., M. Balkanski, R.C.C. Leite and S.P.S. Porto (Flammarion Sciences, Paris) p. 688.
Blinc, R., 1976, Local properties of order–disorder-type ferroelectric phase transitions as studied by double resonance, in: Local Properties at Phase Transitions, Proc. Int. Sch. Phys. Enrico Fermi, Course LIX, Varenna, eds., K.A. Muller and A. Rigamonti (North-Holland, Amsterdam).
Blinc, R., B. Zeks and A.S. Chaves, 1980, Phys. Rev. B **22**, 3486.
Boccara, N., 1968, Ann. Phys. (N.Y.) **47**, 40.
Boissier, M., R. Vacher, D. Fontaine and R.M. Pick, 1978, Brillouin linewidth of phonons in KCN, in: Lattice Dynamics, ed., M. Balkanski (Flammarion Sciences, Paris) p. 641.
Bordeaux, D., J. Bornarel, A. Capiomont, J. Lajzerowicz-Bonneteau, J. Lajzerowicz and J.F. Legrand, 1973, Phys. Rev. Lett. **31**, 314.
Boyer, L., R. Vacher, M. Adam and L. Cecchi, 1971, Rayleigh and Brillouin scattering in succinonitrile in: Light Scattering in Solids, ed. M. Balkanski (Flammarion Sciences, Paris) p. 498.
Brody, E.M. and H.Z. Cummins, 1968, Phys. Rev. Lett. **21**, 1263.
Brody, E.M. and H.Z. Cummins, 1974, Phys. Rev. **B 9**, 179.
Busch, M., 1974, Thesis: Contribution a l'Etude par Diffusion Brillouin des Transitions de Phases dans le Molybdate de Gadolinium et le Niobate de Baryum et de Sodium (Universite de Paris VI, unpublished).
Busch, M., J.C. Toledano and J. Torres, 1974, Opt. Cummun. **10**, 273.
Catlow, C.R.A., R.T. Harley and W. Hayes, 1978, Brillouin scattering and lattice defect energetics of superionics with the fluorite structure, in: Lattice Dynamics, ed., M. Balkanski (Flammarion Sciences, Paris) p. 547.

Chapelle, J.P., C.X. An and J.P. Benoit, 1976, Solid State Commun. **19**, 573.
Cho, M. and T. Yagi, 1980, J. Phys. Soc. Jpn. **49**, 429.
Cho, M. and T. Yagi, 1981, J. Phys. Soc. Jpn. **50**, 543.
Copic, M., M. Zgonik, D.L. Fox and B.B. Lavrencic, 1981, Phys. Rev. **B 23**, 3469.
Courtens, E., 1978, Phys. Rev. Lett. **41**, 17.
Courtens, E. and R. Gammon, 1981, Phys. Rev. **B 24**, 3890.
Courtens, E., R. Gammon and S. Alexander, 1979, Phys. Rev. Lett. **43**, 1026.
Cowley, R.A., 1976, Phys. Rev. **B 13**, 4877.
Cummins, H.Z., 1979, Phil. Trans. Roy. Soc. Lond. **A 293**, 393.
Cummins, H.Z. and P.E. Schoen, 1972, Linear scattering from thermal fluctuations, in: Laser Handbook, eds., F.T. Arecchi and E.O. Schulz-Dubois (North-Holland, Amsterdam) p. 1030.
Daimon, M., S. Nakashima, S. Komatsubara and A. Mitsuishi, 1978, Solid State Commun. **28**, 815.
Damien, J.C. and G. Deprez, 1976, Solid State Commun. **20**, 161.
De Raedt, B., K. Binder and K.H. Michel, 1981, J. Chem. Phys. **75**, 2977.
Dil, J.G., 1982, Rep. Prog. Phys. **45**, 285.
Djabourov, M., C. Levy-Mannheim, J. Leblond and P. Papon, 1977, J. Chem. Phys. **66**, 5748.
Dudnik, E.F., V.V. Gene and I.E. Mnushkina, 1979, Izv. Akad. Nauk. SSSR, Ser. Fiz. **43**, 1723 (Bull. Acad. Sci., USSR, Phys. Ser. **43**, 149).
Dultz, W. and H. Krause, 1978, Phys. Rev. **B 18**, 394.
Dultz, W., H. Krause and J. Ploner, 1979, New high-pressure phase transitions in potassium cyanide, in: High-Pressure Science and Technology, Vol. 1, eds., K.D. Timmerhaus and M.S. Barber (Plenum, New York) p. 441.
Dultz, W., H.H. Otto, H. Krause and J.L. Buevoz, 1980, Elastic neutron scattering investigations of new high pressure phases of KCN (unpublished).
Durvasula, L.N. and R.W. Gammon, 1976, Brillouin scattering in potassium dihydrogen arsenate near $T_C$, in: Light Scattering in Solids, eds., M. Balkanski, R.C.C. Leite and S.P.S. Porto (Flammarion Sciences, Paris) p. 775.
Durvasula, L.N. and R.W. Gammon, 1977, Ferroelectrics, **16**, 199.
Ecolivet, C., 1981a, Thesis: Etude par diffusion Brillouin de la dynamique de reseau et des changements de phase de cristaux moleculaires (Universite de Rennes I, unpublished).
Ecolivet, C., 1981b, Solid State Commun. **40**, 503.
Ecolivet, C. and H. Poignant, 1981, Phys. Status Solidi, **A 63**, K107.
Ecolivet, C. and M. Sanquer, 1980, J. Chem. Phys. **72**, 4145.
Ehrhardt, K.D. and K.H. Michel, 1981, Z. Phys. **B 41**, 329.
Elliott, R.J., S.R.P. Smith and A.P. Young, 1971, J. Phys. J. Phys. **C 4**, L317.
Elliott, R.J., R.T. Harley, W. Hayes and S.R.P. Smith, 1972, Proc. Roy. Soc. **A 328**, 217.
Eremenko, V.V., V.I. Fomin and Yu.A. Popkov, 1976, Soft modes and light inelastic scattering in $KMnF_3$ crystal, in: Light Scattering in Solids, eds., M. Balkanski, R.C.C. Leite and S.P.S. Porto (Flammarion Sciences, Paris) p. 905.
Errandonea, G., 1980, Phys. Rev. B **21**, 5221.
Errandonea, G., 1981, Ferroelectrics **36**, 423.
Errandonea, G. and P. Bastie, 1978, Ferroelectrics, **21**, 571.
Field, R.A., D.A. Gallagher and M.V. Klein, 1978, Phys. Rev. **B 18**, 2995.
Firstein, L.A., G.A. Barbosa and S.P.S. Porto, 1976, Combined Raman–Brillouin scattering in $SrTiO_3$ near 110 K phase transition, in: Light Scattering in Solids, eds., M. Balkanski, R.C.C. Leite and S.P.S. Porto (Flammarion Sciences, Paris) p. 866.
Fleury, P.A., 1971, J. Acoust. Soc. Am. **49**, 1041.
Fleury, P.A., 1976, Recent developments in the spectroscopy of structural phase transitions, in: Light Scattering in Solids, eds., M. Balkanski, R.C.C. Leite and S.P.S. Porto (Flammarion Sciences, Paris) p. 747.

Fleury, P.A., 1980, Interacting collective excitations in crystalline solids, in: Proc. of the VIIth Int. Conf. on Raman Spectroscopy, Ottawa, ed., W.F. Murphy (North-Holland, Amsterdam) p. 2.
Fleury, P.A. and P.D. Lazay, 1971, Phys. Rev. Lett. **26**, 1331.
Fleury, P.A. and K.B. Lyons, 1976, Phys. Rev. Lett. **37**, 1088.
Fleury, P.A. and K.B. Lyons, 1978, Dynamic central peaks and phonon interactions near structural phase transitions, in: Lattice Dynamics, ed., M. Balkanski (Flammarion Sciences, Paris) p. 731.
Fleury, P.A. and K.B. Lyons, 1981, Optic studies of structural phase transitions, in: Structural Phase Transitions I, eds., K.A. Muller and H. Thomas (Springer, 1981) p. 9.
Fleury, P.A., P.D. Lazay and L.G. Van Uitert, 1974, Phys. Rev. Lett. **33**, 492.
Fleury, P.A., S. Chiang and K.B. Lyons, 1979, Solid State Commun. **31**, 279.
Fleury, P.A., K.B. Lyons and R.S. Katiyar, 1982 Phys. Rev. **B 26**, 6397.
Folk, R., H. Iro and F. Schwabl, 1976a, Phys. Lett. **A 57**, 112.
Folk, R., H. Iro and F. Schwabl, 1976b, Z. Phys. **B 25**, 69.
Folk, R., H. Iro and F. Schwabl, 1977, Z. Phys. **B 27**, 169.
Folk, R., H. Iro and F. Schwabl, 1979, Phys. Rev. **B 20**, 1229.
Fomin, V.I. and Yu.A. Popkov, 1976, Sov. Phys. JETP, **43**, 64.
Fontaine, D., M. Krauzman and R.M. Pick, 1976, Plastic phases of NaCN and KCN; a light scattering study, in: Light Scattering in Solids, eds., M. Balkanski, R.C.C. Leite and S.P.S. Porto (Flammarion Sciences, Paris) p. 692.
Fritz, I.J. and P.S. Peercy, 1975, Solid State Commun. **16**, 1197.
Gammon, R.W., 1967, Ph.D. Thesis: Brillouin scattering experiments in the ferroelectric crystal triglycine sulfate (The Johns Hopkins University, Baltimore, Md. unpublished).
Gammon, R.W. and H.Z. Cummins, 1966, Phys. Rev. Lett. **17**, 193.
Gammon, R.W. and L.N. Durvasula, 1973, Brillouin scattering in $CsPbCl_3$ presented at Annual Meeting, Optical Soc. America, Rochester, N.Y. (unpublished).
Garland, C.W. and C.F. Yarnell, 1966, J. Chem. Phys. **44**, 1112.
Ginzburg, V.L., A.P. Levanyuk and A.A. Sobyanin, 1980, Phys. Rep. **57**, 152.
Girard, A., Y. DeLugeard, C. Ecolivet and H. Cailleau, 1982, J. Phys. **C 15**, 2127.
Gorodetsky, G. and B. Luthi, 1971, Solid State Commun. **9**, 2157.
Gross, M. and D. Gerlich, 1978, Investigation of the $\lambda$-type phase transition in ammonium bromide by Brillouin scattering, in: Lattice Dynamics, ed., M. Balkanski (Flammarion Sciences, Paris) p. 500.
Gross, M., D. Gerlich and S. Szapiro, 1979, J. Phys. (Paris) **40**, Colloq, **C 8**, 203.
Gu, B., M. Copic and H.Z. Cummins, 1981, Phys. Rev. **B 24**, 4098.
Harada, J., J.D. Axe and G. Shirane, 1971, Phys. Rev. **B 4**, 155.
Harley, R.T. 1977, J. Phys. **C 10**, L205.
Harley, R.T., K.B. Lyons and P.A. Fleury, 1980, J. Phys. **C 13**, L447.
Hauret, G. and C.X. An, 1977, Rev. Phys. Appl. **12**, 995.
Hauret, G. and J.P. Benoit, 1982, Ferroelectrics, **40**, 1.
Hauret, G., L. Taurel and J.P. Chapelle, 1971, C.R. Acad. Sci., **273**, 627.
Heine, V. and J.D.C. McConnell, 1981, Phys. Rev. Lett. **46**, 1092.
Henkel, W., J. Pelzl, K.H. Hock and H. Thomas, 1980, Z. Phys. **B 37**, 321.
Hikita T. and T. Ikeda, 1977, J. Phys. Soc. Jpn. **42**, 351.
Hikita, T., M. Kitabatake and T. Ikeda, 1981, J. Phys. Soc. Jpn. **50**, 1259.
Hirotsu, S., 1971, J. Phys. Soc. Jpn. **31**, 552.
Hirotsu, S., K. Toyota and K. Hamano, 1979, J. Phys. Soc. Jpn. **46**, 1389.
Hocheimer, H.D., W.F. Love and C.T. Walker, 1977, Phys. Rev. Lett. **38**, 832.
Hocheimer, H.D., W.F. Love and C.T. Walker, 1978, Comparison of the order–disorder phase transition in KCN and NaCN, in: Lattice Dynamics, ed., M. Balkanski (Flammarion Sciences, Paris) p. 638.
Hochli, U.T. and A.D. Bruce, 1980, J. Phys. **C 13**, 1963.

Hochli, U.T. and J.F. Scott, 1971, Phys. Rev. Lett. **26**, 1627.
Hoshizaki, H., A. Sawada, Y. Ishibashi, T. Matsuda and I. Hatta, 1980, Jpn. J. Appl. Phys. **19**, L324.
Ishibashi, Y. and H. Shiba, 1978, J. Phys. Soc. Jpn. **45**, 409.
Itoh, S. and T. Nakamura, 1973, Phys. Lett. **A 44**, 461.
Itoh, S. and T. Nakamura, 1974, Solid State Commun. **15**, 195.
Ivanov, N.R., L.A. Shuvalov, H. Schmidt and E. Stolp, 1975, Izv. Akad. Nauk SSSR, Ser. Fiz. **39**, 933.
Janovec, V., V. Dvorak and J. Petzelt, 1975, Czech. J. Phys. **B 25**, 1362.
Joffrin, C., M. Lambert and G. Pepy, 1977, Solid State Commun. **21**, 853.
Jona, F. and G. Shirane, 1962, Ferroelectric Crystals (MacMillan, New York).
Kaiser, W. and R. Zurek, 1966, Phys. Lett. **23**, 668.
Kanamori, J., 1960, J. Appl. Phys. **31**, 14S.
Kaplyanskii, A.A., Yu.F. Markov and Ch. Barta, 1979, Bull. Acad. Sci. USSR, Phys. Ser. **43**, 77.
Kashida, S. and H. Kaga, 1977, J. Phys. Soc. Jpn. **42**, 499.
Khmel'nitskii, D.E., 1974, Fiz. Tverd. Tela, **16**, 3188 (1975, Sov. Phys. Solid State, **16**, 2079).
Kino, Y. and B. Luthi, 1971, Solid State Commun. **9**, 805.
Kino, Y., B. Luthi and M.E. Mullen, 1972, J. Phys. Soc. Jpn. **33**, 687.
Kobayaski, J., I. Mitzutani, H. Hara, N. Yamada, O. Nakada, A. Kumada and H. Schmid, 1970, J. Phys. Soc. Jpn. **28**, 67.
Krasser, W. and S. Haussuhl, 1976, Solid State Commun. **20**, 191.
Krasser, W., U. Buchenau and S. Haussuhl, 1976, Solid State Commun. **18**, 287.
Krasser, W., B. Janik, K.D. Ehrhardt and S. Haussuhl, 1979, Solid State Commun. **30**, 33.
Kruger, J.K., H.D. Maier, J. Petersson and H.G. Unruh, 1980, Ferroelectrics, **25**, 621.
Kudo, S. and T. Hikita, 1978, J. Phys. Soc. Jpn. **45**, 1775.
Lagakos, N. and H.Z. Cummins, 1974, Phys. Rev. B **10**, 1063.
Lagakos, N. and H.Z. Cummins, 1975, Phys. Rev. Lett. **34**, 883.
Landau, L.D., and E.M. Lifshitz, 1970, Theory of Elasticity, (second edition) (Pergamon Press, Oxford).
Larkin, A.I. and D.E. Khmel'nitskii, 1969, Zh. Eksp. Teor. Fiz. **56**, 2087 (Sov. Phys. JETP, **29**, 1123).
Lavrencic, B.B. and J. Petzelt, 1977, J. Chem. Phys. **67**, 3890.
Lavrencic, B.B., I. Levstek, M. Copic and S. Trost, 1976a, Ferroelectrics, **14**, 637.
Lavrencic, B.B., I. Levstek, M. Copic and S. Trost, 1976b, Brillouin Scattering in $NaH_3(SeO_3)_2$, in: Light Scattering in Solids, ed., M. Balkanskii (Flammarion Sciences, Paris) p. 928.
Lavrencic, B.B., M. Copic, M. Zgonik and J. Petzelt, 1978, Ferroelectrics, **21**, 325.
Lazay, P.D. and P.A. Fleury, 1971, Temperature dependence of the Brillouin–Raman spectrum of $BaTiO_3$, in: Light Scattering in Solids, ed., M. Balkanski (Flammarion Sciences, Paris) p. 406.
Lazay, P.D., J.H. Lunacek, N.A. Clark and G.B. Benedek, 1969, The Rayleigh–Brillouin spectra of ammonium chloride, in: Light Scattering Spectra of Solids, ed., G.B. Wright (Springer, New York) 593.
Levanyuk, A.P. and D.G. Sannikov, 1974, Sov. Phys. Usp. **17**, 199.
Levanyuk, A.P. and D.G. Sannikov, 1976a, Fiz. Tverd. Tela, **18**, 423 (Sov Phys. Solid State, **18**, 245).
Levanyuk, A.P. and D.G. Sannikov, 1976b, Fiz. Tverd. Tela, **18**, 1927 (Sov. Phys. Solid State, **18**, 1122).
Levy-Mannheim, C., M. Djabourov, J. Leblond and P.H.E. Meijer, 1974, Phys. Lett. **A 50**, 75.
Lines, M.E. and A.M. Glass, 1977, Principles and Applications of Ferroelectrics and Related Materials (Clarenden Press, Oxford).

Lockwood, D.J., J.W. Arthur, W. Taylor and T.J. Hosea, 1976, Solid State Commun. **20**, 703.
Lockwood, D.J., A.F. Murray and N.L. Rowell, 1980, J. Phys. **C 14**, 753.
Luspin, Y., 1976, Ph. D. Thesis: Effet Brillouin dans certains cristaux ferroelectriques ou ferroelastiques au voisinage de leur point de transition (Université d'Orleans, France, unpublished).
Luspin, Y. and G. Hauret, 1974, J. Phys. Lett. (Paris) **35**, L-193.
Luspin, Y. and G. Hauret, 1976a, Ferroelectrics, **13**, 347.
Luspin, Y. and G. Hauret, 1976b, Phys. Status Solidi, **B 76**, 551.
Luspin, Y. and G. Hauret, 1977, Ferroelectrics, **15**, 43.
Luspin, Y., G. Hauret and J.P. Chapelle, 1974, C.R. Acad. Sci. **279**, 5.
Luspin, Y., J.L. Servoin and F. Gervais, 1980a, J. Phys. **C 13**, 3761.
Luspin, Y., J.L. Servoin and F. Gervais, 1980b, Ferroelectrics, **25**, 527.
Luthi, B. and W. Rehwald, 1981, Ultrasonic studies near structural phase transitions, in: Topics in Current Physics, vol. 23: Structural Phase Transitions I, eds., K.A. Muller and H. Thomas (Springer, Berlin).
Lyons, K.B. and H.J. Guggenheim, 1979, Solid State Commun. **31**, 285.
Lyons, K.B., T.J. Negran and H.J. Guggenheim, 1980, J. Phys. **C 13**, L415.
Lyons, K.B., R.N. Bhatt, T.J. Negran and H.J. Guggenheim, 1982, Phys. Rev. **B 25**, 1791.
Makita, Y., F. Sakurai, T. Osaka and I. Tatsuzaki, 1977, J. Phys. Soc. Jpn. **42**, 518.
Melcher, R.L., 1979, Research Report, IBM RC 7247 (# 31227), 1.
Melcher, R.L. and B.A. Scott, 1972, Phys. Rev. Lett. **28**, 607.
Melcher, R.L., E. Pytte and B.A. Scott, 1973, Phys. Rev. Lett. **31**, 307.
Michel, K.H., 1976, Nuclear magnetic resonance and incoherent neutron scattering in $NH_4Cl$ (theory and experiment) in: Local Properties at Phase Transitions, Proc. Int. Sch. Phys. Enrico Fermi, Course LIX, Varenna, eds., K.A. Muller and A. Rigamonti (North-Holland, Amsterdam).
Michel, K.H., 1981, Phys. Rev. B **24**, 3998.
Miller, P.B. and J.D. Axe, 1967, Phys. Rev. **163**, 924.
Miyakawa, K. and T. Yagi, 1980, J. Phys. Soc. Jpn. **49**, 1881.
Moudden, A.H., F. Denoyer, J.P. Benoit and W. Fitzgerald, 1978, Solid, Solid State Commun. **28**, 575.
Murray, A.F. and D.J. Lockwood, 1980, Light scattering from excitations in incommensurate $BaMnF_4$: phase and amplitude modes, in: Proc. VIIth Int. Conf. on Raman Spectroscopy, Ottawa, ed., W.F. Murphy (North-Holland, Amsterdam) p. 58.
Naudts, J., 1981, Solid State Commun. **38**, 1233.
Negran, T.J., 1981, Ferroelectrics, **34**, 31.
Newnham, R.E. and L.E. Cross, 1974, Mat. Res. Bull. **9**, 927, 1021.
O'Brien, E.J. and T.A. Litovitz, 1964, J. Appl. Phys. **35**, 180.
O'Keefe, M., 1976, Phase transitions and translational freedom in solid electrolytes, in: Superionic Conductors, eds., G.D. Mahan and W.L. Roth (Plenum, New York) p. 101.
Ohi, K., M. Kimura, H. Ishida and H. Kakinuma, 1979, J. Phys. Soc. Jpn. Lett. **46**, 1387.
Paquet, D. and J. Jerphagnon, 1980, Phys. Rev. B **21**, 2962.
Peercy, P.S., 1973, Phys. Rev. Lett. **31**, 379.
Peercy, P.S., 1976, Soft mode systems: SbSI and rare earth pentaphosphates, in: Proc. Vth Int. Conf. on Raman Spectroscopy, Freiburg, eds., E.D. Schmid, J. Brandmuller, W. Kiefer, B. Schrader and H.W. Schrotter (Hans Ferdinand Schulz, Freiburg) p. 571.
Peercy, P.S., 1978, Pressure dependence of Raman scattering in ferroelectrics and ferroelastics, in: High-Pressure and Low-Temperature Physics, eds., C.W. Chu and J.A. Woollam (Plenum, New York) p. 279.
Peercy, P.S. and I.J. Fritz, 1974, Phys. Rev. Lett. **32**, 466.
Peercy, P.S. and G.A. Samara, 1972, Phys. Rev. B **6**, 2748.

Peercy, P.S., I.J. Fritz and G.A. Samara, 1975, J. Phys. Chem. Solids, **36**, 1105.
Pelous, J. and R. Vacher, 1976, Solid State Commun. **18**, 657.
Pinczuk, A., G. Burns and F.H. Dacol, 1977, Solid State Commun. **24**, 163.
Pinczuk, A., G. Burns and F.H. Dacol, 1978, Linear coupling between soft optical and acoustical modes in ferroelastic $BiVO_4$, in: Lattice Dynamics, ed., M. Balkanski (Flammarion Sciences, Paris) p. 646.
Pinczuk, A., B. Welber and F.H. Dacol, 1979, Solid State Commun. **29**, 515.
Prokhorova, S.D., G.A. Smolensky, I.G. Siny, E.G. Kuzimov, V.D. Mikvabia and H. Arndt, 1980, Ferroelectrics, **25**, 629.
Putley, E.H., 1977, Thermal detectors, in: Optical and Infrared Detectors (Topics in Applied Physics, Vol. 19) ed., R.S. Keyes (Springer, Berlin) p. 91.
Pytte, E., 1971, Phys. Rev. B **3**, 3503.
Quilichini, M., J.F. Ryan, J.F. Scott and H.J. Guggenheim, 1975, Solid State Commun. **16**, 471.
Reese, R.L., I.J. Fritz and H.Z. Cummins, 1973, Phys. Rev. B **7**, 4165.
Rehaber, E. and W. Dultz, 1980, The Phase Diagram in RbCN, in: Proc. VIIth Int. Conf. on Raman Spectroscopy, Ottawa, ed., E.F. Murphy (North-Holland, Amsterdam) p. 58.
Rehwald, W., 1973, Adv. Phys. **22**, 721.
Rehwald, W., 1977, Solid State Commun. **21**, 667.
Rehwald, W., 1978a, J. Phys. C **11**, L157.
Rehwald, W., 1978b, Ultrasonic study of the phase transition in squaric acid, in: Lattice Dynamics, ed., M. Balkanski (Flammarion Sciences, Paris) p. 644.
Rehwald, W., 1980, Ferroelectrics, **24**, 281.
Rehwald, W. and A. Vonlanthen, 1978, Phys. Status Solidi B **90**, 61.
Rehwald, W., J.R. Sandercock and M. Rossinelli, 1977, Phys. Status Solidi, A **42**, 699.
Rehwald, W., A. Vonlanthen, J.K. Kruger, R. Wallerius and H.G. Unruh, 1980a, J. Phys. C **13**, 3823.
Rehwald, W., A. Vonlanthen, E. Rehaber and W. Prettl, 1980b, Z. Phys. B – Condensed Matter, **39**, 299.
Rosasco, G.J., C. Benoit and A. Weber, 1971, Light scattering study of the attenuation and dispersion of hypersound in ammonium bromide, in: Light Scattering in Solids, ed., M. Balkanski (Flammarion Sciences, Paris) p. 483.
Sailer, E. and H.G. Unruh, 1975, Solid State Commun. **16**, 615.
Sailer, E. and H.G. Unruh, 1976, Ferroelectrics, **12**, 285.
Sailer, E., C. Konak, H.G. Unruh and A. Fouskova, 1975, Phys. Status Solidi, A **29**, K73.
Sandercock, J.R., 1980, Light scattering from thermally excited surface phonons and magnons, in: Proc. VIIth Int. Conf. on Raman Spectroscopy, ed., W.F. Murphy (North-Holland) p. 364.
Sandercock, J.R., S.B. Palmer, R.J. Elliott, W. Hayes, S.R.P. Smith and A.P. Young, 1972, J. Phys. **C 5**, 3126.
Sapriel, J., 1975, Phys. Rev. **B 12**, 5128.
Sapriel, J., A. Boudou and A. Perigaud, 1979, Phys. Rev. **B 19**, 1484.
Satija, S.K. and C.H. Wang, 1977, J. Chem. Phys. **66**, 2221.
Satija, S.K. and C.H. Wang, 1978, Solid State Commun. **28**, 617.
Sawada, A., M. Udagawa and T. Nakamura, 1977, Phys. Rev. Lett. **39**, 829.
Sawada, A., A. Hattori and Y. Ishibashi, 1980, J. Phys. Soc. Jpn. **49**, 423.
Sawada, A., J. Sugiyama and Y. Ishibashi, 1981, (Proc. IMF 5) Ferroelectrics, **36**, 385.
Schmid, H., 1970, J. Phys. Soc. Jpn. **28**, 354.
Schneck, J. and F. Denoyer, 1981, Phys. Rev. **B 23**, 383.
Schneck, J., J. Primot, R. Von der Muhll and J. Ravez, 1977, Solid State Commun. **21**, 57.
Schwabl, F., 1980a, Ferroelectrics, **24**, 171.
Schwabl, F., 1980b, Lecture Notes in Physics, **115**, 432 (Springer, Berlin).
Scott, J.F., 1974, Rev. Mod. Phys. **46**, 83.
Scott, J.F., 1979, Rep. Prog. Phys. **42**, 1055.

Scott, J.F., F. Habbal and M. Hidaka, 1981, Bull. Am. Phys. Soc. **26**, 303.
Semmingsen, D. and J. Feder, 1974, Solid State Commun. **15**, 1369.
Shapiro, S.M., 1968, Ph.D. Thesis: Light scattering studies of the alpha beta phase transition in Quartz (Johns Hopkins Univ. Baltimore unpublished).
Shapiro, S.M. and H.Z. Cummins, 1968, Phys. Rev. Lett. **21**, 1578.
Shimizu, H., M. Tsukamoto, Y. Ishibashi and M. Umeno, 1974, J. Phys. Soc. Jpn. **36**, 498.
Shirane, G. and J.D. Axe, 1971, Phys. Rev. Lett. **27**, 1803.
Sinii, I.G., S.D. Prokhorova, E.G. Kuzmimov, V.D. Mikvabia and T.M. Polkholvskaya, 1979, Bull. Acad. Sci. USSR, Phys. Ser. **43**, 93.
Smirnov, P.S., B.A. Strukov, V.S. Gorelik and E.F. Dudnik, 1979, Bull. Acad. Sci. USSR, Phys. Ser. **43**, 1011.
Smolenskii, G.A., S.D. Prokhorova, I.G. Sinii and E.O. Chernyshova, 1971, Bull. Acad. Sci. USSR, Phys. Ser. **41**, # 3, 125.
Smolenskii, G.A., I.G. Sinii, Kh. Arndt, S.D. Prokhorova, E.G. Kuz'minov, V.D. Mikvabiya and N.N. Kolpakova, 1979, Bull. Acad. Sci. USSR, Phys. Ser. **43**, 99.
Stock, M. and W. Dultz, 1979, Phys. Status Solidi, **A 53**, 237.
Sussner, H. and R. Vacher, 1979, Applied Optics, **18**, 3815.
Tanaka, H. and I. Tatsuzaki, 1979, J. Phys. Soc. Jpn. **47**, 878.
Tanaka, H. and I. Tatsuzaki, 1981, J. Phys. Soc. Jpn. **50**, 2006.
Tanaka, H., T. Yagi and I. Tatsuzaki, 1978, J. Phys. Soc. Jpn. Lett. **44**, 2009.
Taylor, W., D.J. Lockwood and H. Vass, 1978, Solid State Commun. **27**, 547.
Tekippe, V.J. and L.L. Abels, 1977, Phys. Lett. **60**, 129.
Timmermans, J., 1961, J. Phys. Chem. Solids, **18**, 1.
Todo, I. and I. Tatsuzaki, 1974, J. Phys. Soc. Jpn. **37**, 1477.
Toledano, J.C., 1974, Ann. Telecommun. **29**, 249.
Toledano, J.C., 1978, Symmetry determined phenomena at crystalline phase transitions, in: Conf. Solid State Chem., Strasbourg (1979, J. Solid State Chem. **27**, 41).
Toledano, J.C., 1979, Phys. Rev. **B 20**, 1147.
Toledano, J.C. and M. Busch, 1975, J. Phys. Lett. (Paris) **36**, L141.
Toledano, J.C. and P. Toledano, 1980, Phys. Rev. **B 21**, 1139.
Toledano, J.C., L. Pateau, J. Primot, J. Aubree and D. Morin, 1975, Mater. Res. Bull. **10**, 103.
Toledano, J.C., M. Busch and J. Schneck, 1976a, Ferroelectrics, **13**, 327.
Toledano, J.C., G. Errandonea and J.P. Jaguin, 1976b, Solid State Commun. **20**, 905.
Toledano, P. and J.C. Toledano, 1976, Phys. Rev. **B 14**, 3097.
Toledano, P. and J.C. Toledano, 1977, Phys. Rev. **B 16**, 386.
Tominaga, Y., M. Udagawa, S. Ushioda, T. Nakamura and H. Urabe, 1980, Hypersonic disperion in tetragonal $BaTiO_3$ measured by double-axis Brillouin spectroscopy and determination of dielectric constant (unpublished).
Torres, J., 1975, Phys. Status Solidi, **B 71**, 141.
Tsujimi, Y., T. Yagi, H. Yamashita and I. Tatsuzaki, 1981, J. Phys. Soc. Jpn. **50**, 184.
Udagawa, M., K. Kohn and T. Nakamura, 1978a, J. Phys. Soc. Jpn. **44**, 1873.
Udagawa, M., K. Kohn and T. Nakamura, 1978b, Ferroelectrics, **21**, 329.
Udagawa, M., Y. Tominaga, K. Kohn, T. Nakamura and M. Maeda, 1979, J. Phys. Soc. Jpn. **47**, 869.
Unruh, H.G., J. Kruger and E. Sailer, 1978a, Ferroelectrics, **20**, 3.
Unruh, H.G., E. Sailer, H. Hussinger and O. Ayers, 1978b, Solid State Commun. **25**, 871.
Uwe, H. and H. Tokumoto, 1979, Phys. Rev. **B 19**, 3700.
Vacher, R., M. Boissier and J. Sapriel, 1981, Phys. Rev. **B 23**, 215.
Wada, M., Y. Nakayama, A. Sawada, S. Tsunekewa and Y. Ishibashi, 1979, J. Phys. Soc. Jpn. **47**, 1575.
Wang, C.H. and S.K. Satija, 1977, J. Chem. Phys. **67**, 851.

Wergin, A., W. Krasser and J.P. Boon, 1977, Mol. Phys. **34**, 1637.
Windsch, W., 1976, The ferroelectric phase transition of tris-Sarcosine calcium chloride (TSCC), in: Local Properties at Phase Transitions, Proc. Int. Sch. Phys. Enrico Fermi, Course LIX, Varenna, eds., K.A. Muller and A. Rigamonti (North-Holland, Amsterdam)
Worlock, J.M., 1971, Light scattering studies of structural phase transitions, in: Structural Phase Transitions and Soft Modes, eds., E.J. Samuelsen, E. Andersen and J. Feder (Universitetsforlaget, Oslo) p. 329.
Yagi, T. and I. Tatsuzaki, 1974, J. Phys. Soc. Jpn. **37**, 1038.
Yagi, T., I. Tatsuzaki and I. Todo, 1974, J. Phys. Soc. Jpn. **37**, 1717.
Yagi, T., H. Tanaka and I. Tatsuzaki, 1976a, J. Phys. Soc. Jpn. **41**, 717.
Yagi, T., M. Tokunaga and I. Tatsuzaki, 1976b, J. Phys. Soc. Jpn. **40**, 1659.
Yagi, T., H. Tanaka and I. Tatsuzaki, 1977, Phys. Rev. Lett. **38**, 609.
Yagi, T., M. Cho and Y. Hidaka, 1979, J. Phys. Soc. Jpn. **46**, 1957.
Yagi, T., Y. Hidaka and M. Miura, 1980, J. Phys. Soc. Jpn. **48**, 2165.
Yamanaka, A., M. Kasahara and I. Tatsuzaki, 1981, J. Phys. Soc. Jpn. **50**, 735.
Yao, W., H.Z. Cummins and R.H. Bruce, 1981, Phys. Rev. B **24**, 424.
Yoshihara, A., T. Fujimura and K.I. Kamiyoshi, 1978, J. Phys. Soc. Jpn. **44**, 1241.
Young, P.W. and J.F. Scott, 1981, Bull. Am. Phys. Soc. **26**, 303.
Zaitseva, M.P., A.I. Krupnyi, Yu.I. Kokorin and V.S. Krasikov, 1975, Izv. Akad. Nauk. SSSR, Ser. Fiz. **39**, 954.

CHAPTER 7

# Central Peaks near Structural Phase Transitions

P.A. FLEURY and K.B. LYONS

*Bell Laboratories*
*Murray Hill, New Jersey 07974*
*USA*

*Light Scattering near Phase Transitions*
*Edited by*
*H.Z. Cummins and A.P. Levanyuk*

© *North-Holland Publishing Company, 1983*

# Contents

## 1. Introduction

The study of dynamical aspects of structural phase transitions entered the modern era in about 1960 when Cochran and Anderson independently introduced the concept of the soft mode in connection with displacive ferroelectric transitions. It was soon realized that this concept is not confined to ferroelectric phase transitions which are associated with soft modes of zero wave vector, but is also applicable to those transitions where the soft mode wave vector lies at other points in the Brillouin zone.

However, even as the application of the soft mode concept widened, it became apparent that it provided an insufficiently detailed description of low-frequency order parameter dynamics close to structural phase transitions. One of the most striking features of the scattered spectrum for either neutrons or light, namely the appearance of a diverging elastic or quasielastic peak centered near zero frequency shift, lies entirely outside the quasiharmonic soft mode description of the dynamics. The first observations of this feature using neutron scattering, in strontium titanate (Riste et al. 1971), precipitated much of the subsequent interest in the so-called "central peak question" over the past decade. However, the observation of a divergence in scattered intensity dates back at least to 1956 when Yakovlev and coworkers (1956) reported a striking increase in the light scattered from quartz near its alpha–beta transition. Other early examples include the observation of so-called critical Rayleigh scattering in lithium tantalate (Johnston and Kaminow 1968) and in ammonium chloride (Lazay et al. 1969). These and similar observations shared the common feature that close to the transition the scattered intensity increased dramatically, sometimes by as much as a factor of 10 000. They also shared with the neutron scattering studies the failure to measure any finite linewidth to the diverging spectral feature. It was only possible to place a maximum value of $\sim 1\ \mathrm{cm}^{-1}$ on that width. This failure, of course, translates into an inability to distinguish whether the scattering arises from a static or dynamic phenomenon.

The shortcoming of the quasiharmonic soft mode model prompted theoretical attempts to resolve the problem. A number of theoretical models appeared, but the observable differences among their predictions consistently lay in the then-inaccessible spectral region below $1\ \mathrm{cm}^{-1}$. Consequently definitive tests of those models were not possible. Expectations grew that when the puzzle was finally solved a single explanation covering all of the systems and observations would suffice. With the development of experimental techniques to probe the previously inaccessible spectral region below $1\ \mathrm{cm}^{-1}$, it has now become clear that this expectation was inappropriate. Rather there are a number of mechanisms responsible for central peak components in various frequency domains, and often two or more of these are active in a single material at the same time.

The proposed central peak mechanisms include: entropy fluctuations, phonon density fluctuations, dielectric relaxations, molecular reorientations, phasons, over-damped soft modes, solitons, degenerate electronic transitions, dynamic and static clusters and domains, as well as a host of phenomena related to impurities, vacancies, strains, dislocations, etc. The question of most fundamental importance was whether the central peaks observed reflected the intrinsic critical dynamics of the phase transitions. If so, then a second characteristic time, not contained in the soft mode picture, would be needed to describe the critical dynamics. On the other hand, features associated with defects or other extrinsic causes, while still intriguing, engage our interest from quite a different perspective. As we shall discuss below, it is now clear that both types of mechanisms play significant roles.

The relative contributions of the various proposed mechanisms in any given instance depend on the quantitative size of certain interaction terms in the Hamiltonian. The present status of lattice dynamical theories in strongly anharmonic systems do not permit *a priori* prediction of which mechanism will dominate any given situation. Therefore, interpretation of experimental data requires careful consideration of a variety of mechanisms, both intrinsic and extrinsic, and, in most cases, is impossible in the absence of detailed and quantitative lineshape information. One purpose of this chapter is to review the recent advances in light scattering techniques which have made such information available.

In the next section, we present a summary of the relevant theoretical ideas. We then discuss in sect. 3 the various experimental techniques relevant to light scattering central peak investigations. Section 4 contains summaries of the experimental results on seven types of prototypical systems – the cooperative Jahn–Teller transition, the order–disorder ferroelectric, the displacive transition, the plastic crystal, the superionic conductor, the incommensurate phase transition, and the defected crystal. In each case we emphasize one or two representative materials which have received the most complete experimental characterization. It will be evident in several cases that more than one central peak mechanism is at work. Finally we summarize in sect. 5 the present state of our understanding of central peak phenomena, and pose some remaining questions which suggest opportunities for future research.

## 2. Theory

In this section we present the theoretical background for central peak studies. We begin with a summary of the light scattering process itself, and then proceed

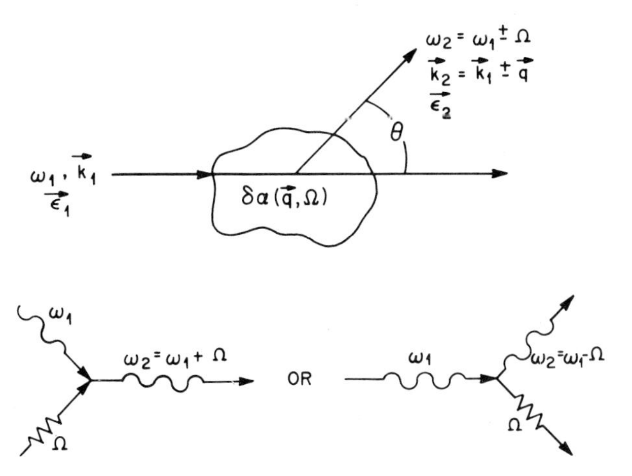


Fig. 1. Geometry of a typical light scattering experiment. The scattering angle $\theta$ determines the scattering momentum transfer $\boldsymbol{q}$. The vector $\boldsymbol{k}_{i,s}$ represent the incident and scattered wavevectors, respectively, $\boldsymbol{q} = \boldsymbol{k}_s - \boldsymbol{k}_i$, while $\omega_{i,s}$ represent the respective frequencies and $\epsilon_{i,x}$ represent the polarization vectors. The inset shows typical excitation diagrams for Stokes and anti-Stokes scattering.

to consider the various proposed central peak mechanisms. The theoretical challenges fall into two categories: (i) to explain the very small central peak linewidths via intrinsic processes involving combinations of the much higher frequencies usually associated with optical phonons and soft modes or (ii) to explicate the possible influences of crystal defects on the dynamics of structural phase transitions.

### *2.1. Light scattering and correlation functions*

Figure 1 shows the geometry of a typical light scattering experiment. The light is incident along a direction $\boldsymbol{k}_\mathrm{i}$, and is scattered along a direction $\boldsymbol{k}_\mathrm{s}$. In most cases, the incident and scattered polarizations are also uniquely defined, usually with reference to the scattering plane formed by $\boldsymbol{k}_\mathrm{i}$ and $\boldsymbol{k}_\mathrm{s}$. In this notation, V refers to a polarization perpendicular to the scattering plane, while H refers to one parallel to it. An alternate notation is a specification such as $\alpha(\beta\gamma)\delta$, where $\alpha$ and $\delta$ refer respectively to the directions of the incident and scattered light, and $\beta$ and $\gamma$ give the respective electric polarization vectors. A light scattering spectrum probes a particular Fourier component of the fluctuations in the polarizability of the medium, that component being $\boldsymbol{q} = \boldsymbol{k}_j - \boldsymbol{k}_i$. For a scattering medium with refractive index $n$, at a scattering angle $\theta$, with incident light of wavelength $\lambda$, then, to a very good approximation, we have $|\boldsymbol{q}| = (4\pi n/\lambda)\sin(\theta/2)$. This relationship is a statement of momentum conservation among the incident and scattered light and the excitation causing the scattering. Similarly, energy conservation requires that $\omega_i - \omega_s = \pm\Omega$, where $\omega_{i,s}$ represent the incident and scattered frequencies and $\Omega$ is the frequency of the excitation. The choice of the + or − sign depends on whether the excitation is created or destroyed, respectively, in the scattering process.

The spectrum of the scattered intensity, $I(\boldsymbol{q}, \omega)$ is given by the Fourier transform of the autocorrelation function of the scattered field. Hence, that spectrum measures the autocorrelation function of the polarizability operator $\alpha(\boldsymbol{r}, t)$. Thus:

$$I(\boldsymbol{q}, \omega) = \mathrm{FT} I(\boldsymbol{r} - \boldsymbol{r}', t - t') = \text{const FT}\langle\alpha(\boldsymbol{r}, t)\alpha(\boldsymbol{r}, t')\rangle, \tag{1}$$

where FT represents the Fourier transform in both space and time. The fluctuation dissipation theorem then relates the spectrum to $\chi$, the dynamic susceptibility, as:

$$I(\boldsymbol{q}, \omega) = -\frac{\hbar}{\pi}(1 - \mathrm{e}^{-\beta\hbar\omega})^{-1} \operatorname{Im} \chi(\boldsymbol{q}, \omega). \tag{2}$$

In general, $\alpha(\boldsymbol{r}, t)$ can be expanded in powers of the dynamic variables $U_i$ of the medium. Expanding to first order in these, we have

$$\alpha(\boldsymbol{r}, t) = \alpha_0 + \sum_i A_i U_i(\boldsymbol{r}, t). \tag{3}$$

Here $U_i(\boldsymbol{r}, t)$ is linear in the creation and annihilation operators for the single elementary excitations. The term $\alpha_0$ has no effect on the scattered spectrum, so we ignore it. Symmetry considerations select the nonzero coefficients $A_i$. For the situations considered in this article, the $U_i$ usually represent atomic displacements. One set of these displacements, and the corresponding $U_s$, will be proportional to the order parameter of a structural phase transition.

Thus, in first order soft mode scattering $\alpha(\boldsymbol{r}, t) = A_s U_s(\boldsymbol{r}, t)$, and the observed spectrum, eq. (2), is quite directly related to the single excitation Green's function:

$$G_s(\boldsymbol{r}, t \,|\, \boldsymbol{r}', t') = G_{UU}(\boldsymbol{r}, t \,|\, \boldsymbol{r}', t') = -\mathrm{i}\langle U_s(\boldsymbol{r}', t) U_s(\boldsymbol{r}', t')\rangle, \tag{4}$$

using eq. (1).

Higher-order terms are always present. When $A_s$ is vanishingly small or identically zero (e.g., via symmetry requirements), $\alpha(\boldsymbol{r}, t)$ is dominantly second order: e.g., $\alpha(\boldsymbol{r}, t) = A_s^{(2)} U_s(\boldsymbol{r}, t) U_s(\boldsymbol{r}, t)$. Here the spectrum is determined by the Fourier transform of the four-point correlation function, as discussed, for example, by Fleury (1978):

$$\chi_{\mathrm{II}} = \chi_{UUUU} = -\mathrm{i}\langle (A_2 U_s(\boldsymbol{r}, t) U_s(\boldsymbol{r}, t) A_2 U_s(\boldsymbol{r}, t') U_s(\boldsymbol{r}, t')\rangle, \tag{5}$$

which, for $A_2 = \text{const}$, is directly proportional to the two particle Green's function: (Fleury 1978) $G_{\mathrm{II}}(\boldsymbol{r}, t \,|\, \boldsymbol{r}, t')$. If the excitations do not interact, $G_{\mathrm{II}}$

is simply a self-convolution of $G_{\mathrm{I}}$. It may, however, be renormalized by interactions among the elementary excitations. Thus, the scattered spectra of two-phonon sum and difference processes may not be simply related to the one-phonon spectrum.

In the more general problem of pairwise interactions among $N$ such excitations (Fleury 1978), the Hamiltonian will contain terms of the form $g_{ij}U_iU_j$. The coupled susceptibilities for this system can be simply derived from the equations of motion $\ddot{U}_i = \Sigma_j g_{ij} U_j = F_i$, where $F_i$ represent the external forces or fields which couple linearly to the amplitude $U_i$. The off-diagonal coupling coefficients $g_{ij}$ are in general complex. The diagonal elements represent the uncoupled mode susceptibilities ('free excitation' Green's functions) $g_{ii} = (G_i^0)^{-1}$. The dynamic susceptibility (2) still gives the scattered spectrum, but with $\chi(\boldsymbol{q}, \omega)$ replaced by $\chi_T$, where

$$\chi_T(\boldsymbol{q}, \omega) = \sum_{ij}^{N} A_i A_j G_{ij}(\boldsymbol{q}, \omega)\,, \tag{6}$$

where $\boldsymbol{G} \equiv \boldsymbol{g}^{-1}$.

When only two modes participate, $\boldsymbol{G}$ takes the simple form

$$G_{ii} = \frac{G_i^0}{1 - g_{12}^2 G_1^0 G_2^0}\,, \qquad G_{12} = \frac{g_{12} G_1^0 G_2^0}{1 - g_{12}^2 G_1^0 G_2^0}\,. \tag{7}$$

Note that not only are the diagonal terms modified, but the interference term $G_{12}$ may introduce additional structure. This simple case often provides an adequate description of a real experimental system. However, we note that the hybridization described by the expressions (7) is not restricted to the case where $G_i^0$ describes a single excitation. In particular, the formalism may describe pair-wise coupling, in which case $G_i^0$ in fact describes a pair excitation. This use of eq. (7) occurs at several places in this chapter, with $G_1^0$ representing a quasiharmonic acoustic phonon response and $G_2^0$ representing a soft mode with different forms of anharmonicity present. This anharmonicity is usually introduced by writing the soft mode response function in the form $G_2^0(\boldsymbol{q}, \omega) = G_{\mathrm{qh}}^0(\boldsymbol{q}, \omega) + \Sigma(\boldsymbol{q}, \omega)$, where $\Sigma$ represents a self-energy correction to the quasiharmonic response function $G_{\mathrm{qh}}$. The relevant forms of $\Sigma$ will be identified where appropriate below. Some examples of this kind of coupling are included in ch. 1 by Ginzburg et al. in this volume (see sect. 3.2).

### 2.2. *Entropy fluctuations and phonon density fluctuations*

In this section we consider the related phenomena of entropy fluctuations and phonon density fluctuations. Since the relation between these has not always been made clear, we shall attempt to explicate it here.

Light scattering by entropy (or temperature) fluctuations in fluids is well known. The heat diffusion equation for a homogeneous medium yields solutions in Fourier transform space ($\boldsymbol{q}$) of the form

$$T_q(t) = T_q^0 \, e^{-\Gamma_q t}, \tag{8}$$

where $\Gamma_q = D_{\rm th} q^2$ and the thermal diffusivity is $D_{\rm th} = \Lambda/\rho C_p$. The symbols $\Lambda$ and $C_p$ represent, respectively, the thermal conductivity and specific heat, while $\rho$ is the mass density. For crystalline solids $\Lambda$ and $D_{\rm th}$ become tensor quantities, and we obtain $\Gamma_q = \boldsymbol{q} \cdot \boldsymbol{D}_{\rm th} \cdot \boldsymbol{q}$. A light scattering experiment probes a particular $\boldsymbol{q}$ component of the temperature fluctuation. Hence, in frequency space, the spectrum should consist of a single Lorentzian centered at $\omega = 0$ with a width $\Gamma_q$. Since fluctuations in the polarizability actually cause the scattering, the intensity of entropy fluctuation scattering is governed by the temperature derivative of the refractive index, $I_{\rm ef} \propto (\mathrm{d}n/\mathrm{d}T)^2$.

At sufficiently high frequencies ($\gg \Gamma_q$), this simple analysis must break down. The concept of a temperature implies an equilibrium distribution of phonons. Entropy fluctuations may be viewed as fluctuations in that phonon density. Those fluctuations, however, are not infinitely fine grained in time. The phonons lifetimes $\Gamma_i^{-1}$ measure the average time a given phonon exists before it decays into one or more other phonons. Therefore, on a time scale comparable to $\Gamma_i^{-1}$, the diffusion equation becomes meaningless. In this regime the polarizability due to a fluctuation in the phonon density will not decay smoothly as indicated in eq. (8), but rather will also contain high-frequency noise which has a characteristic frequency related to some average of the $\Gamma_i$. Thus, even at low $q$, the spectrum will consist of the diffusive Lorentzian of width $D_{\rm th} q^2$ plus a component extending out to much higher frequencies, on the order of an inverse phonon lifetime. Since this intensity results from the breakdown of the hydrodynamic description of the collision dominated regime, its intensity is not included in the integrated intensity calculated on thermodynamic grounds for the heat diffusion peak. In fact, there is no simple relationship between the intensities of these components. Hence, the high-frequency component, which has often been referred to as "phonon density fluctuations" (PDF), may even exceed the integrated intensity of the heat diffusion peak.

This description can be related to the more rigorous and formal treatment of Coombs and Cowley (1973). It is, moreover, of interest to note that the comments above apply equally to any diffusion process where there is a clearly identifiable length or time scale on which the primary event of the diffusion takes place. Thus, for example, quite similar arguments can be applied to particle diffusion in ionic conductors, where $\Gamma_i$ becomes the inverse hopping time. For that case, the approach used by Geisel (see sect. 2.5) again places these ideas on a firm mathematical basis.

It is also important to connect this interpretation of the high-frequency wing on the entropy fluctuation component with the idea of "two-phonon difference" (TPD) scattering (Lyons and Fleury 1976b). The latter attributes the light scattering wing to processes wherein one phonon is created at $\boldsymbol{k}$ while another is annihilated at $\boldsymbol{k} \pm \boldsymbol{q}$. In what follows, we shall use the term PDF to describe *all* such processes, including TPD and others of higher order.

PDF processes scatter light even in the absence of any coupling to phase transition phenomena. For example, in $KTaO_3$ both the entropy fluctuation and PDF components have been observed (Lyons and Fleury 1976b) and spectrally resolved. In an effort to elucidate the properties of PDF in a system amenable to microscopic calculation, Lyons et al. (1980a) have studied the low angle scattering from solid xenon near its melting point, $T_m = 161.5\,K$. At $T \sim 0.97 T_m$, the PDF (or TPD) component exhibits a width of $7.6\,cm^{-1}$, while for $T \sim 0.77 T_m$, the corresponding value is $5.1\,cm^{-1}$. An analytic calculation by Leese and Horton (1979) while exhibiting a component similar to that observed, failed to describe it quantitatively. However, a molecular dynamics calculation by Alder et al. (1976) produced results in considerably better agreement with the observed spectral profile.

Although no such detailed theoretical attack on the PDF problem in the general case is in prospect for the near future, it is clear that, even in the simplest crystals, PDF or two-phonon processes do scatter light observably. Moreover, the characteristic frequencies of a few $cm^{-1}$ make these components excellent candidates for interaction with soft modes as the order-parameter fluctuations slow down near a structural phase transition.

In general, the order parameter may couple linearly to temperature (or entropy) fluctuations below $T_c$ and quadratically above (Schwabl 1974, Enz 1974, Ohnari and Takada 1979, Ohnari 1980). These couplings produce a narrow peak in the structure factor both above and below the phase transitions. The relevant form of the self-energy correction to the soft mode response outside the critical region is $\Sigma(\omega) = [\Gamma - i\omega]^{-1}$. The calculations predict that this component will exhibit critical narrowing near $T_c$. Ginzburg et al. (ch. 1) have given an equivalent discussion of these points in terms of phonon anharmonicities. In sect. 4.2 we shall discuss experimental observations in KDP and TGS which bear on these theoretical results.

The coupling of a soft mode to PDF processes was first considered by Cowley and Coombs (1973). Recently, Bruce and Bruce (1980) have treated explicitly the coupling to PDF processes involving modes on the soft branch. They find a relationship between the linewidths of the one- and two-phonon correlation functions for the soft mode coordinate. They write the width of the $n$-phonon scattering as $\omega^{(n)} = \Omega^{(n)} t^{z\nu}$, with $n = 1, 2$, where $\Omega^{(n)}$ is an amplitude parameter, $t$ is reduced temperature $(T - T_c)/T_c$, and $z$ and $\nu$ are the dynamic and static critical exponents, respectively. The ratio of the amplitudes $\Omega^{(1)}/\Omega^{(2)}$ for a defect-free model then depends only on the universality class of the system and not on the microscopic details. For $SrTiO_3$, the predicted ratio is $\sim 0.06$. This implies that whenever a component due to two-soft-phonon scattering is observed, then a narrower component should exist in the one-phonon spectrum (and vice-versa). Based on the measured width of the broad component in $SrTiO_3$, the narrow component is predicted to have a width of order 1 GHz. This theoretical result may be taken as support for the interpretation of the $SrTiO_3$ spectrum presented in sect. 4.3. Indeed, the analysis of Bruce and Bruce requires that such multiple components exist in a defect-free crystal. It is not clear precisely what effects defects may have on their results, but they state that "defects may control the critical slowing down in real systems and promote quasielastic scattering in both one- and two-phonon spectra."

### *2.3. Degenerate electronic levels*

The entropy fluctuation mechanism considered above relies on a thermodynamic rather than a microscopic description of the system. Hence, entropy fluctuation spectra contain no information which is unobtainable by macroscopic observations (i.e., the thermal diffusivity). The PDF component, which in principle contains information related to the microscopic details of the system, has thus far eluded detailed theoretical description. By contrast, in this section we discuss a central peak mechanism for which a quite thorough microscopic theory is available: the cooperative Jahn–Teller transition.

A cooperative Jahn–Teller (CJT) transition is one caused by the electron–phonon interaction. As the temperature is lowered, this interaction makes it energetically favorable for the crystal to distort so as to lift the degeneracy of electronic levels (or shift nondegenerate levels). The energy cost of the structural distortion is repaid by a lowering of the overall electronic energy. Crystals containing rare earth ions often exhibit CJT transitions. Since the electronic states of rare-earth ions are thoroughly studied, both in isolated atoms and in various crystal fields, there is a large body of knowledge to serve as a basis for a microscopic understanding of the transition mechanism in such crystals.

There is a variety of ground state electronic structures which can lead to a CJT transition. In what follows, we shall consider the particular case of $TbVO_4$, for which the experimental data is presented in sect. 4.1. The low-lying electronic levels of the $Tb^{3+}$ ion in the crystal field of $TbVO_4$ exhibit a degenerate doublet interposed between two closely-spaced singlets (Elliott et al. 1972). At a temperature near $T_D = 32.6$ K, the crystal distorts in such a way that the degeneracy of the doublet lifts and the separation of the singlet states increases. The separation between the two singlet states and of the diverging doublet components can be directly observed in the electronic Raman scattering spectrum. The structural distortion is isomorphic to a transverse acoustic mode polarized in the basal plane, which is the appropriate soft mode for this transition.

A complete dynamic theory applicable above $T_D$ has been derived by Hutchings et al. (1975). They give the following response functions for the electronic and phonon degrees of freedom:

$$G_1^0 = \chi_e^0[1 - J\chi_e^0]^{-1},$$

where

$$\chi_e^0 = \sinh(\beta\epsilon)4\epsilon[4\epsilon^2 - \omega^2 - i\omega\Gamma_1]^{-1} + \beta\Gamma_2[\Gamma_2 - i\omega]^{-1}[1 + \cosh(\beta\epsilon)]^{-1},$$

and

$$G_2^0 = 2\omega_a[\omega_a^2 - \omega^2 - i\omega\gamma_a]^{-1}.$$

where $\beta^{-1} = kT$, while $2\epsilon (= 18\,\mathrm{cm}^{-1})$ and $\Gamma_1$ $(= 10\,\mathrm{cm}^{-1})$ are the frequency and width of the singlet–singlet transition measurable in the Raman spectrum. The quantity $\Gamma_2$ represents the width of the doublet, while $\omega_a$ and $\gamma_a$ are, respectively, the frequency and width of the acoustic phonon. The electronic central peak width $\Gamma_2$ together with $T_c$ determine the maximal coupling $g_{12}$, and $J$ is the exchange coupling for nearest-neighbor ions. The functions $G_i^0$ then couple as described by eq. (7) in sect. 2.1. As we shall see in sect. 4.1, the resulting spectra become quite complex due to the interference of the various terms in the sum (6). Despite that complexity, the theoretical predictions lie in complete and quantitative agreement with the observed spectra.

### 2.4. *Rotational relaxation*

In a plastic crystal, the molecular centers of mass exhibit translational order, but the molecular orientations do not. Relaxation of the quasi-free molecular rotations should produce a central peak scattering similar to the Rayleigh wing scattering observed in fluids composed of anisotropic molecules. Furthermore, in any plastic–solid transition, the fluctuations associated with such relaxation should couple strongly to the order parameter, and singular dynamics might be expected for them. Since the frequencies characteristic of the rotational relaxations are typically a few GHz, these modes may also couple with the shear acoustic modes.

The analysis described in sect. 2.1 is directly applicable to this case, with the uncoupled dynamic susceptibility proposed in the phenomenological theory of Bischofberger and Courtens (1971)

$$G_i^0 = 1/(\omega - i\Gamma), \tag{9}$$

where $\Gamma$ is the inverse relaxation time characteristic of the reorientations. Michel and Naudts (1978) and Wang (1979) have constructed a more microscopic formalism for describing the dynamics of coupled translational and rotational modes in molecular crystals. These models are capable of describing all of the existing experimental data, although, as we shall see in sect. 4.5, they have not been used in the transition region.

*2.5. Particle diffusion*

Another class of systems exhibiting central peaks in the light scattering spectrum are superionic conductors. They share the common feature that one or more types of ions in the lattice are delocalized and capable of diffusion or hopping among a variety of available sites. These motions may arise from thermal activation or may be associated with a structural phase transition.

The simple hopping model (Klein 1976) for superionic dynamics assumes: (i) that each particle may reside only on specific sites; (ii) that its average time between hops is $\tau_h$, the "hopping time"; and (iii) that each particle moves independently of the others. Although assumption (iii) is demonstrably incorrect for a concentrated system, this simple model produces the basic features. Thus, in a time $t$, a particle hops randomly an average of $t/\tau_h$ times. If each hop covers a distance $l$ between sites arranged in a cubic lattice, then this random walk leads to a mean-square displacement $\langle z^2 \rangle = (l^2/3)(t/\tau_h) = (l^2/3\tau_h)t$. We thus identify the diffusion constant $D = l^2/6\tau_h$. For times long compared to $\tau_h$ ($\omega \ll \tau_h^{-1}$) and distances $z \gg l(q \ll l^{-1})$, the diffusion equation is valid, and the expected central peak width is $\Gamma_d = Dq^2$.

This scattering component arises from ion charge density fluctuations, where that density is averaged over a distance much larger than a hopping length, as pointed out by Huberman and Martin (1976). Its line width is very small ($\sim 10^{-4}$ GHz for right angle light scattering). In the regime where the diffusion treatment is invalid, Huberman and Martin recognized that an additional component may result if the ions can occupy (at least) two inequivalent sites in each unit cell. They introduced a pseudospin variable, which decays with a time $\gamma^{-1}$ comparable to a hopping time. They further considered various pairwise coupling limits among the pseudospin density, the charge density, and the acoustic phonons. When the phonons couple to pseudospin density fluctuations, a central peak of width $\gamma$ is predicted in the spectrum.

Thus, Huberman and Martin distinguish two classes of dynamic motion, one associated with the hydrodynamic modes of the charged liquid and the other with the microscopic details of individual sites. A similar distinction results in a different way from an extension of the hopping model based on a particle undergoing Brownian motion in a periodic potential. In this model the particles are no longer confined strictly to the sites. For example, Dieterich et al. (1977) consider inelastic scattering due not only to intersite motion but also to oscillations of the ions within the potential wells. Their model predicts spectral features which are both temperature and wave vector dependent.

Geisel (1977), on the other hand, notes that the spectrum of light scattered by particles in a periodic potential will contain not only terms with a spatial frequency $\boldsymbol{q}$, but also those at $\boldsymbol{K}+\boldsymbol{q}$, where $\boldsymbol{K}$ is a reciprocal lattice vector

($K \sim 2\pi/l$). The modulated polarizability of the ion may be written

$$\alpha(\boldsymbol{r}) = \sum_{\boldsymbol{K}} \alpha(\boldsymbol{K})\, e^{i\boldsymbol{K}\cdot\boldsymbol{r}}. \tag{10}$$

The polarizability density is then the sum of expressions like (10) for $r = r_j$, the position of the $j$th particle. Only the broad nondiffusive component of the spectrum that results from $\alpha(K \neq 0) \neq 0$ is observable in Raman scattering. Figure 2 shows the resulting dynamic structure factor for a particle moving in a sinusoidal potential of height $A$, for various values of the $kT/A$. The evolution from oscillatory motion in the wells to diffusional motion among the wells as $kT$ approaches $A$ is evident.

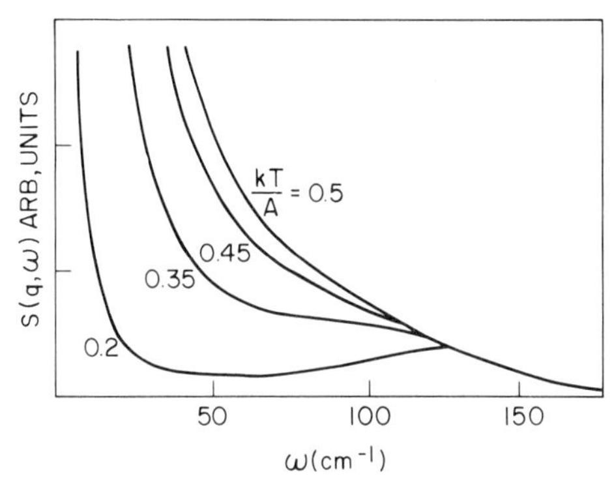


Fig. 2. Dynamic structure factor calculated for a Brownian particle in a sinusoidal potential of amplitude $A$. The numbers labeling the curves represent the quantity $kT/A$. Parameters chosen are typical of $\alpha$-AgI (after Geisel 1977).

We note that these various treatments of the regime where the diffusion equation breaks down differ only in emphasis: there are strong analogies among them. For example the pseudospin density of Huberman and Martin is analogous to the intrasite vibration of Dieterich. We note also that the spectral structure in the regime where the diffusion concept breaks down is completely analogous to the appearance of PDF as a wing on the entropy fluctuation spectrum (sect. 2.2).

Another theoretical attack on ionic dynamics was carried out by Gillan and Dixon (1980) using molecular dynamics techniques. Their model system represents $SrCl_2$, where neutron scattering at large $\boldsymbol{q}$ has revealed quasielastic peaks with measurable linewidths at puzzling locations within the Brillouin zone. The calculated dynamic structure factor (fig. 3) exhibits a quasi-Lorentzian central peak of width $\sim \tau_h^{-1}$ only at temperatures above the onset

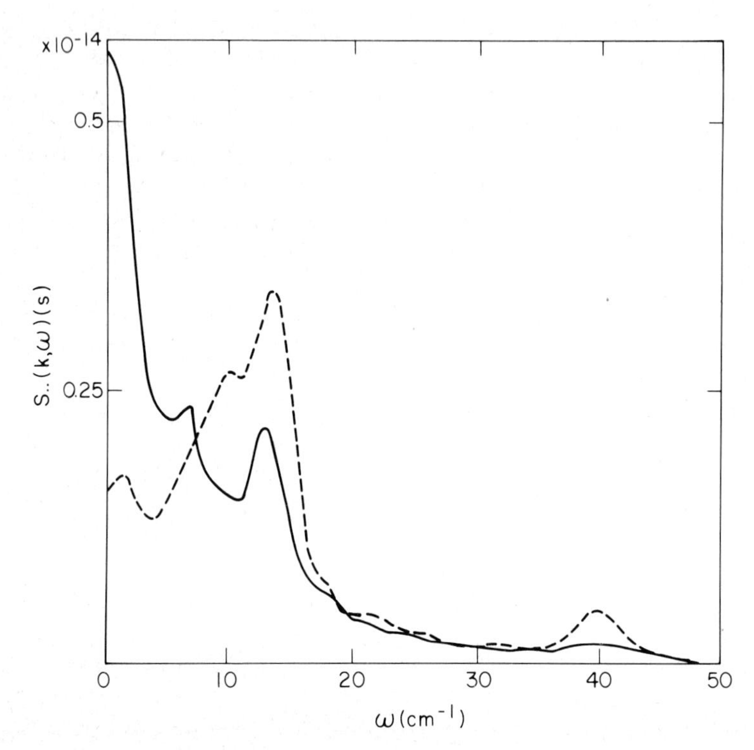


Fig. 3. Partial dynamic structure factor calculated by Gillan and Dixon (1980) by a molecular dynamics simulation of $SrCl_2$ for temperatures of 938 K (broken curve) and 1484 K (solid curve).

of ionic conductivity. The intensity of the peak is strongly dependent upon $q$ and reaches a maximum just beyond the first point in reciprocal space along the (100) direction. Its frequency width approaches $\sim 7\,\text{cm}^{-1}$ for large $\boldsymbol{q}$. Although this value is approximately four times that observed experimentally, the simulation correctly reproduces a number of the previously-puzzling features of the neutron scattering data. The molecular dynamics results naturally do not reflect the very low frequency component $\Gamma \sim Dq^2$, since they are carried out only for times less than 50 ps.

### 2.6. Phasons

In this section we consider a class of excitations peculiar to phases which represent an incommensurate distortion of a parent phase.

An incommensurate phase is defined relative to a parent structure. We may assume that the $l$th parent unit cell contains atoms at a set of equilibrium positions $\{\boldsymbol{u}_\alpha + \boldsymbol{R}_l\}$, where $\alpha$ enumerates the atoms in the cell and $\boldsymbol{R}_l$ is the position of the unit cell (i.e. a lattice vector). An incommensurate distortion of this structure consists of a set of displacements $\{\boldsymbol{\psi}_\alpha^l\}$ having a point symmetry within each unit cell which is a subgroup of the parent lattice point group symmetry. However, these displacements do not have the translational periodicity of the parent lattice. The simplest (but certainly not the only) such displacement field (Bruce and Cowley 1978) is represented by the plane wave approximation

$$\boldsymbol{\psi}_\alpha^l \equiv \psi_0 \boldsymbol{\Delta}_\alpha \cos(\boldsymbol{k}_0 \cdot \boldsymbol{R}_l + \phi)\,, \tag{11}$$

where $\boldsymbol{k}_0$ is a general point in the Brillouin zone of the parent lattice, $\psi_0$ is an overall amplitude, $\phi$ is a phase, and $\boldsymbol{\Delta}_\alpha$ are constant vectors, representing the eigenvector of the distortion.

The incommensurate phase is a curious structure, and requires modification of our usual notion of a crystal, for, in the strictest sense, this "crystal" no longer displays periodicity. Janner and Janssen (1977) have described this situation rigorously in terms of space groups of higher dimensionality, but here we will try to give only an intuitive insight. The experimental observation that $\boldsymbol{k}_0$ is incommensurate with the lattice $\{\boldsymbol{R}_l\}$ can only be made to a finite precision. Likewise, to any desired precision, quasiperiodicity still exists in the (infinite) distorted lattice (the higher the desired precision, the longer the period). Thus, if probed on an atomic scale, the periodicity appears destroyed, and in fact the local symmetry at an arbitrary point $\boldsymbol{R}_l$ may be lower than that of the eigenvector. However, any probe which averages over a finite volume may even exhibit a symmetry higher than that of the eigenvector. By way of illustration, let us assume that the parent phase is centrosymmetric and that $\{\boldsymbol{\Delta}_\alpha\}$ are not. To any desired precision, one may still find a lattice point $\boldsymbol{R}_l$ which exhibits inversion symmetry (i.e., a point for which $\cos(\boldsymbol{k}_0 \cdot \boldsymbol{R}_l) \approx 0$). Thus, to any probe with macroscopic spatial resolution, that crystal still looks centrosymmetric.

These ideas have important consequences for light scattering from the fluctuations of such a distorted phase. Once the eigenvectors $\{\boldsymbol{\Delta}_\alpha\}$ are defined in eq. (11), there are two quantities which define the order parameter below $T_i$: the amplitude $\psi_0$ and the phase $\phi$. In the single plane wave approximation (11) these two quantities may fluctuate independently, yielding two noninteracting branches in the excitation spectrum (Bruce and Cowley 1978), as shown in fig. 4. The upper branch, related to fluctuations in $\psi_0$, is called the amplitude mode, and behaves as a normal soft mode, with its frequency increasing smoothly from zero on both sides of $T_i$. The lower branch represents fluctuations in $\phi$ and is called the phase mode or phason. These fluctuations, indistinguishable from $\delta\psi_0$ when $\psi_0 = 0$ above $T_i$, behave differently below. A fluctuation in $\phi$ which is uniform throughout the crystal requires zero energy below $T_i$ so long as $\boldsymbol{k}_0$ is incommensurate.

It is important to note that a uniform fluctuation in $\phi$ corresponds to an excitation at $\boldsymbol{k}_0$ in the Brillouin zone of the parent phase. A periodic fluctuation

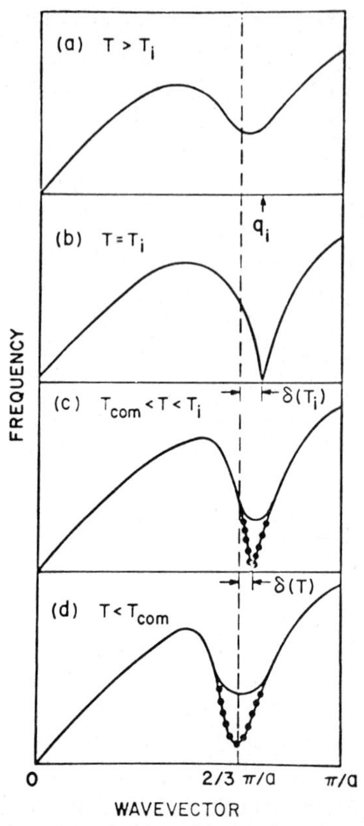


Fig. 4. Representative dispersion curves for the soft mode branch near a transition from a normal structure to an incommensurate phase, in the plane wave approximation. Note the splitting below $T_i$ into two modes, one of which remains at zero frequency (for $\boldsymbol{k} = \boldsymbol{k}_0$) while the other exhibits typical soft mode behavior.

$\delta\phi = A \cos \boldsymbol{q} \cdot \boldsymbol{r}$ (i.e., a "phason at $\boldsymbol{q}$") represents an excitation at $\boldsymbol{k}_0 + \boldsymbol{q}$. In the simplest approximation its frequency will be $\omega_p = c_p q$, where $c_p$ is a phason velocity. Therefore, the dispersion curves (Bruce and Cowley 1978) may be represented as in fig. 4. The frequency of a phason at $\boldsymbol{q}$ must go to zero as $q \to 0$.

The expected low frequency of phase excitations make them attractive candidates for light scattering investigation in those cases where the phason is active in light scattering. We shall now show that the phason is not Raman active within the plane wave approximation in those cases where the parent structure is centrosymmetric. To begin, recall the discussion at the beginning of this section. If the parent phase is centrosymmetric, then the incommensurate phase is centrosymmetric in the limiting sense discussed there. With an appropriate choice of origin, the displacements associated with the amplitude mode, which is always Raman active below $T_i$, are proportional to $\cos(\boldsymbol{k}_0 \cdot \boldsymbol{R}_l)$. Those associated with the phason, on the other hand, go as $\sin(\boldsymbol{k}_0 \cdot \boldsymbol{R}_l)$. Hence the phason parity is opposite that of the amplitude mode, and it will be inactive in first-order light scattering. We note, of course, that the scattering may be Brillouin allowed (i.e., $I \sim q^2$) or may couple to acoustic modes in all cases for appropriate geometries (Golovko and Levanyuk 1981).

Refinements upon the simplest theories of phason dynamics have appeared. Bruce and Cowley (1978) consider breakdown of the plane wave approximation, in which case the amplitude and phase fluctuations are coupled. However, they do not discuss damping. Damping is included in simple models due to Bhatt and McMillan (1975) and to Boriack and Overhauser (1978). In the former case the coupling of an incommensurate CDW excitation to a normal optical vibration is considered. Since their theory applies only for $T > T_i$, they calculate only a single mode. We can extend it easily, however, by noting that the phason dispersion curve should be only weakly temperature dependent below $T_i$ (Bruce and Cowley 1978, Lyons et al. 1982). Therefore, if we write the Bhatt–McMillan results for $T = T_i$, it should represent a reasonable approximation to the phason behavior *below* $T_i$. For a phason at wave vector $\boldsymbol{q}$ one must substitute the wave vector $\boldsymbol{k}_0 + \boldsymbol{q}$ in their equation for the width of the phase mode in the overdamped case (their eq. (20a)), letting $T = T_i$. We obtain

$$\tau_q^{-1} = \frac{2e}{\gamma} q_\parallel^2 + \frac{2f}{\gamma} q_\perp^2 , \tag{12}$$

where $e$ and $f$ are constants in the expression for the free energy, $\gamma$ is the CDW dissipation constant, and $q_{\parallel,\perp}$ represent the components of $\boldsymbol{q}$ parallel and perpendicular, respectively to $\boldsymbol{k}_0$.

This analysis is not limited to overdamped phasons. With appropriate values of the parameters in the Bhatt–McMillan equations, an underdamped phason would result. We have emphasized the overdamped limit here because this case yields a central peak and corresponds to experimental observations in $BaMnF_4$ (sect. 4.6). Levanyuk (1981) has also argued that the phason will always become overdamped as $q \to 0$.

We also note that the plane wave approximation is probably not relevant near a lock-in transition, where the value of $\boldsymbol{k}_0$ moves to a commensurate value. Bruce and Cowley (1978) have argued that in this case the most relevant model is the discommensuration or soliton model of McMillan (1976) in which the average value of $\boldsymbol{k}_0$ deviates from commensuration due to periodic shifts of phase with respect to the parent lattice. In this case, the amplitude and phase modes interact, and Bruce and Cowley (1978) argue that that interaction has the effect of stabilizing $\psi_0$ against fluctuations and the two branches then represent mainly fluctuations in $\phi$ near discommensurations and between them.

### 2.7. Precursor clusters and solitons

In the discommensuration limit, an incommensurate phase can thus be regarded as a periodic array of domains, and the discommensurations become domain walls, or solitons. We consider in this section a very similar picture which may be applicable to normal commensurate phase transitions.

Scattering from static or slowly-moving domain walls is expected and observed to produce a central peak below $T_c$, in the ordered phase. The usual quasiharmonic approximation admits of no such mechanism above the transition temperature. However, as the displacements associated with the collapsing soft mode become large in amplitude, the usual perturbative anharmonic phonon theories break down (see ch. 1). Long-lived regions of the structure appropriate to the low-temperature phase (precursor clusters) may appear above $T_c$. The excitation spectrum should then exhibit low-frequency modes, quite possibly overdamped, associated with fluctuations of the clusters as a whole. Higher-frequency components may result from localized vibrations within the clusters, analogous to the soft mode below $T_c$.

To examine these notions on a firm mathematical base Krumhansl and Schrieffer (1975) constructed a one-dimensional model of a displacive system and investigated its statistical mechanics and dynamics. Above $T_c$, in addition to the usual phonon structure in the dynamic response of this system, they found a central peak due to strongly nonlinear domain wall type of displacement field (a soliton). The authors expressed caution, however, about the extension of these ideas to high-dimensional systems, for it is well known that solitons are a strictly valid concept only in one dimension.

Nevertheless, the molecular dynamics simulations of Schneider and Stoll (1973) suggested the potential relevance of solitons or precursor clusters to systems of higher dimensionality. Their two-dimensional model, constructed to simulate the antiferroelectric transition in strontium titanate, could be adjusted to mimic either an order–disorder or a displacive structural transition. The results of these numerical simulations in the order–disorder limit produced a central peak in the dynamic response which was interpreted as due to the formation and dynamics of precursor clusters. In the displacive limit, the model produced, rather than a central peak, a very low frequency resonant response. Later simulation of a two-dimensional XY model (Schneider and Stoll 1976) produced again a central peak whose origin was associated with damped traveling domain walls bounding clusters.

Analytical support for these two dimensional molecular dynamics results was recently provided by Bruce et al. (1979), using renormalized group methods. A more complete treatment has recently been given by Bruce (1981). Very close to $T_c$ they found a non-Gaussian static displacement distribution corresponding to an effective crossover to order–disorder behavior even for apparently displacive systems. This implies that well-defined clusters of precursor order are

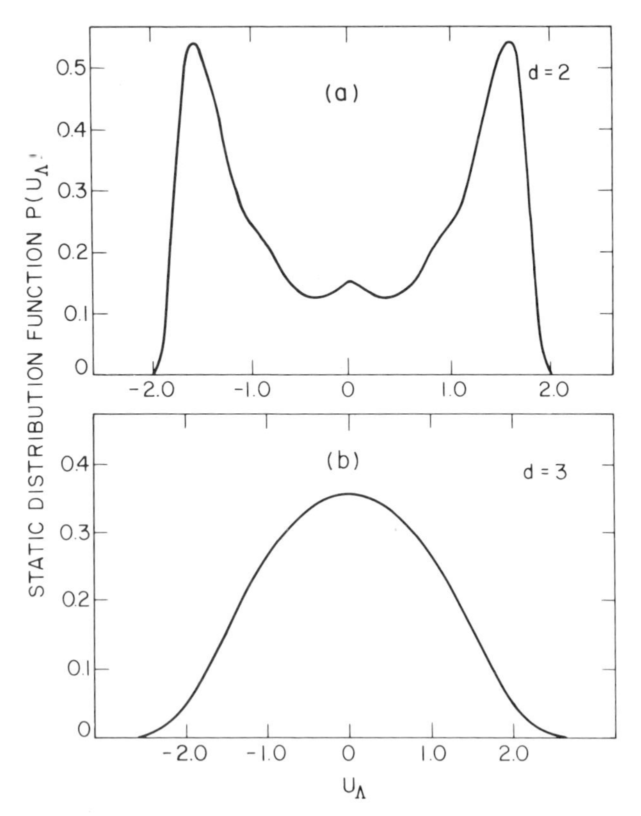


Fig. 5. Static order-parameter distribution function calculated in two and three dimensions by renormalization group techniques (after Bruce et al. 1979).

evident above $T_c$. While no dynamics were calculated in this study, the shape of the static displacement distribution function – particularly for two dimensions – is similar to that derived from molecular dynamics simulations. This result lends credence to the expectation that quasistatic cluster dynamics would cause a central peak in such systems. It is significant, however, that, as seen from fig. 5, the departure from Gaussian behavior is much weaker in three dimensions than in two dimensions. So, as they point out, the relevance of the precursor cluster to central peaks in three dimensions is not clearly demonstrated.

In summary then, the soliton or cluster dynamics mechanism for central peaks is certainly applicable to one dimensional systems, and has been shown by both molecular dynamics and renormalization group analysis to be important in the two-dimensional case. It remains, however, an open question in the three-dimensional case. Because most structural transitions occur in three-dimensional situations, and because a number of alternative mechanisms have been more fully explored theoretically and more clearly verified experimentally, the importance of the precursor cluster or soliton mechanism to existing central peak observations in structural transitions remains an open question.

### *2.8. Transitions in defected systems*

Experimental manifestations of the intrinsic sorts of precursor clusters discussed above are difficult to distinguish from certain possible effects of crystal defects. Defects may influence structural phase transitions in several ways: (i) by shifting or smearing the transition temperature, thereby inducing locally ordered (precursor) clusters; (ii) by altering the critical exponents, causing crossover to a different universality class; and (iii) by changing the dynamic response function. The latter may occur via defect-induced coupling of the soft mode to otherwise orthogonal degrees of freedom, or via the formation of local modes (Schmidt and Schwabl 1977).

Table I
Defect classification near structural phase transitions after Halperin and Varma (1976)

| Defect class | $\Delta T_c$ | $\psi$ Coupling | Crossover | Central peak |
|---|---|---|---|---|
| SB: relaxing | $T_c > T_c^0$ | linear | yes | dynamic<br>strong divergence |
| SB: frozen | $T_c < T_c^0$ | linear | yes | static<br>strong divergence |
| NSB: mobile or quenched | | nonlinear | no | dynamic<br>weak divergence |

SB: symmetry breaking NSB: nonsymmetry breaking

Halperin and Varma (1976) have categorized defects according to symmetry and mobility. Both static and dynamic consequences are indicated in table I. *Symmetry breaking* defects will either increase or decrease $T_c$, depending respectively on whether they are relaxing or frozen. A *relaxing* symmetry breaking defect tends to induce a local nonzero value of $\psi$ above $T_c^0$ in response to the diverging susceptibility. The ability of the defects to relax permits alignment of the locally induced $\psi$'s and hence leads to long-range order at $T_c > T_c^0$. *Frozen* symmetry breaking defects, on the other hand, will result in local nonzero $\psi$'s, randomly directed in space. The resulting spatial *fluctuation* in $\psi(\boldsymbol{r})$ *opposes* attainment of long-range order, and hence lowers $T_c$ below $T_c^0$.

Weakly perturbing *nonsymmetry breaking defects* will couple to $\psi$ only in higher order (through energy terms like $\psi^2$) and will have a relatively weak effect on $T_c$.

The effects on universality class have not yet been thoroughly explored. Obviously, within mean field theory (MFT) there can be no change, since MFT admits only one universality class. Insofar as static phenomena are concerned, alteration of the universality class beyond MFT depends again on defect symmetry and mobility. For completely mobile defects (symmetry breaking or not) Halperin and Varma (1976) expect "no change in asymptotic critical behavior, provided that $(T - T_c)$ is corrected to refer to measurements at constant chemical potential, rather than at constant defect concentration". For frozen symmetry breaking defects, Imry and Ma (1975) predict deviations from pure system asymptotic behavior for any system with spatial dimensionality $d$ less than 6, but explicit calculations for $d \leqslant 3$ systems have not been carried out. For quenched defects (trapped in a given unit cell, but able to relax locally within it) modified static critical behavior is expected, and "crossover" to a different universality class can occur.

While Halperin and Varma have considered the effects of various symmetry defects upon the susceptibilities in more detail, Levanyuk et al. (1976) give an intuitively pleasing treatment of the symmetry breaking case. We shall briefly recount here its major features. The effect of a symmetry breaking defect is to induce, above $T_c$, in its immediate neighborhood, a nonzero value for $\psi$, which decays with a characteristic distance equal to the correlation length, $\xi$. The precise spatial dependence of $\psi(\boldsymbol{r})$ depends on the static universality class. For example, $\psi(\boldsymbol{r})$ falls off exponentially for an isotropic system without long-range forces. For this case, the contribution of a density $N$ of such defects to the optical polarizability may be written

$$\Delta\alpha = N \int (\psi^2(r) - \langle\psi\rangle^2)\,\mathrm{d}V \approx 4\pi\alpha_0\psi_0^2 d_0^2\xi/V\,, \tag{13}$$

where $\alpha_0$ is the coupling between $\alpha$ and $\psi^2$, and $d_0$ is the defect core radius.

So long as $N\xi^3 \ll 1$, the defects scatter light independently and the scattered intensity is proportional to $\langle|\Delta\alpha|_q^2\rangle$ or

$$I_{\mathrm{TOT}} \sim (V/16\pi^2)(\omega_i/c)^4(\Delta\alpha V)^2 N \sim N\xi^2\psi_0^2 \sin^2\phi\,.$$

Since $\xi = \xi_0 t^{\nu}$, this predicts a divergence proportional to $\sim t^{-2\nu}$. In mean field theory $\nu = \frac{1}{2}$. For a uniaxial dipolar system, the prediction is $\sim(\ln\xi)^2$ rather than $\sim\xi^2$. A more complete discussion of these concepts, including more complicated cases, such as ferroelastic transitions, is given by Levanyuk et al. (1976).

The influence of defects on the order parameter dynamics is less clear. For a system with quenched defects, a small departure of the dynamic critical exponent $z$ from its pure system value is expected (Halperin and Varma 1976). For mobile defects the ultimate dynamic critical behavior should be that of the appropriate pure system. More detailed considerations beyond MFT on the one hand and beyond the quasiharmonic phonon approximation on the other await future theoretical effort.

## 3. Experimental techniques

The light scattering techniques applicable to central peak study include grating spectrometry, interferometry, intensity correlation techniques, and speckle photography. In this section we briefly discuss each of these techniques as they apply to the study of central peaks. The interested reader will find relevant background material in, for example, the books by Hayes and Loudon (1978) (for grating and interferometric techniques), by Cummins and Pike (1974) (for correlation spectroscopy), and the article by Durvasula and Gammon (1977) (for speckle photography).

The main experimental problem associated with study of central peak phenomena is the substantial elastic scattering from imperfections unrelated to the phase transitions. That scattering, usually of far greater intensity than the central peak under study (often by factors of $10^3$–$10^7$) can mask features of interest. Even in the case of speckle photography, where the component of interest is elastic, it is important experimentally to minimize the nonsingular contributions, such as might arise from surfaces. For inelastic scattering, the necessity of working very close to the exciting line places stringent requirements on the apparatus. A variety of techniques are necessary, and often even used in concert. We shall begin this section with a discussion of the spectral resolution techniques (grating and interferometric), followed by discussions of the other two.

Table 2
Typical resolution properties of experimental apparatus

| Spectroscopic apparatus | Resolution (HWHM) $\Gamma_{inst}$ | Typical contrast at $5\Gamma_{inst}$ | $20\Gamma_{inst}$ |
|---|---|---|---|
| Grating (double) | 0.2 $cm^{-1}$ | 250 | $10^8$ |
| Fabry Perot | | | |
| planar | 0.1 GHz | 40 | $10^3$ |
| spherical | 0.01 GHz | 40 | $10^3$ |
| multipass | 0.05 GHz | $10^4$ | $10^{10}$ |
| tandem/$I_2$ | 0.1 GHz | $10^{10}$ | $3 \times 10^{11}$ |

(1 $cm^{-1}$ = 30 GHz)

Table 2 summarizes the essential characteristics of the spectral resolution techniques for central peak studies. The concept of contrast has relevance only when specified at a particular frequency separation. Since central peak investigations call for working very close to the exciting line, we have chosen to quote contrast values for a frequency separation of $5\Gamma_{inst}$ (HWHM), which are more relevant for such work than the values typically quoted. This emphasizes two additional problems relevant to the study of central peaks. First, the contrast at $\Delta\nu \sim n\Gamma_{\text{inst}}$, $n \lesssim 5$, is much lower than the values usually given. Second, and really more important, is that the wings of the instrumental function $F(\Delta\nu)$ are not flat in this region, and can thus significantly distort a central peak spectrum. Accurate subtraction of the wings of $F(\Delta\nu)$ presents most severe problems, for the stray light almost always illuminates the pinhole or entrance slit differently from the scattered light, and the exact shape of $F(\Delta\nu)$ in the region $\Delta\nu \sim 5\Gamma_{\text{inst}}$ usually depends substantially on the details of that illumination. Thus, for example, it is rarely reliable to study a central peak a few $\text{cm}^{-1}$ in width with a double grating spectrometer if the stray light is $10^3$ greater than the central peak intensity, although the contrast ratio usually cited for this instrument is $> 10^8$.

The two techniques most commonly used to circumvent these problems are multipass interferometry and resonant ($I_2$) reabsorption. In the remainder of this section we shall discuss these techniques in separate subsections, followed by discussions of correlation spectroscopy and speckle photography.

### *3.1. Multipass Fabry–Perot interferometer*

A single-stage, single-pass Fabry–Perot etalon exhibits a characteristic periodic transmission function, $T(\nu)$. The period is referred to as the "free spectral range", which we denote as $\Omega$, $\Omega = c/2nL$ where $c$ is the velocity of light and $2nL$ is the optical pathlength of the cavity. The etalon resolution is usually specified as a finesse, $F = \Omega/\Gamma_{\text{inst}}$, where $\Gamma_{\text{inst}}$ is the full width at half maximum of the spectrum observed when monochromatic light is incident on the etalon. The free spectral range is inversely proportional to the plate spacing, and values in the range 5–100 GHz are usually employed in studies of solids. The finesse is usually limited by plate surface and coating quality to no more than 75, and a more typical value is 50. The periodic transmission of such an etalon leads to the phenomenon of "order overlap" if the scattering intensity extends significantly into the region $\Delta\nu \gtrsim \Omega/2$. In Brillouin scattering, where sharp acoustic peaks are typically observed, order overlap is not usually a severe problem. However, in the study of the spectral profile of a central peak it must be avoided, unless the detailed characteristic shape of the spectrum is already known from other experiments. Since the latter is rarely the case, a free spectral range several times greater than the central peak width must usually be employed.

Thus, the simple Fabry–Perot etalon is of severely limited utility in the study of central peaks. First, the contrast is simply insufficient. Second, the limited number of resolution elements in the free spectral range (the finesse) imposes the contradictory requirements of resolution sufficient to distinguish a central peak from the instrumental tail of the elastic scattering and the avoidance of order overlap.

One solution to these problems lies in the use of multi-stage Fabry–Perot devices. An apparatus in which the light passes serially through two different etalons is referred as a tandem Fabry–Perot. If the light passes several times through different regions of one etalon, it is called a multi-pass Fabry–Perot. A tandem instrument can substantially reduce the order-overlap problem. The use of a multi-pass etalon provides an extreme improvement in contrast (see table 2) and increases the finesses by a factor of almost two, at the cost of some throughput. Perhaps more important is the very steep drop in the values of $T(\Delta\nu)$ for $\Delta\nu > \Gamma_{\text{inst}}$, which leads to usable contrast even for $\Delta\nu \sim 5\Gamma_{\text{inst}}$. A comparison of typical instrumental functions for single and multi-pass etalons is shown in fig. 6. The multi-pass etalon still has the drawback that its transmission function is periodic, but Sandercock (1978) and Dil et al. (1981) have made recent progress in constructing tandem multi-pass instruments.

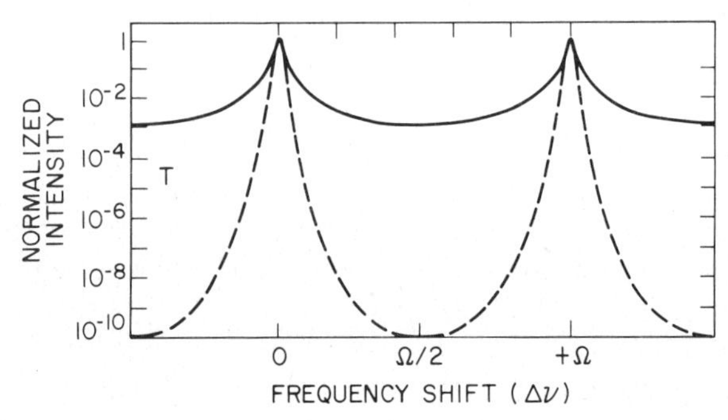


Fig. 6. Transmission functions for one free spectral range of a single and a five-pass Fabry–Perot etalon, shown on a log scale for clarity. Note the very steep drop of the transmission for $\Delta\nu > \Gamma_{\text{inst}}$ in the latter case.

### *3.2. Molecular iodine reabsorption cell*

A different approach to overcoming the limitations of the simple planar Fabry–Perot etalon utilizes a molecular iodine reabsorption cell. This technique is applicable to grating spectroscopy as well, with one drawback, discussed at the end of this subsection.

The possibility of using an $I_2$ filter arises from an accidental overlap between the gain curve of an $Ar^+$ laser operating at 5145 Å and a strong, narrow absorption in molecular iodine. The laser, operated single mode, can be tuned accurately to the iodine absorption frequency. When that tuning is maintained, the elastically scattered light may be attenuated more than eight orders of magnitude, while inelastically scattered light in the range $|\Delta\nu| \gtrsim 0.5$ GHz is transmitted with a throughput of up to 15%. Unfortunately, other iodine

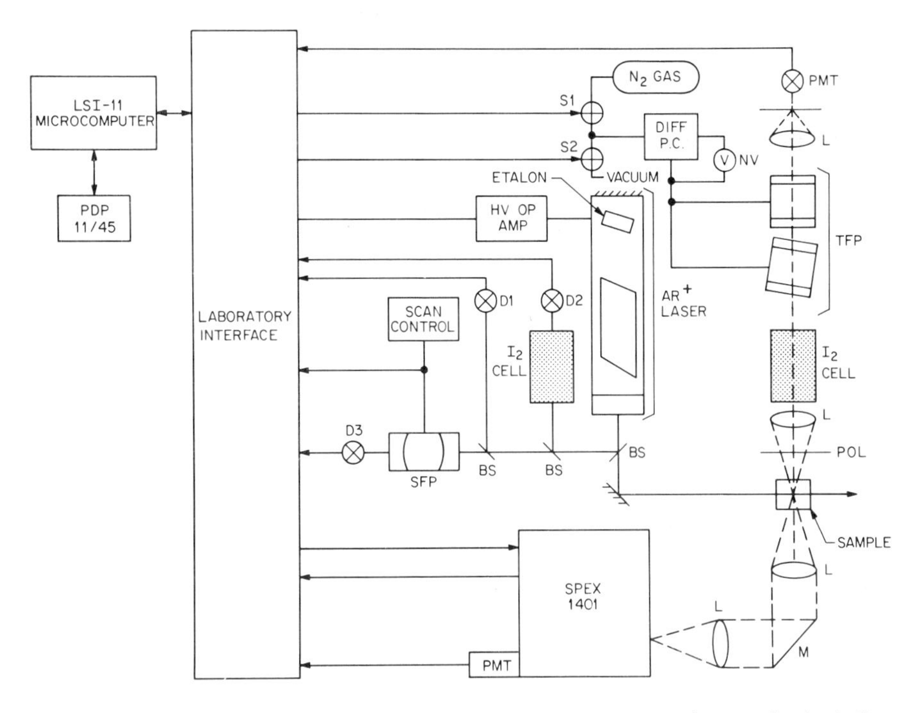


Fig. 7. Schematic of a tandem pressure-scanned Fabry–Perot apparatus using a molecular iodine ($I_2$) reabsorption cell. The $Ar^+$ laser is operated single mode, and tuned by an intracavity etalon manipulated under computer control to maintain its tuning for long periods of time. The auxiliary $I_2$ cell, spherical Fabry–Perot (SFP), and power meters provide continuous monitoring of laser tuning. Spectra are stored on the computer for subsequent normalization analysis. The differential pressure controller (DIFF PC) provides a linear scan of the Fabry–Perot on a time base.

absorption lines introduce considerable structure into the transmitted spectrum. Nevertheless, if the various experimental parameters (scan rate, resolution, iodine cell temperature, etc.) are sufficiently stable, it is possible to normalize spectra so as to recover quantitatively the inelastic spectral profile for $\Delta\nu \gtrsim 0.5$ GHz. Since the iodine cell structure does not exhibit the same periodicity as the etalon, this normalization technique requires use of a tandem (i.e., multi-stage) Fabry–Perot (TFP), as shown schematically in fig. 7, to eliminate the order-overlap problem. By appropriate selection of the free spectral ranges of the two stages, an effective finesse of $\sim 700$ has been obtained (Lyons and Fleury 1976a). The performance of this TFP-$I_2$ system is illustrated in fig. 8, where the raw spectral data obtained through the reabsorption cell and its normalized counterpart are displayed.

The effective contrast of the TFP-$I_2$ system relevant to central peak studies is then the rejection ratio of the iodine cell (up to $10^8$) multiplied by the contrast inherent in the tandem Fabry–Perot. That net value may be as high as $10^{10}$.

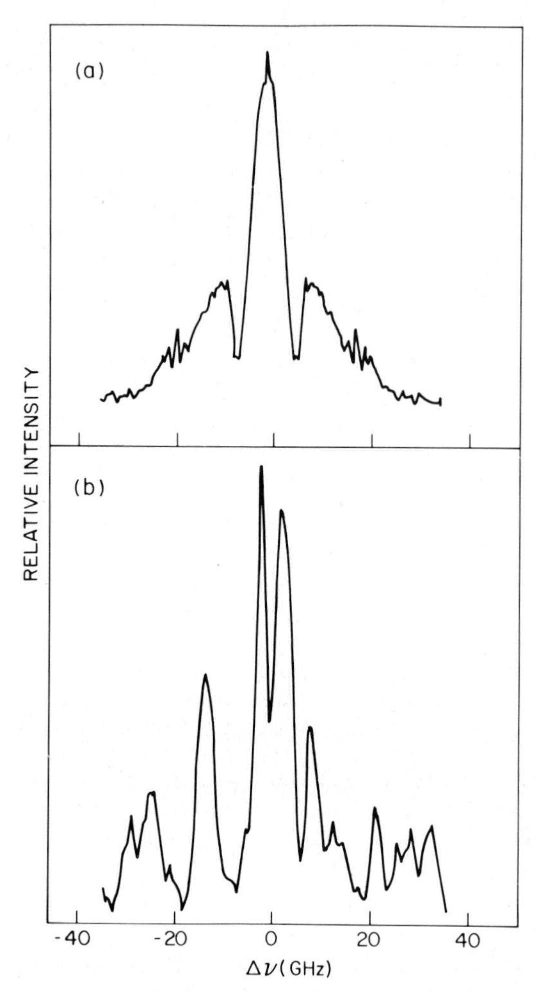


Fig. 8. Typical example of spectra obtained by the TFP-$I_2$ technique (Fleury and Lyons 1981) showing both the normalized data (part a) and the raw data (part b). The expected symmetry of the data has been employed to improve the signal-to-noise in the normalized data. The unfiltered elastically scattered light in this case was about $10^7$ cps. These spectra were obtained in $TbVO_4$.

The $I_2$ filter technique is applicable as well to grating spectroscopy. However, the presence of acoustic (Brillouin) scattering *not* removed by the iodine filter often limits its usefulness in this case. The sharp Brillouin peaks located in the region 0.5 to 1.5 $cm^{-1}$ are quite strong in many materials, often a factor of $10^2$ to $10^3$ stronger than a central peak of interest. The natural width of these Brillouin peaks is typically small ($\Gamma_a \sim 0.05\ cm^{-1}$), but, even with $\Gamma_{inst} = 0.2\ cm^{-1}$, the resulting displaced peaks as seen on a grating instrument exhibit wings which, adding $\Delta\nu = 5\Gamma_{\text{inst}}$ to the already displaced peak center, may obscure central peak scattering as far out as 2.5 cm$^{-1}$ from the exciting line. We should note in passing, however, that, while the preceding statements refer to *instrumental* broadening of the Brillouin lines, an analogous *intrinsic* broadening of those lines is often present near structural phase transitions. Thus, even when Fabry–Perot techniques are used, it may be necessary to apply the analysis of sect. 2.1 in order to describe the entire spectral profile properly.

### *3.3. Correlation spectroscopy*

The various forms of intensity correlation spectroscopy extend the resolution of light scattering techniques to lower frequencies by several orders of magnitude. The techniques rest upon measurement of temporal correlations in the scattered intensity (Cummins and Pike 1974). For example, in the case of autocorrelation or self-beating spectroscopy, the function measured is

$$F_2^0(\tau) \equiv \langle I(t)I(t+\tau)\rangle\,, \tag{14}$$

where $I$ is the scattered intensity, which is proportinal to the photomultiplier current or count rate. The scattered spectrum measured by other techniques represents the Fourier transform of the *field* correlation function, whereas the Fourier transform of the correlation function (14), which is fourth-order in the field, is the *intensity* power spectrum. For a Gaussian light field, the two are simply related (Cummins and Pike 1974).

As in the cases discussed previously, the separation of the scattering of interest from parasitic light is of prime importance. Thus, if the observed intensity is $I(t) = I_B + I_0(t)$ where $I_B$ is a background and $I_0(t)$ is the signal of interest, the correlation function becomes

$$F_2(\tau) = I_B^2 + I_B\langle I_0\rangle + F_2^0(\tau)\,.$$

If $I_B \gg \langle I_0\rangle$, then the contribution of the function of interest will be small.

For diffusive processes, the function $F_2^0(\tau)$ then has the form (Cummins and Pike 1974)

$$F_2^0(\tau) = I_0^2\left(1 + \frac{1}{N}\,e^{-2\Gamma\tau}\right),$$

where $\Gamma$ is an inverse correlation time, and $N \geqslant 1$ is the number of coherence areas subtended by the detector. For a diffusion process with a diffusion constant $D$, the inverse correlation time is $\Gamma = Dq^2$. Note the factor of two in the argument of the exponential, which represents the decay of the $q$th Fourier component of the density–density correlation function $\langle\rho(r,t)\rho(0,0)\rangle$. The diffusing quantity, $\rho$, may be entropy, a concentration of Brownian particles in a liquid, or a concentration of ions in a superionic conductor, to name a few.

There are very few cases where autocorrelation spectroscopy has been successfully applied to the study of a structural phase transition, although at least a couple of important negative results have been reported. The reasons for this are (1) the sensitivity to stray light mentioned above, and (2) the range of times accessible to study. The former prevents application of the technique except in very clean systems or in systems where the scattering of interest rivals the stray light in intensity. The second is a limitation imposed by the speed of the correlators presently available, which presently offer time resolution no better than 5 nsc. The characteristic frequency of a process must thus be lower than about 10 MHZ before it can be measured successfully by autocorrelation. Rather the power of the technique lies in its ability to study very low frequency processes with a limit imposed only by practical considerations such as stability of lasers, temperature control, acoustic vibration, and other such factors. It therefore has potential application to the study of domain wall motion and low-mobility impurities and defects, although these capabilities have not yet been exploited.

### 3.4. Speckle photography

When laser light is incident on any scatterer, the scattered light interferes with itself to produce the speckle patterns familiar to all those who work with lasers. If the scattering objects fluctuate rapidly, the scattered light will appear homogeneous, as in the case of scattering from phonons, entropy fluctuations, diffusing particles, etc. The rapidly fluctuating speckle pattern is in this case only detectable by high-speed electronic techniques. If, on the other hand, the scattering is dominated by that from static defects, surfaces, etc., the pattern is stationary and is not only visible by eye but may be photographed. The appearance of speckle on a photograph implies that the pattern is unchanged over a time at least comparable to the exposure time. Going beyond that, using photographic techniques the patterns may be compared at widely spearated times to check for long-term fluctuations. These techniques are referred to collectively as speckle photography. In essence, speckle photographic techniques represent the extreme case of intensity correlation spectroscopy. They have been used to verify the predominantly static nature of the scattering centers responsible for the large anomaly in the scattered light intensity near structural phase transition such as KDP (Durvasula and Gammon 1977). $SiO_2$ (Shapiro and Cummins 1968), and lead germanate (Taylor et al. 1978).

## 4. Experimental results

In this section we discuss the variety of experimental results which have been obtained in real systems. We make no attempt at exhaustive review, but instead concentrate on prototypical examples to illustrate the mechanisms discussed in sect. 2.

*4.1. Cooperative Jahn Teller transitions*: $TbVO_4$

As discussed in sect. 2.3 the cooperative Jahn teller (CJT) transitions as a class are unique in that they are thoroughly understood microscopically. Hence, they are an ideal place to begin our consideration of experimental central peak systems.

The low-lying electronic levels of the $Tb^{3+}$ ion in the crystal field of the high-temperature phase of $TbVO_4$ exhibit a degenerate doublet interposed between two closely-spaced singlets (Elliott et al. 1972). Both pairs of levels participate in a Jahn–Teller coupling which drives a CJT transition at $T_D = 32.6$ K. Measurement of the splitting of $E_g^1$ phonon mode by Raman scattering provides a direct measure of the order parameter below $T_D$. The accompanying structural distortion is isomorphic to a transverse acoustic mode polarized in the basal plane, which is the appropriate soft mode for this transition. This soft mode is active in light scattering both above and below the transition. The associated acoustic anomaly has been thoroughly studied by ultrasonic and small angle Brillouin scattering experiments (Sandercock et al. 1972).

The doublet and the acoustic mode are only weakly coupled when the acoustic wave propagates in the (110) direction, while the coupling is maximal for propagation in the (100) direction. Furthermore, the coupling parameter is predicted theoretically (Marques 1980) to vary in a prescribed way with wave vector direction, thereby providing an experimental means of turning on smoothly the electron phonon coupling. Harley et al. (1980) investigated the central peak region in detail using the iodine cell technique described in sect. 3.2. The spectra obtained in the uncoupled geometry (fig. 9(a)) provide an experimental determination of the dynamic susceptibility associated with the electronic degrees of freedom. The strong temperature dependence of this linewidth is evident. In contrast, the spectra in fig. 9(b), obtained in the fully coupled geometry, show a far more complex behavior. The highly temperature dependent line shape and its complexity are indicative of the strong interactions and significant interference effects between the various terms contributing to the sum in the coupled mode equation (6). In $TbVO_4$ the individual terms contributing to the sum in (6) are nearly two orders of magnitude larger than their sum, reflecting nearly perfect cancellation of individual contributions in various regions of the spectrum. This accounts for the extreme sensitivity of the spectral lineshape to temperature. It also suggests that the spectra should depend strongly on the direction of $\boldsymbol{q}$ as well. Figure 10 shows a verification of this behavior, where spectra taken at three temperatures in the vicinity of $T_D = 32.6$ K are displayed for different orientations of the $\boldsymbol{q}$ vector. All of these spectral complexities are quantitatively accounted for by the theory described in sect. 2.3, and represented by the solid curves in the figures.

The $TbVO_4$ results represent the first time that any dynamic central peak has been quantitatively described on the basis of a microscopic theory for a

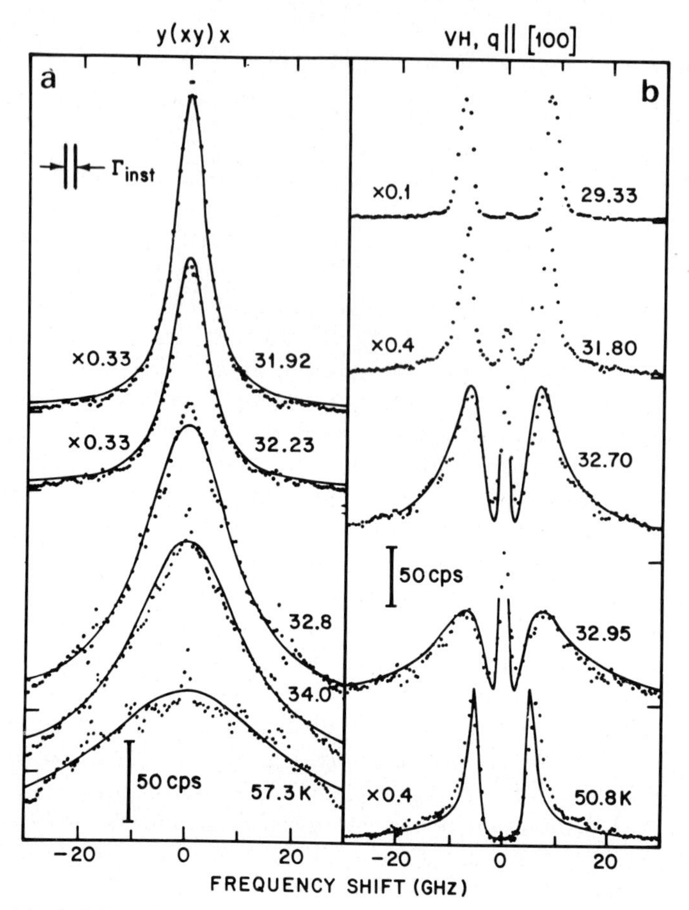


Fig. 9. Quasielastic light scattering of $TbVO_4$ obtained in geometries where the coupling of the electronic and phonon degrees of freedom are minimal (left side) and maximal (right side). In both cases the solid dots represent the normalized experimental data, while the solid lines represent theoretical fits (after Harley et al. 1980).

structural phase transition. The agreement between theory and experiment is essentially perfect in the temperature region ($T > T_D$) where a theory exists. Data obtained below the transition are of equal quality, and should provide considerable stimulus for the development of a dynamical theory for fluctuations in the ordered phase.*

### *4.2. KDP Family–order–disorder ferroelectrics*

Although the structural phase transitions in the materials $AB_2CO_4$ (A = K, Rb, or Cs; B = H or D; and C = P or As) are slightly first order at atmospheric

*Note added in proof: Such an extension has now been made and equally good agreement is obtained below $T_D$ as well (Harley et al. to be published).

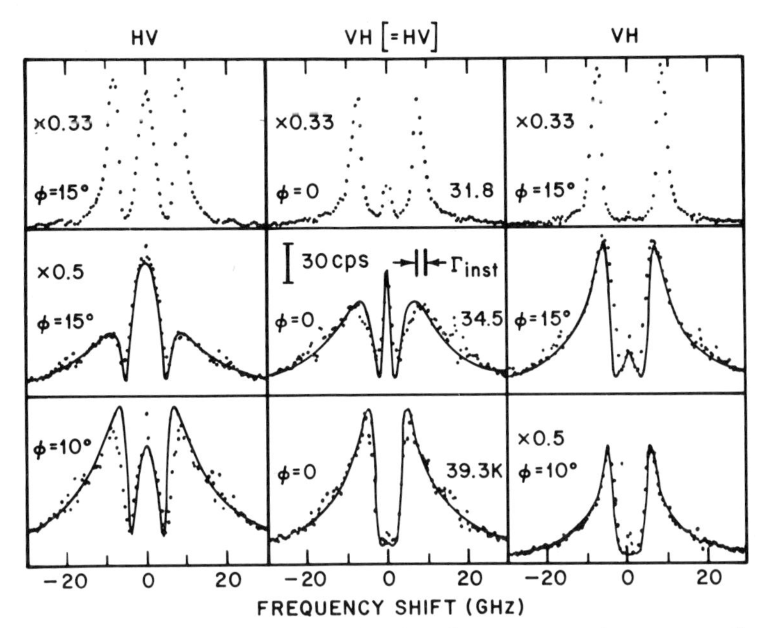


Fig. 10. Central peak spectra of $TbVO_4$ as a function of temperature and wave vector direction. The angle $\phi$ is that formed by $\boldsymbol{q}$ and the $x$-axis. After Harley et al. (1980). The data and lines are as in fig. 9.

pressure, they represent one of the most thoroughly investigated classes from the standpoint of central peaks. Most of the light scattering techniques discussed in sect. 4 have been employed. In this section we shall emphasize mainly results obtained in potassium dihydrogen phosphate (KDP). As we shall see, KDP exhibits a variety of central peak phenomena which are not all amenable to simultaneous characterization with any single experimental technique. It therefore serves to illustrate several mechanisms.

The displacements accompanying the KDP ferroelectric transition involve tunneling of a proton (or deuteron) through the midhump of a double well potential. Above $T_c$, although the crystals are piezoelectric ($D_{2d}^{12}$), the protons are randomly distributed in these wells, while below they order cooperatively. The spontaneous polarization in the low temperature phase ($C_{2v}'$) lies in a direction perpendicular to the direction of the proton motion. Thus the uncoupled proton tunneling mode cannot by itself describe the dynamics of the phase transition.

In fact, the tunneling mode couples strongly to a higher frequency $B_2$ optic mode (Katiyar et al. 1971) to produce the spontaneous polarization observed. This interpretation sparked early attempts to observe a central peak in these materials. Cowley and Coombs (1973) recognized that, within this picture, the

Lyddane–Sachs–Teller relation required the presence of a substantial self-energy term with a low frequency relaxation. Following the failure of early attempts to reveal the expected central peak intensity in CsDA, Lagakos and Cummins (1975) showed that inclusion of an additional strong interaction between the soft (tunneling) mode and a transverse ($C_{66}$) acoustic mode described the spectra correctly, with no self-energy correction, when the difference between the clamped and free response was taken into account.

Nevertheless, experimental observations clearly demonstrated that an anomaly in the scattered intensity in KDP occurred near $T_c$ (Lagakos and Cummins 1975). An investigation using correlation spectroscopy failed to reveal any dynamic component (Lyons et al. 1973). Later, this scattering was shown by speckle photography to be largely the result of scattering by static centers (Durvasula and Gammon 1977). A model based on scattering by deuterium impurities at first seemed capable of describing these results (Courtens 1977). However, subsequent experiments demonstrated that the bulk of this scattering could be removed by an appropriate annealing procedure near $T_c$ (Courtens 1978). Thus, the isotopic impurity mechanism cannot be responsible.

While the nature of the annealable defects responsible for the bulk of the elastic central peak remains unknown, further studies have revealed the presence of additional inelastic components in KDP crystals (Mermelstein and Cummins 1977). The narrowest of these is attributable to entropy fluctuations, and its anomalous intensity is due to the dependence of the polarizability on the order parameter.

It is generally true that in the ordered phase a linear coupling exists between the temperature $T$ and the order parameter, $\psi_0$. In particular, the scattering of light by entropy or temperature fluctuations (sect. 2.2) is proportional to the temperature derivative of the polarizability. This may be written

$$\left(\frac{\partial\epsilon}{\partial T}\right)_P = \left(\frac{\partial\epsilon}{\partial T}\right)_{X,P} + \frac{\partial X}{\partial T}\left(\frac{\partial\epsilon}{\partial X}\right)_{T,P} + \frac{\partial P}{\partial T}\left(\frac{\partial\epsilon}{\partial P}\right)_{T,X} \tag{15}$$

where $X$ and $P$ label strain and polarization variables, respectively. The first term in eq. (15) is quite small for most solids. In KDP, the second and third terms, divergent at $T_c$, describe indirect coupling of light to temperature fluctuations via temperature dependence of the spontaneous strain and polarization below $T_c$. In the absence of dynamic coupling effects, the known thermodynamic parameters for KDP predict a linewidth of 62 MHz for right angle scattering. The observed value of $47 \pm 5$ MHz argues strongly for this identification of the dynamic portion of the central peak (Mermelstein and Cummins 1977). Despite the strongly divergent intensity, the slight first-order nature of the phase transition prevented a quantitative comparison of the observed divergence with that expected theoretically. Experimental problems have thus far prevented verification of the $q$ dependence and critical narrowing of the linewidth. Experiments at higher pressure, where the ferroelectric

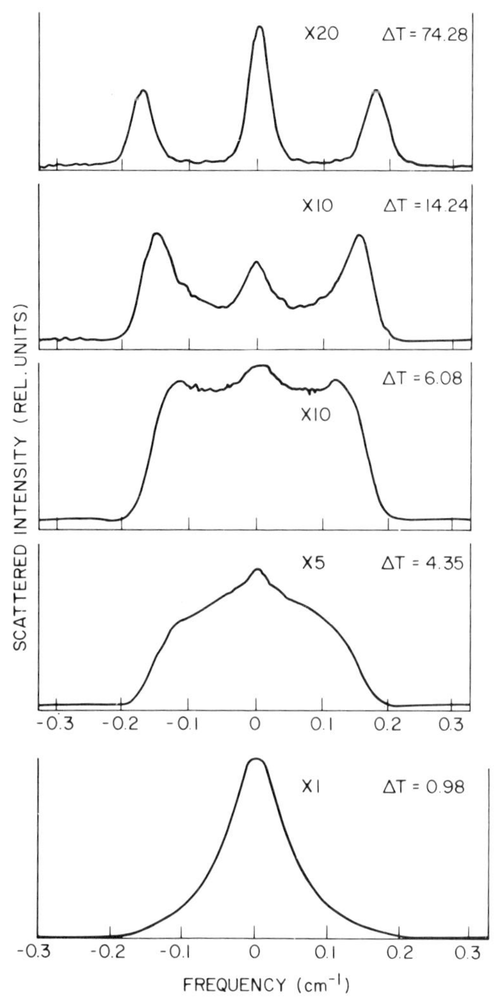


Fig. 11. Five typical $[x+z]([-x+z],y)[-x+z]$ spectra obtained in KD*P, showing the coupled ferroelectric and acoustic soft modes. The temperatures $\Delta T = T - T_0$ and the gain changes are shown for each spectrum. The transition temperature is $T = 220.678$ K. After Reese et al. (1973).

transition in KDP becomes second order ($\sim 2.0 \pm 0.3$ kbar (Peercy 1975a, Peercy 1975b)), as well as at other scattering angles, might permit quantitative verification that this central peak is due to entropy fluctuations.

Early experiments of Reese et al. (1973) had demonstrated the coupling of the transverse ($C_{66}$) acoustic mode to a low frequency relaxation in KD*P ($KD_2PO_4$). Sawafuji et al. (1979a) recently reported a detailed investigation of this spectrum. They observe two well separated frequency scales, including one indistinguishable from zero frequency, which could conceivably contain both defect and entropy fluctuation contributions. The other exhibits a width of several GHz. The two components exhibit different temperature and wave vector dependence.

The transverse ($C_{66}$) acoustic mode observed (Sawafuji et al. 1979a, Reese et al. 1973) for $\boldsymbol{q} \parallel [100]$ exhibits a lineshape very distorted from that of a damped harmonic oscillator (fig. 11). The TA frequency cannot be unambiguously extracted from the spectra, even several degrees from $T_c$. This distortion is due to the presence of the broad central component (HWHM $\sim$ 5 GHz). Reese and Cummins attribute the broad component to scattering from a polarization relaxation mode. These experiments are discussed more fully elsewhere in this volume. Sawafuji et al. (1979a) concluded that the separation of the TA and polarization mode frequencies accounted for the absence of TA softening in the wave vector range typical of Brillouin scattering. However, in a subsequent report, Tanaka and Tatsuzaki (1980) concluded dispersion in the phase velocity of this mode does exist at very small wave vector.

By utilizing an HH scattering geometry (e.g., $x(yx)y$) Sawafuji et al. (1979a) observed the broad KD*P central component with much less interference from the acoustic phonons, and demonstrated that it is indeed due to the Debye relaxation of the polarization. The intensity is dependent on the direction of $\boldsymbol{q}$, but the lineshape is not.

In contrast, they observed that the narrow central peak component in KD*P appears above $T_c$ in a VH scattering geometry as $\boldsymbol{q}$ deviates from [100] in the *ac*-plane (Sawafuji et al. 1979b). This behavior distinguishes it from the broad component. The direction of $\boldsymbol{q}$ for which its intensity is maximal approaches the $c$-axis as $T \to T_c$. They attribute the scattering to charged impurities. By a simple model based on the response of the crystal to such isolated charges, they account completely for the observed $\boldsymbol{q}$ dependence of the elastically scattered intensity. They further make the interesting proposal that the *annealable* defect in KDP may be a strain defect which, by a similar analysis, might be expected to exhibit a different characteristic $\boldsymbol{q}$ dependence.

In summary, then, the KDP family exhibits at least three central peak components, stemming from the following mechanisms: (i) static annealable defects; (ii) entropy fluctuations; and (iii) relaxing polarization (the soft mode). Despite the careful experiments performed to date, certain quantitative aspects of the models relating to these mechanisms remain incompletely explored.

*Triglycine sulfate*

Very recent observations in TGS by Miyakawa and Yagi (1980) demonstrate the presence of a singular central peak with an upper limit on the linewidth of a few hundred MHz. These observations have been interpreted as entropy fluctuation scattering, based on the observed singular behavior of the

Landau–Placzek ratio near $T_c = 322.2$ K, rather than on the basis of direct linewidth measurements. The quantitative consistency between the observed ratios $I_R/2I_B$ and the singular behavior of $(C_P - C_V)/C_V$ argues in favor of the interpretation. The singularity is described by an exponent nearly equal to unity. It should be noted, however, that recent work (Strukov et al. 1980) has shown that the specific heat is nonsingular in sufficiently pure samples. Furthermore, the value of $C_V$, which is the one relevant to the experiments, is not known in any case, and the reported spectra contain no dynamic information. It thus appears likely that defects or impurities may be the source of the central component observed.

### 4.3. *Displacive structural phase transitions*

The soft mode concept was first derived from efforts to understand displacive transitions, where it is most directly applicable. These transitions, including both ferrodistortive and antiferrodistortive transitions, have been thoroughly studied by several experimental techniques. Indeed, the original neutron scattering observation of an anomalous central peak was made in $SrTiO_3$, an antiferrodistortive transition (Riste et al. 1971). In this section we consider the light scattering central peak data available for a representative of each of these two major classes of displacive transitions.

*Lead germanate*

$Pb_5Ge_3O_{11}$ is a uniaxial ferroelectric ($C_3^1$) which undergoes a continuous transition to $C_{3h}^1$ symmetry at $T_c = 451$ K (Iwasaki et al. 1972). Below 390 K the $A_{1g}$ soft mode (polarization fluctuations) appears in the $x(zz)y$ scattering geometry (Hisano and Ryan 1972) ($P_s \| z$) as an underdamped phonon which exhibits a characteristic frequency $\omega_s^2 = a'(T_c - T)$ where $a' = 5850(\mathrm{GHz})^2/\mathrm{K}$.

Between 390 K and 441 K, the overdamped soft mode width obeys $\omega_s^2/\Gamma_s = b'(T_c - T)$ with $b' = 12$ GHz/K. Above about 441 K this temperature dependence slows down, and the linewidth stabilizes above 445 K at $\gamma_s = \omega_s^2/\Gamma_s = 75$ GHz. Above this temperature, an additional quasielastic scattering of unusual lineshape grows in intensity (Lyons and Fleury 1978), reaching a maximum in the vicinity of the transition temperature. This additional scattering exhibits a narrowing linewidth which reaches a minimum value of about 4 GHz (deconvolved half width half maximum) at $T_c$, and couples obviously to the LA Brillouin components. Representative data are shown in fig. 12.

The molecular iodine cell (sect. 3.2) employed in obtaining the spectra in fig. 12 completely removes the elastically scattered light. That scattered intensity increases singularly near $T_c$, as first shown by Lockwood et al. (1976). Both visual inspection of the sample and speckle photography (Taylor et al. 1978) show that the extra intensity emanates from finely distributed centers which

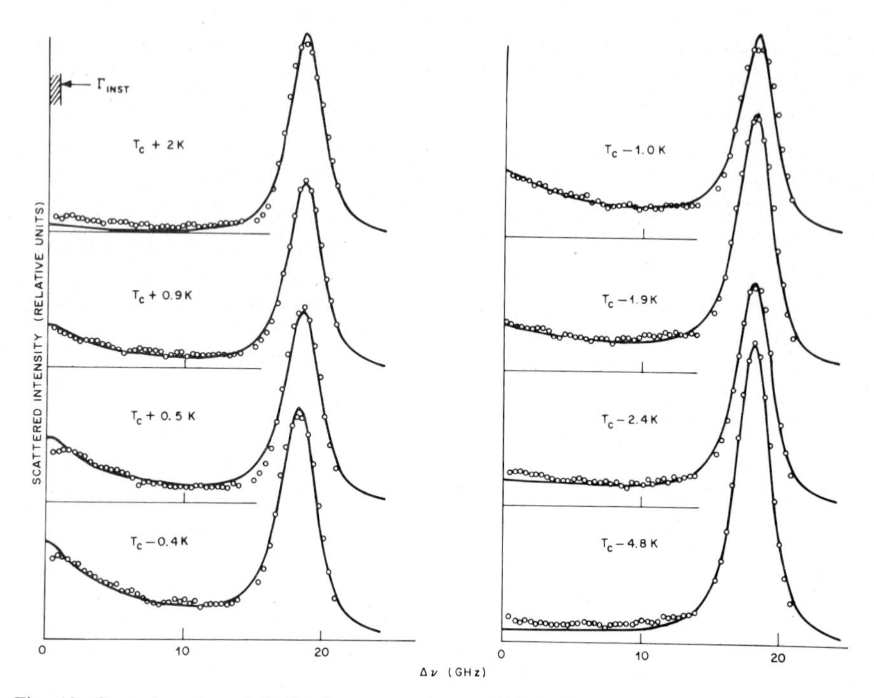


Fig. 12. Central peak and Brillouin spectra (open circles) of lead germanate at various temperatures in a single-domain sample at $E = 40$ V/cm. The solid curves are the result of a simultaneous least-squares fit to all of the data using the formalism described in sect. 2.1. The apparent spatial variation of $T_c$ (about $\pm 0.5$ K) has been taken into account in drawing the theoretical curves (Lyons and Fleury 1978).

appear to be stationary, even over periods of minutes. While this elastic component may be the result of defect scattering, there is to date no evidence that those defects influence the dynamics of the transition (Fleury and Lyons 1976, Lyons and Fleury 1978).

The influence of such defects on the spectra shown in fig. 12 must be small, since the formalism of sect. 2.1 provides a good description of the observed spectral profile, using a completely intrinsic pairwise coupling model (Lyons and Fleury 1978). In this model the acoustic mode susceptibility is represented by a quasiharmonic susceptibility $\chi_1 = [\omega^2 - \omega_1^2 + 2i\omega\Gamma_1]^{-1}$ while the soft optic mode response is assumed to contain a relaxing self-energy term:

$$\chi_2(\omega) = [\omega^2 - \omega_2^2 + 2i\omega\Gamma_2 + i\omega\delta^2\tau/(1 - i\omega\tau)]^{-1}. \tag{16}$$

The parameters $\omega_i$ and $\Gamma_i$ represent the quasiharmonic frequencies and widths of the respective susceptibilities, while $\delta$ and $\tau$ are the self-energy strength and relaxation time, respectively. The relaxing self-energy term was interpreted by Lyons and Fleury (1978) in terms of phonon density fluctuations (sect. 2.2). The parameters for the acoustic mode were extracted from spectra well above $T_c$, where the coupling is zero, and a soft mode (uncoupled) frequency obeying $\omega_2 \propto (T_c - T)^{1/2}$ was assumed in the calculation.

The behavior of the spectrum in the region below $T_c$, where the coupling is nonzero, is more complex and depends on $\tau$, $\delta$, and the ratio of the optical coupling constants, $a_i$, in eq. (6), as well as the coupling constant. The latter is proportional to the piezoelectric coefficient $a_{31}$, and hence were available from independent data as a function of temperature (Klein 1974). The values $\omega_1 = 18.5$ GHz, $\Gamma_1 = 0.5$ GHz, and $(a_2^2 \mathrm{Re}\chi_{22}/a_1^2 \mathrm{Re}\chi_{11}) = 1.8$ were determined from spectral measurements far from $T_c$.

Using these parameters, the theory accounts correctly not only for the overall lineshape of the central peak and Brillouin spectra at all temperatures further than $\sim 1.5$ K from $T_c$, but also for the small anomalies in the LA phonon velocity and attenuation if the values of $\tau^{-1} = 29$ GHz and $\delta = 60$ GHz are taken as constant near $T_c$. Very close to $T_c$ (within about 1.5 K) the dynamic central peak ceases to narrow, presumably because of smearing of $T_c$ over the small $(2 \times 10^{-3}\ \mathrm{mm}^3)$ scattering volume. A smearing of $T_c$ by about 1 K is sufficient to account for the observed behavior and is consistent with the gross variation in $T_c$ observed by varying the location of the scattering volume within the sample. The additional complications introduced by the spatial variation of $T_c$ have thus far prevented a more quantitative test for deviations from mean field behavior predicted for a uniaxial dipolar system by renormalization group theory (Larkin and Khmel'nitskii 1969). While the total scattered intensity in the relaxing soft optic mode exhibits a clear departure from mean field behavior, quantitative assessment of this departure would require the theory to be extended to include effects of symmetry breaking defects.

*Strontium titanate*

The well-known cubic-tetragonal transition at $T_c = 107$ K in $SrTiO_3$ is accompanied by a softening of the $\Gamma_{25}$ symmetry phonon at the R point of the cubic Brillouin zone. Below $T_c$ the order parameter fluctuations appear as zone center Raman active modes of $A_g$ and $E_g$ symmetry in the polarized and depolarized scattering geometries, respectively (Fleury et al. 1968). The polarized spectrum is dominated by central peak components analogous to those observed in $KTaO_3$ (Lyons and Fleury 1976b). These components, representing entropy fluctuations and PDF, as discussed in sect. 2.2, are inelastic, but not detectably singular. Neither of them appears in the depolarized $E_g$ geometry, so that a detailed search for a singular dynamic central peak in that geometry has proven fruitful (Lyons and Fleury 1977). Figure 13 shows the spectral profile as a function of temperature. The facts that the maximum central peak intensity occurs below $T_c$ and that the feature cannot be seen above $T_c$ suggest that the soft phonon PDF appear indirectly via their relaxational contribution to the soft mode self energy, which in turn couples to the TA mode. Such coupling

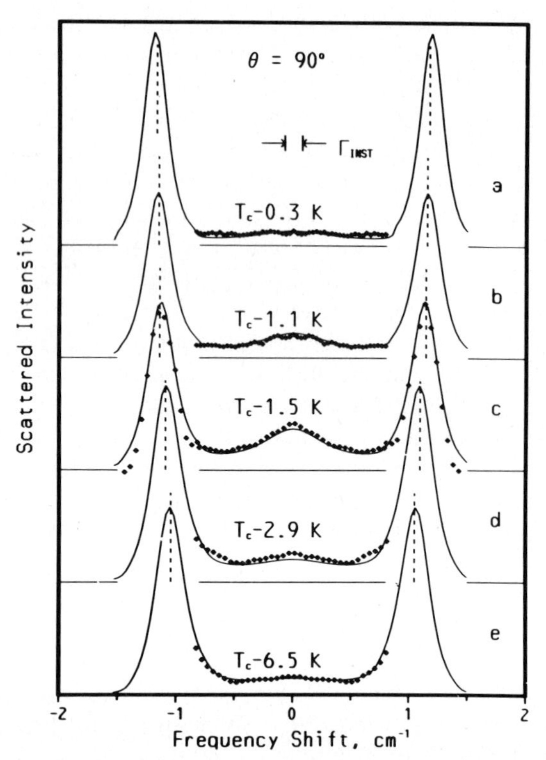


Fig. 13. Depolarized spectra of $SrTiO_3$ near $T_c = 107$ K are shown by the points. Solid lines represent the coupled mode theory described in the text. The vertical dashed lines indicate the TA phonon frequencies determined without the iodine cell (Lyons and Fleury 1977).

affects not only the relaxing soft mode lineshape, but also causes anomalous behavior in both the velocity and attenuation of the TA phonons near $T_c$. Careful application of the formalism of sect. 2.1, similar to that discussed above for lead germanate, leads to the solid curves shown in the figure. Once again, most of the parameters appearing in the expressions are available from other experimental data. The remaining adjustable parameters chosen in the fitting procedure are $a_2^2\mathrm{Re}\chi_{22}(0)/a_1^2\mathrm{Re}\chi_{11}(0)) = 0.12$, $\tau^{-1} = 15$ GHz, and $\delta$, which is temperature dependent. The value of $\delta^2(T_c) = 1.5 \times 10^4$ (GHz)$^2$ joins smoothly with corresponding values inferred from neutron scattering above $T_c$. The value of $\omega_2^2$ inferred from the fit reaches approximately $10^4$ (GHz)$^2$. This is equivalent to the failure of the dynamic central peak to collapse to zero linewidth. Although smearing of $T_c$ could account for this, order parameter measurements give no indication that $T_c$ is appreciably smeared. A more likely explanation, particularly in view of the recent analysis of Bruce and Bruce (1980) (see sect. 2.2) is that there might be an *additional* central peak of intensity comparable to that observed, but of such a width as to lie within the iodine absorption employed to remove the stray light in this experiment. The presence of two singular central peaks in $SrTiO_3$ (the one observed directly with $\tau^{-1} = 15$ GHz, and the second inferred indirectly with $\tau^{-1} < 0.3$ GHz) would account for all observations made to date regarding this transition.

Although the model of Bruce and Bruce provides a mechanism for the presence of two intrinsic central peaks in $SrTiO_3$, a role of defects has been indicated in some neutron scattering experiments. Therefore, the situation is still not fully clarified.

### *4.4. Superionic conductors*

Ionic conduction in superionic conductors stems from the existence of multiple sites for ions of a particular species, usually with many sites available per ion. Diffusion among these sites may be simply thermally activated or may be facilitated as the result of a structural phase transition. In the latter case, the collective phenomena involved and the interaction with the soft mode of the transition provide a complex and interesting dynamical system.

Although a number of Raman and Brillouin investigations have been performed in ionic conductors, the most thoroughly studied from the point of view of central peaks are the silver halides: silver iodide and rubidium silver iodide. These share with copper iodide, copper bromide, and copper chloride the quasielastic tails in the Raman spectrum which extend to several tens of wave numbers.

The most detailed experiments to date are those of Winterling et al. (1977), who studied the spectrum of silver iodide in the temperature range between 178 and 421°C for frequencies between 0.8 and 100 $cm^{-1}$. A typical spectrum is displayed in fig. 14. It has been reconstructed from a composite of the low frequency triple pass Fabry–Perot data at 0.8–2 $cm^{-1}$, and the higher frequency double spectrometer data covering frequencies above 2 $cm^{-1}$. Note the logarithmic intensity scale. The spectrum shown by the solid line can be decomposed into two components: a broad Lorentzian of half width 32 $cm^{-1}$, and a narrower peak with a half width of approximately 4 $cm^{-1}$. These extracted linewidth values are relatively insensitive to the fitting procedures used. However, we should note that the comments in sect. 3.2 relative to contributions from Brillouin scattering are applicable here, and that no allowance was made for this in the data analysis. Therefore the degree to which longitudinal or transverse acoustic phonons contribute to the narrow central peak is not known.

Winterling et al. (1977) interpret the narrow component in terms of the hopping time $\tau_h$ mentioned in sect. 2.5. Using the value $D = 2.3 \times 10^{-5}$ $cm^2/s$ from Funke (1976), simple application of the relationship $6D = l^2/\tau_h$ yields a value of $l$ within 20% of the nearest neighbor site distance. The higher frequency

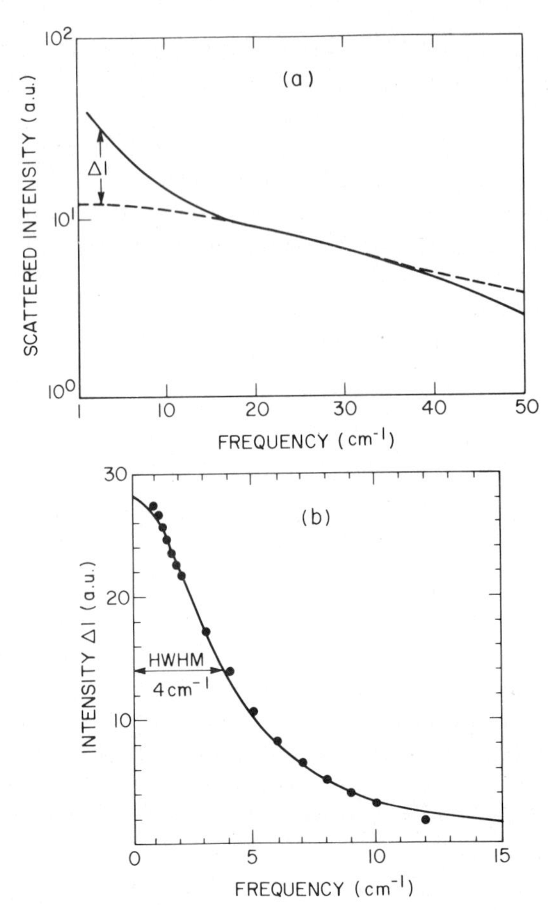


Fig. 14. Typical quasielastic spectrum of α-AgI obtained by superposition of spectra obtained by Fabry–Perot and grating monochromator techniques (Winterling et al. 1977). In part (a) the spectrum (solid line) is shown out to 50 cm$^{-1}$, with a Lorentzian approximation ($\Gamma$ = 32 cm$^{-1}$) to the broad component (dashed line). In part (b) the spectrum is shown over a smaller frequency range, with the Lorentzian fit to the broad component subtracted (points), and the dashed line again represents a Lorentzian ($\Gamma$ = 4 cm$^{-1}$) fit to the narrow component. The scattering geometry is $x(yx)z$, the temperature is 178°C, and the incident wavelength is 647 nm.

component may be related to intrasite vibrations or to the flight time, $\tau_f$. The calculations based on diffusion in periodic potentials (Dieterich et al. 1977, Geisel 1977) automatically include both of these contributions, and we note that those results qualitatively reproduce the observed spectra. However, the predicted temperature evolution of the spectral features has not been observed; the wing is never observed to become resonant below the onset of ionic conductivity; and the diffusive ($q^2$) component has not been observed. In particular, there has not yet been a thorough assessment of the extent to which the observed scattering could be due to disorder-induced first-order Raman scattering or to two-phonon (PDF) scattering (sect. 2.2). Furthermore, the theoretical results have not been compared in detail to the experimental spectra – rather the latter have been analyzed in terms of superimposed Lorentzians. Thus, while the periodic potential models may indeed offer a substantially correct interpretation, there remains a significant need to demonstrate this connection and to achieve a quantitative contact between theory and experiment.

A high resolution study of the Rayleigh–Brillouin spectrum of rubidium silver iodide has recently been reported by Field et al. (1978). Using an iodine reabsorption filter, they demonstrate the presence of a quasielastic scattering feature in the spectral region below 1 $cm^{-1}$, as shown in fig. 15. Although the

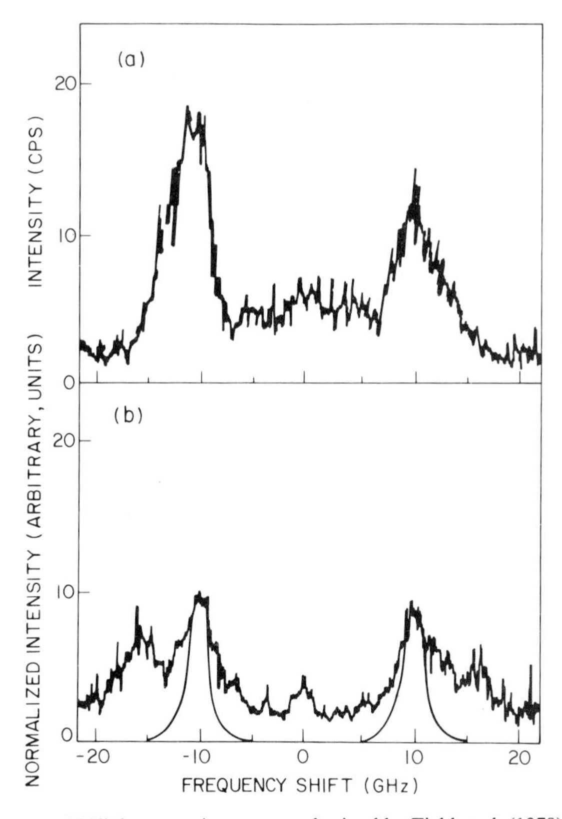


Fig. 15. Low frequency VV light scattering spectra obtained by Field et al. (1978) in $RbAg_4I_5$, using an iodine reabsorption cell. Part (a) shows the raw data, distorted by the subsidiary iodine absorptions, and part (b) shows the same data with corrections made for those absorptions. The solid lines in part (b) show their calculated Brillouin peak lineshapes.

normalized spectra show some effects due to apparatus instabilities, they exhibit a pair of damped acoustic phonon features, with intensity between them extending down to zero frequency. This intensity cannot be clearly distinguished from a flat background. However, the authors were able to estimate its integrated intensity by assuming it to be a Lorentzian roughly 9 GHz in width, and thereby infer an intensity about twice that of the Brillouin peaks. They use this interpretation to extract a dwell time ($\tau_h - \tau_f$) of 18 ps from this feature. As these authors note, a complete and careful study of the inelastic central components in these spectra would be of interest.

For both of the silver ion systems discussed above, the final interpretation must await more detailed data on the temperature and wave vector dependence of the spectra observed. For example, the model of Geisel makes specific predictions as to the temperature evolution of the spectrum, which should be tested explicitly. Furthermore, the temperature evolution of the spectrum predicted near the onset of ionic conduction should be investigated. The fluorite family of materials are clear candidates for a careful quantitative study utilizing experimental techniques discussed elsewhere in this chapter. Some intriguing data from the Brillouin spectra of the fluorites have been obtained by Harley and co-workers on the fluorite family. These have mainly been in the form of elastic constant anomalies preceding the onset of ionic conductivity and have been rather successfully explained by the models of Catlow et al. (1978). Nevertheless, it remains to be seen whether any significant coupling effects similar to those considered by Huberman and Martin (1976) will become evident when the low frequency spectra are studied in detail.

### *4.5. Plastic crystals*

Whereas superionic conductors exhibit a translational disorder in the conducting state, plastic crystals maintain translational order in their plastic phases, but become disordered in the orientation of molecular groups internal to the unit cell. In the plastic phase the molecules undergo reorientation about site centers which remain ordered. As the plastic-to-crystalline transition is approached, it is anticipated that these orientational fluctuations should slow down as a precursor phenomenon to the orientational ordering of the molecules which exists in the crystalline phase.

#### *Succinonitrile*

Plastic crystal phases have been studied by light scattering as well as by neutron scattering, birefringence measurements, and other techniques sensitive to both static and dynamic aspects of the transitions. The studies on succinonitrile exemplify the data available. Birefringence measurements (Bischofberger and Courtens 1971) demonstrated the continuous nature of the solid–plastic transition in this material at 233 K. Bischofberger and Courtens (1971) distinguished two depolarized central components in the Rayleigh–Brillouin spectrum with uncoupled widths of 24 and 42 GHz. The coupled mode spectra containing these modes and the TA phonons exhibit anomalous and strongly angularly-dependent line shapes. Nevertheless, the simple coupled mode theory outlined in sect. 2.1, using the expression (9) for the uncoupled relaxational susceptibility, accounts nicely for the observations. Unfortunately, these experimental observations were confined to a single temperature $\sim 87$ K above the plastic-to-crystalline phase transition, so that no singular behavior could be followed. The situation clearly calls for investigation of the temperature dependence of the coupling and of the central peak linewidths.

*Methane*

To date the only observation of critical slowing down upon approach to a plastic-to-crystalline phase transition has been made in $CD_4$ using inelastic neutron scattering. Press et al. (1974) reported measurements of the quasielastic linewidth using high resolution neutron scattering at a series of temperatures above the 27 K plastic-to-crystalline phase transition in $CD_4$. Within experimental error the central peak linewidth was observed to narrow according to a power law with an exponent equal to 1.13. Because of the large $\boldsymbol{q}$ used in such experiments, no acoustic phonons were observed in the frequency range of interest. Therefore the dynamics could be described solely in terms of the central peak behavior. To date there have been no corresponding studies using optical techniques or inelastic light scattering in either methane or its deuterated relative.

*Ammonium chloride*: $NH_4Cl$

Other experimental studies of the Brillouin spectrum in plastic crystalline phases include KCN and mixed crystals of KCN and KBr. In these materials neutron scattering experiments at larger wave vector have revealed some dynamic central peak in the response. Unfortunately, no detailed light scattering studies of the central peak have appeared.

One closely related system in which dynamic measurements have been reported very recently is $NH_4Cl$ (Andrews and Harley 1981, Hikita et al. 1981). This material exhibits an order–disorder phase transition involving the orientations of the $NH_4$ groups. It is slightly first order at atmospheric pressure but becomes second order at a pressure of 1.5 kbar. As mentioned in the introduction, this is a system which has long been known to exhibit a strong anomaly in the intensity of the scattering at or near zero frequency (Lazay et al. 1969).

Andrews and Harley (1981), using the $I_2$ cell technique, have obtained the spectra shown in fig. 16 for $NH_4Cl$ at a pressure of $p = 0$ bar. They observe a feature centered at zero with a temperature dependent width which ranges from 6 GHz (HWHM) near room temperature down to less than 1 GHz within

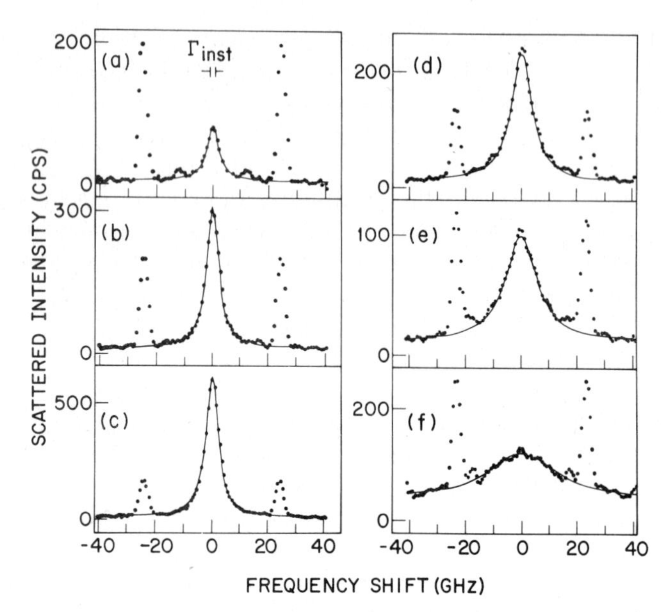


Fig. 16. Spectra of the inealstic component in $NH_4Cl$ obtained in HH ($E_g$) geometry at the following temperatures: (a) 238.2 K; (b) 242.1 K; (c) 248.6 K; (d) 284.5 K; (e) 305.9 K; (f) 323.2 K. The count rate scales are shown for each trace. The curves are Lorentzian fits including a constant background term. After Andrews and Harley (1981).

a few degrees of $T_c = 242.5$ K. Due to the width of the $I_2$ absorption, they could not characterize the spectral profile in detail very near $T_c$, but in the region $T \gtrsim T_c + 5$ K they found a clearly measurable dynamic width, well outside the instrumental resolution. They attribute the peak to fluctuations in the orientational energy density, since its width exhibits no dependence on $\boldsymbol{q}$.

A later experiment (Hikita et al. 1981) utilized a confocal spherical Fabry–Perot interferometer without an $I_2$ cell in an attempt to extend the data to lower frequencies. These workers found that, as in the case of KDP (see sect. 4.2), the anomalous elastic intensity near $T_c$ disappeared after a crystal was extensively annealed. In a sample so treated they report a weak narrow component ($\sim 30$ MHz) within 1 K of $T_c$ at atmospheric pressure. They propose no interpretation. It is not clear whether this is the same feature as seen by Andrews and and Harley (1981), since Hikita et al. report no evolution of the feature as a function of $T$ or $q$. They also report qualitatively similar spectra at lower resolution which are consistent with the observations of Andrews and Harley.

### 4.6. *Incommensurate transitions*

In the last two sections we have considered systems which lose certain aspects of their long-range order at continuous phase transitions. In this section, we consider system exhibiting incommensurate phases, which are unique in that they lose their translational symmetry at the normal-to-incommensurate transition, but maintain their long-range order.

Investigation of incommensurate structures as described in sect. 2.6 dates back to the discovery (Wilson et al. 1975) and elucidation (McMillan 1975) of the charge density wave (CDW) phase transitions in the layered dichalcogenides. More recently, displacive incommensurate phases have been found in a variety of materials (Janner and Janssen 1977). One of the most interesting and thoroughly studied of these materials from the standpoint of light scattering is $BaMnF_4$, which we review in this section.

A detailed neutron scattering (Cox et al. 1979) study provided the first evidence for an incommensurate phase in $BaMnF_4$. Below the transition at $T_i = 247$ K, the incommensurate distortion develops continuously at a wave vector $\boldsymbol{k}_0 = (0.392, 0.5, 0.5)$. The value of $\boldsymbol{k}_0$ is not measurably temperature dependent and, according to these results, there is no lock-in phase transition. However, despite careful investigation, the nature of the soft mode eigenvector has remain unknown. Even so, $BaMnF_4$ is especially interesting, since the high temperature phase lacks inversion symmetry (Keve et al. 1969) and exhibits pyroelectricity (Glass et al. 1979). As noted in sect. 2.6, this may permit direct light scattering from the phason excitation.

The fact that the incommensurate wave vector in the case of $BaMnF_4$ is on the zone boundary presents a complication which was not discussed in sect. 2.6. In fact, there are four equivalent wave vectors (instead of two in the usual incommensurate phase), with the result that there should be four modes associated with the transition (Bruce and Cowley 1978, Cox et al. 1979). Only one of these should exhibit zero frequency at $q = 0$. In the neutron scattering work, only one excitation, presumably the highest lying amplitude mode, was observed. Barring accidental cancellation of large parameters, however, the other modes should lie well outside of the frequency range of the phason below $T_i$. The theory of Bhatt and McMillan as outlined in sect. 2.6 must be extended to include this case. This has been done (Lyons et al. 1982) by identifying $q_{\|,\perp}$ in eq. (12) as the components of $\boldsymbol{q}$ respectively parallel and perpendicular to the *direction of incommensuration*: namely the $a$-axis for $BaMnF_4$. Clearly, the problem of phason damping in this interesting case awaits a more complete theoretical treatment.

Early Raman scattering investigations of $BaMnF_4$ (Ryan and Scott 1974, Popkov et al. 1975) did not disclose the incommensurate nature of the transition. A low angle Brillouin scattering investigation by Bechtle et al. (1978) revealed dispersion in a TA mode at a frequency near 1 GHz. Describing these results by a model which invokes coupling between the TA phonon and the phason, the latter authors predicted an (unobserved) depolarized central component, of width $\sim 0.7$ GHz. They also reported a strongly anomalous elastic (Lockwood et al. 1981) intensity near $T_i = 247$ K. A subsequent inter-

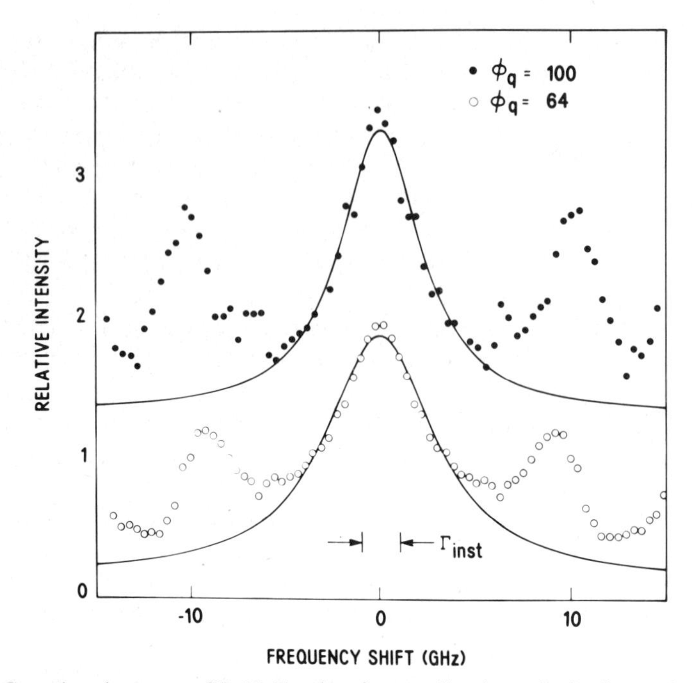


Fig. 17. Central peak spectra of $BaMnF_4$ taken for two directions of $\boldsymbol{q}$ in the *ac*-plane, with $|\boldsymbol{q}|$ held constant. The solid lines are Lorentzians drawn to fit the central peak. The features visible at $\sim \pm 10$ GHz are the TA modes. Instrumental resolution is $\Gamma_{\text{inst}} = 1.9$ GHz. The angle $\phi_q$ is that formed by $\boldsymbol{q}$ and the *a*-axis (Lyons et al. 1980).

pretation by Lockwood and Murray (1981) conjectures that there is an imperfectly resolved, underdamped phason peak in the spectrum, in the vicinity 9–15 GHz. However, the most recent data of Lyons et al. (1980), summarized in fig. 17 and fig. 18, suggest a quite different interpretation.

The spectra shown in fig. 17 clearly exhibit a polarized inelastic central peak. Detailed inspection of the Raman intensity data, obtained simultaneously, showed that this scattering exhibits a maximum intensity at a temperature $T = T_{\text{m}}$ which is about 7 K below $T_{\text{i}} = 254$ K. Moreover, further simultaneous observations (Lyons et al. 1982) demonstrated that the anomaly in the elastic scattered intensity reported by Bechtle et al. (1978) occurs at $T = T_{\text{e}} \approx (T_{\text{m}} - 6\text{ K})$. Pyroelectric measurements (Glass et al. 1979) indicate anomalies near $T_{\text{i}}$ (where $\mathrm{d}P_{\text{s}}/\mathrm{d}T$ changes sign) and at $T \sim T_{\text{i}} - 15\text{ K} \approx T_{\text{e}}$ (where $\mathrm{d}P_{\text{s}}/\mathrm{d}T$ is a maximum). Therefore, there are clearly three characteristic temperatures in the problem.

The inelastic peak observed also exhibits unusual wave vector dependence (Lyons et al. 1980b). For $\boldsymbol{q} \parallel \boldsymbol{c}$, its width scales as $q^2$ for $q$ in the range $2.2 \times 10^5\text{ cm}^{-1}$ to $4.3 \times 10^5\text{ cm}^{-1}$. Furthermore, as $\boldsymbol{q}$ deviates from $\boldsymbol{c}$ in the

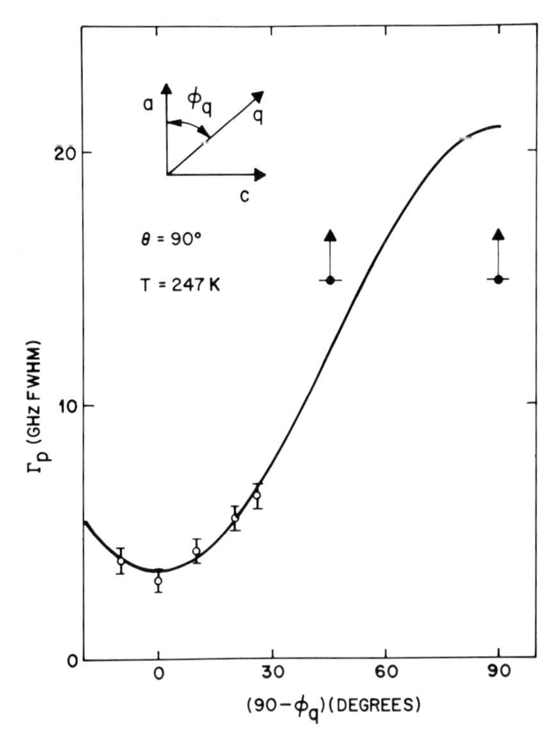


Fig. 18. True linewidth ($\Gamma_p$, full width at half maximum) of the inelastic central peak in $BaMnF_4$ as a function of $\phi_q$ in the *ac*-plane. The temperature and scattering angle are held constant at 247 K and 90°, respectively. The points at $\phi_q = 45°$ and 90° represent negative observations, from which approximate minimum values of $\Gamma_p$ may be inferred (Lyons et al. 1980).

*ac*-plane, the central peak width (at constant $|\boldsymbol{q}|$) increases dramatically. Two sample spectra are shown in fig. 17, and a summary of the observed linewidth behavior is shown in fig. 18. Fitting the data for $\phi_q < 26°$ to the theory of Bhatt and McMillan (eq. (12)) suggests an anisotropy ratio of $e/f \sim 6.8$ for the coefficients in their expression (12) for the dispersion of an incommensurate excitation. The line represents a calculation using $2e/\gamma = 0.98\ \mathrm{cm^2/s}$ and $2f/\gamma = 0.14\ \mathrm{cm^2/s}$ in eq. (12). Thus, all of the data are consistent with a model invoking an overdamped phason with a strongly anisotropic dispersion curve. A simple calculation using the central peak observations fully describes (Lyons et al. 1982) the detailed LA line profile data of Lockwood et al. (1981). This calculation suggests that the main contribution to the phason intensity stems from their interaction with the acoustic modes. This fact is consistent with the assertion by Levanyuk (Golovko and Levanyuk 1981) that the direct phason scattering intensity will be that typical of a Raman mode and, hence, may be far weaker than that typical of Brillouin scattering.

Lack of knowledge of the soft mode eigenvector in $BaMnF_4$ precludes a more detailed theoretical interpretation of these results. Indeed, Scott et al. (1982) have recently presented evidence for a second phase transition occurring 7 K below $T_i$ (which they now identify as 254 K), interpreted as of lock-in character. The discrepancy apparent between these results and the earlier investigation of Cox et al. (1979) remains unresolved at this writing. An alternate recent suggestion (Natterman and Przystawa 1981, Toledano 1981) is that the incommensurate wave vector $\boldsymbol{k}_0$ below the upper phase transition lies off the $a$-axis in the $ab$-plane, and subsequently locks in along $\boldsymbol{a}$ in the phase below 247 K. The precise character of the multiple phase transitions in this material thus remain in considerable doubt. The possibility has also been raised (Scott et al. 1982) that $BaMnF_4$ may exist in different polymorphs, some of which exhibit lock-in while others do not.

### *4.7. Experiments in defected solids undergoing structural phase transitions*

In the foregoing sections we have concentrated on the intrinsic phase transition phenomena and have largely ignored the possible effects of defects. However, all real experimental solid systems contain defects, whether deliberately introduced or not. Although it has been acknowledged from the beginning that defects may be at least partly responsible for the instrumentally narrow central peaks observed, few systematic investigations of their effects have appeared. As it has become clear that multiple central components exist in several cases, it appears likely that certain of those components result from defects. Such defect-related components could conceivably be either static or dynamic. However, of the central peaks observed which might originate from defects, most have been examined under spectral resolution sufficient only to set an upper limit on possible defect mobility. Apparently static components, presumably related to frozen defects, have been observed in lead germanate (sect. 4.3), KDP (sect. 4.2), and in the $\alpha$–$\beta$ transition in $SiO_2$ (Yakovlev et al. 1956).

The narrow component in $Pb_5Ge_3O_{11}$ exhibits a much stronger divergence in its intensity than does the dynamic peak. By a combination of Fabry–Perot spectroscopy (Lyons and Fleury 1978), autocorrelation spectroscopy (Lockwood et al. 1976), and speckle photography (Taylor et al. 1978) its width has been shown to be less than 0.1 Hz. Its behavior is consistent with that expected from *frozen symmetry breaking defects* (Halperin and Varma 1976). However, Lyons and Fleury (1978) find no correlation between the local value of $T_c(r)$ (observed to vary by $\pm$ 3 K in different regions of the crystal) and the elastic central peak intensity. The static central peak appears as a noninteracting addition to the dynamic response, implying that the defects are sufficiently dilute that the condition $N\xi^2 < 1$ is satisfied for $|T_c - T|/T_c$ as small as $5 \times 10^{-4}$.

In the case of $SrTiO_3$, several different spectral probes (light, neutrons, X-rays, $\gamma$-rays, ultrasonics, and EPR) have provided conflicting values for the width of the central peak. At present the evidence suggests that it contains at least two components, one of which remains spectrally unresolved. The analysis of Bruce and Bruce (1980) suggests a possible intrinsic source for such multiple components, but the very small linewidths found by some of the techniques seem to suggest the presence of an extrinsic component as well. A systematic study with defects of known identity and concentration remains to be performed.

The slightly first-order ferroelectric transition at 121 K in KDP is one of the most thoroughly studied of all structural transitions. The structure of its central peak is correspondingly complex, as discussed in sect. 4.2. For the purposes of the present discussion, we take particular note of the predominantly elastic central component observed in speckle photography by Durvasula and Gammon (1977), which was later shown (Courtens 1977) to be the result of annealable defects (see sect. 4.2).

The related $K(H_xD_{1-x})_3(SeO_3)_2$ system has shown evidence of both annealable and persistent defects in the Brillouin spectra (Yagi et al. 1977, Tanaka et al. 1978a, Tanaka et al. 1978b). For $x = 1$, the value of $T_c$ is 212.8 K. For $x = 0$, it is 302 K. Yagi and coworkers (1977) have studied this system for $x = 0$, 0.05, 0.95, and 1.0, and have observed two kinds of apparently elastic central peaks, placing an upper limit of $\sim$ 200 MHz on the central linewidth. For $x = 1$, as shown in fig. 19, temperature cycling and annealing near $T_c$ can reduce the central intensity by nearly an order of magnitude. The annealable central peak is, however, polarized (VV) orthogonally to the scattering from the soft acoustic mode (VH). It has been attributed to strains rather than impurities. For the mixed samples ($x \simeq 0.95$ and 0.05), a depolarized (VH) and nonannealable singular central peak is seen in addition to the annealable one. This has been attributed to scattering from the appropriate (D or H) minority impurity. While measurements to date have not definitely established the degree of mobility (if any), the identify of the defects is rather clear.

## 5. Conclusions and future speculations

Quasielastic light scattering experiments in solids exhibiting structural phase transitions have clarified several important points.

(1) The singular central peaks near $T_c$ have in many cases been distinguished into either static or dynamic phenomena.

(2) Linewidth and lineshape measurements have permitted definitive attribution of central peak mechanisms.

(3) The mechanisms verified to date include: static defects (both annealable and nonannealable); molecular group reorientations, degenerate electronic transitions, entropy fluctuations, dielectric relaxations, phonon density fluctuations, ion diffusion, and phasons. Some of these require further experimental and theoretical work.

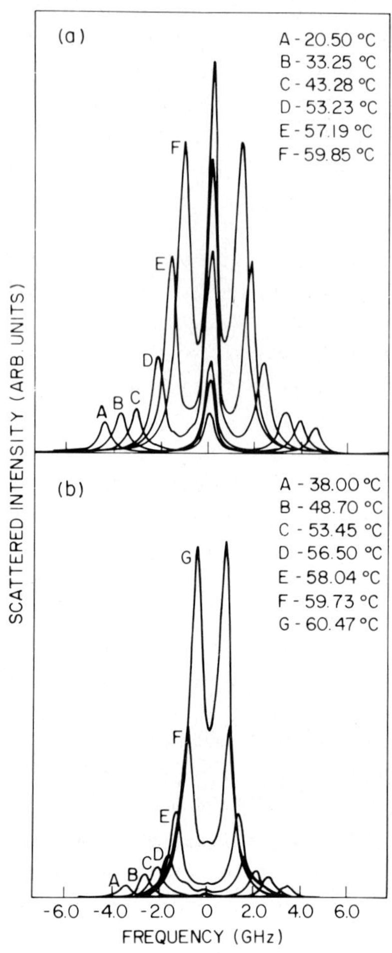


Fig. 19. Unfiltered Brillouin spectra of $KH_3(SeO_3)_2$ from Yagi et al. (1977), showing the effects of annealing near $T_c$. Part (a) shows the spectra of as-grown crystals, and part (b) shows those of crystals after repeated measurements near $T_c$. The temperatures for the various curves are as indicated in the key. $T_c = 60.5$ K.

(4) Any vestige of the notion that central peaks can be attributed to a single phenomenon has been clearly laid to rest. And while the variety and complexity of the many central peak phenomena preclude any simple or general *a priori* explanation of light scattering experiments, the material reviewed in this chapter demonstrates forcefully the necessity of careful and quantitative experiments to provide guidance for, and discrimination among the theoretical approaches to the problem of critical lattice dynamics.

It is worthwhile to consider where the major challenges and opportunities remain in this field. Central peak studies of the plastic–solid transition in molecular crystals directed toward quantitative determination of critical reorientation dynamics and their coupling to translational modes need expanded attention.

More detailed investigation of the complex spectra of ionic conductors as a function of temperature, orientation, and wave vector, including substantiation of the presence of the diffusive $Dq^2$ component, should reveal some novel aspects.

Controlled doping experiments in conjunction with careful spectroscopic studies are clearly needed to elucidate the largely unexplored area of phase transitions and dynamics in defected solids.

The dynamics of the incommensurate phase remain incompletely explored. Experiments in charge density wave systems such as $TaSe_2$ should be carried to the quasielastic regions. Surface wave manifestations of critical dynamics in semiconducting and semimetallic layered compounds deserve examination.

More direct and conclusive evidence is needed to elucidate the applicability of the precursor cluster phenomenon and associated dynamics in three dimensional systems. Careful study of quasi-one- and two-dimensional structural transitions are needed to substantiate experimentally the recently clarified theoretical predictions concerning soliton-like behavior.

Along with these experimental challenges, of course, go the needs for more concrete theoretical predictions. Many of the theoretical treatments relevant to central peak phenomena in solids do not make definitive predictions regarding the crucial characteristic time scale (though clearer predictions regarding its dependence on temperature, wave vector, etc. are often available). Even phenomenological theories can be quite useful, provided that the fundamental parameters they contain are related to some measurable physical quantities. Entropy fluctuation scattering is one good example of this kind. But for several other mechanisms, even the phenomenological theories leave the crucial parameters unrelated to known or measurable quantities. Effort should be expended to elucidate these relationships.

While these theoretical and experimental challenges are formidable indeed, there is reason to hope that many of them will be met within the next few years. The recent past has demonstrated that considerable progress has been made in the extremely difficult area of the nonlinear dynamics of strongly interacting, many body systems – which structural phase transitions represent. The lure of similar, perhaps more subtle physical phenomena, coupled with the momentum generated by recent theoretical and experimental successes, should make such challenges irresistible.

## *References*

Alder, B.J., H.L. Strauss, J.J. Weiss, J.P. Hansen and M.L. Klein, 1976, Physica, **B 83**, 249.
Andrews, S.R. and R.T. Harley, 1981, J. Phys. **C 14**, L207.

Bechtle, D.W., J.F. Scott and D.J. Lockwood, 1978, Phys. Rev. **B 18**, 6213.
Bhatt, R.N. and W.L. McMillan, 1975, Phys. Rev. **B 12**, 2042.
Bischofberger, T. and E. Courtens, 1971, Phys. Rev. Lett. **35**, 1451.
Boriack, M.L. and A.W. Overhauser, 1978, Phys. Rev. **B 17**, 4549.
Bruce, A.D., 1981, J. Phys. **C 14**, 3667.
Bruce, D.A. and A.D. Bruce, 1980, J. Phys. **C** 5871.
Bruce, A.D. and R.A. Cowley, 1978, J. Phys. **C 11**, 3609.
Bruce, A.D., T. Schneider and E. Stoll, 1979, Phys. Rev. Lett. **43**, 1285.
Catlow, C., R.A. Comins, C. Germano, R.T. Harley and W. Hayes, 1978, J. Phys. **C 11**, 3197.
Coombs, G.J. and R.A. Cowley, 1973, J. Phys. **C 6**, 121.
Courtens, E., 1977, Phys. Rev. Lett. **39**, 561.
Courtens, E., 1978, Phys. Rev. Lett. **41**, 1171.
Cowley, R.A. and G.J. Coombs, 1973, J. Phys. **C 6**, 143.
Cox, D.E., S.M. Shapiro, R.A. Cowley, M. Eibschutz and H.J. Guggenheim, 1979, Phys. Rev. **B 19**, 5754.
Cummins, H.Z. and E.R. Pike (editors), 1974, Photon Counting and Light Beating Spectroscopy (Plenum Press, New York).
Dieterich, W., I. Peschel and W.R. Schneider, 1977, in: Lattice Dynamics, ed., M. Balkanski (Flammarion Sciences, Paris) p. 524.
Dil, J.G., N.C.J.A. van Hijningen, F. van Dorst and R.M. Aarts, 1981, Appl. Opt. **20**, 1374.
Durvasula, L.N. and R.W. Gammon, 1977, Phys. Rev. Lett. **38**, 1081.
Elliott, R.J., R.T. Harley, W. Hayes and S.R.P. Smith, 1972, Proc. Roy. Soc. **A 328**, 217.
Enz, C.P., 1974, Helv. Phys. Acta, **47**, 749.
Field, R.A., D.A. Gallagher, and M.V. Klein, 1978, Phys. Rev. **B 18**, 2995.
Fleury, P.A., 1978, Correlation Functions and Quasi Particle Interactions, ed., J.W. Halley (Plenum, New York) p. 325.
Fleury, P.A. and K.B. Lyons, 1976, Phys. Rev. Lett. **37**, 1088.
Fleury, P.A. and K.B. Lyons, 1981, in: Lasers and Applications, eds., W.O.N. Guimaraes, C.-T. Lin and A. Morradian (Springer, New York) p. 16.
Fleury, P.A., J.F. Scott and J.M. Worlock, 1968, Phys. Rev. Lett. **21**, 16.
Funke, K., 1976, Prog. Solid State Chem. **11**, 345.
Geisel, T., 1977, in: Lattice Dynamics, ed., M. Balkanski (Flammarion Sciences, Paris) p. 549.
Gillan, M.J. and M. Dixon, 1980, J. Phys. **C 13**, L835.
Glass, A.M., M.E. Lines, F.S.L. Hsu and H.J. Guggenheim, 1979, Phys. Rev. **B 19**, 5754.
Golovko, V.A. and A.P. Levanyuk, 1981, Ferroelectrics, **36**, 306.
Halperin, B.I. and C.M. Varma, 1976, Phys. Rev. **B 14**, 4030.
Harley, R.T., K.B. Lyons and P.A. Fleury, 1980, J. Phys. **C 13**, L447.
Hayes, W. and R. Loudon, 1978, Scattering of Light by Crystals (Wiley, New York).
Hikita, T., K. Suzuki and T. Ikeda, 1981, Ferroelectrics **39**, 1005.
Hisano, K. and J.F. Ryan, 1972, Solid State Commun. **11**, 1745.
Huberman, B.A. and R.M. Martin, 1976, Phys. Rev. **B 13**, 1498.
Hutchings, M.T., R. Scherm and S.R.P. Smith, 1975, AIP Conf. Proc. **29** Magn. and Magn. Mat. 372.
Imry, Y. and S. Ma, 1975, Phys. Rev. Lett. **35**, 1399.
Iwasaki, H., S. Miyazawa, H. Koizumi, K. Sugii and N. Niizeki, 1972, J. Appl. Phys. **43**, 4907.
Janner, A. and T. Janssen, 1977, Phys. Rev. **B 15**, 643.
Johnston, W.D. and I.P. Kaminow, 1968, Phys. Rev. **168**, 1045.
Katiyar, R.S., J.F. Ryan and J.F. Scott, 1971, Phys. Rev. **B 4**, 2635.
Keve, E.T., S.C. Abrahams and J.L. Bernstein, 1969, J. Chem. Phys. **51**, 4928.
Klein, R., 1974, in: Anharmonic Lattices, Structural Transitions, and Melting, ed. T. Riste (Noordhoff, Leiden) p. 161.

Klein, M.V., 1976, in: Light Scattering in Solids, eds., M. Balkanski, R.C.C. Leite and S.P.S. Porto (Flammarion Sciences, Paris) p. 351.
Krumhansl, J.A. and J.R. Schrieffer, 1975, Phys. Rev. **B 11**, 3535.
Lagakos, N. and H.Z. Cummins, 1975, Phys. Rev. Lett. **34**, 883
Larkin, A.J. and D.E. Khmel'nitskii, 1969, Sov. Phys. JETP, **29** (Zh. Eksp. Teor. Fiz. **56**, 2087).
Lazay, P.D., J.H. Lunacek, N.A. Clark and G.B. Benedek, 1969, Light Scattering Spectra of Solids, ed., G.B. Wright (Springer, New York) p. 593.
Leese, J.G. and G.K. Horton, 1979, J. Low Temp. Phys. **35**, 205.
Levanyuk, A.P., 1981, Ferroelectrics, **36**, 306.
Levanyuk, A.P., V.V. Osipov and A.A. Sobyanin, 1976, in: Theory of Light Scattering in Condensed Matter, eds., B. Bendow, J.L. Birman, V.M. Agranovich (Plenum, New York).
Lockwood, D.J., J.W. Arthur, W. Taylor and T.J. Hosea, 1976, Solid State Commun. **20**, 703.
Lockwood, D.J., A.F. Murray and N.L. Rowell, 1981, J. Phys. **C 14**, 753.
Lyons, K.B. and P.A. Fleury, 1976a, J. Appl. Phys. **47**, 4898.
Lyons, K.B. and P.A. Fleury, 1976b, Phys. Rev. Lett. **37**, 161.
Lyons, K.B. and P.A. Fleury, 1977, Solid State Commun. **23**, 477.
Lyons, K.B. and P.A. Fleury, 1978, Phys. Rev. **B 17**, 2403.
Lyons, K.B., R.C. Mockler and W.J. O'Sullivan, 1973, J. Phys. **C 6**, L420.
Lyons, K.B., P.A. Fleury and H.L. Carter, 1980a, Phys. Rev. **B 21**, 1653.
Lyons, K.B., T.J. Negran and H.J. Guggenheim, 1980b, J. Phys. **C 13**, L415.
Lyons, K.B., R.N. Bhatt, T.J. Negran, H.J. Guggenheim, 1982, Phys. Rev. **B 25**, 1791.
Marques, M.C., 1980, J. Phys. **C 13**, 3149.
McMillan, W.L., 1975, Phys. Rev. **B 12**, 1187.
McMillan, W.L., 1976, Phys. Rev. **B 14**, 1496.
Mermelstein, M.D. and H.Z. Cummins, 1977, Phys. Rev. **B 16**, 2177.
Michel, K.H. and J. Naudts, 1978, J. Chem. Phys. **68**, 216.
Miyakawa, K. and T. Yagi, 1980, J. Phys. Soc. Jpn. **49**, 1881.
Natterman, T. and J. Przystawa, 1981, private communication.
Ohnari, I., 1980, J. Phys. **C 13**, 5911.
Ohnari, I. and S. Takada, 1979, Prog. Theory. Phys. **61**, 11.
Peercy, P.S., 1975a, Phys. Rev. **B 12**, 2725.
Peercy, P.S., 1975b, Solid State Commun. **16**, 439.
Popkov, Yu.A., S.V. Petrov and A.P. Mokhir, 1975, Sov. J. Low Temp. Phys. **1**, 91 (Fiz. Nizk. Temp. **1**, 189).
Press, W., A. Huller, H. Stiller, W. Stirling and R. Currat, 1974, Phys. Rev. Lett. **32**, 1355.
Reese, R.L., I.J. Fritz and H.Z. Cummins, 1973, Phys. Rev. **B 7**, 4165.
Riste, T., E.J. Samuelsen, K. Otnes and J. Feder, 1971, Solid State Commun. **9**, 1455.
Ryan, J.F. and J.F. Scott, 1974, Solid State Commun. **14**, 5.
Sandercock, J., 1978, Bull. Am. Phys. Soc. **23**, 387.
Sandercock, J.R., S.B. Palmer, R.J. Elliott, W. Hayes, S.R.P. Smith and A.P. Young, 1972, J. Phys. **C 5**, 3126.
Sawafuji, M., M. Tokunaga and I. Tatsuzaki, 1979a, J. Phys. Soc. Jpn. **47**, 1860.
Sawafuji, M., M. Tokunaga and I. Tatsuzaki, 1979b, J. Phys. Soc. Jpn. **47**, 1870.
Schmidt, H. and F. Schwabl, 1977, Lattice Dynamics, in: Lattice Dynamics, ed., M. Balkanski, (Flammarion Sciences, Paris) p. 748.
Schneider, T. and E. Stoll, 1973, Phys. Rev. Lett. **31**, 1254.
Schneider, T. and E. Stoll, 1976, Phys. Rev. Lett. **36**, 1501.
Schwabl, F., 1974, in: Anharmonic Lattices, Structural Transitions and Melting, ed., T. Riste (Leiden, Nordhoff) p. 87.
Scott, J.F., F. Habbal and M. Hidaka, 1982, Phys. Rev. **B 25**, 1805.
Shapiro, S.M. and H.Z. Cummins, 1968, Phys. Rev. Lett. **21**, 1578.

Strukov, B.A., S.A. Taraskin, K.A. Minaeva and V.A. Fedorikhin, 1980, Ferroelectrics, **25**, 399.
Tanaka, H. and I. Tatsuzaki, 1980, Solid State Commun. **35**, 285.
Tanaka, H., T. Yagi and I. Taszusaki, 1978a, J. Phys. Soc. Jpn. **44**, 1257.
Tanaka, H., T. Yagi and I. Taszusaki, 1978b, J. Phys. Soc. Jpn. **44**, 2009.
Taylor, W., D.J. Lockwood and H. Vass, 1978, Solid State Commun. **27**, 547.
Toledano, P., 1982, NATO Adv. Study Inst. on Nonlinearities at Phase Transitions and Instabilities, Geilo, Norway, March 29-April 9, 1981 (unpublished).
Wang, C.H., 1979, J. Chem. Phys. **70**, 3796.
Wilson, J.A., F.J. DiSalvo and S. Mahajan, 1975, Adv. in Phys. **24**, 117.
Winterling, G., W. Senn, M. Grimsditch and R. Katiyar, 1977, in: Lattice Dynamics, ed., M. Balkanski (Flammarion Sciences, Paris) p. 553.
Yagi, T., H. Tanaka and I. Tatsuzaki, 1977, Phys. Rev. Lett. **38**, 609.
Yakovlev, I.A., T.S. Velichkina and L.F. Mikheeva, 1956, Sov. Phys. Doklady, **1**, 215.

CHAPTER 8

# Light Scattering Studies of Incommensurate Transitions

M.V. KLEIN

*Department of Physics and Materials Research Laboratory*
*University of Illinois at Urbana-Champaign*
*104 S. Goodwin Ave., Urbana, IL 61801*
*USA*

*Light Scattering near Phase Transitions*
*Edited by*
*H.Z. Cummins and A.P. Levanyuk*

© *North-Holland Publishing Company, 1983*

# Contents

## 1. Introduction

The theory of incommensurate phase transitions is discussed by Golovko and Levanyuk in this volume. In this introduction I shall give only a brief sketch, without attribution, of some of the basic ideas.

The simplest incommensurate phase has primary order parameters

$$Q(\boldsymbol{q}_c) = A\,e^{i\phi}, \tag{1}$$

and $Q(-\boldsymbol{q}_c) = Q(\boldsymbol{q}_c)^*$ with a static distortion that has a primary component of the form

$$\boldsymbol{u}_s(\boldsymbol{R}) = \epsilon A \cos(\boldsymbol{q}_c \cdot \boldsymbol{R} + \phi). \tag{2}$$

When expressed as a linear combination of reciprocal lattice vectors, $\boldsymbol{q}_c$ must have at least one irrational component. To eq. (2) must be added higher harmonics and umklapp terms. If the amplitude $A$ is small and if coupling to secondary order parameters and higher harmonics is neglected, phonon modes of zero wavevector in the presence of the static distortion will consist of an amplitude mode with displacement

$$\delta \boldsymbol{u}_A = \epsilon \delta A \cos(\boldsymbol{q}_c \cdot \boldsymbol{R} + \phi), \tag{3a}$$

and a phase mode with displacement

$$\delta \boldsymbol{u}_\phi = -\epsilon A \delta\phi \sin(\boldsymbol{q}_c \cdot \boldsymbol{R} + \phi). \tag{3b}$$

This represents the so-called "harmonic limit". The phase mode will have zero frequency because the free energy is invariant under a change in $\phi$. It is therefore sometimes referred to as a Goldstone mode. Coupling to other harmonics or other degrees of freedom may give $\delta u_\phi$ a response that is relaxational and/or of a damped oscillator form.

The origin of the Raman activity and the temperature dependence of the amplitude mode frequency will be very much like those properties in a commensurate modulated phase. Whether and to what extent the phase mode is Raman active has been the subject of much recent discussion. It is generally believed to have considerably less Raman activity than the amplitude mode. At wavevector $\boldsymbol{q} = 0$ the phase mode is expected to show linear dispersion in the simplest harmonic case. With relaxation dominating at $\boldsymbol{q} = 0$, this mode will stay overdamped for a finite range of $\boldsymbol{q}$, and the expected Raman response will be in the form of a central peak. Such a mode should couple in some order to acoustic phonons and would therefore be expected to affect their dispersion and damping. Workers who have tried to observe phase modes by light scattering techniques have therefore often used the same Fabry–Perot instruments used for Brillouin studies of acoustic phonons. These attempts will be discussed in the article by Cummins on Brillouin scattering and in the article by Fleury and Lyons on central peaks.

Most of the data to be discussed in this article apply to systems with a two component order parameter ($n = 2$), i.e., the real and imaginary parts of $Q(\boldsymbol{q}_c)$, or $A$ and $\phi$ in eq. (1). Examples include $K_2SeO_4$, $Rb_2ZnCl_4$, "KCP", $Na_2CO_3$, $SC(NH_2)_2$, biphenyl and $ThBr_4$. $BaMnF_4$ has $n = 4$, the 1T polytypes of the transition metal dichalcogenides have $n = 6$; and the 2H polytypes of these materials have $n = 12$. We now discuss the Raman data on these systems in order of increasing $n$. We shall occasionally mention data taken with Brillouin techniques, but shall refer the reader to the articles by Cummins and by Fleury and Lyons for more details.

## 2. Two-component systems

### 2.1. $K_2SeO_4$

At room temperature $K_2SeO_4$ is orthorhombic with the space group Pnam ($D_{2h}^{16}$) (Kálman et al. 1970). This is the P phase – paraelectric. It undergoes two phase transitions (Aiki et al. 1969, Aiki and Hukuda 1969) at $T_i \approx 129$ K and $T_c \approx 93$ K. The intermediate phase (I) was shown to be incommensurate by Iizumi et al. (1977) who performed a neutron-scattering study. They found superlattice reflections characterized by a wavevector $\boldsymbol{q}_c = (1/3 - \delta)\boldsymbol{a}^*$ where $\delta = 0.023$ at $T_i$. Below $T_c$ $\delta$ vanishes, $\boldsymbol{q}_c$ becomes commensurate, and the crystal becomes weakly ferroelectric (F phase).

Iizumi et al. (1977) measured the dispersion curves for the soft optic phonon mode above $T_i$ and found its symmetry to be $\Sigma_2$. The mode showed almost complete softening as $T_i$ was approached from above. The structure of the F phase was also studied in detail. Its space group in $Pna2_1$ ($C_{2v}^9$). Lattice dynamics were studied in detail in the F phase by Axe et al. (1980) using inelastic neutron scattering. They measured the dispersion as a function of $q_x$ of the amplitude mode, the (hardened) phase mode, and a transverse acoustic mode which interacts with the phase mode.

A soft mode was seen by Raman scattering below $T_i$ by Fawcett et al. (1974) and by Caville et al. (1976, 1978) and studied in detail by Wada et al. (1977a). It has $A_1$ symmetry in the F phase and could be followed almost to $T_i$. A much weaker, lower frequency Raman mode of $B_1$ symmetry [$(ac) \equiv (xz)$] was observed by Wada et al. (1977b) in the F phase. (In their notation this was labelled a $B_2$ mode.) Its intensity vanishes with rising temperature somewhat below $T_c$, but its frequency extrapolates to zero near $T_i$. It is believed to be the phase mode, which has a finite frequency in the commensurate F phase and is Raman active with an intensity proportional to the fourth power of the order parameter (Dvorak and Petzelt 1978). As $T$ rises through $T_c$ the frequency of this mode is expected to drop discontinuously to zero.

The two soft modes were studied using Raman and Brillouin Techniques by Fleury et al. (1979), who used an iodine cell to absorb the quasi-elastic component of the scattered light, and by Unruh et al. (1979), who used a double monochromator plus a third monochromator. The most complete Raman study is that of Unruh et al. Their oscillator fits show less of an increase in the width $\gamma$ with increasing temperature than do those of Wada et al. (1977a,b). The results of Fleury et al. seem consistent with those of Unruh et al. Figure 1 shows the spectra taken by Unruh et al., and fig. 2 gives the results of their oscillator fits. Note that the sum of $\gamma$ and $\omega_0$ remains nearly constant.

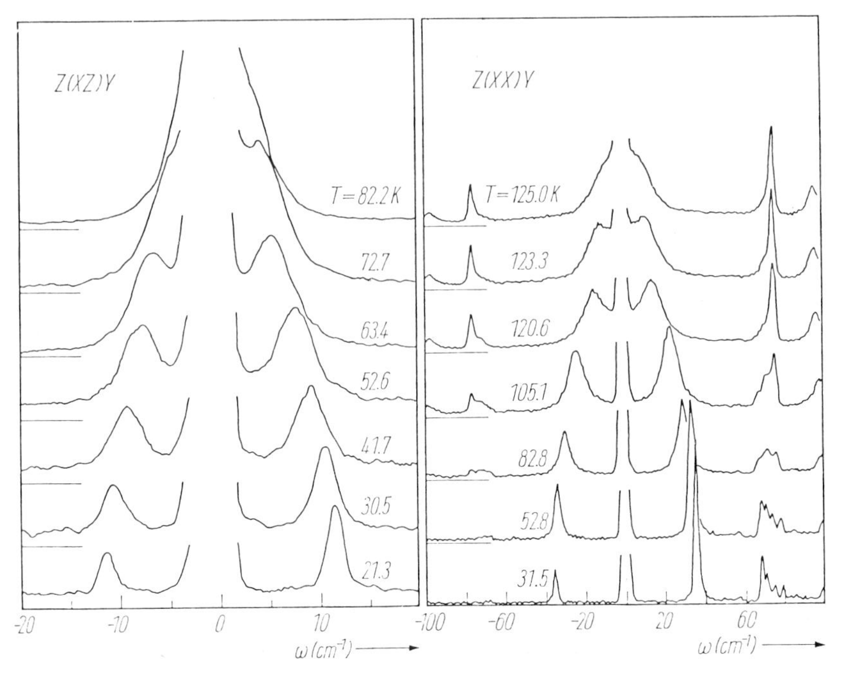


Fig. 1. Stokes and anti-Stokes Raman spectra of $K_2SeO_4$ in the low frequency region at various temperatures (Unruh et al. 1979). Left part: $B_1$ symmetry, showing the soft phase mode in the commensurate (F) state. Right part: $A_1$ symmetry showing the soft amplitude mode in I and F states.

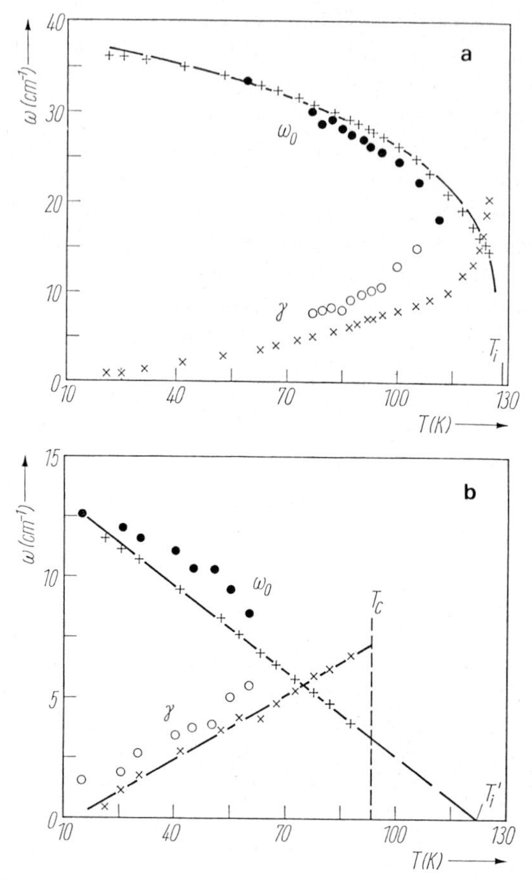


Fig. 2. Results of oscillator fits by Unruh et al. (1979) to their Raman data. Part (a): amplitude mode. Part (b): phase mode. Crosses give their results for mode frequencies and $\times$'s give their results for damping constants. Circles give the results of Wada et al. (1977a, b).

The critical behavior of $\omega_{A1}$ near $T_1$ is of interest. Unruh et al. found that

$$\omega_0^2 \propto (T_i - T)^{2\beta},$$

where $2\beta = 0.70 \pm 0.1$. This result is in good agreement with the result $2\beta = 0.75 \pm 0.05$ obtained by Majkrzak et al. (1980), who measured the intensity of the primary superlattice reflections using neutron diffraction. As discussed by Cowley and Bruce (1978), $K_2SeO_4$ represents the simplest incommensurate phase transition and is in the universality class with the $n = 2$ component Heisenberg ferromagnet in its continuous spin representation. Series expansions (Pfeuty et al. 1974, Ferer et al. 1973) and a scaling relation lead to the theoretical result (Majkrzak et al. 1980) $2\beta = 0.70 \pm 0.04$, in good agreement with the Raman and the neutron results.

Inoue et al. (1980) used a third monochromator with a double monochromator to study the $B_1$ mode at frequencies above about $2.5\,cm^{-1}$ and at temperatures below and above $T_c$. They observed some scattering that looks like the tail of a central peak above $T_c$. This vanished at 121 K and above. This result seems to disagree with the iodine cell results of Fleury et al. (1979), who report no low-frequency inelastic scattering of $B_1$ symmetry above $T_c$, apart from Brillouin components. The results of Inoue et al. (1980) might be explained by stray light from a strong temperature-dependent central peak, static enough to be absorbed by the iodine cell, i.e., $\Delta\nu < 300$ MHz (Lyons and Fleury 1976), and hence not observable by Fleury et al. (1979). Independent evidence of an over-damped infrared-active phase mode in the I phase was provided by the dielectric response measurements of Petzelt et al. (1979).

Thermal hysteresis was observed by Chen et al. (1981) in an X-ray study of the incommensurability parameter $\delta$. On the same sample used for the neutron studies of Iizumi et al. (1977), Chen et al. found that upon rapid cooling to a temperature $T$, $\delta(T)$ agreed with the neutron result, but upon further cooling after several hours of annealing at temperature $T$, $\delta$ remained at $\delta(T)$. Upon heating above $T_c$ the X-ray results showed a superposition of two components, one commensurate and one incommensurate. Neutrons penetrate the entire sample, whereas the X-rays used by Chen et al. (1981) penetrate about 10 $\mu$m. Since the neutron data on the same sample show no evidence of this hysteresis, Chen et al. attribute their X-ray results to a surface layer in which $\delta$ varies with depth. The data shown by Inoue et al. (1980) were taken upon slow cooling, but they state that the same results were obtained upon heating. If they were indeed observing the phase mode in the I phase, and if their sample has surface-related hysteresis, then one would expect their Raman data to show hysteresis.

The lattice dynamices of $K_2SeO_4$ were studied theoretically by Haque et al. (1978) and in more detail by Haque and Hardy (1980). They used a rigid-ion model with Coulomb interactions and a limited number of short-range interactions that could be determined by the equilibrium conditions using the calculated Coulomb coefficients and the observed Raman frequencies. These authors were able to reproduce the measured dispersion curve of the soft $\Sigma_2$ branch and showed that it depended on a delicate balance of the two types of interatomic force. This balance is affected by pressure, as shown experimentally by Press et al. (1980) with neutron scattering and by Kudo and Ikeda (1981) with dielectric measurements. Both groups found that an increase in pressure lowers $T_i$ and $T_c$ but increases $T_i - T_c$. Press et al. (1980) found in addition that pressure increases the value of the incommensurability parameter $\delta$. Haque et al. (1978) found that an increase in pressure lowers the frequency of the $A_1$ amplitude mode. This was studied in more detail by Massa et al. (1982). At a temperature at the sample holder of 82 K they found, for instance, that the amplitude mode softened by 3 $cm^{-1}$ and that the apparent temperature, as determined from the measured anti-Stokes to Stokes intensity, increased to about 100 K. The measurements were made after a ten hour "anneal" at nitrogen temperature before the stress was applied. The anneal reduced the scattered light near zero frequency shift. Its effect may be related to the surface-related hysteresis discovered by Chen et al. (1981).

Massa et al. (1979) observed inelastic scattering in the 24–50 $cm^{-1}$ region at room temperature, which they ascribed to three features that they tracked as a function of temperature from 135 to 573 K. The spectra were clearest in the $a(cc)b$ geometry (Massa et al. 1983). This is a correction; $b(cc)a$ geometry was stated by Massa et al. (1979). The lowest feature appeared at 4 $cm^{-1}$ at 135 K. These features were related by Massa et al. (1979) to features in the phonon dispersion curves measured by Iizumi et al. (1977). Massa et al. explained their results as first-order Raman scattering due to disorder. It was reproducible among their samples; hence they suggested that the disorder is intrinsic. The results of Unruh et al. (1979) are noticeably different from those of Massa et al. It is now believed (Massa et al. 1983) that the disorder is due to slight orientational disordering of the selenate groups in $K_2SeO_4$ crystals grown by slow evaporation of aqueous solutions. Massa et al. (1980) studied low frequency sidebands of an internal mode in crystals doped with about 1% of Na, Rb, or Cs and found no differences from spectra from undoped crystals.

### 2.2. $Rb_2ZnCl_4$ and $Rb_2ZnBr_4$

Dielectric measurements on $Rb_2ZnCl_4$ (Sawada et al. 1977a) and on $Rb_2ZnBr_4$ (Sawada et al. 1977b, De Pater 1978) show that each of these materials undergoes phase transitions analogous to those of $K_2SeO_4$ at 302 and 192 K ($Rb_2ZnCl_4$) and at 291 and 187 K ($Rb_2ZnBr_4$). The crystal structure of the normal phase has not been explicitly studied, but it is presumably the same as that of $Cs_2ZnCl_4$ (Brehler 1957) and $Cs_2ZnBr_4$ (Morosin and Lingafelter 1959), namely Pmcn ($D_{2h}^{16}$). This is the same as the room temperature structure of $K_2SeO_4$, except for the labelling of the axes. Sawada et al. 1977a,b) and those following them use the convention that $a$ (instead of $c$) approximately equals $b\sqrt{3}$.

Neutron scattering studies have been performed on the modulated phases of $Rb_2ZnBr_4$. An incommensurate structure was found by De Pater and Van Dijk (1978) below $T_i = 355$ K. Gesi and Iizumi (1978) found a similar result but with $T_i = 323$ K. They pointed out that $T_i$ is probably quite dependent on small amounts of impurities or lattice strains. They found the incommensurability parameter $\delta$ to be 0.04 just below $T_i$. It remained at that value upon cooling to 199 K, where it started to decrease. At 190 K $\delta$ changed discontinuously to zero. Upon warming there was a few °C of hysteresis. Similar neutron results for $\delta(T)$ were found by De Pater et al. (1979).

A neutron scattering study was performed on $Rb_2ZnCl_4$ by Gesi and Iizumi (1979). They measured the intensity of a superlattice satellite and the value of $\delta(T)$ below $T_i = 302$ K. Just below $T_i$ $\delta$ was 0.028, and it showed a smoother decrease with temperature and less hysteresis than in $Rb_2ZnBr_4$. Hysteresis in $Rb_2ZnCl_4$ was studied in more detail in the dielectric work of Hamano et al. (1980a,b) and in the dielectric and X-ray studies of Mashiyama et al. (1981).

Near $T_i = 353$ K Gesi and Iizumi found for $Rb_2ZnBr_4$ that the intensity of the first order superlattice diffraction peaks had the mean-field temperature dependence $(T_i - T)^{0.5}$, whereas De Pater and Van Dijk found a dependence of $(T_i - T)^{2\beta}$ with $2\beta = 0.60$ with $T_i = 355$ K. Matsuda and Hatta (1980) found that the change in shear elastic constants just below $T_i = 305$ K for $Rb_2ZnCl_4$ gave $2\beta = 0.62$, close to the value for the three-dimensional Ising model. Luspin et al. (1982) found $2\beta = 0.6 \pm 0.1$ from a Brillouin measurement.

Above $T_i$ the soft phonon mode was found by De Pater and Van Dijk (1978) to be close to overdamped. The frequency remained below 0.1 THz up to $T = 475$ K.

Raman measurements were performed on $Rb_2ZnBr_4$ by Takashige et al. (1980a) on a sample which dielectric measurements showed had $T_i = 346$ K and $T_c = 187$ K. In $(aa)$ and $(cc)$ polarizations they found a soft mode which they followed from 20 cm$^{-1}$ at 93 K to 9 cm$^{-1}$ at about 325 K. By analogy with a similar mode in $K_2SeO_4$, this was assigned to the amplitude mode. It interacted weakly with a 24 cm$^{-1}$ mode, which broadened with rising temperature and was difficult to follow above $T_i$. There was some evidence for a phase mode in the $a(ca)b$ geometry, but the temperature was not low enough to be certain. The phase mode was clearly seen in the $c(ac)b$ geometry by Franke et al. (1980b). It moved from 10 cm$^{-1}$ at 90 K to 14 cm$^{-1}$ at 27 K. These authors also found the same two modes in the $c(aa)b$ geometry seen by Takashige et al. (1980a), and they also found a peak at 12 cm$^{-1}$ at 77 K in that geometry and a mode, probably the same mode, in $b(cc)a$ geometry that moved from 8 cm$^{-1}$ at 90 K to 11 cm$^{-1}$ at 49 K. This shift suggested the existence of another phase transition near 140 K.

The soft mode in $Rb_2ZnCl_4$ was first seen below $T_1 \equiv T_i = 302$ K by Wada et al. (1978, 1979). At 77 K Wada et al. (1979) found the amplitude mode at 25 cm$^{-1}$ in the $a(cc)b$ and $c(aa)b$ geometries. Upon warming, it softened and showed evidence of anti-crossing with a non-softening mode at 18 cm$^{-1}$. They also observed a hard mode in the $a(bc)b$ geometry which moved from 16 cm$^{-1}$ at 77 K to 27 cm$^{-1}$ at 350 K. The phase mode was first seen by Franke et al. (1980a) in the $a(ca)b$ geometry. They also observed a new strongly temperature-dependent mode in the $a(cc)b$ geometry whose frequency varied from 15 cm$^{-1}$ at 9 K to about 5 cm$^{-1}$ at 64 K. It remained underdamped and

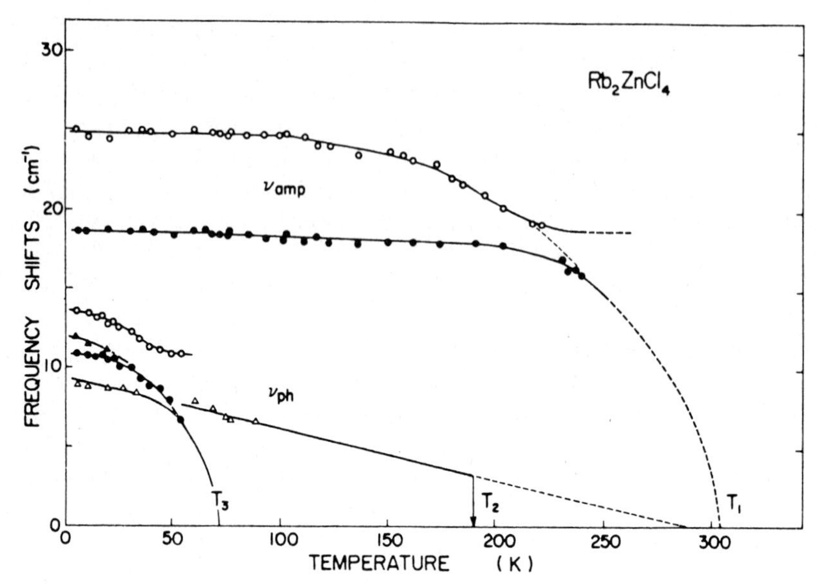


Fig. 3. Temperature dependence of soft mode frequencies in $RbZnCl_4$ (Wada et al. 1981a). See text for further details.

softened according to

$$\omega_0 \propto (T_3 - T)^{1/2},$$

with $T_3 = 72$ K. This was the first evidence of a phase transition in the commensurate state below $T_2 = 189$ K.

Additional evidence for a third phase transition in $Rb_2ZnCl_4$ was provided by the dielectric constant and birefringence measurements of Günter et al. (1981). They found $T_3 = 75$ K and suggested that the phase below $T_3$ is monoclinic with the space group $P2_111$ (using our current assignment of crystal axes).

The most thorough Raman study on $Rb_2ZnCl_4$ is that of Wada et al. (1981a). Their results for the soft mode frequencies are shown in fig. 3. The two modes of highest frequency (open and closed circles in the figure) are the 25 and 18 $cm^{-1}$ amplitude modes seen in $c(aa)b$ and $a(cc)b$ geometries and discussed above. The lower set of open and closed circles starting at 13.5 and 11 $cm^{-1}$, respectively, at low temperature were seen in the $a(cc)b$ geometry. Note the anti-crossing behavior. Oscillator fits were made to the Raman spectra. The damping constants for both modes were approximately proportional to $T$. This pair of modes corresponds to the single mode first observed in the $a(cc)b$ geometry by Franke et al. (1980a). The open triangles indicate the phase mode observed by Wada et al. (1981) in the $a(ca)b$ geometry. For $T < T_3$ another mode appears in the same geometry (solid triangles).

From the doubling of the lowest Raman lines in fig. 3 and from other properties Wada et al. (1981a) suggest that below $T_3$ the structure is commensurate with space group $P2_111$ ($C_2^2$) with twice the unit cell volume as the structure in the commensurate phase between $T_3$ and $T_2$ (where the space group is $P2_1\,cm - C_{2v}^9$).

Unlike the case of $K_2SeO_4$, where pressure lowers the frequency of the amplitude mode, Massa (1981) found no change in this mode with pressure in $Rb_2ZnCl_4$.

This class of crystals is a large one, but few other examples have been studied by Raman scattering. Takashige et al. (1980b) give a brief report on Raman measurements in $[N(CH_3)_4]_2ZnCl_4$ and $[N(CH_3)_4]_2CoCl_4$, which have the same structure as $Rb_2ZnCl_4$ and undergo a series of phase transitions. They found no soft mode in the ferroelectric or incommensurate phases ($T_1 \approx 291$ K).

Some Brillouin results on $Rb_2ZnCl_4$ warrant comment. The elastic constant $C_{33}$ shows dispersion at temperatures near $T_i$ when Brillouin measurements are compared at two scattering angles (Yamanaka et al. 1981) or with ultrasonic data (Luspin et al. 1982). This suggests relaxational behavior of the soft modes near $T_i$. The analysis by Luspin et al. (1982) suggests that it is the amplitude mode that exhibits this behavior. The relaxational processes are probably related to precursor short-range order implied by the tail in the heat capacity anomaly observed above $T_i$ by Hamano et al. (1980a).

### *2.3. Biphenyl*

Biphenyl in its normal phase (I) has space group $P2_1/a$ ($C_{2h}^5$), and the two phenyl groups in each molecule are coplanar. At $T_1 = 38$ K a second-order phase transition occurs to an incommensurate phase (II) with $\boldsymbol{q}_c = \delta_a \boldsymbol{a}^* + \frac{1}{2}(1-\delta_b)\boldsymbol{b}^*$ (Cailleau et al. 1979c). At $T_2 = 21$ K there is a first-order transition to another incommensurate phase (III) with $\boldsymbol{q}_c = \frac{1}{2}(1-\delta_b)\boldsymbol{b}^*$. In the incommensurate phases the planes of the two phenyl groups are thought to have twisted with respect to one another. The soft mode for $T < T_1$ was studied by inelastic neutron scattering by Cailleau et al. (1979a,b). The dispersion of the phase modes was measured by inelastic neutron scattering by Cailleau et al. (1980, 1981). This was the first direct observation of phase modes. They scanned away from $\boldsymbol{q}_c$ in the $a^*$ and $b^*$ directions and observed linear dispersion of the Goldstone mode. A theoretical model of the lattice dynamics of biphenyl in phase I was given by Natkaniec et al. (1981), and a Landau theory for the incommensurate phases and the transition between them was developed by Ishibashi (1981).

A Raman study of the phase transition in biphenyl was first published by Bree and Edelson (1977). They found an activated hard mode at 961 $cm^{-1}$ whose intensity is proportional to $(T_1 - T)$ with $T_1 = 40$ K. At 4.2 K they found four soft modes in an unpolarized spectrum at 33.3, 20, 24.3, and 17.6 $cm^{-1}$. These softened somewhat as the temperature was raised. Polarized spectra were

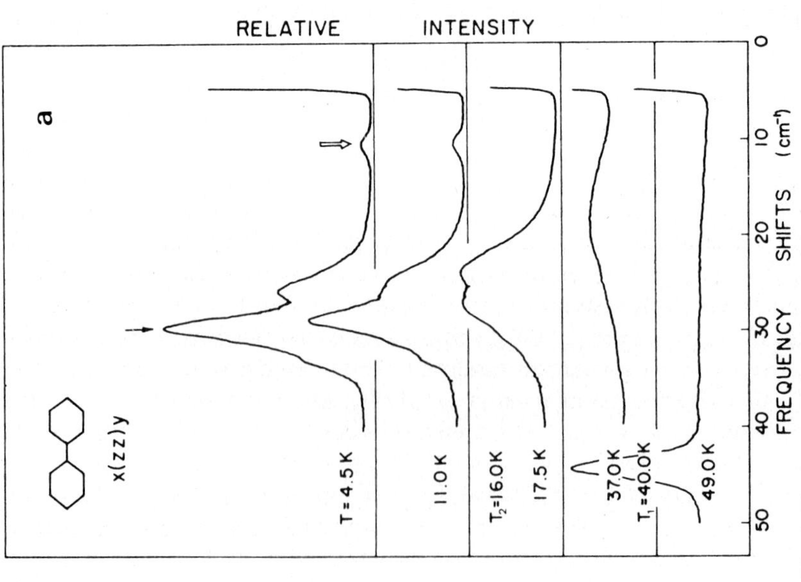


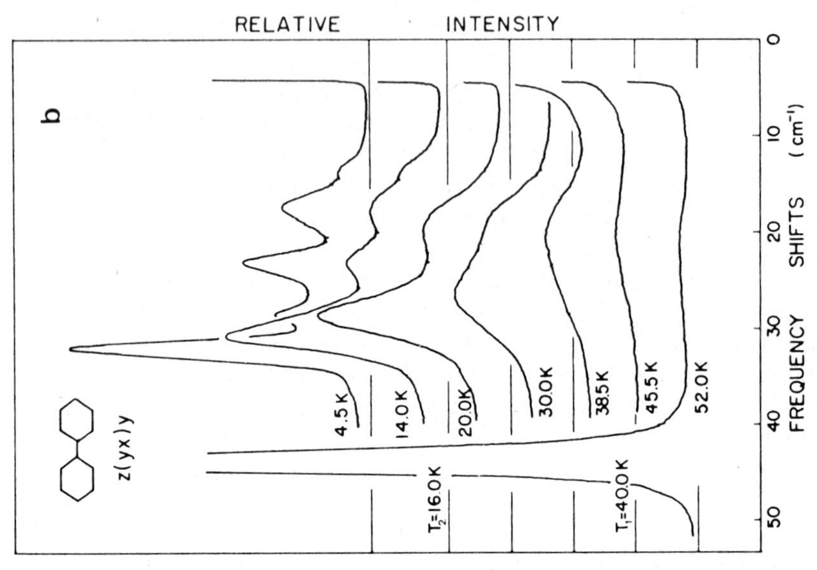


Fig. 4. Temperature-dependent, low-frequency, Raman spectra of biphenyl (Wada et al. 1981b). Solid arrow (left part) denotes soft amplitude mode.

taken by Wada et al. (1981b). Their results are shown in fig. 4. In $x(zz)y$ geometry (solid arrow in fig. 4, left part) at 4.5 K a soft mode can be seen at 30 $cm^{-1}$, which upon warming shows anti-crossing with a mode at about 27 $cm^{-1}$ and can be followed into phase II. In the same geometry a mode is seen at 11 $cm^{-1}$ (open arrow in fig. 4) which softens slightly as the temperature is raised and vanishes above $T_2$.

In $z(yx)y$ geometry (fig. 4, right part) there are five peaks below $T_2$, and two just above $T_2$, which merge as $T_1$ is approached from below and do not soften completely. Just below $T_1$ there is an observable tail of the central component in this geometry.

### 2.4. *Thiourea*

Thiourea, $SC(NH_2)_2$, and its deuterated analogue exist in at least five phases between room temperature and liquid nitrogen temperature (Goldsmith and White 1959). Above $T_4 = 202$ K the structure is orthorhombic and paraelectric with space group Pnma ($D_{2h}^{16}$) and four molecules per unit cell (phase V). Below $T_1 = 169$ K (phase I) the structure is also orthorhombic, but ferroelectric, with the same number of molecules and space group $P2_1ma$ ($C_{2v}^2$). Between these temperatures X-ray satellite reflections imply a modulated structure with period along the $b$-axis of about $8b$ (Futama et al. 1967, Shiozaki, 1971). The incommensurability varies quasi-continuously with temperature (McKenzie 1975a). Lock-in with $\boldsymbol{q}_c = \boldsymbol{b}^*/9$ has recently been shown to occur in $SC(ND_2)_2$ at about 2 K above $T_1$ (Moudden et al. 1979). At $T_1$, $q_c$ drops to zero. The pressure–temperature phase diagram has been determined (Denoyer et al. 1982).

Diffuse scattering was observed above $T_4$ by neutron scattering by McKenzie (1975a), who also found an inelastic component. He measured dispersion curves of acoustic phonons (McKenzie 1975b) along $a^*$, $b^*$, and $c^*$ directions at room temperature and found the frequencies to be low, but he observed no softening as the temperature was decreased.

A soft mode was seen below $T_4$ by Raman scattering (Benoit et al. 1972, Delahaigue et al. 1975, Chapelle and Benoit 1977, Wada et al. 1978b) which is polar below $T_1$. A soft TO mode was studied by infrared reflectivity measurements by Brehat et al. (1976). Its symmetry was $B_{3u}$ in phase V and $A_1$ in phase I. Its frequency extrapolated to zero at $T_4$ in both directions and below $T_1$ it agreed with the Raman results. An analysis by Khelifa et al. (1977) of these and other dielectric results concluded that in addition to the soft polar phonon, inclusion of two Debye relaxations was necessary to explain the data.

An inelastic neutron study was performed on deuterated thiourea by Moudden et al. (1978). Just above $T_4 = 213$ K there is an overdamped mode for $\boldsymbol{q} \approx 0.17\boldsymbol{b}^*$, which gradually becomes underdamped as the temperature is raised to room temperature. There is evidence of coupling between two modes of the same symmetry, an optic mode which becomes $B_{3u}$ at the zone center and a TA mode. In the incommensurate phase a new inelastic neutron peak has been tentatively assigned to the phase mode (Denoyer et al. 1980).

In phase I Wada et al. (1978) assigned the soft Raman mode to a mode that consists primarily of librations of $SC(NH_2)_2$ molecules about the $b$-axis (in our notation) that would have $B_{3u}$ symmetry (in our notation) in phase V. A second Raman peak near 120 $cm^{-1}$ was assigned by them to another mode primarily composed of translations that also would have $B_{3u}$ symmetry in phase V. These are probably the same modes seen in phase V by Moudden et al. (1978).

### 2.5. $Na_2CO_3$

Above $T_1 = 763$ K $Na_2CO_3$ has an hexagonal phase ($\alpha$). Between $T_2 = 620$ K and $T_1$ it is in a monoclinic $\beta$ phase, and between $T_2$ and $T_3 = 130$ K it is in an incommensurate $\gamma$ phase with a primary wavevector $\boldsymbol{q}_c = q_1 a^* + q_3 c^*$, where $q_1 = 0.182$ and $q_3 = 0.318$ at room temperature (van Aalst et al. 1976). Below $T_3$ lock-in occurs to a $\delta$ phase with $q_1 = \frac{1}{6}$ and $q_3 = \frac{1}{3}$ (De Pater and Helmholdt 1979).

A Raman study was made by Maciel and Ryan (1981a,b). In phases $\delta$ and $\gamma$ they observed a splitting of the totally symmetric $\nu_1$ mode of the carbonate ion, which they could follow to about 500 K. Five soft modes were seen in the distorted phases, three in the lower temperature range of the $\gamma$-phase. These were interpreted as the soft and hard mode components of two modes that result from coupling between an acoustic phonon and a librational mode of B symmetry.

### 2.6. KCP

The one-dimensional conductor $K_2Pt(CN)_4Br_{0.3} \cdot 3H_2O$, known as KCP, has a very high room temperature conductivity that is nearly one-dimensional (1D) (Zeller 1973) due to chains of closely-spaced Pt atoms along the $c$-axis. There are two Pt atoms in the average unit cell, which has $C_{4v}$ symmetry (Deiseroth and Schulz 1974) (Williams et al. 1974). Diffuse X-ray measurements (Comès et al. 1973) and inelastic neutron scattering measurements (Renker et al. 1973) show a giant Kohn anomaly at wavevector $2k_F$. Below room temperature, the distortion becomes noticeably 3D and remains incomplete with a new unit cell $2a \times 2a \times 6.7$ $(c/2)$, where $c/2$ is the average Pt–Pt spacing, but with no long-range order transverse to the $c$-axis (Renker et al. 1974).

A low-frequency Raman study by Steigmeier et al. (1975) found a line of $A_1$ symmetry at about 44 $cm^{-1}$. The data were fit well by a damped harmonic oscillator. The damping constant increased by more than a factor of 25 between helium and room temperature. The frequency showed about 25% hardening over the same range. This was interpreted by Steigmeier et al. (1975) as the amplitude mode of the CDW.

Far infrared measurements were performed on KCP by Brüesch et al. (1975). A peak in the optical conductivity was observed near 15 $cm^{-1}$ that was well defined at 4 K, but greatly broadened at room temperature. Oscillator fits agreed well with the data. At 4 K the TO ("bare") phonon frequency was 15 $cm^{-1}$, and the LO–TO splitting implied a plasma frequency of 58 $cm^{-1}$.

The absence of long-range order perpendicular to the $c$-axis will broaden the Raman and infrared peaks but will not eliminate them, provided that the fluctuations into the distorted phase are long lasting on the time scale of the period of a CDW phonon mode. Then an approximate description of the Raman and infrared properties may be given as if the fluctuations were static and exhibited long-range order. Steigmeier et al. (1976a) observed a shift in the frequency of the amplitude mode when $D_2O$ was substituted for $H_2O$ in KCP, evidence of coupling between the amplitude mode and the high-frequency O–H stretching vibration.

### 2.7. $ThBr_4$ and $Sr_2Nb_2O_7$

$ThBr_4$ occupies space group $I4_1/amd$ ($D_{4h}^{19}$) at room temperature and has a Raman-active soft below $T_i = 92$ K, whose frequency obeys

$$\omega_0 \propto (92 - T)^{1/3}$$

(Hubert et al. 1981). $ThCl_4$ shows similar behavior with $T_i = 70$ K. A neutron study (Currat 1981) shows that the modulation wavevector is $\boldsymbol{q}_c = 0.31\boldsymbol{c}^*$. Above $T_i$ soft modes were seen in the $\tau_4$ phonon branch. Below $T_i$ dispersion curves were taken as $\boldsymbol{q}$ moved away from $\boldsymbol{q}_c$. Both an amplitude mode and a Goldstone phase mode were observed.

$Sr_2Nb_2O_7$ has a perovskite slab structure and is ferroelectric with space group $Cmc2_1$ at room temperature (Ishizawa et al. 1975). An elastic anomaly was discovered near 215°C by Ohi et al. (1979), evidence of a phase transition, which was shown by Yamamoto et al. (1980) to be to an incommensurate phase with $\boldsymbol{q}_c = (\frac{1}{2} - \delta)\boldsymbol{a}^*$. Raman measurements were performed by Kojima et al. (1979). They found a soft mode in $b(cc)\bar{b}$ geometry, which anti-crossed with a 32 $cm^{-1}$ mode and softened as $T$ increased towards $T_i = 215$°C according to

$$\omega_0 \propto (T_i - T)^{0.38}\,.$$

## 3. Four-component systems

### 3.1. $Ba_2NaNb_5O_{15}$

$Ba_2NaNb_5O_{15}$ has the tetragonal tungsten bronze structure (space group 4 mm) between 580°C and 300°C, where it becomes orthorhombic and incommensurate with a wavevector $q_c = \pm(1+\delta)(a^*+b^*)/4+c^*/2$ (Schneck and Denoyer 1981). This is an $n = 4$ system. The parameter $\delta$ changes discontinuously upon cooling through 150°C, and by room temperature it equals zero. A detailed neutron study has been made by Schneck et al. (1982). An earlier Raman study by Boudou and Sapriel (1980) found no soft modes in the incommensurate phases. A Brillouin study by Young and Scott (1981) found a shift in the $C_{11}$ and $C_{22}$ elastic constants belowe 300°C that could be described by a Debye-type relaxation process.

### *3.2. $BaMnF_4$*

$BaMnF_4$ is a pyroelectric ferromagnetic with many unusual properties, which have been reviewed by Scott (1979). In its normal phase it is orthorhombic with space group $A2_1am$ ($C_{2v}^{12}$) (Keve et al. 1969). First report of a structural phase transition in $BaMnF_4$ was by Spencer et al. (1970), who found a sharp divergence at 255 K in the ultrasonic attenuation coefficient $\alpha_{cc}$ for the longitudinal phonon propagating along the $c$ direction. Sound velocities were measured ultrasonically by Fritz (1975a), who found dips in $v(T)$ at 248 K, which were largest (24%) and narrowest for $v_{cb}$. (The first index, $c$, gives the direction of the wavevector, and the second, $b$, gives the direction of the displacement.) He also found a broader 13% dip in $v_{cc}$ and dips of the order of 1% in $v_{aa}$ and $v_{bb}$. The corresponding attenuation $\alpha_{cc}$ was found by Fritz (1975b) to be critically diverging above $T_c \approx 250$ K. (He could not measure $\alpha_{cc}$ below $T_c$). He also measured $\alpha_{cb}$ and found it to diverge as $T \to T_c$ from above or below. The critical exponents for $\alpha_{cb}$ were greater (3.9 for $T > T_c$ and 5.7 for $T < T_c$) than for $\alpha_{cc}$ (2.2 for $T > T_c$), qualitatively consistent with the narrower width of the dip in $v_{cb}$. The magnitudes of the exponents were qualitatively consistent with strong two-dimensional correlations, presumably within layered sheets of connected $MnF_6$ octahedra in planes perpendicular to the $b$-axis.

The first confirmation of the phase transition by optical means was provided by the Raman measurements of Ryan and Scott (1974). Below $T_c$ they found that the number of Raman-active modes approximately doubled, suggesting that the phase transition involves condensation of a zone-boundary phonon. Below $T_c$ they found two modes of quasi-transverse, mixed $A_1$–$B_2$ character. One component remained at 31 $cm^{-1}$, and the other softened to 10 $cm^{-1}$ just below $T_c$. Eremenko et al. (1975) published room-temperature Raman spectra over a wide frequency range and listed frequencies of observed peaks at 300 K and 4.2 K. Stray light apparently prevented them from looking much below 35 $cm^{-1}$. Popkov et al. (1975) studied the temperature dependence of one new mode seen below $T_c$ by Eremenko et al. using a $c(b, b+c)a$ geometry. Popkov et al. followed this mode from 153 $cm^{-1}$ at 4.2 K to 138 $cm^{-1}$ just below $T_c$. Lyons et al. (1982) examined a 150 $cm^{-1}$ mode using a geometry that produced a phonon wavevector $q$ parallel to the $a$-axis. The polarization symmetry was not specified. They subtracted the 138 cm$^{-1}$ peak from the spectrum and found that the 150 cm$^{-1}$ peak did not shift. Its intensity extrapolated to zero at a temperature ($T_R$) of 255 K.

Murray and Lockwood (1980) also studied the two low-frequency Raman modes discovered by Ryan and Scott (1974). More data and a coupled-mode analysis were presented by Murray et al. (1981). The $b(aa)c$ spectrum ($A_1$ TO) from this later work differs somewhat from the mixed $A_1$–$B_2$ spectrum of Ryan and Scott. The lines are narrower, and there is less softening as $T \to T_c$ from below. In fact, the coupled mode analysis showed that both mode frequencies stayed above 20 cm$^{-1}$, and the strength remained nearly constant, even at 247 K, the highest temperature studied below $T_c$. The damping increased as $T$ increased towards $T_c$, but the mode remained underdamped, according to this analysis.

The most complete Raman studies are those of Lockwood et al. (1981), who

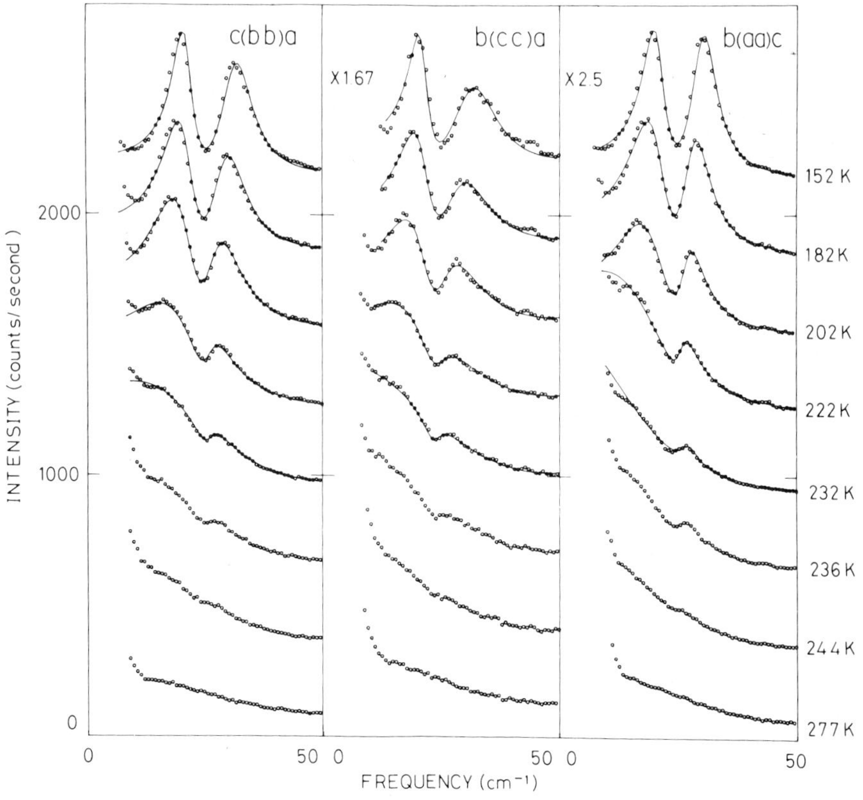


Fig. 5. Polarized Raman spectra of $BaMnF_4$ near $T_c$ and below (Murray et al. 1981). Circles represent data; solid lines represent fits to a coupled-modes model.

studied all four Raman symmetries from about 15 to 500 $cm^{-1}$ over the temperature range from 86 to 257 K, and by Murray et al. (1981), who measured various $A_1$ spectra at 295 K and who tabulated frequencies for all symmetries at the same temperature. Their $A_1$ spectra are shown in fig. 5. These latter authors assigned low-frequency scattering at 295 K to an $A_1$ mode at 21 $cm^{-1}$, and their $c(bb)a$ spectrum is similar to that of Lockwood et al. (1981) below $T_c$ (at 240 K). It is therefore possible that one of the two low-lying 20–30 $cm^{-1}$ peaks observed by Ryan and Scott (1974) is a hard mode, as they suggested; but if this is true, its symmetry must be $A_1$, not $B_2$ as suggested by Ryan and Scott.

The 86 K, $B_2$, spectra of Lockwood et al. show peaks at 25 and 37 $cm^{-1}$. Upon heating these merge and weaken. There is no rise in the scattering near the laser line, as is the case with the $A_1$ spectrum. There is also no low-frequency scattering (apart from apparent polarization leakage) in the $A_2$ and $B_1$ spectra. Since these authors used a triple monochromator, one may conclude that the observed low-frequency $A_1$ scattering is real and that the soft mode or modes have an effective symmetry that is $A_1$ in the distorted phase.

Thus far, there is no unambigious Raman evidence of hardened, i.e., underdamped, soft modes below $T_c$.

Structural information about the distorted phase comes from an extensive neutron study by Cox et al. (1979), which followed a preliminary report by the same group (Shapiro et al. 1976). They found superlattice diffraction peaks at a primary wavevector $\boldsymbol{q}_c = (\xi, \frac{1}{2}, \frac{1}{2})$ where $\xi$ remained almost constant with temperature at a value of 0.392, and also at wavevectors $2\boldsymbol{q}_c$, and $3\boldsymbol{q}_c$ (*modulo* reciprocal lattice vectors). The intensity of the $2\boldsymbol{q}_c$ scattering extrapolated to zero at $T_c = 247$ K, but there was intense critical scattering at $\boldsymbol{q}_c$, extending well above $T_c$. Inelastic neutron scattering measurements found an overdamped phonon for $T > T_c$. This mode remained overdamped to at least 581 K. Its intensity, $I$, divided by $T$ was proportional to $(T - T_c)^{-1}$ with $T_c = 250$ K. High resolution (0.1 meV) scans for $T > T_c$ gave a broad background component (the overdamped soft mode) plus a resolution-limited central peak which diverged as $T \to T_c$. As $\boldsymbol{q}$ was varied away from $\boldsymbol{q}_c$ at 300 K, the soft mode was observed to become hard.

Below $T_c$ an attempt was made to find amplitude or phase modes for wavevectors near $\boldsymbol{q}_c$, but none were found.

Measurements by Scott et al. (1982) of specific heat and paraelectric resonance have shown an anomaly at 255 K (as well as one at 247 K), strongly suggesting a second phase transition. Neutron studies (Cox et al. 1981) on the sample studied at Brookhaven (Cox et al. 1979) rule out two transitions in that sample.

Brillouin scattering measurements on $BaMnF_4$ are reviewed in the chapters by Cummins, by Scott, and by Fleury and Lyons in this volume. Of great interest is the possible identification of the elastic anomalies (Bechtle et al. 1978,

Lyons et al. 1980, 1982) and the central peaks (Lyons et al. (1980, 1982)) with phase modes. Such assignments implicitly assume that the amplitude modes have hardened and are not overdamped. If one accepts the conclusion from the above discussion of the Raman data in the 15 to 30 $cm^{-1}$ spectral region that there is no unambiguous evidence for hard amplitude modes, one is tempted to suggest that the observed anomalies and central peaks may be due to one or two overdamped amplitude modes.

The symmetry of the soft mode warrants discussion. The star of the wavevector $\boldsymbol{q}_c$ contains $\boldsymbol{q}_1 = (\xi, \frac{1}{2}, \frac{1}{2})$ and $\boldsymbol{q}_3 = (\xi, -\frac{1}{2}, \frac{1}{2})$. Phonons at $\boldsymbol{q}_2 = -\boldsymbol{q}_1$ and $\boldsymbol{q}_4 = -\boldsymbol{q}_3$ also condense, giving a $n = 4$ component order parameter (Cox et al. 1979). Dvorak (1975) discussed the case $\xi = 0$ (before it was known that $\xi = 0.39$). The symmetry properties for $\xi \neq 0$ are very similar and are discussed by Dvorak and Fousek (1980). The group of the wavevector contains the identity and the two-fold screw rotation $2_1$ and hence is isomorphic to $C_2$. The irreducible representations of $C_2$ are A (even under $C_2$) and B (odd under $C_2$). From the known properties of the antiferromagnetic phase below $T_N = 26$ K Dvorak and Fousek (1980) conclude that the symmetry of the order parameter of the structural phase transition is B (in our notation).

The Landau theory of Cox et al. (1979) shows that below $T_c$ there are two possibilities for the order parameters $Q(q_1)[= Q(q_2)^*]$ and $Q(q_3)[= Q(q_4)^*]$, depending on the values of the fourth-order Landau parameters: (i) with $Q(q_1) = Q(q_3)$ and (ii) with either $Q(q_1)$ or $Q(q_2) = 0$ and the other non-zero. Solution (i) retains the $C_{2v}$ orthorhombic structure, if the incommensurate modulation is neglected (Dvorak and Fousek 1980). Solution (ii) does not; it describes a monoclinic ($C_2$) incommensurate state, in which the fully symmetric mode ($A_1$) would have *aa*, *bb*, *cc*, and *bc* Raman tensor components (Dvorak 1975). There is no strong experimental evidence that the *bc* spectrum mixes with the diagonal spectra. The Raman data therefore favor solution (i), but arguments such as this based on missing (or very weak) features are always weak.

## 4. *Layered transition-metal dichalcogenides*

The layered dichalcogenides of the group Vb transition metals consist of three-layer "sandwiches" of the form chalcogen-transition metal-chalcogen. They undergo phase transitions of the generalized charge-density-wave (CDW) type (Wilson et al. 1975). The CDW distorts the lattice statically by an amount roughly proportional to the onset temperature for this phase transition. In the 2H polytypes the transition metal atom is coordinated with six chalcogen atoms that form a trigonal prism. There are two transition metal atoms per unit cell, and the space group is hexagonal. These materials are in normal phases at room temperature. The 1T polytype has trigonal symmetry, with the single transition atom in a unit cell octahedrally coordinated with its six chalcogen nearest neighbors. These materials are usually in distorted phases already above room temperature.

The 2H polytype of $TaSe_2$ has been the most extensively studied. The strongest room-temperature Raman peak (Smith et al. 1976) is an overtone of anomalously soft LA-like phonons having wavevectors $\boldsymbol{q}_c$ near $2\Gamma M/3$ in the basal plane of the hexagonal Brillouin zone (Moncton et al. 1975, 1977). The frequency of the Raman peak decreases as the temperature is decreased (Steigmeier et al. 1976b), in agreement with the neutron data (Moncton et al. 1975, 1977). Below the onset temperature $T_0 = 123$ K for the CDW the soft modes "condense" into CDW phonons, and a new low-frequency Raman peak appears and eventually evolves into four peaks at very low temperatures (Steigmeier et al. 1976b, Holy et al. 1976). There are 6 wavevectors in the star of $\boldsymbol{q}_c$. There are two layers per unit cell, which interact only weakly. The LA phonons in both layers go soft. Thus the order parameter has 12 components! The number of these modes and their apparent symmetries ($A_{1g}$ at 44 and 82 $cm^{-1}$ plus $E_{2g}$ at 50 and 65 $cm^{-1}$) in the $3a_0 \times 3a_0$ commensurate phase were shown by Holy et al. (1976) to be consistent with the same space group $P6_3mmc$ ($D_{6h}^4$) that the material has in the normal phase, but independent structural data were lacking.

The symmetry arguments used by Holy et al. (1976) and the microscopic model used by them and later by McMillan (1977) assigned the $A_{1g}$ modes to fully-symmetric linear combinations of LA-like phonons from the 6 wavevectors in the star of $\boldsymbol{q}_c$. The $E_{2g}$ modes corresponded to anti-symmetric linear combinations of these modes. For each wavevector, pair ($\boldsymbol{q}_c$, $-\boldsymbol{q}_c$), one expects one Raman-active CDW phonon in which the amplitudes of the static distortion in adjacent layers oscillate in phase and one Raman-active CDW phonon in which phases of the static distortion in adjacent layers oscillates out of phase. This latter mode, which we shall call the "relative phase mode" is not found in materials with only a single layer per unit cell.

Upon cooling below 123 K 2H–$TaSe_2$ goes into an hexagonal incommensurate phase which locks into the $3a_0 \times 3a_0$ commensurate phase at 85–90 K (Fleming et al. 1980). Upon warming Fleming et al. (1980) found a new incommensurate phase appearing at 93 K, the so called "striped" phase where one of the wavevectors is commensurate and two wavevectors are incommensurate. At 112 K a phase transition occurs to the hexagonal incommensurate phase.

Recent measurements by Fung et al. (1981) using convergent beam electron diffraction have discovered that the commensurate phase of 2H–$TaSe_2$ is orthorhombic (with a most probable space group Cmcm) and not hexagonal. This discovery makes the existence of the striped incommensurate phase easier to understand (Walker and Jacobs 1981, 1982, Littlewood and Rice 1982, McMillan 1982). The effects of the orthorhombic structure must be very subtle, as they have not yet been observed in X-ray diffraction measurements. A Landau theory of the CDW phonon modes in the commensurate orthorhombic state and the origins of their Raman activity were sketched by Klein (1982). For weak interlayer interaction the main effect of the orthorhombic distortion is a splitting of the E modes.

The temperature dependence of the Raman spectra of the CDW phonon modes was first measured by Steigmeier et al. (1976b). It was studied in more detail by Sooryakumar et al. (1979) and in even more detail by Sugai et al. (1981a,c). Results from the latter workers are shown in fig. 6. All the CDW Raman peaks broaden considerably as the incommensurate phases are approached. As the temperature is raised to the commensurate-to-striped phase

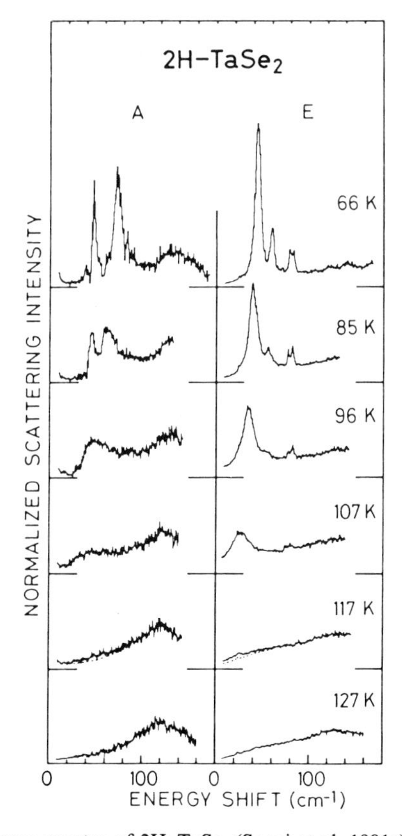


Fig. 6. Low-frequency Raman spectra of 2H–$TaSe_2$ (Sugai et al. 1981c). The Bose-factor $(1+n)$ has been divided out. The E spectra were taken in $z(xy)z$ geometry. The A spectra represent the difference between $z(xx)z$ and $z(xy)z$ spectra. The dotted lines on the 117 K curves duplicate the 127 K spectra.

transition at 93 K, the two A peaks merge. The two E peaks keep their separate identity throughout most of the striped phase, but the weaker, higher frequency, peak essentially vanishes near the stripe-to-incommensurate transition at 112 K. In the incommensurate phase between 112 and 123 K only a single, broad, overdamped feature remains in each spectrum. There are no discontinuities in any spectral features as the two phase transition at 93 and 112 K are traversed from below.

2H–$TaSe_2$ was studied by Bruns (1979) using a grating monochromator and an iodine cell to reject elastically scattered light. He found a central peak in the E spectrum having a width of about one wavenumber. This peak was found over the range from 46 to 98 K and was most intense near 80 K, where it was considerably more intense than the E CDW phonon at about 40 $cm^{-1}$ (Sooryakumar et al. 1979). The origin of this central peak is not known.

In 2H–$NbSe_2$ the soft LA-like phonons condense at $T_0 = 33$ K to form an incommensurate CDW state with a superlattice that is approximately $3a_0 \times 3a_0$ (Moncton et al. 1975, 1977). Below $T_0$ Raman-active CDW modes are seen near 40 $cm^{-1}$ (Tsang et al. 1976b), one each of A and E symmetry (Sooryakumar et al. 1979, Sooryakumar and Klein 1980). These are presumably the in-phase amplitude modes for the two layers in a unit cell. The out-of-phase phase modes for the two layers, which were observed in the commensurate phase of 2H–$TaSe_2$, were not seen.

Below 7 K 2H–$NbSe_2$ becomes superconducting. At 2 K Sooryakumar and Klein (1980) observed one new A mode at 18 $cm^{-1}$ and one extra E mode at 15 $cm^{-1}$. These energies are close to $2\Delta$, where $\Delta$ is the superconducting gap parameter. Behavior in a magnetic field (Sooryakumar and Klein 1980, 1981) and as a function of impurities (Sooryakumar et al. 1981) suggests strongly that these new Raman peaks acquire their Raman activity by coupling quasi-particle pair excitations to the CDW amplitude modes (Balseiro and Falicov 1980), (Littlewood and Varma 1981).

Sugai et al. (1981b) have recently measured the first Raman spectrum on 2H–$TaS_2$. The samples are crinkly and of poor optical quality, and it was necessary to use an iodine filter to remove the strong elastic light signal. At 78 K there is a phase transition to a $3a_0 \times 3a_0$ superlattice state, as determined by electron diffraction measurements by Tidman et al. (1974). Such measurements are not precise enough to rule out a small deviation of the superlattice spots from $a^*/3$. Below this temperature an E Raman peak appears and hardens towards 50 $cm^{-1}$ as the temperature is decreased. An *extremely* weak A peak near 78 $cm^{-1}$ is apparently also present. The presence of only one E CDW phonon suggests that 2H–$TaSe_2$ is more like 2H–$NbSe_2$ than 2H–$TaSe_2$ and therefore is likely to be in an incommensurate CDW state.

1T–$VSe_2$ is unique among the layered transition metal dichalcogenides by existing in a phase that is incommensurate along the $c$-axis. Structural studies are incomplete. Electron diffraction work by Williams (1976) found an incommensurate phase at 140 K with a nearly $4a_0$ by $3c_0$ superlattice. At 40 K he found this to be commensurate $4a_0$ by $3c_0$. Electron diffraction studies by Van Landuyt et al. (1978) found at least three phase transitions between room and liquid nitrogen temperature. More precise X-ray diffraction measurements by Tsutsumi et al. (1981) found that at low temperatures the superlattice wavevector has a component 0.294 $c^*$. This would make the low-temperature unit cell $4a_0$ by approximately 3.3 $c_0$ and hence incommensurate along the $c$-axis. Raman measurements by Smith et al. (1976) found a single peak at 210 $cm^{-1}$ in the undistorted phase at room temperature and four or five additional peaks above 110 $cm^{-1}$ at 2 K. A more detailed Raman study has recently been made by Sugai et al. (1982). In the normal phase they found an $A_{1g}$ mode at 211 $cm^{-1}$ and a weaker $E_g$ mode at 143 $cm^{-1}$, whose intensity grows anomalously strongly with decreasing temperature. At low temperatures new A modes appear at 50, 64, and 174 $cm^{-1}$. The lowest frequency mode softens upon heating towards the phase transition temperature. In the incommensurate phase a central peak was also observed.

The strongest CDWs are found in the 1T polytypes of $TaS_2$ and $TaSe_2$. $TaS_2$ has three distorted phases (Scruby et al. 1975). The crystal forms in the incommensurate $1T_1$ phase. At 350 K the three CDW wavevectors rotate into another incommensurate ($1T_2$) phase, which finally becomes commensurate at 190–205 K ($1T_3$ phase). The latter has a $\sqrt{13}\ a_0 \times \sqrt{13}\ a_0 \times 13c_0$ superlattice containing 169 Ta atoms (Brouwer and Jellinek 1980). Raman measurements on 1T–$TaSe_2$ were made by Smith et al. (1976), Duffey et al. (1976), Sai-Halasz and Perry (1977), and Sugai et al. (1981a, 1982). The most complete studies were those of Duffey et al. and Sugai et al., especially the latter. In the $1T_3$ phase 11 Raman-active modes are seen below 140 $cm^{-1}$. Two $A_g$ modes at 80 and 115 $cm^{-1}$ show the most softening as the temperature is raised (Sugai et al. 1981a, 1982). At the $1T_3$–$1T_2$ phase transition at 205 K Sugai et al. (1981a,b) find large changes in the line width and energy of the $A_g$ modes. The remaining modes broaden considerably and at the transition into the $1T_1$ phase only a broad peak with some structure remains in the 65–80 $cm^{-1}$ range.

Similar results have been obtained on 1T–$TaSe_2$ by Smith et al. (1976), Tsang et al. (1977) and Sugai et al. (1981a, 1982).

## 5. Conclusions

The behavior of the amplitude mode(s) is not very different from that of hardened soft modes seen in commensurate modulated states. In no case have phase modes been seen unambiguously by light scattering techniques in an incommensurate state, but they have been clearly seen by neutron scattering in biphenyl and in $ThBr_4$. Since the origin of any Raman activity by these modes and their couplings to other modes remain topics of great interest, the search for them by light scattering techniques will continue.

## Acknowledgments

Thanks go to K.B. Lyons, N.E. Massa, J.F. Scott, and F.G. Ullman for conversations and preprints. This work was supported by the National Science Foundation through the MRL Grant DMR-80–20250.

## References

Aiki, K. and K. Hukuda, 1969, J. Phys. Soc. Jpn. **26**, 1066.
Aiki, K., K. Hukuda and O. Matamura, 1969, J. Phys. Soc. Jpn. **26**, 1064.
Axe, J.D., M. Iizumi and G. Shirane, 1980, Phys. Rev. **B 22**, 3408.
Balseiro, C.A. and L.M. Falicov, 1980, Phys. Rev. Lett. **45**, 662.
Bechtle, D.W., J.F. Scott and D.J. Lockwood, 1978, Phys. Rev. **B 18**, 6213.
Benoit, J.P., M. Deniau and J.P. Chapelle, 1972, C.R. Acad. Sc. (Paris) **275** B 665.
Boudou, A. and J. Sapriel, 1980, Phys. Rev. **B 21**, 61.
Bree, A., and M. Edelson, 1977, Chem. Phys. Lett. **46**, 500.
Brehat, F., J. Claudel, P. Strimer and A. Hadni, 1976, J. de Phys. Lett. **37**, L-229.
Brehler, B., 1957, Z. Kryst. **109**, 68.
Brouwer, R. and F. Jellinek, 1980, Physica, **99B**, 51.
Brüesch, P., 1975, Optical properties of the one-dimensional Pt-complex compounds, in: One Dimensional Conductors, Proc. GPS Summer School, Saarbrücken, 1974, eds., J. Ehlers, K. Hepp and H.A. Weidenmüller (Springer, Berlin) Lecture Notes, Vol. 34.
Brüesch, Strässler, S. and H.R. Zeller, 1975, Phys. Rev. **B 12**, 219.
Bruns, D.G., 1979, Ph.D. Thesis, University of Illinois at Urbana-Champaign, unpublished.
Cailleau, H., J.L. Baudoir and C.M.E. Zeyen, 1979a, Acta Crystall. **B 35**, 426.
Cailleau, H., A. Girard, F. Moussa and C.M.E. Zeyen, 1979b, Solid State Commun. **29**, 259.
Cailleau, H., F. Moussa and J. Mons, 1979c, Solid State Commun. **31**, 521.
Cailleau, H., F. Moussa, C.M.E. Zeyen and J. Bouillot, 1980, Solid State Commun. **33**, 407.
Cailleau, H., F. Moussa, C.M.E. Zeyen and J. Bouillot, 1981, Proc. Int. Conf. on Phonon Physics, Bloomington, 1981, ed., W. Bron, J. de Phys., **42**, C6–704.
Carneiro, K., G. Shirane, S.A. Werner and S. Kaiser, 1976, Phys. Rev. **B 13**, 4258.
Caville, C., V. Fawcett and D.A. Long, Proc. 5th Int. Conf. Raman Spectroscopy, Freiburg, 1976, ed., E.D. Schmid (Flamarion, Paris) p. 626.
Caville, C., V. Fawcett and D.A. Long, 1978, J. Raman Spectrosc. **7**, 43.
Chapelle, J.P. and J.P. Benoit, 1977, J. Phys. **C 10**, 145.
Chen, C.-E., Y. Schlesinger and A.J. Heeger, 1981, Phys. Rev. **B 24**, 5139.
Comès, R., M. Lambert, H. Launois and H.R. Zeller, 1973, Phys. Rev. **B 8**, 571.
Cowley, R.A. and A.D. Bruce, 1978, J. Phys. **C 11**, 3577.
Cox, D.E., S.M. Shapiro, R.A. Cowley, M. Eibschütz and H.J. Guggenheim, 1979, Phys. Rev. **B 19**, 5754.
Cox, D.E., S.M. Shapiro, M. Eibschutz and H.J. Guggenheim, 1981, Bull. Am. Phys. Soc. **26**, 303.
Currat, R., 1981, Proc. Int. Conf. Phonon Physics, Bloomington, 1981, ed., W. Bron, J. de Phys. **42**, C6-693.
Deiseroth, H.J. and H. Schulz, 1974, Phys. Rev. Lett. **33**, 963.
Delahaigue, A., B. Khelifa and P. Jouve, 1975, Phys. Status Solidi (b) **72**, 585.
Denoyer, F., A.H. Moudden and M. Lambert, 1980, Ferroelectrics, **24**, 43.
Denoyer, F., A.H. Moudden, R. Currat, C. Vettier, A. Bellamy and M. Lambert, 1982, Phys. Rev. **B 25**, 1697.
De Pater, C.J., 1978, Phys. Status Solidi (a) **48**, 503.

De Pater, C.J. and C. van Dijk, 1978, Phys. Rev. **B 18**, 1281.
De Pater, C.J. and R.B. Helmholdt, 1979, Phys. Rev. **B 19**, 5735.
De Pater, C.J., J.D. Axe and R. Currat, 1979, Phys. Rev. **B 19**, 4684.
Duffey, J.R., R.D. Kirby and R.V. Coleman, 1976, Solid State Commun. **20**, 617.
Dvorak, V., 1975, Phys. Status Solidi (b) **71**, 269.
Dvorak, V. and J. Petzelt, 1978, J. Phys. **C 11**, 4827.
Dvorak, V. and J. Fousek, 1980, Phys. Status Solidi (a) **61**, 99.
Eremenko, V.V., A.P. Mokhir, Yu. A. Popkov and O.L. Reznikskaya, 1975, Ukr. Fiz. Zh. **20**, 144.
Fawcett, V., R.J.B. Hall, D.A. Long and V.N. Sankaranayanan, 1974, J. Raman Spectrosc. **2**, 629.
Ferer, M., M.A. Moore and M. Wortis, 1973, Phys. Rev. **B 8**, 5205.
Fleming, R.M., D.E. Moncton, D.B. McWhan and F.J. DiSalvo, 1980, Phys. Rev. Lett. **45**, 576.
Fleury, P.A., S. Chiang and K.B. Lyons, 1979, Solid State Commun. **31**, 279.
Franke, E., M. LePostollec, J.P. Mathieu and H. Poulet, 1980a, Solid State Commun. **33**, 155.
Franke, E., M. LePostollec, J.P. Mathieu and H. Poulet, 1980b, Solid State Commun. **35**, 183.
Fritz, I.J., 1975a, Phys. Lett. **51A**, 219.
Fritz, I.J., 1975b, Phys. Rev. Lett. **35**, 1511.
Fung, K.K., S. McKernan, J.W. Steeds and J.A. Wilson, 1981, J. Phys. **C 14**, 5417.
Futama, H., Y. Shiozaki, A. Chiba, E. Tanaka, T. Mitsui and J. Furuichi, 1967, Phys. Lett. **25A**, 8.
Gesi, K. and M. Iizumi, 1978, J. Phys. Soc. Jpn. **45**, 1777.
Gesi, K. and M. Iizumi, 1979, J. Phys. Soc. Jpn. **46**, 697.
Goldsmith, G.J. and J.G. White, 1959, J. Chem. Phys. **31**, 1175.
Günter, P., R. Santuary, H. Arend and W. Seidenbusch, 1981, Solid State Commun. **37**, 883.
Hamano, K., Y. Ikeda, T. Fujimoto, K. Ema and S. Hirotsu, 1980a, J. Phys. Soc. Jpn. **49**, 2278.
Hamano, K., Y. Ikeda, T. Fujimoto, K. Ema and S. Hirotsu, 1980b, Proc. 2nd Japan–Soviet Symp. Ferroelectricity, Kyoto, 1980, J. Phys. Soc. Jpn. **49**, Suppl. B, 10.
Hague, M.S. and J.R. Hardy, 1980, Phys. Rev. **B 21**, 245.
Hague, M.S., J.R. Hardy, Q. Kim and F.G. Ullman, 1978, Solid State Commun. **27**, 813.
Holy, J.A., M.V. Klein, W.L. McMillan and S.F. Meyer, 1976, Phys. Rev. Lett. **37**, 1145.
Hubert, S., P. Delamoge, S. Lefrant, M. Lepostollec and M. Hussonnois, 1981, J. Solid State Chem. **36**, 36.
Iizumi, M., J.D. Axe, G. Shirane and K. Shimaoka, 1977, Phys. Rev. **B 15**, 4392.
Inoue, K., S. Koiwai and Y. Ishibashi, 1980, J. Phys. Soc. Jpn. **48**, 1785.
Ishibashi, Y., 1981, J. Phys. Soc. Jpn. **50**, 1255.
Ishizawa, N., F. Marumo, T. Kawamura and M. Kimura, 1975, Acta Crystallogr. **B 31**, 1912.
Kálman, A., J.S. Stephens and D.W.J. Cruickshank, 1970, Acta Crystallogr. **B 26**, 1451.
Keve, E.T., S.C. Abrahams and J.L. Bernstein, 1969, J. Chem. Phys. **51**, 4928.
Khelifa, B., A. Delahaigue and P. Jouve, 1977, Phys. Status Solidi (b) **83**, 139.
Klein, M.V., 1982, Phys. Rev. **B 25**, 7192.
Kojima, S., K. Ohi, M. Takashige, T. Nakamura and H. Kakinuma, 1979, Solid State Commun. **31**, 755.
Kudo, S. and T. Ikeda, 1981, J. Phys. Soc. Jpn. **50**, 733.
Littlewood, P.B. and C.M. Varma, 1981, Phys. Rev. Lett. **47**, 811.
Littlewood, P.B. and T.M. Rice, 1982, Phys. Rev. Lett. **48**, 7.
Lockwood, D.J., A.F. Murray and N.L. Rowell, 1981, J. Phys. **C 14**, 753.
Luspin, Y., M. Chabin, G. Hauret and F. Giletta, 1982, J. Phys. **C 15**, 1581.
Lyons, K.B. and P.A. Fleury, 1976, J. Appl. Phys. **47**, 4898.
Lyons, K.B., T.J. Negran and H.J. Guggenheim, 1980, J. Phys. **C 13**, L415.
Lyons, K.B., R.M. Bhatt, T.J. Negran and H.J. Guggenheim, 1982, Phys. Rev. **B 25**, 1794.
Maciel, A. and J.F. Ryan, 1981a, J. Phys. **C 14**, L509.
Maciel, A. and J.F. Ryan, 1981b, Proc. Int. Conf. on Phonon Physics, Bloomington, 1981, ed., W. Bron, J. de Phys. **42**, C-6, 725.

Majkrzak, C.F., J.D. Axe and A.D. Bruce, 1980, Phys. Rev. **B 22**, 5278.
Mashiyama, H., S. Tanisaki and K. Hamano, 1981, J. Phys. Soc. Jpn. **50**, 2139.
Massa, N.E., 1981, (Proc. Int. Conf. on Phonon Physics, Bloomington, 1981, ed., W. Bron) J. de Phys. **42**, C6-593.
Massa, N.E., F.G. Ullman and J.R. Hardy, 1979, Solid State Commun. **32**, 1005.
Massa, N.E., F.G. Ullman and J.R. Hardy, 1980, Proc. 4th European Conf. on Ferroelectricity, Portoroz, Yugoslavia, 1979, Ferroelectrics, **25**, 601.
Massa, N.E., F.G. Ullman and J.R. Hardy, 1982, Solid State Commun. **42**, 175.
Massa, N.E., F.G. Ullman and J.R. Hardy, 1983, Phys. Rev. **B 27**, 1523.
Matsuda, T. and I. Hatta, 1980, J. Phys. Soc. Jpn. **48**, 157.
McKenzie, D.R., 1975a, J. Phys. **C 8**, 1607.
McKenzie, D.R., 1975b, J. Phys. **C 8**, 2003.
McMillan, W.L., 1977, Phys. Rev. **B 16**, 1977.
McMillan, W.L., 1982, unpublished work.
McWhan, D.B., R.M. Feming, D.E. Moncton and F.J. DiSalvo, 1980, Phys. Rev. Lett. **45**, 269.
Moncton, D.E., J.D. Axe and F.J. DiSalvo, 1975, Phys. Rev. Lett. **34**, 734.
Moncton, D.E., J.D. Axe and F.J. DiSalvo, 1977, Phys. Rev. **B 16**, 801.
Morosin, B. and E.C. Lingafelter, 1959, Acta Crystallogr. **12**, 744.
Moudden, A.H., F. Denoyer, J.P. Benoit and W. Fitzgerald, 1978a, Solid State Commun. **28**, 575.
Moudden, A.H., F. Denoyer and M. Lambert, 1978b, J. de Phys. **39**, 1323.
Moudden, A.H., F. Denoyer, M. Lambert and W. Fitzgerald, 1979, Solid State Commun. **32**, 933.
Murray, A.F. and D.J. Lockwood, 1980, Proc. 7th Int. Conf. Raman Spectroscopy, ed., W.F. Murphy (North-Holland, Amsterdam) pp. 58–9.
Murray, A.F., G. Brims and S. Sprunt, 1981, Solid State Commun. **39**, 941.
Natkaniec, I., A.V. Bielushkin and T. Wasiutynski, 1981, Phys. Status Solidi (b) **105**, 413.
Ohi, K., M. Kimura, H. Ishida and H. Kakinuma, 1979, J. Phys. Soc. Jpn. **46**, 1397.
Petzelt, J., G.V. Kozlov, A.A. Voklov and Y. Ishibashi, 1979, Z. Phys. **B 33**, 369.
Pfeuty, P., D. Jasnow and M.E. Fisher, 1974, Phys. Rev. **B 10**, 2088.
Popkov, Yu.A., S.V. Petrov and A.P. Mokhir, 1975, Fiz. Nizkikh Temp. **1**, 189 (Trans: Sov. J. Low Temp. Phys. **1**, 191).
Press, W., C.F. Majkrzak, J.D. Axe, J.R. Hardy, N.E. Massa and F.G. Ullman, 1980, Phys. Rev. **B 22**, 332.
Renker, B., H. Rietschel, L. Pintschovius, W. Gläser, P. Brüesch, D. Kuse and M.J. Rice, 1973, Phys. Rev. Lett. **30**, 1144.
Renker, B., L. Pintschovius, W. Gläser, H. Rietschel, R. Comes, L. Liebert and W. Drexel, 1974, Phys. Rev. Lett. **32**, 836.
Ryan, J.F. and J.F. Scott, 1974, Solid State Commun. **14**, 5.
Sai-Halasz, G.A. and P.B. Perry, 1977, Solid State Commun. **21**, 995.
Sawada, A., Y. Makita and Y. Takagi, 1976, J. Phys. Soc. Jpn. **41**, 174.
Sawada, S., Y. Shiroishi, A. Yamamoto, M. Takashige and M. Matsuo, 1977a, J. Phys. Soc. Jpn. **43**, 2099.
Sawada, S., Y. Shiroishi, A. Yamamoto, M. Takashige and M. Matsuo, 1977b, J. Phys. Soc. Jpn **43**, 2101.
Schneck, J. and F. Denoyer, 1981, Phys. Rev. **B 23**, 383.
Schneck, J., J.C. Toledano, C. Joffrin, J. Aubree, B. Joukoff and A. Gabelotaud, 1982, Phys. Rev. **B 25**, 1766.
Scott, J.F., 1979, Rep. Prog. Phys. **12**, 1055.
Scott, J.F., F. Habbal and M. Hidaka, 1982, Phys. Rev. **B 25**, 1805.
Scruby, C.B., P.M. Williams and G.S. Parry, 1975, Phil. Mag. **31**, 255.
Shiozaki, Y., 1971, Ferroelectrics, **2**, 245.
Shiozaki, S., A. Sawada, Y. Ishibashi and Y. Takagi, 1977, J. Phys. Soc. Jpn. **43**, 1314.

Shapiro, S.M., R.A. Cowley, D.E. Cox, M. Eibschütz and H.J. Guggenheim, 1976, in: Proc. Conf. on Neutron Scattering, ed., R.M. Moon (NTIS, Springfield, Virginia) p. 399.
Smith, J.E., J.C. Tsang and M.W. Shafter, 1976, Solid State Commun. **19**, 283.
Sooryakumar, R. and M.V. Klein, 1980, Phys. Rev. Lett. **45**, 660.
Sooryakumar, R. and M.V. Klein, 1981, Phys. Rev. **B 23**, 3213.
Sooryakumar, R., D.G. Bruns and M.V. Klein, 1979, Raman scattering from charge density waves and superconducting gap excitations in 2H–$TaSe_2$ and 2H–$NbSe_2$, in: Light Scattering in Solids, Second USA-USSR Symposium, New York, 1979, eds., J.L. Birman, H.Z. Cummins and K.K. Rebane (Plenum New York).
Sooryakumar, R., M.V. Klein and R.F. Frindt, 1981, Phys. Rev. **B 23**, 3222.
Spencer, E., H.J. Guggenheim and G.J. Kominiak, 1970, Appl. Phys. Lett. **17**, 300.
Steigmeier, E.F., R. Loudon, G. Harbeke and H. Auderset, 1975, Solid State Commun. **17**, 1447.
Steigmeier, E.F., D. Baeriswyl, G. Harbeke and H. Auderset, 1976a, Solid State Commun. **20**, 661.
Steigmeier, E.F., G. Harbeke, H. Auderset and F.J. DiSalvo, 1976b, Solid State Commun. **20**, 667.
Sugai, S., K. Murase, S. Uchida and S. Tanaka, 1981a, Proc. Yamada Conf. IV on Physics and Chemistry of Layered Materials, Sendai, 1980, eds., Y. Nishina, S: Tanuma and H.W. Myron, Physica, **105B**, 405.
Sugai, S., K. Murase, S. Uchida and S. Tanaka, 1981b, Solid State Commun. **40**, 399.
Sugai, S., K. Murase, S. Uchida and S. Tanaka, 1981c, Proc. Int. Conf. on Phonon Physics, Bloomington, 1981, ed., W. Bron, J. de Phys. **42C**–6, 728.
Sugai, S., K. Murase, S. Uchida and S. Tanaka, 1982, unpublished work.
Takashige, M., T. Nakamura, M. Udagawa, S. Kojima, S. Hirotsu and S. Sawada, 1980a, J. Phys. Soc. Jpn. **48**, 150.
Takashige, M., T. Nakamura and S. Sawada, 1980b, Ferroelectrics, **24**, 143.
Tidman, J.P., O. Singh, A.E. Curzon and R.F. Frindt, 1974, Phil. Mag. **30**, 1191.
Tsang, J.C., J.E. Smith and M.W. Shafer, 1976, Phys. Rev. Lett. **37**, 1407.
Tsang, J.C., J.E. Smith, M.W. Shafer and S.F. Meyer, 1977, Phys. Rev. **B 16**, 4239.
Tsutsumi, K., T. Sambongi, A. Toriumi and S. Tanaka, 1981, Proc. Yamada Conf. IV on Physics and Chemistry of Layered compounds, Sendai, 1980, eds., Y. Nishina, S. Tanuma and H.W. Myron, Physica, **105B**, 419.
Unruh, H.-G., W. Eller and G. Kirt, 1979, Phys. Status Solidi (a) **55**, 173.
Van Aalst, V., J. den Hollender, W.J.A.M. Peterse and P.M. de Wolff, 1976, Acta Crystallogr. **32**, B47.
Van Landuyt, J., G.A. Wiegers and S. Amelinckx, 1978, Phys. Status Solidi (a) **46**, 479.
Wada, M., A. Sawada, Y. Ishibashi and Y. Takagi, 1977a, J. Phys. Soc. Jpn. **42**, 1229.
Wada, M., H. Uwe, A. Sawada, Y. Ishibashi, Y. Takagi and T. Sakudo, 1977b, J. Phys. Soc. Jpn. **43**, 544.
Wada, M., A. Sawada and Y. Ishibashi, 1978a, J. Phys. Soc. Jpn. **45**, 1429.
Wada, M., A. Sawada, Y. Ishibashi and Y. Takagi, 1978b, J. Phys. Soc. Jpn. **45**, 1905.
Wada, M., A. Sawada and Y. Ishibashi, 1979, J. Phys. Soc. Jpn. **47**, 1185.
Wada, M., A. Sawada and Y. Ishibashi, 1981a, J. Phys. Soc. Jpn. **50**, 531.
Wada, M., A. Sawada and Y. Ishibashi, 1981b, J. Phys. Soc. Jpn. **50**, 737.
Walker, M.B. and A.E. Jacobs, 1981, Phys. Rev. **B 24**, 6770.
Williams, J.M., J.L. Petersen, H.M. Gerdes and S.W. Peterson, 1974, Phys. Rev. Lett. **33**, 1079.
Wilson, J.A., F.J. DiSalvo and S. Mahajan, 1975, Adv. Phys. **24**, 117.
Yamanaka, A., M. Kasahara and I. Tatsuzaki, 1981, J. Phys. Soc. Jpn. **50**, 735.
Yamamoto, H., K. Yagi, G. Honjo, M. Kimura and T. Kawamura, 1980, J. Phys. Soc. Jpn. **48**, 185.
Young, P.W. and J.F. Scott, 1981, Bull. Am. Phys. Soc. **26**, 303.
Zeller, H.R., 1973, in: Festkörperprobleme, Vol. 13, ed., H.J. Queisser (Pergamon, New York) p. 31.

CHAPTER 9

# Light Scattering Investigations of the Critical Region in Fluids

WALTER I. GOLDBURG

*Department of Physics Astronomy*
*University of Pittsburgh*
*Pittsburgh, PA 15260*
*USA*

*Light Scattering near Phase Transitions*
*Edited by*
*H.Z. Cummins and A.P. Levanyuk*

© *North-Holland Publishing Company, 1983*

# Contents

## 1. Introduction

Light scattering is in certain respects the ideal method for studying critical phenomena in fluids. In the thermodynamic range where the laws of critical behavior are applicable, the correlation length of the fluctuations is much larger than the atomic size and, very near the critical point, becomes comparable to the wavelength of light. As a result, an incident photon beam is strongly scattered and easily detected. Moreover, the light source, which is usually a low-intensity laser beam, can be highly collimated and transmitted with negligible attenuation through the walls of the sample cell and the temperature-stabilizing bath surrounding it. Finally, the coherence properties of the laser allow for convenient measurement of the lifetime of the fluctuations.

At the beginning of the last decade the door to our understanding of critical phenomena was opened by Wilson's renormalization group idea (Wilson and Kogut 1974). Wilson's ideas in turn, were built on the pivotal scaling ideas of Kadanoff (1966) and of Widom (1965a and b). Since that time, the theory and experiment have become increasingly refined, and interest has broadened to include first-order transitions and the dynamics of phase separation itself, i.e., nucleation and spinodal decomposition (Riste 1975). In addition, theorists and experimentalists have begun to identify intriguing properties of steady-state systems which support temperature gradients or shear flow (Beysens et al. 1979a,b, Onuki and Kawasaki 1979). Here again, dramatic effects have been observed and quantified by light scattering.

This chapter is a review of selected experiments which, in the author's view, have especially contributed to advancing our understanding of critical phenomena. Sections 2–4 are concerned with experiments on fluids in thermodynamic equilibrium. Even though this review covers the last decade only, the body of literature is so large that aspirations to completeness must be abandoned. The reader should therefore regard the experiments referred to here as no more than illustrative. For a more comprehensive recent review of the equilibrium properties of systems, but with less emphasis on experimental techniques, an excellent source is the review article by Sengers and Sengers (1978). Spinodal decomposition and nucleation studies are covered in sects. 5 and 6. Here the body of recent literature on fluid behavior near the critical point is smaller and the discussion is more complete.

To conform to the notation used in most of the references cited in this article, we depart in several important instances from that used in ch. 1. These notational changes are as follows:

| Present chapter | Ch. 1 | Meaning of symbol |
|---|---|---|
| $\epsilon$ | $\tau$ | $(T - T_c)/T_c$ (and dielectric constant) |
| $\Omega$ | $V$ | volume |
| $\omega$ | $\Omega$ | frequency change |
| $\eta$ | $\eta$ | order parameter (no change) |
| $\xi(\xi_0)$ | $r_c(r_{c0})$ | correlation length (its amplitude) |
| $I_1$ | $I(q, \Omega)$ | scattered intensity |
| $S(q)$ | – | static structure factor |
| $S(q, \omega)$ | – | dynamic structure factor |

The dynamic structure factor which was not introduced in ch. 1, is the time and space Fourier transform of $\langle\eta(0, 0)\eta(r, t)\rangle$, and $S(q)$ is its frequency integral.

In most of the experiments to be discussed in subsequent sections, the measured parameter is the dynamic structure factor $S(q, \omega)$ or its frequency integral, the static structure factor $S(q)$. Recently, however, interferometric techniques have provided some of the most accurate values of critical exponents and amplitudes in fluids (Wilcox and Balzarini 1968, Hocken and Moldover 1976). The microscope too has been used effectively in the study of non-equilibrium aspects of critical behavior (Kim et al. 1978). The term "light scattering" in the title of this chapter will therefore be interpreted broadly enough to include discussion of all these experiments.

Within the above stated limitations, the aim of this chapter is to identify some recent high-precision light scattering experiments, as well as experimental methods which hold great promise. The limitation of various types of experiment will also be identified wherever possible.

The fundamental theory of light scattering is reviewed in ch. 1. For a summary of research prior to 1970, the article by Cummins and Swinney (1970) is highly recommended; their article also contains a lucid discussion of the theory of light-beating spectroscopy, sometimes called photon correlation spectroscopy. The same volume contains several other articles which, taken together, provide an excellent introduction to the theory of critical phenomena (Berne and Pecora 1976, Stanley 1971, and Chu 1974). Another useful reference is the continuing series of monographs entitled "Phase Transitions and Critical Phenomena", edited until now by C. Domb and M.S. Green.

## 2. Systems in thermal equilibrium

### 2.1. Relating $I(q)$ to $\chi(q)$

In a typical light scattering experiment, the fluid sample is exposed to a beam of monochromatic light, usually provided by a laser whose output is attenuated to a milliwatt or less to limit local heating (Sorensen et al. 1977, 1978). The angular distribution of the scattered light is recorded with a photomultiplier, usually operating in the photon counting mode. If the product of scattering cross section and sample thickness is small enough so that multiple scattering may be neglected, the frequency-resolved signal, $I(q,\omega)$ may be simply related to the dynamic structure factor $S(q,\omega)$ (Berne and Pecora 1976). In the absence of frequency resolution, the scattered intensity $I(q)=\int I(q,\omega)\,\mathrm{d}\omega$ is proportional to the static structure factor $S(q)$. The constant of proportionality depends, of course, on the quantum efficiency of the detector and other parameters characterizing the fluid and the geometry of the scattering arrangement. Since these factors are often difficult to measure, there exist very few light scattering measurements of the absolute value of $S(q,\omega)$. This is regrettable, since such measurements would provide valuable thermodynamic parameters, viz the isothermal compressibility of simple fluids and the osmotic susceptibility of binary mixtures. The latter is especially difficult to determine with precision near the critical point by conventional techniques of physical chemistry.

To measure the frequency dependent function, $I(q,\omega)$, one may, in principle, spectrally analyze the fluctuating light signal or alternatively measure the intensity correlation function $\overline{\langle I(q,t)I(q,t+\tau)\rangle}=G(\tau)$, or rather the photocurent correlation function $\langle \mathrm{i}(q,t)\mathrm{i}(q,t+\tau)\rangle$. (The angular bracket and bar indicate statistical and temporal averages respectively.) As will be discussed further below, $I(q,\omega)$ is simply related to $G(\tau)$ under usual experimental conditions. Typically the lifetime of the critical fluctuations conveniently falls in the kHz range. If the scattered light is weak, i.e., if there are no more than a few photons per correlation time, it becomes desirable to make measurements in the photon counting mode. For a full discussion of the technical aspects of self-beat and photon correlation spectroscopy, see Pike (1974).

Since visible light probes fluctuations of wavelength much larger than the range of atomic interactions, the scattering medium may be treated as a continuum, in which case,

$$I(q)=\langle|\epsilon(q)|^2\rangle_\Omega\,, \tag{2.1-1}$$

where $\Omega$ is the volume of sample exposed to the light, and $\epsilon(\boldsymbol{q})$ is the Fourier component of the local dielectric constant, $\epsilon(\boldsymbol{r})$. This quantity also depends on the average value of the order parameter $\eta$ as well as other thermodynamic variables such as temperature and pressure. (We defer consideration of the temporal fluctuations in $\epsilon$.)

In a simple fluid one may write (Landau and Lifshitz 1958):

$$\langle|\epsilon(q)|^2\rangle_\Omega=\left(\frac{\partial\epsilon}{\partial n}\right)_T^2\langle|n(q)|^2\rangle+\left(\frac{\partial\epsilon}{\partial T}\right)_n^2\langle|T(q)|^2\rangle\,, \tag{2.1-2}$$

with

$$\langle n(q)T(-q)\rangle = 0\,.$$

The variables chosen here are atomic density $n$ and temperature.

Since $n$ may be regarded as the order parameter for a simple fluid, the first term will dominate near the critical point. To obtain the absolute value of $S(q)$ from a scattering measurement, it is obviously necessary to measure $(\partial\epsilon/\partial n)_T$ or to calculate it using the Clausius–Mossoti relation (Kittel 1966). In the limit $q\to 0$, $\langle|n(q)|^2\rangle \to \Omega^{-1}n^2 k_\mathrm{B}T\kappa_T(0)$ where $\kappa_T(0)=\kappa_T$ is the isothermal compressibility (Berne and Pecora 1976).

In a binary mixture, the situation is far more complicated. Mountain and Deutch (1969) and Cohen et al. (1971) have derived an analog of eq. (2.1-2) for a mixture at constant temperature and pressure rather than constant volume (see also Chang et al. 1979). They expand $\langle\epsilon|(q)|^2\rangle$ in terms of a set of orthogonal thermodynamic variables and find:

$$\langle|\epsilon(q)|^2\rangle = \left(\frac{\partial\epsilon}{\partial X}\right)^2_{T,P}\langle|X(q)|^2\rangle + \left(\frac{\partial\epsilon}{\partial T}\right)^2_{P,X}\langle|\Phi(q)|^2\rangle + \left(\frac{\partial\epsilon}{\partial P}\right)^2_{\Phi,X}\langle|P(q)|^2\rangle\,, \tag{2.1-3}$$

where $X$ is the mole fraction of one of the components and

$$\mathrm{d}\Phi = \mathrm{d}T - \frac{T}{C_{P,X}}\left(\frac{\partial\Omega}{\partial T}\right)_{P,X}\mathrm{d}P\,.$$

Near the critical point the second and third terms are only weakly singular (Griffiths and Wheeler 1970) and their sum is expected to be less than 10% of the first term (Chang et al. 1979). Therefore the scattered intensity may be approximated by

$$I(q) = I_0\left(\frac{\partial\epsilon}{\partial X}\right)^2_{T,P}\langle|X(q)|^2\rangle + \Delta I\,,$$

where $\Delta I$ is treated as a temperature-independent background term over a limited temperature range, say $|T-T_\mathrm{c}| \lessapprox 1\,\mathrm{K}$. In the limit $q\xi\to 0$ $\langle|X(q)|^2\rangle$ in the first term becomes:

$$\lim_{q\to 0}\langle|X(q)|^2\rangle = \frac{k_\mathrm{B}T}{N}\left(\frac{\partial X}{\partial\mu}\right)_{P,T} \equiv \frac{k_\mathrm{B}T}{\Omega}\chi_\eta(0)\,, \tag{2.1-4}$$

where $\mu \equiv \mu_\mathrm{A} - \mu_\mathrm{B}$ is the difference in chemical potential of the two species comprising the mixture. The order parameter susceptibility $\chi_\eta(q)$ (sometimes called the osmotic compressibility) is defined here to have the same dimensions as the isothermal compressibility of a simple fluid. With this notation, then, the scattered intensity in both fluids and binary mixtures can be written:

$$I(q) = I_0\left(\frac{\partial\epsilon}{\partial\eta}\right)^2_\theta \chi_\eta(q) + \Delta T\,, \tag{2.1-5}$$

where $\theta = T$ in a simple fluid and $T, P$ in a binary mixture. The correction term, $\Delta I$, contains weakly singular contributions, varying as $|T - T_c|^{-\alpha}$, and an analytic background contributions as well (see ch. 1 for a definition of critical exponents).

The above representation of $\langle|\epsilon(q)|^2\rangle$ is not unique; for example, this function could be expressed in terms of the chemical potentials $\nu \equiv \frac{1}{2}(\mu_A + \mu_B)$ and $\Delta \equiv \frac{1}{2}(\mu_A - \mu_B)$ or in terms of the chemical potentials $\mu_A$ and $\mu_B$ of the two species themselves (Jasnow and Goldburg 1972). If one uses the variable $\nu$ and $\Delta$, then $I(q)$ will contain two strongly divergent contributions. In the limit $q\xi \to 0$, these terms are proportional to $\chi_\eta$ and to the compressibility $\chi_{T,N} = (\partial n/\partial \nu)_{T,\Delta}$. These two terms are of the same order of magnitude and cannot be separated. In mixtures with more than three components, additional chemical potentials appear and the order parameter becomes difficult to identify (Griffiths 1974). Finally it should be noted that the temperature dependence of the prefactor $(\partial\epsilon/\partial\eta)_\theta$ may sometimes be important to take into account, especially in light scattering measurements aimed at obtaining critical amplitudes or their ratios (Kim et al. 1979, 1980), as will be seen in sect. 4.4.

### 2.2. *Approximants of the static correlation function*

A central problem in the theory of critical phenomena is the calculation of the wavenumber-dependent susceptibility $\chi_\eta(q)$. According to the now well-established scaling hypothesis (Widom 1965a,b) this function is expected to be of homogeneous form. On the critical isochore ($\eta = \eta_c$),

$$\chi(q) \equiv \chi_\eta(q) = \Gamma\epsilon^{-\gamma} g(q\xi), \tag{2.2-1}$$

with $\Gamma$ so chosen (Chang et al. 1979) that

$$\lim_{q\xi \to 0} g(q\xi) = 1 .$$

Table 1 contains definitions of the most commonly encountered critical exponents and amplitudes and their typical values in simple fluids and mixtures. The exponents are connected by scaling relations and by the hyperscaling relation (Stanley 1971) discussed in ch. 1 and listed here:

$$\begin{aligned} &\gamma = \beta(\delta - 1), \\ &2 - \alpha = \beta(\delta + 1) = 2\beta + \gamma, \\ &\alpha = \alpha', \\ &\gamma = \nu(2 - \hat{\eta}), \\ &\dim \nu = 2 - \alpha, \quad (\text{"hyperscaling"}). \end{aligned} \tag{2.2-2}$$

In the hyperscaling relation, dim is the dimensionality of the system. The scaling relations were predicted by the homogeneity hypothesis of Widom

Table 1
Definitions of critical exponents and amplitudes

| Quantity | Path in density–temperature plane | Equation | Typical experimental values* |
|---|---|---|---|
| Coexistence curve | $\rho = \rho_{cx}, \quad \epsilon < 0$ | $\frac{\lvert\rho - \rho_c\rvert}{\rho_c} = B\lvert\epsilon\rvert^{\beta}$ | $\beta = 0.326 \pm 0.00$ |
| Susceptibility | $\rho = \rho_c, \quad \epsilon > 0$<br>$\rho = \rho_{cx}, \quad \epsilon < 0$ | $\chi = \Gamma\epsilon^{-\gamma}$<br>$\chi = \Gamma'\epsilon^{-\gamma}$ | $\gamma \simeq \gamma' = 1.21 \pm 0.04$ with $\gamma'$ significantly less than $\gamma$ in pure fluids |
| Specific heat | $\rho = \rho_c, \quad \epsilon > 0$<br>$\rho = \rho_{cx}, \quad \epsilon < 0$ | $C_v = \frac{A}{\alpha}\lvert\epsilon\rvert^{-\alpha} + B_0$<br>$C_v = \frac{A'}{\alpha'}\lvert\epsilon\rvert^{-\alpha'} + B_0$ | $\alpha = 0.113 \pm 0.005$ |
| Critical isotherm | $\epsilon = 0$ | $\frac{\mu - \mu_c}{\mu_c} = D\left(\frac{\rho - \rho_c}{\rho_c}\right)\left(\frac{\rho - \rho_c}{\rho_c}\right)^{\delta - 1}$ | $\delta = 4.35 \pm 0.10$ |
| Correlation length at $T - T_c$ | $\rho = \rho_c, \quad \epsilon = 0$ | $\chi(q) = C_1^{-1}q^{2-\hat{\eta}}$ | $\hat{\eta} = 0.02 \pm 0.02$ |
| Correlation length | $\rho = \rho_c, \quad \epsilon > 0$<br>$\rho = \rho_{cx}, \quad \epsilon < 0$ | $\xi = \xi_0^+\epsilon^{-\nu}$<br>$\xi = \xi_0^-\epsilon^{-\nu'}$ | $\nu = 0.625 \pm 0.01$ |
| Correction to scaling term | Critical isochore or coexistence curve ($f$ is any thermodynamic quantity) | $f - f_c = A_i\epsilon^{\lambda}(1 + a_i\lvert\epsilon\rvert^{\Delta} + b_i\epsilon^{2\Delta} + \ldots)$ | $\Delta = 0.50$ |

* From compendia, by A. Kumar et al. (1980) and P.C. Hohenberg (1978).

(1965a,b) and confirmed by renormalization group (RG) theory (Wilson and Kogut 1974). They are supported by experiment, though controversy still surrounds the hyperscaling relation. A more exact verification of these relations remains a continuing challenge for the experimentalist. In this connection it should be noted that even the best scattering experiments cannot measure the asymptotic ($T \to T_c$) or "true" values of the critical exponents; rather they only reveal "effective values" over a limited range of reduced temperatures. There is, however, good theoretical and experimental evidence that the effective exponents obey the scaling relations (Aharony and Ahlers 1980).

Theoretically, one expects that $\chi(q)$ in eq. (2.2-1) has a simple algebraic form only in the limits of very small or very large $q\xi$; at intermediate values, the numerical calculations of $g(q\xi)$ is often approximated by expressions (approximants) which fit theoretically expected forms for $g(x)$ in the limits $x \simeq 0$, $x \gg 1$. We will see, however, that even the best experimental data can be very satisfactorily fitted to the simple Ornstein–Zernike (OZ) result, $g(x) = g_{OZ}(x) = (1 + x^2)^{-1}$ over a wide range of $x$, even though it should hold in principle for $x \ll 1$ only (Fisher 1964).

In the limit $x \gg 1$, $g(x)$ is expected to be of the Fisher–Langer form, $g = g_{FL}(x)$ (Fisher and Langer 1968, Bray 1976a,b and references therein). Hence a satisfactory approximant must satisfy the conditions (Chang et al. 1979):

$$g_{OZ}(x) = 1/(1 + x^2), \qquad x \ll 1\,, \tag{2.2-3}$$

$$g_{FL}(x) = (C_1/x^{2-\hat{\eta}})(1 + C_2/x^{(1-\alpha)/\nu} + C_3/x^{1/\nu}), \qquad x \gg 1\,, \tag{2.2-4}$$

where the constants $C_i$ depend on the universality class of the Hamiltonian. For fluids, this symmetry is not known with certainty, though it is widely accepted that fluids and mixtures have the symmetry of the Ising model. It is worth noting that the approximant used by Chang et al. (1979) in the analysis of their light scattering data, fits the Ising model calculations *and* the OZ form to within 2% out to $q\xi = 30$.

## 2.3. *Critical exponents, amplitudes, and corrections to scaling*

To obtain the true values of critical exponents from thermodynamic measurements, it is necessary to approach very close to the critical point in order to reduce the value of the so-called corrections-to-scaling terms (Sengers and Sengers 1980), first identified experimentally by Greywall and Ahlers (1972) and by Ahlers (1973). These terms are relatively large in magnitude and slowly converging, according to RG theory (Wegner 1972). Singular thermodynamic functions, $f_i$, such as the order parameter susceptibility, etc., are expected to be of the form

$$f_i - f_{ic} = A_i|\epsilon|^{\lambda}(1 + a_i|\epsilon|^{\Delta} + \ldots)\,. \tag{2.3-1}$$

In fluids the exponent $\Delta$ is estimated to be $0.498 \pm 0.020$ (Le Guillou and Zinn-Justin 1980) and the system-dependent coefficients $a_i$ should be of order unity (Ley-Koo and Green 1977). To assure that the second term is, say, no more than 1% of the first one, the reduced temperature, $\epsilon$, must be kept below $5 \times 10^{-4}$ (Sengers and Sengers 1980). For at least some thermodynamic functions, the coefficient $a$ is smaller in binary mixtures than in simple fluids (Kumar et al. 1980).

It now seems likely that the neglect of scaling corrections have contributed to the discrepancy between measured critical exponents in fluids and the theoretically expected Ising model values. The first clear evidence of this came from the very precise experiments of Hocken and Moldover (1976). Using an interferometric technique of Wilcox (Wilcox and Balzarini 1968), they measured critical parameters in Xe, $SF_6$, and $CO_2$. With a temperature stability of $\pm 20\,\mu K$ they were able to confine the observation range to $-1.5 \times 10^{-5} < \epsilon < 5 \times 10^{-5}$, where correction to scaling terms should be unimportant. These experiments yielded values of $\delta$, $\hat{\eta}$, $\alpha$, $\beta$, $\gamma$, and $\Gamma/\Gamma'$ that are consistent with Ising model predictions (see table 2), whereas some of the earlier experiments did not, presumably because the data were fitted to a simple power-law form over an excessive temperature range.

There now exist thermodynamic measurements of sufficient precision that the second term in eq. (2.3-1) must be retained, as for example, in measurements of the coexistence curve in simple fluids (Ley-Koo and Green 1977, Hayes and Carr 1977, Pittman et al. 1979). In binary mixtures, on the other hand, the correction is smaller than in simple fluids (see for example Nagaranjan et al. (1980), Kumar et al. (1980)). A number of scattering experiments are consistent with a finite value of $a_\Gamma$, but there exist no data which require the presence of this correction term. Thermodynamic measurements are consistent with $a_\Gamma \simeq 1$ in steam (Balfour et al. 1978, Sengers and Levelt Sengers 1978) and in $^3$He (Pittman et al. 1979). Its presence is also suggested by the light scattering experiments discussed in the next section.

All systems within a given unversality class are expected to have the same value of certain critical amplitude ratios (see table 2). The universality of $R_\xi^+$ follows from the assumption of "two-scale factor universality (Stauffer et al. 1972) or alternatively from RG calculations based on a Hamiltonian of restricted symmetry (Hohenberg et al. 1976). This symmetry implies that thermodynamic functions of scaled form, are characterized by two system-dependent amplitudes rather than three. Whereas some of these amplitude ratios, e.g., $\xi_0^+/\xi_0^-$, can be verified by light scattering measurements alone, others, such as $R_\xi^+$ obviously cannot.

In table 2 are listed some critical exponents and amplitude ratios obtained from Ising model series expansions and RG calculations.

Table 2
From Le Guillou and Zinn-Justin (1980), Hohenberg (1980) and Brezin et al. (1976) unless otherwise noted

| Theoretical critical exponent values | | |
|---|---|---|
| | Ising model series expansions | RG theory (Ising symmetry) |
| $\alpha$ | $0.125 \pm 0.020$ | $0.109 \pm 0.004$ |
| $\beta$ | $0.312 \pm 0.005$ | $0.325 \pm 0.001$ |
| $\gamma$ | $1.250 \begin{smallmatrix} +0.003 \\ -0.007 \end{smallmatrix}$ | $1.241 \pm 0.002$ |
| $\nu$ | $0.638 \begin{smallmatrix} +0.002 \\ -0.008 \end{smallmatrix}$ | $0.630 \pm 0.0015$ |
| $\delta$ | $5.00 \pm 0.05$† | $4.82 \pm 0.02$ |
| $\Delta$ | $0.50 \pm 0.05$ | $0.496 \pm 0.004$ |
| $\hat{\eta}$ | $0.041 \pm 0.006$* | $0.0315 \pm 0.0025$ (see Chang et al. 1979) |
| $\Gamma/\Gamma'$ | 5.07 | 4.80 |
| $R_\xi^+ \equiv \xi_0^+ A^{1/d}$*** | 0.26 | $0.2696 \pm 0.0008$** |
| $\xi_0^+/\xi_0^-$ | $1.96 \pm 0.03$ | 1.91 |
| $\Gamma D B^{\delta-1}$ | 1.75 | 1.6 |
| | †Bervillier (1976); Bervillier and Godrèche (1980) | |
| $A\Gamma/B^2$ | 0.059 | 0.066 |
| $A/A'$ | 0.51 | 0.48 |

†Sengers and Sengers (1978).
*Sengers (1980).
**Bervillier and Godrèche (1980).
***$d$ is the system dimensionality.

## 3. Measurements of $S(q)$

### 3.1. Gravitational effects and multiple scattering

Of the many high-precision light scattering measurements made within the last decade, we need only review a few in order to appreciate the level of precision presently achievable, and the systematic errors typically encountered. The accuracy of these measurements is usually determined by multiple scattering and gravity, both effects limiting the closeness of approach to the critical point.

The gravitational field produces a vertical density or composition gradient and suppresses fluctuations in the order parameter. To minimize the effect of the former, the beam diameter must be small enough so as to produce a small fractional variation of $S(q, t)$ over its diameter. The latter effect cannot be controlled except by staying sufficiently far from $T_c$ that the gravitational energy of a fluctuation is small compared to its thermal energy. A thorough discussion of gravitational effects is to be found in the review article by

Moldover et al. (1979) (see also Hohenberg and Barmatz 1972). Gravitational effects become important, at the 1% level of precision when $\epsilon$ is below $10^{-4}$ or $10^{-5}$, if the laser beam is focussed to a diameter of 100 μm.

Gravitational effects can be suppressed by thermal or mechanical stirring, and perhaps by using a vertical incident beam rather than a horizontal one (Spittorff and Miller 1974, Leung and Miller 1975, Cannell 1977a, Cannell 1977b), though the latter technique may have its own limitations (Cannell 1977a, Leung and Miller 1977, Kumar et al. 1980).

In binary mixtures the influence of gravity is felt at comparable values of $\epsilon$. However, diffusion is so slow in these systems that the experiment can sometimes be completed before the vertical concentration gradients have time to develop (Greer et al. 1975, Chang et al. 1979). With sample cells 1 cm high, the equilibration time can be days or even weeks. For this reason almost all experiments carried out in binary mixtures have not been under conditions of thermodynamic equilibrium.

In mixtures, the influence of gravity may be further reduced by choosing components whose densities are closely matched. Practically speaking, however, multiple scattering usually imposes a more restrictive limitation on the choice of components (Moldover et al. 1979, Cannell 1977b). While double scattering can be corrected for analytically (Moldover et al. 1979, Beysens and Zalczer 1977, Sorensen et al. 1977, Trappeniers 1977), some mixtures and almost all pure fluids become so turbid near the critical point (even for samples of small optical path length), that higher-order scattering cannot be ignored. It has been frequently observed that linewidth measurements are less affected by multiple scattering (Ferrell and Bhattacharjee (1979), Schroeter et al. (1979) and references therein).

To obtain the susceptibility and correlation length from scattering measurements, it is sometimes desirable to measure the turbidity $\tau$ of the fluid rather than the angular distribution of the light. The turbidity is defined by

$$I_F = I_0 \, e^{-\tau(\epsilon)w}, \tag{3.1-1}$$

where $I_F$ is the power of the unscattered beam, $w$ is its path length in the fluid, and $I_0$ is the incident beam power. Since $\tau$ is proportional to the angular integral of $S(q)$, one must assume a form for this function to extract the parameters $\chi(0)$ $(\equiv \chi)$ and $\xi$. Taking $g(x)$ to be of OZ form gives

$$\tau = A_0 \chi f(\alpha), \tag{3.1-2}$$

with

$$f(\alpha) = \frac{2\alpha^2 + 2\alpha + 1}{\alpha^3} \ln(1 + 2\alpha) - \frac{2(1+\alpha)}{\alpha^2}, \tag{3.1-3}$$

where $\alpha = 2(q_0 \xi)^2$. Here $q_0 = 2\pi/\Lambda_0$, the wavenumber of the incident light in the fluid. For a one component system,

$$A_0 = \left(\pi k_B \frac{T_0}{\Lambda_0^4}\right)\left(\frac{\partial n'^2}{\partial n}\right)^2, \tag{3.1-4}$$

where $n$ is the number density and $n'^2 = \epsilon$ is the refractive index (Puglielli and Ford 1970).

Multiple scattering does not influence turbidity measurements, and therefore should not interfere with an accurate determination of $\xi(\epsilon)$. However, the determination of this parameter requires considerable extrapolation toward small values of $|T - T_c|$.

### 3.2. *A measurement of $S(q)$ in a simple fluid*

In this section and the next one we discuss careful measurements of the static structure factor in a simple fluid ($SF_6$) and in a binary mixture (3-methyl pentane-nitroethane) by Cannell (1975, 1977b) and Chang et al. (1979) respectively.

Cannell employed a differential measuring scheme (Lunacek and Cannell 1971) to determine $\xi(\epsilon)$ along the critical isochore and in the liquid phase along the coexistence curve. To minimize the effect of drift and other systematic errors, the scattered intensity was recorded at a forward (F) and backward (B) angle ($\theta_F = 8.5^\circ$ and $\theta_B = \pi - \theta_F$). The two signals $I_F$ and $I_B$ were alternatively sent to a single photomultiplier, using a system of mirrors and lenses and a rotating chopper wheel. The periodic signal driving the chopper also provided a reference signal to a lock-in amplifier which recorded the photomultiplier output. The phase of the input signals to the lock-in was chosen so that its output, $V$, is $V = G_F I(q_F) - G_B I(q_B)$, where $G_F$ and $G_B$ are gain parameters. In addition, the forward signal was measured with the backward beam blocked. At each temperature the ratio

$$S = 1 - \frac{G_B\, I(q_B)}{G_F\, I(q_F)}$$

was formed. Assuming that $I(q) \propto \epsilon^{-\gamma} g_{OZ}(q\xi)$, the above equation is easily inverted to give values of $\xi_0$, $\nu$, and $\gamma$.

By using nested temperature baths, Cannell achieved a temperature stability of 50 μK over 72 h. However, gravitational effects and multiple scattering, rather than temperature stability, limited the approach to the critical point to $\epsilon_{min} \simeq 1.5 \times 10^{-4}$.

Near the critical point, a gravitationally induced density variation bends the incident laser beam sharply downward, producing obviously undesirably geometrical effects. To eliminate them Cannell convectively stirred the fluid by applying a transient temperature gradient. For about 30 minutes after this stirring, the density gradient (i.e., the beam bending) disappeared, but of necessity this produces a vertical temperature gradient $\mathrm{d}T/\mathrm{d}z = (\partial P/\partial T)_{\Omega}^{-1}\rho g$, where $\rho$ is the mass density. Using parameters appropriate to this experiment, $(\partial P/\partial T)_{\Omega} = 10^6$ dyn cm$^{-2}$ K$^{-1}$, laser beam diameter $h = 0.05$ cm, the variation of the reduced temperature across the beam turns out to be very small, viz, $\sim 50\ \mu\mathrm{K}/T_c \simeq 10^{-7}$. Stirring was therefore used in the turbidity measurements. With gravitational effects attenuated, the turbidity experiments were extended down to $\epsilon = 4 \times 10^{-6}$ $(T > T_c)$ and $1.25 \times 10^{-6}$ $(T < T_c)$, with $\epsilon_{max} \simeq 5 \times 10^{-2}$.

Other effects taken into account in these experiments were loss of transmitted light by reflection from the cell windows and multiple scattering, which can be corrected for when small ($\lesssim 5\%$). Since one aim of the $SF_6$ experiment was to determine the absolute value of $\chi$, it was necessary to measure independently the temperature dependence of $n'(T)$ and hence $(\partial n'^2/\partial n)_T$ in eq. (3.1-4).

Because gravity causes the fluid density to vary with height $z$ above the liquid–vapor interface, the vertical position of the laser beam was varied to explore points in the $\rho$–T plane which are off the coexistence curve in $SF_6$. Carrying out such measurements in the liquid phase and fitting them to a parametric equation of state, enabled all the parameters in this equation to be determined. With these additional data, the deduced value of $\chi(z=0) \propto (\partial\rho/\partial\mu)_T$ required the inclusion of the first correction-to-scaling in eq. (2.3-1):

$$\frac{\partial\rho}{\partial\mu} = A|\epsilon|^{-\gamma}(1 + a|\epsilon|^{x}) .$$

Fixing $\gamma_L'$ (liquid phase) at 1.223, the scattering data were best fitted with $x = 0.222$ rather than the theoretically expected value, $x = \Delta = 0.50$ (see table 1). On the other hand, with $\gamma_L'$ and other parameters permitted to float, the result was: $A_L' = 1.824 \times 10^{-10}$ g$^2$/erg cm$^3$, $\gamma_L' = 1.200$, $a' = 0.834$, and $x = 0.533$. These results differ slightly from the Ising values for both $\gamma$ and $\Delta$ quoted in table 2. A correction to scaling term was not required for the differential measurement of the correlation length, which turned out to be $\xi_L(z=0) = 1.26|\epsilon|^{-0.596}$. Cannell's findings are also slightly inconsistent with the scaling predition, $\gamma = \gamma_L'$, $\nu = \nu_L'$. With the additional freedom resulting from inclusion of a correction-to-scaling term in the correlation length, equality of these exponents could almost certainly have been achieved (see, e.g., Ahlers 1973).

These measurements may be summarized with the observation that they cannot be fitted to a simple power law alone below $T_c$, and that the measured or "effective exponents" (Aharony and Ahlers 1980) did not seem to obey scaling in the two-phase region. This may mean that correction-to-scaling terms are larger below $T_c$ than above.

Cannell's numerical results are quoted here to give an indication of the precision attainable in a careful experiment and to point out a provocative disagreement between Ising model predictions and the scattering measurements. Cannell (1975) estimates that $\xi$ and $(\partial\rho/\partial\mu)_T$, $\gamma$ and $\nu$ were measured to within no better than a few percent in these experiments, even though the directly measured variables were known to a much higher precision.

### *3.3. A measurement of $S(q)$ in a binary mixture*

Chang et al. (1979) set for themselves the task of measuring the very small critical exponent $\hat{\eta}$ along the critical isochore ($c = c_c$, $T > T_c$) of a binary mixture. Prior to this work, the most precise measurements of this exponent come from Cannell's $SF_6$ measurements discussed in sect. 3.2. From his determination of $\gamma$ and $v$, Cannell used the scaling relation $\gamma = (2 - \hat{\eta})v$ to deduce $\hat{\eta} = 0.03 \pm 0.03$ (Cannell 1975).

An accurate determination of $\hat{\eta}$ requires measurements at the largest possible values of $q\xi$. Hence observation at large scattering angles and small values of $\epsilon$ are required. In this so called "critical regime" the effects of multiple scattering are correspondingly large. To minimize multiple scattering, a mixture was chosen whose components, trimethylpentane and nitroethane, were very closely matched in refractive index. All measurements were made at a fixed scattering angle of 90°, where unwanted reflections from the cell walls are at a minimum. Also, gravitational effects should be smaller at this angle (Cannell 1977a, Leung and Miller 1977, Moldover et al. 1979).

The temperature stability was 0.3 mK over 7 h, and the incident laser intensity was servo-controlled. Furthermore, the photodetector periodically monitored a small fraction of the incident beam, so that slight drifts in the electronics could be corrected for. Because the mixture was a weak scatterer, the photodetector operated in the photon counting mode, the dark-current count rate being only 0.1 counts/s. With these precautions, the standard deviation of the intensity measurements at various temperatures averaged 0.17%.

To deduce $\chi(q)$ from the measured intensity, the Clausius–Mossotti equation, along with the ideal fluid assumption, was used to calculate the temperature dependence of $(\partial\epsilon/\partial\eta)_{P,T}$ in eq. (2.1-5). The order parameter was chosen to be the volume fraction rather than the mole fraction (Greer 1978, Kumar et al. 1980). Finally eq. (2.2-1) was applied to obtain $g(q\xi)$. The parameters $I_0$ and $\Delta I$ in eq. (2.1-5) and the values of $\eta$, $v$, and $\xi_0$ were then extracted from a least-square fit to various approximants for $g(x)$. For example, using an approximant due to Bray (1976a,b), the following results were obtained when the data were analyzed in the range $10^{-6} < \epsilon < 2.7 \times 10^{-3}$.

$$\hat{\eta} = 0.017 \pm 0.015, \qquad v = 0.625 \pm 0.006\,,$$
$$\xi_0 = 2.29 \pm 0.10\ \text{Å}, \qquad \gamma = 1.240 \pm 0.017\,.$$

Two other approximants yielded $\hat{\eta} = 0.030 \pm 0.025$ and $\hat{\eta} = 0.020 \pm 0.017$, the quoted errors being two standard deviations. All three exponents are consistent with the theoretical values quoted in table 1.

The critical parameters were then recalculated with a correction-to-scaling term included. The exponent $\Delta$ was fixed at 0.5, but the amplitude of the first correction was adjustable, along with other critical amplitudes and exponents. The results for $\hat{\eta}$, $\nu$, $\xi_0$, and $\gamma$ differed little, however, from the previous fits, where no such term was included. To illustrate, this procedure gave $\hat{\eta} = 0.016 \pm 0.016$.

In summary, these experiments measured $\gamma$ and $\nu$ to 1% or better and the various estimates of $\hat{\eta}$ differed by no more than 0.015.

### 3.4. *Amplitude ratios*

With the predictions of two-scale factor universality, the measurement of critical amplitude ratios has assumed increased importance. There is interest not only in the amplitude ratios of the leading singular term, i.e., $A_i$ in eq. (2.3-1) (Aharony and Hohenberg 1976) but also in the ratios of the $a_i$ in that same equation (Aharony and Ahlers 1980). Light scattering should play an increasingly important role in obtaining these ratios, though few such measurements now exist. For example Beysens et al. (1979c) have determined the ratio $R_\xi^+ \equiv \xi_0^+ A^{1/3}$ (see table 2) using photon correlation spectroscopy to measure $\xi_0^+$, and density measurements to find the heat capacity amplitude $A$ in the same sample. Measurements in three binary mixtures gave $R_\xi^+ = 0.29 \pm 0.04$ (nitroethane–isooctane), $0.280 \pm 0.011$ (isobutyric acid and water), and $0.255 \pm 0.013$ (triethylamine–water). Though these results are in good agreement with the theoretical prediction, $R_\xi^+ = 0.27 \pm 0.02$ in table 2, the measurements may contain systematic errors. One possible source of error relates to proper identification of the background contribution to the heat capacity; another is connected with the determination of $\xi_0$ from linewidth measurements. Whereas Beysens et al. equate the relaxation rate $\Gamma(q\xi \to 0)$ to $\xi_0$ through the equation, $\Gamma = k_B T q^2/5\pi\eta\xi_0^+ t^{-\nu}$, recent measurements suggest that the factor of 5 should actually be 6 (see sect. 4.3).

Other studies aimed at verifying two-scale factor universality are the experiments of Thoen et al. (1978), and a reanalysis of earlier experimental data in simple fluids carried out by Sengers and Moldover (1978).

The measurement of an amplitude ratio in an asymmetric tricritical system has raised some puzzling theoretical questions (Kim et al. 1979, Kaufman et al. 1980, Kim et al. 1980), as we will now see. Before proceeding, however, it should be mentioned that light scattering experiments in asymmetric systems can be somewhat difficult to interpret; the order parameter is an unknown linear combination of the composition variables, and changing the sample temperature only, moves one along a thermodynamic path which is not easily characterized. Nevertheless one might expect that in a mixture of tricritical composition, $I(q)$ would diverge as $\epsilon^{-\gamma_t/(1-\alpha_t)}$, where $\gamma_t$ and $\alpha_t$ should have their mean-field values, 1 and $\frac{1}{2}$, respectively (Griffiths 1974). Along this same path the scaling relation, $\gamma_t = (2 - \hat{\eta}_t)\nu_t$ implies that $\nu_t = 1$, since $\hat{\eta}_t = 0$ (Griffiths 1974, Gollub et al. 1976, Riedel and Wegner 1972, Bausch 1972).

There have been few light scattering experiments in tricritical systems (Griffiths 1974). Leiderer et al. have made measurements in $^3He$–$^4He$ near its tricritical point (Leiderer et al. 1975), and the asymmetric multicomponent mixture, ethanol–water–bromobenzene–ammonium sulfate (EWBA) has been studied by Gollub et al. (1976) and by Wu (1978). The experiments yield critical exponents $\gamma_t = 2\nu_t = 1$, but the EWBA experiments do not constitute a severe test of mean-field theory because they span a relatively small range of reduced temperature.

In a multicomponent system like EWBA this range is limited by the extreme difficulty of preparing a sample whose composition is close to the tricritical value. However, by measuring the ratio of intensities scattered by the three coexisting phases, the problem can be circumvented, since an unambiguous prediction of mean-field theory can be tested at each temperature. This prediction, which is implicit in Griffiths' pioneering analysis of tricritical behavior of asymmetric systems (Griffiths 1974), expresses itself in the form of various sum rules governing the suceptibility of three coexisting phases ($\alpha$, $\beta$, $\gamma$). Griffiths' "first sum rule" (Kaufman et al. 1980) is:

$$R_\chi = \frac{\chi_\alpha^{1/2} + \chi_\gamma^{1/2}}{\chi_\beta^{1/2}} = 1\,. \qquad (3.4\text{-}1)$$

The three static susceptibilities in the above equation refer to the upper ($\alpha$), middle ($\beta$), and lower ($\gamma$) phases respectively. Griffiths' second sum rule is

$$\chi_\alpha^{-1/2} + \chi_\gamma^{-1/2} - \chi_\beta^{-1/2} \propto T - T_t \qquad (3.4\text{-}1a)$$

where $T_t$ is the tricritical temperature.

Kim et al. (1979, 1980) determined the susceptibilities in eq. (3.4-1) by measuring the scattered intensity in all three phases of EWBA. The phase diagram of this system was first mapped out by Lang and Widom (1975). The angular distribution measurements of Kim et al. were fitted to the OZ equation,

$$I(q) \propto \left(\frac{\partial n'}{\partial \eta}\right)^2_{T,P} (\xi^{-2} + q^2)^{-1}\,, \qquad (3.4\text{-}2)$$

with $I(0) \propto \chi$. Since mean field theory predicts that $\chi \propto \xi^2$ (Stanley 1971), eq. (3.4-1) can also be written:

$$R_\xi = (\xi_\alpha + \xi_\gamma)/\xi_\beta = 1\,, \qquad (3.4\text{-}3)$$

and Griffith's second sum rule, eq. (3.4-1a), becomes

$$\xi_\alpha^{-1} + \xi_\gamma^{-1} - \xi_\beta^{-1} \propto T - T_t\,. \qquad (3.4\text{-}4)$$

The applicability of these sum rules requires only that three phases be present and, as always, that the system is reasonably near the critical point so that $\xi \gg \xi_0$.

In the experiments of Kim et al., eq. (3.4-3) was confirmed, but significant deviations from eqs. (3.4-1) and (3.4-3) were seen. Taking $(\partial n'/\partial \eta)_{pT}$ to have the same value in all three phases, the measurements gave $R_I = 1.18 \pm 0.02$ to $1.27 \pm 0.05$. Inserting reasonable estimates of $(\partial n'/\partial \eta)^{\alpha,\beta,\gamma}_{P,T}$ reduced these ratios somewhat but still left a large discrepancy between experiment and theory. The correlation length ratio, which is immune to this type of correction, was $R_\xi = 1.3 \pm 0.08$. While logarithmic corrections to mean-field behavior are expected near the tricritical point, an RG calculation shows that they do not modify eqs. (3.4-1) and (3.4-3) in first order (Stephen 1975, Stephen et al. 1975), Kaufmann et al. (1980) offer some suggestions to resolve the theoretical dilemma posed by these amplitude ratio measurements, but so far no calculations exist.

It seems likely that the measurement of amplitude ratios near ordinary and multicritical points may provide the most severe tests of theories of critical behavior.

## 4. Critical dynamics

### 4.1. Signal detection

While light scattering is a useful probe of the equilibrium properties of critical systems, it is also valuable for the study of fluctuations about the equilibrium state. By measuring the spectrum of these fluctuations, one can determine diagonal transport coefficients, such as the thermal diffusivity of simple fluids, or the concentration diffusivity in mixtures. In addition, off-diagonal coefficients such as the thermal diffusion ratio (Landau and Lifshitz 1959), can also be obtained as in the Soret effect measurements of Giglio and Vendramini (1975a, 1975b, 1977, 1978).

In this chapter we consider only those transport properties which are related to the slow fluctuations of the order parameter. Their spectra typically fall in the range 10 Hz to $10^4$ Hz and are therefore readily measured by photon correlation spectroscopy or by rapid Fourier analysis of the fluctuating output signal of the photodetector. These techniques are discussed in many books and review articles (see for example, Cummins and Swinney (1970)). It will suffice here to note that the physical quantity of interest is the order parameter correlation function $\langle \eta(q,0)\eta^*(q,t)\rangle$, whereas a photon correlation measurement gives $G_I \equiv \langle I(q,0)I(q,t)\rangle$. To connect these two correlation functions one must assume a linear relation between $\delta\eta(r,t)$ or $\delta\epsilon(r,t)$, and the electric field, $E_s(r,t)$, scattered by a small volume of fluid surrounding the point $r$. Finally, the electric *field* correlation function $G_E \equiv \langle E(00)E^*(q,t)\rangle$ and $G_I$

are related by the equation,

$$G_I(q,\tau) = \langle I(q)\rangle^2\left[1+\beta\frac{|\langle E^*(q,t)E(q,t+\tau)\rangle|^2}{\langle I(q)\rangle^2}\right], \tag{4.1-1}$$

where $E$ is the scattered electric field, $\langle I\rangle = \langle |E|^2\rangle$, and $\beta$ is a geometric factor which is of order unity if the area of the photodetector and source are sufficiently small (Jakeman 1974). Equation (4.1-1) holds if the field fluctuation $E$ is a Gaussian random variable. The central limit theorem assures that this restriction is met if the laser beam encompasses many correlation volumes, $\xi^3$. Since $E(q,t) \propto \eta(q,t)$, the Fourier transform of the electric field correlation function is proportional to $I(q,\omega)$ (Cummins and Swinney 1970).

To an excellent approximation, the order parameter correlation function decays exponentially, i.e.,

$$\frac{|\langle \eta^*(q,t)\eta(q,t+\tau)\rangle|}{\langle|\eta(q)|^2\rangle} = \frac{|\langle E^*(q,t)E(q,t+\tau)\rangle}{\langle I(q)\rangle} = e^{-\Gamma(q)\tau}, \tag{4.1-2}$$

with

$$\Gamma(q) = D(q)q^2\,. \tag{4.1-3}$$

Here $D(q)$ is the wavenumber-dependent diffusity and $D(0) \equiv D$ is its static value.

While the above equations predict an exponential decay of the order parameter correlation function, recent experiments (Burstyn et al. 1980a,b) have detected a small non-exponential contribution very near the critical point in the binary mixture 3-methylpentane–nitroethane. This effect is connected with the frequency dependence of the viscosity (Perl and Ferrell 1972, Lo and Kawasaki 1973, Ohta and Kawasaki 1976, Ohta 1980).

### 4.2. Theory

According to the fluctuation–dissipation theorem (Martin 1968), the diffusivity of a fluid may be written as the ratio of a conductivity $\lambda$ and the susceptibility $\chi$:

$$D \equiv \lim_{q\xi\to 0} D(q,\xi) = \lambda/\chi\,. \tag{4.2-1}$$

In a simple fluid, $D$ is the thermal diffusivity, and in mixtures it is the concentration diffusivity.

Near the critical point, the singular temperature dependence of transport coefficients is determined by a few slowly relaxing modes, whose coupling is crucial. In the absence of this coupling $D \sim \chi^{-1} \sim \epsilon^{\gamma}$, whereas it is experimentally observed that $D \sim \epsilon^{z_{\bar\eta}+\nu}$ with $z_{\bar\eta} \ll \nu$. Together with eqs. (4.2-1) and (2.2-1), this implies that $\lambda$ diverges strongly at the critical point.

This singular behavior of $\lambda$ was first predicted by Kawasaki and taken into account in theories of the so-called mode-coupling type (Kawasaki 1970, Kadanoff and Swift 1968). Mode-coupling (MC) theory is equivalent to the decoupled-mode theory of Ferrell (Ferrell 1970, Perl and Ferrell 1972). The approximations which enter these theories have been supported by an RG analysis (Siggia 1976, Halperin and Hohenberg 1974, Ohta and Kawasaki 1976, Garisto and Kapral 1976). The theory of dynamic critical phenomena has been recently reviewed by Hohenberg and Halperin (1977), and we will draw heavily from this reference. The present discussion however, will be limited to light scattering, i.e., to calculations and measurements of $\Gamma$ in eq. (4.1-2).

According to MC and RG theory, the diffusivity at finite $q$ may be written in the scaling form:

$$D(q, \xi) = \frac{\lambda(q)}{\chi(q)} = \left(\frac{R}{k_B T \bar{\eta} \xi}\right) \bar{\Omega}(q\xi), \tag{4.2-2}$$

where $R$ is a universal constant and $\bar{\eta}$ is the shear viscosity. Mode coupling and RG calculations give $R = (6\pi)^{-1}$ and $R \simeq (5\pi)^{-1}$ respectively. According to MC theory (Kawasaki 1970),

$$\bar{\Omega}(x) = \tfrac{3}{4} x^{-2} [1 + x^2 + (x^3 - x^{-1}) \arctan x], \tag{4.2-3}$$

and

$$\bar{\eta} = \bar{\eta}_0 [1 + (8/15\pi^2) \ln \xi] \equiv \bar{\eta}_0 + \bar{\eta}_s . \tag{4.2-4}$$

In the temperature range accessible in most experiments, the background contribution to shear viscosity $\bar{\eta}$ is much larger than the singular part, $\bar{\eta}_s$.

Both $\lambda(q)$ and $\bar{\eta}(q)$ are predicted to be of scaling form, i.e.,

$$\lambda(q) = q^{-z_\lambda} L(q\xi) = \xi^{z_\lambda} \tilde{L}(q\xi) = \lambda_0 + \lambda_s , \tag{4.2-5}$$

$$\bar{\eta}(q) = q^{-z_{\bar{\eta}}} E(q\xi) = \xi^{z_{\bar{\eta}}} \tilde{E}(q\xi) = \bar{\eta}_0 + \bar{\eta}_s , \tag{4.2-6}$$

with

$$z_\lambda + z_{\bar{\eta}} = 4 - d - \hat{\eta} . \tag{4.2-7}$$

From an RG analysis (Siggia 1976),

$$z_\lambda = 0.916, \qquad z_{\bar{\eta}} = 0.065 \tag{4.2-8}$$

A background term appears in both $\lambda$ and $\chi$ (Sengers and Keyes 1971) Chang et al. 1971, Oxtoby and Gelbart 1974), and hence in $\Gamma(q) = D(q) q^2$. Taking this into account in the MC approximation (Hohenberg and Halperin 1977), one finds,

$$\Gamma(q) = \frac{\lambda_0 q^2}{\chi(q)} + \frac{k_B T}{6\pi \bar{\eta} \xi^3} \bar{\Omega}(q\xi) H(q\xi) = \Gamma_B + \Gamma_s , \tag{4.2-9}$$

where $\bar{\eta}$ is given by eq. (4.2-6). The background conductivity, $\lambda_0$, like $\bar{\eta}_0$, is system dependent. The factor $H(q\xi)$ is unity if vertex corrections are ignored (Kawasaki and Lo 1972, Lo and Kawasaki 1973, Swinney and Henry 1973) and if $\chi(q)$ is taken to be of OZ form in the derivation of $\Gamma_s$. In calculating the background contribution it will be assumed that $\chi(q)$ in the denominator of eq. (4.2-9) is also of OZ form. In a simple fluid,

$$\chi(q) \propto C_p/(1 + q^2\xi^2), \tag{4.2-10}$$

where $C_p$ is the specific heat. (In binary mixtures, $\Gamma_B$ is observed to be negligibly small.)

Many scattering experiments, a few of which will be discussed below, provide striking confirmation of MC theory and are accurate enough to reveal the predicted weak singularity in the viscosity, eq. (4.2-8).

### *4.3. Linewidth measurements*

In the usual experimental arrangement, the relaxation rate $\Gamma(q)$ is measured in the homodyne or "self-beat" mode (Jakeman 1974). With this scheme, the only signal received by the photodetector is that which is scattered by the source so that $G_I(q, \tau)$ in eq. (4.1-1) becomes

$$G_I(q, \tau) = \langle I(q)\rangle^2[1 + \beta\, e^{-2\Gamma(q)\tau}]. \tag{4.3-1}$$

In an alternative measuring scheme (heterodyne spectroscopy), the scattered light is coherently mixed with a small portion of the incident laser beam. In this case the factor of two in the exponent of the above equation disappears (Chu 1974). Sometimes it is experimentally difficult to fully exclude elastically scattered stray light in a homodyne experiment, with the result that both terms, $\exp[\Gamma_\eta(q)\tau]$ and $\exp[2\Gamma_\eta(q)\tau]$ will appear on the right of eq. (4.3-1). Obviously this unwanted mixture of homodyne and heterodyne terms complicates the data analysis.

As noted in the previous section, there now exist many measurements in binary mixtures and simple fluids which confirm the MC prediction, eqs. (4.2-9). Figure 1 is a plot of $\Gamma^* \equiv \Gamma_s(6\pi\bar{\eta})/k_BT)q^{-3}$ vs $q\xi$ for eight such systems; it was taken from the review article by Swinney and Henry (1973). According to eq. (4.2-9), $\Gamma^* = H(q\xi)\bar{\Omega}(q\xi)q^{-3}$, with $\bar{\Omega}(x)$ given by eq. (4.2-3). This function appears as the dashed line in the figure. The parameters $C_p$ and $\lambda_0$ which enter the subtracted term $\Gamma_B$ in eq. (4.2-9), were separately measured, as was the viscosity $\bar{\eta}$. (In some cases it was necessary to take into account the small, singular contribution to $\bar{\eta}$ in eq. (4.2-4).) Thus there are no adjustable parameters to affect the excellent agreement between experiment and MC theory which is seen in this figure.

Another test of the MC theory has come from scattering experiments on the binary mixture, methanol–cyclohexane. In these experiments of Sorensen et al.

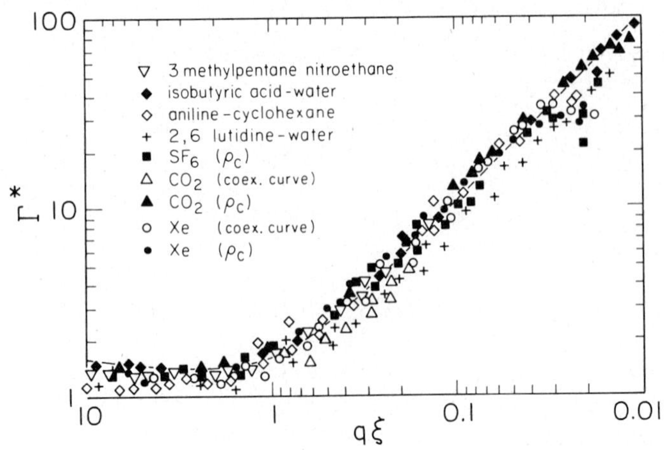


Fig. 1. $\Gamma^*$ as a function of $q\xi$ for various simple fluids and binary mixtures. A background subtraction has been made for the pure fluids only. All experiments were on the critical isochore except the following, where the decay rate was measured along the coexistence curve: xenon (○) and $CO_2$ (△). (After Swinney and Henry (1973).)

(1977), the temperature stability was $\pm 15\,\mu K$, enabling much closer approach to the critical point than the data of fig. 1. Also investigated was the linewidth of depolarized, and hence multiply scattered light. The authors conclude that while multiple scattering has a small effect on the linewidth, it did influence the agreement between experiment and theory when $q\xi \gtrsim 1$. For $q\xi \lesssim 10$ their measurements were consistent with MC theory and with the predictions of the dynamic droplet model (Sorensen et al. 1976). This latter model, however, contains an adjustable parameter not present in the MC theory and does not appear to rest on as firm as theoretical foundation (see also Schroeter et al. (1979) and references therein).

The experiments of Sorensen et al. (1977) underline the importance of understanding multiple scattering effects on linewidth measurements and of choosing very well index-matched mixtures in linewidth measurements at large $q\xi$.

While mode-coupling theory predicts that $\Gamma(q\xi \gg 1) = k_B T/6\pi\bar{\eta}\xi$, recent renormalization group calculations indicate that the factor, 6, in the denominator is closer to 5. Recent experiments in a binary mixture (Burstyn et al. 1980) and a single fluid (Guttinger and Cannell 1980) are in much better agreement with the MC value but this discrepancy should not be given undue emphasis, since both the MC and RG calculations are subject to appreciable uncertainty. More important is the agreement between these two measurements of $\Gamma$; this is the best existing evidence of the expected universality of dynamic behavior in two widely different fluid systems.

## 4.4. The viscosity singularity

Though the shear viscosity in fluids has a divergent part, it is very difficult to observe because, the exponent $z_{\bar{\eta}}$ associated with this divergence is very small (see eq. (4.2-8)), and because the background part, $\bar{\eta}_0$ in eq. (4.2-5) is much larger than the singular contribution at ordinarily accessible values of $\epsilon$. Nevertheless, careful light scattering measurements of $\Gamma(q)$ can detect its presence through the scaling relation,

$$\Gamma_s(q) = q^z \bar{\Omega}(q\xi)\,, \tag{4.4-1}$$

with

$$z = 3 + z_{\bar{\eta}}\ . \tag{4.4-2}$$

These equations follow from the relation $\Gamma(q) = D(q)q^2$, together with eqs. (4.2-2), (4.2-3) and (4.2-6).

Since $\Omega(q\xi)\to$const as $q\xi\to\infty$, the linewidth in the extreme critical limit is given by,

$$\lim_{q\xi\to\infty} \Gamma_s(q) = \text{const} \times q^z\,, \tag{4.4-3}$$

with $z = 3.065$ (see eq. (4.2-8)). To approach this limiting value using an optical source, it is necessary to observe the scattering at large angles (and hence large $q$) and to approach the critical point very closely in order to maximize $\xi$ and to minimize corrections-to-scaling terms.

The first experiment to reveal a value of $z$ larger than 3 was reported by Chu and Lin (1974), their sample being a ternary liquid mixture. More recently, Burstyn and Sengers (1980) have obtained $z$ from a linewidth measurement in 3-methylpentane-nitroethane. After subtraction of a small background they fitted their data to the equation

$$\Gamma_s(q, T) \propto q^{z_{\text{eff}}(T)}\,. \tag{4.4-4}$$

Figure 2 is a plot of $z_{\text{eff}}$ vs $T - T_c$ over the range $0.5\,\text{mK} \leqslant T - T_c \leqslant 10\,\text{mK}$. An extrapolation of $z_{\text{eff}}(t)$ to $T - T_c = 0$ gives $z = 3.06 \pm 0.024$, the error being two standard deviations. The measurements were also fitted to a functional form postulated by Ohta (1977), and this procedure yielded a similar value of $z$.

The above value of $z$, while consistent with the theoretical value, disagrees with the measurements of Sorensen et al. (1978), who found $z = 2.992 \pm 0.014$ in the same binary mixture. The discrepancy between these two experiments is due in part to an acknowledged small error in the measurement of the viscosity of the mixture, but it is also indicative of the difficulties encountered in reducing systematic errors in line width measurements below $\frac{1}{2}\%$.

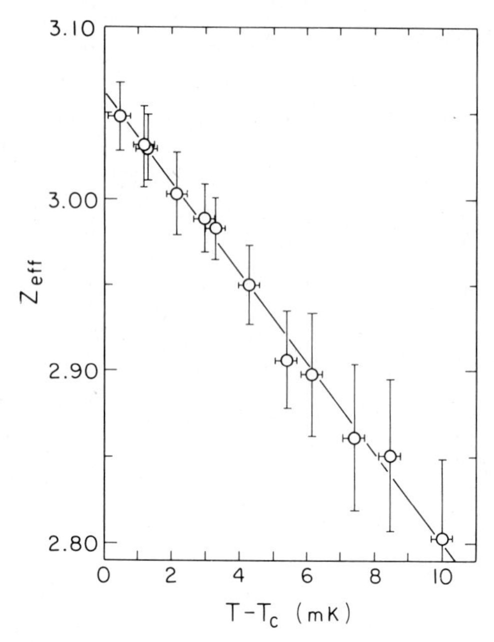


Fig. 2. Temperature dependence of the decay-rate exponent $z_{\mathrm{eff}}$ in eq. (4.4-4) measured by Burstyn and Sengers (1980). The system is a critical mixture of 3-methylpentane-nitroethane; the critical temperature is approached along the critical ioschore.

## 5. Light scattering from phase-separating fluids

### 5.1. Nucleation vs spinodal decomposition

The rest of this chapter is concerned with the dynamics of phase separation in fluids. It is useful to view this process as proceeding either by nucleation or by spinodal decomposition. Nucleation is associated with metastability and the presence of an energy barrier which must be surmounted before a droplet of the new phase can develop. In homogeneous nucleation, the only type of interest here, a local thermodynamic fluctuation produces a droplet which exceeds the critical size; droplets smaller than this do not develop because the cost in surface energy is too great.

In contrast, phase separation is said to proceed by spinodal decomposition when this energy barrier is absent. To produce spinodal decomposition, the system must be so deeply quenched that the diffusion constant $D(q)$ in eq. (4.2-1) must be negative. When this condition is satisfied even the smallest fluctuation will grow according to the equation

$$\eta(q, t) = \eta(q, 0)\exp[-D(q)q^2 t] \tag{5.1-1a}$$

where $\eta(q, t)$ is the $q$th Fourier component of density or composition.

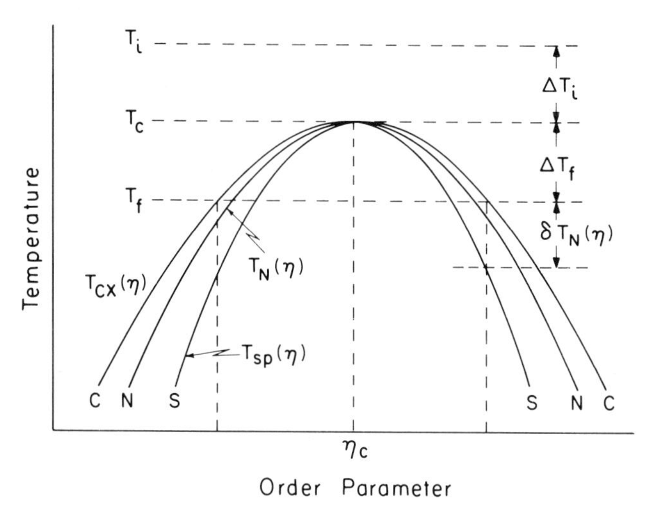


Fig. 3. Idealized coexistence curve, $T_{cx}(\eta)$ appropriate to a simple fluid ($\eta$ = density) or binary mixture ($\eta$ = molecular density of one of the components). Also shown are the spinodal line, $T_{sp}(\eta)$ and nucleation line, $T_N(\eta)$. Other parameters are defined in sect. 6.

According to eq. (4.2-1), $D(q)$ is the ratio of an intrinsically positive function $\lambda(q)$, to the order parameter susceptibility $\chi(q)$. In fig. 3 the curve CC is the coexistence curve, $T_{cx}(\eta)$, and the line SS defines the locus of points $T_{sp}(\eta)$ where $\chi = 0$ and below which $\chi < 0$. This so-called spinodal line is, of course a theoretical construct; its definition requires that the Helmholtz free energy density $\phi$ be extended to include both metastable states – those between the coexistence and spinodal lines – as well as the unstable states within the spinodal region.

Since the spinodal and coexistence curves touch each other at the critical point ($T_c$, $\eta_c$), the quenching of a critical system will of necessity proceed by spinodal decomposition. To observe this phenomenon in an "off-critical sample" ($\eta \neq \eta_c$) the quench must be so rapid that insipient droplets have inadequate time to nucleate before the temperature of the system has reached its final value $T_f < T_{cx}(\eta)$.

Between the curves CC and SS in fig. 3, is a third line, NN that might be called "the fuzzy line of nucleation", $T_N(\eta)$. A sample of fixed $\eta$ can be supercooled below this line (by an amount $\delta T_N$) before nucleation is observed to occur; if the system is supercooled by a smaller or larger amount, nucleation proceeds at an unmeasurably slow or unmeasurably rapid rate. Theoretically, the nucleation rate $J$ (in droplets/cm$^3$ s) is a very strong function of $T - T_{cx}$, so $J$ increases from say $10^{-2}$/cm$^3$ s to $10^2$/cm$^3$ s with a very small increase in supercooling. Therefore the fuzzy line $T_N(\eta)$ is, from the experimental point of view, a very sharp one. We might arbitrarily define it as that value of $T(< T_{cx})$ at which $J = 1/\text{cm}^3\,\text{s}$. Since nucleation arises from spontaneous local fluctuations in $\eta$, and these fluctuations increase near the critical point, $\delta T_N$ is a function of $T_c - T_{cx}(\eta)$ in fig. 3.

Using the above operational definition, $T_N(\eta)$ can be theoretically calculated – but not necessarily measured, since the presence of 1–$10^3$ droplets in 1 $\text{cm}^3$ of fluid may not in fact be detectable by light scattering or by any other means. A better definition of $\delta T_N$ might therefore be that value of $T_{cx} - T$ such that an appreciable fraction $f$, of the background phase be converted to droplets of the new phase (one would hope that $\delta T$ will depend very weakly on $f$). As we will see, the nucleation observability problem is a crucial one, especially in experiments carried out close to the critical point. In this region of the $T$–$\eta$ phase diagram, droplets grow so slowly that minutes to hours may elapse before they reach detectable size.

The spinodal line, like the nucleation line, must also be a blurred one; physically there is no reason to expect that phase separation will proceed differently just above it where an energy barrier exists, then just below, as long as this barrier energy is no larger than the average energy $k_B T$ of thermal fluctuations.

The next three sections (5.2–5.4) will be concerned with the spinodal decomposition phenomenon, i.e., the time evolution of the structure factor, $S(q, t)$, for a fluid which is quenched (at $t = 0$) from an initial temperature $T_i(\eta)$ to a final temperature $T_f(\eta) < T_N(\eta)$. The most thoroughly explored situation is that of a "critical quench", where $\eta = \eta_c$. The behavior of the system under this condition is relatively simple because no metastable states are encountered along the vertical path from the initial temperature $T_i$ to the final value $T_f$, through the intermediate temperature $T_c$ (see fig. 3).

Section 6.1 takes up a review of nucleation theory, setting the stage for a discussion of recent anomalous supercooling experiments carried out near the critical point. As will be seen in sect. 6.2, earlier experimental evidence suggesting the failure of nucleation theory, was probably misinterpreted. Subsequent experimental and theoretical investigations indicate that these studies provided evidence about the *growth* of droplets rather than their *birth*. It now appears that nucleation theory has yet to be tested near the critical point.

### 5.2. *Spinodal decomposition: theory*

The basic idea of spinodal decomposition (and the term as well), is due to Cahn (1968). It was offered to explain the spatially periodic composition variation that was observed in certain metallic alloys that were deeply quenched from the melt. The origin of this periodicity is contained in Cahn's linear theory which starts with the diffusion equation for the local composition,

$$\frac{\partial c}{\partial t}(\boldsymbol{r}, t) = D\nabla^2 c(\boldsymbol{r}, t)\,, \tag{5.2-1b}$$

and the phenomenological equation relating the interparticle flux $\boldsymbol{J}(\boldsymbol{r}, t)$ and the chemical potential gradient,

$$\boldsymbol{J} - -\lambda \nabla \mu(c(\boldsymbol{r}, t)) . \tag{5.2-2}$$

The transport coefficient $\lambda$ (whose $q$ dependence we ignore) was introduced in eq. (4.2-1). In the present context it is usually referred to as the mobility and denoted as $M_0$.

Cahn assumed $\mu(c(\boldsymbol{r}, t))$ to be of the Ginsburg–Landau form discussed in ch. 1 (see also Langer (1975)),

$$\mu = \left(\frac{\partial \phi'}{\partial c}\right)_{P,T} + 2K(\nabla c)^2 , \tag{5.2-3}$$

where $\phi'(c)$ is the free energy density of a uniform system and

$$\chi^{-1} = \left(\frac{\partial^2 \phi'}{\partial c^2}\right)_{P,T} = \left(\frac{\partial \mu}{\partial c}\right)_{P,T} . \tag{5.2-4}$$

On combining the above equation with the conservation law,

$$\nabla \cdot \boldsymbol{J} + \frac{\partial c}{\partial t} = 0 , \tag{5.2-5}$$

and dropping all terms non-linear in $c$, one obtains the basic result,

$$\frac{\partial c}{\partial t} = M\nabla^2[\chi^{-1} + 2K\nabla^2 c] . \tag{5.2-5a}$$

The solution to this equation is

$$c(q, t) = c(q, 0) \exp[-D(q)q^2 t] ,$$

where

$$D(q) = M(\chi^{-1} + 2Kq^2) .$$

From the last pair of equations, it follows that when the fluid is quenched into the spinodal region ($\chi < 0$), composition fluctuations of long wavelength, $q < q_c$, will grow exponentially, with

$$q_c = (2K\chi)^{-1} = \xi^{-1}(T_f) .$$

The second equality follows from the relation, $\chi = \xi^2/2K$, given by the Ginsburg–Landau theory. The growth of the unstable modes is directly observable by light scattering, since $S(q, t) \propto \langle |c(q, t)|^2 \rangle$.

According to this linear theory, a fluid which is quenched into the spinodal region at $t = 0$, will produce a scattering pattern in the form of a circular ring. After a short time, the scattering is dominated by the most rapidly growing Fourier component $q_m = q_c/\sqrt{2}$. This ring will brighten at an exponential rate and its diameter will be independent of time. Contrary to this prediction, experiments show that the ring collapses with time. This is illustrated in fig. 4, which shows a sequence of photographs of the scattering pattern produced by

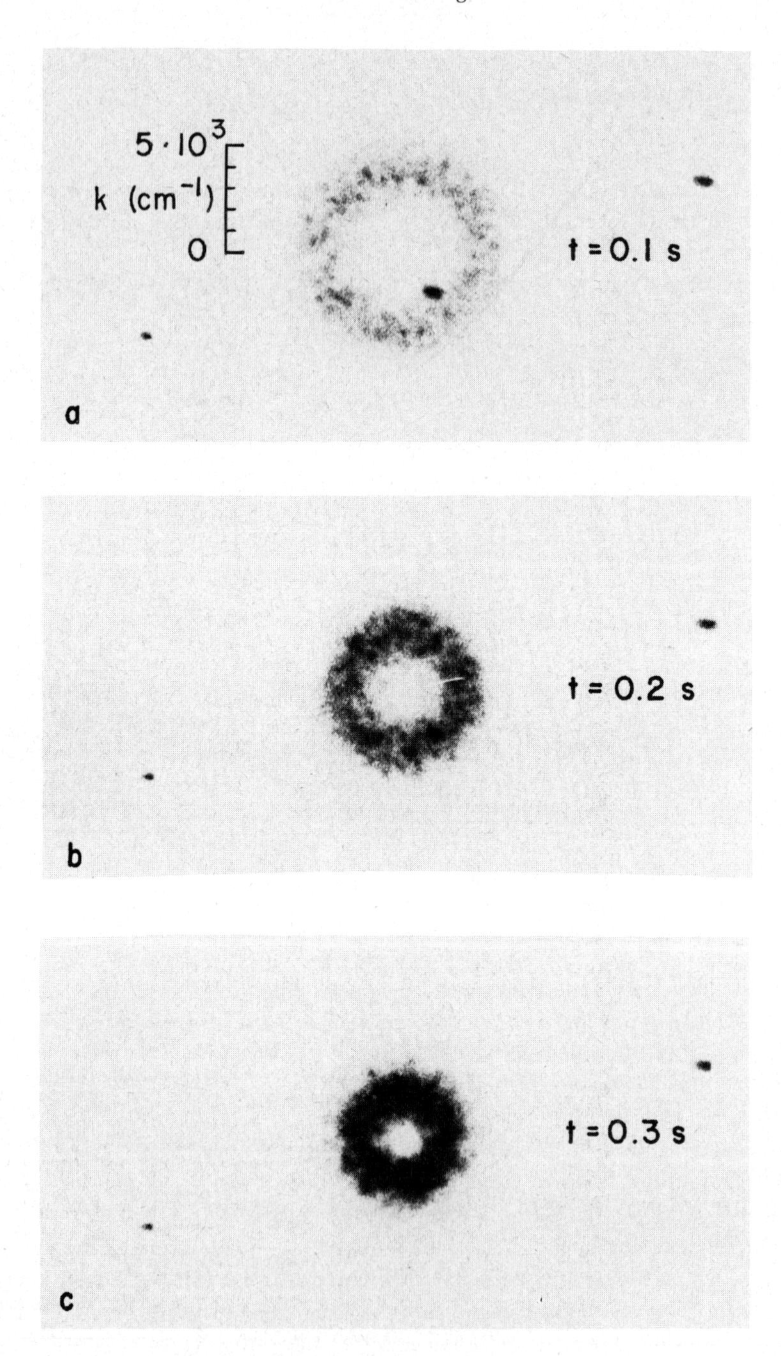


Fig. 4. Sequence of photographs showing spinodal decomposition in a near-critical mixture of $^3$He and $^4$He (see Hoffer et al. 1980).

a quenched mixture of $^3$He–$^4$He of near-critical composition (Hoffer et al. 1980). The measurements were made at $t = 0.1$ s, 0.2 s and 0.3 s. The "quench" which was accomplished by a sudden reduction of the pressure on the fluid (Wong and Knobler 1977), corresponds to $\Delta T_f = 0.037$ K, with $T_c = 0.867$ K. The granular structure of the pattern is a diffraction effect. The speckles flicker relatively slowly, revealing potentially interesting information about the dynamics of the coarsening process (see sect. 5.3). The collapse of the halo, as well as its growth in intensity, are clearly in evidence in fig. 4.

The collapse of the ring is due to the non-linear terms which have been omitted in the above analysis. Also omitted was an important spontaneous thermal fluctuation term. (Such fluctuations are, of course, entirely responsible for nucleation.) Finally the formulation contains no room for the influence of fluid flow. All these effects are taken into account, at least partially, in subsequent theories, to be summarized briefly below. First, however, it is useful to express the results of the linear theory in dimensionless units of wavenumber and time through the definitions:

$$\bar{q} \equiv q\xi \,, \tag{5.2-6}$$

$$\bar{\tau} = D(q = 0)/\xi^2 \,. \tag{5.2-7}$$

In the Born approximation $S(q, t) \propto \langle |c(q, t)|^2 \rangle$, with $\langle |c(q, 0)|^2 \rangle \propto \chi(q)$, evaluated at the quench temperature $T_f$. In the Ginsburg Landau theory $\chi(q) = \text{const} \times \epsilon^{-\gamma}/(1 + q^2\xi^2)$. Combining this last equation with results given above, and writing the result in dimensionless units, one finds

$$\frac{S(\bar{q}, \bar{\tau})}{S(0, 0)} \epsilon^{\gamma} = \frac{1}{1 + \bar{q}^2} \exp[-2\bar{q}^2(-1 + \bar{q}^2)] \,. \tag{5.2-8}$$

There is a $-1$ on the right because $x < 0$ within the spinodal region.

A questionable step in deriving this equation was the replacement of $S(0, 0)$ by its thermal equilibrium value at $T = T_f$. In actuality the quenched system will remember its initial state at $T = T_i$ for the short time, $\bar{\tau} \lesssim 1$.

It follows from this last equation that scattering measurements in all systems should be characterized by a single characteristic curve, if the intensity is plotted in units of $\epsilon^{-\gamma}$ and reduced units of wavenumber and time are used. This simple function also appears in non-linear theories of spinodal decomposition if the role of surface tension is neglected; its inclusion brings a new parameter into the problem.

It is tempting to treat mode coupling terms, ignored so far, as a small perturbation – at least when $t$ is small, but this approach seems unsatisfactory. Early evidence for the failure of perturbation theory came from the computer experiments of Bortz et al. (1974) and Marro et al. (1975). In these Monte Carlo simulations, and those that followed (Marro et al. 1975, Rao et al. 1976, Sur et al. 1977), an Ising A–B lattice was quenched from the completely disordered state (initial temperature $T_i = \infty$), to a final temperature ($T_f$) within the miscibility gap. Contrary to the linear theory, the ring intensity, $S(\bar{q} = \bar{q}_m, \tau)$ grew algebraically rather than at an exponential rate, and the ring diameter decreased in the same way. Specifically, the computer simulations could be fitted to the equation

$$S(q_m, t) \underset{\sim}{\propto} (t + 10)^{\theta}, \tag{5.2-9}$$

with $\theta = 0.70 \pm 0.05$, and

$$q_m \underset{\sim}{\propto} (t + 10)^{-\phi}, \tag{5.2-10}$$

with $\phi = 0.22 \pm 0.03$. Here $t$ and $q_m$ are in units of the spin exchange time and inverse lattice spacing respectively. Both $\theta$ and $\phi$ were almost independent of quench depth over the wide reduced temperature interval $0.6 \leqslant (T_c - T_f)/T_c \leqslant 0.9$, and the dimensionless time interval $1 \leqslant t \leqslant 30$. Only nearest neighbor spin exchanges are allowed, and the kinetics of the model is of the Kawasaki type, previously used by Flinn (1974) and Binder (1974). The computer experiments show, in addition, that when the lattice is quenched to a final temperature that also lies in the one phase region, a collapsing and brightening ring also appears. (This phenomenon was subsequently treated analytically by Binder (1977) and observed by Wong and Knobler (1979)). Almost coincident with the computer experiments were a number of analytical attempts to solve the spinodal decomposition problem. It was quickly recognized that the inclusion of thermal fluctuations played a crucial role in spinodal decomposition as well as nucleation (Binder 1974, Langer 1973). Langer et al. (1975) devised a very successful computational scheme for calculating $S(q, t)$, starting with a Markovian master equation for the probability density functional, $\rho([a(r)], t)$. As in Cahn's theory, the composition flux is proportional to a chemical potential gradient calculated from a free energy of $G$–$L$ form. The fluctuations are driven by thermal noise, which is included by addition of a Langevin term noise term to eq. (5.2-5a).

The main simplification required in the approach of Langer, Bar-on and Miller (LBM) was the self-consistent introduction of a single characteristic length $A(t)$, whose functional form could be solved for. Partly because of this assumption, the theory cannot describe the later stage of phase separation, where at least two lengths obviously appear, viz the correlation length $\xi$ and the size, $L(t) \simeq q_m^{-1}(t)$, of a typical cluster or domain (Billotet and Binder 1979). Like the theory of Cahn, the LBM theory implicitly assumes the viscosity of the system be infinite.

An entirely different approach to the spinodal decomposition problem has been taken by Binder and his associates (Binder (1977), Binder et al. (1978) and references contained therein). The starting point is a kinetic equation for the growth of clusters of B atoms surrounded by A atoms. The clusters grow by nucleation condensation and coagulation according to the Kawasaki spin-exchange model (1966a,b,c). With this approach there is no reference to a spinodal region ($\chi < 0$). The barrier to cluster growth by spin exchange never goes to zero; it merely drops well below $k_B T$ deep inside the coexistence curve. On making various approximations one recovers the results of Cook (1970) (who included fluctuations in Cahn's linearized theory), of LBM, and also the cluster-growth predictions of Lifshitz and Slyozov (1961). This latter model is very important and will be further discussed below. The structure factor $S(q, t)$ does not emerge in a simple way in the Binder approach, and it can only be evaluated qualitatively. The analysis is in a sense complimentary to Langer's approach, since its approximations work best far from the critical point and at large times (Binder et al. 1978), rather than at small $t$, where $\bar{q}_m \lessapprox 1$.

So far we have been discussing models which apply to solids only. In fluids, cluster coagulation is speeded up by diffusion, and the domain growth exponent $\phi$, defined in eq. (5.2-10), should no longer be given from the computer simulations of Lebowitz et al. According to the following dimensional argument, due to Binder and Stauffer (1974), the growth of a domain $L(t)$, expressed in reduced units, should be given by

$$L(t)/\xi \sim \bar{q}_m^{-1} \simeq \bar{\tau}^{\phi} = \bar{\tau}^{1/3} . \tag{5.2-11}$$

It is assumed that the system is deeply quenched, forming an interconnected (percolating) structure, so that the size of the domains, $L$, is roughly equal to their spacing. Thus only one length enters the growth-rate problem. The volume $\Omega \simeq L^3$ of a cluster should increase by a coalescence according to the simple equation, $d\Omega/dt \sim \Omega/\tau_L$ where $\tau_L$ is the time for an adjacent domain to diffuse through the intervening distance, which is also $L$ according to Stoke's law, $\tau_L^{-1} \sim D/L^2$, with $D$ given by $D = k_B T/6\pi\eta L$. Combining these equations gives $dL^3/dt \sim k_B T/6\pi\eta$. Integration of this equation leads to eq. (5.2-11) if one expresses $t$ in dimensionless units of eq. (5.2-6).

As further step toward inclusion of fluid degrees of freedom in spinodal decomposition, Kawasaki (1977) and Kawasaki and Ohta (1978) took into account the flow-induced coupling of distant composition fluctuations. This flow was in turn generated by pressure gradients produced by local composition gradients. Their calculation of $S(q, t)$ utilized the computational scheme of LBM and is therefore subject to the same limitation, viz that it becomes untrustworthy when $\bar{q}_m$ falls much below unity.

Furukawa (1979) has devised another method for approximating $S(q, t)$ in the spinodal region. His analysis, which makes strong use of scaling arguments, appears limited in applicability to off-critical quenches. Its starting point is an equation of the Langevin type. As in the Binder cluster model (1977), no distinction is drawn between metastable and unstable states (Furukawa 1978). At the outset, the structure factor is assumed to have the simple scaling form,

$$S(q, t) = L^3(t)F(qL) , \tag{5.2-12}$$

where the scaling function $F(x)$ contains no explicit time dependence. This type of scaling form for $S$ was first introduced by Binder and Stauffer (1976) and appears in subsequent theoretical studies of spinodal decomposition by Binder (1977) and Binder et al. (1978). It is born out by computer simulations (Marro et al. 1979) and by scattering experiments described in the next section.

It was first pointed out by Cahn and Moldover (1977) that phase separation in a fluid should also be influenced by surface tension, which has not been taken into account in the theories discussed above. In systems which are quenched into the spinodal region near the critical point, the phase-separating domains will be interconnected. If we imagine the domains as tube-like in form, we can expect these tubes to become pinched off to reduce the surface energy of the system. This squeezes fluid into remaining pockets causing them to grow. Invoking dimensional analysis to find the growth rate one finds,

$$L \propto t . \tag{5.2-13}$$

Siggia (1979) has investigated in some detail, the hydrodynamic effects associated with this late-stage phase separation mechanism, and the results will be compared with experiment in the next section.

We conclude this discussion by returning to the fundamental question: Does the spinodal line really exist, i.e., can an experiment determine whether a fluid has been quenched to a temperature just above the spinodal line to a temperature just below it? While theories and computer experiments discussed so far, imply that the answer is no, a different view is expressed in the interesting review article by Abraham (1979). The author considers the problem analytically and also presents molecular dynamics calculations for liquid argon which show that the linear theory of Cahn is indeed valid for $\bar{\tau} \lesssim 1$. The number of atoms in his system is not extremely large (1000 atoms), but this limitation aside, the calculations should reflect the behavior of a real fluid, including viscous flow and surface tension effects. Additional molecular dynamics calculations, carried closer to the critical point, may prove as stimulating as the Monte Carlo simulations of Lebowitz, summarized above.

### *5.3. Spinodal decomposition: experiments*

Binary mixtures are especially suitable for the study of spinodal decomposition (SD) because the diffusion constant is roughly one hundred times smaller than in simple fluids, so the phenomenon can be conveniently studied at small $\bar{\tau}$. Mixtures can be quenched into the spinodal by quickly lowering the temperature of the surrounding bath (as with simple fluids), or by suddenly changing the pressure so as to shift the critical temperature (Wong and Knobler 1977, 1978). Applying the pressure jump method to the isobutyric acid and water system produces an effective temperature change of $\sim -60$ mK/atm. This shift of the coexistence curve (at fixed $T$) may also be accompanied by

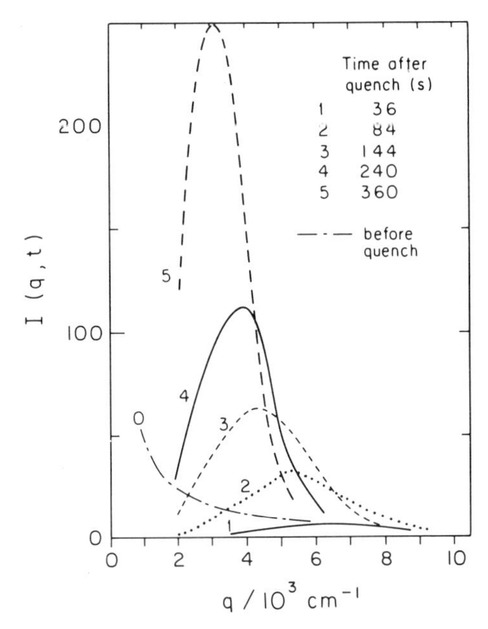


Fig. 5. Scattered intensity as a function of $q$ and $t$ in a critically quenched mixture of isobutyric acid and water. The quench depth is 0.9 mK (from Wong and Knobler (1978)).

adiabatic heating or cooling. Therefore the pressure quench scheme is effective only with those mixtures, where the second effect is relatively small.

In mixtures with an inverted coexistence curve, such as 2,6-lutidine and water (Stein et al. 1972, Gulari et al. 1972), a rapid quench into the two-phase region has been effected by Joule heating produced by driving a current pulse through the fluid (Goldburg et al. 1978a) and by dielectric heating with a pulse of microwave radiation (Goldburg et al. 1978b). Both schemes have the disadvantage of changing the sample temperature relative to that of the surrounding bath. To avoid this, the bath temperature must be independently changed as well.

Figure 5 shows the angular distribution of light scattered from a critically-quenched mixture of isobutyric acid and water (IW) (Wong and Knobler 1978). The curves are numbered by the observation times following the quench. Even though the pressure quench took only a small fraction of a second, a well-defined ring did not emerge until approximately 30 s; its collapse and decrease in half-width with time are clearly seen. In the decade of time spanned by the experiments, the domain size $q_m^{-1}$ increased from 1.6 μm to 3.3 μm. In dimensionless units this corresponds to a decrease in $\bar{q}_m$ from 0.5 to 0.25. So far the most *stringent* tests of SD theory have come from measurements of $q_m$ and the ring intensity $I(q_m, t)$ vs $t$.

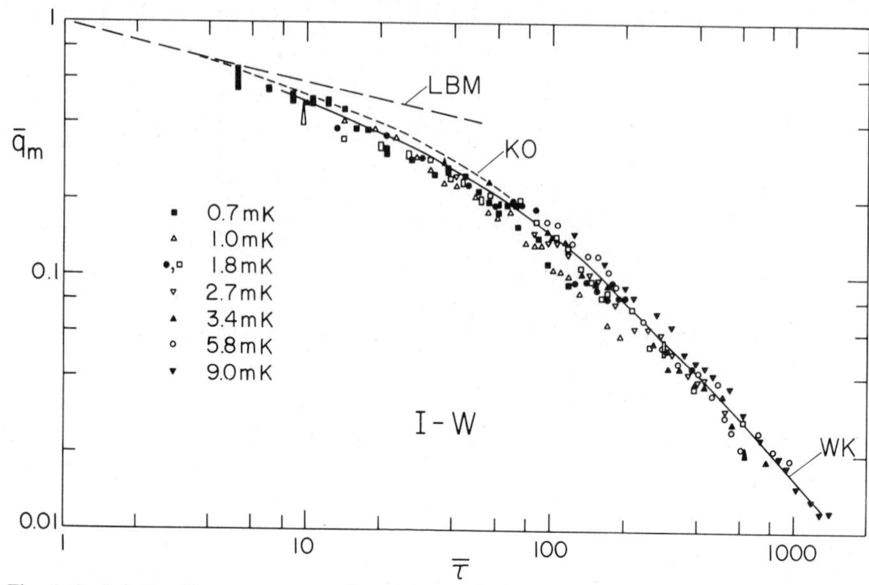


Fig. 6. Scaled ring diameter versus reduced time in isobutyric acid and water. The curved solid line (WK) summarizes similar measurements on the same system by Wong and Knobler (1978). The broken and dashed lines are respectively the theoretical calculations of Langer et al. (1975) and of Kawasaki and Ohta (1978) (from Chou and Goldburg (1979)).

Figure 6 shows $\bar{q}_m$ vs $\bar{\tau}$ in a critically quenched mixture of isobutyric acid and water (IW). The data points are those of Chou and Goldburg (1979) and the solid line summarizes similar experiments by Wong and Knobler (1978). The quench depths used by Chou and Goldburg (CG) are as indicated, and the Wong–Knobler experiments covered an almost identical range. The excellent agreement between the two sets of results, shows that the method of quench is irrelevant; CG quenched by raising the temperature, and WK used the pressure jump method. The dashed lines in fig. 6 show the theoretical results of LBM and of Kawasaki and Ohta (KO). To express $q_m$ and $t$ in reduced units, it is of course necessary to have independent measurements of the temperature dependence of $\xi$ and $D$. These parameters were known from previous experiments of Stein et al. (1972) and Gulari et al. (1972).

Chou and Goldburg (1979) also measured $\bar{q}_m$ vs $\bar{\tau}$ in a critical mixture of 2,6-lutidine and water. Superposition of the LW and IW curves shows them to be in excellent agreement with each other, as scaling arguments suggest. The measurements depart from the LBM theory at relatively small $\bar{\tau}$ but agree with the results of Kawasaki and Ohta (KO) out to $\bar{\tau} \simeq 100$. It appears then that the hydrodynamic effects, partially included by KO, are important.

Though the measurements in fig. 6 cannot be fitted to a simple algebraic form over the full range of $\bar{\tau}$, one can extract from the curves the limiting slopes

$\phi(\bar{q}_m \simeq 1)$ and $\phi(\bar{q}_m \ll 1)$ in the power law,

$$\bar{q}_m = A\bar{\tau}^{-\phi}. \qquad (5.3\text{-}1)$$

The results are:

$$\phi = 0.3 \pm 0.1, \qquad 0.6 \geqslant \bar{q}_m \geqslant 0.3, \qquad (5.3\text{-}2a)$$

$$\phi = 1.1 \pm 0.1, \qquad 0.1 \geqslant \bar{q}_m \geqslant 0.08. \qquad (5.3\text{-}2b)$$

These results support the coalescence model of Binder and Stauffer, eq. (5.2-11), at small $\tau$ and confirm eq. (5.2-13) at large $\tau$.

Siggia (1979) has further studied domain growth in the early and late stages of domain growth and obtains a semi-quantitative estimate of the amplitude $A$ in these two regimes. His estimates are (Chou and Goldburg 1979):

$$\bar{q}_m \simeq (5/36)\bar{\tau}^{-1/3}, \qquad 1 \geqslant \bar{q}_m \geqslant \tfrac{1}{10}, \qquad (5.3\text{-}3)$$

$$\bar{q}_m \simeq \bar{\tau}^{-1}/\pi, \qquad \tfrac{1}{10} \geqslant \bar{q}_m \geqslant q_g. \qquad (5.3\text{-}4)$$

Gravitational effects enter when $q_m \lessapprox q_g \equiv \xi g \Delta\rho/\sigma$, where $\rho$ is the average mass density of the fluid, $\Delta\rho(\tau)$ is the miscibility gap, and $\sigma$ is the surface tension. In LW and IW, $q_g < 7 \times 10^{-3}$ when $\Delta T_f > 1$ mK, and therefore lies outside the range of the measurements.

In the coalescence regime, the measured and calculated values of $q_m$ agree within a factor of two, but in the flow-driven stage of phase separation, the grain size calculated from eq. (5.3-4) exceeds the measured values of $q_m^{-1}$ by roughly 100. The problem of obtaining a quantitative solution to the late-stage coarsening problem remains an open one.

While ring diameter measurements in critically quenched mixtures support the LBM–KO theory at small $\tau$, the ring intensity measurements do not. The disagreement is apparent in fig. 7, which shows $\mathscr{I}_m(\bar{\tau}) = I(q_m, \bar{\tau})\epsilon^{\gamma}$ in I–W (Chou and Goldburg 1979). The intensity has been normalized as in eq. (6.2-8) and a correction has been made for multiple scattering, which unfortunately was severe at large $\bar{\tau}$. The graph also shows the calculation of $S(\bar{q}_m, \bar{\tau})$ by LBM and KO. Since the scattered intensity was measured in relative units only, it was necessary to normalize the theoretical and experimental results at an arbitrarily chosen time, viz $\bar{\tau} = 20$.

While data in fig. 6 scale with quench depth, i.e., all data points fall on a smooth curve, this curve does not coincide with a similar one obtained from measurements in LW. This may be an experimental artifact or the result of system-dependent cross over from coalescence to flow driven behavior. The form of $\mathscr{I}_m(\bar{\tau})$ is, however, qualitatively the same as that measured in the same system by Wong and Knobler (1978). Taken together, the two experiments leave little doubt that existing structure factor calculations are inadequate at large $\bar{\tau}$ and require extension to include surface tension.

Though calculation of $S(q, t)$ in the late stage of spinodal decomposition

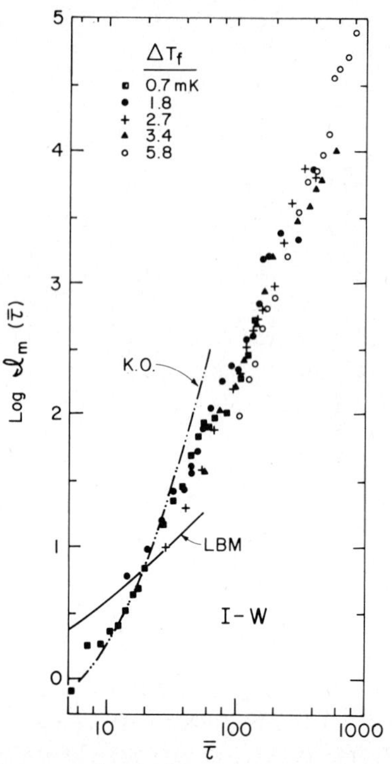


Fig. 7. Scaled ring intensity versus reduced time in isobutyric acid and water. The theoretical calculations of Langer et al. and of Kawasaki and Ohta have been normalized to fit the data at the arbitrarily chosen time, $\tau = 20$ (from Chou and Goldburg (1979)).

must be a formidable problem, the following argument suggests that $S(q_m, t)$ and $q_m(t)$ are simply related (Siggia 1979, Chou and Goldburg 1979): Suppose that each nucleating domain of size $L$, contributes independently to the scattering, that these domains are closely packed, and that the scattering from each one is proportional to $L^6$, in the Rayleigh–Gans limit of scattering theory. Then $S(q, t) \sim$ (number of domains per unit volume) $\times L^6 \sim L^{-3} \times L^6 = L^3 \sim q^{-3}$, so that product $q_m^3 S(q_m, t)$ is independent of time. This result is roughly in accord with the critical-quench experiments of Chou and Goldburg (1978), but should apply equally well to the off-critical-quench experiments of Wong and Knobler (1978). It is in fact but a special case of the more general scaling expression for $S(q, t)$ given in eq. (5.2-12). To compare this equation with experiment, we write it in slightly different form:

$$\bar{S}(q, t)K^3(t) = F(q/K(t)), \tag{5.3-5a}$$

where

$$\bar{S} = S(q, t) \Big/ \int S(q, t) q^2 \, dq. \tag{5.3-5b}$$

Here $K^{-1}$ is some measure of the inverse domain size $L$. In the computer simulations of Marro et al. (1979) and in the scattering experiments to be discussed next, the limits on the integral in eq. (5.3-5b) are of course finite, but the integral included most of the area under the collapsing ring (see fig. 5).

To verify eq. (5.3-5), Chou and Goldburg (1980) measured $I(q, t)$ vs $q$ at various times $t$ and various quench depths in both IW and LW. Figure 8 summarizes their observations for IW. Each of the four curves refers to a series of angular distribution measurements, made at various times but a single quench depth, $\Delta T_{\mathrm{f}}$. There the inverse domain size $K$ is identified with the "ring diameter" $q_{\mathrm{m}}$, so that the scaling hypothesis was tested by plotting $q_{\mathrm{m}}^3 I(q, t)$ as a function of $q/q_{\mathrm{m}}(t)$ (in the computer simulations of Marro et al. (1979) the structure factor best conformed to eq. (5.3-5) when $K$ was set equal to the second moment of $S(q, t)$).

Since the structure factor $S$ is normalized according to eq. (5.3-5b), there are no adjustable parameters in the figure. One therefore might hope that an

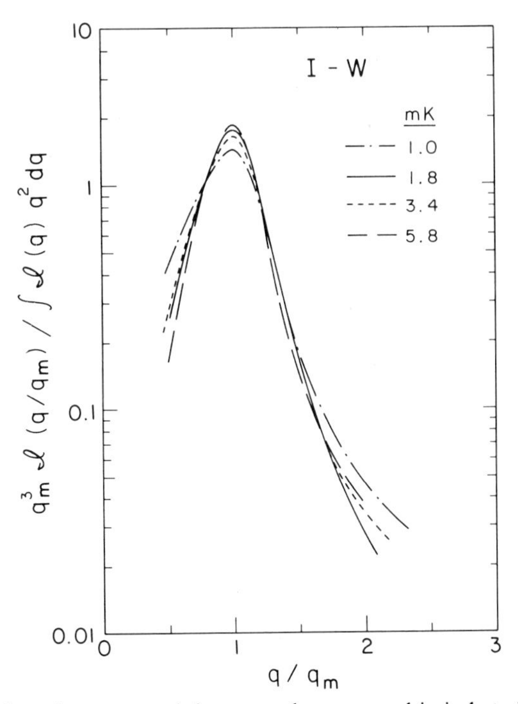


Fig. 8. The scaled intensity versus scaled wavenumber measured in isobutyric acid and water at various quench depths. Each curve refers to a different quench depth and summarizes a series of angular distribution measurements made at various times. The vertical coordinate is the function $F(x)$ defined in eq. (5.3-5), with $K = q_{\mathrm{m}}$ (Chou and Goldburg 1980).

identical study in a critical LW mixture, would yield the same function $F(x)$; this was indeed found (Chou and Goldburg 1980). Furthermore, a reanalysis of the angular distribution measurements of Wong and Knobler (1979) yields a function $F(x)$ whose form is qualitatively similar to that of fig. 8.

The confirmation of eq. (5.3-5) which the above experiments provide, is perhaps surprising, in light of the large time interval which the data span; there was no evidence of a change in the form of $F(x)$ as the system traversed the boundary between diffusion-driven coarsening and domain growth driven by surface tension.

### 5.4. *Temporal fluctuations*

In the experiments and theories discussed so far, no attention has been given to the temporal fluctuations of the scattering intensity discussed in sect. 5.2, i.e., to the lifetime of the speckles seen in fig. 4. The only reported investigations of this phenomenon are light scattering measurements made in the late stage of phase separation (Kim et al. 1978). In this study a photomultiplier measured the power spectrum $S_I(f)$ of the temporally fluctuating intensity $I(q, t)$ at various small scattering angles in the vicinity of $\theta = \theta_m$. Figure 9 shows a typical intensity trace in a critically quenched LW mixture at $\Delta T_f = 2.1$ mK, $q = 8.7 \times 10^3\ \text{cm}^{-1}$. During this run the ring is collapsing, with $q_m$ decreasing from $\sim 10^4\ \text{cm}^{-1}$ to $10^3\ \text{cm}^{-1}$ in the first hundred seconds. Roughly one coherence area of the photodetector was exposed, so that the amplitude of the fluctuations was not averaged out.

The two most notable features of fig. 9 are the very large amplitude of the fluctuations in $I(q, t)$ and their long lifetime. In this particular run the lifetime, $f_0^{-1}$, of the fluctuations was comparable to the diffusional relaxation rate, $\Gamma = Dq^2 \simeq 2\pi \times 50$ mHz. Measurements at larger scattering angles revealed, however, that $f_0$ was weakly dependent on $q$, though there was a clear increase in $f_0$ with increasing quench depth.

Further evidence for the non-equilibrium nature of these large-amplitude fluctuations was found in their non-Lorentzian power spectrum. Surprisingly, the spectrum could be well fitted to the form, $S_I(f) \propto \exp - (|f|/f_0)$, in the frequency range $10^{-2}\ \text{Hz} < f < 10^{-1}\ \text{Hz}$ and at various values of $q$ and $\Delta T_f$ in the interval $3 \times 10^3\ \text{cm}^{-1} \leqslant q \leqslant 2 \times 10^4\ \text{cm}^{-1}$ and $2\ \text{mK} \leqslant \Delta T_f \leqslant 5\ \text{mK}$. By definition $S_I(f)$ is the Fourier transform of $I(q, t)I(q, t + \tau)$ averaged over $t$.

While there is no theory to account for the above observations, data of this sort contain very detailed information about the coarsening process. With further experimentation, one should be able to determine, for example, if the noise is generated by oscillatory modes of the sponge-like nucleating structure or whether it comes from phase fluctuations of the electric field resulting from a coherent coarsening of this sponge.

The average intensity in the above late-stage measurements varied little over

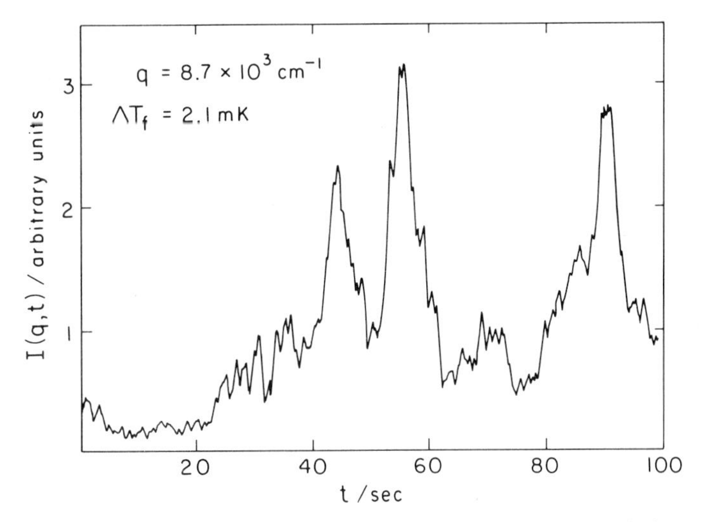


Fig. 9. Temporal fluctuations in the scattering intensity during the late stage of phase separation in LW (from Kim et al. (1978)).

the full measurement interval, $1\text{ s} \lesssim t < 400\text{ s}$. Therefore one could safely imagine $I(q, t)$ to be a stationary random variable and obtain $S_I(f)$ by splitting up a long record into smaller time intervals. In the absence of this approximate stationarity, $S_I(f)$ could only be deduced from an ensemble of measurements at each value of $q$ and $\Delta T_{\mathrm{f}}$.

Recently Billotet and Binder (1980) have used the LBM computational scheme to calculate the time-dependent structure factor $S(q, t_1, t_2)$ in the early stage of phase separation. Their results cannot be applied to the late-stage measurements described above but are very interesting in their own right. For one thing they reveal a damped oscillatory form of this function in a certain range of quench depth and $t_1$ and $t_2$. Their findings could be checked by measurements of "nucleation noise" in the early stage, but so far no such measurements exist.

## 6. Nucleation

### 6.1. Review of nucleation theory

It has been known for many years that supercooled liquids behave anomalously near the critical point. From visual observation and heat capacity measurements it is observed that binary mixtures (Sundquist and Oriani 1962, Heady and Cahn 1973, Huang et al. 1974) and simple fluids (Dahl and Moldover 1971, Huang et al. 1975) can be supercooled more than expected on the basis of classical nucleation theory (Frenkel 1955, Zettlemoyer 1969). In its simplest form, this theory predicts that the rate, $J$, at which droplets of a new phase nucleate out of the parent phase is given by

$$J = J_0 \exp(-\Delta\tilde{\Phi}^*/k_B T), \tag{6.1-1}$$

where $J$ has the units of droplets/cm$^3$ s. The prefactor $J_0$ is an attempt rate and $\Delta\tilde{\Phi}^*$ is an energy barrier that prevents the formation of droplets whose radius is less than the critical value, $R^*$. The growth of a droplet is inhibited by the cost in surface energy $\sigma(\sigma \sim \epsilon^{\gamma'})$ and is compensated for by a gain in volume energy, so both terms are contained in $\Delta\Phi^*$. This theory is obviously an oversimplification, since it includes only one variable (the radius $R$ of the embryo) and ignores other "coordinates" of the problem (Binder and Stauffer 1976). The value of $J_0$ depends only weakly on the amount of supercooling, $\delta T$, defined in fig. 10, but $\Delta\Phi^*$ depends strongly on this variable. Various other parameters are also defined in fig. 10, which is the phase diagram for a mixture such as LW, which has a lower critical point.

Because $J_0$ is very large (in fluids $J_0 \sim 10^{34}\,\mathrm{cm}^{-3}\,\mathrm{s}^{-1}$) and the exponential term in eq. (6.1-1) is such a strong function of $\delta T$, $J$ increases from a very small to a very large number with only a slight increase in supercooling or, equivalently, supersaturation, $\delta c_1$ in fig. 10. (Experimentally $\delta T$ is the parameter directly measured, but $\delta c_1$ enters more naturally into the theory).

At any finite value of $\delta T$, a metastable system will, in time, separate into two phases. Nevertheless it is sensible to define an admittedly fuzzy nucleation limit as determined by that value of $\delta T = \delta T^*$ (for a sample of fixed composition $c$) such that $J \simeq 1$ droplet/cm$^3$ s, i.e., where $J$ falls in a conveniently measurable range.

As the critical point is approached, $\delta T$ must necessarily decrease to zero,

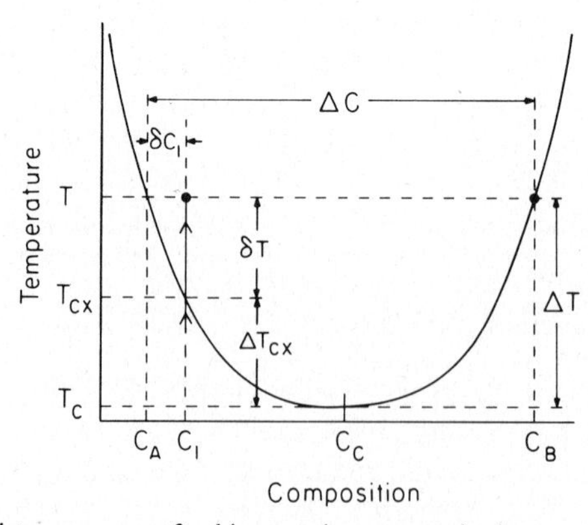


Fig. 10. Idealized coexistence curve of a binary mixture near its lower critical point. The figure defines various parameters appearing in the text.

since the fluctuations, which are required to surmount the energy barrier are infinite there. From scaling arguments as well as detailed calculations (Langer and Turski 1973), one finds that

$$\Delta\tilde{\Phi}^*/k_B T = (\Delta T/\delta T)^{\lambda}(\tau_0/\tau_c)^2 , \qquad (6.1\text{-}2)$$

where $\lambda = 3\gamma' - 2\beta - \gamma = 2$. The parameter, $\tau_0/T_c \equiv x_0$ is a non-universial combination of critical amplitudes; in $CO_2$, $x_0 = 1.1$ (Langer and Turski 1973, Goldburg and Huang 1975) and in binary mixtures it is also near unity (Howland et al. 1980).

Neglecting the unimportant temperature dependence of $J_0$, it follows from eqs. (6.1-1) and (6.1-2) that along the nucleation line, i.e., when $\delta T = \delta T^*$, $\delta T^*(c)/\Delta T(c) = \text{const.}$, i.e., the nucleation line has the same shape as the coexistence curve and is tangent to it at the critical point. Inserting the known values of $\tau_0$ and $J_0$ into eqs. (6.2-1) and (6.2-2) and setting $J = 1\ \text{cm}^{-3}$ gives $\delta T^*/\Delta T = 0.13$ (Goldburg and Huang 1975).

Classical nucleation theory predicts, then, that a $CO_2$ sample of arbitrary density can be supercooled by 13%. This value appears to vary little from one simple fluid or binary mixture to another (Schwartz et al. 1980).

### 6.2. *Anomalous supercooling experiments*

Nucleation is usually detected visually; a sealed sample, i.e., one of fixed composition or density ($\eta$), is cooled continuously or in steps until cloudiness appears at a temperature $T_N(\eta)$. The experiment is then repeated with a series of samples to plot out the entire nucleation line (by definition, $\delta T_N = |T_N - T_{cx}|$).

To inhibit surface nucleation, the wall-wetting phase is usually selected as the majority, or initial one. With this choice an incipient droplet of the new phase will not form at the wall where the nucleation barrier is reduced. In $CO_2$ for example, the initial phase must be the more dense one if the cell is of pyrex. Therefore it is difficult if not impossible to obtain the nucleation line $T_N(\rho)$ on the vapor side of the coexistence curve, $\rho < \rho_c$.

If the diffusion constant is very small, as in binary mixtures, nucleation can be studied in either phase or in two coexisting phases, provided the sample is cooled rapidly enough that inhomogeneous nucleation at the walls does not have time to propagate through an appreciable volume of the fluid (Wong and Huang et al. 1974).

The apparent failure of nucleation theory near the critical point is evident in fig. 11, which shows reduced supercooling ($\delta T_N/\Delta T$) vs $T_{cx}/T_c$ (see fig. 10). The dashed horizontal line is the prediction of classical nucleation theory, using the value of $\tau_0$ for $CO_2$ (Schwartz et al. 1980, Langer and Turski 1973). The diamonds, triangles, crosses, circles and squares refer respectively to measurements of $\delta T_N/\Delta T$ in $C_7H_{14}$–$C_7F_{14}$ (Heady and Cahn 1973), $CO_2$

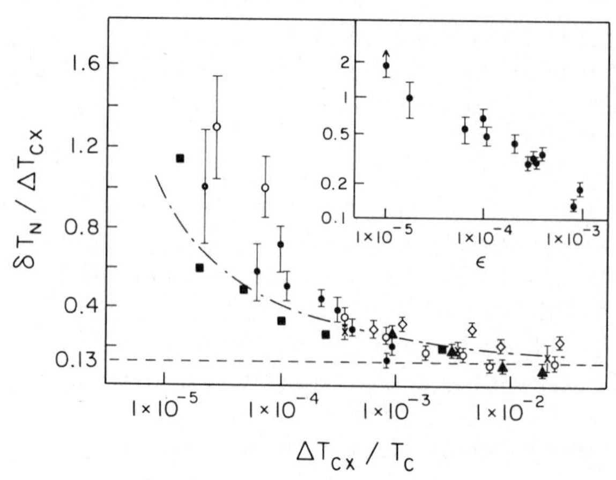


Fig. 11. Reduced supercooling versus $\log(\Delta T_{cx}/T_c)$ as measured in various binary mixtures and simple fluids. The LW measurements (solid circles) are replotted on a log–log scale in the inset. The horizontal dashed line shows the prediction of classical nucleation theory, using parameters appropriate to $CO_2$. The curved broken line is the theoretical result of Binder and Stauffer (1976).

(Huang et al. 1975), $He^3$ (Dahl and Moldover 1971), LW (Schwartz et al. 1980), and IW (Howland et al. 1980). The data exhibit a strong dependence on $\Delta T_{cx}$ in obvious contradiction to the theoretical prediction. This point is more dramatically made in the inset figure, which shows the LW measurement in a log–log plot; it appears that the supercooling is diverging algebraically at the critical point!

To assess the implications of fig. 11 one must be aware of the conditions under which the experiments were carried out. In every study the sample was supercooled or superheated (LW sample only) into the metastable region in steps or continuously, until nucleation occurred. The temperature interval from $T_{cx}$ to $T_N$ was usually traversed in minutes though in one set of experiments on LW, this "completion time", $t_c$, was several hours.

### 6.3. *The role of critical slowing down*

To explain the apparent failure of nucleation theory, Binder and Stauffer (1976) advanced the following argument. Because $D$ approaches zero at the critical point, nucleating droplets (of radius $R \geqslant R^*$) may grow so slowly that they leave no signature of their presence for a surprisingly long time. Therefore the experimentalist who pauses only seconds or minutes between each temperature step, may falsely conclude that nucleation has not taken place. He will then proceed to a deeper quench before reaching the cloud point. This experimental procedure will lead him to a falsely large value of $\delta T^*$.

Binder and Stauffer therefore suggested the more realistic definition of the supercooling, $\delta T_N$, namely that value of $\delta T$, such that a fraction $f = 1$, say, of the mother phase has been converted to the new phase after an experimental observation time $t_c$. As it turns out $\delta T_N$ depends weakly on $f$, but the dependence on $t_c$ is crucial.

The dashed line in fig. 11 shows the results of Binder and Stauffer, assuming $t_c = 1$ s. This completion time corresponds closely to the conditions under which the L–W experiments were carried out (Schwartz et al. 1980).

The work of Binder and Stauffer has since been followed by a more detailed analysis of the nucleation-growth problem by Langer and Schwartz (1980). They start with a differential equation for the time rate of change of the droplet size distribution, $\nu(R, t)$, assuming it to be the sum of a droplet *birth* term and a *growth* term previously studied by Lifshitz and Slyozov (1961).

At the heart of the Lifshitz–Slyozov theory is the equation,

$$\frac{\mathrm{d}R}{\mathrm{d}t} = \frac{D}{R}\left[\frac{\delta c}{\Delta c} - \frac{2d_0}{R}\right], \tag{6.3-1}$$

together with an equation which assures that solute molecules are conserved. The miscibility gap $\Delta c$ in eq. (6.3-1) is defined in fig. 10; the capillary length $d_0 = \xi/6$ (Langer and Schwartz 1980).

Making certain simplifying assumptions, Langer and Schwartz determine the time dependence of the supersaturation $y(\tau)$, the mean droplet radius $\rho(\tau)$ and the droplet density $n(\tau)$. All of these quantities, including time ($\tau$) are expressed in terms of scaled, dimensionless units. Referring to fig. 10, the scaling constants include the supercooling $\delta c_1$, the temperature difference $\Delta T = \Delta T_{cx} + \delta T$ and various critical amplitudes characterizing the fluid.

This theory permits determination of the initial supersaturation,

$$y(0) \equiv y_1 ,$$

as a function the reduced temperature,

$$\epsilon = \Delta T/T_c = (\Delta T_{cx} + \delta T)/T_c , \tag{6.3-2}$$

assuming that the reaction has gone half way to completion in $t_c$ seconds. (In conventional nucleation experiments, $t_c$ cannot be varied over a wide range.) It is this function $y_1(\epsilon, t_c)$ that should really be compared with experiment rather than $\delta T/\Delta T$ vs $\epsilon$ or $\Delta T_{cx}/T_c$.

To express the supersaturation in system-independent units, one must introduce another reduced temperature

$$\epsilon_c = G\epsilon , \tag{6.3-3}$$

where $G$ is a complicated function of various critical amplitudes. Figure 12 (solid lines) show $y_1(\epsilon_c)$ for the various completion times indicated. As expected, far from the critical point, i.e., when $\epsilon_c$ is large, the supersaturation $y_1$ is small, especially for large completion times.

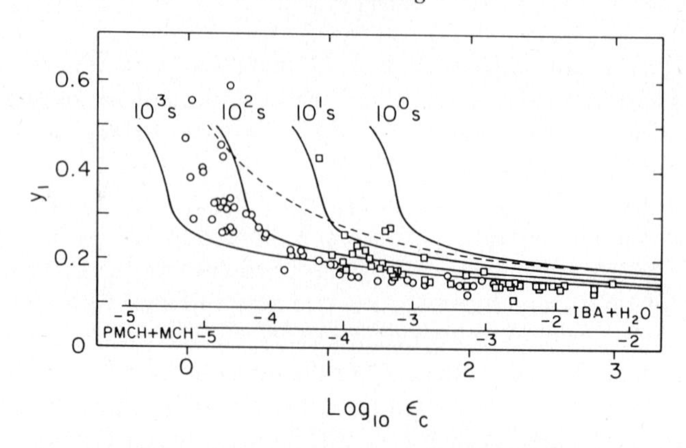


Fig. 12. Initial supersaturation ($y_I$) as a function of scaled temperature at four completion times ($t_c$). The data are those of Howland et al. (1980). The solid lines are the theoretical results of Langer and Schwartz (1980); the dashed curve is the prediction of Binder and Stauffer for $t_c = 1$ s. The measurements were made at $t_c = 200$–$400$ s.

The data in fig. 12 are the cloud point observations of Howland et al. (1980) in the two binary mixtures isobutyric acid and water (squares) and $C_7H_{14}$–$C_7F_{14}$, here labelled PMCH (circles). The horizontal scales designated IBA–$H_2O$ and PMCH–MCH are labelled with the values of $\epsilon$ that correspond to the values of $\epsilon_c$ on the scale below. The dashed line was calculated from a Binder–Stauffer model (1976) with $t_c$ taken to be 1 s.

In the experiments of Howland et al., the two mixtures were continuously supercooled by lowering the bath temperature and by changing the pressure on the mixture. The initial state of the system was one of two phase coexistence. (In another set of experiments on PMCH–MCH in which only one phase was present, greater values of supercooling could be achieved.) The total cooling time $t_c$ was 200–400 s. As theoretically expected the I–W measurements fall between the curves labelled $t_c = 10^3$ s and $10^2$ s.

The above experimental results leave little doubt that excess supercooling near the critical point results from slow growth of the new phase rather than from the failure of nucleation theory itself.

Krishnamurthy and Goldburg (1980) have used a microscope to measure the growth of nucleating droplets in a quenched (superheated) mixtures of 2,6-lutidine and water. Their observations, which we now summarize, strongly suggest that all the anomalous supercooling experiments discussed above, did not test nucleation theory itself; rather the experiments reveal information about the growth of droplets after they have already been formed.

Each of their samples consisted of a single water-rich phase of L–W. The experiment itself consisted of measuring the radius of the individual droplets as they slowly formed after a quench of 5 to 97 mK. Since the relevant critical parameters of LW are known, the reduced variables $y(\tau)$, $n(\tau)$ and

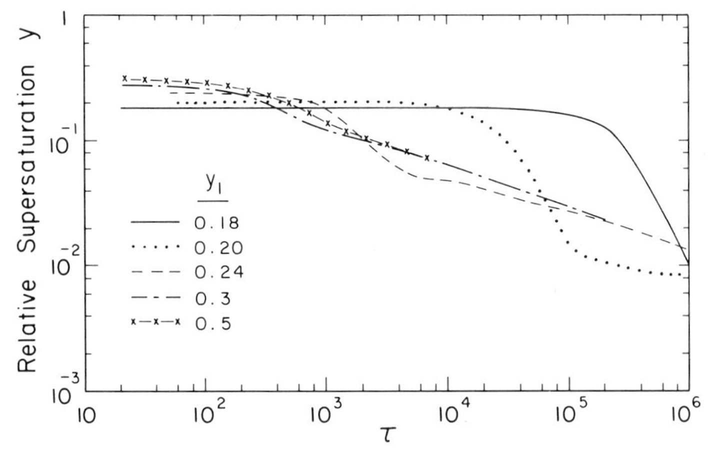


Fig. 13. Reduced supersaturation ($y$) versus reduced time at various values of initial saturation, $y = y_{\mathrm{I}}$. The calculations are by Langer and Schwartz (1980).

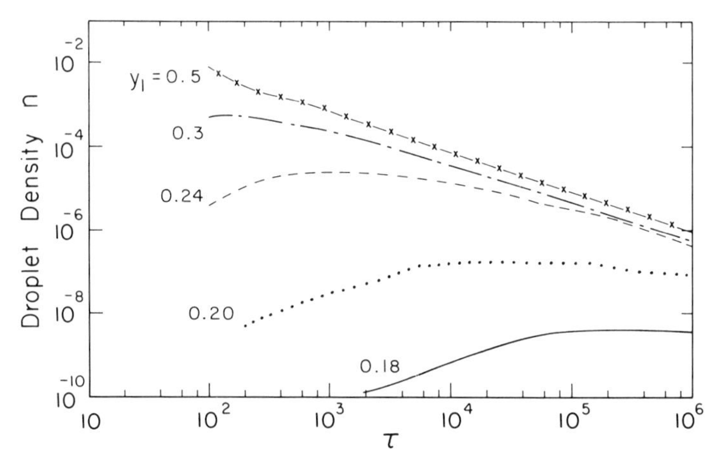


Fig. 14. Calculated reduced droplet density versus reduced time at various initial supersaturations (from Langer and Schwartz (1980)).

$\rho(\tau)$ could be obtained from the measured droplet distribution together with the usual assumption that the droplets were of equilibrium composition, i.e., $c = c_{\mathrm{B}}$ in fig. 10.

Figures 13 and 14, taken from Langer and Schwartz show $y(\tau)$ and $n(\tau)$, respectively, for various initial supersaturations, $y_{\mathrm{I}}$. Observe that when the initial supersaturation is small, e.g., $y_{\mathrm{I}} = 0.18$, many decades of time elapse before the supersaturation $y(\tau)$ is appreciably depleted and before the droplet density becomes large. During the interval in which $n(\tau)$ is increasing, nucleation dominates, but at subsequent times the behavior of the system is dictated by diffusive growth and simultaneous depletion of the supersaturation. The latter phenomenon is accompanied by an increase in $R^*$ so that droplets of radius $R < R^*$ dissolve. The net result is a fall in $n(\tau)$, accompanied by an increase in the average size, $\rho$.

According to fig. 14, the droplet density at the early (nucleation) stage of phase separation is a very strong function of initial quench depth, $y_1$. This manifested itself in the experiments as a lack of reproducibility of the initially observed droplet density for a fixed quench depth, $\delta T$ (or $y_1$). To eliminate this experimental uncertainty would have required an unattainably high temperature stability and quench depth reproducibility. Therefore it was not possible to check the Langer–Schwartz theory in full, so an alternative strategy was pursued; the measured droplet density at small $t$ was taken as the input information for a numerical integration of the Lifshitz–Slyozov Equation (6.3-1) (with the constraint that solute is conserved). The integration was started at a droplet size large enough to assure that only the growth mechanism was operative.

Figure 15 shows the measurements of $\rho(\tau)$ at three values of $\Delta T_{cx}$ (i.e. at three

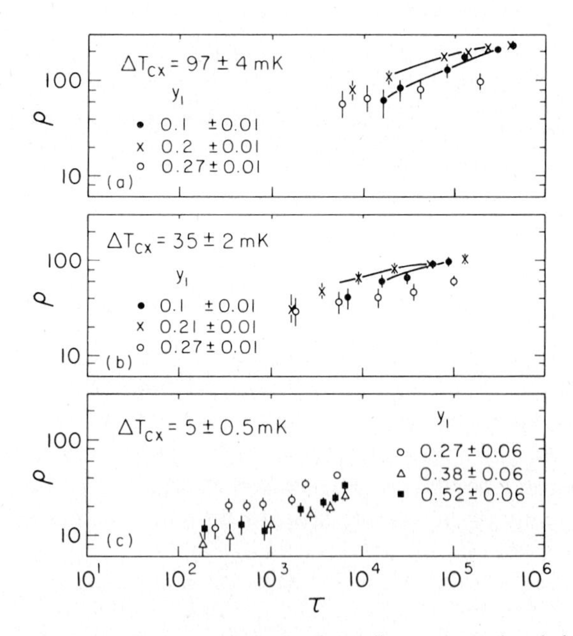


Fig. 15. Reduced supersaturation and mean droplet radius versus reduced time obtained from microphotographic measurements of droplet growth. The solid lines come from numerically integrating the Lifshitz–Slyozov equations for diffusive droplet growth (from Krishnamurthy and Goldburg 1980).

sample compositions) and at three values of $y_1$ for each sample of composition. The solid lines represent the numerical integration of eq. (6.3-1) for the two sets of measurements at $y_1 = 0.1$ and 0.21. The data corresponding to the deepest quench ($y_1 = 0.27$), were not analyzed, because coalescence of the droplets intervened at relatively small values of $T_{cx}$. Agreement between experiment and the Lifshitz–Slyozov droplet growth theory (solid lines) is seen to be satisfactory.

The origin of the apparent failure of nucleation theory becomes evident if one translates the dimensionless units of fig. 15 into dimensional ones: At the quench depth, $y_1 = 0.1$, $\delta T/\Delta T_{cx}$ in fig. 11 is 0.11. Even though droplets will ultimately emerge at this small supercooling, they are not observable with a microscope until 300 s after the quench. What is more, even one hour later ($t = 2800$ $\tau = 10^5$) the droplets occupy only 1.6‰ of the sample volume! (At $t = 300$ s, this volume fraction is 0.13‰.) It is no wonder, then, that an experimenter who quenches to this depth and pauses only a few seconds there, concludes that no nucleation has occurred and proceeds to a deeper quench in order to observe cloudiness in the fluid or appreciable light scattering.

When the droplets first became visible in the experiments of Krishnamurthy et al., their radii were calculated to be much larger than the critical value $R^*$. From this observation together with a comparison of the measured and calculated values of $n(\tau)$, $y(\tau)$ and $\rho(\tau)$, one concludes that these experiments explored the growth regime only. In retrospect the same can be said of all the measurements in fig. 11.

The study of the droplet birth process near the critical point, will require the observation of droplets whose size is 1 μm or smaller and whose volume density is very small. The solution of this problem falls to the next generation of nucleation experiments.

## *References*

Abraham, F.F., 1979, Phys. Rep. **53**, 93.
Aharony, A. and P.C. Hohenberg, 1976, Phys. Rev. B **13**, 3081.
Aharony, A. and G. Ahlers, 1980, Phys. Rev. Lett. **44**, 782.
Ahlers, G., 1973, Phys. Rev. A **7**, 2145.
Balfour, F.W., J.V. Sengers, M.R. Moldover and J.M.H. Levelt Sengers, 1978, Phys. Lett. **65** A, 223.
Bausch, R., 1972, Z. Phys. **254**, 81.
Berne, B.J. and R. Pecora, 1976, Dynamic Light Scattering (Wiley, New York).
Bervillier, C., 1976, Phys. Rev. B **14**, 4964.
Bervillier, C. and C. Godreche, 1980, Phys. Rev. B **21**, 5427.
Beysens, D. and G. Zalczer, 1977, Phys. Rev. **A 15**, 765.
Beysens, D., M. Gbadamassi and L. Boyer, 1979a, J. de Phys. Lett. **40**, L-623.
Beysens, D., M. Gbadamassi and L. Boyer, 1979b, Phys. Rev. Lett. **43**, 1253.
Beysens, D., R. Tufeu and Y. Garrabos, 1979c, J. de Phys. Lett. **40**, 623.
Billotet, C. and K. Binder, 1979, Z. Phys. **B 32**, 195.
Billotet, C. and K. Binder, 1980, Physica, **103A**, 99.

Binder, K., 1974, Z. Phys. **267**, 313.
Binder, K., 1977, Phys. Rev. **B 15**, 4425.
Binder, K. and D. Stauffer, 1974, Phys. Rev. Lett. **33**, 1006.
Binder, K. and D. Stauffer, 1976, Adv. Phys. **25**, 343.
Binder, K., C. Billotet and P. Mirold, 1978, Z. Phys. **B 30**, 183.
Bortz, A.B., M.H. Kalos, J.L. Lebowitz and J.L. Zendejas, 1974, Phys. Rev. B **10**, 535.
Bray, A.J., 1976a, Phys. Rev. Lett. **36**, 285.
Bray, A.J., 1976b, Phys. Rev. **B 14**, 1248.
Brezin, E., J.C. Le Guillou and J. Zinn-Justin, 1976, Field Theoretic Approach to Critical Phenomena in Phase Transitions and Critical Phenomena, eds., C. Domb and M.S. Green (Academic, London) Vol. 6.
Burstyn, H.C. and J.V. Sengers, 1980, Phys. Rev. Lett. **45**, 259.
Burstyn, H.C., R.F. Chang and J.V. Sengers, 1980a, Phys. Rev. Lett. **44**, 410.
Burstyn, H.C., J.V. Sengers and P. Esfandiari, 1980b, Phys. Rev. **A 22**, 282.
Cahn, J.W., 1968, Trans. Metall. Soc. AIME **242**, 166.
Chan, J.W. and Moldover, M.R., 1977 (private communication).
Cannell, D.S., 1975, Phys. Rev. **A 12**, 225.
Cannell, D., 1977a, Phys. Rev. **A 16**, 431.
Cannell, D., 1977b, Phys. Rev. **A 15**, 735.
Chang, M-c. and A. Haughton, 1980, Phys. Rev. Lett. **44**, 785.
Chang, R.F., P.H. Keyes, J.V. Sengers and C.O. Alley, 1971, Phys. Rev. Lett. **27**, 1706.
Chang, R.F., H. Burstyn and J.V. Sengers, 1979, Phys. Rev. **A 19**, 866.
Chou, Y.C. and W.I. Goldburg, 1979, Phys. Rev. **A 20**, 2105.
Chou, Y.C. and W.I. Goldburg, 1980, Phys. Rev. **A 23**, 858.
Chu, B., 1974, Laser Light Scattering (Academic, New York).
Chu, B. and F.L. Lin, 1974, J. Chem. Phys. **61**, 5132.
Cohen, C., J.W.H. Sutherland and J.M. Deutch, 1971, Phys. Chem. Liq., **2**, 213.
Cook, H.E., 1970, Acta Met. **18**, 297.
Cummins, H.Z., 1971, Light Scattering Spectroscopy of Critical Phenomena, in: Proc. Sch. Phys. Enrico Fermi, Course LI, Varenna, 1970, ed., M.S. Green (Academic, New York) p. 381.
Cummins, H.Z. and H.L. Swinney, 1970, Light Beating Spectroscopy in: Prog. in Optics, Vol. 8, ed., E. Wolf (North-Holland, Amsterdam) p. 133.
Cummins, H.Z. and E.R. Pike, eds., 1974, Photon Correlation and Light Beating Spectroscopy (Plenum, New York).
Dahl, D. and M.R. Moldover, 1971, Phys. Rev. Lett. **27**, 1421.
Ferrell, R.A., 1970, Phys. Rev. Lett. **24**, 1169.
Ferrell, R.A. and J.K. Battacharjee, 1979, Phys. Rev. A **19**, 348.
Fisher, M.E., 1964, J. Math. Phys. **5**, 944.
Fisher, M.E. and J.S. Langer, 1968, Phys. Rev. Lett. **20**, 665.
Flinn, P.A., 1974, J. Stat. Phys. **10**, 89.
Frenkel, J., 1955, Kinetic Theory of Liquids (Dover, New York) Ch. VII.
Furukawa, H., 1978, Prog. Theor. Phys. **59**, 1072.
Furukawa, H., 1979, Phys. Rev. Lett. **43**, 136.
Garisto, F. and R. Kapral, 1976, Phys. Rev. **A 14**, 884.
Giglio, M. and A. Vendrimini, 1975a, Phys. Rev. Lett. **34**, 561.
Giglio, M. and A. Vendrimini, 1975b, Phys. Rev. Lett. **35**, 168.
Giglio, M. and A. Vendrimini, 1977, Phys. Rev. Lett. **38**, 26.
Giglio, M. and A. Vendrimini, 1978, J. Chem. Phys. **68**, 2016.
Goldburg, W.I. and Huang, J.S., 1975, Phase Separation Experiments Near the Critical Point, in: Fluctuations, Instabilities, and Phase Transitions, Geilo, 1975, ed., T. Riste (Plenum, New York).

Goldburg, W.I., C-H. Shaw, J.S. Huang and M.S. Pilant, 1978a, J. Chem. Phys. **68**, 484.
Goldburg, W.I., A.J. Schwartz and M.W. Kim, 1978b, Prog. Theor. Phys. Suppl. No. 64, 477.
Gollub, J.P., A.A. Koenig and J.S. Huang, 1976, J. Chem. Phys. **65**, 639.
Greer, S.C., T.E. Block and C.M. Knobler, 1975, Phys. Rev. Lett. **34**, 250.
Greer, S.C., 1978, Acc. Chem. Res. **11**, 427.
Greywall, D.S. and G. Ahlers, 1972, Phys. Rev. Lett. **28**, 1251.
Griffiths, R.B. and J.C. Wheeler, 1970, Phys. Rev. A **2**, 1047.
Griffiths, R.B., 1974, J. Chem. Phys. **60**, 195.
Gulari, E., A.F. Collings, R.L. Schmidt and C.J. Pings, 1972, J. Chem. Phys. **56**, 6169.
Guttinger, H. and D. Cannell, 1980, Phys. Rev. **A 22**, 285.
Halperin, B.I. and P.C. Hohenberg, 1974, Phys. Rev. Lett. **32**, 1289.
Hayes, C.E. and H.Y. Carr, 1977, Phys. Rev. Lett. **39**, 1558.
Heady, R.B. and J.W. Cahn, 1973, J. Chem. Phys. **58**, 896.
Hocken, R. and M.R. Moldover, 1976, Phys. Rev. Lett. **37**, 29.
Hoffer, J.K., L.J. Campbell and R.J. Bartlett, 1980, Phys. Rev. Lett. **45**, 912.
Hohenberg, P.C., 1978, Critical Phenomena in Fluids, in: Microscopic Structure and Dynamics of Liquids (Plenum, New York) p. 333.
Hohenberg, P.C. and M. Barmatz, 1972, Phys. Rev. **A 6**, 289.
Hohenberg, P.C. and B.I. Halperin, 1977, Rev. Mod. Phys. **49**, 435.
Hohenberg, P.C. and D.R. Nelson, 1980, Phys. Rev. **B 20**, 2665.
Hohenberg, P.C., A. Aharony, B.I. Halperin and E.D. Siggia, 1976, Phys. Rev. B **13**, 2986.
Howland, R.G., N-C.W. Wong and C.M. Knobler, 1980, J. Chem. Phys. **73**, 522.
Huang, J.S., S. Vernon and N.C. Wong, 1974, Phys. Rev. Lett. **33**, 140.
Huang, J.S., W.I. Goldburg and M.R. Moldover, 1975, Phys. Rev. Lett. **34**, 639.
Jacobs, D.T., D.J. Anthony, R.C. Mockler and W.J. O'Sullivan, 1977, Chem. Phys. **20**, 219.
Jakeman, E., 1974, Photon Correlation, in: Photon Correlation and Light Beating Spectroscopy, eds., H.Z. Cummins, and F.R. Pike (Plenum, New York) p. 75.
Jasnow, D. and W.I. Goldburg, 1972, Phys. Rev. **A 6**, 2492.
Kadanoff, L.P., 1966, Physics **2**, 263.
Kadanoff, L.P. and J. Swift, 1968, Phys. Rev. **166**, 89.
Kaufman, M., K.K. Bardham and R.B. Griffiths, 1980, Phys. Rev. Lett. **44**, 77.
Kawasaki, K., 1966a, Phys. Rev. **145**, 224.
Kawasaki, K., 1966b, Phys. Rev. **148**, 375.
Kawasaki, K., 1966c, Phys. Rev. **50**, 285.
Kawasaki, K., 1970, Ann. Phys. (New York) **61**, 1.
Kawasaki, K., 1977, Prog. Theor. Phys. **57**, 826.
Kawasaki, K. and S.M. Lo, 1972, Phys. Rev. Lett. **29**, 48.
Kawasaki, K. and T. Ohta, 1978, Prog. Theor. Phys. **59**, 362.
Kim, M.W., A.J. Schwartz and W.I. Goldburg, 1978, Phys. Rev. Lett. **41**, 657.
Kim, M.W., W.I. Goldburg, P. Esfandiari and J.M.H. Levelt Sengers, 1979, J. Chem. Phys. **71**, 4888.
Kim, M.W., W.I. Goldburg, P. Esfandiari, J.M.H. Levelt Sengers and E.-S. Wu, 1980, Phys. Rev. Lett. **44**, 80.
Kittel, C., 1966, Introduction to Solid State Physics (John Wiley, New York) 3rd ed., Ch. 12.
Krishnamurthy, S. and W.I. Goldburg, 1980, Phys. Rev. **A 22**, 2147.
Kumar, A., E.S.R. Gopal and H.R. Krishnamurthy, 1980, (unpublished).
Kwon, O., D.M. Kim and R. Kobayashi, 1977, J. Chem. Phys. **66**, 4925.
Landau, L.D. and E.M. Lifshitz, 1958, Statistical Physics (Addison Wesley, Reading MA) Ch. XII.
Landau, L.D. and I.M. Lifshitz, 1959, Fluid Mechanics (Addison Wesley, Reading, MA) Ch. VI.
Lang, J.C., Jr. and B. Widom, 1975, Physica, **81** A, 190.

Langer, J.S., 1973, Acta Met. **21**, 1649.
Langer, J.S., 1975, Spinodal Decomposition, in: Fluctuations, Instabilities, and Phase Transitions, Geilo, 1975, ed., R. Riste (Plenum, New York).
Langer, J.S. and A.J. Schwartz, 1980, Phys. Rev. **A 21**, 948 (1980).
Langer, J.S. and L.A. Turski, 1973, Phys. Rev. **A 8**, 3230.
Langer, J.S., M. Bar-on and Harold D. Miller, 1975, Phys. Rev. A **11**, 1417.
LeGuillou, J.C. and J. Zinn-Justin, 1980, Phys. Rev. **B 21**, 3976.
Leiderer, P., D.R. Nelson, D.R. Watts and W.W. Webb, 1975, Phys. Rev. Lett. **34**, 1080.
Leung, H.K. and B.N. Miller, 1975, Phys. Rev. **A 12**, 2162.
Leung, H.K. and B.N. Miller, 1977, Phys. Rev. **A 16**, 435.
Ley-Koo, M. and M.S. Green, 1977, Phys. Rev. **A 16**, 2483.
Lifshitz, I.M. and V.V. Slyozov, 1961, J. Phys. Chem. Solids, **19**, 35.
Lo, S.M. and K. Kawasaki, 1973, Phys. Rev. **A 8**, 2176.
Lunacek, J.H. and D.S. Cannell, 1971, Phys. Rev. Lett. **27**, 841.
Marro, J., A.B. Bortz, M.H. Kalos and J.L. Lebowitz, 1975, Phys. Rev. **B 12**, 2000.
Marro, J., J.L. Lebowitz and M.H. Kalos, 1979, Phys. Rev. Lett. **43**, 282.
Martin, P.C., 1968, in: Many Body Physics, eds.; C. De Witt and R. Balian (Gordon and Breach, New York).
Moldover, M.R., J.V. Sengers, R.W. Gammon and R.J. Hocken, 1979, Rev. Mod. Phys. **51**, 79.
Mountain, R.D. and J.M. Deutch, 1969, J. Chem. Phys. **50**, 1103.
Nagarajan, N., A. Kumar, E.S.R. Gopal and S.C. Green, 1980, J. Phys. Chem. **84**, 2883.
Ohta, T., 1977, J. Phys. **C 10**, 791.
Ohta, T., 1980 Prog. Theor. Phys. **64**, 536.
Ohta, T. and K. Kawasaki, 1976, Prog. Theor. Phys. **55**, 1384.
Onuki, A. and K. Kawasaki, 1979, Ann. Phys. (New York) **121**, 456.
Oxtoby, D.W. and W.H. Gelbart, 1974, J. Chem. Phys. **61**, 2957.
Perl, R. and R.A. Ferrell, 1972, Phys. Rev. **A 6**, 2358.
Pittman, C., T. Doiron and H. Meyer, 1979, Phys. Rev. **B 20**, 3678.
Puglielli, V.G. and N.C. Ford Jr., 1970, Phys. Rev. Lett. **25**, 143.
Rao, M., M.H. Kalos, J.L. Lebowitz and J. Marro, 1976, Phys. Rev. **B 13**, 7325.
Riedel, E.K. and F.J. Wegner, 1972, Phys. Rev. Lett. **29**, 349.
Riste, T. (ed.), 1975, Fluctuations, Instabilities, and Phase Transitions, Geilo, 1975 (Plenum, New York).
Schroeter, J.D., D.M. Kim and R. Kobayashi, 1979, Phys. Rev. **A 19**, 2402.
Schwartz, A.J., 1979, (unpublished).
Schwartz, A.J., S. Krishnamurthy and W.I. Goldburg, 1980, Phys. Rev. **A 21**, 1331.
Sengers, J.M.H.L. and J.V. Sengers, 1980, How close is "close to the critical point"? in: Perspectives in Statistical Physics, ed., H.J. Raveche (North-Holland, Amsterdam) Ch. 14.
Sengers, J.V. and P.H. Keyes, 1971, Phys. Rev. Lett. **26**, 70.
Sengers, J.V. and M.R. Moldover, 1978, Phys. Lett. **66 A**, 44.
Sengers, J.V. and J.M.H. Levelt Sengers, 1978, Critical Phenomena in Classical Fluids, in: Progress in Liquid Physics, ed., C.A. Croxton (Wiley, Chichester, U.K.) p. 103.
Sengers, J.S., 1980, Lecture at M.S. Green Symposium, Washington, D.C. (unpublished).
Siggia, E.D., 1976, Phys. Rev. B **13**, 3218.
Siggia, E.D., 1979, Phys. Rev. A **20**, 595.
Sorensen, C.M., B.J. Ackerson, R.C. Mockler and W.J. O'Sullivan, 1976, Phys. Rev. A **13**, 1593.
Sorenson, C.M., R.C. Mockler and W.J. O'Sullivan, 1977, Phys. Rev. **A 16**, 365.
Sorenson, C.M., R.C. Mockler and W.J. O'Sullivan, 1978, Phys. Rev. Lett. **40**, 777.
Spittorf, O. and B.N. Miller, 1974, Phys. Rev. **A 9**, 550.
Stanley, H.E., 1971, Introduction to Phase Transitions and Critical Phenomena (Oxford, New York).
Stauffer, D., M. Ferrer and M. Wortis, 1972, Phys. Rev. Lett. **29**, 345.

Stein, A., S.J. Davidson, J.C. Allegra and G.F. Allen, 1972, J. Chem. Phys. **56**, 6164.
Stephen, M.J., 1975, Phys. Rev. **B 12**, 1015.
Stephen, M.J., E. Abraham and J.P. Straley, 1975, Phys. Rev. **B 12**, 256.
Sundquist, B.E. and R.A. Oriani, 1962, J. Chem. Phys. **36**, 2604.
Sur, A., J.L. Lebowitz, J. Marro and M.L. Kalos, 1977, Phys. Rev. **B 15**, 535.
Swinney, H.L. and D.L. Henry, 1973, Phys. Rev. **A 8**, 2586.
Thoen, T., E. Bloemen and W. Van Dael, 1978, J. Chem. Phys. **68**, 735.
Trappeniers, N.J., R.H. Huijser and A.C. Michels, 1977, Chem. Phys. Lett. **48**, 31.
Wegner, F., 1972, Phys. Rev. **B 5**, 4529.
Widom, B., 1965a, J. Chem. Phys. **43**, 3892.
Widom, B., 1965b, J. Chem. Phys. **43**, 3898.
Wilcox, L.R. and D. Balzarini, 1968, J. Chem. Phys. **48**, 753.
Wilcox, L.R. and W.T. Estler, 1971, J. Phys. (Paris), Colloq. **32**, C5-A-175.
Wong, N-C. and C.M. Knobler, 1977, J. Chem. Phys. **66**, 4707.
Wong, N-C. and C.M. Knobler, 1978, J. Chem. Phys. **69**, 725.
Wong, N-C. and C.M. Knobler, 1979, Phys. Rev. Lett. **43**, 1733.
Wu, E.-S., 1978, Phys. Rev. **A 18**, 1641.
Wilson, K.G. and J. Kogut, 1974, Phys. Rep. **12** C, 75.
Zettlemoyer, A.C., ed., 1969, Nucleation (M. Dekker, New York).

CHAPTER 10

# Scattering Spectroscopy of Liquid Crystals

J.D. LITSTER

*Center for Materials Science and Engineering and*
*Department of Physics*
*Massachusetts Institute of Technology*
*Cambridge, MA 02139*
*USA*

*Light Scattering near Phase Transitions*
*Edited by*
*H.Z. Cummins and A.P. Levanyuk*

© *North-Holland Publishing Company, 1983*

# Contents

## *1. Introduction*

One of the most interesting research areas in physics today is the study of which condensed phases can exist in matter, what are their properties, and an explanation for these properties on as fundamental basis as possible. One of the fundamentally important concepts in statistical mechanics is that of spontaneously broken symmetry and the concomitant appearance of new hydrodynamic (Goldstone) modes; thermally excited fluctuations of these modes, the symmetry of ordered and disordered phases, and the dimensionality of space determine which ordered phases can and cannot exist. For second-order or weakly first-order phase changes we now believe that if the space dimensionality is greater than an upper critical value, fluctuations do not play a significant role and one can calculate correctly using the mean field approximation. (For most phase changes in most isotropic materials the upper critical dimension is four.) In lower dimensional spaces the fluctuations modify the observed behavior and the development of renormalization group techniques was an important breakthrough to enable calculations of the properties of materials near critical points in three dimensions. There is also a lower critical dimension for most phases, when thermal fluctuations become so strong as to prevent the establishment of order which the interactions between the atoms or molecules would otherwise favor; for example we believe that it is not possible for solids to exist in two or fewer dimensions. These concepts have come as the result of studies of fairly simple phase changes, such as fluids near their critical point, consolute mixtures, and simple magnetic transitions. In this chapter, I shall discuss how they have been helpful to understand more complex phase changes, such as those which occur in liquid crystals, and how the study of the rich variety of phase changes that occur in liquid crystals can be used to extend the ideas to understand more complex phases.

To discuss the condensed phases of matter, it is useful to introduce a density function $\rho(\boldsymbol{r})$ defined so $\rho(\boldsymbol{r})\,\mathrm{d}\boldsymbol{r}$ gives the probability of finding an atom in volume $\mathrm{d}\boldsymbol{r}$. In liquids $\rho(\boldsymbol{r})$ is a constant and therefore has continuous translational symmetry as well as complete isotropy; it also has inversion symmetry if the liquid molecules have no stereo-isomers or if it is a racemic mixture. In crystalline solids $\rho(\boldsymbol{r})$ has the translational periodicity of the lattice as well as rotational anisotropy. Landau (1937) pointed out that $\rho(\boldsymbol{r})$ could be anisotropic in materials which possessed continuous translational symmetry, and he felt that liquid crystals (which had been known since 1890) could be characterized in this way.

There are two ways for $\rho(\boldsymbol{r})$ to be anisotropic and yet have continuous translational symmetry. The first way is observed in the nematic phase of liquid crystals; anisotropic molecules are orientationally ordered but with no positional ordering. The second way is more subtle. Imagine a crystal in which the positional ordering of the atoms on the lattice has been lost but the underlying orientational anisotropy of the lattice remains. This means that an imaginary line joining the centers of two neighboring molecules will have long-range orientational order even though there is no positional ordering of the molecules. This is commonly called "bond orientational" order, although I emphasize that no physical bonds are involved.

## 2. The phases of thermotropic liquid crystals

Thermotropic liquid crystal phases (de Gennes 1974) occur on changing the temperature of many pure organic compounds; we believe that the ideas of molecular orientational order, bond orientational order, and molecular positional order can serve as a basis for describing the variety of thermotropic liquid crystal phases that are known. There are two phases which have only molecular orientational order. The *nematic*, mentioned above, derives its name from the Greek word for thread since defects in the nematic state appear like threads running through the sample. Nematics are optically uniaxial and the ordered state has quadrupolar symmetry; no ferroelectric nematics are known. The optic axis is usually represented by a unit vector $\hat{\boldsymbol{n}}$ called the director and the Goldstone modes (analogous to spin waves in a ferromagnet) are fluctuations in the director. Many chiral molecules (such as esters of cholesterol) form phases similar to nematics except that the optic axis twists slowly (with a period of about $0.5\,\mu$m typically); this phase is commonly called *cholesteric*. The refractive index of cholesterics is therefore modulated at just the right period to scatter visible light, and these materials are attractively colorful. A racemic mixture of cholesteric liquid crystals forms a nematic phase.

All other thermotropic liquid crystal phases are called smectics, from the Greek word for soap, since many soaps have smectic phases when dissolved in water. Smectic phases have some positional order of the molecules and are commonly referred to as "layered" phases. They have been alphabetically labelled approximately in order of their discovery, and phases smectic A through smectic I are known. Not all are truly liquid crystals and the nature of ordering in some of them is not well known. In this chapter, I shall concentrate on the smectic phases A and C. Illustrated in fig. 1, these phases have nematic order combined with a density wave. If $\hat{\boldsymbol{n}}$ lies along $\hat{z}$, then in a smectic A phase one can write the density (De Gennes 1972a) as

$$\rho = \rho_0[1 + \Psi \exp(iq_0 z)]\,. \tag{1}$$

Here $q_0$ is the wave vector of the density wave ($2\pi/q_0$ is the layer spacing) and $\Psi = |\Psi|\,e^{i\phi}$ is the smectic A order parameter. Experimentally the higher harmonics of the density wave are extremely weak, so the layers are not so well defined as fig. 1 suggests. The smectic C phase is like a smectic A, except that the density wave is tilted with respect to the director. A more detailed discussion, as needed, will be given when presenting results of experimental studies of these phases.

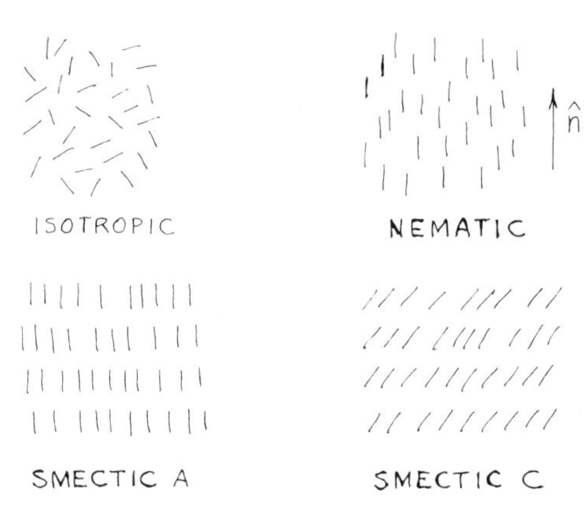


Fig. 1. A sketch of the molecular arrangement in isotropic, nematic, smectic A, and smectic C phases of thermotropic liquid crystals.

## 3. *Interaction of electromagnetic radiation with liquid crystals*

The interaction of radiation with matter is given by the term $(\boldsymbol{p} - (e/c)\boldsymbol{A})^2$ in the Hamiltonian; the precise form it takes depends upon the frequency of the radiation compared to the absorption frequencies of the atom. Far away from atomic resonances, as for X-rays away from absorption edges, the term proportional to $A^2$ dominates and we get Thomson scattering from the individual electrons in the molecules. Thus X-ray scattering will give us information on the electron, or equivalently, the mass distribution in the liquid crystal. The energy of the X-ray photons, typically 8 keV, is so much greater than the energy of thermal fluctuations that one always integrates over the energy spectrum of the scattered radiation; this means that only the equal time correlation function $\langle \rho(\boldsymbol{r}, t)\rho(\boldsymbol{r}, t)\rangle$ can be measured. For visible light, the term

$\boldsymbol{p} \cdot \boldsymbol{A}$ taken to second order dominates the scattering. This is Rayleigh scattering and the scattered light gives information on fluctuations in the dielectric constant tensor. Energy changes of $10^{-13}$ eV are readily detected so light scattering permits us to study the dynamics of thermal fluctuations as well as the statics.

The most intense light scattering in a liquid crystal comes from the nematic phase Goldstone modes or director fluctuations; since they involve reorientation of the optic axis, they cause a change in the polarization of the scattered light. For convenience, let us take $\hat{\boldsymbol{n}}$ to lie along $\hat{z}$. Then the director modes scatter light by fluctuations $\delta\epsilon_{xz}$ and $\delta\epsilon_{yz}$ in the dielectric constant. If we take the wave vector $\boldsymbol{q}$ for the fluctuation to be in the $xz$-plane (this may be done without loss of generality as $\hat{z}$ is an axis of symmetry) the two Goldstone modes of a nematic are $\delta n_1 = \delta n_x$ and $\delta n_2 = \delta n_y$. The corresponding mean squared fluctuations in $\epsilon$ are

$$\langle \delta\epsilon_{xz}^2(\boldsymbol{q}) \rangle = \epsilon_a^2 \langle \delta n_1^2(\boldsymbol{q}) \rangle = \frac{\epsilon_a^2 kT}{K_1 q_x^2 + K_3 q_3^2}, \tag{2a}$$

and

$$\langle \delta\epsilon_{yz}^2(\boldsymbol{q}) \rangle = \epsilon_a^2 \langle \delta n_2^2(\boldsymbol{q}) \rangle = \frac{\epsilon_a^2 kT}{K_2 q_x^2 + K_3 q_z^2}. \tag{2b}$$

Here $K_1$, $K_2$, and $K_3$ are phenomenological elastic constants (De Gennes 1974) for splay, twist, and bend distortions in the director field, respectively. The difference in dielectric constant for light polarized along, $\epsilon_\parallel$, and transverse, $\epsilon_\perp$, to $\hat{\boldsymbol{n}}$ is given by $\epsilon_a = \epsilon_\parallel - \epsilon_\perp$; this quantity is proportional to the orientational order parameter of the nematic phase. The elastic constants $K_i$ are of order $10^{-6}$ dynes and $\epsilon_a$ is typically 0.5; these combine to give rather large fluctuations $\delta\epsilon$ and intense scattering of light. The mean free path of a photon is about 1 mm in most nematics before scattering by director fluctuations.

## 4. *Light scattering from the isotropic phase*

Fluctuations of the nematic order parameter cause fluctuations in the magnitude of the birefringence and also scatter light. They are completely masked by the much more intense director fluctuations in the ordered phase, but fluctuations in short-range order are readily observable in the isotropic phase of nematics. They were thoroughly studied some years ago (Stinson 1972) and I shall briefly summarize the results. The nematic order parameter is

$$S = \tfrac{1}{2} \langle 3\cos^2\theta - 1 \rangle, \tag{3}$$

where the average is over a small but macroscopic volume and $\theta$ is the angle between the long molecular axis and the local optic axis. In the isotropic phase there is no preferred orientation and fluctuations $\delta S$ cause fluctuations in both diagonal and off-diagonal components of $\epsilon$; the polarized-to-depolarized intensity ratio is 4 : 3. The quadrupolar symmetry of $S$ leads to a cubic invariant in a Landau expansion of the free energy near the nematic–isotropic transition (De Gennes 1969) and so the transition must be weakly first order. The intensity of light scattered by $\langle \delta S^2(\boldsymbol{q}) \rangle$ measures the susceptibility for $S$ and on cooling towards the nematic phase shows a divergence which is cut off by the first-order phase change; from the spectrum of the scattered light one also observes (Stinson 1972) a critical slowing down of the fluctuations. The short-range order is spatially correlated: $\langle \delta S(0) \delta S(r) \rangle \sim \mathrm{e}^{-r/\xi}$ and the correlation length $\xi$ can be measured from the $\boldsymbol{q}$ dependence of the intensity (Stinson 1973); it diverges from about 20 Å to 200 Å before the phase change intervenes. Light scattering was a powerful tool to elucidate the nature of the nematic–isotropic transition; it appears to pose no fundamental unsolved questions and is not much studied today.

## *5. The smectic A phase and the smectic A to nematic transition*

### *5.1. Properties of the smectic A phase*

The simplest smectic phase is the A, whose order parameter was introduced and discussed in sect. 2. The phase of the order parameter may be written $\phi = q_0 u$ where $u(\boldsymbol{r})$ is the displacement along $\hat{z}$ of the smectic layer from its equilibrium

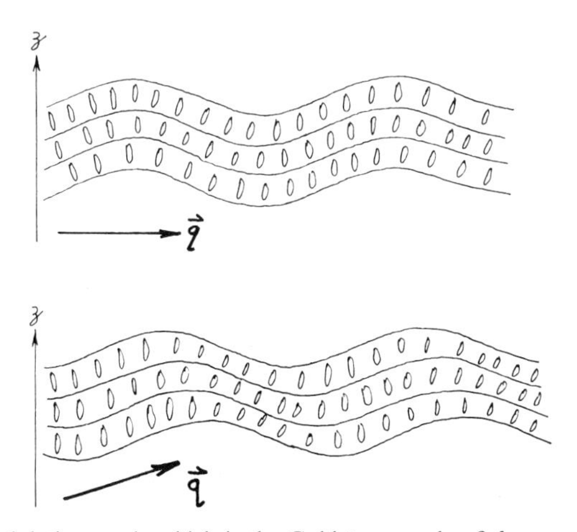

Fig. 2. The layer undulation mode which is the Goldstone mode of the smectic A phase. Layer compression occurs when $q_z \neq 0$, but only bending or molecular splay if $q_z = 0$.

position. The Goldstone mode of the smectic A phase consists of fluctuations in $u(\boldsymbol{r})$ or, equivalently, the phase of the order parameter. The elastic properties are highly anisotropic for reasons which are illustrated in fig. 2. The nature of the mode is sketched in the upper half of the figure when $\boldsymbol{q}$ is accurately in the plane of the smectic layers, and in the lower half with a component along $\hat{z}$. In the latter case, it can be seen that the smectic layers are compressed and expanded along the $\hat{z}$ direction. As for a solid, this gives a contribution $\frac{1}{2}Bq_z^2$ to the elastic free energy, where $B$ is a compressibility. Typical values of $B$ are $10^8$ erg cm$^{-3}$. When $q_z = 0$, there is no compression of the layers and the only resistance to the distortion arises from curvature of the layers. Since the molecules remain normal to the layers, this corresponds to a splay distortion of the director field $\hat{\boldsymbol{n}}$ and the elastic free energy is $\frac{1}{2}K_1q_x^4$. Applying the equipartition theorem to the smectic A Goldstone mode, one calculates

$$\langle u^2(\boldsymbol{q})\rangle = kT/(Bq_z^2 + K_1q_\perp^4)\,. \tag{4}$$

Because of its coupling to $\hat{\boldsymbol{n}}$, this mode can be studied by light scattering. It can be readily shown (Brochard 1973) that in the smectic A phase, eq. (2a) becomes

$$\langle\delta\epsilon_{xz}^2(\boldsymbol{q})\rangle = \frac{\epsilon_a^2kT[1 + (B/D)(q_z/q_x)^2]}{K_1q_x^2 + K_3q_z^2 + (B/D)(D + K_1q_x^2 + K_3q_z^2)(q_z/q_x)^2} \tag{5a}$$

$$\simeq \epsilon_a^2kTq_x^2/(K_1q_x^4 + Bq_z^2)\,, \tag{5b}$$

and eq. (2b) becomes

$$\langle\delta\epsilon_{yz}^2\boldsymbol{q}\rangle = \epsilon_a^2kT/(D + K_2q_x^2 + K_3q_z^2)\,. \tag{5c}$$

In these equations $D$ is an elastic constant which provides a restoring force to maintain the molecules normal to the smectic layers; for a tilt $\theta$ between the molecules and the layer normal, the free energy is $\frac{1}{2}D\theta^2$. Usually $D$ is about 10 times smaller than $B$.

As I shall explain presently, the phenomenological model for a smectic A is analogous to the Landau–Ginzburg model (Ginzburg and Landau 1950) for a charged superconductor and $B$ and $D$ are the analogs of the superfluid density $\rho_s$. The length $\lambda = (K_1/B)^{1/2}$ is the liquid crystal equivalent to the penetration depth in a superconductor.

### *5.2. The smectic A to nematic transition*

In the phenomenological model for the smectic A to nematic transition (De Gennes 1972a, McMillan 1971, Kobayashi 1970) the free energy near the transition is written,

$$\begin{aligned}\Phi &= \Phi_0 + a|\Psi|^2 + \tfrac{1}{2}b|\Psi|^4 + (1/2M_v)|\partial_z\Psi|^2 \\ &\quad + (1/2M_t)\{|(\partial_x + \mathrm{i}q_0n_x)\Psi|^2 + |(\partial_y + \mathrm{i}q_0n_y)\Psi|^2\} + \Phi_n\,,\end{aligned} \tag{6}$$

where $\Phi_n$ is the nematic phase elastic free energy (De Gennes 1974)

$$\Phi_n = \tfrac{1}{2}K_1[\partial_x n_x + \partial_y n_y]^2 + \tfrac{1}{2}K_2[\partial_x n_y - \partial_y n_x]^2 + \tfrac{1}{2}K_3[(\partial_z n_x)^2 + (\partial_z n_y)^2]. \tag{7}$$

Equation (2) was derived by applying the equipartition theorem to the Fourier transform of eq. (7). In eq. (6) if $b > 0$, a second-order phase change occurs when $a$ passes through zero. One writes $a = a_0(T - T_c)^\gamma$, where $\gamma = 1$ in the mean field approximation. If the Ginzburg criterion (Ginzburg 1960) is applied to eq. (6) we find that the mean field approximation should only be valid in greater than four dimensions. Thus it is necessary to make use of scaling laws (Jähnig and Brochard 1974) in order to use eq. (6) to discuss the nematic–smectic A transition; this requires $\gamma \simeq \frac{4}{3}$, and that $b$ vary as $(T - T_c)^{\gamma - 2\beta}$ where the exponent $\beta$ describes how the order parameter $\Psi$ vanishes for $T < T_c$. If we consider fluctuations one may readily calculate (Lifshitz and Pitaevskii 1980) that

$$\langle \delta\Psi(0)\delta\Psi(\boldsymbol{r})\rangle = \frac{kT}{a}\exp\left(-\frac{z}{\xi_\parallel} - \frac{\sqrt{x^2+y^2}}{\xi_\perp}\right). \tag{8}$$

The correlation lengths are given by

$$\xi_\parallel^2 = 1/(2aM_v) \sim |T - T_c|^{-\nu_\parallel}, \tag{9a}$$

$$\xi_\perp^2 = 1/(2aM_t) \sim |T - T_c|^{-\nu_\perp}, \tag{9b}$$

The scaling laws require $\nu_\parallel = \nu_\perp$ since they are based on the physical assumption that one diverging length determines all of the anomalous thermodynamic behavior for $T \simeq T_c$. It is also quite straightforward to calculate that the elastic constants $B$ and $D$ of the smectic A phase should vanish as $B \sim \xi_\parallel^{-1}$, $D \sim \xi_\perp^{-1}$.

### *5.3. Smectic phase elastic constants*

By examing eqs. (5b) and (5c) we see that the elastic constants $B$ and $D$ can be measured by light scattering from the director modes. When the first results were reported (Birecki et al. 1976) the results were surprising. Instead of the anticipated $(T_c - T)^{2/3}$ dependence for $B$ and $D$, it was found that $B \sim (T_c - T)^{1/3}$ and $D \sim (T_c - T)^{1/2}$ in cyanobenzylidene-octyloxyanilene. This anomalous behavior has since been confirmed in other materials. In figs. 3 and 4 the temperature dependence of $B$ and $D$ in the smectic A phase are shown for butoxybenzylidene-octylaniline. This behavior is not yet quantitatively understood, but it is widely believed to be the consequence of the Landau–Peierls instability which I shall discuss in the following section. As a result of this instability, the smectic A phase lacks long range order in three dimensions so that in the limit of very long wavelengths $B$ and $D$ should vanish. A recent theory (Grinstein and Pelcovits 1981) predicts that $B$ should vanish logarithmically as $q \to 0$, but it will be very diffcult to observe the predicted behavior.

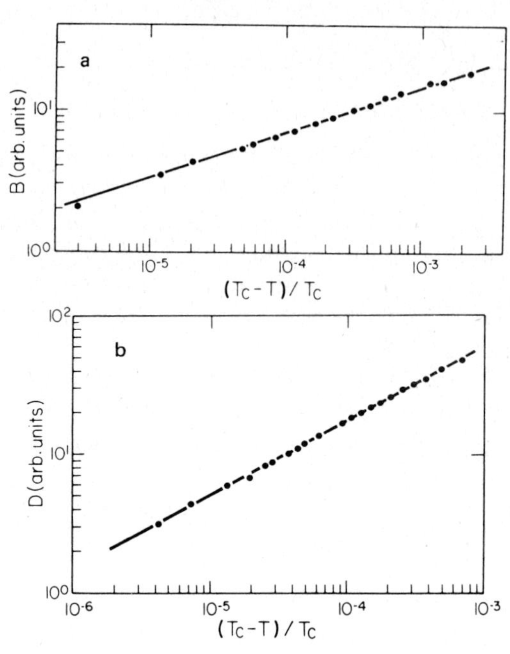


Fig. 3. The smectic A elastic constants $B$ and $D$, defined in the text, just below the nematic to A transition in butoxybenzylidene octylaniline (40.8). From von Känel and Litster (1981).

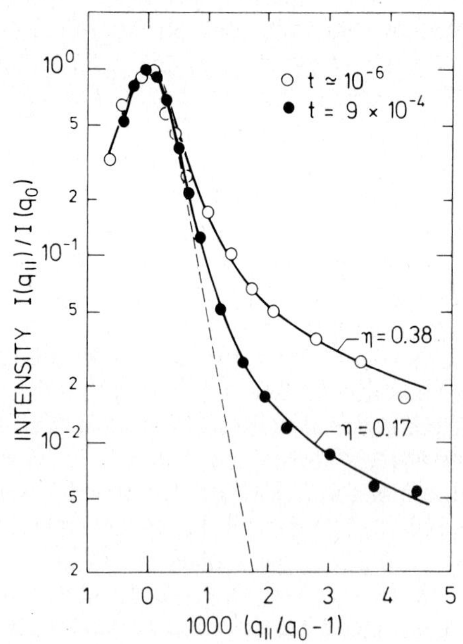


Fig. 4. Data showing algebraic decay of correlations in the phase of the smectic A density wave in octyloxy cyanobiphenyl. The parameter $\eta$ is discussed in the text. From Als-Nielsen et al. (1980).

### *5.4. Landau–Peierls instability in the smectic A phase*

It has been known theoretically since the 1930s (Landau 1937, Peierls 1934) that the translational order of a solid cannot exist in two dimensions because it is destroyed by thermally excited fluctuations. A closely related phenomenon occurs in smectic A and smectic C liquid crystal phases in three dimensions (De Gennes 1974). If we use the free energy for the smectic A Goldstone mode, eq. (3), it is straightforward to calculate the mean squared fluctuations in the position of the smectic layers

$$\langle u^2(r)\rangle = \frac{kT}{(2\pi)^3}\int \frac{\mathrm{d}^3q}{Bq_z^2 + K_1 q_\perp^4} = \frac{kT}{4\pi(BK_1)^{1/2}}\ln(q_0 L),$$

where I have cut off the momentum space integral at $q_0$ and $2\pi/L$, where $L$ is the sample size. The divergence as $L\to\infty$ indicates there can be no long-range order.

The way to test this situation experimentally is to carry out X-ray scattering from the smectic density wave. It is convenient to introduce a correlation function

$$G(\boldsymbol{r}) = \langle \exp\{\mathrm{i}q_0[u(0) - u(\boldsymbol{r})]\}\rangle, \tag{10}$$

to discuss the X-ray scattering. If we let $S(\boldsymbol{q})$ be the Fourier transform of $G(\boldsymbol{r})$, then $S(\boldsymbol{q})$ gives the X-ray scattering intensity with $\boldsymbol{q}$ being the momentum transfer with respect to $q_0\hat{z}$. In a system with long-range order $G(\boldsymbol{r})$ assumes a constant value, the Debye–Waller factor, as $r\to\infty$; the Fourier transform, a delta function, corresponds to a Bragg peak. In the case of a smectic A liquid crystal, and also a two-dimensional solid, where there is no long-range order $G(r)\to 0$ as $r\to\infty$. However $G(r)$ does not go rapidly to zero like $\mathrm{e}^{-r/\xi}$ as it would if only short-range order were present; rather it decays algebraically as $r^{-\eta}$, where $\eta$ is somewhat smaller than unity. The precise form of $G(\boldsymbol{r})$ for a smectic A (Caillé 1972) decays as $z^{-\eta}$ and $(x^2+y^2)^{-\eta}$ where $\eta = kT(q_0^2/8\pi)(Bk_{\|})^{-1/2}$. On taking the Fourier transform, one finds the power law singularity $|q-q_0|^{-2+\eta}$ instead of a Bragg peak. To distinguish this from a Bragg peak is, however, a delicate experimental problem. The details are discussed by Als-Nielsen et al. (1980) and I show the results of their experiment in fig. 5. The theoretical expression for $G(r)$ involved only two parameters: the penetration depth $\lambda$ (see sect. 5.1), which was measured by light scattering as in eq. (5b), and the exponent $\eta$. The experimental data were fit to $S(q)$, convoluted with the measured X-ray spectrometer resolution function, with $\eta$ as an adjustable parameter. In fig. 5 the dashed curve shows the expected scattering for a Bragg peak along with the values of $\eta$, which is temperature dependent, for two different measurements. From $\eta$ it is possible to deduce values of the

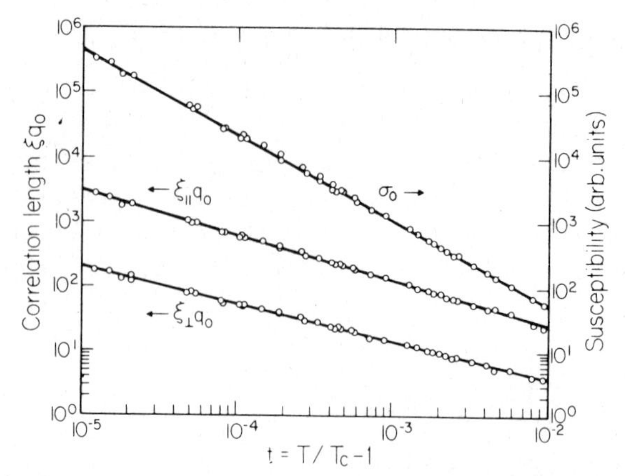


Fig. 5. The result of X-ray scattering measurements of the susceptibility and correlation lengths in nematic 40.8 just above the smectic A phase. From Birgeneau et al (1981).

splay elastic constant $K_1$ which are in good agreement with independent measurements.

This experiment in a smectic A liquid crystal is the first experimental verification of the algebraic decay of correlation functions predicted to occur at the lower marginal space dimension when thermal fluctuations just prevent long-range order. Because of this subtle behavior, a quantitative theory of the smectic A to nematic phase change has yet to be formulated. This includes a quantitative description of the pretransitional nematic phase behavior I discuss in the next section.

### 5.5. *Smectic A short-range order in the nematic phase*

In the nematic phase near the smectic A phase there can be thermally excited short-range fluctuations in the smectic order parameter. These are described by eq. (8). As can be seen from eqs. (4) and (5), bend and twist director modes, which involve curl $\hat{\boldsymbol{n}}$, are excluded from the smectic phase; thus $\hat{\boldsymbol{n}}$ plays a role analogous to the magnetic vector potential in a superconductor and the exclusion of bend and twist is a "Meissner effect" in the smectic A phase. The short-range smectic A order also causes a nematic phase analogue of fluctuation diamagnetism and a divergence of the elastic constants $K_2$ and $K_3$. The divergent contributions at long wavelengths are (Jähnig and Brochard 1974)

$$\tilde{K}_2 = kT(q_0^2/24)\xi_\perp^2/\xi_\parallel\,, \tag{11a}$$

and

$$\tilde{K}_3 = kT(q_0^2/24\pi)\xi_\parallel\,. \tag{11b}$$

If the wave vector $q$ of the nematic director mode becomes comparable to $\xi^{-1}$, the divergences of eq. (11) are reduced. This non-hydrodynamic behavior is well known near critical points (Halperin and Hohenberg 1969). A coupled-mode calculation for $\boldsymbol{q} = q_z\hat{z}$ (Jähnig and Brochard 1974) gives the result

$$\tilde{K}_3 = kT(q_0 q_z/8\pi)\left[\left(1+\frac{1}{X^2}\right)\tan^{-1}X - \frac{1}{X}\right], \tag{12}$$

where

$$X = \tfrac{1}{2}q_z\xi_\parallel .$$

The effects of smectic A short-range order can be studied in detail by X-ray scattering and by light scattering. The X-rays probe mass density fluctuations and measure the Fourier transform of eq. (8). In practice the X-ray scattering intensity is fit to

$$I(\boldsymbol{q}) = \chi[1 + \xi_\parallel^2(q_\parallel - q_0)^2 + \xi_\perp^2 q_\perp^2 + c\xi_\perp^4 q_\perp^4]^{-1}, \tag{13}$$

convoluted with the instrumental resolution function, where $\chi \sim 1/a = 1/a_0(T - T_c)^\gamma$. The Fourier transform of eq. (8) would not contain the $\xi_\perp^4 q_\perp^4$ term. The fact that it is required by the experiments indicates that the coefficient of the $\nabla_\perp^2$ terms in the free energy (6) is sufficiently small that higher order $\nabla_\perp^4$ terms are also needed. The X-ray scattering experiments provide a direct measurement of the susceptibility $\chi$ and the two correlation lengths $\xi_\parallel$ and $\xi_\perp$. In fig. 5 the results of measurements (Birgeneau et al. 1981) for butoxybenzylidene octylaniline (40.8) are shown. The susceptibility diverges

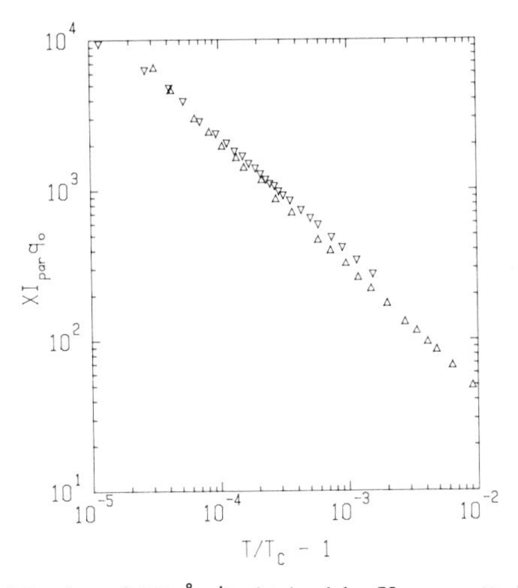


Fig. 6. A comparison of $\xi_\parallel q_0$($q_0 = 0.222$ Å$^{-1}$) obtained by X-ray scattering (triangles) and light scattering (inverted triangles) just above the N–A transition in 40.8.

with an exponent $\gamma = 1.31 \pm 0.02$ and the correlation lengths diverge with $\nu_\parallel = 0.70 \pm 0.01$ and $\nu_\perp = 0.57 \pm 0.01$. That $\nu_\parallel \neq \nu_\perp$ contradicts the simple scaling hypothesis since apparently two lengths are involved; however $\nu_\parallel \simeq \nu_\perp + 0.13$ has been observed in a number of different materials. As an additional complication, the observed values of $\nu_\parallel$ appear to vary from one material to another, although the results are quite reproducible for a given material.

Light scattering by director fluctuations may also be studied to test the model of eq. (6). By carrying out experiments in the non-hydrodynamic region, eq. (12) can be used to obtain an absolute value for $\xi_\parallel$ from light scattering. The data are well fit by this equation (Von Känel and Litster 1981) and the resulting values for $\xi_\parallel$ are shown in fig. 6 along with those measured directly by X-ray scattering. The agreement is excellent close to the phase transition, but the effect of a non-divergent background to $K_3$ and possibly different corrections to scaling give somewhat larger values for the light scattering data well above $T_c$. The much larger background contribution and the small value of $\xi_\perp/\xi_\parallel$ combine to make light scattering determinations of $\xi_\perp$ very difficult. The two measurements reported in the literature (Delaye et al. 1973, Chu and McMillan 1975) are not in agreement. The experimental situation for the nematic to smectic A transition is qualitatively similar to the simpler theoretical predictions (of eq. (6) and the three dimensional X–Y model) but there are serious quantitative discrepancies, particularly with isotropic scaling laws. The hypothesis of two scale factor universality (Stauffer et al. 1972, Hohenberg et al. 1976) states that the free energy per correlation volume is independent of temperature near the phase transition. If the specific heat diverges as $|T - T_c|^{-\alpha}$, then this requires $\nu_\parallel + 2\nu_\perp = 2 - \alpha$. A similar prediction comes from an anisotropic scaling model (Lubensky and Chen 1978) and appears to be verified by experiment (Birgeneau et al. 1981). A recent defect mediated model for the transition (Nelson and Toner 1981) predicts that $\xi_\parallel = 2\xi_\perp$ asymptotically close to $T_c$ as well as a discontinuous jump in the elastic constant $B$; the latter may have been observed (Fisch et al. 1982) but it appears that the Nelson–Toner predictions for $\xi_\parallel = \xi_\perp^2$ will only occur much closer to $T_c$ than current experiments can approach. We are still lacking, therefore, a theory which can quantitatively describe behavior observed over the temperature region as $\xi_\parallel$ diverges from 5 Å to 5 $\mu$m; this difficulty probably also arises from the fact that the smectic A phase lacks true long-range order. An additional complication can arise when the smectic order is coupled to other properties of the material. An example is the reentrant nematic behavior (Guillon et al. 1981) that results when the smectic order is coupled to the density. In this case a simple model using standard multicritical theory (Kortan et al. 1981) shows that the observed exponents can be quite different from the intrinsic divergences of the phase transition. It is likely that similar coupling to other quantities (e.g., the nematic order parameter) will be able to explain the apparent material dependence of the observed divergences near the nematic to smectic A transition.

## 6. The smectic C phase

As shown in fig. 1, the smectic C phase is similar to the smectic A but the density wave is not parallel to the director, alternatively one might say the director is tilted with respect to the smectic "layers". The component of the director in the plane of the layers behaves similar to a nematic with elastic restoring forces coming only from the curvature of the vector field. Thus director fluctuations in the C phase can scatter light strongly in many directions and the smectic C phase is turbid like the nematic one. The elastic free energy is quite complicated (De Gennes 1974) and, largely due to the difficulty of preparing good monodomain samples, only one Rayleigh scattering study has been reported (Galerne et al. 1971). The C phase should also have the same algebraic decay of correlations for the phase of the density wave as the smectic A, however this has yet to be observed.

### 6.1. The smectic A to C transition

In many liquid crystals there is a high temperature smectic A phase which on cooling transforms to a C phase by a second-order phase change (no latent heat). Since the C phase results from a tilt of the molecules with respect to the layers, it can be described phenomenologically by letting the elastic constant $D$ of eq. (5c) go to zero; in the C phase then $D$ would be negative and a higher-order positive term would be added for stability (Chen and Lubensky 1976). In the A phase one may write correlation functions for the tilt $\omega$ by using eq. (6). One obtains (Safinya et al. 1980):

$$\langle \omega_1^2(\boldsymbol{q}) \rangle = kT/(D + K_1 q_x^2 + K_3 q_z^2)\,, \tag{14}$$

and

$$\langle \omega_2^2(\boldsymbol{q}) \rangle = kT/(D + K_2 q_x^2 + K_3 q_z^2)\,, \tag{15}$$

where $\omega_1$ is the tilt along $\boldsymbol{q}$ and $\omega_2$ is transverse to $\boldsymbol{q}$. The tilt can be taken as the magnitude of the C order parameter, which is actually a two component quantity that can be written as

$$\psi_c = \omega\, e^{i\phi}\,, \tag{16}$$

where the phase $\phi$ gives the azimuthal direction of the tilt. Equations (14) and (15) can also be derived by a Landau expansion for the free energy near the AC transition temperature (De Gennes 1972b). Because $\psi_c$ has two degrees of freedom we expect that mean field behavior should occur only in greater than four dimensions and the observed critical behavior at the AC transition should be that calculated for the three dimensional X–Y model which is observed for superfluid helium.

The experimental situation has, until very recently, been unclear. The first optical measurements (Taylor et al. 1970) of the tilt angle in the C phase of

TBBA (terephthal-bis-p-butylaniline) suggested an order parameter that vanished as $(T_c - T)^\beta$ with $\beta \simeq 0.3$ as expected for the X–Y model. Experiments by Delaye and Keller (1976) on undecylazoxymethylcinnamate (AMC11) seemed to suggest mean field behavior at the AC transition in that compound, but later studies (Delaye 1979) of nonyloxy-benzoate-butyloxyphenol (9OBO4) yielded critical (3-dimensional X–Y exponents).

The tilt angle may be directly measured by X-ray scattering, if the director is held fixed by a magnetic field, as the angle between the director and density wave vector in reciprocal space. Safinya et al. (1980) found mean field behavior for the vanishing of the order parameter in pentylphenylfhiol-octyloxybenzoate ($\overline{8}$ S5). They also found the period of the density wave (i.e., the "layer" thickness) scaled as the square of the tilt angle, indicating it is a secondary-order parameter and that the primary mechanism for the AC transition is a simple tilt of the molecules. As de Vries (1979) has pointed out, the nematic order parameter is not saturated, so attempts to deduce the molecular length from the density wave period and the tilt angle will be misleading. Safinya et al. also pointed out the explanation for observing mean field behavior for the AC transition in three dimensions. It is the same reason as for superconductors (Ginzburg 1960). The fluctuations do not become sufficiently strong to affect the transition unless

$$\frac{|T - T_{AC}|}{T_{AC}} < \frac{k^2}{32\pi^2(\Delta C)^2(\xi_\parallel^0)^2(\xi_\perp^0)^4}, \tag{17}$$

where $\Delta C$ is the heat capacity jump at the transition and $\xi_\parallel^0$ and $\xi_\perp^0$ are the bare correlation lengths along and transverse to the director. From eqs. (14) and (15) we obtain

$$\xi_\parallel = \xi_\parallel^0(T/T_{AC} - 1)^{-\nu_\parallel} = (K_3/D)^{1/2}, \tag{18}$$

and

$$\xi_\perp = \xi_\perp^0(T/T_{AC} - 1)^{-\nu_\perp} = [(K_1 + K_2)/2D]^{1/2}. \tag{19}$$

From estimates of $D$ far away from $T_{AC}$, Safinya et al. calculated that $\xi^0$ could be as large as 70 Å. Subsequent measurements of $\xi_\parallel^0 \simeq 20$ Å by light scattering (Schaetzing 1980) were still large enough when combined with the known value for $\Delta C$ (Schantz and Johnson 1978) to explain why mean field behavior should be observed. Such was the somewhat confused situation discussed in more detail by Safinya et al., two years ago. It was not improved when Galerne (1981) reported the tilt angle in AMC11 vanished with a critical, or not mean field, exponent in apparent contradiction to Delaye's results for $T > T_{AC}$.

The clue to understanding the behavior near the AC transition was provided by specific heat measurements of the nearly unpronounceable material methylbutylphenyl-nonyloxybiphenyl-carboxylate (2M4P9OBC) by Huang and Viner (1982). They found the heat capacity anomoly to be that expected for a mean field model in which the sixth power invariant of the order parameter in the Landau expansion plays a significant role. That is, a free energy of the form

$$F = a\psi_c^2 + b\psi_c^4 + c\psi_c^6 + \cdots + l^2(\nabla\psi)^2, \tag{20}$$

with $a = a_0(T/T_{AC} - 1) = a_0 t$. This predicts a tilt angle

$$\omega^2 = (b/3c)(\sqrt{1 - 3t/t_0} - 1), \tag{21}$$

and specific heat

$$C = \frac{T}{2T_{AC}^2}\left(\frac{a_0^3}{c}\right)^{1/2}(t_0 - 3t)^{-1/2}$$
$$= 0, \quad \text{if} \quad t > 0. \tag{22}$$

Huang and Viner found that eq. (22) represented their data very well for 2M4P9OBC with a value of $t_0 = b^2/a_0 c \simeq 5.5 \times 10^{-3}$. This means that the order parameter shows crossover behavior for $t \simeq t_0$. If $|t| \ll t_0$ one finds $\omega \sim |t|^\beta$ with $\beta = \frac{1}{2}$, while $|t| \gg t_0$ will show $\beta \simeq \frac{1}{4}$. In practice, depending on the value of $t_0$, one may observe over two decades of $|t|$, which is all the experiments usually cover, effective values of $\beta$ from $\sim 0.30$ to $\sim 0.50$.

Recent experiments at M.I.T. (Meichle 1982) show similar behavior for AMC11 with $t_0 \simeq 1.6 \times 10^{-3}$, which reconciles the difference between Galerne's

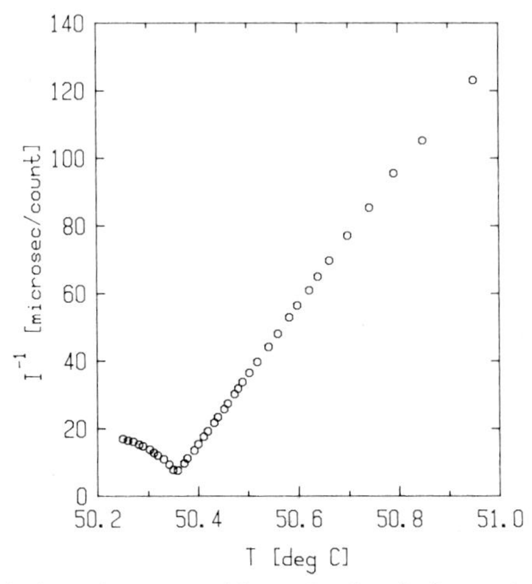


Fig. 7. Reciprocal of light intensity scattered by molecular tilt fluctuations near the smectic A to smectic C phase transition in butoxy-benzylidene-heptylaniline (40.7). The data clearly show mean field divergences for susceptibility and correlation lengths. From L-J. Yu and J.D. Litster, unpublished.

and Delaye's observations of this compound. In fig. 7 I show some measurements made recently at M.I.T. by L.J. Yu of the light scattered by tilt fluctuations near the AC transition. The reciprocal of the intensity is plotted and shows a minimum at $T_{AC}$. The data are for $\boldsymbol{q}$ along $\hat{z}$ and one can see clearly that $D$ vanishes as $T - T_{AC}$ while $\xi_\parallel^2 \sim (T_{AC} - T)^{-1}$; analysis gives $\xi_\parallel^0 = 20.4 \pm 0.7$ Å. It now appears that most AC transitions show mean field behavior over the experimentally accessible range of $T_{AC} - T$ because the base correlation lengths are large. The importance of the $c\psi_c^6$ term in eq. (20) causes the tilt angle to saturate quickly for $T < T_{AC}$ and yield effective exponents $\beta$ in some materials which are close to those predicted by the three dimensional X–Y model. The crossover temperature $t_0$ is not a universal parameter and it is also conceivable that not all materials will show mean field behavior. Delaye's data (1979) for nonyloxybenzoate-butoxyphenol are internally consistent with critical (not mean field) behavior, but a definite answer will require careful heat capacity measurements.

### 6.2. *The NAC multicritical point*

By mixing different members of a homologous series of liquid crystals it is possible to vary the microscopic interactions and, unlike the situation in solid phases, to have truly homogeneous materials. Some very interesting phase diagrams result. One discovered by David Johnson and his colleagues at Kent State University (Brisbin et al. 1969) is shown in fig. 8. The region in mixtures of $\bar{8}$ S5 and $\bar{7}$ S5 where the N, A, and C phases are all in equilibrium is especially interesting.

A number of theoretical models have been proposed to explain behavior near this point (Chu and McMillan 1977, Benguigi 1979, Van der Meer and Vertogen 1979). The statistical mechanics is most intriguing for the Chen–Lubensky model since it should be an example of an $m = 2$ Lifshitz model which, like the smectic A phase itself, marginally lacks long-range order in three dimensions (Hornreich et al. 1975). Being thus doubly marginal, the NAC point should show very interesting fluctuation effects. Most of the NAC models are simply a mean field treatment with coupled order parameters. In the Chen–Lubensky model there is only one order parameter and the AC transition occurs when the coefficient of the transverse gradient term in the free energy, $1/2M_t$ in eq. (6), passes through zero. In its simplest mean field approximation, the model predicts that the X-ray line shapes scattered by short-range order in the nematic phase should have a non-Lorentzian $q_\perp^{-4}$ form in some regions of the phase diagram. None of the other theories predict results in agreement with recent X-ray scattering experiments (Safinya et al. 1981) but it is not yet established if the Chen–Lubensky model is the correct one. Further experiments and a calculation which properly includes fluctuations are required. The nematic to smectic C transition may be second order, by symmetry,

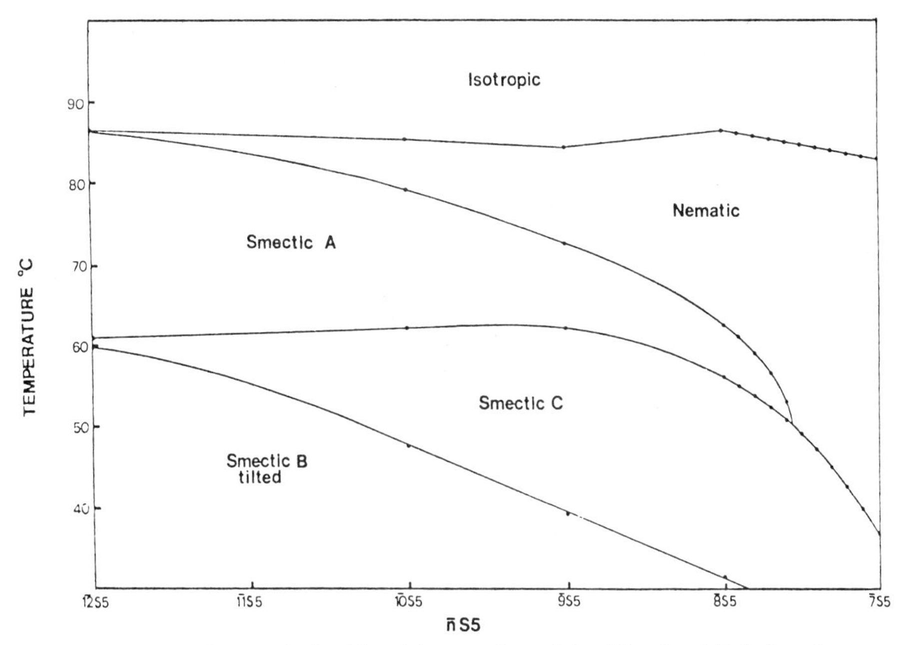


Fig. 8. The phase diagram obtained by mixing members of the alkly-phenylthiol-alkoxybenzoate (n̄ S5) homologous series. From Brisbin et al. (1979).

in mean field models. Experimentally it is always first order. Brazovskii (1975) has shown that if a density wave with an infinite number of characteristic wave vectors is established in a material, fluctuations will always cause the transition to be first order. As Swift (1976) has pointed out, the degeneracy of the smectic C density wave in $\boldsymbol{q}$ space places the N–C transition in the Brazovskii class.

The behavior of phases obtained by mixing liquid crystals offers many examples of fundamental interest in statistical mechanics, and there are many experiments involving light scattering and X-ray scattering that have yet to be done.

## 7. Conclusions

By discussing some experiments in the better known smectic A and C phases I have tried to illustrate some of the most pressing intellectual questions in statistical mechanics, how they can be elucidated by studying liquid crystals, and the power of combining light scattering with high resolution X-ray scattering to perform these studies.

There are many other phases of thermotropic liquid crystals, labelled smectic

B and smectics D through I, which have yet to be studied as intensely. Some are not truly liquid crystals (Moncton and Pindak 1979, Pershan et al. 1981) yet they manifest interesting structural phase changes. Other phases show entirely new concepts of ordering, such as bond-orientational long-range order (Birgeneau and Litster 1978) to be explored. One of the chief obstacles has been the difficulty of preparing well-aligned samples. There are grounds to hope that these problems may be overcome by the use of submicron structures (Von Känel et al. 1981) or freely suspended films (Moncton and Pindak 1979) of liquid crystals. One may also expect that similar experimental methods and theoretical tools will be successfully applied to increase our understanding of the many lyotropic liquid crystal phases.

## *Acknowledgements*

It is a pleasure to thank my colleague Robert Birgeneau for a fruitful collaboration which produced much of the work discussed in this article. Important financial assistance came from the National Science Foundation under grants DMR78–23555 and DMR78–24185.

## *References*

Als-Nielsen, J., J.D. Litster, R.J. Birgeneau, M. Kaplan, C.R. Safiny, A. Lindegaard-Andersen and S. Mathiesen, 1980, Phys. Rev. **B 22**, 312.
Benguigi, L., 1979, J. Phys. (Paris) **40**, C3–419.
Birecki, H., R. Schaetzing, F. Rondelez and J.D. Litster, 1976, Phys. Rev. Lett. **36**, 1376.
Birgeneau, R.J. and J.D. Litster, 1978, J. Phys. Lett. (Paris) **39**, L-399.
Birgeneau, R.J., C.W. Garland, G.B. Kasting and B.M. Ocko, 1981, Phys. Rev. A **24**, 2624.
Brazovskii, S.A. 1975, Sov. Phys. JETP, **41**, 85.
Brisbin, D., R. DeHoff, T.E. Lockhart, and D.L. Johnson, 1979, Phys. Rev. Lett. **43**, 1171.
Brochard, F., 1973, J. Phys. (Paris) **34**, 411.
Caillé, A., 1972, C.R. Acad. Sci. Ser. B **274**, 891.
Chen, J.H. and T.C. Lubensky, 1976, Phys. Rev. **A 14**, 1202.
Chu, K.C. and W. McMillan, 1975, Phys. Rev. **A 11**, 1059.
Chu, K.C. and W.L. McMillan, 1977, Phys. Rev. **A 15**, 1181.
De Gennes, P.G., 1969, Phys. Lett. **A 30**, 454.
De Gennes, P.G., 1972a, Solid State Commun. **10**, 753.
De Gennes, P.G., 1972b, Comptes rendus **B 274**, 758.
De Gennes, P.G., 1974, The Physics of Liquid Crystals (Oxford, London).
Delaye, M., 1979, J. Phys. (Paris) **40**, C3–350.
Delaye, M. and P. Keller, 1976, Phys. Rev. Lett. **37**, 1065.
Delaye, M., R. Ribotta and G. Durand, 1973, Phys. Rev. Lett. **31**, 443.
De Vries, A., 1979, J. Chem. Phys. **71**, 25.
Fisch, M.R., L.B. Sorensen and P.S. Pershan, 1982, Phys. Rev. Lett. **48**, 943.
Galerne, Y., 1981, Phys. Rev. **A 24**, 2284.
Galerne, Y., J.L. Martinand, G. Durand and M. Veyssié, 1971, Phys. Rev. Lett. **29**, 561.

Ginzburg, V.L., 1960, Sov. Phys. Solid State, **2**, 1824.
Ginzburg, V.L. and L.D. Landau, 1950, J.E.T.P. (USSR) **20**, 1064.
Grinstein, G. and R. Pelcovits, 1981, Phys. Rev. Lett. **47**, 856.
Guillon, D., P.E. Cladis and J. Stamatoff, 1981, Phys. Rev. Lett. **41**, 1598.
Halperin, B.I. and P.C. Hohenberg, 1969, Phys. Rev. **177**, 952.
Hohenberg, P.C., A. Aharony, B.I. Halperin and E.D. Siggia, 1976, Phys. Rev. **B 13**, 2986.
Hornreich, R., M. Luben and S. Shtrikman, 1975, Phys. Rev. Lett. **35**, 1678.
Huang C.C. and J.M. Viner, 1982, Phys. Rev. **A 25** (issue of June 1982).
Jähnig, G. and F. Brochard, 1974, J. Phys. (Paris) **35**, 301.
Kobayashi, K.K., 1970, J. Phys. Soc. Jpn. **29**, 101.
Kortan, R., H. von Känel, R.J. Birgeneau and J.D. Litster, 1981, Phys. Rev. Lett. **47**, 1206.
Landau, L.D., 1937. English translation in Collected Papers of L.D. Landau, ed., D. ter Haar (Gordon and Breach, New York, 1965) p. 209.
Lifshitz, E.M. and L.P. Pitaevskii, 1980, Statistical Physics (Pergamon, New York).
Lubensky, T.C. and J.H. Chen, 1978, Phys. Rev. **B 17**, 366.
McMillan, W.L., 1971, Phys. Rev. **A 4**, 1238.
Meichle, M., 1982, private communication.
Moncton, D.E. and R. Pindak, 1979, Phys. Rev. Lett. **43**, 701.
Nelson, D.R. and J. Toner, 1981, Phys. Rev. **B 24**, 363.
Peierls, R.E., 1934, Helv. Phys. Acta. Suppl. **7**, 81.
Pershan, P.S., G. Aeppli, J.D. Litster and R.J. Birgeneau, 1981, Mol. Cryst. Liq. Cryst. **67**, 205.
Safinya, C.R., R.J. Birgeneau, J.D. Litster and M.E. Neubert, 1981, Phys. Rev. Lett. **47**, 668.
Safinya, C.R., M. Kaplan, J. Als-Nielsen, R.J. Birgeneau, D. Davidov and J.D. Litster, 1980, Phys. Rev. B **21**, 4149.
Schaetzing, R. 1980, Ph.D. Thesis, M.I.T., unpublished.
Schantz, C.A. and D.L. Johnson, 1978, Phys. Rev. **A 17**, 1504.
Stauffer, D., M. Ferer and M. Wortis, 1972, Phys. Rev. Lett. **29**, 345.
Stinson, T.W. and J.D. Litster, 1973, Phys. Rev. Lett. **30**, 688.
Stinson, T.W., J.D. Litster and N.A. Clark, 1972, J. Phys. (Paris) **33**, C1–69.
Swift, Jack, 1976, Phys. Rev. **A 14**, 2274.
Taylor, T.R., J.L. Fergason and S.L. Arora, 1970a, Phys. Rev. Lett. **24**, 359.
Taylor, T.R., J.L. Fergason and S.L. Arora, 1970b, Phys. Rev. Lett. **25**, 722.
Van der Meer, B.W. and G. Vertogen, 1979, J. Phys. (Paris) **40**, C3–222.
Von Känel, H. and J.D. Litster, 1981, Phys. Rev. **A 23**, 3251.
Von Känel, H., J.D. Litster, J. Melngailis and H.I. Smith, 1981, Phys. Rev. **24**, 2713.

CHAPTER 11

# Light Scattering in Quartz and Ammonium Chloride and its Peculiarities in the Vicinity of Phase Transition of Crystals A Retrospective View and Recent Results

I.A. YAKOVLEV and O.A. SHUSTIN

*Department of Physics*
*Moscow State University II7234, Moscow*
*USSR*

*Light Scattering near Phase Transitions*
*Edited by*
*H.Z. Cummins and A.P. Levanyuk*

© *North-Holland Publishing Company, 1983*

# Contents

## 1. Introduction

The molecular mechanism of phase transitions is currently a topic of great interest. The light scattering phenomena are clearly needed to provide a sufficient insight into the problem. One of the optically most perfect crystals that have a phase transition at a temperature of 573°C, quartz, is very attractive for an optical investigation of second-order (and close to second-order) phase transitions in solids.

The investigation of light scattering in quartz has a long and interesting history. In 1927, Landsberg revealed, for the first time, the molecular light scattering in crystals. In 1928, Landsberg and Mandelstam observed the combination light scattering in quartz. In 1930, Gross examined the fine structure of the spectral lines of scattered light in an experiment based on the idea of Mandelstam.

A detailed experimental and theoretical study of the molecular light scattering in quartz up to 250°C was made by Motulevich (1950). To characterize the quartz transition, we have plotted in fig. 1 the temperature dependence of the heat capacity of quartz around phase transition (Sinel'nikov 1953). On the other hand, some properties of quartz suggest the existence of some peculiarities of the scattered light intensity under these conditions. According to Perrier and Mandrot (1922) the isothermal moduli of elasticity of quartz pass through a sharp minimum (fig. 2). This leads to an increase in the isothermal compressibility of the crystal, to a rise of density fluctuations and the optical inhomogeneity of the crystal at the transition. It is also worth noting that the refractive indices of quartz have a strong temperature dependence at the transition point (fig. 3) (Rinne and Kolb 1910, Yakovlev and Velichkina 1957). Therefore, Yakovlev et al. (1956) considered that in the formula for the intensity of light scattering,

$$I \sim I_0 \overline{\Delta\epsilon^2} = I_0 \left[ \left( \frac{\partial \epsilon}{\partial \rho} \right)_T^2 \overline{\Delta\rho^2} + \left( \frac{\partial \epsilon}{\partial T} \right)_\rho^2 \overline{\Delta T^2} \right],$$

both the terms should be taken into account and not only the first one, as is commonly done.

The case is that in the small volumes of light scattering $V = (0.1\lambda)^3$ the temperature fluctuations $\overline{\Delta T^2} = kT^2/C_V \simeq 0.01$°C where $C_V$ is the heat capacity of the crystal in the fluctuation volume. The calculation with the concrete data

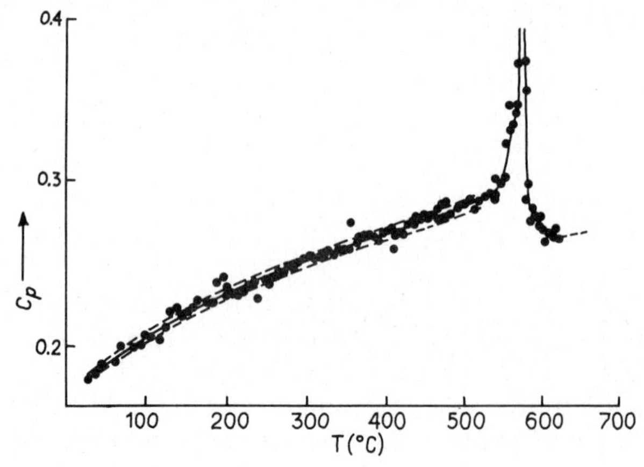


Fig. 1. Specific heat of quartz near the phase transition point vs temperature.

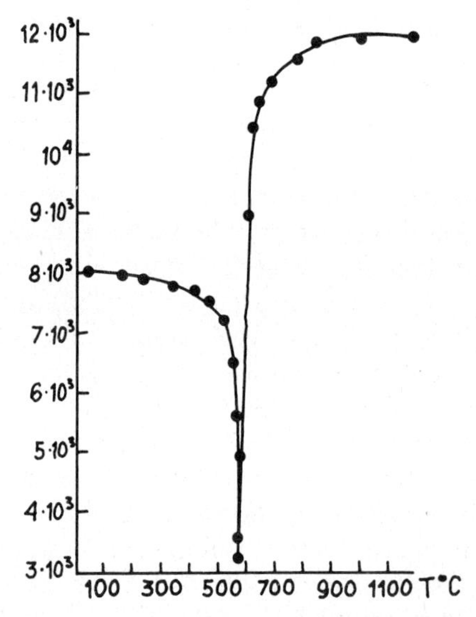


Fig. 2. Young's modulus of quartz vs temperature (⊥ to the optical axis in kg/mm²).

for quartz shows that the derivative,

$$\left(\frac{\partial \epsilon}{\partial T}\right)_\rho = \left(\frac{\partial \epsilon}{\partial T}\right)_p + \left(\frac{\partial \epsilon}{\partial p}\right)_T \frac{\alpha_p}{\beta_T} \approx \left(\frac{\partial \epsilon}{\partial T}\right)_p,$$

and the magnitude of $(\partial \epsilon/\partial T)_p$ is large (fig. 3). In the last relation $\rho$ is the density of a medium, $p$ is the pressure, $\alpha_p$ is the coefficient of volume expansion and $\beta_T$ is the isothermal compressibility. This suggests a unique possibility of connecting the temperature fluctuations and an optically observable light

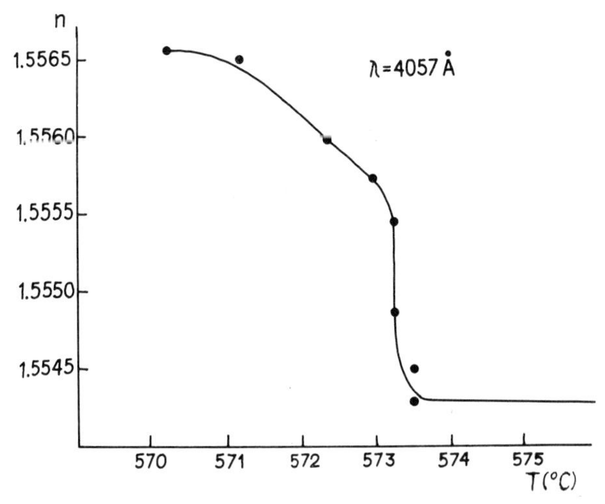


Fig. 3. Refraction index of quartz for a specific wavelength vs temperature.

scattering effect in quartz. If, following this suggestion, we calculate the relation of the intensity $I_\theta$ of light scattering by the fluctuations $(\partial\epsilon/\partial T)^2_\rho \overline{\Delta T^2}$ at the transition point to the intensity $I_{\rho 0}$ of light scattering by the density fluctuations $\rho$ at room temperature $T_0$, then:

$$I_\theta/I_{\rho 0} = 4n^2\theta^2\left(\frac{\partial n}{\partial T}\right)^2_\rho \bigg/ \rho^2\left(\frac{\partial \epsilon}{\partial \rho}\right)^2_T \beta_T C_V T_0 = 9 \times 10^3 . \tag{1}$$

Here, $n^2 = \epsilon$, and $(\partial n/\partial T)^2_\rho$ is from the article of Yakovlev and Velichkina (1957).

These considerations were used by Yakovlev et al. (1956) on a basis for the experiments on light scattering at the phase transition of quartz.

Ginzburg (1955) who used a different approach in estimating the scattered light intensity in second-order phase transitions, considered possible phenomena in terms of the Landau scheme of phase transition theory.

The calculation of Ginzburg gives the following relation of the intensity $I_\eta$ of light scattering by the fluctuations $\Delta\epsilon$ connected with the fluctuations of a characteristic transition parameter $\eta$ at the transition temperature, to the intensity $I_\rho$ of light scattering by the density fluctuations also at the transition,

$$\frac{I_\eta}{I_\rho} = \frac{4n^2(\Delta n)^2\theta}{\Delta C_p(\Delta T)^2(\rho(\partial\epsilon/\partial\rho))^2\beta_T} \simeq 10^4 . \tag{2}$$

In this formula $\Delta n$ is the magnitude of the sharp change of the refractive index at the transition temperature and $\Delta T = 0.1$ K is the temperature width of this "jump" (Yakovlev and Velichkina 1957), $\Delta C_p$ is a "jump" of the heat capacity at the phase transition (Sinel'nikov 1953). Later it was shown (see, for example, Ginzburg et al. (1980)) that the taking into account of the shear modulus in a solid changed radically all the estimations and gave no reason to expect the intensity of molecular scattering at the $\alpha \rightleftarrows \beta$ transition in quartz to be increased by more than several times. Fortunately, that the above-given estimations were exaggerated was not revealed until the experiment was performed.

## *2. Discovery of the anomalous light scattering in quartz*

(A) Let us describe the experiments on light scattering in quartz in the 20–600°C temperature interval including the vicinity of the phase transition (Yakovlev et al. 1956).

Layout of the experimental unit is given in fig. 4. The light beam from a mercury lamp traversed the quartz crystal placed in an oven. The light scattered by quartz at a right angle to the primary light beam, was registered by a photoelectric multiplier.

The oriented polished-quartz parallelepipeds of 20 × 20 × 40 mm were made of the best samples of optical quartz having no twins. Yet, the final criterium of suitability of the samples for further investigations was the temperature dependence of the light scattering intensity up to 250°C. According to experimental and theoretical data, the dependence should be linear and could be compared, in the course of measurements, with the thorough measurements of Motulevich (1950). The 120 × 120 mm square oven was made of steel of 8 mm thickness. The heat-protective glass windows of a 20 mm diameter were

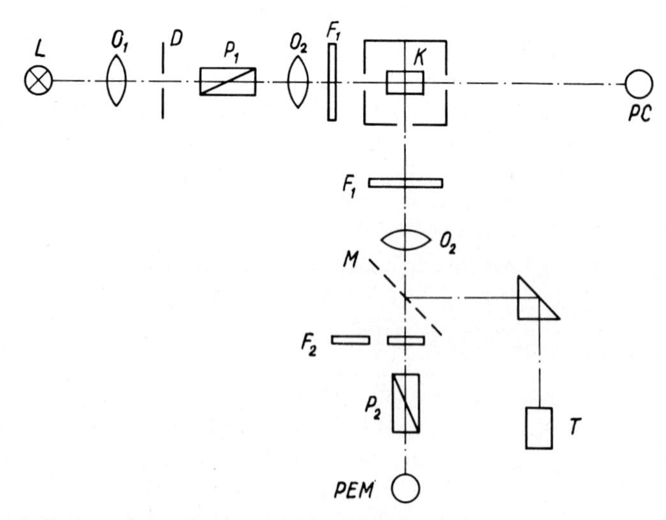


Fig. 4. Layout of the experimental unit for the investigation of scattered light.

at a distance of 160 mm from the oven base. The light-protective metallic tubes were laid from the crystal to the windows inside the oven. The investigated crystal K was placed on the metallic table fixed on a thin porcelain tube. After the crystal was fixed, a removable cover of the furnace and the lower perimeter were hermetically sealed. Thus, the oven had a stagnation thermal zone. The heating coils of the oven were supplied from an accumulator voltage. Temperature was measured by a thermocouple Pt, Pt, Rh with a potentiometer. The thermocouple and the crystal had vertical, horizontal and rotational micrometric displacements in the hermetically sealed oven. The temperature gradients were measured by shifting the thermocouple at strictly constant conditions in the oven. The gradients observed in the operating zone were compensated by changing currents in the coils of the bottom and cover. Near 573°C the vertical gradient in the vicinity of the crystal was lowered to 0.01°C/mm. The horizontal gradient in the direction of the light beam was reduced to 0.03°C/mm. Thus the isothermal layers of the crystal were near perpendicular to the light beam passing through the crystal. Special experiments have established the possible time dependence of temperature at which the temperature of the thermocouple corresponded to the temperature of the closely located crystal. The oven was heated rather slowly: the complete cycle (20–600–20)°C lasted more than 72 hours. Near the transition point the temperature changed at a rate of approximately 0.3°C/h.

(B) The final measurements of the temperature dependence of the scattered light intensity in quartz were made with three samples of quartz, two of them belonging to the same quartz deposit. Several experiments were made for each of the three samples. The analyzed combinations of directions of the primary and scattered light beams in quartz and of the polarization states for which the light scattering intensities were measured, are tabulated.

The table presents the summed values of $I_x$ and $I_y$ (in the second case) because in the propagation of the scattered light along the $z$-axis the optical activity of quartz provides an ambiguity to separate values of $I_x$ and $I_y$. $I_x$, $I_y$, $I_z$ were

Table 1
Primary and scattered beam

| No. | Primary light | | Scattered light | |
|---|---|---|---|---|
| | Direction of propagation | Intensity and state of polarization | Direction of propagation | Intensity and state of polarization |
| 1 | $x$ | $J_z$ | | $J_x, J_z$ |
| 2 | $x$ | $J_y$ | $y$, $z$ } samples 2 and 3 | $J_x + J_y$ |
| 3 | $y$ | $J_z$ | $x$ sample 1 | $J_z, J_y$ |

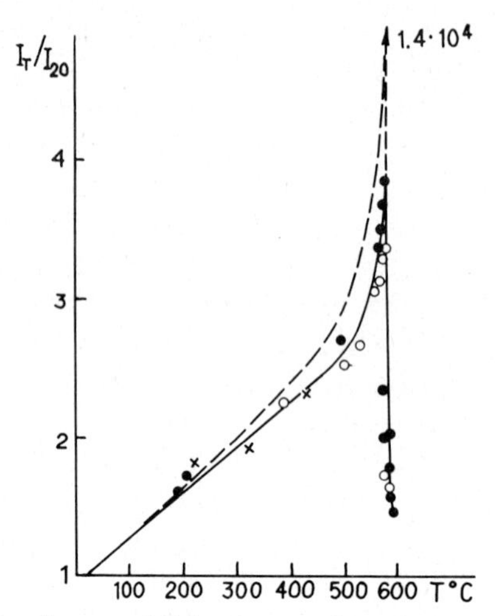


Fig. 5. Intensity of scattered light vs temperature (case 1 in the table 1).

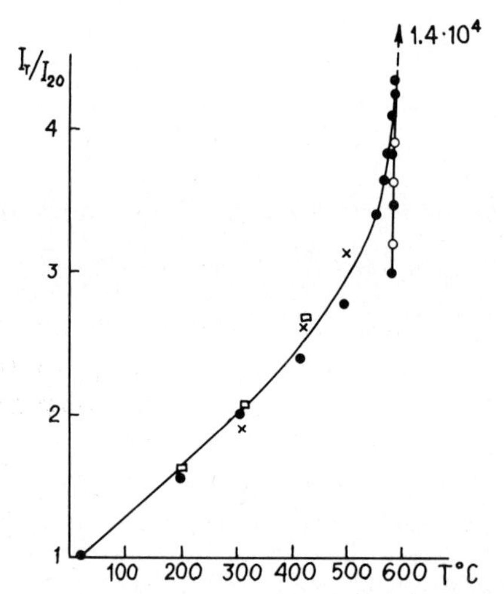


Fig. 6. The same as in fig. 5 for case 2 in the table 1.

measured for the total light of a mercury lamp, for a group of the mercury lines of 4360, 4078 and 4047 Å and for the line 5460 Å. The dependence of the intensity of scattered light of the wavelength permits the estimation of the size of optical inhomogeneities. Percentage of the parasitic light in the total flux of registered light will be smaller for the short-wave part of the spectrum than for white light. The results of the measurements with a violet light filter are, therefore, more reliable and are given in the figures. Incidentally, owing to constancy of a weak parasite light, the obtained temperature dependences of

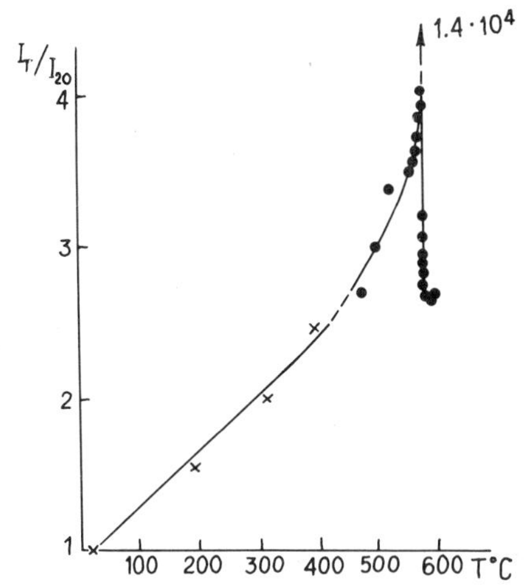


Fig. 7. The same as in fig. 5 for case 3 in the table 1.

the scattered light intensity for blue, white and green light were in perfect agreement.

The sequence of the measurement results in figs. 5, 6 and 7 is the same as in table 1.

In each figure, temperature is plotted along the abscissa and the ratio $I_T/I_{20^\circ}$ (the light scattering intensity at a temperature $T$ to the intensity at room temperature) along the ordinate. A correction for the temperature independent parasite light making up no more than 0.3 $I_{20^\circ}$, was introduced by extrapolation of the linear part of the curve* to the absolute zero. The slopes of linear sections of the curves coincide with an accuracy of 4%, with the results of Motulevich (1950) obtained by photographic photometry in the 20–250°C range.

The characteristic results of examination of the temperature dependence of the depolarization of light scattering are given in fig. 8. The abscissa is the temperature; the ordinate, the depolarization $\Delta = I_x/I_z$. The results pertain to case 1, table 1, and are the same for the other cases.

At the tops of the curve peaks in figs. 5, 6 and 7 arrows are placed with $1.4 \times 10^4$. This mark corresponds to a particular increase in the scattered light intensity, $10^4$ times as compared with $I_{20^\circ}$ in the temperature range of 0.1°C near 573°C. The increase is associated with a specific phenomenon of the quartz opalescence near the transformation temperature. As is stated in the visual and photographic observations of the crystal made in parallel with the light scattering intensity measurements, the boundary region between the α- and

*Corrections for the combination light scattering were not introduced. Due to the heat-protective filters at 600°C, the thermal radiation of the oven is 5% of the intensity of light scattering in quartz at the same temperature.

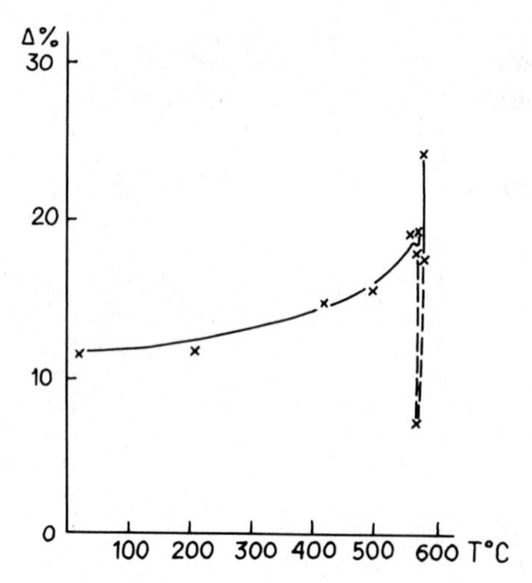


Fig. 8. Temperature dependence of the scattered light depolarization.

$\beta$-phases is a region of strong optical inhomogeneity of the crystal. At a temperature of 573°C a "fog" band strongly scattering the light started moving inside the crystal* from the warmer end of the crystal to the colder one. On both sides of the "fog" zone the light was scattered weakly. On one side of the band there was the $\alpha$-phase and on the other one, the $\beta$-phase. As the temperature increased, the "fog" band reached the opposite edge of the crystal and the crystal became again transparent and weakly scattered the light.

In cooling, the "fog" band appeared once again at the colder end and passed through the crystal to the warmer end. The transverse cross section of the fog zone sweeped the entire crystal; the linear thickness of the band was 0.5 to 3 mm. It could be varied by changing a horizontal temperature gradient in quartz. The temperature width of the band was 0.1°C. By stabilizing the oven temperature, the fog zone was kept at the central part of the crystal during 3 h. Figures 9, 10 and 11 give a succession of photographs: the primary light beam in quartz before the appearance of a fog zone at the phase boundary (1 h exposure), the intersection of the fog band by the primary light beam at a horizontal temperature gradient of 0.3°C/cm (1 s exposure) and 1°C/cm. The contours of the primary light beam on the two last photographs are given in a broken line.

The phenomenon was observed in all investigated quartz and could be repeatedly reproduced in each sample. An X-ray examination of the crystal before and after the heating failed to reveal any changes in the structure. Thus, irreversible changes accompanying the revealed phenomenon in the crystal are

*It will be recalled that there was actually only a horizontal temperature gradient along the primary light beam in the crystal.

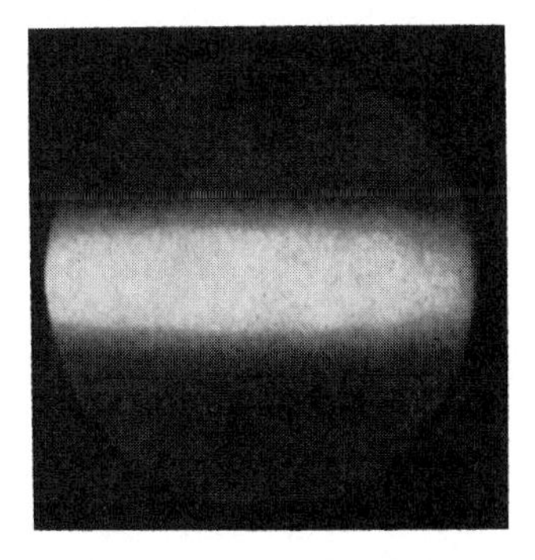

Fig. 9. Photograph of the primary beam of light before and after the phase transition.

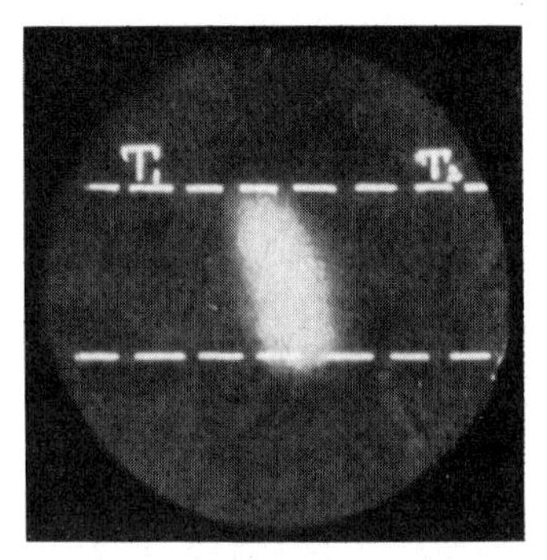


Fig. 10. Photograph of the fog zone during the phase transition (temperature gradient 0.3°C/cm).

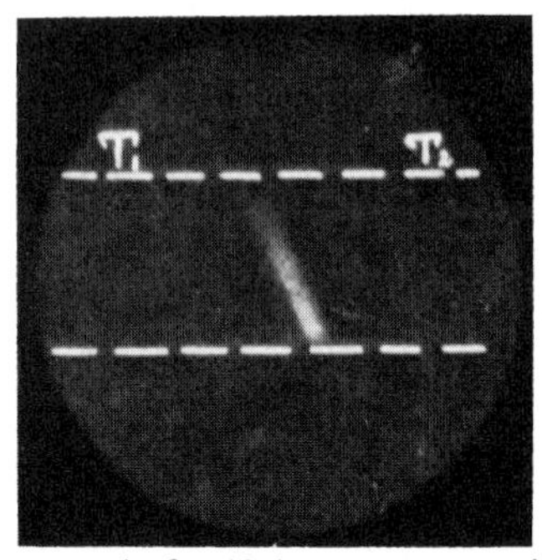


Fig. 11. The same as in fig. 10 (temperature gradient 1°C/cm).

nonexistent. The depolarization of light scattered by the fog band in the crystal makes up 6% whereas the depolarization of light scattered by quartz under normal conditions is 12% which indicates a weak anisotropy of the fog structure (see fig. 8) as compared with quartz crystal at other temperatures.

Visual observations of the fog band in a white light show that the blue light scattering is dominant on the inlet side of fog to the primary light beam, and the yellow light scattering, on the outlet side. This phenomenon is typical of the light passage through a strongly scattering medium. The ratio of the intensities in a blue and green light was controlled throughout the experiment, starting at room temperature. The value of the ratio was 5. The same ratio for the initial light of a mercury lamp was 1.6. For the light scattered by the fog band, it reached 10. The ratio of the intensities of blue and green light scattering was also measured for pure benzole in which the molecular character of light scattering is indisputable. The searched-for value for benzole was 4.2 which permitted a suggestion of the existence of the optical inhomogeneities smaller than the light wavelength in the fog band. The increased value of the ratio of the intensity for blue and green light scattering in fog, as compared with benzole, is, seemingly, due to multiple light scattering.

(C) It is convenient to discuss the experimental results following the natural separating of the whole temperature range into several intervals characteristic of the phenomena described.

The 20–400°C interval corresponds to a linear dependence of the light scattering intensity $I_T$ on temperature $T$. In view of the fact that the theoretical calculations also establish a linear function for the explicit dependence of $I_T$ on $T$, the obtained result indicates that up to 400°C the elastic and elasto-optical parameters of quartz determining, along with temperature, the scattered light intensity, are weakly dependent on temperature.

The 400–573°C and 573–600°C intervals correspond to two regions of nonlinear dependence of $I_T$ on $T$. The strong increase and also the decrease in the light scattering intensity in these intervals agree with other experimental facts pertaining to quite different properties of quartz in the vicinity of the phase transition. It is exactly a sharp decrease of the modulus of elasticity of quartz $E$ (by 1.7 times for $E_{\parallel z}$ and by 2.5 times for $E_{\perp z}$) revealed by Perrier (1922) that indicates an increase in compressibility of quartz near 573°C, which leads to increasing density fluctuations in a medium. The theoretical curve of the dependence of $I_T$ on $T$ taking into account the changes in compressibility of quartz, according to the temperature dependence of the Young modulus only, is plotted in fig. 5 (a broken line). The trend of the broken line curve is in good agreement with the experimental results. The temperature dependence of the refractive index ought have also been taken into account, in addition to the temperature dependence of compressibility, when constructing the theoretical curve. The measurements of Baransky (Yakovlev and Velichkina 1957) establish the dependence of $n$ on $T$. However, in view of natural conditions of experiment the dependence is measured at constant pressure whereas the scattered light intensity is determined by $(\partial\epsilon/\partial T)_\rho$. The conversion from $(\partial\epsilon/\partial T)_p$ to $(\partial\epsilon/\partial T)_\rho$ is mathematically described above.

Let us now discuss the opalescence phenomenon in quartz in the immediate vicinity of the phase transition. The reversible opalescence in a solid was first revealed in the above-described paper. The opalescence in quartz proved to be stronger than in some liquids in the critical state. The very fact of the presence of opalescence speaks in favour of sharp small-size optical inhomogeneities in quartz. The quantitative data on the light scattering intensity at the quartz transition are in good agreement with the calculated results (Ginzburg 1955). Yet, the experiments to be described in what follows have shown that the inhomogeneities are quasistatic and the observed light scattering is essentially elastic. As has already been noted (Ginzburg 1955, Ginzburg et al. 1980) the approximate estimations of the intensity of scattered light from the dynamic fluctuations of the refractive index proved to be overrated and the light scattering by the dynamic fluctuations is "overwhelmed" by the more intense elastic scattering. Especially as the calculation employed an assumption that at the transition temperature the quartz sample remained an optically homogeneous medium and no relatively large statical inhomogeneities of the refractive index, revealed later, arose. Nevertheless, the anomalous light scattering is, seemingly, due to the large values of $(\partial \epsilon / \partial T)_\rho$, and also the increase in compressibility of the crystal is connected with its elastic moduli which decrease at the phase transition.

## *3. Subsequent investigations of the light scattering in quartz*

(A) To elucidate the origin of opalescence in quartz, Shapiro and Cummins (Shapiro and Cummins 1968, Shapiro 1969, Cummins 1976) made thorough investigations of the light scattering in quartz at phase transition.

Ginzburg and Levanyuk (1960) suggested that the frequency of the 207 $cm^{-1}$ combination spectral line should tend to zero and the intensity should increase substantially as the crystal temperature reached the phase transition. Therefore, to verify a conception soft mode the authors (Shapiro and Cummins 1968, Shapiro 1969) have studied the combination scattering spectra in the temperature range close to the phase transition. It was shown that the frequency of the 207 $cm^{-1}$ line is actually invariable as the crystal reaches the transition temperature. However, the frequency of the weak 147 $cm^{-1}$ line at the phase transition reaches 30 $cm^{-1}$ and the line disappears when opalescence arises. At the minimal frequency the width of the line reaches 80 $cm^{-1}$. The intensity of this line at the transition point goes to zero and, consequently, the concept of the soft optical mode cannot explain the opalescence.

The authors also studied the temperature dependence of the Mandelstam–Brillouin components at the transformation temperature of the crystal. The spectral line components corresponding to the light scattering by elastic waves propagating along the [100], [010], [001] and [110] directions, were investigated. It was shown that the fine structure components at the phase transition undergo substantial changes. Figure 12 illustrates the temperature dependence of the

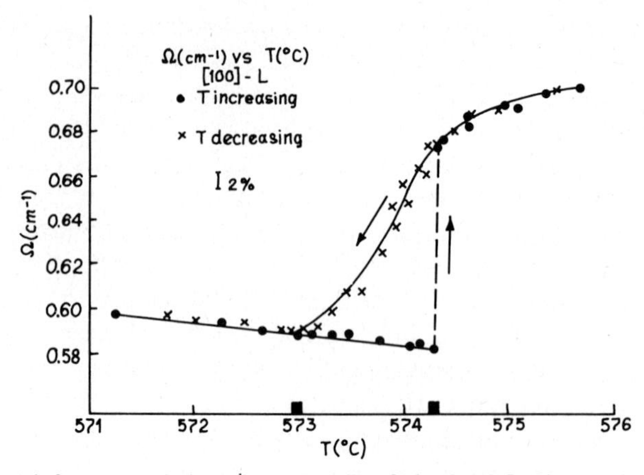


Fig. 12. Measured frequency $\Omega$ ($cm^{-1}$) vs $T$ (°C) of the [100]–L phonon in the transition temperature region.

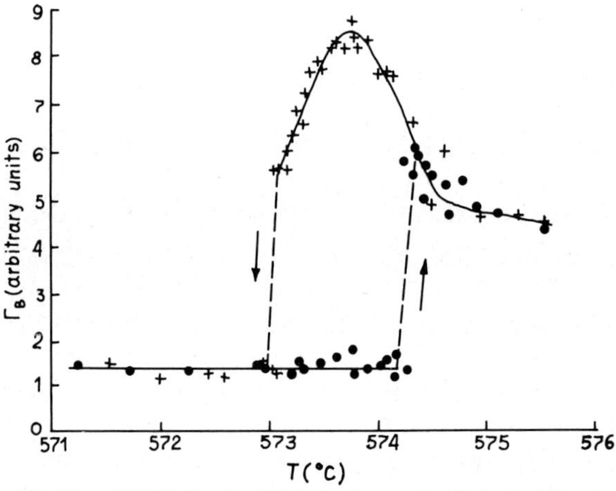


Fig. 13. The linewidth $\Gamma$ vs $T$ (°C) for the [100]–L phonon in the transition temperature region (arrows indicate direction of the temperature change).

Mandelstam–Brillouin component frequency corresponding to the light scattering by a longitudinal sound wave propagating along [100]. The frequency increases strongly in heating, at $T = 574.3$°C, and mildly declines in cooling. At 573.0°C the $\beta$–$\alpha$ phase transition occurs. Opalescence is observed at these two temperatures. Figures 13 and 14 show the behaviour of the width and intensity of the same fine structure component. The temperatures at which the noted characteristics change appreciably, correspond to the opalescence temperature. Similar changes in the spectra are also observed in the case of scattering by elastic waves propagating in different directions. Following the

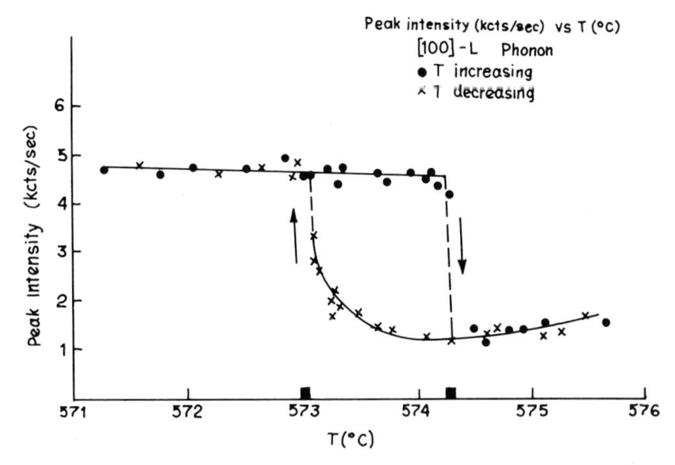


Fig. 14. Peak intensity vs $T$ (°C) of the [100]–L phonon in the transition temperature region.

obtained data, the authors conclude that no fine-structure line frequency goes to zero and the opalescence cannot be accounted for by a soft acoustic mode. A study of the fine structure of the scattered light spectral line has shown that a strong increase in the scattering intensity (by $10^4$ times) is connected with a central nonshifted component of the spectral line. With an accuracy up to $0.06\,cm^{-1}$, the spectral width of this component is equal to the width of a laser line. Visual observations have shown that in the presence of opalescence the scattering volume has a grannular structure which, according to the authors, is connected with the static inhomogeneities responsible for the light scattering. The authors have provided two possible explanations of the existence of the static inhomogeneities in the crystal. First, the strong light scattering can occur on the stressed surface between the Dauphine twins produced at the transition temperature. Second, the anomalous light scattering can be connected with the formation of a heterophase region of the coexistence of the $\alpha$ and $\beta$ phase in the crystal. Yet, the second assumption is not in line with the results described below.

(B) Dolino and Bachheimer (1977) and Dolino (1979) described the polarizing-microscope observations of the phase transformation of a quartz crystal in the immediate vicinity of the phase transition. The light scattering at a right angle to the direction of the exciting light beam and at small angles to the optical axis of the crystal was also studied. The samples used are $1\,cm^3$ natural quartz cubes. The $x$-, $y$-, $z$-axes are perpendicular to the faces. The temperature was regulated with an accuracy of $\pm 0.01$°C. A collimated white light beam was directed along the $z$-axis of the crystal. The samples placed between two polarizing prisms were observed with a microscope giving a resolution of $\sim 5\ \mu m$.

Small-angle scattering and right-angle scattering of light were registered simultaneously. For this purpose, the light beam from a He–Ne laser was divided into two beams. The first beam was sent along $y$ to observe the scattered light along $x$. The second beam was sent at a small-angle to $z$ to produce small-angle scattering. In some experiments the small-angle scattering was observed in the case when the laser light beam propagated along $y$.

It was established that the maxima of small-angle and right-angle scattering near $z$ are observed at somewhat different temperatures. Figure 15 presents the temperature dependence of the light-scattering intensity at a fixed point of the sample. Figures 15(a) and 15(b) correspond to the crystal in heating and in cooling, respectively. It is seen that as the temperature of the sample increases there appears, first, the peak of the right-angle intensity and, then, the peak of small-angle scattering (the full line and the broken line in fig. 15(a), respectively). The case is opposite in cooling.

If the sample has a temperature gradient in the $xy$-plane, the regions of the anomalous light scattering at the 90° angle and near $z$ have different locations in the crystal. It is established that the small-angle scattering is induced by cylinders with a cross section of some tens of microns which are elongated in the $z$ direction. The scattered light is concentrated in a cone with the angle of several degrees near $z$. According to the authors, this type of scattering is connected with the presence of a heterophase region of coexistence of the $\alpha$- and $\beta$-phases in the crystal. The cylindrical form of nuclei of one phase in the

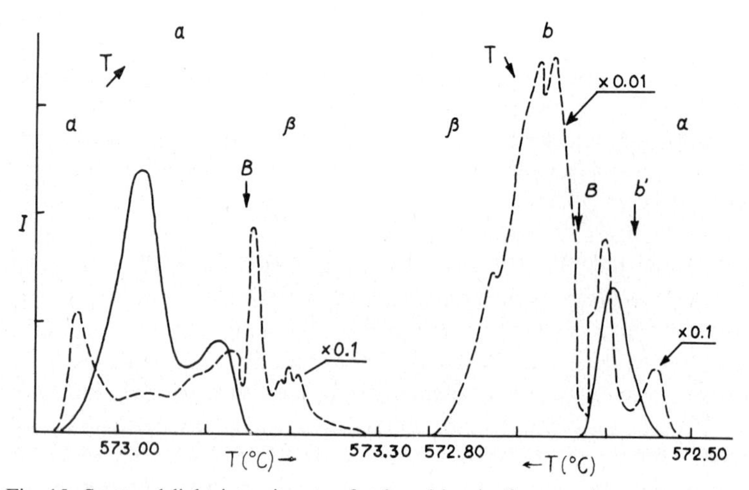


Fig. 15. Scattered light intensity at a fixed position in the crystal vs temperature.

other one is energetically more favourable. Right-angle scattering is only observed when the scattering vector $\Delta K$ is in the plane perpendicular to $z$. This scattering is of static character and, according to the authors, is caused by Dauphine twins whose characteristic size is 0.5 μm and the walls are parallel to $z$.

The intensity of small-angle scattering is about two hundred times greater than for right-angle scattering. The microscopic observations in the crossed polaroids have shown that the temperature stresses appear in the vicinity of the phase transition in the presence of a temperature gradient in the sample. When the $\beta$-phase region becomes larger than the $\alpha$-phase region, the field of vision assumes colour. A faint line (the B-line) perpendicular to the temperature gradient appears in the sample as the volume of the $\beta$-phase increases. As the temperature of the crystal increases, this line moves rapidly through the crystal to the lower temperature region and is followed at some distance by a second more contrasted line (the B-line). Behind the B-line in the $\beta$-phase region there is a granular structure responsible for the small-angle scattering. The right-angle scattering is observed between these two lines in the crystal. A polarizing microscope observation of the granular structure responsible for the small-

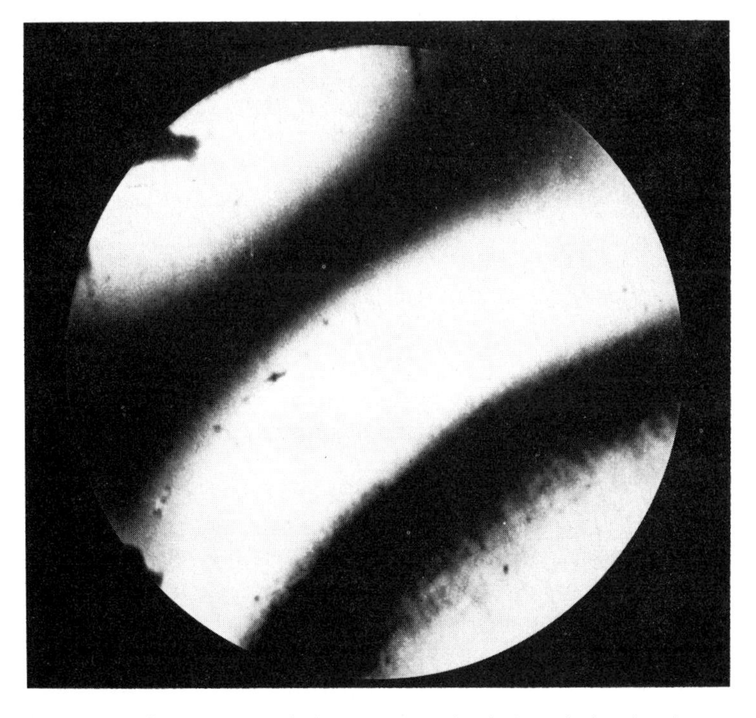

Fig. 16. Photograph of a quartz sample between crossed polarisers during the phase transition.

angle scattering, has shown that this structure sweeps the whole sample along $z$.

(C) Similar polarization investigations of the quartz crystal in the vicinity of the phase transition were made by Mikheeva and Shustin (1964) and Shustin (1964). The samples used were natural quartz discs 5 mm in diameter and 1 mm in thickness. An optical axis of the crystal was perpendicular to the disc plane. A near parallel linearly polarized monochromatic beam of light from a mercury lamp was sent along $z$. In the immediate vicinity to the phase transition the light transmitted through the plate was not linearly polarized. Figure 16 is a photographic picture of the surface of the investigated quartz sample in this temperature range. An analyzing prism is adjusted so as to transmit the minimum light intensity.

The photographic picture shows three regions in a narrow temperature interval at the phase transition where the linearly polarized light propagating along the optical axis of the crystal is no longer linealry polarized in leaving the sample. In the picture, these regions correspond to the transparent regions of the field of vision. The transmitted light proved to be eliptically polarized in these regions. The axial ratio (amplitude) for the wavelength $\lambda = 5780$ Å is 10 for the central part of elipticity between two black bands. For two other regions, the elipse is still more extended. The black bands in the picture correspond to the regions of the crystal where the light remains to be linearly polarized after leaving the crystal. As the temperature of the crystal is varied the observed picture shifts along the direction of the temperature gradient and the surface of the crystal becomes equally dark both above and below the phase transition.

To clarify the origin of ellipticity of the light passing through the crystal, the conoscopic observations were made. In a special oven employed, the aperture angles of a light beam and a beam forming conoscopic figures, made up 90°. A free quartz sample was a disc of 5 mm diameter and 0.95 mm thickness, cut perpendicular to the optical axis of the crystal. The conoscopic figures were formed by the rays passing through a small region of the crystal, 1 mm in diameter, that was not covered by the walls of the oven which permitted one to observe successively the conoscopic figures corresponding to different regions of ellipticity.

The conoscopic investigations have shown that in the temperature region closely adjacent to the transformation temperature the quartz crystal becomes optically biaxial. For the central region located between black bands, the plane of the optical axes contains the direction of the temperature gradient. For the two other regions this plane is perpendicular to the gradient direction. The black bands correspond to a one-axial quartz crystal. The calculation has shown that the angle between the optical axes for the central region of the maximum ellipticity is 5° for $\lambda = 5780$ Å and for the side regions 3°15.

Thus, in the immediate vicinity to the transformation temperature the ellipsoid of optical-dielectric permeability of the quartz crystal ceases to be an ellipsoid of rotation corresponding to a one-axial crystal. In this case, the quartz crystal is characterized by a three-axial ellipsoid of dielectric permeability with the difference of length of axes in the elliptic cross section perpendicular to the longer axis of ellipsoid equal to $5 \times 10^{-5}$.

Apparently, the above-described phenomenon is similar to that observed later (Dolino and Bachheimer 1977, Dolino 1979) and is connected with the elastic stresses in the crystal sample arising from the temperature gradient that is inevitable in the sample. Differences in the observed patterns may be due to different size and shape of the quartz samples used.

(D) To understand the nature of inhomogeneities induced at the phase transition of the crystal and leading to a strong increase in the light scattering intensity, Shustin et al. (1978, 1981) investigated the small-angle light scattering in a quartz single crystal. The experimental unit designed was aimed at making measurements of the intensity of light scattering, fixing a pattern of Fraunhofer diffraction and, using the Töpler method, at making photographic pictures of the sample examined at the same time.

Samples of natural and synthetic quartz used for this study are discs of a 5 mm diameter, 1–4 mm thick, cut either perpendicular or parallel to the optical axis of the crystal.

The experiments indicate that relatively large inhomogeneities appear in the vicinity of the phase transition of the crystal in a narrow temperature range (from several hundredths of a degree to 1°C) in various samples, which leads to intensive light scattering, at small angles in particular.

Figure 17(a) shows a picture of Fraunhofer diffraction in a sample of 1 mm

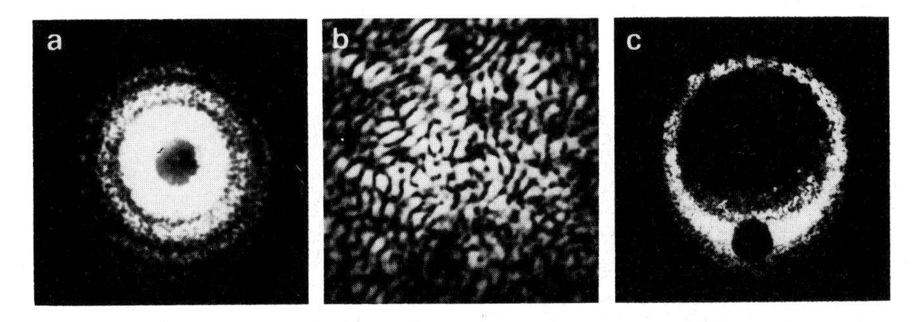


Fig. 17. Diffraction of light on inhomogeneities of a quartz crystal: (a) the incident beam of light is propagating along the optical axis of the crystal; (b) is a photographic picture of inhomogeneities made by the Töpler method; (c) is the incident beam of light propagation at a small angle to the optical axis.

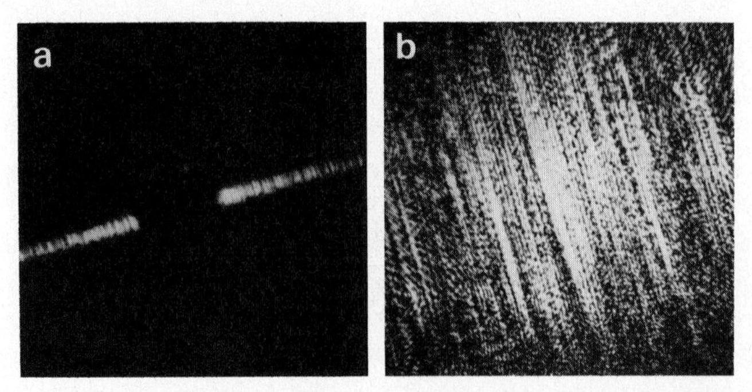


Fig. 18. Diffraction of light on inhomogeneities of a quartz crystal: (a) is the incident beam of light propagating perpendicular to the optical axis of the crystal; (b) is a photographic picture of inhomogeneities made by the Töpler method.

thick cut perpendicular of the optical axis of the crystal (the incident beam propagates along the optical axis). In this case the diffracted light is practically linearly polarized. One can see diffracted rings corresponding to the diffraction pattern on inhomogeneities distributed at random. Such a diffraction picture could not be observed (Yakovlev et al. 1956) due to different geometric parameters used in that experiment (different sizes of samples and different angular divergence of the incident beam). Figure 17(b) shows the picture of optical inhomogeneities inside the crystal obtained by the Töpler method. If the optical axis of the crystal is at some angle to the direction of propagation of the incident beam the diffraction picture will be similar to the one shown in fig. 17(c). Such a picture corresponds to diffraction patterns appearing on phase inhomogeneities elongated along the optical axis of the crystal. (A similar ring diffraction picture can be easily reproduced in the model experiment when the laser beam is obliquely incident to the wire diffraction grating in the plane perpendicular to the plane of the grating and parallel to the wire axes of the grating). The conclusion is supported by the photographic pictures shown in fig. 18 where the sample is cut parallel to the optical axis of the crystal (the incident light is perpendicular to the optical axis). Figure 18(a) shows the picture of Fraunhofer diffraction which represents a system of diffracted maxima positioned along the straight line perpendicular to the optical axis of the crystal. In some cases the diffraction picture is more or less periodic whereas in other occasions there is no periodicity at all. Figure 18(b) shows the photographic picture of the crystal inhomogeneities for this particular case of observations.

It is evident that there is some state of the crystal that is characterized with rather sharp stationary inhomogeneities of the refraction index in a narrow temperature range in the vicinity of the phase transition of the crystal. The inhomogeneities induce intensive light scattering especially at small scattering angles. The results of the diffraction measurements indicate that inhomogeneities represent small columns with a mean cross section of 20–30 μm elongated along the optical axis of the crystal. The diffraction shown in fig. 17(a) is similar to that which is produced in diffracted light on a multitude of round apertures or opaque discs arranged at random. This seems to be associated with the fact that the cross sections of inhomogeneities produced are very close to circles in configuration and the scattering in the values of their diameters is rather small. As it was described earlier small-angle scattering light at the phase transition of quartz was also studied by Dolino and Bachheimer (1977) and Dolino (1979). They believe that some heterophase structure produced by the coexistence of the $\alpha$ and $\beta$ phases appears in the vicinity of the phase transition of the crystal, which leads to anomalous light scattering. However, as is known (Yakovlev and Velichkina 1957), the change in the refraction index of quartz at the transition of the crystal from one phase to the other is $\sim 10^{-3}$ and when the thickness of the crystal is 1 mm the optical pass difference for light waves passed through various thermodynamic phases of the crystal at its transition changes by the value of the order of the light wavelength. Consequently, if some heterophase structure resulted from the coexistence of $\alpha$ and $\beta$ phases appeared in the vicinity the phase transition of the crystal the value of spatial modulation of the phase at the output of the sample should also reach values close to $\pi$. This could have led to a much higher intensity of the diffracted light than that of the experiment. By evaluating the modulation of the light wave phase it is assumed that inhomogeneities transfix the whole sample examined, and the evidence is supported both by microscopic examinations (Dolino 1979) and the measurements of the dependence of the diffracted light intensity on the size of the sample along the $z$-axis which will be described below.

To elucidate the value of alterations in the refraction index of inhomogeneities induced in the crystal interferometric investigations of the crystal was made and also measurements of the dependence of the diffracted light intensity on the diffraction angle.

To evaluate the spatial modulation of the refraction index to the tenfold multiplicated image of the crystal produced by a light wave passed directly through the crystal (the light propagated along the optical axis of the crystal) a coherent plane light wave was superimposed. The direction of the former was at the $\sim 10°$ angle from that of the latter. The samples used in the study were polished discs 1–2 mm thick with a diameter of 5 mm cut perpendicular to the optical axis of the crystal. Samples of both natural and synthetic quartz were studied. The interferometric picture obtained in the plane of the crystal image (practically straight equidistantly arranged fringes) were microscopically observed. Special attention was paid to elimination of air convection flows. The interferometric picture was quite stable at a constant temperature of the sample. It is ascertained that no local distortions in the interferometric fringes occur when inhomogeneities are produced in the vicinity of the phase transition. Consequently, the refraction index of inhomogeneities differs insignificantly from the mean refraction index of the crystal and the optical pass difference in inhomogeneities changes by a value which is much smaller than that of the light wavelength.

To determine a spatial modulation value of the refraction index the dependence of the diffracted light intensity on the diffraction angle was measured. The layout of the experiment unit is shown in fig. 19. A narrow collimated beam of light from the helium–neon laser propagated in a sample along its optical axis (the diameter of the beam in the crystal was equal to 0.3 mm). Samples of the crystal were the same as in the interference experiments. The straight light passing through the sample is stopped at an opaque screen S of a 2 mm diameter positioned 100 mm apart from the sample. The screen stops also the light scattered by the sample in the range from 0 to 0.6°C. The lense O accumulates the diffracted light at the input diaphragm of the photoelectric multiplier. The diaphragm of a variable diameter limits the cone of rays falling on the photoelectric multiplier.

Experimental results on the dependence of the diffracted light intensity on inhomogeneities of the crystal at the phase transition on the angle $\phi$ of the cone of rays falling on the photoelectric multiplier are shown by dots in fig. 20. The angle $\phi$ is dependent upon the diameter of the diaphragm D. The temperature of the crystal is stabilized in such a manner that during the measurements of the dependence the total intensity of diffracted light remains practically unchanged. The solid line indicates an approximate theoretical curve expressed analytically as:

$$I_\phi = I[1 \quad J_0^2(b\phi) \quad J_1^2(b\phi)] \quad I' . \tag{3}$$

Here $b = 1.22\pi/\phi_{\min}$, $\phi_{\min}$ is an angle at which the first minimum was detected in the diffraction picture. $J_0$ and $J_1$ are Bessel functions. The physical sense of $I$ and $I'$ is as follows: $I$ is the total intensity of diffracted light, $I'$ is an intensity of light that was subject to diffraction at small angles and was stopped at the screen S. $I$ and $I'$ are determined by the measured diffraction light intensities at two values of the angle $\phi$.

To obtain the approximating function (3) we used the known result of the

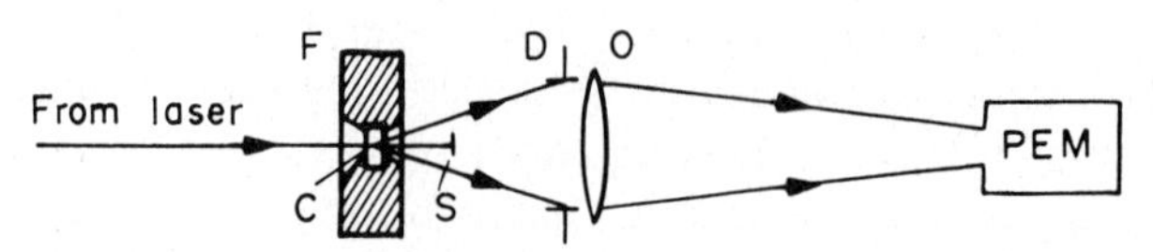


Fig. 19. Layout of the experimental unit for measuring the angular dependence of the intensity of diffracted light.

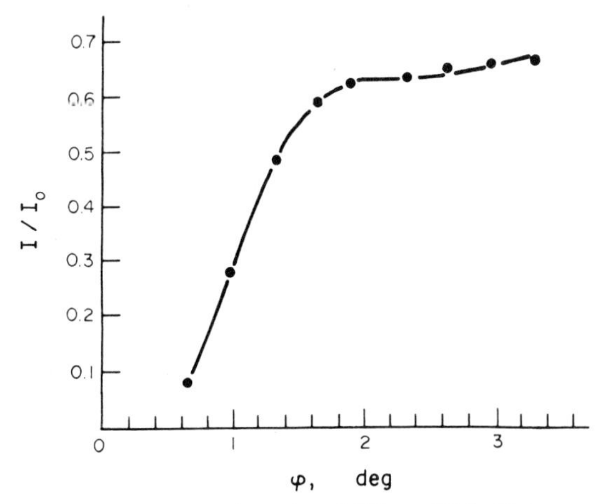


Fig. 20. Angular dependence of a relative intensity of diffracted light (dots are experimental results, the solid line is a theoretical curve).

diffraction of light on the round aperture (Born and Wolf 1964) proceeding from the fact that some optical phase inhomogeneities distributed at random are induced in the sample at the thermodynamic phase transition, which yields a picture similar to one obtained on a multitude of round apertures or opaque discs distributed at random, except for the area of the incident light beam. Besides, a low intensity of diffracted light indicates that when the sample is 1 or 2 mm thick the spatial modulation range of the light wave phase induced by inhomogeneities will be much lower than 1 radian on the output face of the crystal. As one can see in fig. 20, the approximative curve describes well the diffraction picture experimentally observed. Calculating the total intensity of diffracted light it is assumed that changes in the refraction index $\Delta n$ on inhomogeneities are small and the phase difference resulted from them in the output of the crystal will be $\Delta\varphi = (2\pi/\lambda)l\Delta n \ll 1$ where $l$ is the thickness of the sample. Besides, it is assumed that the sample is illuminated by a plane light wave.

For the electric field strength of the light wave passing directly through the sample (zero order) we shall obtain:

$$E_0 = \frac{E_{\mathrm{inc}}}{S}\, e^{i\langle\varphi\rangle} \iint_S e^{i\Delta\varphi(x,y)}\, dx\, dy\,,$$

where $\langle\varphi\rangle = (2\pi/\lambda)l\langle n\rangle$ and $\langle n\rangle$ is a mean refraction index of the crystal, $n(x, y)$ is a refraction index in a given point of the sample and $S$ is a cross-section area of the illuminating beam. Expanding the exponential factor

$\exp[i\Delta\varphi(x, y)]$ in series along $\Delta\varphi$ and integrating the expression we shall obtain:

$$E_0 = E_{\mathrm{inc}}\, e^{i\langle\varphi\rangle}[1 - \tfrac{1}{2}\langle(\Delta\varphi)^2\rangle],$$

where

$$\langle(\Delta\varphi)^2\rangle = \frac{1}{S}\iint_S [\Delta\varphi(x, y)]^2\, dx\, dy .$$

Consequently, the intensity of the light wave passing directly through the crystal will be equal to $I_0 = I_{\mathrm{inc}}[1 - \langle(\Delta\varphi)^2\rangle]$ and the total intensity of diffracted light will be $I = I_{\mathrm{inc}}\langle(\Delta\varphi)^2\rangle$.

Assuming that the change in the refraction index on inhomogeneities is equal to $\Delta n$ and remains constant over the whole area of the inhomogeneity we shall obtain the final formula:

$$I = I_{\mathrm{inc}}\left(\frac{2\pi}{\lambda}\, l\Delta n\right)^2 \frac{s}{S}\left(1 - \frac{s}{S}\right), \tag{4}$$

where $s$ is an area occupied by inhomogeneities in the cross section of the illuminating light beam.

Using formula (4) the modulation range of the refraction index $\Delta n$ was calculated for several samples of synthetic quartz. As the intensity of diffracted light in the vicinity of the phase transition of the crystal is dependent upon its temperature the maximum intensity value of scattered light of each sample was accepted as $I$. It is also assumed that the area occupied by inhomogeneities is equal to a half of that of the cross section of the illuminating light beam. When calculating $\Delta n$ the maximum value of the diffracted light intensity obtained in each experiment in cooling were used. In heating the maximum diffracted light intensity was usually several times lower than in cooling. The calculation indicates that $\Delta n$ is subject to changes from sample and amounts to $(0.7\text{–}1.5) \times 10^{-5}$.

One of the samples of 2 mm thick was repolished to the thickness of 1 mm after being used in the previous experiment and the intensity of diffracted light was measured again. It proved to be four times as low as in the 2 mm sample. This supports the evidence that inhomogeneities transfix the whole sample along the optical axis of the crystal.

It is worth noting one more circumstance associated with the preheating treatment of the crystal. If the sample is heated to 900°C the intensity of diffracted light increases by 2–5 times at the phase transition as compared to the values measured prior to heating.

To examine the problem of a possible low-frequency time modulation of scattered light which can be produced by some movement of inhomogeneities appearing in the crystal the holographic method of measuring small widenings of spectra lines suggested earlier (Shustin et al. 1977) was used. The results of the experiments indicate that when the exposition time was changed from 0.04 to 5 s, the diffraction efficiency of holograms of scattered light was fairly high and remained constant on a 10% accuracy level. Thus inhomogeneities appearing in the crystal do not induce time modulation of scattered light with frequencies higher than $10^{-2}$ c.p.s., or the role of modulated light is very insignificant compared to the light diffracted from stationary inhomogeneities.

The stationary optical phase inhomogeneities with a cross section of $\sim 20\,\mu$m elongated along the optical axis of the crystal are, therefore, responsible for diffraction of light at small angles at the phase $\alpha$–$\beta$ transition of quartz. The modulation range of the refraction index is roughly lower by two orders than the difference of the refraction index of $\alpha$ and $\beta$ quartz on both sides of the phase transition of the crystal. Thus, the state of the crystal in the vicinity of the phase transition cannot be considered as a heterophase state of coexistence of $\alpha$ and $\beta$ phases. The assumption that the presence of significant elastic stresses induced in the crystal in the vicinity of the phase transition is responsible for the decline in the difference of the refraction indices of supposedly-existent thermodynamic phases fails to explain the diminution of $\Delta n$ by two orders of the value as compared to its value for the $\alpha$ and $\beta$ phases. The appearance of the optical phase structure of the crystal seems to be associated with impurities and defects of the crystal producing an insignificant heterogeneous deformation which is most clearly displayed in the vicinity of the phase transition. This is also emphasized by the fact that the intensity of diffracted light and temperature variations where the heterogeneous state of the crystal is observed change substantially from sample to sample.

## 4. *Light scattering at phase transition of ammonium chloride crystals*

(A) The $NH_4Cl$ crystal with a phase transition at a temperature of $-31°C$ is one more crystal revealing anomalous light scattering. Below the transition temperature (in a less symmetric phase) the investigated crystals belong to the cubic system (class $T_d$) and above the transition temperature, to class $O_h$. The phase transition in the ammonium chloride crystal is a first-order transition of the "order–disorder" type and is close to the Curie critical point. Below the transition temperature, the orientation of the $NH_4$ groups permits them to be referred to the one definite system of spatial diagonals of the cubic lattice cell. In the symmetric phase, the orientation of the $NH_4$ groups is equally probable relative to two possible systems of spatial diagonals of a unit cell. The reasons why the authors investigated the light scattering at the ammonium chloride transition are the same as in the analysis of the phase transition in quartz. It was assumed that the fluctuations of the refractive index of the crystal leading

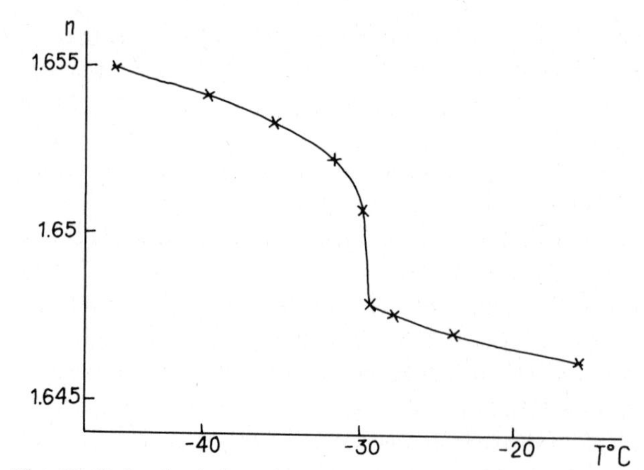


Fig. 21. Refractive index of ammonium chloride vs temperature.

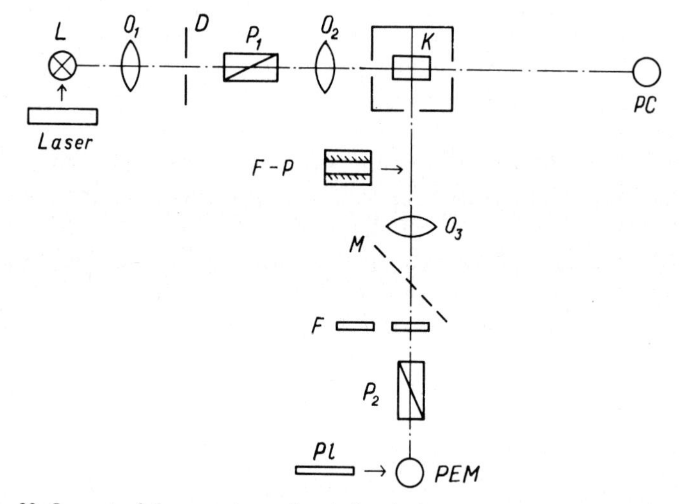


Fig. 22. Layout of the experimental unit for the investigation of scattered light.

to anomalously large light scattering, should also arise at first-order phase transition close to the Curie critical point. Indirect evidence for this assumption is an appreciable change of the refractive index at the phase transition of ammonium chloride (Barbaron 1951). Figure 21 is the temperature dependence of the refractive index of the crystal according to the present data. It follows from this dependence that the derivative $(\partial n/\partial T)_p$ in the vicinity of the phase transition reaches the value of $0.01°C^{-1}$.

(B) Further we describe the experiments (Shustin 1966). Figure 22 gives an optical scheme of the experimental arrangement. The discharge column of a super-high pressure mercury lamp N is projected by a lens O on the diaphragm D which cuts out the central part of the discharge picture. Behind the diaphragm is a polarizor H. The second lens $O_2$ forms a narrow weakly convergent light beam passing through a thermostat and a crystal K contained in it. A photocell is to continuously measure the intensity of the primary light beam. A constant spectral composition of emission of the mercury lamp is maintained by strictly constant discharge conditions. The light scattered by an ammonium chloride crystal at the 90° angle to the primary light beam is incident on the lens $O_3$. With the aid of the analyzer $H_2$ and changeable spectra light filters F this lens focusses the scattered light to the cathode of a photoelectron multiplier (PEM). The cathode carries a diaphragm that permits a small fraction of the scattering volume ($1 \times 1 \times 1\ mm^3$) can be cut from the image of an investigated crystal. The ammonium chloride samples were placed inside the thermostat which was cooled and thermally stabilized with an accuracy of 0.01–0.02°C by means of a basic Wobser thermostat.

(C) 4.3. The ammonium chloride crystals used in the light scattering measurements were cut perpendicular to edges of the cubic unit cell ($10 \times 7 \times 2\ mm^2$). The primary light beam propagated along one edge, and the scattered light along the perpendicular edge. The electric vector in the primary light beam was perpendicular to the plane of scattering. The measurements were made for the wavelength 4358 Å and 5460 Å of the mercury spectrum. Two runs of measurements were made simultaneously. One run of experiments measured

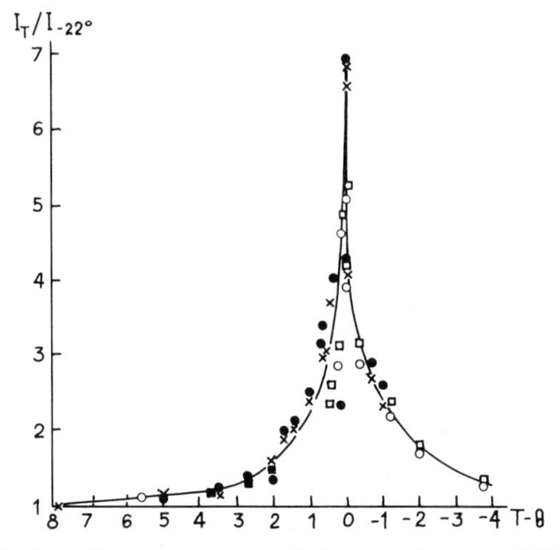


Fig. 23. Scattered light intensity vs temperature (primary and scattered light beams polarized perpendicular to the scattering plane).

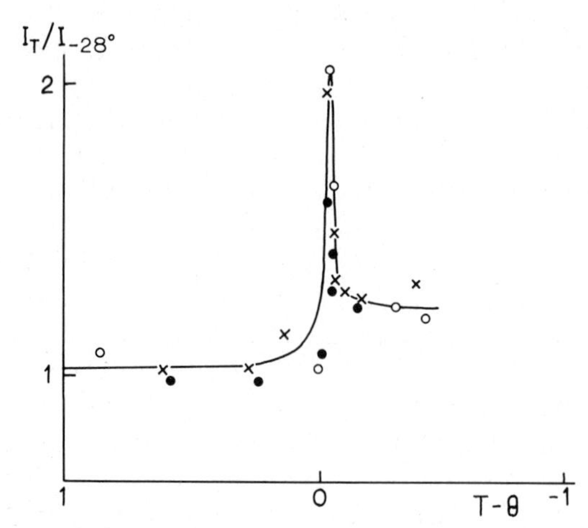


Fig. 24. The same as in fig. 23, but the scattered beam polarized in the scattering plane.

the scattered light with the same linear polarization as in the case of the primary light beam (fig. 23). On the abscissa is plotted the temperature counted off the temperature corresponding to the maximum of the intensity of scattered light. Along the ordinate is the ratio of the intensity of scattered light at the temperature of experiment to the same one at $-22°C$. In constructing the curve, the light scattered by microscopic inhomogeneities and crystal defects, was not excluded. Different signs for the experimental dots correspond to different experiments and wavelengths.

The second run of experiments measured the light scattering intensity with an electric vector lying in the plane of scattering (fig. 24). Here, the intensity of scattered light at $-28°C$ is noted as 1. The results of the two runs of experiments show strong evidence for the presence of the expected effect; the increase of the light scattering at the phase transition of an ammonium chloride crystal.

The figures clearly show a number of differences in the temperature behaviour of the light scattering for different polarization. The temperature half width of a peak corresponding to the scattered light with the same state of polarization as in primary beam is 0.5°C, whereas the temperature halfwidth for the scattered light with an electric vector lying in the plane of scattering is only 0.03°C. It will be also noted that the maxima of the intensity of scattered light in the first and second runs of experiments occurred at somewhat different temperatures ($\Delta t = 0.02°C$). The maximum of the intensity of scattered light for the component polarized perpendicular to the plane of scattering is at a lower temperature.

(D) To estimate the size of optical inhomogeneities responsible for the light scattering, the dependence of the light scattering intensity on the wavelength

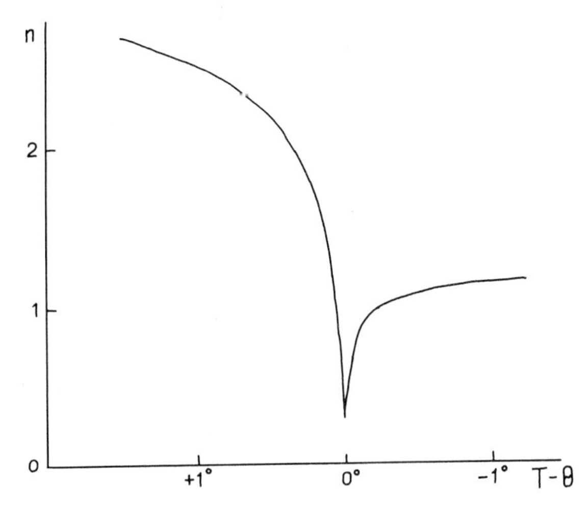


Fig. 25. Temperature dependence of the power $n$ in the relation $J_{scat} \sim 1/\lambda^n$.

was investigated. As noted above, for this purpose the intensity of scattered light was measured for two wavelengths 4358 Å and 5460 Å. These measurements gave the index $n$ describing the dependence of the intensity of scattered light on the wave length ($J_{scat} \sim 1/\lambda^n$). Figure 25 illustrates the dependence of $n$ on the temperature counted off the phase transition point $\theta$, for the component of scattered light that is polarized perpendicularly to the scattering plane. Not all the experimentally measured intensity was recognized as the intensity of scattered light, but only its increase connected with the approach to the transition since the light scattering far from the transformation temperature is completely determined by the scattering by the macroscopic inhomogeneities of the investigated samples. The index $n$ is seen to decrease while approaching the phase transition and drops to zero in the close proximity to the phase transition. This indicates an increasing size of the optical inhomogeneities responsible for the scattering.

The dependence of the intensity of scattered light on the wavelength for the scattered light component polarized in the scattering plane was measured only at the maximal intensity of scattered light in view of an extremely small temperature width of the scattering peak. In this case, it proved to be 2.5. Thus, the scattering light component at the phase transition of the crystal is characterized by a stronger dependence on the wavelength than the scattered light polarized perpendicular to the scattering plane.

(E) In addition to the measurement of the total intensity of scattered light, a study was also made of the fine structure of the Rayleigh scattering line. The experiments used a helium–neon laser as a light source. The laser light propagated along one edge of the cubical crystal lattice, the scattered light along another edge, i.e., at 90° to the exciting light beam. The electric vector in the incident light beam was perpendicular to the scattering plane. The scattered light spectra were taken with a Fabry–Perot interferometer with a 3 mm distance between mirrors.

The Mandelstam–Brillouin components corresponding to the scattering both by the longitudinal and transversal Debye waves propagating along the diagonal of a face of the cubical crystal lattice, were registered at room temperature. Analysis of the obtained spectrograms yielded the following values for the shifting $\Delta\lambda$ of the Mandelstam–Brillouin components: $0.228 \pm 0.001$ Å for the longitudinal wave and $0.117 \pm 0.001$ Å for the transversal wave. The velocities of a longitudinal and transversal hypersonic wave were calculated using these values of $\Delta\lambda$: $4650 \pm 25$ m/s and $2380 \pm 25$ m/s.

The spectrograms of the scattered light of the same polarization as in the case of the exciting light beam were obtained at temperatures of the crystal close to the phase transition. Analysis of the obtained spectrograms has shown that an increase of the intensity of scattered light at the phase transition of the crystal is connected with a strong increase of the intensity of the central nonshifted component of the spectral line. The intensity and location of the Mandelstam–Brillouin components at the phase transition change insignificantly.

The obtained scattering spectra permitted one to exclude the spurious scattering connected with the macroscopic crystal defects and to estimate the ratio of the intensity at the phase transition of the crystal to the intensity of molecular light scattering at room temperature. The calculations employed the following assumptions: (1) at room temperature, i.e., far from the transition, the whole intensity of the central nonshifted component of the spectral line is connected with macroscopic inhomogeneities of the crystal since the central components should be absent for an ideal crystal; (2) the intensity of light scattered by the crystal defects does not undergo any appreciable change while approaching the phase transition. Analysis of the experimental results on these assumptions has shown that at the phase transition the intensity of scattered light is $\sim 10^3$ times the intensity of molecular light scattering at room temperature.

The increase of the intensity of scattered light at the phase transition of ammonium chloride was theoretically analyzed in complete analogy with the calculation performed for the quartz crystal. The analysis is made taking it into account that, according to Barbabon (1951) $(\partial\epsilon/\partial T)_p = -2.5 \times 10^{-2}$ $I$/degree at the phase transition and is by three orders larger than the value of the noted derivative far from the phase transition.

The calculations have shown that the intensity of scattered light at the phase transition should be 600 times the intensity of molecular light scattering at room temperature which agrees with the experimental results of Shustin (1966) in the order of magnitude.

(F) The Mandelstam–Brillouin scattering in ammonium chloride crystals in the – 50 to + 50°C interval has been thoroughly investigated by Lazar et al. (1968). The temperature dependence of frequencies and intensities of the Mandelstam–Brillouin components connected with the longitudinal and transversal hypersonic waves propagating in the [100] and [110] directions, was discussed. The velocity of hypersonic waves was obtained from the frequency measurements. It was shown that the velocity of a transversal hypersonic wave at 9 Gc/s increased strongly (by 4%) in the transition to the low-temperature phase of crystal which is connected, according to the authors, with a strong decrease in the volume of a unit cell. The intensity of the Mandelstam–Brillouin component linearly decreased with temperature.

The velocity of longitudinal waves at 18 Gc/s has a deep minimum at the phase transition of the crystal and then increases rapidly with decreasing temperature. A change in the velocity near the transformation temperature is 5% of its mean value. The intensity of light scattered by the longitudinal waves has a maximum at 35°C and decreases when approaching the phase transition where it takes the minimal value, and then gradually increases below the phase transition.

The temperature dependence of the intensity of the central unshifted component of the spectral line of scattered light has also been studied. The results are analogous to the data described above (Shustin 1966). However, for scattering plane the authors have not revealed an intensity maximum at the phase transition but observed a sharp increase in the intensity of scattered light in the transition to the low-temperature phase of the crystal. This is believed to be due to an extremely small temperature width of the scattering peak for this component of scattered light and, therefore, the temperature gradients are required to be minimal for the scattering peak to be observed. The authors lay emphasis on the circumstance that in the vicinity of the phase transition the illuminated region of the crystal nonuniformly scattered the light due to macroscopic static inhomogeneities.

(G) Pique et al. (1977) described the results of the polarizing microscope observations of the phase transition of an ammonium chloride crystal. The samples were prepared either by slow cooling (0.1°C/day near the temperature 45°C) or by slow evaporation at a stabilized temperature (25°C). Practically all samples at room temperature exhibited internal stresses induced by the slip bands perpendicular to the [110] direction of the crystal.

In studying the phase transition in the crystal, a light beam from the He–Ne laser or a white light beam propagated in the [110] direction. In cooling down to – 28°C, the above described slip bands only were observed. At – 28°C, thin needles perpendicular to [111] and [111] which first arose in the most compressed regions of the slip bands, were observed. As the temperature is lowered, new nucleations of the low-temperature phase appear and the needles grow throughout the crystal volume. The angles between needles is $109 \pm 2°$ and corresponds to the angle between [111] and [111]. At the same time there is such an intense light scattering that virtually no light remains in the [110] direction. Near the forward direction the scattered light forms a cross with arms perpendicular to the needle directions on a screen perpendicular to the beam. The authors believe that the intense light scattering is connected with the heterophase region of coexistence of two phases in the crystal. With further cooling, the heterophase structure disappears near $-31°C$ but some traces visible in polarized light remain showing plastic deformations of the crystal. This suggests a rather low limit of elasticity of the ammonium chloride crystal. On heating, the same phenomena are observed between $-30$ and $-27°C$.

The whole body of experimental data shows strong evidence for the increase of light scattering at the phase transitions of crystals that are close to second order. The experiments have shown however that the effect is a more complicated process than was expected. In the case of quartz, there are two types of light scattering – diffuse and small angle stimulated by different inhomogeneities arising in the crystal at the phase transition but at somewhat different temperature. Both the types are induced by quasistatic inhomogeneities. The polarization observations of free quartz samples at the minimal thermal gradients in the vicinity of the phase transition emphasize how compliant the crystal lattice is under the rearrangement conditions.

All the revealed effects show that the optical methods of studying the phase transition are very helpful in providing much new evidence. However, new experiments and also the theoretical development are clearly needed for a complete explanation of the phenomena.

## References


Barbaron, M., 1951, Ann. de Phys. **6**, 899.

Born, M. and E. Wolf, 1964, Principles of Optics (Pergamon, New York).

Cummins, H.Z., 1976, The Theory of Light Scattering in Solids, in: Proc. First Sov.-Amer. Symp., Moscow, 1975, eds., V.M. Agranovich and J.L. Birman (Nauka, Moscow).

Dolino, G. and J.P. Bachheimer, 1977, Phys. Status Solidi (a) **41**, 673.

Dolino, G., 1979, J. Phys. Chem. Solids, **40**, 121.

Ginzburg, V.L., 1955, Dokl. Akad. Nauk SSSR, **105**, 240.

Ginzburg, V.L. and A.P. Levanyuk, 1960, Zh Exp. Teor. Fiz. **39**, 192.

Ginzburg, V.L., A.P. Levanyuk and A.A. Sobyanin, 1980, Phys. Rep. **57**.

Lazay, P.D., J.H. Lunacek, N.A. Clark and G.B. Benedek, 1969, in: Light Scattering Spectra of Solids, New York, 1969, ed., G.W. Wright (Springer, Berlin).

Mikheeva, L.F. and O.A. Shustin, 1964, Kristallografiya, **9**, 423.

Motulevich, G.P., 1950, Trudy FIAN SSSR, **V**, 12.

Perrier, A. and R. Mandrot, 1922, Comp. rend. **175**, 622.

Pique, J.P., G. Dolino and M. Vallade, 1977, de Phys. **38**, 1527.

Rinne, P. and R. Kolb, 1910, Neues Jahrbuch Mineralogie, **2**, 138.

Shapiro, S.M., 1969, Thesis, The John Hopkins University.
Shapiro, S.M. and H.Z. Cummins, 1968, Phys. Rev. Lett. **21**, 1578.
Shustin, O.A., 1964, Kristallografiya, **9**, 925.
Shustin, O.A., 1966, Pis'ma Zh. Exp. Theor. Fiz. **12**, 491.
Shustin, O.A., T.G. Chernevich and S.A. Ivanov, 1977, Pis'ma Zh. Exp. Theor. Fiz. **26**, 669.
Shustin, O.A., T.G. Chernevich, S.A. Ivanov and I.A. Yakovlev, 1978, Pis'ma Zh. Exp. Theor. Fiz. **7**, 349.
Shustin, O.A., T.G. Chernevich, S.A. Ivanov and I.A. Yakovlev, 1981, Solid State Commun. **37**, 65.
Sinel'nikov, N.N., 1953, Dokl. Akad. Nauk SSSR, **92**, 369.
Yakovlev, I.A., T.S. Velichkina and L.F. Mikheeva, 1956, Kristallografiya, **1**, 91.
Yakovlev, I.A. and T.S. Velichkina, 1957, Usp. Fiz. Nauk, **63**, 411.

# AUTHOR INDEX

# SUBJECT INDEX

# MATERIALS INDEX

MARSTON SCIENCE LIBRARY

MARSTON SCIENCE LIBRARY
Date Due

MAY
AP